Methods in Enzymology

Volume 312

SPHINGOLIPID METABOLISM AND CELL SIGNALING

Part B

METHODS IN ENZYMOLOGY

EDITORS-IN-CHIEF

John N. Abelson   Melvin I. Simon

DIVISION OF BIOLOGY
CALIFORNIA INSTITUTE OF TECHNOLOGY
PASADENA, CALIFORNIA

FOUNDING EDITORS

Sidney P. Colowick and Nathan O. Kaplan

Methods in Enzymology

Volume 312

Sphingolipid Metabolism and Cell Signaling

Part B

EDITED BY

Alfred H. Merrill, Jr.

EMORY UNIVERSITY CENTER FOR NUTRITION AND HEALTH SCIENCES,
ATLANTA, GEORGIA

Yusuf A. Hannun

MEDICAL UNIVERSITY OF SOUTH CAROLINA
CHARLESTON, SOUTH CAROLINA

ACADEMIC PRESS
San Diego London Boston New York Sydney Tokyo Toronto

This book is printed on acid-free paper. ♾

Academic Press
A Harcourt Science and Technology Company
525 B Street, Suite 1900, San Diego, California 92101-4495, USA

http://www.academicpress.com

Academic Press Limited
32 Jamestown Road, London NW1 7BY, UK

International Standard Book Number: 0-12-182213-3

PRINTED IN THE UNITED STATES OF AMERICA
00 01 02 03 04 05 06 MM 9 8 7 6 5 4 3 2 1

Table of Contents

Section I. Methods for Analyzing Sphingolipids

Section II. Methods for Analyzing Aspects of Sphingolipid Metabolism in Intact Cells

Section III. Sphingolipid–Protein Interactions and Cellular Targets

Section IV. Sphingolipid Transport and Trafficking

Section V. Other Methods

Contributors to Volume 312

Article numbers are in parentheses following the names of contributors.
Affiliations listed are current.

DOMENICO ACQUOTTI (21), *Interfaculty Center for Measurements, University of Parma, Parma 43100, Italy*

VERED AGMON (23), *Department of Biochemistry, Hebrew University, Hadassah School of Medicine, Jerusalem 91120, Israel*

TOSHIO ARIGA (10), *Clinical Research Center, Eisai Co., Ltd., Tokyo 120, Japan*

AHMET AYAR (31), *Department of Pharmacology, Firat University Medical School, Elazig, Turkey*

DAGMAR BACIKOVA (25), *Department of Biochemistry and Molecular Biology, Uniformed Services University of the Health Sciences, Bethesda, Maryland 20814*

MANJU BASU (17), *Department of Chemistry and Biochemistry, University of Notre Dame, Notre Dame, Indiana 46556-5670*

SHIB S. BASU (17), *Department of Chemistry and Biochemistry, University of Notre Dame, Notre Dame, Indiana 46556-5670*

SUBHASH BASU (17), *Department of Chemistry and Biochemistry, University of Notre Dame, Notre Dame, Indiana 46556-5670*

NINA BEROVA (19), *Department of Chemistry, Columbia University, New York, New York 10027*

ERHARD BIEBERICH (27), *Department of Biochemistry and Molecular Biophysics, Medical College of Virginia Campus, Virginia Commonwealth University, Richmond, Virginia 23298-0614*

ALICJA E. BIELAWSKI (4, 34), *Department of Biochemistry and Molecular Biology, Medical University of South Carolina, Charleston, South Carolina 29425*

JACQUES BODENNEC (9), *Laboratory of Tumor Glycobiology, University Claude Bernard-Lyon I, Oullins, Cedex 69921, France*

BETH BOYD (38), *Departments of Laboratory Medicine and Pathobiology and Biochemistry, University of Toronto and Research Institute, Hospital for Sick Children, Toronto, Ontario M5G 1X8, Canada*

GÉRARD BRICHON (9), *Institute Michel Pacha, University Claude Bernard-Lyon I, La Seyne sur Mer 83500, France*

DEBORAH A. BROWN (22), *Department of Biochemistry and Cell Biology, State University of New York at Stony Brook, Stony Brook, New York 11794-5215*

JOSEF BRUNNER (35), *Department of Biochemistry, Swiss Federal Institute of Technology, Zurich ETHZ, Switzerland*

THOMAS B. CALIGAN (1), *Department of Biochemistry, Emory University School of Medicine, Atlanta, Georgia 30322-3050*

GEORGE M. CARMAN (29), *Department of Food Science, Rutgers University, New Brunswick, New Jersey, 08901*

CHARLES E. CHALFANT (34), *Department of Biochemistry and Molecular Biology, Medical University of South Carolina, Charleston, South Carolina 29425*

BRIAN E. COLLINS (36), *Department of Pharmacology and Molecular Science, The Johns Hopkins Medical School, Baltimore, Maryland 21205*

ANDREAS CONZELMANN (42), *Institute of Biochemistry, University of Fribourg, CH-1700 Fribourg, Switzerland*

ARIE DAGAN (23, 26), *Institute for Gene Therapy and Molecular Medicine, Mount Sinai School of Medicine, New York, New York 10029*

SARA DASTGHEIB (17), *Department of Chemistry and Biochemistry, University of Notre Dame, Notre Dame, Indiana 46556-5670*

TAMA DINUR (23, 26), *Institute for Gene Therapy and Molecular Medicine, Mount Sinai School of Medicine, New York, New York 10029*

MICHEL DOMINGUEZ (44), *Department of Biochemistry and Molecular Biology, Thoracic Diseases Research Unit, Mayo Clinic and Foundation, Rochester, Minnesota 55905-0001*

TERESA M. DUNN (25), *Department of Biochemistry and Molecular Biology, Uniformed Services University of the Health Sciences, Bethesda, Maryland 20814*

LISA EDSALL (2), *Laboratory of Molecular and Cellular Regulation, National Institute of Mental Health, Bethesda, Maryland 20892*

SHAI ERLICH (26), *Institute for Gene Therapy and Molecular Medicine, Mount Sinai School of Medicine, New York, New York 10029*

JACQUES FANTINI (41), *Laboratoire de Biochimie et Biologie de la Nutrition, UPRESA-CNRS 6033, Faculté des Sciences de St. Jérôme, Marseille, Cedex 20, 13397, France*

ANTHONY H. FUTERMAN (15, 16), *Department of Biological Chemistry, Weizmann Institute of Science, Rehovot 76100, Israel*

KEN GABLE (25), *Department of Biochemistry and Molecular Biology, Uniformed Services University of the Health Sciences, Bethesda, Maryland 20814*

SHIMON GATT (23, 26), *Institute for Gene Therapy and Molecular Medicine, Mount Sinai School of Medicine, New York, New York 10029*

ISABELLE GUILLAS (42), *Institute of Biochemistry, University of Fribourg, CH-1700 Fribourg, Switzerland*

SEN-ITIROH HAKOMORI (30), *Pacific Northwest Research Institute, Seattle, Washington 98122-4327*

SEN-ITIROH HAKOMORI (37, 40), *Departments of Pathobiology and Microbiology, University of Washington, Seattle, Washington 98122*

AKIKAZU HAMAGUCHI (30), *Pacific Northwest Research Institute, Seattle, Washington 98122-4327*

DJILAI HAMMACHE (41), *Laboratoire de Biochimie et Biologie de la Nutrition, UPRESA-CNRS 6033, Faculté des Sciences de St. Jérôme, Marseille, Cedex 20, 13397, France*

KENTARO HANADA (24), *Department of Biochemistry and Cell Biology, National Institute of Infectious Diseases, 1-23-1, Toyama, Shinjuku-ku, Tokyo 162-8649, Japan*

KAZUKO HANDA (37, 40), *Pacific Northwest Research Institute, Seattle, Washington 98122-4327*

YUSUF A. HANNUN (4, 28, 33, 34), *Department of Biochemistry and Molecular Biology, Medical University of South Carolina, Charleston, South Carolina 29425*

MICHAEL HEINRICH (35), *Institute of Immunology, University of Kiel, D-24105 Kiel, Germany*

TOMOKO HIRAMA (18), *Institute for Biological Sciences, National Research Council Canada, Ottawa, Ontario K1A OR6, Canada*

DAI ISHIKAWA (12, 13), *Molecular Medical Science Institute, Otsuka Pharmaceutical Co., Ltd., Kawauchicho, Tokushimashi, Tokushima 771-0192, Japan*

SERGE IVALDI (41), *Laboratoire de Biochimie et Biologie de la Nutrition, UPRESA-CNRS 6033, Faculté des Sciences de St. Jérôme, Marseille, Cedex 20, 13397, France*

KAZUHISA IWABUCHI (30, 40), *Pacific Northwest Research Institute, Seattle, Washington 98122-4327*

REIJI KANNAGI (14), *Program of Experimental Pathology, Research Institute, Aichi Cancer Center, Chikusaku, Nagoya 464-8681, Japan*

AKIRA KAWAMURA (19), *Department of Chemistry, Columbia University, New York, New York 10027*

KATSUYA KISHIKAWA (34), *Department of Biochemistry and Molecular Biology, Medical University of South Carolina, Charleston, South Carolina 29425*

NAOYA KOJIMA (37), *Faculty of Engineering, Tokai University, Kanagawa, Japan*

MARTIN KRÖNKE (35), *Institute of Medical Microbiology and Hygiene, University of Cologne, D-50935 Köln, Germany*

STEPHAN LADISCH (11), *Center for Cancer and Transplantation Biology, Children's Research Institute, and Departments of Pediatrics and Biochemistry/Molecular Biology, George Washington University School of Medicine, Washington, DC 20010-2970*

RUIXIANG LI (11), *Center for Cancer and Transplantation Biology, Children's Research Institute, and Departments of Pediatrics and Biochemistry/Molecular Biology, George Washington University School of Medicine, Washington, DC 20010-2970*

ZHIXIONG LI (17), *Department of Chemistry and Biochemistry, University of Notre Dame, Notre Dame, Indiana 46556-5670*

CLIFFORD A. LINGWOOD (38, 39), *Division of Immunity, Infection, Injury and Repair, Research Institute, and Departments of Laboratory Medicine and Pathobiology and Biochemistry, University of Toronto and Research Institute, Hospital for Sick Children, Toronto, Ontario M5G 1X8, Canada*

SEAN S. LIOUR (27), *Department of Biochemistry and Molecular Biophysics, Medical College of Virginia Campus, Virginia Commonwealth University, Richmond, Virginia 23298-0614*

ERWIN LONDON (22), *Department of Biochemistry and Cell Biology, State University of New York at Stony Brook, Stony Brook, New York 11794-5215*

CHIARA LUBERTO (33) *Department of Biochemistry and Molecular Biology, Medical University of South Carolina, Charleston, South Carolina 29425*

ROGER C. MACKENZIE (18), *Institute for Biological Sciences, National Research Council Canada, Ottawa, Ontario K1A OR6, Canada*

MARC MARESCA (41), *Laboratoire de Biochimie et Biologie de la Nutrition, UPRESA-CNRS 6033, Faculté des Sciences de St. Jérôme, Marseille, Cedex 20, 13397, France*

TAMAR MEGIDISH (30), *Pacific Northwest Research Institute, Seattle, Washington 98122-4327*

RICHARD MENDELSOHN (20), *Department of Chemistry, Rutgers University, Newark College of Arts and Science, Newark, New Jersey 07102*

ALFRED H. MERRILL, JR. (1, 28, 47), *Department of Biochemistry, Emory University School of Medicine, Atlanta, Georgia 30322-3050*

SHELDON MILSTIEN (2), *Laboratory of Molecular and Cellular Regulation, National Institute of Mental Health, Bethesda, Maryland 20892*

SILVIA R. P. MIRANDA (26), *Institute for Gene Therapy and Molecular Medicine, Mount Sinai School of Medicine, New York, New York 10029*

WIEBKE MÖBIUS (45), *Kekulé-Institut für Organische Chemie und Biochemie, Universität Bonn, D-53121 Bonn, Germany*

ERIN MONAGHAN (25), *Department of Biochemistry and Molecular Biology, Uniformed Services University of the Health Sciences, Bethesda, Maryland 20814*

DAVID J. MOORE (20), *International Specialty Products, Wayne, New Jersey 07470*

JOHANNES MÜTHING (6), *Technical Faculty, Institute for Cell Culture Technology, University of Bielefeld, D-33501 Bielefeld, Germany*

MURUGESAPILLAI MYLVAGANAM (39), *Division of Immunity, Infection, Injury and Repair, Research Institute, Hospital for Sick Children, Toronto, Ontario M5G 1X8, Canada*

KOJI NAKANISHI (19), *Department of Chemistry, Columbia University, New York, New York 10027*

MASAHIRO NISHIJIMA (24), *Department of Biochemistry and Cell Biology, National Institute of Infectious Diseases, 1-23-1, Toyama, Shinjuku-ku, Tokyo 162-8649, Japan*

ANITA NUTIKKA (38), *Departments of Laboratory Medicine and Pathobiology and Biochemistry, University of Toronto and Research Institute, Hospital for Sick Children, Toronto, Ontario M5G 1X8, Canada*

LINA M. OBEID (28), *Department of Biochemistry, Medical University of South Carolina, Charleston, South Carolina 23425*

JOYCE OU (1), *Department of Biochemistry, Emory University School of Medicine, Atlanta, Georgia 30322-3050*

RICHARD E. PAGANO (44), *Department of Biochemistry and Molecular Biology, Thoracic Diseases Research Unit, Mayo Clinic and Foundation, Rochester, Minnesota 55905-0001*

DAVID K. PERRY (4), *Department of Biochemistry and Molecular Biology, Medical University of South Carolina, Charleston, South Carolina 29425*

KATHERINE PETERS (1), *Department of Biochemistry, Emory University School of Medicine, Atlanta, Georgia 30322-3050*

MARTINE PFEFFERLI (42), *Institute of Biochemistry, University of Fribourg, CH-1700 Fribourg, Switzerland*

GÉRARD PIÉRONI (41), *INSERM U470, Marseille, France*

JAMIE POLLOCK (31), *Department of Biomedical Sciences, Institute of Medical Sciences, Aberdeen University, Aberdeen AB25 2ZD, Scotland*

JACQUES PORTOUKALIAN (9), *Laboratory of Tumor Glycobiology, University Claude Bernard-Lyon I, Oullins, Cedex 69921, France*

RENÉ J. RAGGERS (46), *Department of Cell Biology and Histology, Academic Medical Center, University of Amsterdam, Amsterdam 1100 DC, The Netherlands*

RONALD L. SCHNAAR (36), *Departments of Pharmacology and Neuroscience, The Johns Hopkins Medical School, Baltimore, Maryland 21205*

EDWARD H. SCHUCHMAN (26), *Department of Human Genetics and Institute for Gene Therapy and Molecular Medicine, Mount Sinai School of Medicine, New York, New York 10029*

STEFAN SCHÜTZE (35), *Institute of Immunology, University of Kiel, D-24105 Kiel, Germany*

ANDREAS SCHWARZ (15, 16), *Kekulé-Institute, Universität Bonn, D-53121 Bonn, Germany*

GÜNTER SCHWARZMANN (45), *Kekulé-Institut für Organische Chemie und Biochemie, Universität Bonn, D-53121 Bonn, Germany*

RODERICK H. SCOTT (31), *Department of Biomedical Sciences, Institute of Medical Sciences, Aberdeen University, Aberdeen AB25 2ZD, Scotland, United Kingdom*

DAN J. SILLENCE (46), *Department of Cell Biology and Histology, Academic Medical Center, University of Amsterdam, Amsterdam 1100 DC, The Netherlands*

E. R. SMITH (28), *Department of Biochemistry, Emory University School of Medicine, Atlanta, Georgia 30322-3050*

SANDRO SONNINO (21), *Department of Medical Chemistry and Biochemistry-LITA, University of Milan, Medical School, Segrate, Milan 20090, Italy*

SARAH SPIEGEL (2, 32), *Department of Biochemistry and Molecular Biology, Georgetown University Medical Center, Washington, DC 20007*

EIKO SUGIYAMA (8), *Research Center on Aging and Adaptation, Shinshu University School of Medicine, Matsumoto 390-8621, Japan*

M. CAMERON SULLARDS (5), *Department of Biochemistry, Emory University School of Medicine, Atlanta, Georgia 30322-3050*

TAMOTSU TAKETOMI (8), *Research Center on Aging and Adaptation, Shinshu University School of Medicine, Matsumoto 390-8621, Japan*

TAKAO TAKI (12, 13), *Molecular Medical Science Institute, Otsuka Pharmaceutical Co., Ltd., Kawauchicho, Tokushimashi, Tokushima 771-0192, Japan*

ANNEMIEK D. TEPPER (3), *Laboratory for Genetic Metabolic Diseases, Departments of Clinical Chemistry and Pediatrics, Academic Medical Center, University of Amsterdam, 1100 DE Amsterdam, The Netherlands*

GEORG C. TERSTAPPEN (16), *Medicines Research Centre, Glaxo Wellcome, Verona I-37135, Italy*

NICOLA M. THATCHER (31), *Discovery Biology Department, Central Research, Pfizer Limited, Kent CT13 9NJ, England, United Kingdom*

WIM J. VAN BLITTERSWIJK (3), *Division of Cellular Biochemistry, The Netherlands Cancer Institute, 1066 CX Amsterdam, The Netherlands*

JAMES R. VAN BROCKLYN (32), *Department of Pathology, College of Medicine and Public Health, The Ohio State University, Columbus, Ohio 43210-1239*

GERHILD VAN ECHTEN-DECKERT (7), *Kekulé-Institut für Organische Chemie und Biochemie, Universität Bonn, D-53121 Bonn, Germany*

GERRIT VAN Meer (46), *Department of Cell Biology and Histology, Academic Medical Center L3, University of Amsterdam, Amsterdam 1105 AZ, The Netherlands*

LEWIS VANN (2), *Laboratory of Molecular and Cellular Regulation, National Institute of Mental Health, Bethesda, Maryland 20892*

ALEXANDER VON COBURG (45), *Kekulé-Institut für Organische Chemie und Biochemie, Universität Bonn, D-53121 Bonn, Germany*

ELAINE WANG (1), *Department of Biochemistry, Emory University School of Medicine, Atlanta, Georgia 30322-3050*

RIKIO WATANABE (44), *Department of Biochemistry and Molecular Biology, Thoracic Diseases Research Unit, Mayo Clinic and Foundation, Rochester, Minnesota 55905-0001*

THOMAS WEBER (35), *Memorial Sloan-Kettering Cancer Center, New York, New York 10021*

CHRISTINE WHEATLEY (44), *Department of Biochemistry and Molecular Biology, Thoracic Diseases Research Unit, Mayo Clinic and Foundation, Rochester, Minnesota 55905-0001*

MARC WICKEL (35), *Institute of Immunology, University of Kiel, D-24105 Kiel, Germany*

ELIZABETH M. WILSON-KUBALEK (43), *Department of Cell Biology, MB25, The Scripps Research Institute, La Jolla, California 92037*

SUPANDI WINOTO-MORBACH (35), *Institute of Immunology, University of Kiel, D-24105 Kiel, Germany*

WEN-I. WU (29), *Lord and Taylor Laboratory for Lung Biochemistry and Anna Perahia Adatto Clinical Research Center, National Jewish Medical and Research Center, Denver, Colorado 80206*

XIAOLIAN XU (22), *Department of Biochemistry and Cell Biology, State University of New York at Stony Brook, Stony Brook, New York 11794-5215*

NOUARA YAHI (41), *Laboratoire de Virologie, Hôpital de la Timone, Marseille, France*

LYNDA J.-S. YANG (36), *Department of Pharmacology and Molecular Science, The Johns Hopkins Medical School, Baltimore, Maryland 21205*

ROBERT K. YU (10), *Institute of Molecular Medicine and Genetics, Medical College of Georgia, Augusta, Georgia 30912*

ROBERT K. YU (27), *Department of Biochemistry and Molecular Biophysics, Medical College of Virginia Campus, Virginia Commonwealth University, Richmond, Virginia 23298-0614*

URI ZEHAVI (31), *Institute of Biochemistry, Food Science and Nutrition, Hebrew University of Jerusalem, Rehovot 76100, Israel*

GEORGES ZWINGELSTEIN (9), *Institute Michel Pacha, University Claude Bernard-Lyon I, La Seyne sur Mer 83500, France*

Preface

Sphingolipids are a highly diverse family of compounds constructed from more than 60 different sphingoid base backbones, numerous long-chain fatty acids, and hundreds of different headgroups, which include highly complex carbohydrates and even some covalently linked proteins. They are found in all eukaryotic organisms, as well as in some prokaryotes and viruses, primarily as components of membranes, but, also, of other lipid-rich structures such as lipoproteins and skin.

Research during the past decades has provided a conceptual framework and methodologies for elucidation of the structures, biophysical properties, biosynthesis and turnover, trafficking, and biological functions of these complex molecules as determinants of specialized membrane structures, modulators of growth factor receptors and extracellular matrix proteins, and intracellular mediators for a growing list of agonists and toxins. As this knowledge base has grown, investigators from a wide range of scientific disciplines have become "sphingolipidologists," or at least have begun to explore the role of sphingolipids in their experimental system.

This volume of *Methods in Enzymology* and its companion, Volume 311, present techniques that are useful for such investigations, including assays of enzymes of sphingolipid biosynthesis and turnover, plus their purification, cloning, expression, and characterization; approaches for the selection of genetic mutations; preparation of structurally defined sphingolipids, radiolabeled compounds, analogs, and inhibitors by chemical, enzymatic, and microbial syntheses; analysis of sphingolipid structures and biophysical properties; quantitation of sphingolipids and bioactive sphingolipid metabolites by a variety of techniques; and characterization of protein–sphingolipid interactions, especially with respect to cell regulation and signal transduction.

These methods, plus the increasing availability of sphingolipids, sphingolipid analogs, antibodies to sphingolipids, and other tools from commercial sources, should improve the ease and sophistication of research on the structures, metabolism, and functions of sphingolipids, as well as on their roles in the etiology, prevention, and treatment of disease.

ALFRED H. MERRILL, JR.
YUSUF A. HANNUN

METHODS IN ENZYMOLOGY

VOLUME I. Preparation and Assay of Enzymes
Edited by SIDNEY P. COLOWICK AND NATHAN O. KAPLAN

VOLUME II. Preparation and Assay of Enzymes
Edited by SIDNEY P. COLOWICK AND NATHAN O. KAPLAN

VOLUME III. Preparation and Assay of Substrates
Edited by SIDNEY P. COLOWICK AND NATHAN O. KAPLAN

VOLUME IV. Special Techniques for the Enzymologist
Edited by SIDNEY P. COLOWICK AND NATHAN O. KAPLAN

VOLUME V. Preparation and Assay of Enzymes
Edited by SIDNEY P. COLOWICK AND NATHAN O. KAPLAN

VOLUME VI. Preparation and Assay of Enzymes (*Continued*)
Preparation and Assay of Substrates
Special Techniques
Edited by SIDNEY P. COLOWICK AND NATHAN O. KAPLAN

VOLUME VII. Cumulative Subject Index
Edited by SIDNEY P. COLOWICK AND NATHAN O. KAPLAN

VOLUME VIII. Complex Carbohydrates
Edited by ELIZABETH F. NEUFELD AND VICTOR GINSBURG

VOLUME IX. Carbohydrate Metabolism
Edited by WILLIS A. WOOD

VOLUME X. Oxidation and Phosphorylation
Edited by RONALD W. ESTABROOK AND MAYNARD E. PULLMAN

VOLUME XI. Enzyme Structure
Edited by C. H. W. HIRS

VOLUME XII. Nucleic Acids (Parts A and B)
Edited by LAWRENCE GROSSMAN AND KIVIE MOLDAVE

VOLUME XIII. Citric Acid Cycle
Edited by J. M. LOWENSTEIN

VOLUME XIV. Lipids
Edited by J. M. LOWENSTEIN

VOLUME XV. Steroids and Terpenoids
Edited by RAYMOND B. CLAYTON

VOLUME XVI. Fast Reactions
Edited by KENNETH KUSTIN

Volume LV. Biomembranes (Part F: Bioenergetics)
Edited by Sidney Fleischer and Lester Packer

Volume LVI. Biomembranes (Part G: Bioenergetics)
Edited by Sidney Fleischer and Lester Packer

Volume LVII. Bioluminescence and Chemiluminescence
Edited by Marlene A. DeLuca

Volume LVIII. Cell Culture
Edited by William B. Jakoby and Ira Pastan

Volume LIX. Nucleic Acids and Protein Synthesis (Part G)
Edited by Kivie Moldave and Lawrence Grossman

Volume LX. Nucleic Acids and Protein Synthesis (Part H)
Edited by Kivie Moldave and Lawrence Grossman

Volume 61. Enzyme Structure (Part H)
Edited by C. H. W. Hirs and Serge N. Timasheff

Volume 62. Vitamins and Coenzymes (Part D)
Edited by Donald B. McCormick and Lemuel D. Wright

Volume 63. Enzyme Kinetics and Mechanism (Part A: Initial Rate and Inhibitor Methods)
Edited by Daniel L. Purich

Volume 64. Enzyme Kinetics and Mechanism (Part B: Isotopic Probes and Complex Enzyme Systems)
Edited by Daniel L. Purich

Volume 65. Nucleic Acids (Part I)
Edited by Lawrence Grossman and Kivie Moldave

Volume 66. Vitamins and Coenzymes (Part E)
Edited by Donald B. McCormick and Lemuel D. Wright

Volume 67. Vitamins and Coenzymes (Part F)
Edited by Donald B. McCormick and Lemuel D. Wright

Volume 68. Recombinant DNA
Edited by Ray Wu

Volume 69. Photosynthesis and Nitrogen Fixation (Part C)
Edited by Anthony San Pietro

Volume 70. Immunochemical Techniques (Part A)
Edited by Helen Van Vunakis and John J. Langone

Volume 71. Lipids (Part C)
Edited by John M. Lowenstein

Volume 72. Lipids (Part D)
Edited by John M. Lowenstein

VOLUME 108. Immunochemical Techniques (Part G: Separation and Characterization of Lymphoid Cells)
Edited by GIOVANNI DI SABATO, JOHN J. LANGONE, AND HELEN VAN VUNAKIS

VOLUME 109. Hormone Action (Part I: Peptide Hormones)
Edited by LUTZ BIRNBAUMER AND BERT W. O'MALLEY

VOLUME 110. Steroids and Isoprenoids (Part A)
Edited by JOHN H. LAW AND HANS C. RILLING

VOLUME 111. Steroids and Isoprenoids (Part B)
Edited by JOHN H. LAW AND HANS C. RILLING

VOLUME 112. Drug and Enzyme Targeting (Part A)
Edited by KENNETH J. WIDDER AND RALPH GREEN

VOLUME 113. Glutamate, Glutamine, Glutathione, and Related Compounds
Edited by ALTON MEISTER

VOLUME 114. Diffraction Methods for Biological Macromolecules (Part A)
Edited by HAROLD W. WYCKOFF, C. H. W. HIRS, AND SERGE N. TIMASHEFF

VOLUME 115. Diffraction Methods for Biological Macromolecules (Part B)
Edited by HAROLD W. WYCKOFF, C. H. W. HIRS, AND SERGE N. TIMASHEFF

VOLUME 116. Immunochemical Techniques (Part H: Effectors and Mediators of Lymphoid Cell Functions)
Edited by GIOVANNI DI SABATO, JOHN J. LANGONE, AND HELEN VAN VUNAKIS

VOLUME 117. Enzyme Structure (Part J)
Edited by C. H. W. HIRS AND SERGE N. TIMASHEFF

VOLUME 118. Plant Molecular Biology
Edited by ARTHUR WEISSBACH AND HERBERT WEISSBACH

VOLUME 119. Interferons (Part C)
Edited by SIDNEY PESTKA

VOLUME 120. Cumulative Subject Index Volumes 81–94, 96–101

VOLUME 121. Immunochemical Techniques (Part I: Hybridoma Technology and Monoclonal Antibodies)
Edited by JOHN J. LANGONE AND HELEN VAN VUNAKIS

VOLUME 122. Vitamins and Coenzymes (Part G)
Edited by FRANK CHYTIL AND DONALD B. MCCORMICK

VOLUME 123. Vitamins and Coenzymes (Part H)
Edited by FRANK CHYTIL AND DONALD B. MCCORMICK

VOLUME 124. Hormone Action (Part J: Neuroendocrine Peptides)
Edited by P. MICHAEL CONN

VOLUME 141. Cellular Regulators (Part B: Calcium and Lipids)
Edited by P. MICHAEL CONN AND ANTHONY R. MEANS

VOLUME 142. Metabolism of Aromatic Amino Acids and Amines
Edited by SEYMOUR KAUFMAN

VOLUME 143. Sulfur and Sulfur Amino Acids
Edited by WILLIAM B. JAKOBY AND OWEN GRIFFITH

VOLUME 144. Structural and Contractile Proteins (Part D: Extracellular Matrix)
Edited by LEON W. CUNNINGHAM

VOLUME 145. Structural and Contractile Proteins (Part E: Extracellular Matrix)
Edited by LEON W. CUNNINGHAM

VOLUME 146. Peptide Growth Factors (Part A)
Edited by DAVID BARNES AND DAVID A. SIRBASKU

VOLUME 147. Peptide Growth Factors (Part B)
Edited by DAVID BARNES AND DAVID A. SIRBASKU

VOLUME 148. Plant Cell Membranes
Edited by LESTER PACKER AND ROLAND DOUCE

VOLUME 149. Drug and Enzyme Targeting (Part B)
Edited by RALPH GREEN AND KENNETH J. WIDDER

VOLUME 150. Immunochemical Techniques (Part K: *In Vitro* Models of B and T Cell Functions and Lymphoid Cell Receptors)
Edited by GIOVANNI DI SABATO

VOLUME 151. Molecular Genetics of Mammalian Cells
Edited by MICHAEL M. GOTTESMAN

VOLUME 152. Guide to Molecular Cloning Techniques
Edited by SHELBY L. BERGER AND ALAN R. KIMMEL

VOLUME 153. Recombinant DNA (Part D)
Edited by RAY WU AND LAWRENCE GROSSMAN

VOLUME 154. Recombinant DNA (Part E)
Edited by RAY WU AND LAWRENCE GROSSMAN

VOLUME 155. Recombinant DNA (Part F)
Edited by RAY WU

VOLUME 156. Biomembranes (Part P: ATP-Driven Pumps and Related Transport: The Na,K-Pump)
Edited by SIDNEY FLEISCHER AND BECCA FLEISCHER

VOLUME 157. Biomembranes (Part Q: ATP-Driven Pumps and Related Transport: Calcium, Proton, and Potassium Pumps)
Edited by SIDNEY FLEISCHER AND BECCA FLEISCHER

VOLUME 158. Metalloproteins (Part A)
Edited by JAMES F. RIORDAN AND BERT L. VALLEE

VOLUME 159. Initiation and Termination of Cyclic Nucleotide Action
Edited by JACKIE D. CORBIN AND ROGER A. JOHNSON

VOLUME 160. Biomass (Part A: Cellulose and Hemicellulose)
Edited by WILLIS A. WOOD AND SCOTT T. KELLOGG

VOLUME 161. Biomass (Part B: Lignin, Pectin, and Chitin)
Edited by WILLIS A. WOOD AND SCOTT T. KELLOGG

VOLUME 162. Immunochemical Techniques (Part L: Chemotaxis and Inflammation)
Edited by GIOVANNI DI SABATO

VOLUME 163. Immunochemical Techniques (Part M: Chemotaxis and Inflammation)
Edited by GIOVANNI DI SABATO

VOLUME 164. Ribosomes
Edited by HARRY F. NOLLER, JR., AND KIVIE MOLDAVE

VOLUME 165. Microbial Toxins: Tools for Enzymology
Edited by SIDNEY HARSHMAN

VOLUME 166. Branched-Chain Amino Acids
Edited by ROBERT HARRIS AND JOHN R. SOKATCH

VOLUME 167. Cyanobacteria
Edited by LESTER PACKER AND ALEXANDER N. GLAZER

VOLUME 168. Hormone Action (Part K: Neuroendocrine Peptides)
Edited by P. MICHAEL CONN

VOLUME 169. Platelets: Receptors, Adhesion, Secretion (Part A)
Edited by JACEK HAWIGER

VOLUME 170. Nucleosomes
Edited by PAUL M. WASSARMAN AND ROGER D. KORNBERG

VOLUME 171. Biomembranes (Part R: Transport Theory: Cells and Model Membranes)
Edited by SIDNEY FLEISCHER AND BECCA FLEISCHER

VOLUME 172. Biomembranes (Part S: Transport: Membrane Isolation and Characterization)
Edited by SIDNEY FLEISCHER AND BECCA FLEISCHER

VOLUME 173. Biomembranes [Part T: Cellular and Subcellular Transport: Eukaryotic (Nonepithelial) Cells]
Edited by SIDNEY FLEISCHER AND BECCA FLEISCHER

VOLUME 174. Biomembranes [Part U: Cellular and Subcellular Transport: Eukaryotic (Nonepithelial) Cells]
Edited by SIDNEY FLEISCHER AND BECCA FLEISCHER

VOLUME 175. Cumulative Subject Index Volumes 135–139, 141–167

VOLUME 227. Metallobiochemistry (Part D: Physical and Spectroscopic Methods for Probing Metal Ion Environments in Metalloproteins)
Edited by JAMES F. RIORDAN AND BERT L. VALLEE

VOLUME 228. Aqueous Two-Phase Systems
Edited by HARRY WALTER AND GÖTE JOHANSSON

VOLUME 229. Cumulative Subject Index Volumes 195–198, 200–227

VOLUME 230. Guide to Techniques in Glycobiology
Edited by WILLIAM J. LENNARZ AND GERALD W. HART

VOLUME 231. Hemoglobins (Part B: Biochemical and Analytical Methods)
Edited by JOHANNES EVERSE, KIM D. VANDEGRIFF, AND ROBERT M. WINSLOW

VOLUME 232. Hemoglobins (Part C: Biophysical Methods)
Edited by JOHANNES EVERSE, KIM D. VANDEGRIFF, AND ROBERT M. WINSLOW

VOLUME 233. Oxygen Radicals in Biological Systems (Part C)
Edited by LESTER PACKER

VOLUME 234. Oxygen Radicals in Biological Systems (Part D)
Edited by LESTER PACKER

VOLUME 235. Bacterial Pathogenesis (Part A: Identification and Regulation of Virulence Factors)
Edited by VIRGINIA L. CLARK AND PATRIK M. BAVOIL

VOLUME 236. Bacterial Pathogenesis (Part B: Integration of Pathogenic Bacteria with Host Cells)
Edited by VIRGINIA L. CLARK AND PATRIK M. BAVOIL

VOLUME 237. Heterotrimeric G Proteins
Edited by RAVI IYENGAR

VOLUME 238. Heterotrimeric G-Protein Effectors
Edited by RAVI IYENGAR

VOLUME 239. Nuclear Magnetic Resonance (Part C)
Edited by THOMAS L. JAMES AND NORMAN J. OPPENHEIMER

VOLUME 240. Numerical Computer Methods (Part B)
Edited by MICHAEL L. JOHNSON AND LUDWIG BRAND

VOLUME 241. Retroviral Proteases
Edited by LAWRENCE C. KUO AND JULES A. SHAFER

VOLUME 242. Neoglycoconjugates (Part A)
Edited by Y. C. LEE AND REIKO T. LEE

VOLUME 243. Inorganic Microbial Sulfur Metabolism
Edited by HARRY D. PECK, JR., AND JEAN LEGALL

VOLUME 244. Proteolytic Enzymes: Serine and Cysteine Peptidases
Edited by ALAN J. BARRETT

VOLUME 320. Cumulative Subject Index Volumes 290–319 (in preparation)

VOLUME 321. Numerical Computer Methods (Part C)
Edited by MICHAEL L. JOHNSON AND LUDWIG BRAND

VOLUME 322. Apoptosis
Edited by JOHN C. REED

VOLUME 323. Energetics of Biological Macromolecules (Part C)
Edited by MICHAEL L. JOHNSON AND GARY K. ACKERS

VOLUME 324. Branched-Chain Amino Acids (Part B)
Edited by ROBERT A. HARRIS AND JOHN R. SOKATCH

VOLUME 325. Regulators and Effectors of Small GTPases (Part D: Rho Family)
Edited by W. E. BALCH, CHANNING J. DER, AND ALAN HALL

VOLUME 326. Applications of Chimeric Genes and Hybrid Proteins (Part A: Gene Expression and Protein Purification)
Edited by JEREMY THORNER, SCOTT D. EMR, AND JOHN N. ABELSON

VOLUME 327. Applications of Chimeric Genes and Hybrid Proteins (Part B: Cell Biology and Physiology)
Edited by JEREMY THORNER, SCOTT D. EMR, AND JOHN N. ABELSON

VOLUME 328. Applications of Chimeric Genes and Hybrid Proteins (Part C: Protein–Protein Interactions and Genomics) (in preparation)
Edited by JEREMY THORNER, SCOTT D. EMR, AND JOHN N. ABELSON

VOLUME 329. Regulators and Effectors of Small GTPases (Part E: GTPases Involved in Vesicular Traffic) (in preparation)
Edited by W. E. BALCH, CHANNING J. DER, AND ALAN HALL

VOLUME 330. Hyperthermophilic Enzymes (Part A) (in preparation)
Edited by MICHAEL W. W. ADAMS AND ROBERT M. KELLY

VOLUME 331. Hyperthermophilic Enzymes (Part B) (in preparation)
Edited by MICHAEL W. W. ADAMS AND ROBERT M. KELLY

VOLUME 332. Regulators and Effectors of Small GTPases (Part F: Ras Family I) (in preparation)
Edited by W. E. BALCH, CHANNING J. DER, AND ALAN HALL

VOLUME 333. Regulators and Effectors of Small GTPases (Part G: Ras Family II) (in preparation)
Edited by W. E. BALCH, CHANNING J. DER, AND ALAN HALL

VOLUME 334. Hyperthermophilic Enzymes (Part C) (in preparation)
Edited by MICHAEL W. W. ADAMS AND ROBERT M. KELLY

VOLUME 335. Flavonoids and Other Polyphenols (in preparation)
Edited by LESTER PACKER

Section I

Methods for Analyzing Sphingolipids

[1] Analysis of Sphingoid Bases and Sphingoid Base 1-Phosphates by High-Performance Liquid Chromatography

By ALFRED H. MERRILL, JR., THOMAS B. CALIGAN, ELAINE WANG, KATHERINE PETERS, and JOYCE OU

Introduction

Long-chain bases (sphingosine, sphinganine, phytosphingosine, and homologs of these compounds) and their derivatives (sphingoid base 1-phosphates, psychosines, etc.) are highly bioactive compounds that appear as intermediates of sphingolipid metabolism and cell signaling, and are elevated on disruption of sphingolipid metabolism in disease. Among the methods that are available for quantitation of sphingoid bases, reversed-phase high-performance liquid chromatography (HPLC) of the fluorescent, *o*-phthalaldehyde derivatives remains one of the most sensitive and informative since it is sensitive and able to resolve individual molecular species.[1,2] This chapter describes these procedures as well as their application for analysis of sphingoid bases[1] and sphingoid base 1-phosphates[2] as substrates and products of sphingosine kinase.

Measurement of Free Sphingoid Bases

Free sphingoid bases are relatively easy to analyze because they can be extracted in high yield using standard organic solvents and, once glycerolipids are removed by base treatment, there are relatively few contaminants that elute on reversed-phase HPLC in the vicinity of the *o*-phthalaldehyde derivatives of most naturally occurring sphingoid bases.[1] Because there is considerable interest in analyzing the amounts of sphinganine in blood, tissues, and urine from animals exposed to naturally occurring inhibitors of ceramide synthase (fumonisins), many methods have been developed for sphingoid base extraction and analysis,[3–7] most of which are modifications of the method described here.

[1] A. H. Merrill, Jr., E. Wang, R. E. Mullins, W. C. Jamison, S. Nimkar, and D. C. Liotta, *Anal. Biochem.* **171,** 373 (1988).

[2] T. B. Caligan, K. Peters, J. Ou, E. Wang, J. Saba, and A. H. Merrill, Jr., *Anal. Biochem.* **281,** 36 (2000).

[3] R. T. Riley, E. Wang, and A. H. Merrill, Jr., *J. AOAC Int.* **77,** 533 (1994).

[4] M. Castegnaro, L. Garren, I. Gaucher, and C. P. Wild, *Nat. Toxins* **4,** 284 (1996)

0076-6879/00 $30.00

Lipid Extraction

Sphingoid bases can be extracted from most mammalian samples (tissues, cells in culture, serum, etc.) as follows. Tissues are usually homogenized in 4 volumes (w/v) of ice-cold 0.05 *M* potassium phosphate buffer (pH 7.0) and 0.1 ml of the homogenate is used. Cells in culture are chilled, washed with cold buffered saline, and scraped from the dishes (10^6 cells are usually used). Blood, serum, or plasma (0.1–0.5 ml) can be used directly. The samples are placed in 13 × 100-mm screw-cap test tubes (standard borosilicate tubes with Teflon caps); 1.5 ml of $CHCl_3$: methanol (1 : 2, v/v) is added (more of this solvent can be added if phase separation occurs); an internal standard of 50–400 pmol of eicosasphinganine (C_{20}-sphinganine, which is available commercially from a number of suppliers, such as Matreya, Pleasant Gap, PA) is added; and the extract components are mixed vigorously (for example, by sonication in a bath-type sonicator). Next, 2 ml each of $CHCl_3$ and water are added, and the two phases are separated by centrifugation using a tabletop centrifuge. The upper phase is discarded, and the $CHCl_3$ phase is washed 2–3 times with water, drained through a small column (Pasteur pipette) containing anhydrous sodium sulfate (granular), and dried *in vacuo*.

The extracts are resuspended in 1 ml of 0.1 *M* KOH in methanol and $CHCl_3$ (4 : 1, v/v), incubated at 37° for 1 hr, and the sphingoid bases are extracted by adding 2 ml each of $CHCl_3$ and water and the $CHCl_3$ phase is washed with water, dried over sodium sulfate, and the solvent removed *in vacuo*.

Formation and Analyses of o-Phthalaldehyde Derivatives of Sphingoid Bases

The *o*-phthalaldehyde (OPA) derivatives are prepared by a dissolving the base-treated extracts in 50 μl of methanol, and adding (with rapid mixing) 50 μl of OPA reagent. The OPA reagent is prepared by mixing (a) 99 ml of 3% (w/v) boric acid in water (pH adjusted to 10.5 with KOH), (b) 1 ml of ethanol containing 50 mg of OPA, and (c) 50 μl of 2-mercaptoeth-

[5] M. Solfrizzo, G. Avantaggiato, and A. Visconti, *J. Chromatogr. B Biomed. Sci. Appl.* **692,** 87 (1997).

[6] M. Castegnaro, L. Garren, D. Galendo, W. C. A. Gelderblom, P. Chelule, M. F. Dutton, and C. P. Wild, *J. Chromatogr. B Biomed. Sci. Appl.* **720,** 15 (1998).

[7] G. S. Shephard and L. Van der Westhuizen, *J. Chromatogr. B Biomed. Sci. Appl.* **710,** 219 (1998).

anol. The OPA reagent is stable for approximately 1 week when stored in the refrigerator.

After incubation for approximately 15 min at room temperature, 250–500 μl of the mobile phase solvent for HPLC (methanol : 5 m*M* potassium phosphate buffer, pH 7.0, usually in a ratio of 90 : 10, v/v) is added. After a few minutes, the samples are centrifuged briefly in a microcentrifuge to clarify, and aliquots are analyzed by HPLC. Until injection, the OPA derivatives are kept at $\leq 8°$; the derivatives are usually stable for up to 48 hr when kept cold.

HPLC analyses are conducted using a C_{18} reversed-phase column (Waters, Milford, MA, Radial Pak C_{18} column: 5 μm, type 8NVC18) with a small C_{18} guard column (2.5 cm; Universal Scientific, Atlanta, GA), and an isocratic solvent system (methanol : 5 m*M* potassium phosphate, pH 7.0, at 90 : 10, v/v) at a flow rate of 2 ml/min. The fluorescence of the OPA derivatives is measured with an excitation wavelength of 340 nm and an emission wavelength of 455 nm (or a 418-nm cutoff filter).

Under these conditions, the retention times of sphingosine, sphinganine, and 4-hydroxysphinganine (phytosphinganine) are approximately 11, 15, and 7 min, respectively. The C_{20}-sphinganine internal standard elutes at approximately 20 min. The sensitivity of the method will depend on the fluorescence detector; with a Shimadzu (Kyoto, Japan) RF-535 spectrofluorometer, we have found that fluorescence is proportional to the amount of sphingosine from at least 10–1000 pmol of injected sphingoid base.

Comments

Some samples also contain fluorescent contaminants that may be mistaken for sphingoid bases, but this can be ascertained by examining the extract on HPLC without OPA derivatization.

Sphingoid bases of 18-carbon atoms predominate in most mammalian samples; however, other homologs (including 20-carbon species) are found in milk sphingolipids and brain gangliosides, for example, and are common in other organisms, such as yeast, fungi, and plants. Hence, the HPLC profile for a given sample should be examined without adding the C_{20}-sphinganine internal standard to ensure that it does not interfere with the analysis. When other sphingoid bases are present, the HPLC solvent can be varied (including the use of a gradient system) to achieve the desired separation. Other alkylamines can also be used as internal standards, but the fluorescence yield versus that of sphingoid bases must be determined.

This procedure is also able to analyze psychosine and presumably other lysosphingolipids, although these will probably not be extracted efficiently by this procedure. For more polar sphingoid bases and derivatives, the solid-phase extraction described below can be substituted.

Other Modifications to This Method

A number of simplified versions of this method have been published.[3–8] One that we have used omits the first extraction step by first mixing the sample with 1 ml of $CHCl_3$:methanol (1:4, v/v) containing 0.1 *M* KOH, incubating it for 1 hr at 37°, then continuing the two-phase extraction described above to obtain a dry extract for OPA derivatization and HPLC. In most cases, this gives a slightly higher yield of the sphingoid bases.

Other methods for analyzing sphingoid bases have been summarized in Ref. 1. There is also a method that uses sphingosine kinase to quantify sphingosine.[8]

Measurement of Sphingoid Base 1-Phosphates

The analysis of sphingoid base 1-phosphates presents two additional challenges: their high water solubility results in partial extraction by organic solvents, and the OPA derivatives of sphingoid base 1-phosphates are more unstable. These problems were solved by the following modifications.[2]

Solid-Phase Extraction of Sphingoid Base 1-Phosphates

Each sample is mixed with an equal volume of methanol, then gravity loaded onto a small RP-18 column. The column is prepared by suspending LiChroprep RP-18 (40- to 63-μm particle size, EM Science, Gibbston, NJ) in methanol:H_2O (1:1, v/v), allowing the matrix to settle, then aspirating off the fines. The column (approximately 5 mm × 4 cm) is prepared by placing a small amount of coarse glass wool in the neck of a Pasteur pipette, adding the RP-18 slurry, then washing the resin (the RP-18 is sometimes washed first with hexane to remove fluorescent contaminants) with at least 3 ml of methanol:H_2O (1:1, v/v) before addition of the sample. If necessary, precipitated material in the sample–methanol mixture can be removed by centrifugation in a tabletop centrifuge.

As soon as the last portion of the sample is loaded, the column is washed with 2 ml methanol:H_2O (1:1, v/v) followed by 4 ml of methanol:H_2O (3:1, v/v) containing 0.1% acetic acid. Sphingoid base 1-phosphates are next eluted with 4 ml of methanol:H_2O (9:1) containing 10 m*M* potassium phosphate buffer (pH 7.2), and the solvent is removed under vacuum,

[8] A. Olivera, J. Rosenthal, and S. Spiegel, *Anal. Biochem.* **223,** 306 (1994).

with care to analyze the samples as soon as possible after drying to minimize decomposition. Using this procedure, recoveries are typically ~85%.

Preparation and Analysis of o-Phthalaldehyde Derivatives

The OPA derivatives are prepared by a minor modification of the method described above for free sphingosine. This modification is the addition of 100 μl of 10 m*M* EDTA to the redissolved RP-18 eluate, which increase the stability of the OPA derivative. Next, 100 μl of 3% (w/v) boric acid in water (pH adjusted to 10.5 with KOH) is added, followed by 50 μl of the OPA reagent (prepared as described above). After incubation for 20 min at room temperature and centrifugation to remove any precipitated material, the derivatives are analyzed by HPLC. Until injection, the derivatives are stored at 4° to 8° to increase their stability (losses begin to be significant after approximately 10 hr).

The HPLC analyses are conducted using a C_{18} reversed-phase column (Radial Pak C_{18} column, 5 μm type 8NVC18 from Waters) with a small guard column. The solvent system is methanol : 10 m*M* potassium phosphate, pH 7.2 : 1 *M* tetrabutylammonium dihydrogen phosphate (TBAP in water) in the ratio 83 : 16 : 1 (v/v/v), and a flow rate of 2 ml/min. The OPA derivatives are detected using an excitation wavelength of 340 nm and an emission wavelength of 455 nm. Under these conditions, the retention times for the OPA derivatives of sphingosine 1-phosphate and sphinganine 1-phosphate are 13.8 and 21.5 min, respectively, and 8.6 min for 4-hydroxysphinganine 1-phosphate. Quantitation is achieved by comparison of the fluorescence versus that of an internal standard (such as sphinganine 1-phosphate) or by addition of a known amount of sphingosine 1-phosphate to half of the extract and comparison of the spiked and unspiked results. Using sphingosine 1-phosphate, the fluorescence has been shown to be proportional to the amount analyzed over at least 2 orders of magnitude (5–500 pmol).

Comments

Use gravity flow for the RP-18 column to optimize recovery.

There can be overlap between some sphingoid bases and 1-phosphates. For example, sphingosine 1-phosphate and 4-hydroxysphinganine are separated by only 0.1 min under these conditions. When samples contain both compounds, they can be resolved by changing the mobile phase to pH 5.5, which slows the elution of the 1-phosphates by ~2 min. This pH shift can also be used to verify the identities of peaks in a complicated chromatogram. At pH 5.5, the retention times for 4-hydroxysphinganine, sphingosine 1-

phosphate, sphingosine, sphinganine 1-phosphate, and sphinganine are 13.7, 16.3, 22.6, 26, and 37.5 min, respectively. Changing the pH of the mobile phase (5.5 or 7.2) does not appear to affect the stability of the OPA derivatives.

This method has been useful for the analysis of sphingoid base 1-phosphates from serum, plasma, and platelets as well as for sphingosine kinase assays (see below), but many tissues and mammalian cell lines have other amines that complicate the HPLC profile. In those cases, other methods that can be used include the dephosphorylation of sphingoid base 1-phosphates and analysis of the sphingoid bases as described above,[9] or by other methods.[10] It is also possible to convert sphingoid base 1-phosphates to radiolabeled derivatives (using radiolabeled acetic anhydryde),[11] although this is a more time-consuming procedure and does not resolve analogs that have similar structures. The most definitive method is mass spectrometry[12] (see chapter by Sullards in this volume[12a]).

Application of This Method for Assay of Sphingosine Kinase

This assay method has been used to assay sphingosine kinase from a number of sources, including platelets and a number of cell lines. For these analyses, the cells are collected by centrifugation, the pellet resuspended in 0.1 *M* potassium phosphate buffer, pH 7.4, 1 m*M* 2-mercaptoethanol, 1 m*M* EDTA, 1 m*M* sodium orthovanadate, 1 m*M* phenylmethylsulfonyl fluoride (PMSF), 15 m*M* sodium fluoride, and 10 μg/ml leupeptin and aprotinin, then the cells are lysed by sonication on ice in 5 bursts of 10 sec each, with at least 10-sec intervals between bursts. The cytosol is recovered by ultracentrifugation [for example, using a Beckman (Palo Alto, CA) Optima TL ultracentrifuge with a TLA 100.3 rotor at 50,000 rpm at 4° for 15 min], and stored frozen at −80° until use (activity appears to be stable for one to two months).

The assay mixture for sphingosine kinase (based on Refs. 13 and 14) contains 20 μM sphingosine and 55 m*M* octyl-β-D-glucopyranoside (added from a 10-fold concentrated stock; to prepare this stock, 66.7 μl of a 3 m*M*

[9] E. R. Smith and A. H. Merrill, Jr., *J. Biol. Chem.* **270,** 18749 (1995).

[10] L. C. Edsall and S. Spiegel, *Anal. Biochem.* **272,** 80 (1999).

[11] Y Yatomi, F. Ruan, H. Ohta, R. J. Welch, S. Hakomori, and Y. Igarashi, *Anal. Biochem.* **230,** 315 (1995).

[12] P. P. Van Veldhoven, P. De Ceuster, R. Rozenberg, G. P. Mannaerts, and E. De Hoffmann, *FEBS Lett.* **350,** 91 (1994).

[12a] M. C. Sullards, *Methods Enzymol.* **312,** [5], (2000) (this volume).

[13] A. Olivera, J. Rosenthal, and S. Spiegel, *Anal. Biochem.* **223,** 306 (1994).

[14] B. M. Buehrer and R. M. Bell, *J. Biol. Chem.* **267,** 3154 (1992).

solution of sphingosine in methanol is added to a borosilicate tube, dried under a stream of nitrogen, 1 ml of 550 m*M* octyl-β-D-glucopyranoside is added, and the mixture is sonicated), 0.1 *M* potassium phosphate buffer (pH 7.2), 2 m*M* $MgCl_2$, 1 m*M* EDTA, 1 m*M* 2-mercaptoethanol, 0.5 m*M* 4-deoxypyridoxine, 40 μ*M* β-glycerophosphate, 20% glycerol, and 50 μg of protein, with the reaction being initiated by addition of 10 μl of 10 m*M* adenosine triphosphate for a total volume of 0.1 ml. The assay tubes are incubated in a shaking water bath at 37° for 30 min, the reaction is stopped by addition of 750 μl of methanol, and then brought to 1 : 1 methanol : H_2O (v/v) and loaded onto RP-18 columns and analyzed by HPLC as described in the preceding section.

[2] Enzymatic Method for Measurement of Sphingosine 1-Phosphate

By LISA EDSALL, LEWIS VANN, SHELDON MILSTIEN, and SARAH SPIEGEL

Introduction

Sphingosine 1-phosphate (SPP) is a bioactive sphingolipid metabolite that functions both as an intracellular second messenger and extracellularly as a ligand for specific G-protein-coupled receptors.[1] These dual roles make SPP a unique and pivotal lipid in several signal transduction pathways.

Endogenous SPP biosynthesis is mediated by sphingosine kinase.[2] SPP levels are also regulated via metabolism by an endoplasmic reticulum-associated pyridoxal phosphate-dependent lyase, and also by specific phosphatases, which convert it back to sphingosine.[3–5] Various physiologic stimuli activate sphingosine kinase and concomitantly increase cellular levels of SPP.[1] This increased intracellular SPP in turn has been associated with regulation of a wide variety of cellular processes, including calcium mobilization, cell proliferation, and survival (reviewed in Ref. 6). Moreover,

[1] S. Spiegel, O. Cuvillier, L. C. Edsall, T. Kohama, R. Menzeleev, Z. Olah, A. Olivera, G. Pirianov, D. M. Thomas, Z. Tu, J. R. Van Brocklyn, and F. Wang, *Ann. N.Y. Acad. Sci.* **845,** 11 (1998).

[2] A. Olivera, J. Rosenthal, and S. Spiegel, *J. Cell. Biochem.* **60,** 529 (1996).

[3] P. P. Van Veldhoven and G. P. Mannaerts, *J. Biol. Chem.* **266,** 12502 (1991).

[4] P. P. Van Veldhoven and G. P. Mannaerts, *Biochem. J.* **299,** 597 (1994).

[5] S. Mandala, R. Thornton, Z. Tu, M. Kurtz, J. Nickels, J. Broach, R. Menzeleev, and S. Spiegel, *Proc. Natl. Acad. Sci. U.S.A.* **95,** 150 (1998).

[6] S. Spiegel, D. Foster, and R. N. Kolesnick, *Curr. Opin. Cell Biol.* **8,** 159 (1996).

0076-6879/00 $30.00

competitive inhibitors of sphingosine kinase block formation of SPP and inhibit these biologic effects,[7–15] thus emphasizing the importance of sphingosine kinase.

SPP binds with a high degree of specificity and affinity to a family of orphan G-protein-coupled cell surface receptors belonging to the endothelium differentiation gene (*edg*) family.[16–18] The cellular events initiated through SPP binding to EDG-1, EDG-3, and EDG-5 are distinct from those that occur when SPP acts as an intracellular second messenger[17] and result in platelet activation,[19] inhibition of melanoma cell motility,[20] activation of G_i-protein-gated inward rectifying K^+ channels in atrial myocytes,[21] cytoskeletal changes in N1E-115 neurons,[22] and induction of neurite retraction and soma rounding in differentiated PC12 cells.[18]

As the biologic scope of SPP has broadened, the usefulness of a rapid, routine, and accurate method to quantitate both intra- and extracellular levels of SPP has become apparent. Previously, SPP mass levels have been determined by mass spectrometry[23] and liquid chromatography/ion spray

[7] A. Olivera and S. Spiegel, *Nature* **365,** 557 (1993).
[8] C. S. Rani, A. Berger, J. Wu, T. W. Sturgill, D. Beitner-Johnson, D. LeRoith, L. Varticovski, and S. Spiegel, *J. Biol. Chem.* **272,** 10777 (1997).
[9] L. C. Edsall, G. G. Pirianov, and S. Spiegel, *J. Neurosci.* **17,** 6952 (1997).
[10] B. Kleuser, O. Cuvillier, and S. Spiegel, *Cancer Res.* **58,** 1817 (1998).
[11] O. Cuvillier, G. Pirianov, B. Kleuser, P. G. Vanek, O. A. Coso, S. Gutkind, and S. Spiegel, *Nature* **381,** 800 (1996).
[12] M. Machwate, S. B. Rodan, G. A. Rodan, and S. I. Harada, *Mol. Pharmacol.* **54,** 70 (1998).
[13] O. Choi, J.-H. Kim, and J.-P. Kinet, *Nature* **380,** 634 (1996).
[14] A. Melendez, R. A. Floto, D. J. Gillooly, M. M. Harnett, and J. M. Allen, *J. Biol. Chem.* **273,** 9393 (1998).
[15] D. Meyer zu Heringdorf, H. Lass, R. Alemany, K. T. Laser, E. Neumann, C. Zhang, M. Schmidt, U. Rauen, K. H. Jakobs, and C. J. van Koppen, *EMBO J.* **17,** 2830 (1998).
[16] M.-J. Lee, J. R. Van Brocklyn, S. Thangada, C. H. Liu, A. R. Hand, R. Menzeleev, S. Spiegel, and T. Hla, *Science* **279,** 1552 (1998).
[17] J. R. Van Brocklyn, M. J. Lee, R. Menzeleev, A. Olivera, L. Edsall, O. Cuvillier, D. M. Thomas, P. J. P. Coopman, S. Thangada, T. Hla, and S. Spiegel, *J. Cell Biol.* **142,** 229 (1998).
[18] J. R. Van Brocklyn, Z. Tu, L. Edsall, R. R. Schmidt, and S. Spiegel, *J. Biol. Chem.* **274,** 4626 (1999).
[19] Y. Yatomi, S. Yamamura, F. Ruan, and Y. Igarashi, *J. Biol. Chem.* **272,** 5291 (1997).
[20] S. Yamamura, Y. Yatomi, F. Ruan, E. A. Sweeney, S. Hakomori, and Y. Igarashi, *Biochemistry* **36,** 10751 (1997).
[21] C. J. van Koppen, D. Meyer zu Heringdorf, K. T. Laser, C. Zhang, K. H. Jakobs, M. Bünnemann, and L. Pott, *J. Biol. Chem.* **271,** 2082 (1996).
[22] F. R. Postma, K. Jalink, T. Hengeveld, and W. H. Moolenaar, *EMBO J.* **15,** 2388 (1996).
[23] P. P. Van Veldhoven, P. De Ceuster, R. Rozenberg, G. P. Mannaerts, and E. de Hoffmann, *FEBS Lett.* **350,** 91 (1994).

ionization tandem mass spectrometry.[24] Measurement of SPP by selective extraction and derivatization with [^{3}H]acetic anhydride is another widely used method.[25] However, these methods require highly specialized instrumentation and involve complicated procedures. Utilizing partially purified[26] or recombinant sphingosine kinase,[27] we have developed a simple, specific, and sensitive enzymatic assay that can be used to measure SPP in cells, tissues, and serum.

Principle

The determination of SPP is based on selective alkaline solvent extraction to completely separate SPP from sphingosine, dephosphorylation of the extracted SPP to sphingosine with alkaline phosphatase, and rephosphorylation by recombinant sphingosine kinase in the presence of [γ-^{32}P]ATP to yield [^{32}P]SPP, which is then separated by thin-layer chromatography (TLC) and quantitated.

Materials and Methods

Sphingosine Kinase

Sphingosine kinase of sufficient specific activity can be conveniently isolated from several sources. Purification of rat kidney sphingosine kinase is described in detail elsewhere.[26] Another convenient source is cytosolic extracts from Swiss 3T3 fibroblasts.[28] We have cloned and expressed high levels of murine sphingosine kinase in HEK293 cells. This preparation is described in detail below.

Preparation of Recombinant Sphingosine Kinase

Sphingosine kinase (SPHK1a; GenBank accession number AF068748) is subcloned into a pcDNA vector (Invitrogen, San Diego, CA) by PCR using a 5′ primer with a *Bam*HI restriction site (5′-GAGGGATCCATG-GAACCAGAATGCCCTCGAGGA-3′), and as the 3′ primer, the last

[24] N. Mano, Y. Oda, K. Yamada, N. Asakawa, and K. Katayama, *Anal. Biochem.* **244,** 291 (1997).

[25] T. Yatomi, F. Ruan, H. Ohta, R. J. Welch, S. Hakomori, and Y. Igarashi, *Anal. Biochem.* **230,** 315 (1995).

[26] A. Olivera, T. Kohama, Z. Tu, S. Milstien, and S. Spiegel, *J. Biol. Chem.* **273,** 12576 (1998).

[27] T. Kohama, A. Olivera, L. Edsall, M. M. Nagiec, R. Dickson, and S. Spiegel, *J. Biol. Chem.* **273,** 23722 (1998).

[28] A. Olivera, J. Rosenthal, and S. Spiegel, *Anal. Biochem.* **223,** 306 (1994).

21 nucleotides of the SPHK1a sequence with an *Eco*RI overhang (5′-GAGGAATTCTTATGGTTCTTCTGGAGGTGG-3′).[27] Human embryonic kidney 293 cells (HEK293, ATCC CRL) are plated on polylysine-coated 100-mm dishes (5×10^6 cells per plate). After 24 hr, cells are transfected with 4 μg of vector and 20 μl of Lipofectamine PLUS reagent and 30 μl of Lipofectamine reagent (Life Technologies Inc., Gaithersburg, MD) according to the manufacturer's directions. Cells are harvested after 48 hr and lysed by freeze-thawing in buffer A [20 m*M* Tris (pH 7.4), 20% glycerol, 1 m*M* 2-mercaptoethanol, 1 m*M* EDTA, 1 m*M* sodium orthovanadate, 40 m*M* β-glycerophosphate, 15 m*M* NaF, 10 μg/ml leupeptin, aprotinin, and soybean trypsin inhibitor, 1 m*M* phenylmethylsulfonyl fluoride (PMSF), and 0.5 m*M* 4-deoxypyridoxine], yielding a cytosolic extract with sphingosine kinase activity of approximately 0.023 μmol/min/mg protein.

Extraction of Lipids

Five to 10 million cells, either grown in 100-mm culture dishes or in suspension, are used for the determination of cellular SPP levels. Cells are washed with phosphate-buffered saline (PBS) and harvested in 1 ml of methanol containing 2.5 μl concentrated HCl. Lipids are extracted by addition of 2 ml chloroform/1 *M* NaCl (1:1) plus 100 μl 3 *N* NaOH and phases separated by centrifugation. SPP (90–95%) partitions into the aqueous upper phase at alkaline pH,[29] with the added advantage of excluding the majority of cellular phospholipids and sphingolipids, including sphingosine (>99%), which remain in the organic phase. The basic aqueous phase is transferred to a siliconized glass tube, the organic phases reextracted with 1 ml methanol/1 *M* NaCl (1:1) plus 50 μl 3 *N* NaOH, and the aqueous fractions combined. Various amounts of standard SPP (1–200 pmol) are carried through the procedure to correct for losses and to generate a standard curve.

Extractions of lipids from serum and tissues are performed similarly, after adjusting the sample volumes to 1 ml with 1 *M* NaCl. Lipids are then extracted with 2 ml of chloroform:methanol (1:1) plus 100 μl 3 *N* NaOH and phases separated. Samples are reextracted as described above. Tissue extracts are prepared by homogenization in 25 m*M* HCl/1 *M* NaCl.

Measurement of SPP

To the aqueous fraction of lipid extracts (3 ml total), 450 μl of buffer B (200 m*M* Tris-HCl, pH 7.4, 75 m*M* $MgCl_2$ in 2 *M* glycine, pH 9.0) are

[29] H. Zhang, N. N. Desai, A. Olivera, T. Seki, G. Brooker, and S. Spiegel, *J. Cell Biol.* **114,** 155 (1991).

added. Samples are vortexed and 50 units of alkaline phosphatase (bovine intestinal mucosa, Type VII, Sigma, St. Louis, MO) are added and samples incubated at 37° for 30 min. Then 50 μl concentrated HCl are added and lipids are extracted twice with 1.5 ml $CHCl_3$. The $CHCl_3$ layer is transferred to siliconized glass tubes, evaporated under a nitrogen stream in a fume hood, and the residue resuspended in 175 μl of buffer A containing 0.25% Triton X-100. After vortexing and brief sonication, recombinant or partially purified sphingosine kinase (10 μg) is added and sphingosine kinase reactions started by addition of 10 μl of a solution containing [γ^{32}-P]ATP (10 μCi), 20 m*M* ATP, and 200 m*M* $MgCl_2$ as described.[30] After 30 min at 37°, 50 μl 2 *N* HCl and 0.5 ml $CHCl_3$/methanol/concentrated HCl (100 : 200 : 1) are added. Samples are mixed by vortexing, and phases separated by addition of 250 μl $CHCl_3$ and 250 μl 1 *M* NaCl. After centrifugation, an aliquot of the organic phase is spotted on silica gel G60 TLC plates and separated using $CHCl_3$/methanol/acetone/acetic acid/water (10 : 5 : 3 : 2 : 1, v/v) as solvent. [^{32}P]SPP is visualized and quantified with a phosphorimager or, alternatively, the radioactive spots corresponding to authentic SPP are scraped from the plates and counted in a scintillation counter.[27] Mass level amounts of SPP are calculated from SPP standards.

Measurement of Phospholipids

To control for differences in recoveries, levels of SPP in cell and tissue extracts are normalized to total cellular phospholipids in the organic phase of the initial extract. Phospholipids in lipid extracts are quantified as previously described.[9] Briefly, to an aliquot of cellular lipid extract (usually 1% of the initial organic extract), 40 μl of a mixture of 10 *N* H_2SO_4/70% perchloric acid (1 : 3, v/v) is added, and samples are heated for 30 min at 210°. After cooling, 75 μl of water and 400 μl of a mixture of 1 part 4.2% ammonium molybdate in 4 *N* HCl and 3 parts 0.045% malachite green (Sigma) are added. Samples are incubated at 37° for 15 min, and absorbances measured at 660 nm. Standards ranging from 1–10 nmol of Na_2HPO_4 are used to generate a standard curve, together with a blank for background subtraction.

Discussion

Quantitation of SPP

The formation of [^{32}P]SPP is directly proportional to the amount of SPP standard added using partially purified rat kidney sphingosine kinase

[30] L. C. Edsall and S. Spiegel, *Anal. Biochem.* **272,** 80 (1999).

as a source of enzyme (Fig. 1A). Linear standard curves are also generated using recombinant sphingosine kinase or Swiss 3T3 fibroblast extracts (data not shown). The ^{32}P-labeled product prepared by the method described here has the same mobility on TLC as standard SPP (R_f of 0.25 ± 0.01). Recovery of SPP is linear over a wide concentration range, up to at least 5 nmol of SPP (Fig. 1A and data not shown). Furthermore, using recombinant sphingosine kinase, amounts of SPP as low as 250 fmol can readily be measured,[30] supporting the validity of this method for measurement of low levels of SPP in biologic samples.

To verify that the presence of sphingosine does not interfere with SPP measurements, equal molar amounts of standard SPP and sphingosine have

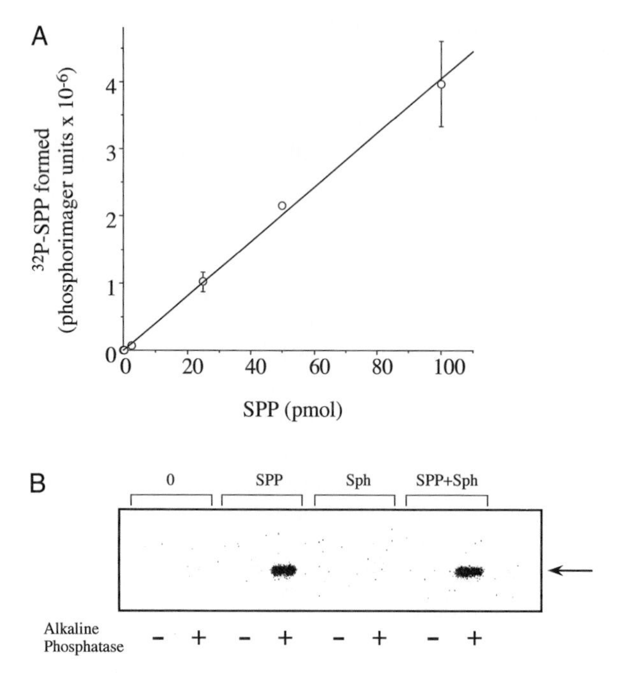

Fig. 1. (A) Standard curve for analysis of SPP. Different amounts of SPP standards are extracted as described in the Materials and Methods section. SPP is determined using partially purified sphingosine kinase from rat kidney as a source of kinase activity[26] and radioactivity in spots corresponding to [^{32}P]SPP quantified with a phosphorimager. Data are means ±SD of duplicate values. (B) Autoradiogram of a TLC demonstrating lack of effect of added sphingosine on SPP determination. Samples containing either 200 pmol SPP, 200 pmol sphingosine (Sph), or a mixture of equal amounts are extracted as described. Samples are treated without (−) or with (+) alkaline phosphatase (50 units) followed by phosphorylation with [γ-^{32}P]ATP catalyzed by recombinant sphingosine kinase and TLC analysis. Arrow indicates the location of SPP visualized with molybdenum blue spray (R_f = 0.25 ± 0.01).

TABLE I
LEVELS OF SPHINGOSINE 1-PHOSPHATE IN RAT TISSUES AND CELLS[a]

Source	SPP (pmol/nmol phospholipid)
Tissues	
Brain	0.394 ± 0.100
Liver	0.103 ± 0.012
Spleen	0.462 ± 0.070
Kidney	0.065 ± 0.006
Cells	
HL60	0.015 ± 0.026
NIH 3T3	0.047 ± 0.009
Swiss 3T3	0.038 ± 0.002
C6 glioma	0.053 ± 0.010
PC12	0.013 ± 0.003

[a] SPP and phospholipids were measured in various rat tissues and cell types as described. The data are the means ±SD of triplicate determinations.

been mixed and extracted as described in Materials and Methods, and SPP measured with and without alkaline phosphatase treatment. No SPP is detected in samples incubated without alkaline phosphatase digestion, indicating that carryover of sphingosine during lipid extraction is negligible (Fig. 1B). Moreover, the presence of sphingosine in the lipid extracts does not affect the determination of SPP (Fig. 1B), since the standard curve is the same in the absence or presence of sphingosine. There is also no carryover of sphingosine into alkaline lipid extracts from horse and fetal bovine serum, which contain significant endogenous levels of both SPP and sphingosine.[30–32] Other phospholipids in extracts prepared from various types of cells also do not significantly affect the recovery of [^{32}P]SPP.

SPP in Rat Tissues and Cultured Cells

Levels of SPP in various rat tissues measured by this enzymatic method are shown in Table I. High levels are found in brain and spleen and somewhat lower levels in kidney and liver. These values are in excellent

[31] Y. Yatomi, Y. Igarashi, L. Yang, N. Hisano, R. Qi, N. Asazuma, K. Satoh, Y. Ozaki, and S. Kume, *J. Biochem.* **121,** 969 (1997).

[32] M. Castegnaro, L. Garren, D. Galendo, W. C. Gelderblom, P. Chelule, M. F. Dutton, and C. P. Wild, *J. Chromatogr. Biomed. Sci. Appl.* **720,** 15 (1998).

agreement with previously reported SPP mass levels determined by other methods.[33,34] Although sphingosine levels have been determined in a number of cell types,[28,35–37] there are only a few reports of cellular SPP measurements. The cellular levels of SPP measured by the recombinant sphingosine kinase method (Table I) are almost identical to values obtained by mass spectrometry methods.[24,36] However, it should be noted that the levels of SPP determined in HL60 and PC12 cells are only 1/10 of those previously reported by Yatomi *et al.* using the acetic anhydride method.[9–11,25,38] It is possible that this difference results from a contaminating lipid that can be acetylated and has the same mobility on TLC as phosphorylated C_2-ceramide. Because the SPP assay described here utilizes sphingosine kinase that selectively phosphorylates only D-*erythro*-sphingosine and D-*erythro*-sphinganine, it should represent true endogenous levels of phosphorylated long-chain sphingoid bases.

Acknowledgments

This work was supported by research grants GM43880 and CA61774 from the National Institutes of Health.

[33] T. Kobayashi, K. Mitsuo, and I. Goto, *Eur. J. Biochem.* **172,** 747 (1988).
[34] A. H. Merrill, E. Wang, R. E. Mullins, W. C. Jamison, S. Nimkar, and D. C. Liotta, *Anal. Biochem.* **171,** 373 (1988).
[35] E. Wilson, E. Wang, R. E. Mullins, D. J. Uhlinger, D. C. Liotta, J. D. Lambeth, and A. H. Merrill, *J. Biol. Chem.* **263,** 9304 (1988).
[36] P. P. Van Veldhoven, W. R. Bishop, and R. M. Bell, *Anal. Biochem.* **183,** 177 (1989).
[37] H. Ohta, F. Ruan, S. Hakomori, and Y. Igarashi, *Anal. Biochem.* **222,** 489 (1994).
[38] K. Sato, H. Tomura, Y. Igarashi, M. Ui, and F. Okajima, *Biochem. Biophys. Res. Commun.* **240,** 329 (1997).

[3] Ceramide Mass Analysis by Normal-Phase High-Performance Liquid Chromatography

By ANNEMIEK D. TEPPER and WIM J. VAN BLITTERSWIJK

Introduction

Ceramide (Cer) and its generation from sphingomyelin by sphingomyelinase has received a lot of attention during the past few years, mainly because its formation in cells is associated with stress responses and

0076-6879/00 $30.00

apoptosis induction (see Ref. 1 for a review). Assays employed to quantitate cellular Cer levels are often indirect, depending on enzyme activities (such as via the conversion to ceramide phosphate by *Escherichia coli* diacylglycerol kinase) or based on metabolic labeling using radioactive precursors (e.g., serine or palmitate), which do not provide real mass data. In this chapter, we describe a nonenzymatic, nonradioactive mass assay to quantitate Cer by normal-phase high-performance liquid chromatography (HPLC) after its conversion to nonpolar, chromophoric benzoate derivatives. The assay is essentially based on published methods by Blank *et al.*[2] for the analysis of diradylglycerols as benzoate derivatives and Previati *et al.*[3] for analysis of fluorescent diradylglycerol and ceramide derivatives. We have adapted these methods to measure the accumulation of Cer during CD95-(Apo-1/Fas)-induced apoptosis in Jurkat cells.[4]

Lipid Extraction

Cell samples (approximately 5×10^6 cells or one 10-cm dish of confluent cells) are collected, washed twice with phosphate-buffered saline (PBS), and total lipids are extracted according to standard procedures.[5] Briefly, for each milliliter of cell suspension, 3.75 ml of chloroform/methanol [1:2 (v/v) containing 0.001% butylated hydroxytoluene (BHT) as antioxidant] is added. After vigorous vortexing, the mixture is allowed to stand at room temperature for at least 30 min to permit complete extraction of the lipids. The extract is then partitioned by the addition of 1.25 ml of chloroform and 1.25 ml of 1 *M* NaCl. After centrifugation (725*g* for 2 min at room temperature), the lower (organic) phase, containing most of the cellular polar and neutral lipids, is transferred to a clean 12-ml glass tube with a Pasteur pipette. Care should be taken to leave behind the protein that remains in the interphase because this will interfere with the derivatization. The upper methanol/water phase is washed with 2 ml of chloroform and the lower phases are pooled and dried under nitrogen.

Benzoylation

Total lipid extracts are subjected to a derivatization reaction using benzoic anhydride in toluene as a solvent with 4-(dimethylamino)pyridine

[1] Y. Hannun, *Science* **274,** 1855 (1996).
[2] M. L. Blank, E. A. Cress, and F. Snyder, *J. Chromatogr.* **392,** 421 (1987).
[3] M. Previati, L. Bertolaso, M. Tramarin, V. Bertagnolo, and S. Capitani, *Anal. Biochem.* **233,** 108 (1996).
[4] A. D. Tepper, J. G. R. Boesen-de Cock, E. de Vries, J. Borst, and W. J. van Blitterswijk, *J. Biol. Chem.* **272,** 24308 (1997).
[5] E. G. Bligh and W. J. Dyer, *J. Biochem. Physiol.* **37,** 911 (1959).

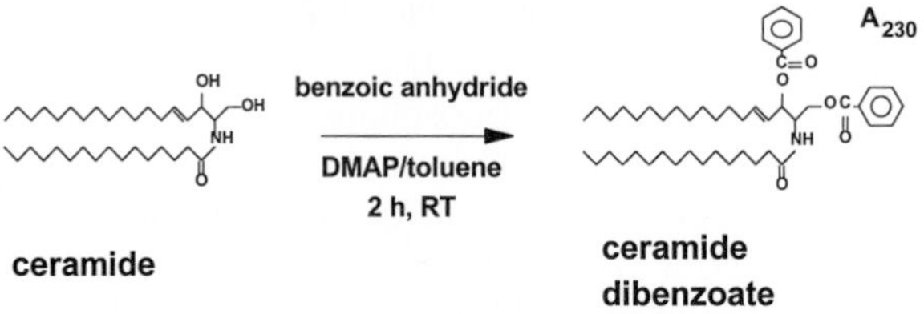

FIG. 1. Scheme for the derivatization of ceramide with benzoic anhydride at room temperature (RT) to form ceramide dibenzoate, which is readily detectable by UV absorption (230 nm).

(DMAP) as a catalyst, in which any free hydroxyl group is quantitatively converted into a chromophoric, nonpolar benzoyl ester (Fig. 1). Under these conditions, only O- but not N-benzoylation (at the amide nitrogen) occurs, in contrast to benzoylation using benzoyl chloride in pyridine.[6]

The derivatization reagent should be prepared fresh each time before use. To the dried lipid extract, 0.25 ml of toluene containing 8 mg of benzoic anhydride and 3 mg of 4-(dimethylamino)pyridine is added. After vortexing, the mixture is incubated at room temperature for 2 hr. To avoid evaporation of the solvent, the tubes should be closed with screw caps or covered with glass marbles. After cooling on ice, 2 ml of 25–28% NH_4OH is added dropwise to neutralize the mixture and stop the reaction. The samples are then warmed to room temperature and 2 ml of *n*-hexane is added. After vigorous vortexing and centrifugation (725*g* for 2 min), the upper *n*-hexane/toluene layer containing the benzoate derivatives is transferred to a clean glass tube. If the two phases are turbid and/or insoluble material remains in the interphase, more NH_4OH should be added. Also, it may help to warm the tubes in a 37° water bath. The lower phase is extracted two more times with 2 ml of *n*-hexane and the pooled fractions are dried under nitrogen.

Dissolve the benzoates in 1 ml of chloroform and filter the sample through a 0.2-μm HPLC polyvinylidene fluoride (PVDF) sample filter (Chrompack, The Netherlands) attached to a 1-ml plastic syringe. Collect the filtrate in a 2-ml screw-capped glass vial (Brown Chromatography Supplies, Wertheim, Germany) and evaporate the solvent under nitrogen. The benzoates, which should appear as a white residue, can be stored under nitrogen at −20° or dissolved in a small volume (usually 50 μl) of carbon tetrachloride. At this stage, the sample is ready for HPLC analysis.

Separation of Ceramide Benzoates by HPLC

Ceramide derivatives can be separated from other (benzoylated) lipids by isocratic normal-phase HPLC with a ChromSpher silica analytical col-

[6] K. Gross and R. H. McCluer, *Anal. Biochem.* **102,** 429 (1980).

umn (4.6 × 250 mm, 5-μm pore size; Chrompack) and a mobile phase consisting of cyclohexane/0.45% (v/v) 2-propanol at a flow rate of 1.5 ml/min. The solvent mixture should be degassed (either by sonication or under vacuum) before use.

A sample of 10–15 μl is injected onto the column and the eluates are quantitatively measured at 230 nm (ε_{max} 26,000 for dibenzoates) using an on-line absorbance detector. The attenuation should usually be set at a full scale of 0.05–0.1 absorbance unit. Data can be collected and further processed using a conventional integrator or specialized HPLC software such as Millenium (Waters, Milford, MA) or EZChrom (Scientific Software, Inc., Pleasanton, CA).

The Cer benzoates elute as a doublet as is shown in Fig. 2 where Cer from bovine brain sphingomyelin (NFA-Cer; e.g., Sigma, St. Louis, MO) containing primarily C18:0 and C24:1 fatty acids was used as a standard. The first peak contains Cer molecular species with long-chain (C_{22}–C_{24})

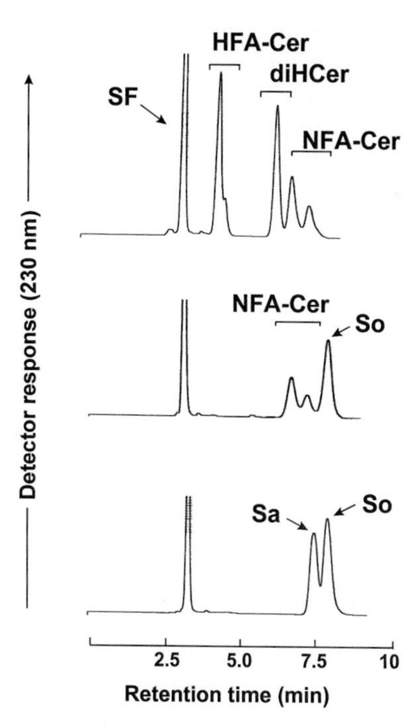

FIG. 2. Separation of benzoylated ceramides and related lipids by normal-phase HPLC. Mixtures of various lipid standards are benzoylated, injected onto a 4.6 × 250 mm, 5-μm pore size ChromSpher silica analytical column and eluted with cyclohexane/0.45% (v/v) 2-propanol at a flow rate of 1.5 ml/min. Detection is by absorbance at 230 nm. SF, solvent front; HFA–Cer, hydroxy fatty acid ceramide; diHCer, dihydroceramide (C24:1); NFA–Cer, nonhydroxy fatty acid ceramide; So, sphingosine; Sa, sphinganine.

fatty acids, being more hydrophobic and more soluble in the mobile phase, whereas those with intermediate-chain fatty acids (C_{14}–C_{18}) elute in the second peak. The retention time of the Cer doublet will depend on the particular column used and the precise amount of 2-propanol in the mobile phase, but a complete separation can be achieved within 10 min. In addition to nonhydroxy fatty acid (NFA)–Cer, some cells or tissues contain 2′-hydroxy fatty acid (HFA)–Cer molecular species. After benzoylation, HFA–Cer is more hydrophobic than NFA–Cer since it carries three instead of two hydroxy groups, allowing HFA–Cer species to elute at approximately one-third of the relative retention time of NFA–Cer.

An important advantage of the system described here is that besides separating Cer species differing in fatty acid chain length and degree of hydroxylation, it also resolves dihydroceramide (diHCer), the direct precursor of Cer in the *de novo* biosynthetic pathway, as well as the long-chain base sphingosine. After benzoylation, Cer and related molecules elute in the following order: HFA-Cer, diHCer, NFA-Cer, sphingosine. For each subclass of Cer, species containing long- and intermediate-chain fatty acids will elute as a doublet. Sphinganine can be separated from sphingosine but, because it elutes just before sphingosine, the low cellular sphinganine levels are usually masked by the more abundant intermediate-chain Cer peak. Quantitation of sphingosine and sphinganine is better achieved by reversed-phase HPLC as described by Merrill *et al.*[7] (and [1] of this volume[8]).

Figure 3 shows the separation of benzoylated total lipid extracts of Jurkat cells which were left untreated, induced to undergo apoptosis by anti-CD95 monoclonal antibody, or treated with *Bacillus cereus* sphingomyelinase. In total lipid samples, derivatized neutral lipids (mainly sterols, tri-, di-, and monoacylglycerols), which are less polar than Cer, will elute with the solvent front, together with excess derivatization reagent. Reduction of the isopropanol content to 0.1–0.2% (v/v) would enable the separation and quantitation of these lipid classes within the same sample.

A standard curve of bovine brain Cer can be used to determine the relationship between peak area and Cer amount, which should be linear in the range between 100 and 5000 pmol. In addition, a lipid standard that is absent or present in only trace amounts in the sample of interest can be used as an internal standard for quantitation of Cer. Estrone (Sigma) is very suitable for this purpose because benzoylated estrone elutes between HFA- and NFA-Cer. A known amount of (benzoylated) estrone can be added before lipid extraction, before derivatization, or before HPLC analy-

[7] A. H. Merrill, Jr., E. Wang, R. E. Mullins, W. C. L. Jamison, S. Nimkar, and D. C. Liotta, *Anal. Biochem.* **171,** 373 (1988).

[8] A. H. Merrill, Jr., *Methods Enzymol.* **312,** [1], (2000) (this volume).

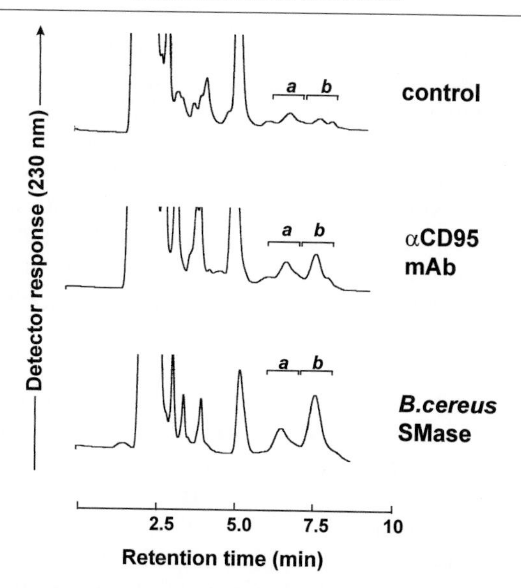

FIG. 3. HPLC profiles of benzoylated total lipid extracts from Jurkat cells (volumes injected correspond to 1.5×10^6 cells) that were either left untreated (control) or were treated with the anti-CD95 monoclonal antibody CH-11 (1 μg/ml; 5 hr) or *B. cereus* sphingomyelinase (300 mU/ml; 15 min). Conditions for separation were as described in Fig. 2. Regions indicated by *a* and *b* represent Cer species containing long-chain and intermediate-chain fatty acids, respectively.

sis. When the estrone peak area is used to calculate the amount of Cer in the sample, it should be taken into account that benzoylated Cer carries two absorbing benzoate groups, whereas derivatized estrone contains only one. To evaluate the overall recovery of derivatized Cer, trace amounts of [^{3}H]- or [^{14}C]Cer can be added (to the bovine brain Cer or to cellular lipid extracts) before the derivatization procedure. After HPLC separation, collection of the peak corresponding to Cer and counting recovered radioactivity will indicate the final yield, which usually ranges from 80 to 90%.

General Comments

1. Accumulation of excess derivatization reagents and polar lipids on the HPLC column will ultimately alter the chromatographic characteristics and reduce the performance of the column. Routine injection of approximately five sample volumes of 100% 2-propanol at the end of each set of cell samples is therefore recommended to prevent continuous accumulation

of polar material. Alternatively, prefractionation by standard column chromatography (see, e.g., Ref. 9) to obtain the neutral (glyco)lipid fraction will considerably lengthen the lifespan of the HPLC column.

2. It is commonly observed that normal-phase silica HPLC is very sensitive to ambient conditions such as temperature and humidity, as is reflected by changes in peak symmetry and appearance of (front) shoulders. Also, during the analysis of a large set of samples, such small deviations are sometimes observed but can be resolved by subtle changes in flow rate and/or 2-propanol content. It is recommended that a standard Cer sample be injected after every set of approximately five samples to monitor possible changes in the elution profiles.

3. A typical setup for a pilot experiment to quantitate (changes in) cellular Cer levels will include the following samples: (a) reaction blank (i.e., derivatization reagents only in a glass tube); (b) different amounts of standard bovine brain Cer; (c) total lipid extracts of cells of interest (under normal and stimulated conditions); (d) total lipid extracts of cells of interest after treatment with bacterial sphingomyelinase (*B. cereus* or *Staphylococcus aureus*) to reveal the approximate Cer species pattern that will result from cellular sphingomyelin hydrolysis by endogenous sphingomyelinase(s).

4. Under certain cell stimulation conditions, preparation containing glycerol (e.g., monoclonal antibodies, purified enzymes) are used. It is difficult to completely wash away the glycerol from the cells prior to lipid extraction and derivatization. A major peak eluting right after the solvent front corresponding to glycerol tribenzoate will appear in the chromatogram of these samples.

[9] M. Iwamori, C. Costello, and H. W. Moser, *J. Lipid Res.* **20,** 86 (1979).

[4] Quantitative Determination of Ceramide Using Diglyceride Kinase

By DAVID K. PERRY, ALICJA BIELAWSKA, and YUSUF A. HANNUN

Introduction

During the last few years it has become apparent that ceramide, an intermediary in sphingolipid metabolism, is an important intracellular mediator of stress responses. This has led to widespread interest in its function by investigators from diverse fields of study and has underscored

0076-6879/00 $30.00

the importance of a rapid and quantitative assay for ceramide determination.

Several methods have been developed for quantifying ceramide including mass spectrometry and HPLC, both of which require lengthy periods of processing and/or analysis.[1,2] This chapter focuses on an enzymatic assay using diglyceride kinase (DG kinase) that uses ceramide and [γ-^{32}P]ATP as substrates to obtain a radiolabeled and quantifiable product, ceramide 1-[^{32}P]phosphate. The primary advantages of this assay are its rapidity, sensitivity, reliability, and ease of use in processing large numbers of samples.

History

DG kinase was originally described in animal tissues by Hokin and Hokin.[3] The enzyme was validated as an analytical tool in measuring diglyceride levels by the demonstration of a linear relationship between the amount of diglyceride added to an *in vitro* assay and the amount of product (phosphatidic acid) formed.[4] In this report the level of sensitivity of the assay was in the low picomole range. The use of DG kinase to determine diglyceride levels was refined and was demonstrated to give quantifiable conversion of substrate to product over a range of 25 pmol to 25 nmol.[5]

Ceramide shares structural similarities with diglyceride, and Schneider and Kennedy reported that DG kinase can utilize it as a substrate with a K_m nearly five times greater than that for diglyceride.[6] As with diglyceride, early attempts to use DG kinase to quantify ceramide revealed a linear but nonquantitative relationship between substrate added and product formed.[7] Further modifications of the assay demonstrated that DG kinase could also be used for quantitative conversion of ceramide to ceramide 1-phosphate over a range of 25 pmol to 2 nmol, and it is this study that provides a basis for this chapter.[8]

[1] J. D. Watts, M. Gu, A. J. Polverino, S. D. Patterson, and R. Aebersold, *Proc. Natl. Acad. Sci. U.S.A.* **94,** 7292 (1997).

[2] M. Iwamori, C. Costello, and H. W. Moser, *J. Lipid Res.* **20,** 86 (1979).

[3] L. E. Hokin and M. R. Hokin, *Biochim. Biophys. Acta* **31,** 285 (1959).

[4] D. A. Kennerly, C. W. Parker, and T. J. Sullivan, *Anal. Biochem.* **98,** 123 (1979).

[5] J. Priess, C. R. Loomis, W. R. Bishop, R. Stein, J. E. Niedel, and R. M. Bell, *J. Biol. Chem.* **261,** 8597 (1986).

[6] E. G. Schneider and E. P. Kennedy, *J. Biol. Chem.* **248,** 3739 (1973).

[7] M. Jones and R. W. Keenan, *J. Neurosci. Methods* **5,** 383 (1982).

[8] P. P. Van Veldhoven, W. R. Bishop, D. A. Yurivich, and R. M. Bell, *Biochem. Mol. Biol. Int.* **36,** 21 (1995).

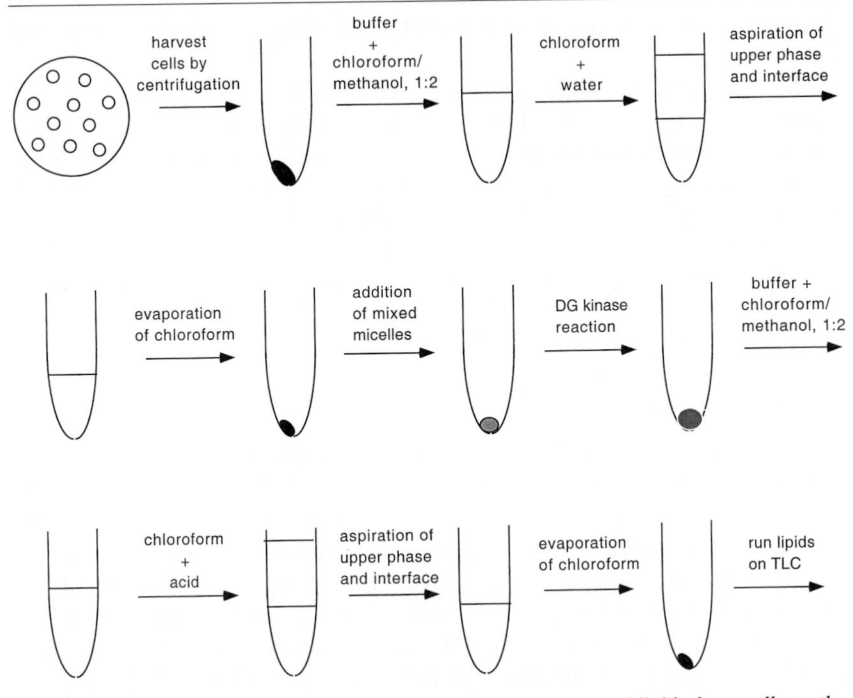

FIG. 1. Scheme for the DG kinase assay from the extraction of lipids from cells to the fractionation of lipids by thin-layer chromatography (TLC).

Procedure

Throughout this procedure refer to scheme in Fig. 1.

Lipid Extraction

Description. The nonpolar properties of ceramide require that it be extracted from cells in organic solvents. This is accomplished by using a modification of the Bligh and Dyer method that involves the lysis of cells with an organic solvent followed by dilution with chloroform and water to obtain phase separation.[9] Approximately 100% of the biologic ceramide species are extracted into the organic phase under these conditions.

The critical parameters of the lipid extraction are the ratios of chloroform, methanol, and buffer/water before and after dilution (Fig. 1). Although the absolute volumes may be changed to reflect the amount of tissue being extracted, it is important that the ratios be maintained at

[9] E. G. Bligh and W. J. Dyer, *Can. J. Biochem. Physiol.* **37,** 911 (1959).

0.5 : 1.0 : 0.4 and 1.0 : 1.0 : 0.9 (v/v/v) before and after dilution, respectively. Additionally, when the biologic material is a limiting factor, preliminary studies may be performed to determine the minimal amount of tissue needed to obtain a control level of ceramide that is within the detection limits of the assay. In the following method, we use 10^6 MOLT-4 human leukemia T cells that were treated with the chemotherapy agent etoposide. Similar procedures for the extraction of ceramide from primary tissues may be used.[10]

Reagents. Phosphate-buffered saline (PBS), pH 7.4, chloroform/methanol (1 : 2, v/v), chloroform, and water.

Method. After treatment of cell cultures, cells are harvested by centrifugation at 4° and residual serum lipids are removed by rinising once in ice-cold PBS, pH 7.4. Then 10^6 cells are suspended in 0.4 ml of ice-cold PBS and added to 1.5 ml of chloroform/methanol (1 : 2, v/v) in a 13-mm × 100-mm glass test tube followed immediately by vortexing for 15 sec. Chloroform (0.5 ml) and water (0.5 ml) are then added and each addition is followed by additional vortexing for 15 sec.

The biphasic mixture that is observed is composed of a lower lipid-containing chloroform phase and an upper phase consisting of methanol and water. The upper phase, and any resulting precipitate at the interface, is removed by aspiration and the resulting lipids in the organic phase are concentrated by evaporation of the chloroform.

Lipid Solubilization with Mixed Micelles

Description. In the DG kinase reaction, the presentation of the substrate in a soluble form to the enzyme is critical for its optimal conversion to product. Mixed micelles containing a nonionic detergent and phospholipid are utilized for this purpose.

Reagents. n-Octyl-β-D-glucopyranoside (β-OG) (Calbiochem, San Diego, CA), 1,2-dioleoyl-*sn*-glycero-3-[phospho-*rac*-(1-glycerol)] (DOPG) (Avanti Polar Lipids, Alabaster, AL), acetone, and diethyl ether

*Method: Recrystallization of Detergent.** Five grams of β-OG is added to 20 ml of acetone and solubilized with heating at 40°. The solution is filtered through a Pyrex sintered glass funnel, 100 ml of diethyl ether are added to the filtrate, and crystals are allowed to form by cooling at −20° overnight. The crystals are collected over a Buchner glass funnel with a fixed perforated plate and are washed with 500 ml of ice-cold diethyl ether. After drying at 4°, the crystals are stored in a sealed glass vial at −20°.

[10] A. E. Bielawska, J. P. Shapiro, L. Jiang, H. S. Melkonyan, C. Piot, C. L. Wolfe, L. D. Tomei, Y. A. Hannun, and S. R. Umansky, *Am. J. Pathol.* **151,** 1257 (1997).

* If an ultrapure form of β-OG is purchased, the recrystallization procedure is not required.

Formation of Mixed Micelles. A 20 mg/ml solution of DOPG in chloroform is added in 0.97-ml aliquots to screw-top glass tubes and concentrated under nitrogen gas as a thin film around the bottom of the tube. A 7.5% (≈250 m*M*) solution of β-OG in water is made and added in 1-ml aliquots to each tube containing DOPG for a final DOPG concentration of 25 m*M*. The DOPG can be solubilized with either sonication or by allowing to stand overnight at 4° followed by vigorous vortexing. The resulting micelles are stored at −20°.

Solubilization of Lipid. Twenty microliters of the micelle solution is added to the lipid residue and vortexed briefly followed by incubation for 30 min at 37°. After additional vortexing, the solubilized lipids are subjected to the DG kinase reaction. Alternatively, the lipids may be solubilized with the micelles by water bath sonication.

Preparation of a Standard Curve

Description. The quantification of ceramide can be determined from the slope of a standard curve using known amounts of ceramide once total or near total conversion of ceramide is established. More rigorous results are obtained by calculating the mass of ceramide 1-phosphate (the product) using the known specific activity of adenosine triphosphate (ATP) (Table I).

Reagents. Naturally occurring ceramides, such as those derived from phospholipase C hydrolysis of sphingomyelin (Avanti Polar Lipids, Alabaster, AL) or type III ceramide from Sigma (St. Louis, MO), serve as the

TABLE I
QUANTIFICATION OF CERAMIDE[a]

Calculations of ceramide in standard curve from lanes 1–6			Calculations of ceramide in experimental samples in lanes 8–12			
					Ceramide (pmol) determined from	
Ceramide (pmol)	Ceramide (cpm) (corrected for background)	Ceramide (pmol) determined by specific activity of ATP	Etoposide treatment (hr)	Ceramide (cpm)	Specific activity of ATP	Slope of standard curve
0	0	0	0	2604	35.7	39.8
30	2192	30.0	1	3252	44.5	49.7
60	3982	54.5	2	4026	55.2	61.5
120	8116	111.2	4	3592	49.2	54.9
300	19626	268.8	6	6640	91.0	101.5
600	39416	539.9				

[a] Determined from the experiment performed in Fig. 2.

best substrates because they are quantitatively converted to product. We have observed that some synthetic ceramides are not quantitatively converted to product and, therefore, these should be avoided as standards in the DG kinase assay.

Method. A solution of ceramide standard in chloroform is aliquoted into glass test tubes for a standard curve range of 0–600 pmol. The lipid is concentrated by the evaporation of the chloroform and solubilized in micelles as previously described.

DG Kinase Reaction

Description. The micelle-solubilized lipid extracts from the experimental samples and the ceramide standard curve are subjected to the DG kinase reaction wherein ceramide is converted to a quantifiable product by the transfer of [^{32}P]phosphate from [γ-^{32}P]ATP to ceramide.

Reagents. Membranes from an *Escherichia coli* strain overexpressing DG kinase; reaction mixture containing 100 m*M* imidazole, pH 6.6, 100 m*M* LiCl, 25 m*M* $MgCl_2$, 2 m*M* EGTA, and 4 m*M* dithiothreitol (DTT); initiation mixture containing 10 m*M* imidazole, pH 6.6, 1 m*M* diethylenetriaminepentaacetic acid, and 5 m*M* ATP [γ-^{32}P]ATP (spec. act. $\approx$ 50,000–100,000 cpm/nmol).

Method: Preparation of DG Kinase Membranes.[11] Ten milliliter of LB-ampicillin media is inoculated with *E. coli* strain N4830/pJW10 containing the structural gene for bacterial DG kinase under control of the λp_L promoter and are grown for 18 hr at 32°. The 10-ml culture is added to 500 ml of LB-ampicillin media and grown at 32° until the A_{600} is approximately 1.0. Expression of DG kinase is then induced by heat induction at 42° for 4 hr. Bacteria are harvested by centrifugation at 50,000*g* for 20 min and the cell pellet is suspended in 150 ml of TNE buffer (50 m*M*, Tris-HCl, pH 7.5, 100 m*M* NaCl, 1 m*M* EDTA, and 5 m*M* 2-mercaptoethanol). Following centrifugation at 5000*g* for 20 min, the resulting pellet is suspended in 150 ml of high salt buffer (50 m*M* KH_2PO_4, pH 7.5, 150 m*M* KCl, 50 m*M* sodium-pyrophosphate, 5 m*M* EDTA, and 5 m*M* 2-mercaptoethanol). The bacteria are lysed by two passages through a French press at 16,000 psi and the homogenate is centrifuged at 5000*g* for 15 min. The supernatant is retained and DG kinase-containing membranes are isolated by centrifugation at 200,000*g* for 90 min. The resulting pellet is suspended in 5 ml of TNE buffer and disrupted by passing through an 18-gauge needle. The volume is brought to 100 ml with TNE buffer and the membranes are pelleted at 200,000*g* for 60 min. The pellet is then suspended in 5 ml of

[11] C. R. Loomis, J. P. Walsh, and R. M. Bell, *J. Biol. Chem.* **260,** 4091 (1985).

storage buffer (50 mM Tris-HCl, pH 7.5, 1 mM EDTA, 5 mM 2-mercaptoethanol, and 10% (v/v) glycerol) and homogenized with a Teflon pestle. The protein concentration (typically 5–10 mg/ml) is determined on the membrane preparation and aliquots are frozen at −80°.

Reaction Constituents. To the solubilized experimental samples and ceramide standards in 20 μl of micelles are added 50 μl of reaction mixture, 5 μg of the membrane protein from the bacterial DG kinase preparation, and water for a total volume of 80 μl. The reaction is begun by the addition of 20 μl of the initiation mixture and is allowed to proceed for 30 min at 26°. The reaction is terminated by the addition of 1.5 ml of chloroform/methanol (1 : 2, v/v) and 0.3 ml of water followed by vigorous vortexing. An additional 0.5 ml of chloroform and 0.5 ml of 1% perchloric acid are added followed by vortexing. The upper aqueous phase is removed by aspiration and the remaining lipid-containing chloroform phase is concentrated by the evaporation of chloroform.

Comments. To obtain quantitative conversion of ceramide to ceramide 1-phosphate, it is imperative that the assay not be performed under Michaelis–Menten conditions where enzyme concentration is a limiting factor but rather under conditions of excess enzyme. A detailed discussion of these considerations has been previously summarized.[12]

Analysis of Reaction Products by Thin-Layer Chromatography (TLC)

Description. Several substrates in lipid extracts are commonly phosphorylated by DG kinase. Therefore, liquid scintillation counting of the organic extract obtained after the DG kinase reaction does not provide an accurate determination of ceramide 1-[^{32}P]phosphate content. Consequently, the products of the DG kinase reaction are chromatographed by thin-layer chromatography, and ceramide 1-[^{32}P]phosphate is identified by comigration with phosphorylated ceramide standards (Fig. 2).

Reagents. Chloroform, acetone, methanol, acetic acid, and water.

Method. The lipid residue from the DG kinase reactions is solubilized in 40 μl of chloroform/methanol (4 : 1, v/v) and one-half of this amount is aliquoted at the origin of a prescored silica gel 60 plate (Whatman, Clifton, NJ). A solvent system of chloroform/acetone/methanol/acetic acid/water (10 : 4 : 3 : 2 : 1, v/v) is added to a TLC chamber, which is allowed to saturate with vapors by using a sheet of filter paper as a wick. The silica gel plate is then placed into the chamber and the solvent front is allowed to migrate to the top of the plate. The plate is then removed and allowed to air dry. Radioactive lipids are visualized by exposure (typically for 12 hr) to X-ray

[12] D. K. Perry and Y. A. Hannun, *Trends Biochem. Sci.* **24,** 226 (1999).

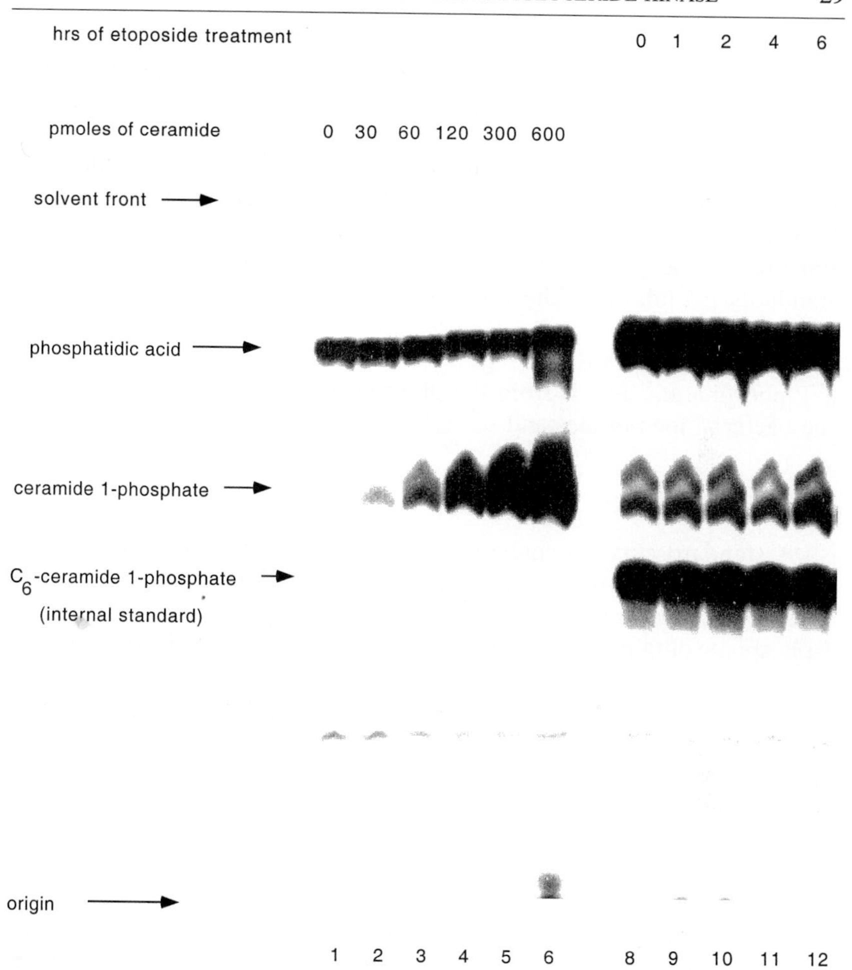

FIG. 2. Autoradiograph of lipids resulting from the DG kinase assay that were fractionated by TLC.

film (Fig. 2). Radioactivity comigrating with ceramide 1-[^{32}P]phosphate ($R_f \approx 0.55$–0.58) is scraped from the plate and quantified by liquid scintillation counting. Alternatively, if the assay has been validated (see discussion below), the radioactivity can be quantified by densitometry.

Comments. Two bands of ceramide 1-[^{32}P]phosphate can be observed in both the standard and experimental samples. It is presumed, though not verified, that these bands represent ceramide species containing fatty acids of different acyl chain length. The radioactivity in both of these bands is used in determining the total mass of ceramide. In the autoradiograph shown in Fig. 2, a band ranging in R_f from 0.50 to 0.60, which encompasses the R_f values of both bands of ceramide, was scraped from the plate and quantified by liquid scintillation counting.

The [^{32}P]phosphatidic acid that is observed in Fig. 2 is a result of the phosphorylation of diglyceride in the assay and differs in migration ($R_f \approx$ 0.75) from ceramide 1-[^{32}P]phosphate sufficiently as to not present distortion in ceramide quantification. In lanes 1–6 (Fig. 2) containing the ceramide standards, [^{32}P]phosphatidic acid is from the phosphorylation of diglyceride in the assay present in the bacterial membranes used as the source of DG kinase. In lanes 8–12 (Fig. 2) containing the experimental samples, [^{32}P]phosphatidic acid is from the phosphorylation of diglyceride in both the bacterial membranes and the cellular lipid extract.

Data Interpretation

A standard curve is constructed using picomoles of ceramide on the abscissa and counts per minute of ceramide 1-[^{32}P]phosphate on the ordinate. After determining the slope of this curve, the amount of ceramide 1-phosphate obtained in the experimental samples is determined by dividing the cpm in the sample by the slope of the standard curve. The moles of ceramide 1-phosphate are equivalent to the moles of ceramide in the assay (Table I).

Ceramide is commonly measured in cells that are undergoing apoptosis. Because apoptosis is often a lengthy process, normalizing ceramide mass to starting cell number may be misleading. For example, if an apoptotic experiment is allowed to proceed for 24–48 hr as is typically the case for agents such as tumor necrosis factor (TNF), proliferation of cells in a control sample may greatly exceed the starting cell number. Likewise, there may be a loss of cells in the later time periods of the treated samples. For this reason, ceramide mass is most appropriately normalized to a parameter whose value is determined at the time the cell treatment is ended. We routinely normalize ceramide mass to cellular lipid–phosphate content. If this normalization parameter is used, the starting cell number needs to be doubled to 2×10^6 and one-half of the organic phase obtained after the initial Bligh and Dyer extraction should be set aside for the lipid–phosphate determination.

Validation of the Assay

To ensure accurate determination of ceramide in the experimental samples using the slope of the standard curve, the quantitative conversion of ceramide standards to ceramide 1-phosphate should be validated. This is accomplished by dividing the cpm obtained in the phosphorylated standards by the specific activity of the $[^{32}P]$ATP (7.3×10^4 cpm/nmol in this experiment) in the DG kinase reaction. This calculation assumes 100% purity of the ATP, and ceramide mass determined in this manner should approach 100% of the mass of ceramide standard in the reactions. The data calculations at left in Table I, obtained from the radioactive ceramide 1-phosphate standards in Fig. 2, demonstrate that this method resulted in ceramide measurements ranging from approximately 90% to 100% of the added ceramide. Moreover, the quantification of mass levels of ceramide determined from cells treated with the chemotherapy agent etoposide (Fig. 2) is shown at right in Table I. In this experiment, the mass values obtained using the specific activity of the ATP differed in value by approximately 10% from those obtained using the slope of the standard curve.

A recent report has demonstrated that under specific biologic conditions, a perceived increase in ceramide by DG kinase analysis was due to a factor in the lipid extract that activated the enzyme rather than an actual increase in ceramide.[1] The influence of DG kinase-stimulating factors on the quantification of ceramide would have been avoided by using an excess of enzyme in the assay.

Additionally, variations of enzyme activity in the assay may be monitored by the use of an internal standard. The internal standard cannot be a physiologic species of ceramide because these will have a similar R_f to the ceramide in the experimental samples and will be indistinguishable by thin-layer chromatography. Ceramide species containing short-chain fatty acids (i.e., C_{6-12}-ceramide) are useful for this purpose as demonstrated in Fig. 2 where C_6-ceramide is used as an internal standard. In lanes 8–12 (Fig. 2) there was less than 10% variation in C_6-ceramide 1-$[^{32}P]$phosphate levels between any two of the experimental samples, indicating that enzyme activity was not significantly altered among these samples.

The shorter acyl chain ceramide 1-phosphates are not efficiently extracted in a normal Bligh and Dyer protocol. However, protonation of the phosphate group significantly enhances their extraction, and it is therefore important to acidify the extraction in order to gain optimal usage of the internal standard. This was accomplished in the current protocol by substituting 1% perchloric acid for water during the extraction of lipids after the DG kinase reaction (Fig. 1).

Acknowledgment

This work is supported in part by NIH grant GM 43825 to Y.A.H.

[5] Analysis of Sphingomyelin, Glucosylceramide, Ceramide, Sphingosine, and Sphingosine 1-Phosphate by Tandem Mass Spectrometry

By M. Cameron Sullards

Introduction

Since the discovery of "sphingosin" by Thudichum,[1] sphingolipids have been shown to be a broad and diverse class of molecules.[2] Sphingolipids help define structural properties of membranes and lipoproteins and also participate in a wide variety of biological functions including cell growth, development, and cell death (apoptosis) through participation in signaling, membrane trafficking, and other behaviors.[3]

The bioactivities of sphingolipids are of great interest and depend on their structure. For example, the free sphingoid base sphingosine (*trans*-4-sphingenine, So) inhibits protein kinase C *in vitro* and in intact cells, and has been found to affect more than 100 different cellular systems.[3] Ceramide (Cer) is a building block in the biosynthesis of more complex sphingolipids and an intermediate in degradation, but is also a lipid second messenger[4,5] via activation of protein phosphatase(s), kinase(s), and other signal transduction pathways. Likewise, complex sphingolipids having differing polar headgroups are involved in a myriad of biological functions.[6–10]

[1] J. L. W. Thudichum, "A Treatise on the Chemical Constitution of Brain," Bailliere, Tindall, and Cox, London, 1884.

[2] Free sphingoid bases share the core structure 2-amino-1,3-dihydroxyoctadecane (named sphinganine or d18:0). The core structure may be substituted with an additional hydroxyl group at position 4 (4-hydroxy sphinganine or t18:0, where "t" denotes a trihydroxy base) and/or has double bonds at positions 4 (commonly called sphingosine or *trans*-4-sphingenine, d18:$1^{\Delta 4}$), 8 (8-sphingenine or d18:$1^{\Delta 8}$), 4 and 8 (4,8-sphingadiene or d18:$2^{\Delta 4,\Delta 8}$), or at position 8 with a hydroxyl at position 4 (4-hydroxy-8-sphingenine or t18:$1^{\Delta 8}$). Complex sphingolipids, however, are comprised of three basic components: a sphingoid base, a polar headgroup linked at the 1-hydroxy position, and a long-chain fatty acid amide linked at the 2-amino position.

[3] A. H. Merrill, Jr., D. C. Liotta, and R. T. Riley, *in* "Handbook of Lipid Research, Volume 8: Lipid Second Messengers" (R. M. Bell, J. H. Exton, and J. M. Prescott, eds.), Chap. 6, p. 205. Plenum Press, New York, 1996.

[4] Y. A. Hanun, *J. Biol. Chem.* **269,** 3125 (1994).

[5] R. N. Kolesnick and M. Krönke, *Annu. Rev. Physiol.* **60,** 643 (1998).

[6] Y. Barenholz and T. E. Thompson, *Biochim. Biophys. Acta* **604,** 129 (1980).

[7] S. I. Hakomori, *J. Biol. Chem.* **265,** 18713 (1991).

[8] S. Spiegel and A. H. Merrill, Jr., *FASEB J.* **10,** 1388 (1996).

0076-6879/00 $30.00

Commonly used methods of identification and quantification of sphingolipids, such as gas chromatography (GC), thin-layer chromatography (TLC), high-performance liquid chromatography (HPLC), and enzyme- or antibody-based assays, do not readily provide complete details regarding the molecular species being analyzed. For example, they require extraction, hydrolysis, and chemical modification, which are not only laborious, but may introduce artifacts into the analyses. More importantly, these methods do not yield critical information about the pairing of specific headgroups with various combinations of long-chain bases and fatty acids in the intact molecular species.

Electron impact mass spectrometry (EIMS) was initially used to elucidate the structures of ceramides[11,12] and neutral glycolipid species.[13,14] These early experiments permitted the analysis of sphingolipids as intact molecular species, and yielded diagnostic fragmentations that could distinguish isomeric sphingolipid structures.[15–18] Use of EI, however, required derivatization of the sphingolipids to trimethylsilyl or permethyl ethers to increase their volatility to enter the gas phase for ionization. Furthermore, since EI is a high-energy ionization method that causes extensive fragmentation of molecular species, observation of molecular ions of larger species is precluded. Additionally, it is difficult to resolve complex mixtures of sphingolipids containing various headgroup/sphingoid base/fatty acid combinations.

The development of "softer" ionization techniques such as fast atom bombardment (FAB) and liquid secondary ionization mass spectrometry (LSIMS) facilitated the generation of intact molecular ions without prior derivitization, yielding numerous sphingolipid species in the resulting mass spectra. Structural information regarding individual molecular species could then be obtained by using multiple stages of mass analysis (i.e., tandem mass spectrometry, MS-MS) to select an ion of interest, collisionally dissociate it, and detect the subsequently formed product ions.

When either $(M + H)^+$ or $(M - H)^-$ precursor ions fragment, they do

---

[9] T. Harder and K. Simons, *Curr. Opin. Cell Biol.* **9,** 534 (1997).

[10] L. Riboni, P. Viani, R. Bassi, A. Prinetti, and G. Tettamanti, *Prog. Lipid Res.* **36,** 153 (1997).

[11] B. Samuelsson and K. Samuelsson, *Biochem. Biophys. Acta* **164,** 421 (1968).

[12] B. Samuelsson and K. Samuelsson, *J. Lipid Res.* **10,** 41 (1969).

[13] K. Samuelsson and B. Samuelsson, *Biochem. Biophys. Res. Commun.* **37,** 15 (1969).

[14] C. C. Sweeley and G. Dawson, *Biochem. Biophys. Res. Commun.* **37,** 6 (1969).

[15] K. Samuelsson and B. Samuelsson, *Chem. Phys. Lipids* **5,** 44 (1970).

[16] S. Hammarstrom, B. Samuelsson, and K. Samuelsson, *J. Lipid Res.* **11,** 150 (1970).

[17] S. Hammarstrom, *J. Lipid Res.* **11,** 175 (1970).

[18] A. Hayashi and F. Matsura, *Chem. Phys. Lipids* **10,** 51 (1973).

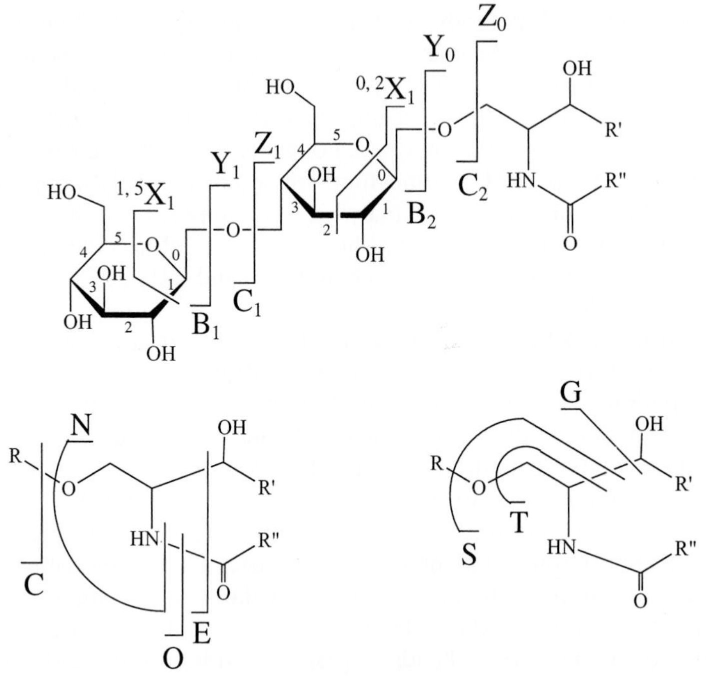

FIG. 1. Mass spectrometry nomenclature for cleavages of complex sphingolipids.

so in specific positions to yield product ions distinctive of the headgroup, sphingoid base, or fatty acid.[19] Additionally, mass spectral fragmentation pathways may be directed with ionization via alkali metal ions [$(M + Me)^+$ in which Me = Li^+, Na^+, K^+, Rb^+, or Cs^+] to improve structural determination of sphingolipids.[20,21] These collections of work resulted in a system of nomenclature descriptive of the product ions observed (Fig. 1) that has been thoroughly reviewed elsewhere.[19]

FAB mass spectrometry and LSIMS have limitations with regard to sphingolipid analysis. For example, both methods require a matrix to solubilize and subsequently ionize the analyte. This results in a significant degree of background chemical noise arising from the matrix, which serves to limit sensitivity. Additionally, quantitating samples can be difficult because analyte is often introduced via a solid probe. Dynamic FAB[22,23] and dynamic

[19] J. Adams and Q. Ann, *Mass Spectrom. Rev.* **12,** 51 (1993).

[20] Q. Ann and J. Adams, *J. Am. Soc. Mass Spectrom.* **3,** 260 (1992).

[21] Q. Ann and J. Adams, *Anal. Chem.* **65,** 7 (1993).

[22] M. Suzuki, M. Sekine, T. Yamakawa, and A. Suzuki, *J. Biochem.* **105,** 829 (1989).

[23] M. Suzuki, M. Sekine, T. Yamakawa, and A. Suzuki, *J. Biochem.* **108,** 92 (1990).

LSIMS attempt to address these issues by continuously infusing a mixture of solvent and matrix onto the probe tip. Unfortunately, even when diluted by a factor of ~100, chemical noise from the matrix is still a problem. An additional limitation of instrumental vacuum systems is that the rate of sample introduction is limited (≤ 10 μl min^{-1}) with respect to typical HPLC flow rates (≥ 200 μl min^{-1}). Thus, the eluent must be split, reducing the amount of sample reaching the probe tip.

In recent years, the field of mass spectrometry has undergone a revolution with the advent of electrospray ionization (ESI). This technique allows an analyte in solution to be infused directly into a specialized ion source, which consists of a metalized needle held at a high potential. At the needle tip, highly charged droplets containing both solvent and analyte are formed. The charged droplets are subsequently drawn into the orifice of the mass spectrometer via a potential difference. In the transition from atmosphere to vacuum, neutral solvent is pumped away, resulting in the soft ionization of the analyte.

Initially, ESI required low flow rates and multiple stages of pumping to remove excess solvent. This resulted in greatly reduced chemical noise and yielded sensitivity orders of magnitude lower than FAB. Subsequent improvements in pumping speed, heated ion sources, and the addition of nebulizing gas have allowed flow rates to be increased greatly. Now it is possible to directly connect eluents from liquid chromatography (LC) columns to a mass spectrometer without splitting for the analysis of complex biological materials.

Liquid chromatography–tandem mass spectrometry (LC-MS/MS) is emerging as a method of choice to obtain unambiguous data with regard to both structure elucidation and quantitation of specific sphingolipids. Recently, some reports have been generated regarding the analyses of sphingolipids in biological materials by nano-[24] and LC-MS/MS.[25] This is often accomplished by using known structure specific fragmentations and MS/MS to identify classes of sphingolipids (e.g., SM or Cer) present in complex mixtures. Additionally, quantitation may be determined with the judicious selection of internal standards. It should be noted that issues regarding sampling efficiency, gas phase basicity, and kinetics of dissociation are of critical importance with regard to accurate quantitation of these species. Therefore, any quantitative data reported not taking these factors into consideration must be considered highly suspect.

[24] B. Brugger, G. Erben, R. Sandhoff, F. T. Wieland, and W. D. Lehmann, *Proc. Natl. Acad. Sci. U.S.A.* **94,** 2339 (1997).

[25] N. Mano, Y. Oda, K. Yamada, N. Asakawa, and K. Katayama, *Anal. Biochem.* **244,** 291 (1997).

It is the primary goal of this chapter to identify and clarify various mass spectral methods with regard to their role in identification and/or quantitation of sphingolipids. This will involve utilizing ESI and a triple quadrupole mass spectrometer. This arrangement was chosen because of the wide range of flow rates amenable to ESI, and the versatility of tandem mass spectrometric methods such as neutral loss, precursor ion, and product ion scans available with this type of instrument. Furthermore, new HPLC-MS/MS methods incorporating multiple reaction monitoring (MRM) experiments will be discussed with regard to automation, high-throughput, and sample screening; and metabolism and quantitation of both naturally occurring and synthetic sphingolipid analogs.

Mass Spectrometric Scanning Methods

Some of the basic principles and vocabulary of biomolecular mass spectrometry are summarized here.

Mass Spectrometry

In a single MS scan, ions over a user-defined m/z range are sequentially passed to the detector. This yields a simple mass spectrum of species ionized from a crude sample and may provide some clues to sphingolipids present. However, interferences from other species present can either suppress ionization of sphingolipids or be of such great abundance (chemical noise) relative to the analyte that its detection is precluded. Furthermore, this type of scan can only provide information regarding molecular mass; no information regarding structure or quantitation can be accurately determined.

Tandem Mass Spectrometry: MS-MS

Product Ion Scan. In a product ion scan, the first mass analyzer (MS1 or Q1) is set to pass a single ion of interest (m/z), which is then collisionally induced to decompose in the next quadrupole (Q2) (Fig. 2, top). In the second stage of mass analysis, MS2 (Q3) is scanned across a range of m/z values to pass the subsequently formed product ions to the detector. The fragmentation pattern in the resulting mass spectrum, thus, provides structural information about the specific precursor ion selected in the form of both product ions detected, and neutral species lost. Additionally, the relative abundances of the product ions observed are reflective of the kinetics of various dissociation pathways and vary with collision energy.

Precursor Ion Scan. In a precursor ion scan, MS2 (Q3) is set to pass an m/z value that corresponds to a structurally specific product ion. MS1 is

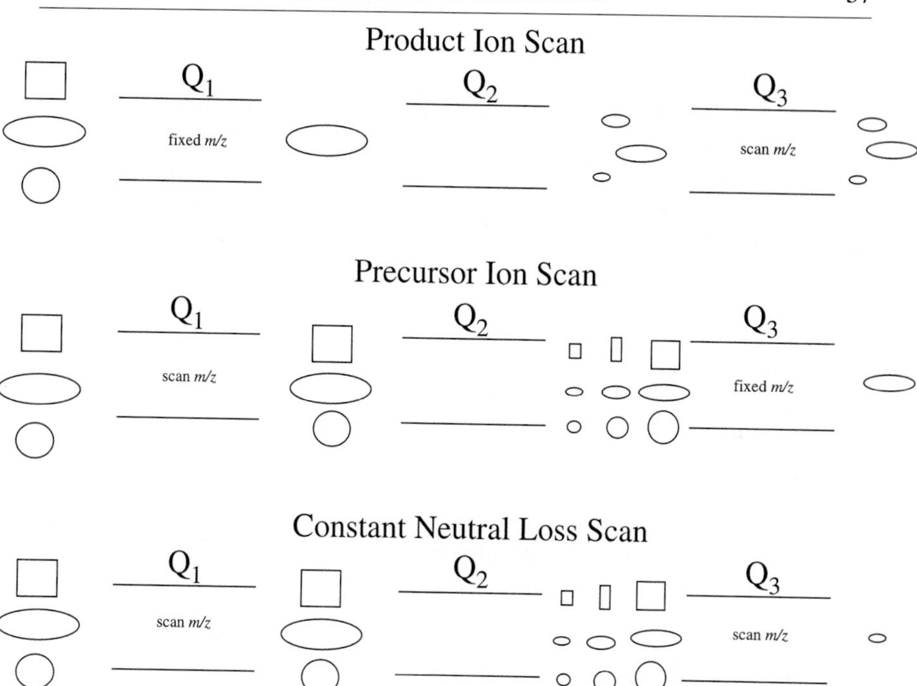

FIG. 2. Illustrations of tandem mass spectrometric scanning methods [product ion (top), precursor ion (middle), and constant neutral loss (bottom)] using a triple quadrupole (Q1–Q2–Q3) mass spectrometer. The large square, oval, and circle on the far left represent intact molecular ions, which can be mass analyzed by the first quadrupole (Q1). These subsequently undergo collisionally activated dissociations in the second quadrupole to yield their corresponding fragment ions of various m/z values (smaller squares, circles, and ovals), which may then be mass analyzed by the third quadrupole (Q3).

then scanned across a range of m/z values sequentially allowing precursor ions to be fragmented into product ions (Fig. 2, middle). Only those precursor ions that decompose to the selected product ion of interest will be passed by Q3 to the detector. This has the net effect of eliminating chemical noise and identifying specific species from complex mixtures. Note that the relative abundances of the precursor ions detected may be dependent on collision energy, and are governed by several kinetic factors.

Constant Neutral Loss (CNL) Scan. In a neutral loss scan, MS1 (Q1) is offset from MS2 (Q3) by fixed m/z, which corresponds to a structure specific neutral loss. Both MS1 and MS2 are scanned in concert across a range of m/z values (Fig. 2, bottom). Thus, only those precursor ions that decompose to lose the correct mass fragment as a neutral will be passed to the detector. This also has the effect of identifying specific species from complex mixtures and eliminating chemical noise. Likewise, the kinetics of dissociation are determined by collision energy and other factors.

Multiple Reaction Monitoring (MRM). In an MRM experiment, MS1 (Q1) is set to pass a specific precursor ion (m/z), and MS2 (Q3) is set to pass a specific product ion (m/z). Only those ions that meet both conditions will be passed to the detector. Thus, individual molecular species are identified from complex mixtures not only by mass, but also by structure. This makes MRM experiments ideal for use with HPLC, quantitation, and drug/metabolite identification for two reasons. First, these types of experiments maximize the amount of time detecting specific ion transitions instead of wasting time scanning regions that have no ions of interest. This serves to increase signal generated and lower the limit of detection. Furthermore, multiple transitions may be monitored sequentially, providing analyses of several analytes in a single HPLC run. Second, each individual transition may be optimized with regard to ion formation and decomposition. This yields maximum sensitivity and eliminates any discrimination that may arise from differences in kinetics of dissociation, thus providing an accurate representation of ion abundances for quantitation.

Free Sphingoid Bases: So, Sa, So-1-P, Sa-1-P

Sphingosine (d18:$1^{\Delta 4}$) and sphinganine (d18:0) readily protonate to form $(M + H)^+$ ions of m/z 300.3 and 302.3, respectively. Product ion scans show that they both dissociate primarily by neutral loss of H_2O to yield carbocations. This decomposition is optimal at a collision energy of 25 eV. The single dehydration product ions are much more abundant than the double dehydration products over a range of collision energies. Furthermore, precursor ion scans of m/z 282/284 and m/z 264/266 show that the single dehydration products yield approximately 10$\times$ more signal and suppress chemical noise to a greater extent than the double dehydration products.

Optimization of ionization and collision conditions for sphingosine and sphinganine reveal two important points regarding accurate quantitation of these species. First, the free sphingoid bases are extremely susceptible to dehydration in the ion source. Decomposition of the analyte in the ion source may adversely affect any subsequent quantitation. Thus, atmospheric voltages must be kept low to avoid dissociation of $(M + H)^+$ ions prior to mass analysis. Second, species containing a $\Delta 4$ double bond yield much more abundant dehydration products than do saturated "dihydro" species of similar abundance. The kinetic difference in reaction rate is a result of dehydration allylic to the double bond, forming a stable conjugated carbocation. This difference in reaction rate requires an internal standard for saturated and unsaturated species so that they may be accurately quantitated.

Phosphorylated sphingoid bases yield either $(M + H)^+$ or $(M - H)^-$ ions when analyzed in positive or negative ion mode, respectively. The $(M + H)^+$ ions of sphingosine 1-phosphate (m/z 380.3), for example, dissociate to form highly abundant product ions of m/z 264. Precursor ion scans of m/z 264, however, reveal only those phosphorylated sphingoid bases that have the specific structure d18:1 or t18:0. Saturation of the $\Delta 4$ double bond, chain length, or other structural modifications, therefore, will result in a corresponding shift in m/z of product ions of this type, as well as changes in the kinetics of dissociation. $(M - H)^-$ ions, however, dissociate to form abundant product ions of m/z 79, corresponding to PO_3^-. Precursor ion scans of this m/z yield all free sphingoid bases that have been phosphorylated, regardless of substitution or structural modification. Note that although the positive ions do not appear as revealing as the negative ions, they are often more sensitive.

Ceramides

Ceramides (Cer), like the free sphingoid bases, easily protonate to form $(M + H)^+$ species generally detected between m/z 500 and 700 (for d18:1 bases and c16:0-c24:0 fatty acids). Product ion scans of protonated ceramides show that these ions undergo cleavage of the amide bond and dehydration to yield structurally distinctive O″ ions (Fig. 1). The O″ ions are characteristic of the sphingoid base and thus may vary in m/z depending on degree of unsaturation, chain length, or other structural modifications of the base. For example, ceramides having d18:1, d18:0, t18:0, or d20:1 sphingoid bases will give rise to O″ ions of m/z 264, 266, 264, and 292, respectively.

Optimization of ionization conditions for ceramides showed that they are also susceptible to dehydration in the ion source. Therefore, atmospheric voltages must be kept low to avoid dissociation of $(M + H)^+$ ions prior to mass analysis in order to yield accurate quantitative results. Likewise, ceramides containing a $\Delta 4$ double bond yield much more abundant O″ ions (m/z 264) than do dihydro species (m/z 266) at similar concentrations. As before, the difference in reaction rate requires an internal standard for either saturated or unsaturated species so that they may be quantitated accurately.

The optimal collision energy for O″ ion formation from ceramides is variable and directly proportional to its degrees of freedom. For example, a small C_2-ceramide (d18:1/c2:0) has an optimal collision energy of 25 eV whereas 45 eV is optimal for a large ceramide (e.g., d18:1/c24:0). Thus, a precursor ion scan performed at 25 eV (Fig. 3A) will yield different relative

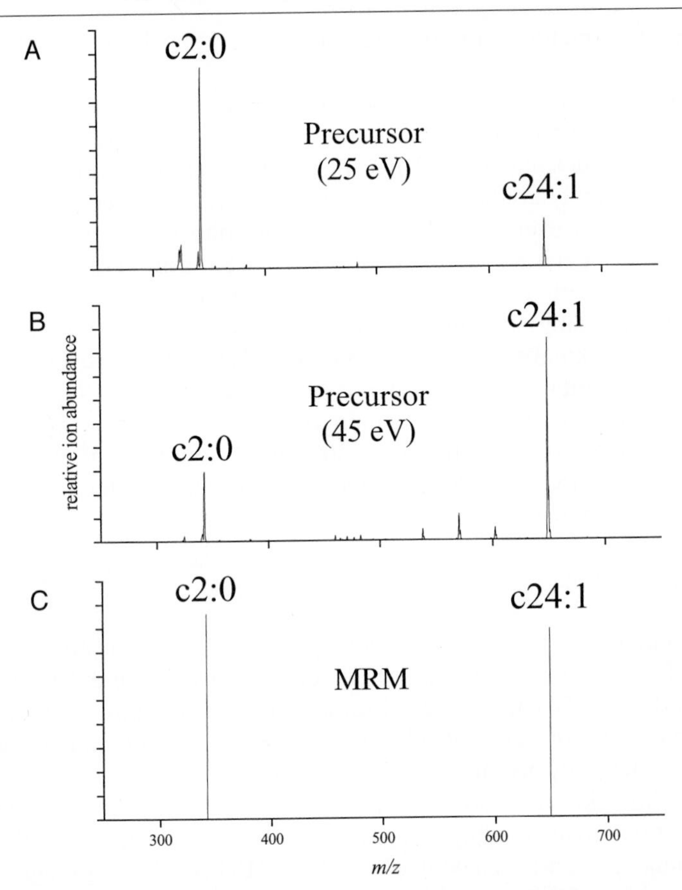

FIG. 3. Tandem mass spectra of an equimolar solution of c2:0- and c24:1-ceramide acquired using (A) precursor ion scans at 25 eV, (B) precursor ion scans at 45 eV, and (C) MRM experiments.

ion abundances than one performed at 45 eV (Fig. 3B). It is for this reason that precursor ion scans do not provide accurate quantitation of ceramides even when internal standards are used, but rather reflect the kinetic differences in dissociation related to the mass of the ions. MRM scans, however, allow the collision energy to be optimized for each individual species monitored and are, therefore, ideal for providing accurate quantitative data (Fig. 3C).

Glucosylceramides

The $(M + H)^+$ ions of glucosylceramides (GluCer) are usually detected in the *m/z* 675–875 range. Product ion scans of these ions show that they dissociate via two pathways. At low collision energies bond cleavage occurs at the glycosidic linkage. The sugar headgroup is lost as a neutral species with charge retention on the remaining ceramide moiety forming the Y_0/Z_0 type ions (Fig. 4A). At higher collision energies both the sugar headgroup and the fatty acid acyl chain are cleaved with the charge retained on the dehydrated sphingoid base resulting in N″ ions (Fig. 4B). (*Note:* The N″ ions are structurally identical to the O″ ions of ceramides. The difference in nomenclature is the result of the glucosylceramides having a headgroup other than a hydrogen atom.)

Ionization conditions required for formation of $(M + H)^+$ ions of glucosylceramides were not as sensitive as with either free sphingoid bases or

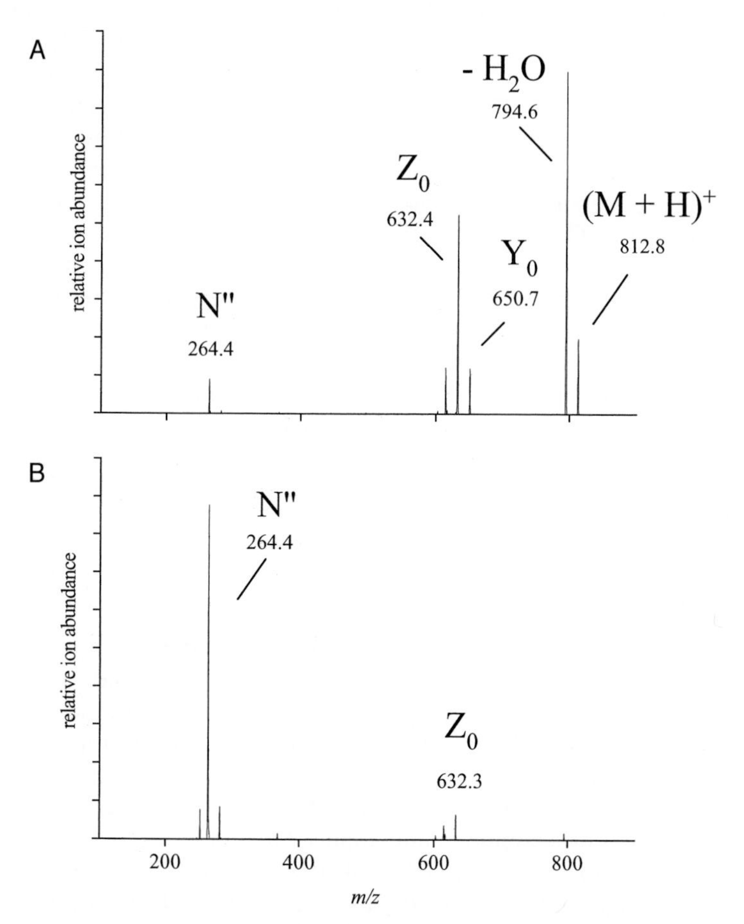

FIG. 4. Product ion spectra of the $(M + H)^+$ ions of d18:1/c24:0 glucosylceramide (*m/z* 812.8) acquired at a collision energy of (A) 25 eV and (B) 50 eV.

ceramides. Dehydration of these ions in the source occurred only at high voltages. It was again observed that there were kinetic differences in the formation of N″ ions depending on the degree and position of saturation in the sphingoid base. Likewise, the collision energy required to maximize formation of N″ ions from various glucosylceramides is directly proportional to degree of freedom. Accurate quantitation of these species can therefore be acquired by using internal standards having the same sphingoid base backbone(s) as the analyte, and by using MRM scans individually optimized for each glucosylceramide present.

Sphingomyelin

Sphingomyelin (SM) forms highly abundant $(M + H)^+$ ions typically between *m/z* 650 and 850 (for d18:1 bases and c16:0-c24:0 fatty acids). These ions are distinctive from the other sphingolipids studied for several reasons. First, they will have odd masses as a result of having an even number of nitrogen atoms and being an even electron ion (nitrogen rule). Second, on collisional activation the phophorylcholine headgroup is cleaved with the charge retained on this moiety yielding highly abundant and structurally distinctive C ions of *m/z* 184 (Fig. 5A).

Optimization of ionization conditions for sphingomyelin shows little propensity for decomposition in the ion source. Dissociation of these ions prior to mass analysis is observed only at the highest potential differences. As with the other sphingolipids, optimal collision energy for formation of C ions is mass dependent. However, we have found no difference in the relative abundance of C ions for sphingomyelins that contain a saturated sphingoid base or a Δ4 double bond (Fig. 5B). This indicates that a single internal standard can be used in conjunction with individually optimized MRM scans to accurately quantitate these species.

Summary

Free sphingoid bases such as sphingosine, sphinganine, and the respective phosphorylated bases, as well as the complex sphingolipids ceramides, glucosylceramide, and sphingomyelin, all dissociate to form structurally distinctive product ions. For sphingomyelin these ions are characteristic of their phosphorylcholine headgroup and are observed at *m/z* 184. The other sphingolipids dissociate to form carbocations characteristic of their sphingoid base. For common mammalian sphingoid bases such as d18:1 or d18:0 these product ions are detected at *m/z* 264 or 266, respectively. However, changes in the sphingoid base chain length, degree of unsaturation, or other modifications may correspondingly result in a shift in *m/z* of

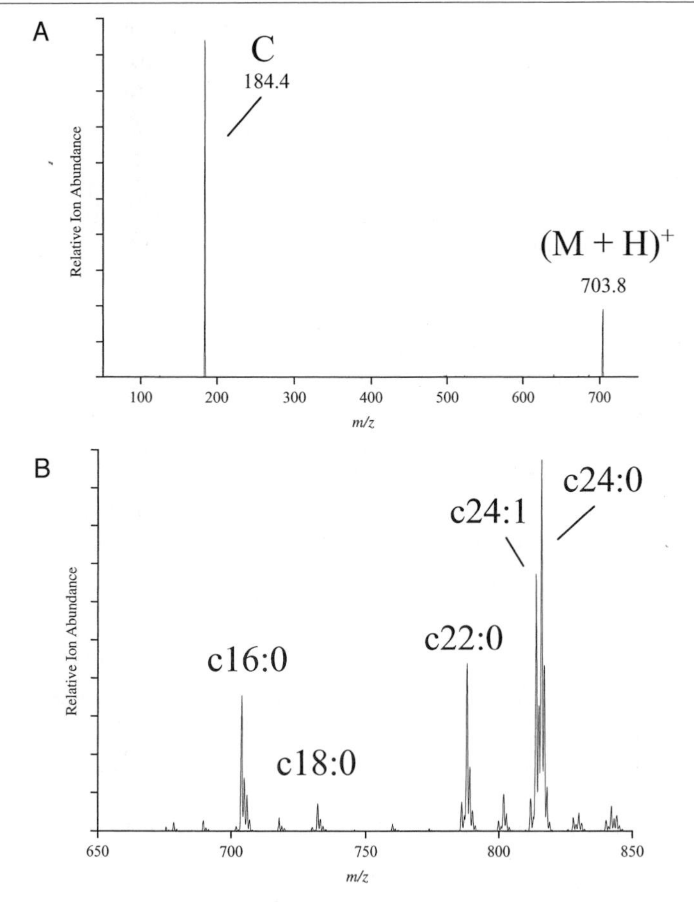

FIG. 5. Tandem mass spectra of bovine sphingomyelin using (A) product ion scans of *m/z* 703.8 (d18:1/c16:0) and (B) precursor ion scans of *m/z* 184.4.

this product ion. Additionally, the kinetics that govern the formation of these product ions is affected by the presence of a Δ4 double bond. Thus, internal standards for each type of sphingoid base are required for quantitative data.

Structurally distinctive product ions, when used with either precursor ion or constant neutral loss scans allow highly specific and sensitive methods for sphingolipid analysis. They serve to greatly reduce background chemical noise, and enhance detection of sphingolipids at very low concentrations. This occurs by allowing only those ions that dissociate to yield a specific product ion or neutral loss to be passed to the detector. Additionally, these scans reveal the exact combinations of headgroup, sphingoid base, and fatty acid in a complex mixture by mass. The free sphingoid bases and Cer readily decompose in the ion source, whereas GlcCer and SM do not. Finally, each individual sphingolipid species fragmented optimally at a different collision energy, precluding the use of either precursor ion or neutral loss scans for quantitation.

Multiple reaction monitoring (MRM) experiments directly address the issues regarding accurate quantitation of sphingolipids that precursor ion and neutral loss scans cannot. In these experiments both ionization and dissociation parameters are optimized for each individual species. By detecting only specific precursor and product ion pairs instead of scanning wide *m/z* ranges maximum sensitivity is attained. Furthermore, relative ion abundance data are not biased with regard to instrumental parameters.

At this point simple loop injections can be used with the MRM scanning methods developed to observe changes in sphingolipid type and quantity in crude extracts on a class-by-class basis. This, however, is labor intensive requiring multiple injections and multiple runs for each class in order to obtain a complete picture of all sphingolipids present. As an alternative

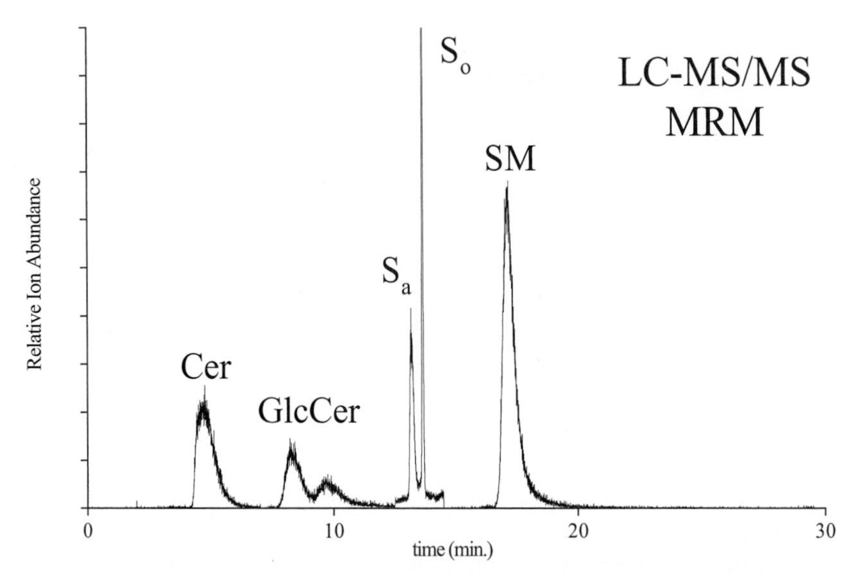

FIG. 6. LC-MS/MS run acquired using MRM experiments on a mixture containing Cer, GlcCer, So, Sa, and SM on a 2.1-mm × 25-cm silica column at a flow rate of 200 μl min^{-1}. Mobile phase A was 97:2:1 (v/v) acetonitrile (AcCN)/methanol/acetic acid with 5 m*M* ammonium acetate, and mobile phase B was 64:15:20:1 (v/v) methanol/water/butanol/acetic acid with 5 m*M* ammonium acetate. The elution profile was 100% A for 5 min, a linear gradient to 100% B over 5 min, and 15 min with 100% B. A total of 27 individual sphingolipids was monitored.

to loop injections, HPLC-MS/MS methods are being developed. In these methods sphingolipids are eluted by class, thus, each individually optimized MRM method can be used at specific times in an LC run. This provides a highly sensitive and accurate quantitation as well as a complete picture of all sphingolipids in a single run (Fig. 6). Additionally, this methodology is amenable to automation and can be used for high-throughput screening of multiple samples.

Acknowledgments

This work was supported by NIH grants NCI-CA73327 and ES09204.

[6] Analyses of Glycosphingolipids by High-Performance Liquid Chromatography

By JOHANNES MÜTHING

Introduction

The German clinician J. L. W. Thudichum, born in the year 1829 in Büdingen (Hessen, Germany), was the first scientist to isolate and name several lipids such as cerebroside, sphingosine, ceramide, sphingomyelin, and cephalin, and thus can be considered the father of sphingolipid research.[1] Since their discovery more than 100 years ago, a large number of glycosphingolipid (GSL) structures has been recorded. They have been found in every vertebrate tissue studied so far[2] and the qualitative and quantitative GSL content has been determined for many organs and tissues. According to a review published over a decade ago,[3] more than 250 different GSLs have been reported and the number is still increasing.

GSLs are amphipathic molecules consisting of two structural elements: a lipophilic membrane anchor, the ceramide portion, which is formed by a long-chain amino alcohol and a fatty acid, and a hydrophilic carbohydrate moiety. In most vertebrate GSLs the long-chain aminodiol sphingosine (4-sphingenine) serves as the lipid backbone. Fatty acids of varying lengths are attached via amide bonds to the amino group of sphingosine, forming

[1] T. Yamakawa, *Glycoconjugate J.* **13,** 123 (1996).

[2] R. K. Yu and M. Saito, *in* "Neurobiology of Glycoconjugates" (R. U. Margolis and R. K. Margolis, eds.), Chap. 1, p. l. Plenum Press, New York, 1989.

[3] C. L. M. Stults, C. C. Sweeley, and B. A. Macher, *Methods Enzymol.* **179,** 167 (1989).

0076-6879/00 $30.00

TABLE I
STRUCTURES OF GANGLIO-, GLOBO-, LACTO-, AND NEOLACTO-SERIES NEUTRAL GSLs

Structure	Trivial name	Symbol[a]	Abbreviation
Glcβ1-1Cer	Glucosylceramide	GlcCer	—
Galβ1-4Glcβ1-1Cer	Lactosylceramide	LacCer	—
GalNAcβ1-4Galβ1-4Glcβ1-1Cer	Gangliotriaosylceramide	$GgOse_3Cer$	Gg_3
Galβ1-3GalNAcβ1-4Galβ1-4Glcβ1-1Cer	Gangliotetraosylceramide	$GgOse_4Cer$	Gg_4
Galα1-4Galβ1-4Glcβ1-1Cer	Globotriaoslyceramide	$GbOse_3Cer$	Gb_3
GalNAcβ1-3Galα1-4Galβ1-4Glcβ1-1Cer	Globotetraosylceramide	$GbOse_4Cer$	Gb_4
GlcNAcβ1-3Galβ1-4Glcβ1-1Cer	Lactotriaosylceramide	$LcOse_3Cer$	Lc_3
Galβ1-3GlcNAcβ1-3Galβ1-4Glcβ1-1Cer	Lactotetraosylceramide	$LcOse_4Cer$	Lc_4
Galβ1-4GlcNAcβ1-3Galβ1-4Glcβ1-1Cer	Neolactotetraosylceramide	$nLcOse_4Cer$	nLc_4
Galβ1-4GlcNAcβ1-3Galβ1-4GlcNAcβ1-3Galβ1-4Glcβ1-1Cer	Neolactohexaoylceramide	$nLcOse_6Cer$	nLc_6

[a] Nomenclature follows the IUPAC-IUB recommendations.[10]

the GSL's hydrophobic ceramide moiety. The oligosaccharide units are bound to the primary hydroxyl group of ceramide via glycosidic linkages. GSL biosynthesis starts in the endoplasmic reticulum with the formation of ceramide,[4] which then is glycosylated stepwise in the Golgi complex.[5,6] GSLs are divided into three main classes: (1) noncharged neutral GSLs, (2) sulfate containing sulfatides, and (3) sialic acid containing gangliosides. Gangliosides are characterized by the presence of one or more sialic acid units in the oligosaccharide chain. Their parent compounds are *N*-acetylneuraminic acid (Neu5Ac) and *N*-glycolylneuraminic acid (Neu5Gc), which are known to play crucial roles in various biologic functions.[7,8] Additionally, the sialic acid can be chemically modified, e.g., by deacetylation or by *O*-acetylation.[9]

Chemical Structure and Supramolecular Organization

The classification of GSLs is based on the oligosaccharide backbone, and most GSLs can be grouped into one of the four main structural families: the ganglio-, globo-, lacto-, and neolacto-series (Table I). The GSL nomen-

[4] E. C. Mandon, I. Ehses, J. Rother, G. van Echten, and K. Sandhoff, *J. Biol. Chem.* **267,** 11144 (1992).
[5] G. Pohlentz, D. Klein, G. Schwarzmann, D. Schmitz, and K. Sandhoff, *Proc. Natl. Acad. Sci. U.S.A.* **85,** 7044 (1988).
[6] G. van Echten and K. Sandhoff, *J. Biol. Chem.* **268,** 5341 (1993).
[7] A. Varki, *Glycobiology* **2,** 25 (1992).
[8] R. Schauer, S. Kelm, G. Reuter, P. Roggentin, and L. Shaw, *in* "Biochemistry and Role of Sialic Acids" (A. Rosenberg, ed.), Chap. 2, p. 7. Plenum Press, New York, 1995.
[9] G. Reuter and R. Schauer, *Methods Enzymol.* **230,** 168 (1994).

clature used in this article follows the recommendations of the IUPAC-IUB Nomenclature Commission[10] and sialic acids are designated according to the suggestions of Reuter and Schauer.[11] Structures of gangliosides from mouse brain and human granulocytes, which were used for preparative high-performance liquid chromatography (HPLC) (see below), are listed in Tables II and III, respectively. Most frequently, the designation of the ganglio-series gangliosides according to Svennerholm[12] is used (see Table II).

Concerning their supramolecular organization, sphingomyelin and GSLs have been proposed to be assembled together with cholesterol to form *GSL–cholesterol rafts* that move within the fluid bilayer.[13] The GSL assembly, termed *GSL-enriched microdomain* (GEM) according to Hakomori *et al.*,[14] indicates a cluster of GSLs in the membrane that plays a functional role in cell interaction and cell adhesion. The term *glycosignaling domain* (GSD) indicates a GEM with emphasis on the physical and functional connection of GSLs that initiate binding to ligands and also initiate signaling through inhibition or activation of transducer molecules. An increasing number of publications indicate the importance of the self-assembling ability of GSLs in cell surface microdomains, in defining antigenicity, ligand-binding, and GSL-dependent cell adhesion.

Biologic Functions

GSL oligosaccharide chains spread out in the aqueous environment at the cell surface, and this makes them excellent candidates for cell surface and cell–cell recognition molecules.[15,16] GSL-mediated mechanisms of action are likely to converge on two general schemes[17]: interactions of GSLs with complementary receptors in apposing membranes (*trans* recognition) and modulation of activities of proteins in the same membrane (*cis* regulation). A combination of these mechanisms may link GSL recognition via microdomains, organized as glycosignaling domains, to regulation of ligand binding and cellular signaling through inhibition or activation of transducer molecules.[14]

[10] IUPAC-IUB Commission on Biochemical Nomenclature, *Eur. J. Biochem.* **79,** 11 (1977).
[11] G. Reuter and R. Schauer, *Glycoconjugate J.* **5,** 133 (1988).
[12] L. Svennerholm, *J. Neurochem.* **10,** 613 (1963).
[13] K. Simons and E. Ikonen, *Nature* **387,** 569 (1997).
[14] S.-I. Hakomori, K. Handa, K. Iwabuchi, S. Yamamura, and A. Prinetti, *Glycobiology* **8,** xi (1998).
[15] R. L. Schnaar, *Glycobiology* **1,** 477 (1991).
[16] P. R. Crocker and T. Feizi, *Curr. Opin. Struct. Biol.* **6,** 679 (1996).
[17] S.-I. Hakomori, *J. Biol. Chem.* **265,** 18713 (1990).

Structures, metabolism, and functions of GSLs have been widely reviewed.[18–20] They act as receptors for toxins, bacteria, and viruses[21] and provide biologic specificity for numerous cellular functions such as cell growth regulation, cell–cell adhesion, development, and differentiation.[22–25]

GSLs reside primarily in the outer leaflet of the plasma membrane. There are, however, several indications of intracellular pools as well,[26–30] and the association of GSLs with intermediate filaments of a variety of cell types has been well documented.[31,32] Therefore, intracellular pools of GSLs, while quantitatively less than the plasma membrane, may also serve important cellular functions.

Principles of Isolation and Structural Characterization

Due to the important biologic role of GSLs, much effort has been spent on their isolation and structural characterization.[33–38] GSLs are usually extracted with chloroform–methanol–water or 2-propanol–hexane–water mixtures with a recommended tissue-to-solvent ratio of 1:20 to 1:40 (w/v).

[18] G. Schwarzmann and K. Sandhoff, *Biochemistry* **29,** 10866 (1990).
[19] C. B. Zeller and R. B. Marchase, *Am. J. Physiol.* **262,** C1341 (1992).
[20] L. D. Bergelson, *Immunol. Today* **16,** 483 (1995).
[21] K. A. Karlsson, *Curr. Opin. Struct. Biol.* **5,** 622 (1995).
[22] T. Feizi, *Trends Biochem. Sci.* **16,** 84 (1991).
[23] T. Feizi, *Curr. Opin. Struct. Biol.* **3,** 701 (1993).
[24] E. G. Bremer, *in* "Current Topics in Membranes" (D. Hoekstra, ed.), Vol. 40, Chap. 15, p. 387. Academic Press, San Diego, 1994.
[25] S.-I. Hakomori and Y. Igarashi, *J. Biochem.* **118,** 1091 (1995).
[26] F. W. Symington, W. A. Murray, S. I. Bearman, and S.-I. Hakomori, *J. Biol. Chem.* **262,** 11356 (1987).
[27] V. Chigorno, M. Valsecchi, D. Acquotti, S. Sonnino, and G. Tettamanti, *FEBS Lett.* **263,** 329 (1990).
[28] K.-F. J. Chan and Y. Liu, *Glycobiology* **1,** 193 (1991).
[29] B. Kniep and K. M. Skubitz, *J. Leukoc. Biol.* **63,** 83 (1998).
[30] J. Müthing, U. Maurer, U. Neumann, B. Kniep, and S. Weber-Schürholz, *Carbohydr. Res.* **307,** 135 (1998).
[31] B. K. Gillard, J. P. Heath, L. T. Thurmon, and D. M. Marcus, *Exp. Cell Res.* **192,** 433 (1991).
[32] B. K. Gillard, L. T. Thurmon, and D. M. Marcus, *Glycobiology* **3,** 57 (1993).
[33] L. Svennerholm and P. Fredman, *Biochim. Biophys. Acta* **617,** 97 (1980).
[34] M. C. Byrne, M. Sbaschnig-Agler, D. A. Aquino, J. R. Sclafani, and R. W. Ledeen, *Anal. Biochem.* **148,** 163 (1985).
[35] S. Ladisch and B. Gillard, *Methods Enzymol.* **138,** 300 (1987).
[36] H. Waki, K. Kon, Y. Tanaka, and S. Ando, *Anal. Biochem.* **222,** 156 (1994).
[37] D. Heitmann, M. Lissel, R. Kempken, and J. Müthing, *Biomed. Chromatogr.* **10,** 245 (1996).
[38] H. Dreyfus, B. Guérold, L. Freysz, and D. Hicks, *Anal. Biochem.* **249,** 67 (1997).

Details concerning the isolation and purification procedures have been reviewed by Ledeen and Yu,[39] Kniep and Ebel[40] and Schnaar.[41]

After extraction from biologic sources, anion-exchange chromatography is generally the method of choice as the first purification step of crude GSL extracts, resulting in separation of neutral GSLs and gangliosides. Since the first report,[42] the separation of gangliosides by DEAE-linked matrices has become a well-established standard procedure.[43–47] Further improvements in separation were the application of DEAE-Fractogel[48] as well as the strong anion-exchangers Q-Sepharose[49] and Mono Q.[50]

Because of its resolving power and easy handling, high-performance thin-layer chromatography (HPTLC) has become a standard tool for separation of neutral GSL and ganglioside mixtures for analytical and preparative applications.[51–55] The oligosaccharide portions of GSLs can be visualized by conventional staining with orcinol,[56] and sialic acid with resorcinol.[57] TLC immunostaining (overlay technique) enables sensitive detection of distinct GSLs in complex mixtures directly on the TLC plate by use of specific anti-carbohydrate antibodies and other carbohydrate-specific reagents, such as toxins and lectins.[54,58–61] For the structural determination

[39] R. W. Ledeen and R. K. Yu, *Methods Enzymol.* **83,** 139 (1982).

[40] B. Kniep and F. Ebel, *in* "Methods of Immunological Analysis" (R. F. Masseyeff, W. H. Albert, and N. A. Staines, eds.), Vol. 2, p. 90. VCH Verlagsgesellschaft, Weinheim, 1993.

[41] R. L. Schnaar, *Methods Enzymol.* **230,** 348 (1994).

[42] C. C. Winterbourn, *J. Neurochem.* **18,** 1153 (1971).

[43] T. Momoi, S. Ando, and Y. Nagai, *Biochim. Biophys. Acta* **441,** 488 (1976).

[44] M. Iwamori and Y. Nagai, *Biochim. Biophys. Acta* **528,** 257 (1978).

[45] P. Fredman, O. Nilsson, J.-L. Tayot, and L. Svennerholm, *Biochim. Biophys. Acta* **618,** 42 (1980).

[46] S. K. Kundu, *Methods Enzymol.* **72,** 174 (1981).

[47] S. Ando, H. Waki, and K. Kon, *J. Chromatogr.* **408,** 285 (1987).

[48] T. Tanaka, Y. Arai, and Y. Kishimoto, *J. Neurochem.* **52,** 1931 (1989).

[49] Y. Hirabayashi, T. Nakao, and M. Matsumoto, *J. Chromatogr.* **445,** 377 (1988).

[50] J.-E. Månsson, B. Rosengren, and L. Svennerholm, *J. Chromatogr.* **322,** 465 (1985).

[51] J. Müthing and D. Heitmann, *Anal. Biochem.* **208,** 121 (1993).

[52] J. Müthing, *J. Chromatogr. B* **657,** 75 (1994).

[53] R. L. Schnaar and L. K. Needham, *Methods Enzymol.* **230,** 371 (1994).

[54] J. Müthing, *J. Chromatogr. A* **720,** 3 (1996).

[55] J. Müthing and S. E. Kemminer, *Anal. Biochem.* **238,** 195 (1996).

[56] L. Svennerholm, *J. Neurochem.* **1,** 42 (1956).

[57] L. Svennerholm, *Biochim. Biophys. Acta* **24,** 604 (1957).

[58] J. L. Magnani, D. F. Smith, and V. Ginsburg, *Anal. Biochem.* **109,** 399 (1980).

[59] J. Müthing and P. F. Mühlradt, *Anal. Biochem.* **173,** 10 (1988).

[60] B. Kniep and P. F. Mühlradt, *Anal. Biochem.* **188,** 5 (1990).

[61] J. Müthing, *in* "Methods in Molecular Biology" (E. F. Hounsell, ed.), Vol. 76, p. 183. Humana Press, Totowa, NJ, 1998.

of GSLs, a multidisciplinary approach is needed, including preparative purification as well as analytical determination of GSLs by high-performance liquid chromatography (HPLC),[62,63] oligosaccharide sequencing with specific exoglycosidases and endoglycoceramidases,[64,65] determination of sialic acids,[66,67] fast atom bombardment mass spectrometry (FAB-MS),[68–70] methylation analysis and gas chromatography mass spectrometry (GC-MS)[71–73] and nuclear magnetic resonance (NMR) spectroscopy.[74,75]

High-Performance Liquid Chromatography of Glycosphingolipids

Growing interest in the numerous and important biologic functions of GSLs, particularly of gangliosides, has engendered the need for capable analytical and preparative HPLC separation techniques. GSLs do not possess a characteristic chromophore that permits their quantitative detection when high sensitivity is required. However, they can be derivatized with various chromophores (see below) to form stable products, which then can be quantitatively measured with high sensitivity by their absorption of UV light. Therefore, several precolumn derivatization reactions prior to separation by HPLC have already been reported.

HPLC of Derivatized GSLs

For analytical purpose, GSLs can be converted with benzoyl chloride or benzoic anhydride to stable O-, N-, or O-benzoylated products. These derivatives can be quantitatively measured with high sensitivity by their

[62] R. Kannagi, K. Watanabe, and S.-I. Hakomori, *Methods Enzymol.* **138,** 3 (1987).
[63] R. H. McCluer, M. D. Ullman, and F. B. Jungalwala, *Methods Enzymol.* **172,** 538 (1989).
[64] G. S. Jacob and P. Scudder, *Methods Enzymol.* **230,** 280 (1994).
[65] H. Higashi, M. Ito, N. Fukaya, S. Yamagata, and T. Yamagata, *Anal. Biochem.* **186,** 355 (1990).
[66] S. Hara, Y. Takemori, M. Yamaguchi, M. Nakamura, and Y. Ohkura, *Anal. Biochem.* **164,** 138 (1987).
[67] A. E. Manzi, S. Diaz, and A. Varki, *Anal. Biochem.* **188,** 20 (1990).
[68] J. Peter-Katalinić and H. Egge, *Methods Enzymol.* **193,** 713 (1990).
[69] A. Dell, A. J. Reason, K.-H. Khoo, M. Panico, R. A. McDowell, and H. R. Morris, *Methods Enzymol.* **230,** 108 (1994).
[70] J. Peter-Katalinić, *Mass Spectrom. Rev.* **13,** 77 (1994).
[71] S. B. Levery and S.-I. Hakomori, *Methods Enzymol.* **138,** 13 (1987).
[72] R. Geyer and H. Geyer, *Methods Enzymol.* **230,** 86 (1994).
[73] F.-G. Hanisch, *Biol. Mass Spectrom.* **23,** 309 (1994).
[74] J. Dabrowski, *Methods Enzymol.* **179,** 122 (1989).
[75] C. E. Costello and J. E. Vath, *Methods Enzymol.* **193,** 738 (1990).

absorption of UV light[76–80] (for a review, see McCluer *et al.*[63]). GSL mixtures can be also separated into their molecular species by HPLC as *O*-acetyl-*N*-*p*-nitrobenzoyl derivatives.[81,82] Furthermore, derivatives of *p*-nitrobenzyloxyamine-[83] and 2,3-dichloro-5,6-dicyanobenzoquinone/$NaBH_4$-treated gangliosides[84] have been reported. Moreover, highly sensitive methods for quantitative HPLC analyses of *p*-bromophenylacyl- and 2,4-dinitrophenylhydrazine-derivatized gangliosides[85–87] and *o*-phthalaldehyde-coupled lysogangliosides[88] are in use. The enumerated methods are reasonably simple to perform, highly sensitive, reproducible, and enable the measurement of low quantities of neutral GSLs and/or gangliosides (for details see cited references).

HPLC of Underivatized GSLs

HPLC has been applied in the separation of native GSLs for preparative purposes as originally described by several research groups. The material of the stationary phase, the quantity of the GSL mixture to be separated, the size of the column, the conditions of elution, particularly the composition and slope of the gradient elution, and the speed of elution can be varied from one case to another. The absence of a significant UV absorption of GSLs hampers detection using commonly available UV detectors. The maximum absorption of gangliosides is at 197 nm, a wavelength at which solvents frequently used in gradients (e.g., chloroform) may interfere.

To overcome these disadvantages, Tjaden *et al.*[89] employed a flame ionization detector for monitoring GSL separation with chloroform–methanol–water mixtures using silica gel as the stationary phase. This gradient system was applied for the isolation of pure mono- and disialoganglio-

[76] M. D. Ullman and R. H. McCluer, *J. Lipid Res.* **18,** 371 (1977).
[77] E. G. Bremer, S. K. Gross, and R. H. McCluer, *J. Lipid Res.* **20,** 1028 (1979).
[78] S. K. Gross and R. H. McCluer, *Anal. Biochem.* **102,** 429 (1980).
[79] W. M. F. Lee, M. A. Westrick, and B. A. Macher, *Biochim. Biophys. Acta* **712,** 498 (1982).
[80] M. D. Ullman and R. H. McCluer, *J. Lipid Res.* **26,** 501 (1985).
[81] A. Suzuki, S. Handa, and T. Yamakawa, *J. Biochem* **82,** 1185 (1977).
[82] A. Suzuki, S. K. Kundu, and D. M. Marcus, *J. Lipid Res.* **21,** 473 (1980).
[83] T. D. Traylor, D. A. Koontz, and E. L. Hogan, *J. Chromatogr.* **272,** 9 (1983).
[84] S. Sonnino, R. Ghidoni, G. Gazzotti, G. Kirschner, G. Galli, and G. Tettamanti, *J. Lipid Res.* **25,** 620 (1984).
[85] H. Nakabayashi, M. Iwamori, and Y. Nagai, *J. Biochem.* **96,** 977 (1984).
[86] K. Miyazaki, N. Okamura, Y. Kishimoto, and Y. C. Lee, *Biochem. J.* **235,** 755 (1986).
[87] H. Kadowaki, K. E. Rys-Sikora, and R. S. Koff, *J. Lipid Res.* **30,** 616 (1989).
[88] T. Kobayashi, and I. Goto, *Biochim. Biophys. Acta* **1081,** 159 (1991).
[89] U. R. Tjaden, J. H. Krol, R. P. van Hoeven, E. P. M. Oomen-Meulemans, and P. Emmelot, *J. Chromatogr.* **136,** 233 (1977).

sides on a Hypersil silica gel column[90] and for one-step fractionation of neutral GSLs and gangliosides on serially connected DEAE-derivatized and underivatized controlled-pore glass columns.[91]

Alternative HPLC separations can be performed with UV transparent elution solvent mixtures like acetonitrile–phosphate buffer by employing LiChrosorb-NH_2,[92] aminopropyl-silica gel,[47] Zorbax-NH_2[93] or Spherisorb-NH_2 columns.[94] Starting from single gangliosides, molecular species homogenous in their oligosaccharide and ceramide moieties can be obtained by reversed-phase chromatography with acetonitrile–phosphate buffer[95] and methanol–water gradient elution systems.[96] The solvent system, composed of *n*-hexane, 2-propanol, and water, originally developed for the preparative HPLC of phospholipids on silica gel columns,[97] was successfully applied for the preparative purification of neutral GSLs on Iatrobeads[98] and gangliosides on Zorbax Sil[99] by use of a mixture of 2-propanol–hexane–water with increasing water and decreasing hexane gradient (for a review, see Kannagi *et al.*[62]).

To monitor the purification of nonderivatized GSLs, several postcolumn detection strategies have been established. Enzyme-linked immunosorbent assay (ELISA) and TLC immunostaining,[100] *in situ* fluorescence labeling on TLC layers,[101] and use of enzymes[102,103] for sensitive and specific GSL identification in combination with previous HPLC separation have found application for particular approaches. Furthermore, the on-line combination of HPLC and mass spectrometry technologies bears potential advantage over postcolumn probing techniques[104–107] and this connection may

[90] J. Gottfries, P. Davidsson, J.-E. Månsson, and L. Svennerholm, *J. Chromatogr.* **490,** 263 (1989).

[91] K. Watanabe and Y. Tomono, *Anal. Biochem.* **139,** 367 (1984).

[92] G. Gazzotti, S. Sonnino, and R. Ghidoni, *J. Chromatogr.* **348,** 371 (1985).

[93] R. F. Menzeleev, Y. M. Krasnopolsky, E. N. Zvonkova, and V. I. Shvets, *J. Chromatogr. A* **678,** 183 (1994).

[94] R. Wagener, B. Kobbe, and W. Stoffel, *J. Lipid Res.* **37,** 1823 (1996).

[95] G. Gazzotti, S. Sonnino, R. Ghidoni, G. Kirschner, and G. Tettamanti, *J. Neurosci. Res.* **12,** 179 (1984).

[96] H. Kadowaki, J. E. Evans, and R. H. McCluer, *J. Lipid Res.* **25,** 1132 (1984).

[97] W. M. A. Hax and W. S. M. G. van Kessel, *J. Chromatogr.* **142,** 735 (1977).

[98] K. Watanabe and Y. Arao, *J. Lipid Res.* **22,** 1020 (1981).

[99] S. K. Kundu and D. D. Scott, *J. Chromatogr.* **232,** 19 (1982).

[100] J. Buehler, U. Galili, and B. A. Macher, *Anal. Biochem.* **164,** 521 (1987).

[101] Y. Tomono, K. Abe, and K. Watanabe, *Anal. Biochem.* **184,** 360 (1990).

[102] S. W. Johnson, M. Masserini, and J. A. Alhadeff, *Anal. Biochem.* **189,** 209 (1990).

[103] R. L. Myers, M. D. Ullman, R. F. Ventura, and A. J. Yates, *Anal. Biochem.* **192,** 156 (1991).

[104] Y. Kushi, C. Rokukawa, Y. Numajir, Y. Kato, and S. Handa, *Anal. Biochem.* **182,** 405 (1989).

contribute greatly to the detailed structural characterization of GSLs. Merging of HPLC with the mentioned techniques enhances its power as a tool for a wide range of applications.

Preparative Anion-Exchange HPLC of Gangliosides

The objective of the following paragraphs is to provide the general rules and particularly practical advice for preparative HPLC separations of different types of ganglioside mixtures using the strong anion-exchanger trimethylaminoethyl(TMAE)-Fractogel. Ganglioside fractions used were (1) ganglio-series brain gangliosides (see Table II) to demonstrate separation not only according to their degree of sialylation into mono-, di-, tri-, and tetrasialogangliosides, but also in separation of positional isomers within the four elution domains, e.g., GD1a and GD1b of the disialoganglioside fraction[108]; (2) neolacto-series monosialogangliosides from human granulocytes (see Table III) to show the separation of isomeric compounds, which only differ in their type of sialylation (α2-3, α2-6)[108]; and (3) ganglioside fractions mainly composed of GM3(Neu5Ac) and GM3(Neu5Gc) to demonstrate the separation of closely allied GM3 specimens, which only differ in one hydroxyl group at position 5 of the neuraminic acid (Neu5Ac versus Neu5Gc).[109]

Materials and Methods

Materials

Chloroform and methanol of HPLC grade are purchased, or analytical grade solvents are distilled before use.

Gangliosides that have been used in these studies are prepared as follows: gangliosides from female CBA/J mouse brains (Table II) are isolated according to standard procedures.[39] Gangliosides from human granulocytes are isolated as described by Müthing *et al.*[110,111] with final purification by

---

[105] M. Suzuki, M. Sekine, T. Yamakawa, and A. Suzuki, *J. Biochem.* **105,** 829 (1989).

[106] M. Suzuki, T. Yamakawa, and A. Suzuki, *J. Biochem.* **108,** 92 (1990).

[107] M. Suzuki, T. Yamakawa, and A. Suzuki, *J. Biochem.* **109,** 503 (1991).

[108] J. Müthing and F. Unland, *J. Chromatogr. B* **658,** 39 (1994).

[109] D. Heitmann, H. Ziehr, and J. Müthing, *J. Chromatogr. B* **710,** 1 (1998).

[110] J. Müthing, F. Unland, D. Heitmann, M. Orlich, F.-G. Hanisch, J. Peter-Katalinić, V. Knäuper, H. Tschesche, S. Kelm, R. Schauer, and J. Lehmann, *Glycoconjugate J.* **10,** 120 (1993).

[111] J. Müthing, R. Spanbroek, J. Peter-Katalinić, F.-G. Hanisch, C. Hanski, A. Hasegawa, F. Unland, J. Lehmann, H. Tschesche, and H. Egge, *Glycobiology* **6,** 147 (1996).

Iatrobeads 6RS-8060 chromatography (Macherey-Nagel, Düren, Germany) according to Ueno *et al.*[112] The whole ganglioside fraction (composed of gangliosides enumerated from 1 to 8, see Table III, and designated with HGG1) is eluted with chloroform–methanol (1:2, v/v); alternatively, stepwise elution is performed with chloroform–methanol (85:15), (3:1), (2:1), (1:2), each by volume, and finally methanol. A neolacto-series ganglioside fraction without GM3, composed of gangliosides 3 to 8 (GM3 depleted; see Table III) and designated with HGG2, is obtained in the chloroform–methanol (1:2, v/v) eluate.

To obtain the gangliosides from mouse–mouse hybridoma cells, the cells are homogenized, lyophilized, and threefold extracted with methyl isobutyl ketone–methanol–water (40:80:30, v/v/v) and further purified according to a procedure recently published.[37] The ganglioside fraction, mainly composed of GM3(Neu5Ac) and GM3(Neu5Gc) is finally purified by adsorption chromatography on Iatrobeads 6RS-8060 (Macherey-Nagel) according to Ueno *et al.*[112] The chloroform–methanol (1:2, v/v) eluate contains purified whole gangliosides.

Thin-Layer Chromatography

The gangliosides obtained by HPLC are also examined by thin-layer chromatography. High-performance thin-layer chromatography plates (HPTLC plates, size 10 cm × 10 cm, glass backed and precoated with 0.2-mm silica gel 60, No. 5633, E. Merck, Darmstadt, Germany) are used. Gangliosides are usually applied to the HPTLC plate using an automatic sample applier AS30 (Desaga, Heidelberg, Germany) equipped with a high-precision 10-μl syringe. HPTLC separation of gangliosides is performed in the neutral solvent 1 [chloroform–methanol–water (120:85:20, v/v/v)] and the alkaline solvent 2 [chloroform–methanol–2.5 *M* NH_4OH (120:85:20, v/v/v)], each with 2 m*M* $CaCl_2$.[54] Gangliosides are visualized with resorcinol according to Svennerholm[57] and lipid-bound sialic acid estimated by densitometry. Chromatograms can be scanned densitometrically with band intensities measured in reflectance mode at 580 nm with a light beam slit of 0.1 mm × 2 mm.

Anion-Exchange HPLC of Gangliosides

A Superformance universal glass cartridge system from Merck can be used for anion-exchange HPLC of gangliosides. Two glass cartridges of different size [150 mm × 10 mm and 110 mm × 30 mm with column volumes

[112] K. Ueno, S. Ando, and R. K. Yu, *J. Lipid Res.* **19,** 863 (1978).

of 12 ml and 78 ml, respectively] filled with Fractogel EMD TMAE-650(S) are fitted into a HPLC system (Gilson Abimed, Langenberg, Germany) consisting of three M303 HPLC pumps, a high-pressure mixer M811, and a fraction collector M202.

The following washing, regeneration, and equilibration steps are described for a small column (150 mm × 10 mm) using a flow rate of 0.5 ml/min. The procedure for a large column (110 mm × 30 mm) is identical, but the flow rate is set to 3 ml/min resulting in a sixfold volume increase.

The anion exchanger is converted into the acetate form by successive rinsing with Milli-Q water (Millipore, Bedford, MA; Milli-Q water purification system), NaCl–Tris buffer (0.5 *M* NaCl, 0.25 *M* Tris, pH 8.5), Milli-Q water, 1 *M* acetic acid, Milli-Q water and methanol, each for 200 min. Finally, the column is equilibrated with chloroform–methanol–water (30:60:8, v/v/v) and gangliosides are applied in this solvent onto the column via a Rheodyne 7125 manual injector connected with a 5-ml sample loop. The details on the HPLC program for gradient elution with increasing concentrations of ammonium acetate in methanol and the chromatographic parameters are given below. The relative elution volume is calculated as the quotient of the elution volume to the bed volume of the column. After the HPLC run, the column is regenerated by successive rinsing with the solvents described above for converting the anion exchanger into the acetate form. After HPTLC evaluation, gangliosides containing samples are pooled, evaporated, and resuspended in water for subsequent dialysis. Alternatively, the residues can be taken up in 0.88% KCl and desalted by reversed-phase chromatography on octadecyl columns (0.5 g; Burdick and Jackson, Inert SPE System, Muskegon, MI) according to Williams and McCluer.[113]

HPLC of Mouse Brain Ganglio-Series Gangliosides on TMAE-Fractogel

A thin-layer chromatogram of unfractionated mouse brain gangliosides is shown in Fig. 1 (lane a). The major gangliosides from mouse brain are the ganglio-series gangliosides GM1, GD1a, GD1b, GT1b, and GQ1b (see Table II). The ganglioside mixtures can be applied to TMAE-Fractogel at up to 300 nmol/ml (about 0.5 mg/ml) without any loss of monosialogangliosides. The maximum loading capacity of TMAE-Fractogel tested is at least about 1 μmol per milliliter of resin.

The HPLC program for gradient elution of bound gangliosides with increasing concentrations of ammonium acetate in methanol is displayed in Table IV. The HPLC elution profile of mouse brain gangliosides and the corresponding TLC pattern of single fractions are shown in Fig. 2.

[113] M. A. Williams and R. H. McCluer, *J. Neurochem.* **35,** 266 (1980).

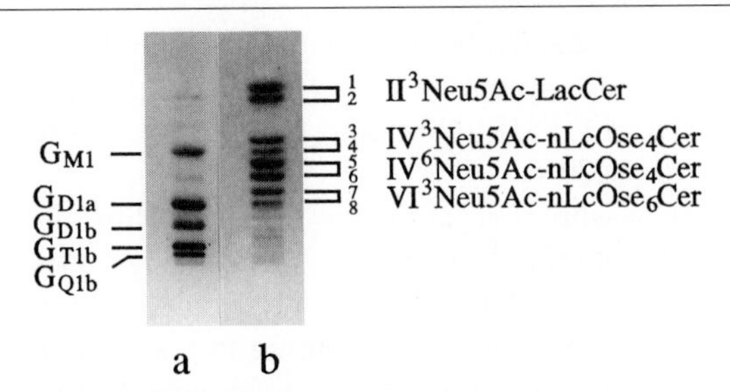

FIG. 1. Thin-layer chromatogram of 10 μg of gangliosides, each, of mouse brain (a) and human granulocytes (b, HGG1). After chromatography in the neutral solvent 1, gangliosides are visualized by spraying the plate with resorcinol–hydrochloric acid reagent. The structures of gangliosides from mouse brain and human granulocytes are listed in Tables II and III, respectively.

Gangliosides elute according to their degree of sialylation into mono-, di-, tri-, and tetrasialogangliosides, and separation is complete after seven relative column volumes, indicating the high-resolution capacity of TMAE-Fractogel. In all four elution peaks, less polar gangliosides elute prior to the more polar gangliosides. Thus, fractions of pure gangliosides are obtained with good separation of isomers, e.g., GD1a and GD1b. The higher resolution capacity of stronger anion exchangers compared to DEAE-coupled matrices has also been reported by Hirabayashi *et al.*[49] and Månsson *et al.*,[50] who used Q-Sepharose for large-scale fractionation of bovine brain gangliosides and Mono Q for small-scale separation of human brain gangliosides, respectively.

In the case of the weak DEAE-anion exchangers, the number of sialic acids is the dominant parameter that determines the retention of gangliosides. In addition to separation according to the degree of sialylation, the

TABLE II
STRUCTURES OF MOUSE BRAIN GANGLIO-SERIES GANGLIOSIDES

Structures[a]	Svennerholm designation[b]
II^3Neu5Ac-GgOse$_4$Cer	GM1
IV^3Neu5Ac,II^3Neu5Ac-GgOse$_4$Cer	GD1a
II^3(Neu5Ac)$_2$-GgOse$_4$Cer	GD1b
IV^3Neu5Ac,II^3(Neu5Ac)$_2$-GgOse$_4$Cer	GT1b
IV^3(Neu5Ac)$_2$,II^3(Neu5Ac)$_2$-GgOse$_4$Cer	GQ1b

[a] Only gangliosides with Neu5Ac substitution are displayed. For core structure of GgOse$_4$ oligosaccharide, see Table I.

[b] Nomenclature follows the designations of Svennerholm[12] (per 1997 IUPAC revisions).

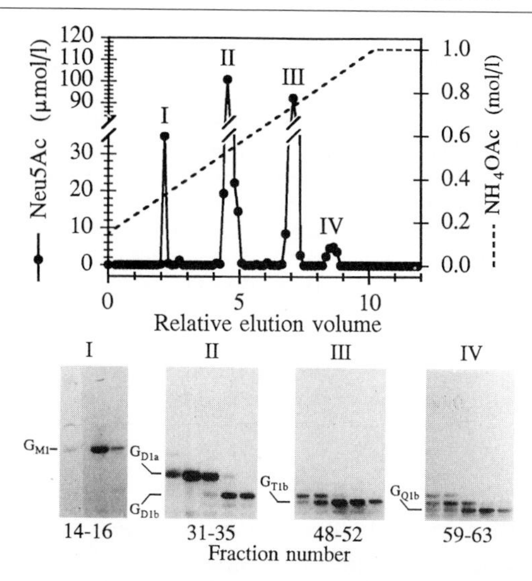

FIG. 2. TMAE-Fractogel HPLC of 20 mg of mouse brain gangliosides. Gangliosides are dissolved in 40 ml chloroform–methanol–water (30:60:6, v/v/v), applied to a 150-mm × 10-mm column (for details of the gradient, see Table IV) and collected in 1.5-ml fractions (= 3-min fractions). One percent aliquots of single fractions can be analyzed by TLC in the neutral solvent 1 and stained with resorcinol–hydrochloric acid reagent. I, Mono-; II, di-; III, tri-; and IV, tetrasialogangliosides. The structures of mouse brain gangliosides are listed in Table II.

strong anion-exchanger TMAE-Fractogel enables the separation of positional isomers within the ganglioside fractions.

HPLC of Granulocyte Neolacto-Series Gangliosides on TMAE-Fractogel

The thin-layer chromatogram of total human granulocyte gangliosides (HGG1) is shown in Fig. 1 (lane b). In a previous paper[110] we showed that, besides GM3, the neolacto-series gangliosides IV^3Neu5Ac-nLcOse$_4$Cer, IV^6Neu5Ac-mLcOse$_4$Cer, and VI^3Neu5Ac-nLcOse$_6$Cer are the major gangliosides isolated from a pool of human granulocytes, according to results of Fukuda *et al.*[114] The characteristic pattern of four ganglioside double bands (Fig. 1, lane b) is caused by substitution of fatty acids with different chain length (mainly C_{24} and C_{16} fatty acids) in the ceramide portions (see Table III). For HPLC, the GM3 depleted ganglioside mixture HGG2 is applied on the TMAE-Fractogel column at 470 nmol/ml (about 0.75 mg/ml).

[114] M. N. Fukuda, A. Dell, J. E. Oates, P. Wu, J. C. Klock, and M. Fukuda, *J. Biol. Chem.* **260,** 1067 (1985).

TABLE III
STRUCTURES OF NEOLACTO-SERIES MONOSIALOGANGLIOSIDES FROM HUMAN GRANULOCYTES

Band[a]	Fatty acid	Structure	Abbreviation
1	24:1, 22:0	II^3Neu5Ac-LacCer	GM3
2	16:0	II^3Neu5Ac-LacCer	GM3
3	24:1, 22:0	IV^3Neu5Ac-nLcOse$_4$Cer	IV^3nLc4
4	16:0	IV^3Neu5Ac-nLcOse$_4$Cer	IV^3nLc4
5	24:1	IV^6Neu5Ac-nLcOse$_4$Cer	IV^3nLc4
6	16:0	IV^6Neu5Ac-nLcOse$_4$Cer	IV^6nLc4
7	24:1	VI^3Neu5Ac-nLcOse$_6$Cer	VI^3nLc6
8	16:0	VI^3Neu5Ac-nLcOse$_6$Cer	VI^3nLc6

[a] According to Fig. 1.

The highest loading capacity that we have tested with this ganglioside mixture is 1.6 μmol per milliliter of resin.

The HPLC program for gradient elution of bound gangliosides with increasing concentrations of ammonium acetate in methanol is identical to that for separation of mouse brain gangliosides (see Fig. 2) and is displayed in Table IV.

TABLE IV
HPLC PROGRAM FOR SEPARATION OF MOUSE BRAIN AND HUMAN GRANULOCYTE GANGLIOSIDES ON TMAE-FRACTOGEL[a]

		Eluent[c]		
Time[b] (min)	Flow (ml/min)	A (%)	B (%)	C (%)
0	0.0	100	0	0
3	0.5	100	0	0
210	0.5	100	0	0
220	0.5	0	100	0
247	0.5	0	100	0
267	0.5	0	90	10
447	0.5	0	60	40
487	0.5	0	0	100
546	0.5	0	0	100
547	0.0	0	0	100

[a] See Figs. 2 and 3. 150-mm × 10-mm column.
[b] Fractionation is started at t = 277 min.
[c] A: chloroform–methanol–water (30:60:8, v/v/v), B: methanol, C: 1 M ammonium acetate in methanol.

HPLC results in two baseline separated monosialoganglioside peaks (Fig. 3, Ia and Ib) and a double peaked disialoganglioside fraction (Fig. 3, II) within a relative elution volume of six column volumes. Isomeric IV^3 Neu5Ac-nLcOse$_4$Cer (compounds 3 and 4) and IV^6Neu5Ac-nLcOse$_4$Cer (compounds 5 and 6) are clearly separated in this system. Separation of α2-3 sialylated IV^3Neu5Ac-nLcOse$_4$Cer (bands 3 and 4) and the Galβ1-4GlcNAc elongated α2-3 sialylated VI^3Neu5Ac-nLcOse$_6$Cer (compounds 7 and 8) is also achieved, as shown by the TLC control in Fig. 3. Fractions 13 and 15, each, contain single purified isomers. Thus, all structurally well-established major gangliosides of elution peak Ia contain a Neu5Acα2-3 terminus, whereas a Neu5Acα2-6 terminus of neolacto-series is characteristic for the main components of elution peak Ib.

From these data, the minor ganglioside of peak Ia (to be seen in fractions 16 and 17 of the TLC in Fig. 3) is predicted as VIII^3Neu5Ac-nLcOse$_8$Cer, and the minor ganglioside of peak Ib (to be seen in fractions 28 and 29) is speculated to represent VI^6Neu5Ac-nLcOse$_6$Cer, which is supported by data of Fukuda *et al.*,[114] who have shown the expression of VI^6Neu5Ac-nLcOse$_6$Cer in human granulocytes.

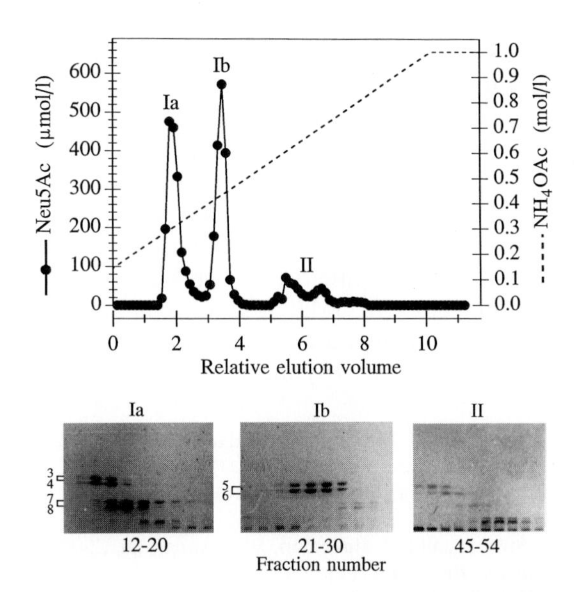

FIG. 3. TMAE-Fractogel HPLC of 30 mg amount of the GM3-depleted ganglioside mixture (HGG2) of human granulocytes. Gangliosides are dissolved in 40 ml chloroform–methanol–water (30:60:6, v/v/v), applied to a 150-mm × 10-mm column (for details of the gradient, see Table IV) and collected in 1.5-ml fractions (= 3-min fractions). One percent aliquots of single fractions can be analyzed by TLC in the neutral solvent 1 and stained with resorcinol–hydrochloric acid reagent. I, mono-; and II, disialogangliosides. The structures of human granulocyte gangliosides are listed in Table III.

Concerning the disialogangliosides separated by TMAE-Fractogel chromatography (Fig. 3, peak II), two pairs of disialogangliosides of yet unknown structures were detected (Fig. 3, fractions 45 to 54). Fukuda *et al.*[114] have noted the presence of a disialoganglioside within a purified monosialoganglioside fraction, but no further structural data are available due to low amounts present in granulocytes.

In conclusion, the strong anion-exchanger TMAE-Fractogel separates isomeric compounds that differ only in their type of sialylation (α2-3, α2-6) as well as gangliosides that have the same sialic acid substitution, but differ in the neutral oligosaccharide moieties. Thus, beside improved separation of gangliosides, extended structural information about unknown gangliosides can be obtained by the strong anion exchanger. Furthermore, the described technique offers the opportunity to accumulate minor components for other studies, such as receptors for carbohydrate binding ligands, e.g., in cell–cell recognition[111,115–117] or interaction with microorganisms.[118–120]

HPLC of GM3(Neu5Ac) and GM3(Neu5Gc) on TMAE-Fractogel

The thin-layer chromatogram of the whole ganglioside fraction from hybridoma cells chromatographed in the alkaline solvent 2 is shown in Fig. 4 (lane R). Chromatography in the alkaline solvent reveals two GM3 double bands, the upper pair representing GM3(Neu5Ac) and the lower pair GM3(Neu5Gc), each substituted with C_{24} (upper band) and C_{16} fatty acid (lower band).[121] Figure 4 depicts HPLC of this fraction on a small TMAE-Fractogel column (150 mm $\times$ 10 mm) (loaded with 20 mg of whole gangliosides from hybridoma cells). The gangliosides are eluted by ammonium acetate gradient chromatography (Fig. 4A) according to the HPLC program depicted in Table V. GM3(Neu5Ac) containing fractions [with GM3(Neu5Gc) contaminants] and GM3(Neu5Gc) harboring fractions can be combined in three pools (Fig. 4B) and, after desalting by reversed-phase

[115] M. Tiemeyer, S. S. Swiedler, M. Ishihara, M. Moreland, H. Schweingruber, P. Hirtzer, and B. K. Brandley, *Proc. Natl. Acad. Sci. U.S.A.* **88,** 1138 (1991).

[116] M. R. Stroud, K. Handa, M. E. K. Salyan, K. Ito, S. B. Levery, and S.-I. Hakomori, *Biochemistry* **35,** 770 (1996).

[117] K. Handa, M. R. Stroud, and S.-I. Hakomori, *Biochemistry* **36,** 12412 (1997).

[118] J. Müthing, *Carbohydr. Res.* **290,** 217 (1996).

[119] M. Matrosovich, H. Miller-Podraza, S. Teneberg, J. Robertson, and K.-A. Karlsson, *Virology* **223,** 413 (1996).

[120] H. Miller-Podraza, M. A. Milh, S. Teneberg, and K.-A. Karlsson, *Infect. Immun.* **65,** 2480 (1997).

[121] J. Müthing, H. Steuer, J. Peter-Katalinić, U. Marx, U. Bethke, U. Neumann, and J. Lehmann, *J. Biochem.* **116,** 64 (1994).

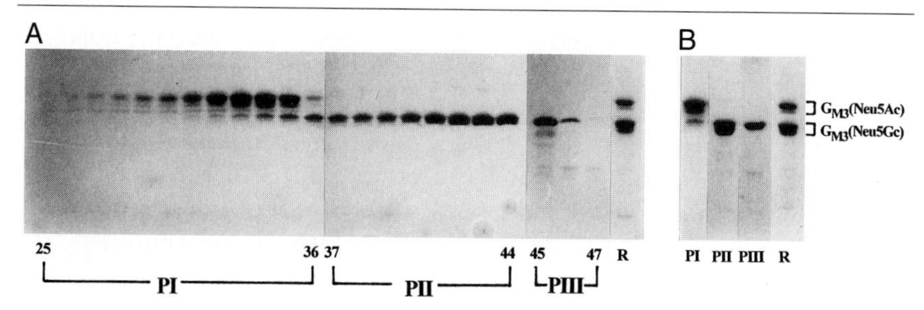

FIG. 4. TMAE-Fractogel HPLC of 20 mg of hybridoma gangliosides (A). Gangliosides are dissolved in 40 ml chloroform–methanol–water (30:60:8, v/v/v), applied to a 150-mm × 10-mm column (for details of the gradient, see Table V), and collected in 1.5-ml fractions (= 3-min fractions). Aliquots of 10 μl (0.67%) of single fractions were analyzed by TLC. Fractions can be pooled as indicated (PI to PIII) and separated by TLC (B). Gangliosides were chromatographed in alkaline solvent 2 and stained with resorcinol–hydrochloric acid reagent. The amount applied to TLC from each pool were PI, 10 μg; PII, 10 μg; PIII, 4 μg; R (reference hybridoma whole gangliosides), 10 μg.

TABLE V
HPLC PROGRAM FOR GANGLIOSIDE SEPARATION OF GM3(Neu5Ac) AND GM3(Neu5Gc) FROM HYBRIDOMA CELLS ON TMAE-FRACTOGEL[a]

		Eluent[d]		
Time[b] (min)	Flow[c] (ml/min)	A (%)	B (%)	C (%)
0	0.0	100	0	0
3	0.5	100	0	0
210	0.5	100	0	0
220	0.5	0	100	0
247	0.5	0	100	0
370	0.5	0	90	10
450	0.5	0	75	25
490	0.5	0	0	100
546	0.5	0	0	100
547	0.0	0	0	100

[a] See Figs. 4, 5, and 6. 150-mm × 10-mm column.
[b] Fractionation is started at $t = 247$ min.
[c] Using the 110-mm × 30-mm column with a flow rate of 3.0 ml/min.
[d] A: chloroform–methanol–water (30:60:8, v/v/v), B: methanol, C: 1 *M* ammonium acetate in methanol.

chromatography, 6.5, 10.5, and 2.1 mg gangliosides is recovered in pools I, II, and III, respectively, corresponding to a yield of 95.5%. Pure GM3(Neu5Gc) with differing ceramide portions is obtained in pools II and III, whereas the GM3(Neu5Ac) enriched pool I contains minor quantities of GM3(Neu5Gc). Therefore, pool I can be reapplied to the column and repeatedly chromatographed under the same conditions described above (see Table V and Fig. 5) to result in pure GM3(Neu5Ac) (pool IV) and GM3(Neu5Gc) (pool V).

A large-scale purification procedure of 500 mg whole gangliosides of hybridomas can also be performed. A large TMAE-Fractogel column (110 mm × 30 mm) is loaded with 500 mg of whole gangliosides of hybridoma cells, and gangliosides are eluted by an ammonium acetate gradient chromatography according to the HPLC program displayed in Table V. Deduced from the HPLC elution profile (Fig. 6A), pure GM3(Neu5Ac) and GM3(Neu5Gc) fractions are in pool I [GM3(Neu5Ac), 14.9 mg] and pools III and IV [GM3(Neu5Gc), 135.8 mg + 64.4 mg = 200.2 mg] (Fig. 6B). The intermediate collection (pool II) contains equal amounts of GM3(Neu5Ac) and GM3(Neu5Gc) (184.4 mg in total) and a combined fraction of gangliosides with higher molecular masses than GM3 (not shown in Fig. 6) made up 27.1 mg. The pooled fractions can be desalted by dialysis for a total recovery of 426.6 mg (85.3%).

In view of the biologic importance of GM3 in various biologic events

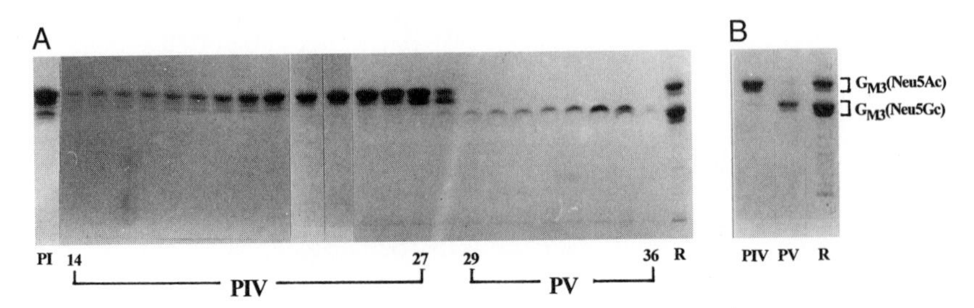

FIG. 5. TMAE-Fractogel HPLC of 6.5 mg of gangliosides from pool I of Fig. 4 [GM3(Neu5Ac)-enriched fraction] (A). Gangliosides are dissolved in 8 ml chloroform–methanol–water (30:60:8, v/v/v), applied to a 150-mm × 10-mm column (for details of the gradient, see Table V), and collected in 1.5-ml fractions (= 3-min fractions). Aliquots of 20 μl (1.33%) of single fractions are analyzed by TLC (shown as 14 to 36) in comparison to 10 μg of PI (starting material for the HPLC run). Fractions can be pooled as indicated (PIV and PV) and separated by TLC (B). Panel B shows TLC of PIV (4 μg), PV (2 μg), and reference hybridoma whole gangliosides (R, 10 μg). Gangliosides were chromatographed in the alkaline solvent 2 and stained with resorcinol–hydrochloric acid reagent.

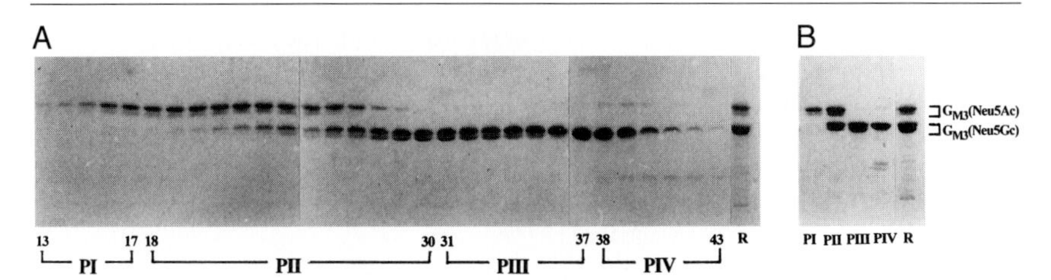

FIG. 6. TMAE-Fractogel HPLC of 500 mg of hybridoma gangliosides (A). Gangliosides are dissolved in 260 ml chloroform–methanol–water (30 : 60 : 8, v/v/v), applied to a 110-mm × 30-mm column (for details of the gradient, see Table V), and collected in 9-ml fractions (= 3-min fractions). Aliquots of 2 μl (0.022%) of single fractions are analyzed by TLC. Fractions can be pooled as indicated (PI to PIV) and separated by TLC (B). Panel B shows PI (2 μg), PII (10 μg), PIII (6 μg), PIV (4 μg) and reference hybridoma whole gangliosides (10 μg). Gangliosides were chromatographed in the alkaline solvent 2 and stained with resorcinol–hydrochloric acid reagent.

(such as immunosuppression and immunomodulation,[20,122,123] signal transduction,[24,124] and modulation of Ca^{2+} flux[125–127]), there is a need for GM3 as a starting material[128] for various GM3 derivatives.[129] Therefore, this improved method enables convenient scale-up isolation of bulk quantities (20–500 mg) of GM3 gangliosides. Hybridoma cells from industrial monoclonal antibody production are a readily available source for sialylated lactosylceramide and for preparative isolation of pure GM3(Neu5Ac) and GM3(Neu5Gc) specimens in high milligram quantities.

Acknowledgments

The author is indebted to Prof. Dr. P. F. Mühlradt, Prof. Dr. J. Lehmann, and all colleagues and coauthors of past and present cooperative projects as cited in the joint research papers and quoted in the reference list, for their continuous support. The author also thanks Dr. B.

[122] S. Ladisch, A. Hasegawa, R. Li, and M. Kiso, *Biochemistry* **34,** 1197 (1995).

[123] M. Sorice, A. Pavan, R. Misasi, T. Sansolini, T. Garofalo, L. Lenti, G. M. Pontieri, L. Frati, and M. R. Torrisi, *Scand. J. Immunol.* **41,** 148 (1995).

[124] A. Rebbaa, J. Hurh, H. Yamamoto, D. S. Kersey, and E. G. Bremer, *Glycobiology* **6,** 399 (1996).

[125] L. H. Wang, Y. P. Tu, X. Y. Yang, Z. C. Tsui, and F. Y. Yang, *FEBS Lett.* **388,** 128 (1996).

[126] Y. Yatomi, Y. Igarashi, and S.-I. Hakomori, *Glycobiology* **6,** 347 (1996).

[127] J. Müthing, U. Maurer, and S. Weber-Schürholz, *Carbohydr. Res.* **307,** 147 (1998).

[128] M. Kiso and A. Hasegawa, *Methods Enzymol.* **242,** 173 (1994).

[129] G. A. Nores, N. Hanai, S. B. Levery, H. L. Eaton, M. E. K. Salyan, and S.-I. Hakomori, *Methods Enzymol.* **179,** 242 (1989).

Kniep for valuable advice and critical reading of the manuscript. The financial support of the Deutsche Forschungsgemeinschaft (DFG) (grants Mu 845/1-2, SFB 223-C06, and SFB 549-B07) and the Ministry of Science and Research (Nordrhein-Westfalen; AZ IV A6-108 411 89) is gratefully acknowledged.

[7] Sphingolipid Extraction and Analysis by Thin-Layer Chromatography

By GERHILD VAN ECHTEN-DECKERT

Introduction

Most biologic membranes are lipid bilayers in which several membrane proteins are embedded. The major classes of membrane lipids are phospholipids, sphingolipids, and cholesterol.

Like all membrane lipids, complex sphingolipids (sphingomyelins and glycosphingolipids) are amphipathic molecules with a nonpolar aliphatic "tail," ceramide, and a polar headgroup. Ceramides are *N*-acyl (fatty acid) derivatives of long-chain "sphingoid" bases. They occur only in small amounts in tissues but form the parent compounds of more abundant complex sphingolipids. The hydrophilic headgroup, phosphocholine and a mono- or oligosaccharide for sphingomyelins and for glycosphingolipids, respectively, is linked to the C-1 position of ceramide. Today more than 300 different glycosphingolipid structures are known that can be classified in a few families based on the structure of their "core" carbohydrates (usually a neutral tetrasaccharide linked to ceramide).[1]

Cerebrosides, the simplest glycosphingolipids, are ceramides with a single sugar residue. Galactocerebrosides (galactosylceramides) are prevalent in the white matter of the brain, whereas glucocerebrosides (glucosylceramides) are ubiquitous membrane components. Like other neutral glycosphingolipids, cerebrosides are nonionic compounds. The galactose residues of some galactocerebrosides, however, are sulfated at their C-3 positions to form ionic compounds known as sulfatides.

Gangliosides, the most complex group of glycosphingolipids, contain at least one sialic acid residue (*N*-acetylneuraminic acid and its derivatives) in their oligosaccharide chain. Thus, gangliosides are negatively charged at physiologic pH and hence anionic glycosphingolipids. Along with the development of sophisticated methods for detection and structural charac-

[1] G. van Echten-Deckert and K. Sandhoff, *in* "Comprehensive Natural Products Chemistry" Vol. 3 (B. M. Pinto, ed.), Chapter 05. Elsevier Science, Amsterdam, 1999.

0076-6879/00 $30.00

terization of glycolipids, numerous novel minor components of this lipid class have been described in both neural and extraneural tissues.[2] Gangliosides represent 6% of brain lipids and are especially abundant on neuronal cell surfaces. It is well known that the cell surface carbohydrate profile is cell and species specific and characteristically changes in development, differentiation, organ regeneration, and oncogenic transformation, suggesting its significance for cell–cell interactions and cell adhesion.[2,3]

The oligosaccharide chains of glycosphingolipids are binding sites for lectins, specific carbohydrate recognizing proteins such as bacterial toxins, binding proteins of viruses and antibodies, which by means of their binding to cell surfaces might influence cellular activity.[2] During the last few years sphingolipid metabolites have been identified as endogenous signal transducing molecules. Sphingomyelin, the major membrane sphingolipid, can be hydrolyzed by sphingomyelinases to form ceramide, which stimulates differentiation, inhibits proliferation, and has also been associated with apoptosis.[4–6] Moreover, sphingosine and sphingosine 1-phosphate, originally proposed as negative regulators of protein kinase C,[7] were shown to play alternative signaling roles as mitogenic second messengers.[6]

Whereas analysis of sphingoid bases and of sphingosine 1-phosphate is described in detail elsewhere in this volume,[8,9] we focus our attention on extraction and analysis of complex sphingolipids and of ceramides. The extraction and separation procedures described here take into account both the hydrophobic nature of ceramide and the amphipathic character of glycosphingolipids and of sphingomyelin. The isolation of glycosphingolipids was described in several prior reviews in this series.[10–13]

Sphingolipid purification involves three steps: (1) extraction of lipids from a biologic source, using organic solvents; (2) separation of lipid and nonlipid contaminants; and (3) chromatographic resolution of individual species.

[2] L. Svennerholm, A. K. Asbury, R. A. Reisfeld, K. Sandhoff, K. Suzuki, G. Tettamanti, and G. Toffano, eds., "Biological Functions of Gangliosides," Progress in Brain Research, Vol. 101. Elsevier, Amsterdam, 1994.

[3] S. Hakomori, *J. Biol. Chem.* **265,** 18715 (1990).

[4] C. C. Kan and R. Kolesnick, *TIGG* **5,** 99 (1993).

[5] Y. A. Hannun and L. Obeid, *TIBS* **20,** 73 (1995).

[6] S. Spiegel, D. Foster, and R. Kolesnick, *Curr. Opin. Cell Biol.* **8,** 159 (1996).

[7] Y. A. Hannun and R. M. Bell, *Science* **234,** 500 (1989).

[8] A. H. Merrill, Jr., T. B. Caligan, E. Wang, K. Peters, and J. Ou, *Methods Enzymol.* **312,** [1], (2000) (this volume).

[9] L. Edsall, L. Vann, S. Milstien, and S. Spiegel, *Methods Enzymol.* **312,** [2], (2000) (this volume).

[10] W. J. Esselmann, R. A. Laine, and C. C. Sweely, *Methods Enzymol.* **28,** 140 (1972).

[11] R. W. Ledeen and R. K. Yu, *Methods Enzymol.* **83,** 139 (1982).

The methods described in this chapter deal with analytical amounts of sphingolipids isolated either from different types of cultured cells (50–500 μg of cellular protein) or from various animal tissues (150–400 mg wet weight).

Extraction

Chloroform–methanol mixtures have long been used for the extraction of lipids from various tissues,[14,15] and are still most commonly utilized. However, many variations of extraction procedures have been published. The most widely used were summarized in this series.[13] Recently six different solvent mixtures in which chloroform was replaced with less harmful solvents were shown to give good yields, at least for glycosphingolipids.[16]

Organic solvents employed for lipid extraction should be of analytical grade. In some cases distillation of solvents is recommended to obtain the required quality. All solvents used in our laboratory are of analytical grade (minimum purity 99.8%) from Riedel de Haen (Seelze, Germany) or Merck (Darmstadt, Germany).

Prior to extraction, tissues or cells are kept at reduced temperature (below 0°). After addition of water (see below) tissues are homogenized for several minutes with a Potter–Elvehjem glass–Teflon homogenizer, whereas cell pellets are suspended in the respective amount of water by repeated passage through plastic pipette tips and then sonicated for 1–2 min on cool water (Branson sonifier 250 with water-cooled cuphorn, Branson, Danbury, CT). At this step aliquots for other measurements (e.g., protein determination) can be taken. Then organic solvents are added to give a final ratio of chloroform–methanol–water–pyridine of 60:30:6:1 (v/v/v/v) (total volume never exceeds 10 ml). The large proportion of water improves the extraction of gangliosides.[17,18] Extraction is performed for 48 hr at 48° with continuous stirring in Reacti-Therm heating and stirring modules from Pierce (Rockford, IL). To avoid solvent loss by evaporation during extraction, screw-capped Pyrex tubes with Teflon inlays are used throughout.

Depending on cell type and amount of material, extraction temperature

---

[12] R. Kannagi, K. Watanabe, and S. Hakomori, *Methods Enzymol.* **138,** 3 (1987).
[13] R. Schnaar, *Methods Enzymol.* **230,** 348 (1994).
[14] J. Folch, M. Lees, and G. H. Sloane-Stanley, *J. Biol. Chem.* **226,** 497 (1957).
[15] E. G. Bligh and W. J. Dyer, *Can. J. Biochem. Physiol.* **37,** 911 (1959).
[16] D. Heitmann, M. Lissel, R. Kempken, and J. Müthing, *Biomed. Chromatogr.* **10,** 245 (1996).
[17] G. Tettamanti, F. Bonali, S. Marchesini, and V. Zambotti, *Biochim. Biophys. Acta* **296,** 160 (1973).
[18] L. Svennerholm and P. Fredman, *Biochim. Biophys. Acta* **617,** 97 (1980).

and time can be reduced without significant loss of efficiency. Thus extraction of neuronal cell pellets containing about 200 μg of protein gave similar results when performed at room temperature or at 48° from 6 to 48 hr with continuous stirring. However, in some cases more drastic extraction conditions are recommended (see above).

After extraction, denaturated protein particles are removed by passing the samples through cotton wadding. Small pieces of wadding are introduced with a glass stick into glass Pasteur pipettes, which are used as filtration columns (final height of the wadding column is 0.5 cm). Prior to application of the lipid extract, the wadding filter is rinsed with 1 ml of extraction solvent. The filtered sample is collected in a new screw-capped Pyrex tube with Teflon inlay placed under the pipette tip. Finally the wadding filter is rinsed with 2 ml of extraction solvent that is collected in the same tube. The solvent is then evaporated under a stream of nitrogen or alternatively in a Speed-Vac AES 2000 (Savant, Life Sciences International, Egelsbach, Germany).

Removal of Lipid Contaminants

Phospholipids are the major lipid components of biologic membranes. Like glycosphingolipids, phospholipids are amphiphilic molecules and are efficiently extracted along with sphingolipids. In solvent systems used for sphingolipid separation by thin-layer chromatography (TLC, see below) some phospholipids comigrate with different sphingolipid species. Therefore, they should be removed from the cellular lipid extracts. The esters of phospholipids (as well as those of triglycerides) are cleaved by treatment with mild alkali under conditions in which sphingolipids are stable. Therefore, mild alkaline methanolysis (saponification) represents an efficient and simple possibility to remove these major lipid contaminants. Extracts are dissolved in 2.5 ml of methanol and sonicated for 5 min in a sonifier (Sonorex RK 100, Bandelin, Berlin, Germany). Then sodium hydroxide (4 *M* stock solution in water) is added to a final concentration of 100 m*M*. After shaking for 2 hr at 37°, samples are neutralized by addition of about 10 μl of concentrated acetic acid. Finally the solvent is evaporated as described above.

Note that *O*-acyl groups of glycosphingolipids are also alkali labile moieties. Therefore, when the structures of interest contain such groups alkali treatment must be avoided.

Removal of Salts and Other Small Hydrophilic Contaminants

Crude lipid extracts always contain small amounts of nonlipid, water-soluble, low molecular weight contaminants. Small hydrophilic molecules such as salts, amino acids, sugars, small peptides extracted along with sphingolipids, as well as salts formed by addition of sodium hydroxide (see alkali treatment above), interfere with lipid behavior during separation by TLC. An optimal chromatographic resolution of different sphingolipid species therefore requires removal of these contaminants from the lipid extract. This can be achieved by dialysis,[13] by chromatography on a dextran gel column[19] or Sephadex G-50.[20]

A rapid and less elaborate method for the separation of nonpolar lipids from polar nonlipid contaminants is reversed-phase chromatography. Reversed-phase chromatography is a form of liquid–liquid partition chromatography in which the polar character of the phases is reversed: The stationary phase consists of a nonpolar liquid immobilized on an inert solid. The mobile phase is a more polar liquid. The hydrophobic interactions in reversed-phase chromatography are strong, so that the eluting mobile phase must be highly nonpolar (chloroform or other organic solvents). Williams and McLuer[21] used reversed-phase chromatography during isolation and purification of gangliosides. In this method the solvent composition was that of a theoretical upper phase (chloroform–methanol–aqueous salt solution, 3:48:47, v/v/v). Such a polar solvent dissolves gangliosides readily, but not neutral glycosphingolipids.

Kyrklund[22] then reported two different procedures to remove polar contaminants from a crude brain lipid extract by reversed-phase chromatography. In this method the complete lipid sample can be bound to the column while polar contaminants remain in the eluate. Based on this method, we developed the following procedure for desalting lipid extracts.

Preparation of Gel

1. Suspend (1:1, v/v) silica gel RP18 (silica gel LiChroprep RP18, particle size 40–63 μm from Merck, Darmstadt, Germany) in chloroform–methanol (2:1, v/v) and cautiously shake for 30 min.
2. Allow the silica gel to settle and remove (pour) the supernatant.
3. Resuspend the silica gel in methanol (1:1, v/v), shake carefully for 30 min, allow the gel to settle, and remove the supernatant. Repeat this step another 3–4 times with fresh methanol.
4. Store the suspended gel in methanol at 4°.

[19] A. Wells and J. C. Dittmer, *Biochemistry* **2,** 1259 (1963).
[20] K. Ueno, S. Ando, and R. K. Yu, *J. Lipid. Res.* **19,** 863 (1978).
[21] A. Williams and R. H. McLuer, *J. Neurochem.* **35,** 266 (1980).
[22] T. Kyrklund, *Lipids* **22,** 274 (1987).

Preparation of Columns

1. Introduce small pieces of silanized glass fiber wadding (from Macherey-Nagel, Düren, Germany) into glass Pasteur pipettes and add 2 ml of the silica gel RP18 suspension.
2. Rinse these columns two times with 1 ml of chloroform–methanol–0.1 *M* potassium chloride (6 : 96 : 94, v/v/v) at any time.

Preparation of Samples

1. Dissolve the samples (lipid extracts) in 1 ml of methanol (sonicate and warm with a hair dryer if necessary).
2. Add 1 ml of ammonium acetate (300 m*M* in H_2O) to each sample.

Chromatography

1. Apply the samples to the columns
2. Rinse the empty sample tubes two times with 0.5 ml of ammonium acetate (200 m*M* in methanol–water, 1 : 1, v/v) at any time and apply to the column. The bulk of polar contaminants remains in the flow-through fraction.
3. Wash the columns with 6 ml of water (doubly distilled quality) to elute all polar contaminants.
4. Place new tubes under each column for collection of the lipid fraction.
5. Elute lipids with 1 ml of methanol and with 8 ml of chloroform–methanol (1 : 1, v/v).
6. Dry samples by evaporating the solvent under a stream of nitrogen or in a Speed-Vac.

Separation of Anionic and Neutral Sphingolipids

Separation of anionic glycosphingolipids, such as gangliosides and sulfatides, from neutral glycosphingolipids, sphingomyelin, and ceramide can be easily achieved by anion-exchange chromatography. This is useful since comigration on TLC plates of anionic and neutral glycosphingolipids can be avoided. For example, the resolution of ganglioside GM3 and sphingomyelin as well as of sulfatide and the neutral sphingolipid lactosylceramide is very poor.

Anion-exchange chromatography or high-performance liquid chromatography (HPLC) is often used for isolation of gangliosides as reviewed in this volume.[23] For this reason I will only briefly describe the method used in our laboratory for separation of anionic and neutral sphingolipids.

[23] J. Müthing, *Methods Enzymol.* **312,** [6], (2000) (this volume).

Preparation of Resin

In our hands DEAE-Sephadex A-25 from Pharmacia LKB Biotechnology (Uppsala, Sweden) turned out to be the best resin for anion-exchange chromatography. For stability reasons this resin is delivered in its chloride form and has to be converted to its acetate form.

1. Let the DEAE-Sephadex swell overnight in distilled water (10 g in 150 ml).
2. Discard the supernatant.
3. Wash the resin with distilled water. Discard the water from the settled resin.
4. Add solvent A, 1 *M* sodium acetate in water.
5. Allow the resin to settle and remove the supernatant.
6. Add fresh solvent A.
7. Repeat steps 5 and 6 until the supernatant is devoid of chloride anions. This can be tested by acidification of the discarded supernatant by addition of concentrated nitric acid and subsequent addition of silver nitrate. Conversion of DEAE-Sephadex to the acetate form is completed when no silver chloride precipitate is formed in the supernatant.
8. Store the acetate form of DEAE-Sephadex in methanol (1:1, v/v) at 4°.

Preparation of Columns

1. Introduce small pieces of silanized glass fiber wadding (from Macherey-Nagel) into glass Pasteur pipettes and add 2 ml of the resin suspension.
2. Wash the column with 1 ml of methanol.
3. Wash the column three times with 1 ml of solvent B, chloroform–methanol–water (3:7:1, v/v/v) at any time.
4. Place a tube under each column for collection of the neutral lipids, which are eluted in the flow-through fraction.

Preparation of Samples

Dissolve each sample (dried, desalted lipid extracts) in 1 ml of solvent B and sonicate if necessary.

Chromatography

1. Apply the samples to the columns. Rinse each empty sample tube two times with solvent B (2 ml at any time) and apply to the column.

2. Elute neutral lipids with 3 ml of solvent B. After evaporation of the solvent this lipid fraction is ready for TLC.
3. Place new tubes under each column for collection of the anionic lipid fraction.
4. Elute anionic lipids with 8 ml of solvent C, chloroform–methanol–0.8 *M* ammonium acetate (3:7:1, v/v/v).
5. Dry the anionic lipid fractions and remove the salts by reversed-phased chromatography as described above.

Thin-Layer Chromatography

TLC has proved to be an invaluable tool for the qualitative and quantitative study of glycosphingolipids. However, a single TLC band, even in different solvent systems, cannot guarantee the absence of multiple species. Many closely related glycosphingolipids comigrate and are difficult to separate. Therefore, TLC alone cannot be accepted for structural identification of sphingolipids. Structural determinations often require a certain amount of highly purified material, which is not available from cell culture experiments. In this case additional evidence for the identity of certain sphingolipid species can be obtained from enzymatic degradation experiments.[24]

Due to its superior resolving power, high-performance thin-layer chromatography (HPTLC) is recommended for diverse applications.[25] However, HPTLC plates have the disadvantage that the bands, although clearly resolved, are often very closely spaced. Thus, scraping of individual bands from these plates is almost impossible.

A basic principle of separation by TLC is that resolution can be increased by decreasing the solvent strength (equivalent to solvent polarity).[26] However, a decrease of solvent strength is accompanied by an exponential decrease in R_f. To overcome this difficulty, migration mobility can be increased by high solvent velocities achieved by continuous development in a short-bed chamber.[27,28]

For more details concerning general application of TLC in lipid analysis the reader is encouraged to consult a practical approach by Henderson and

[24] G. van Echten and K. Sandhoff, *J. Neurochem.* **52,** 207 (1989).
[25] S. Ando, N.-C. Chang, and R. K. Yu, *Anal. Biochem.* **89,** 512 (1978).
[26] A. Perry, *J. Chromatogr.* **165,** 117 (1979).
[27] W. W. Young, Jr. and C. A. Borgman, *J. Lipids Res.* **27,** 120 (1986).
[28] W. W. Young, Jr. and C. A. Borgman, *Methods Enzymol.* **138,** 125 (1987).

Tocher.[29] Moreover, prior reviews dealing with TLC of glycosphingolipids[30,31] and of gangliosides[32] are recommended.

Application of Samples on TLC Plates

Glass-backed silica gel 60 precoated TLC plates from Merck are usually employed for sphingolipid TLC. Aluminum-backed plates are used only in experiments that require dipping of the plate, e.g., in sodium borate buffer for the separation of ceramide and dihydroceramide[33] or in primuline, the most sensitive nondestructive dye for lipids[34] (see below).

Samples (maximal volume 50 μl) are applied as 0.5- to 1-cm bands on the TLC plate either using small glass capillaries (length 50 mm, diameter 0.5 mm) from Hilgenberg (Malsfeld, Germany) or automatically with a Linomat (Camag, Berlin, Germany). To avoid edge effects, it is important to leave sufficient margins (2.0–2.5 cm) on each side of the plate. TLC plates are always dried overnight in a desiccator under vacuum before development.

Development of Chromatograms

Most commonly used solvent systems for sphingolipid TLC are mixtures of chloroform–methanol–water (or aqueous salts), which form complex vapor–liquid equilibria in the development tank. Therefore many factors such as tank geometry and conditions during running (temperature, unintended shaking, etc.) alter chromatographic migration and resolution. We always use rectangular glass separating chambers with polished glass lids from Desaga (Heidelberg, Germany). To improve closure, we always place weights on the lid. To avoid shaking, the chambers are placed in a cupboard. The solvents used are always freshly mixed and added to the tank to a depth of 0.5–1 cm. (Note that the margin of the TLC plate is at least 2 cm.) Prior to initiating chromatography, we allow the formation of the vapor–liquid equilibrium for at least 60 min. Depending on the solvent system used, we place filter paper on the inside glass of the chamber to help equilibrate the vapor phase. This is advantageous for the resolution of neutral glycosphingolipids, which migrate to the upper half of the TLC plate.

[29] R. J. Henderson and D. R. Tocher, *in* "Lipid Analysis: A Practical Approach" (R. J. Hamilton and S. Hamilton, eds.), p. 65. Oxford University Press, Oxford, 1992.

[30] R. L. Schnaar and L. K. Needham, *Methods Enzymol,* **230,** 348 (1994).

[31] J. Müthing, *J. Chromatogr.* **720,** 3 (1996).

[32] J. Müthing, *in* "Methods in Molecular Biology" (E. F. Hounsell, ed.), Vol. 76, p. 183. Humana Press, Totowa, NJ, 1998.

[33] H. Schulze, C. Michel, and G. van Echten-Deckert, *Methods Enzymol.* **311,** 22 (2000).

[34] V. P. Skipiski, *Methods Enzymol.* **35,** 396 (1975).

Multiple development carried out in the same or different solvent systems is often used for lipid separation.[35,36] In some cases two-dimensional chromatography can improve resolution.[29] However, two-dimensional TLC is less reproducible than are one-dimensional systems. For the solvent systems described here, except for system D (see below), a single run in ascending direction is sufficient for the desired resolution.

After running, TLC plates are removed and solvents are allowed to evaporate. Drying can be accelerated by mild heating and blowing with a hair dryer or by placing under vacuum.

Solvent Systems

As mentioned above, chloroform–methanol–aqueous mixtures of different polarities are most widely used as TLC developing solvents for sphingolipids. A good starting system for a variety of glycosphingolipids is solvent mixture A (chloroform–methanol–0.22% aqueous $CaCl_2$, w/v, 60:35:8, v/v/v). Because salts alter ganglioside mobility, thus improving their resolution, it is advantageous to replace water with aqueous salt. As illustrated in Fig. 1, highly hydrophobic lipids like fatty acids, ceramide, and cholesterol all run to the front and are thus not resolved. Also, the resolution of sphingomyelin and ganglioside GM3 as well as of lactosylceramide and sulfatide is not always satisfactory in this solvent system, especially when lipids extracted from cells or tissues are applied on TLC (not authentic lipids as in the chromatograms shown in Figs. 1–4).

One possibility to overcome this drawback is to separate anionic and neutral lipids by anion-exchange chromatography (see above). This chromatographic step allows separation of ganglioside GM3, of sulfatide, and of fatty acids (all anionic) from sphingomyelin, lactosylceramide, ceramide, and cholesterol (all neutral), respectively.

A second possibility is to use solvent system B (chloroform–methanol–2 *M* aqueous ammonia, 65:25:4, v/v/v). Ammonia in the aqueous phase changes the relative mobility of anionic lipids compared to a neutral aqueous phase and is recommended for resolving neutral sphingolipids (Fig. 2). In this system resolution of sphingomyelin and of ganglioside GM3 is better and also that of fatty acid and of ceramide. However, resolution of ceramide and of cholesterol is still not satisfactory.

To separate ceramide from cholesterol, we use solvent system C (chloroform–methanol–water, 80:10:1, v/v/v). Whereas ceramide and cholesterol are clearly resolved in this system, the R_f values of ceramide and of fatty acids are quite similar depending on the chain length of the molecular

[35] S. Harth, H. Dreyfus, P. F. Urban, and P. Mandel, *Anal. Biochem.* **86,** 543 (1978).

[36] J. K. Yao and G. M. Rastetter, *Anal Biochem.* **150,** 11 (1985).

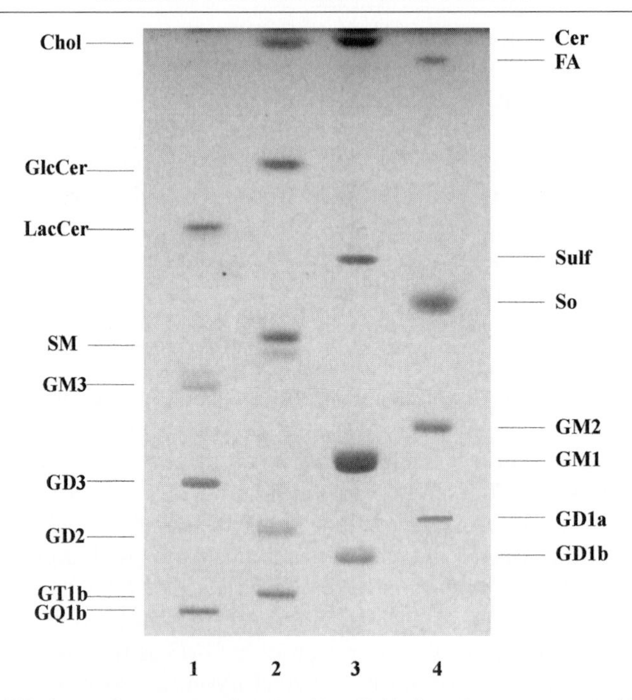

FIG. 1. Thin-layer chromatography of sphingolipids in solvent system A (chloroform–methanol–0.22% aqueous $CaCl_2$, w/v, 60 : 35 : 8, v/v/v). Standard lipids (3–18 nmol) dissolved in methanol or chloroform–methanol (1 : 1, v/v) were applied with a linomat on 1-cm lines on a 20- × 20-cm glass silica gel TLC plate as follows: lane 1, GQ1b, GD3, GM3, and lactosylceramide (LacCer); lane 2, GT1b, GD2, sphingomyelin (SM), GlcCer, and cholesterol (Chol); lane 3, GD1b, GM1, sulfatide (Sulf), and ceramide (Cer); lane 4, GD1a, GM2, sphingosine (So), and oleic acid (FA). Spingolipids were visualized with cupric sulfate in aqueous phosphoric acid as described in the text.

species investigated (Fig. 3). Therefore, we usually apply this solvent system after anion-exchange chromatography.

In solvent system D water is replaced with concentrated acetic acid (chloroform–methanol–acetic acid, 190 : 9 : 1, v/v/v). The chromatogram obtained after two consecutive runs (note that between the runs plates must be dried) in the same direction in this solvent system is shown in Fig. 4. Acetic acid reduces the mobility of ceramide and thus improves the resolution of this sphingolipid and both fatty acid and cholesterol. In both solvent systems C and D, glycosphingolipids as well as long-chain bases have a very reduced mobility and are found close to the origin after development of the plate (Figs. 3 and 4). Furthermore, different ceramide species can be resolved in these solvent systems. Therefore, solvent systems C and

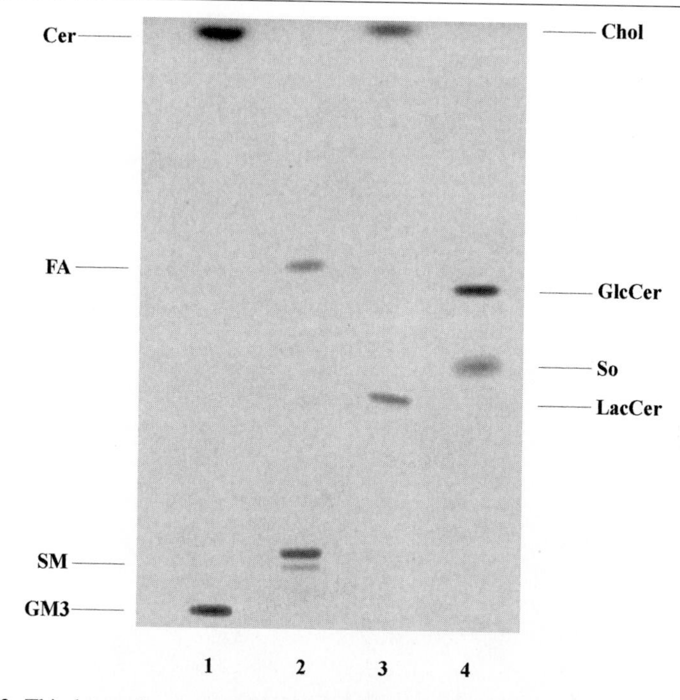

FIG. 2. Thin-layer chromatography of sphingolipids in solvent system B (chloroform–methanol–2 *M* aqueous ammonia, 65 : 25 : 4, v/v/v). Authentic lipids (6–18 nmol) were applied and visualized as described in the legend to Fig. 1. Lane 1, GM3 and ceramide (Cer); lane 2, sphingomyelin (SM) and oleic acid (FA); lane 3, lactosylceramide (LacCer) and cholesterol (Chol); lane 4, glucosylceramide (GlcCer) and sphingosine (So).

D are recommended for direct qualitative and quantitative determination of ceramide in the desalted lipid extract.

The separation of ceramide from dihydroceramide using borate-impregnated TLC plates is described in detail in another chapter of volume 311.[33] The chemical structures of the two molecules differ only in the 4,5-*trans* double bond in the long-chain base moiety of ceramide.

Visualization and Quantitative Determination of Sphingolipids on TLC

Autoradiography

Autoradiography is a very simple procedure for detection of radioactive lipids after TLC. Cellular sphingolipids can be biosynthetically labeled with radioactive precursors like L-[3-^{14}C]serine, L-[^{3}H]serine, ^{3}H- or ^{14}C-labeled

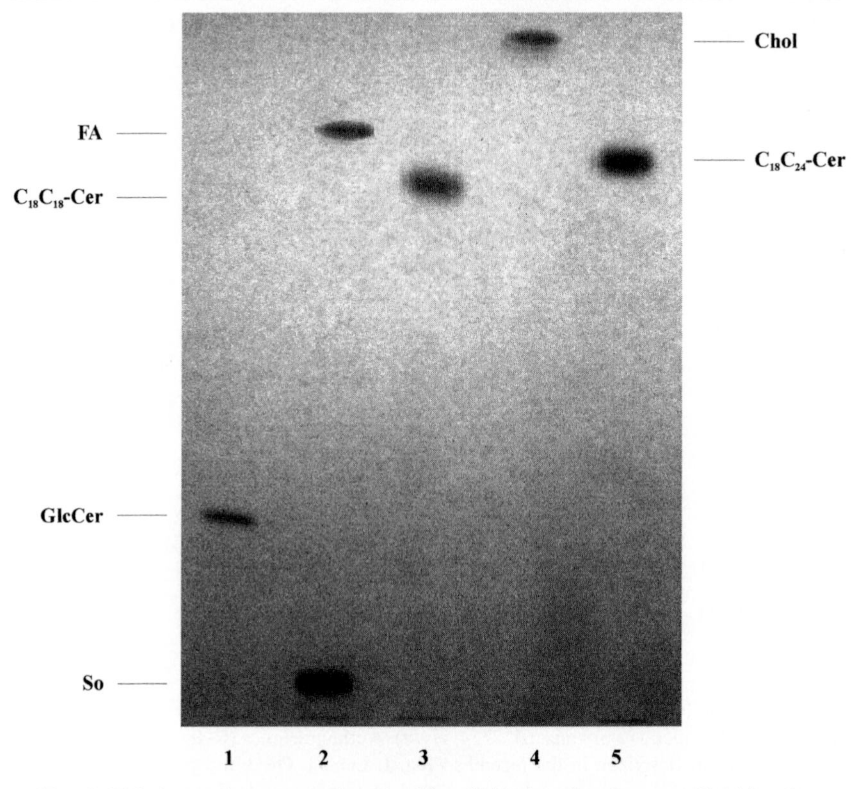

FIG. 3. Thin-layer chromatography of sphingolipids in solvent system C (chloroform–methanol–water, 80:10:1, v/v/v). Authentic lipids (6–18 nmol) were applied and visualized as described in the legend to Fig. 1. Lane 1, glucosylceramide (GlcCer); lane 2, sphingosine (So) and oleic acid (FA); lane 3, ceramide, *N*-stearoylsphingosine ($C_{18}C_{18}$-Cer); lane 4, cholesterol (Chol); lane 5, ceramide, *N*-nervonoylsphingosine ($C_{18}C_{24}$-Cer).

fatty acids, long-chain bases tritiated either at the 4,5 double bond or at carbon atom 3 and for glycosphingolipids also [^{3}H]- or [^{14}C]galactose or other sugars. Autoradiography can be achieved with different techniques. The simplest method is to cover the TLC plate with X-ray film (Kodak, Rochester, NY; X-Omat AR). To ensure good contact between the plate and the film, we use an autoradiograph cassette. Cassettes also have the advantage of being light-tight. After exposure for an appropriate time at $-70°$ the film is developed following the manufacturer's instructions. The exposure time must be determined empirically because it depends on the isotope used, the amount of radioactivity in each sphingolipid species, and also the requirement of the experiment. Because the energy of the tritium β particle is very low and has a short path length, fluorography is recom-

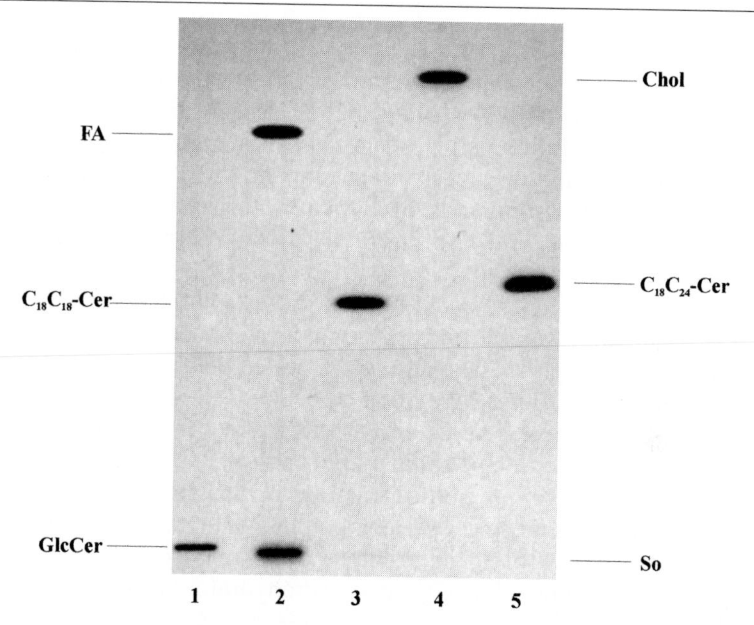

FIG. 4. Thin-layer chromatography of sphingolipids in solvent system D (chloroform–methanol–acetic acid, 190:9:1, v/v/v). Authentic lipids were applied exactly as described in the legend to Fig. 3 and visualized as described in the legend to Fig. 1. Two consecutive runs in the same direction were performed as described in the text.

mended for detecting tritium on TLC. The developed and dried TLC plate is dipped in a scintillator (12% 2,5-diphenyloxazole in ethyl ether, w/v) prior to autoradiography. Note that evaporation of the solvent must occur before covering the plate with a film.

The developed film can then be used to mark the radioactive spots on the TLC plate. Regions of interest can be scraped off the plate and counted in a scintillation counter. Furthermore, the spots on the film can be quantified by densitometric scanning (densitometer from Shimadzu, dual-wavelength TLC scanner CS-910, Duisburg, Germany).

Alternatively, radioactive spots on the TLC plates can be directly detected and quantified with a TLC linear analyzer, Tracemaster 40 (Berthold, Wildbad, Germany). A more elegant method is detection and evaluation of radioactive spots on TLC plates with the bioimaging analyzer Fujix Bas 1000 using software TINA 2.08 from Raytest (Straubenhardt, Germany). The high sensitivity of the latter two methods allows for quick detection of tritium on TLC without prior fluorography.

Staining

Numerous spray reagents of different specificity and sensitivity are used for lipid visualization on TLC.[37] Many of these reagents are quite aggressive. Therefore the dried plate is placed upright in a protective frame (spray box from Desaga) and the dye solution is applied with a fine mist sprayer (Desaga) moving the spray in a zigzag pattern from one edge of the plate to the other until the whole surface is covered. Alternatively, plates can be dipped in the dye solution (e.g., in the case of primuline). Usually sprayed plates have to be heated in an oven for color development.

A very simple, yet unspecific procedure is to expose the chromatogram to iodine vapors in a closed chamber. Unsaturated lipids and some saturated lipids as well as many other compounds appear as yellow or brown fractions on a light background. This coloring is reversible and therefore recommended for first detection of sphingolipid spots.

The most selective and sensitive spray for gangliosides is resorcinol, which is specific for sialic acid.[38] All compounds containing sialic acid appear as blue–violet zones. Both gangliosides and neutral glycosphingolipids can be visualized with a number of general carbohydrate stains (for review see Ref. 30).

For qualitative and quantitative determination of sphingolipids, we recommend cupric sulfate in aqueous phosphoric acid.[36] It is a sensitive char reagent that can detect 250 pmol of GM1 and 1 nmol of sulfatide. Prepare the reagent by mixing 15.6 g of cupric sulfate pentahydrate with 9.4 ml of phosphoric acid (85% in water, w/v) and 100 ml of distilled water. Spray the TLC plate heavily (see above) until it appears wet. Drain the plate for about 5 min. Then heat it in an oven at 150°–180° for about 15 min. In addition to sphingolipids, cholesterol and phospholipids appear brown against a white background (see Figs. 1–4). Avoid color fade by keeping the plate light protected and cool. The spots are quantified with a Shimadzu dual-wavelength flying spot scanning densitometer (CS-9301PC), using software Quanta Scan 2D(9301) from Shimadzu. Although very sensitive, a certain disadvantage of charring is that samples are destroyed by the visualization process.

An even more sensitive and nondestructive method is fluorescent staining with primuline.[34] It reveals nanogram amounts of sphingolipids (corresponding to about 50 pmol). The blue fluorescence quality of primuline was first described by Pick when he used the dye as a vital stain.[39] Primuline

[37] E. Vioque, *in* "Handbook of Chromatography" (H. K. Mangold, ed.), Vol. II, p. 309. CRC Press, Boca Raton, Florida, 1984.

[38] L. Svennerholm, *Biochim. Biophys. Acta* **4,** 405 (1957).

[39] J. Pick, *J. Microsc. Acta* **51,** 338 (1935).

is an anionic thiazole dye that is moderately soluble in water. It is a yellow–gold powder and its aqueous solutions display a pale yellow color when observed in visible light. Under ultraviolet light a blue fluorescence is apparent.

Prepare a stock solution by dissolving 100 mg of primuline (from Sigma, St. Louis, MO, or Aldrich, Milwaukee, WI) in 10 ml of acetone and add distilled water to a final volume of 100 ml. Store this solution at 4° and protect it from light. Before use, 1 ml of the stock solution is added to 100 ml of phosphate buffered saline. The developed and dried aluminum-backed TLC plate is first dipped for 5 min in phosphate-buffered saline (PBS). Then the wet plate is dipped in primuline for 10 min in the dark. Finally the plate is washed for 10 min in PBS and then carefully dried by mild blowing and heating with a hair dryer. The washing steps prior and after primuline staining reduce the background coloring. Lipid bands appear as light bands against a dark background under ultraviolet light. Fluorescent lipid bands can be determined analogous to the quantification of lipid spots after charring using the Shimadzu densitometer (see above) equipped with a fluorometry attachment (xenon lamp). Because this stain is reversible, primuline can be removed by silicic acid chromatography[40] after recovery of lipids from the TLC plate by extraction with organic solvents.

Acknowledgments

The help of colleagues and technicians during my work in the field of sphingolipids is gratefully acknowledged. I thank Prof. Konrad Sandhoff (Universität Bonn, Germany) for kind support through all my years of working in his laboratory. This chapter is dedicated to Prof. Sandhoff on the occasion of his 60th birthday. The financial support of the Deutsche Forschungsgemeinschaft (grant SFB 284) and the German–Israeli Foundation for Scientific Research and Development is greatly acknowledged.

[40] G. K. Ostrander, S. B. Levery, H. L. Eaton, M. E. K. Salyan, S. Hakomori, and E. H. Holmes, *J. Biol. Chem.* **263,** 18716 (1988).

[8] Extraction and Analysis of Multiple Sphingolipids from a Single Sample

By TAMOTSU TAKETOMI and EIKO SUGIYAMA

Introduction

A long and fascinating history of sphingolipid research, beginning with the discovery of brain sphingomyelin and cerebroside containing sphingosine as the characteristic unit of the sphingolipids in 1884 by Thudichum,[1] has established the chemical structures of numerous sphingolipids in animal tissues and cells. Particularly during the decades after World War II, improved methods of isolation and structure determination have revealed an astonishing variety of oligosaccharide structures that occur in the 300 or more glycosphingolipids.[2] Biologic functions of trace amounts of exogenous and endogenous sphingolipids in many vertebral and invertebral cells now command a great deal of attention. The chemical identification of the sphingolipids and related materials in the cells or tissues always should be prerequisite to studies of their function and undertaken easily, quickly, and accurately.

Fortunately, an excellent method, matrix-assisted laser desorption ionization time-of-fight mass spectrometry with delayed ion extraction (DE MALDI-TOF MS), can be applied to the identification and analysis of sphingolipids extracted from individual animal cells or tissues. In this chapter we describe the application of DE MALDI-TOF MS to gala-, globo-, and ganglio-series glycosphingolipids and sphingomyelin as well as some examples of experimental procedures and practical methods for simultaneous analysis of multiple sphingolipids in individual animal cells or tissues.

Principle of DE MALDI-TOF MS

In matrix-assisted laser desorption ionization (MALDI), the sample is embedded in a low molecular weight, UV-absorbing matrix [2,5-dihydroxybenzoic acid (2,5-DHB) or α-4-hydroxycinnamic acid (α-4-CHCA)] that

[1] J. L. W. Thudichum, "A Treatise on the Chemical Constitution of the Brain." Ballière, Tindall & Cox, London, 1884.

[2] R. W. Ledeen, *in* "Sphingolipids as Signaling Modulators in the Nervous System" (R. W. Ledeen, S. Hakomori, A. J. Yates, J. S. Schneider, and R. K. Yu, eds.), *Ann. N.Y. Acad. Sci.* **845,** xi (1998).

0076-6879/00 $30.00

enhances sample ionization. While the exact role of the matrix is not completely defined, it appears to transfer enough energy to the sample to ionize it to a molecular ion, but not enough energy to fragment the ions into atomic components.

Time-of-flight mass spectrometry works on the principle that if ions are accelerated with the same kinetic energy from a fixed point and at a fixed initial time, the ions will separate according to their mass/charge ratios. Light ions travel more quickly to the analyzer, heavier ions more slowly. The time required for ions to reach the detector and the ion intensity (abundance) are measured. Drift time is proportional to the square root of the mass:

$$t = s(m/2KE)^{1/2}$$

where t is the drift time, s is the drift distance, m is mass, and $EK = 1/2m(s/t)^2$.

Furthermore, MALDI-TOF MS with the delayed ion extraction (DE) technique has the advantages of a broad mass range (>100 kDa), high sensitivity (~100 fmol), high accuracy (~0.1%), and high resolution (5000 full-width at half-maximum intensity).

Identification of Individual Molecular Species of Sphingolipid by DE MALDI-TOF MS

Because any one class of sphingolipid consists of many individual molecular species by the combination of long-chain bases (d18:1 and d20:1 sphingosine, d18:0 and d20:0 dihydrosphingosine, and t18:0 and t20:0 phytosphingosine, etc.) and a variety of fatty acids (unbranched, monounsaturated and α-hydroxy alkyl chain lengths from 14 to 28 carbon atoms), the isolation and identification of individual molecular species of sphingolipid have previously been impossible. DE MALDI-TOF MS is capable of giving rise to molecular weight-related ions corresponding to the molecular species, and the numerous molecular ions can be identified in combination of these results with other analyses of the long-chain bases in the form of the lysosphingolipids prepared by deacylation of the starting sphingolipid.

Preparation of Lysosphingolipids

Because our usual method[3–5] for the preparation of lysosphingolipids was not optimized for microanalysis, a new method has been invented using

[3] T. Taketomi and T. Yamakawa, *J. Biochem.* **54**, 444 (1963).
[4] T. Taketomi and N. Kawamura, *J. Biochem.* **68,** 475 (1970).
[5] T. Taketomi, N. Kawamura, A. Hara, and S. Murakami, *Biochim. Biophys. Acta* **424,** 106 (1970).

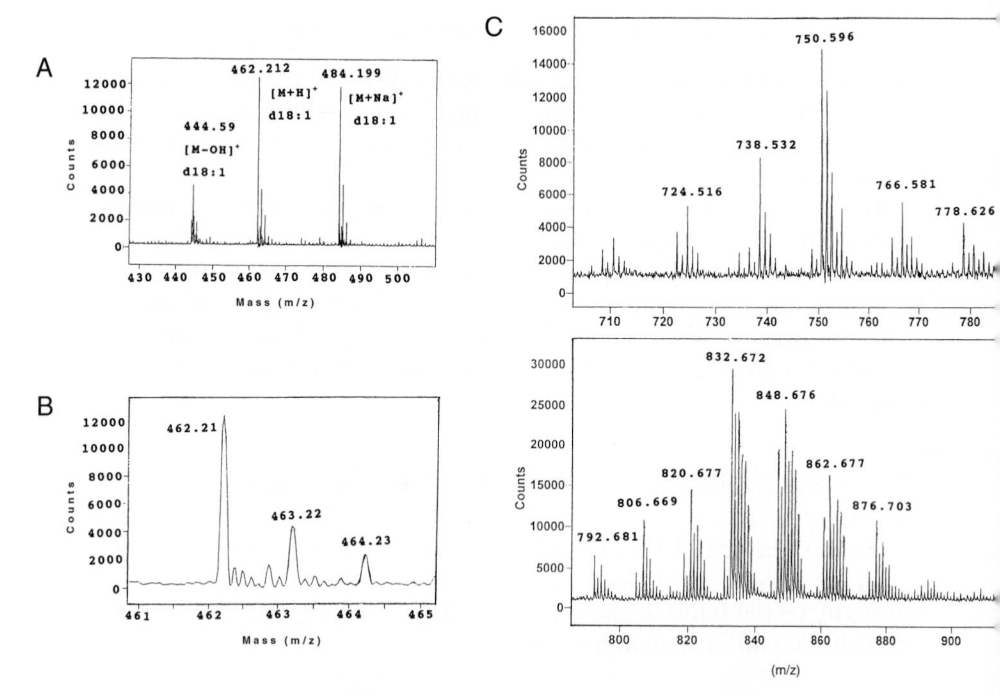

FIG. 1. (A) DE MALDI-TOF mass spectrum of psychosine deacylated from galactosylceramide (cerebroside of monkey brain) in the positive ion mode. (B) Enlarged high-resolution spectrum of isotopically resolved peaks of protonated psychosine (d18:1); $[M + H]^+$ at m/z 462.21 in (A) for the determination of psychosine (d18:0). (C) DE MALDI-TOF mass spectrum of intact cerebroside (monkey brain).

Mass (m/z)	Ions	LCB	FA	Mass (m/z)	Ions	LCB	FA
710	$[M + Na]^+$	d18:0	C15:0	820	$[M + Na]^+$	d18:1	C23:0
722	$[M + Na]^+$	d18:1	C16:0	822	$[M + Na]^+$	d18:0	C23:0
724	$[M + Na]^+$	d18:0	C16:0		$[M + Na]^+$	d18:1	C22:0h
736	$[M + Na]^+$	d18:1	C17:0	832	$[M + Na]^+$	d18:1	C24:1
738	$[M + Na]^+$	d18:0	C17:0	834	$[M + Na]^+$	d18:1	C24:0
750	$[M + Na]^+$	d18:1	C18:0	836	$[M + Na]^+$	d18:0	C24:0
754	$[M + H]^+$	d18:1	C20:1		$[M + Na]^+$	d18:1	C23:0h
764	$[M + Na]^+$	d18:1	C85:0h	846	$[M + Na]^+$	d18:1	C25:1
766	$[M + Na]^+$	d18:0	C18:0h	848	$[M + Na]^+$	d18:1	C25:0
768	$[M + H]^+$	d18:1	C20:0h		$[M + Na]^+$	d18:1	C24:1h
778	$[M + Na]^+$	d18:1	C20:0	850	$[M + Na]^+$	d18:1	C24:0h
780	$[M + H]^+$	d18:1	C22:0		$[M + Na]^+$	d18:0	C24:1h
782	$[M + H]^+$	d18:0	C22:0	860	$[M + Na]^+$	d18:1	C26:1
792	$[M + Na]^+$	d18:1	C20:1h	862	$[M + Na]^+$	d18:1	C26:0
794	$[M + Na]^+$	d18:1	C20:0h	864	$[M + Na]^+$	d18:0	C26:0
804	$[M + Na]^+$	d18:1	C22:1	874	$[M + Na]^+$	d18:1	C27:1
806	$[M + Na]^+$	d18:1	C22:0	876	$[M + Na]^+$	d18:1	C27:0
818	$[M + Na]^+$	d18:1	C23:1	878	$[M + Na]^+$	d18:0	C27:0

LCB, long-chain base; FA, fatty acid; h, hydroxy fatty acid.

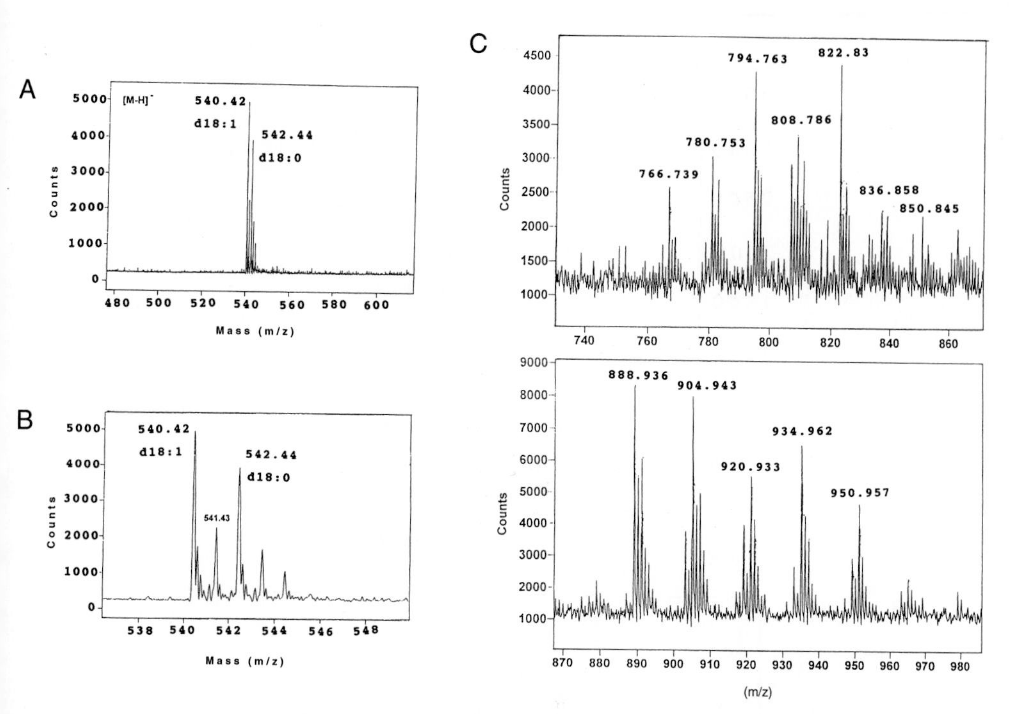

FIG. 2. (A) DE MALDI-TOF mass spectrum of lysosulfatide deacylated from sulfatide (monkey brain) in the negative ion mode. (B) Enlarged high-resolution spectrum of isotopically resolved peaks of deprotonated lysosulfatide (d18:1), $[M - H]^-$ at m/z 540.42 in (A) for the determination of lysosulfatide (d18:0). (C) DE MALDI-TOF mass spectrum of intact sulfatide (monkey brain).

Mass (m/z)	Ions	LCB	FA	Mass (m/z)	Ions	LCB	FA
765	$[M - H]^-$	d18:1	C15:0	889	$[M - H]^-$	d18:1	C24:1
767	$[M - H]^-$	d18:0	C15:0	891	$[M - H]^-$	d18:1	C24:0
779	$[M - H]^-$	d18:1	C16:0	903	$[M - H]^-$	d18:1	C25:1
781	$[M - H]^-$	d18:0	C16:0	905	$[M - H]^-$	d18:1	C25:0
783	$[M - H]^-$	d18:0	C15:0h		$[M - H]^-$	d18:1	C24:1h
793	$[M - H]^-$	d18:1	C17:0	907	$[M - H]^-$	d18:0	C25:0
795	$[M - H]^-$	d18:1	C16:0h		$[M - H]^-$	d18:1	C24:0h
797	$[M - H]^-$	d18:0	C16:0h	919	$[M - H]^-$	d18:1	C26:0
807	$[M - H]^-$	d18:1	C18:0		$[M - H]^-$	d18:1	C25:1h
809	$[M - H]^-$	d18:0	C18:0	921	$[M - H]^-$	d18:0	C26:0
811	$[M - H]^-$	d18:0	C17:0h		$[M - H]^-$	d18:1	C25:0h
823	$[M - H]^-$	d18:1	C18:0h	933	$[M - H]^-$	d18:1	C27:0
825	$[M - H]^-$	d18:0	C18:0h		$[M - H]^-$	d18:1	C26:1h
833	$[M - H]^-$	d18:1	C20:0	935	$[M - H]^-$	d18:0	C27:0
837	$[M - H]^-$	d18:0	C20:0		$[M - H]^-$	d18:1	C26:0h
839	$[M - H]^-$	d18:0	C19:0h	949	$[M - H]^-$	d18:0	C28:0
851	$[M - H]^-$	d18:1	C20:0h	951	$[M - H]^-$	d18:1	C27:0h
879	$[M - H]^-$	d18:0	C23:0	965	$[M - H]^-$	d18:0	C28:0h
	$[M - H]^-$	d18:1	C22:0h				

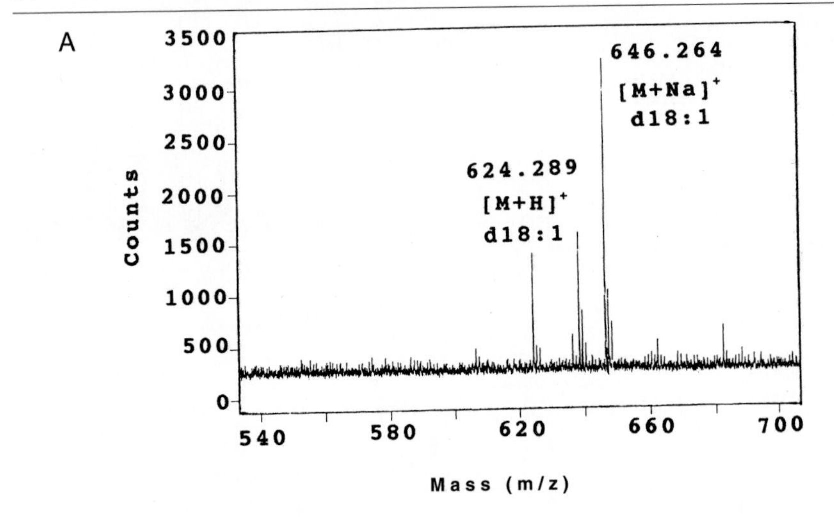

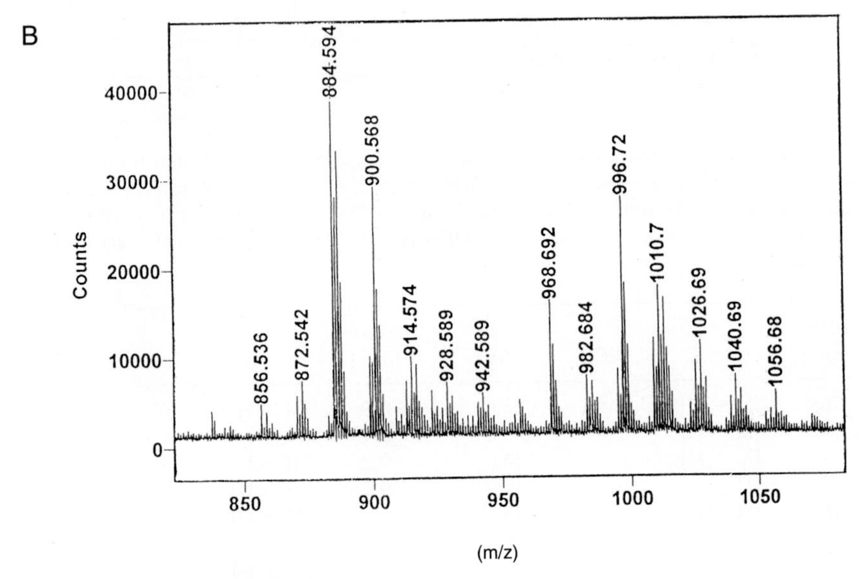

FIG. 3. (A) DE MALDI-TOF mass spectrum of lysolactosylceramide deacylated from lactosylceramide (porcine erythrocytes) in the positive ion mode. Protonated and sodium-adducted lysolactosylceramide (d18:1), $[M + H]^+$ at m/z 624.29 and $[M + Na]^+$ at m/z 646.26. (B) DE MALDI-TOF mass spectrum of intact lactosylceramide (porcine erythrocytes) in the positive ion mode.

microwave-mediated saponification of glycosphingolipids except sphingomyelin.[6–8] This is carried out as follows: about 1 mg of glycosphingolipids is dissolved in 0.5 ml of 0.1 *M* NaOH in methanol in a thick-walled Pyrex glass centrifuge tube (10 ml) with a Teflon-lined screw cap. The atmosphere is replaced by argon (or nitrogen) and the tubes are irradiated in a microwave oven (500 W, Toshiba model ER-V-11) for 2 min. For some sphingolipids, the yield is improved by conducting the microwave irradiation while the sample is under an atmosphere of argon (personal communication with Dr. Alfred H. Merrill). The volume of the alkaline methanol solution must not exceed 0.5 ml and other safety precautions should be taken because the tube may explode. A sealed glass tube must not be used owing to the possibility of explosion of the tube.

After saponification, the slightly turbid solution is cooled to room tem-

[6] T. Taketomi, A. Hara, K. Uemura, and E. Sugiyama, *J. Biochem.* **120,** 573 (1996).
[7] T. Taketomi, A. Hara, K. Uemura, H. Kurahashi, and E. Sugiyama, *Biochem. Biophys. Res. Commun.* **224,** 462 (1996).
[8] T. Taketomi, A. Hara, K. Uemura, H. Kurahashi, and E. Sugiyama, *J. Biochem.* **121,** 264 (1997).

FIG. 3. (*continued*)

Mass (m/z)	Ions	LCB	FA	Mass (m/z)	Ions	LCB	FA
856	$[M + Na]^+$	d18:1	C14:0	984	$[M + Na]^+$	d18:0	C23:0
872	$[M + Na]^+$	d18:0	C15:0		$[M + Na]^+$	d18:1	C22:0h
884	$[M + Na]^+$	d18:1	C16:0	986	$[M + Na]^+$	d18:0	C22:0h
886	$[M + Na]^+$	d18:0	C16:0	994	$[M + Na]^+$	d18:1	C24:1
898	$[M + Na]^+$	d18:1	C17:0	996	$[M + Na]^+$	d18:1	C24:0
900	$[M + Na]^+$	d18:1	C16:0h	1008	$[M + Na]^+$	d18:1	C25:1
	$[M + Na]^+$	d18:0	C17:0	1010	$[M + Na]^+$	d18:1	C25:0
912	$[M + Na]^+$	d18:1	C18:0	1012	$[M + Na]^+$	d18:0	C25:0
914	$[M + Na]^+$	d18:0	C18:0		$[M + Na]^+$	d18:1	C24:0h
	$[M + Na]^+$	d18:1	C17:0h	1022	$[M + Na]^+$	d18:1	C26:1
916	$[M + Na]^+$	d18:0	C17:0h	1024	$[M + Na]^+$	d18:1	C26:0
928	$[M + Na]^+$	d18:1	C18:0h	1026	$[M + Na]^+$	d18:0	C26:0
	$[M + Na]^+$	d18:0	C19:0		$[M + Na]^+$	d18:1	C25:0h
930	$[M + Na]^+$	d18:0	C18:0h	1028	$[M + Na]^+$	d18:0	C25:0h
940	$[M + Na]^+$	d18:1	C20:1	1038	$[M + Na]^+$	d18:1	C27:0
942	$[M + Na]^+$	d18:1	C20:0	1040	$[M + Na]^+$	d18:0	C27:0
	$[M + Na]^+$	d18:0	C20:1		$[M + Na]^+$	d18:1	C26:0h
944	$[M + Na]^+$	d18:0	C20:0	1042	$[M + Na]^+$	d18:0	C26:0h
968	$[M + Na]^+$	d18:1	C22:0	1052	$[M + Na]^+$	d18:1	C28:0
	$[M + Na]^+$	d18:0	C22:1	1054	$[M + Na]^+$	d18:0	C28:0
982	$[M + Na]^+$	d18:1	C23:0	1056	$[M + Na]^+$	d18:0	C27:0h

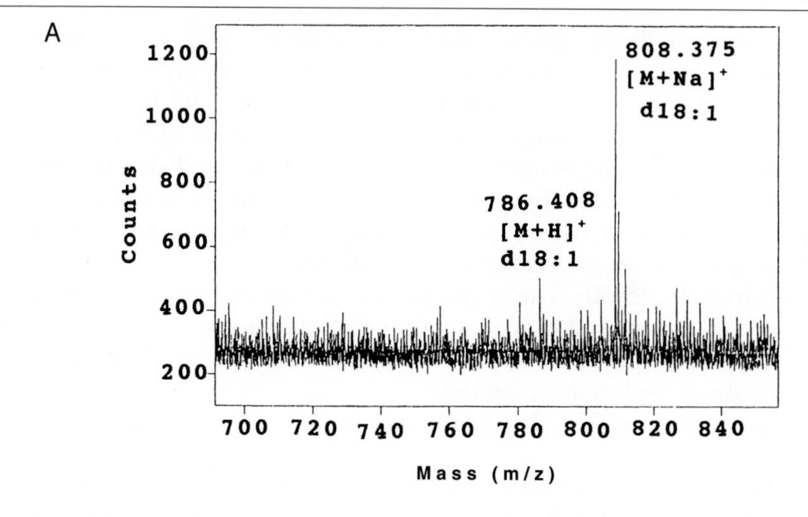

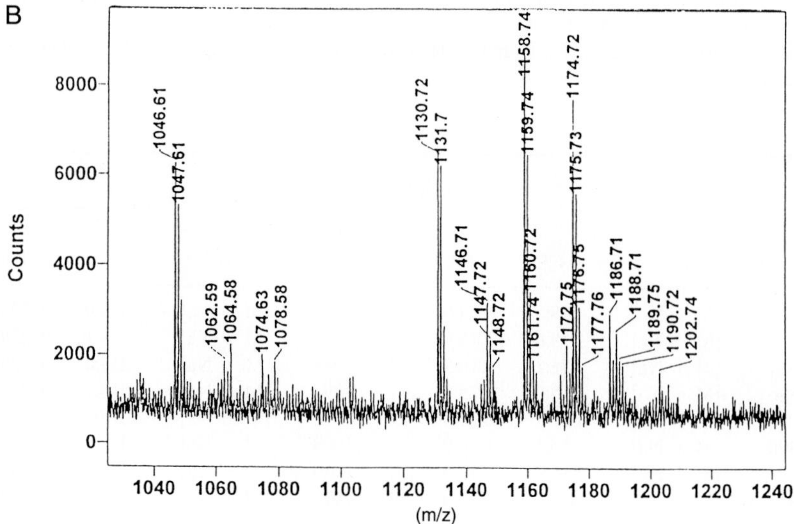

FIG. 4. (A) DE MALDI-TOF mass spectrum of lysoglobotriaosylceramide deacylated from globotriaosylceramide (porcine erythrocytes) in the positive ion mode. Protonated and sodium-adducted lysoglobotriaosylceramide (d18:1), $[M + H]^+$ at m/z 786.41 and $[M + Na]^+$ at m/z 808.37. (B) DE MALDI-TOF mass spectrum of globotriaosylceramide (porcine erythrocytes) in the positive ion mode.

perature, and acidified with one drop of 3 *N* HCl together with one drop of distilled water. The reaction mixture is vigorously shaken with 1 ml of hexane using a vortex mixer and is allowed to separate into an upper phase and a lower phase. The upper phase contains free fatty acids and is removed carefully with a micropipette. After the lower phase is extracted with hexane, it is applied to a Sep-Pak C_{18} cartridge (Waters Associates, Milford, MA) to remove salt. The cartridge is washed with about 5 ml of methanol–water (3:7, v/v), 5 ml of water, and then the lysoglycosphingolipids (and occasionally unreacted glycosphingolipids) are eluted with about 5 ml of ethanol–water (8:2, v/v) and 5 ml of chloroform–methanol–water (60:30:4.5, v/v/v). After the solvent has been evaporated under N_2, the lyso compound is subjected to DE MALDI-TOF MS analysis for the confirmation of lysoglycosphingolipid as well as the determination of long-chain base composition.

Analysis of Lysosphingolipids and Sphingolipids by DE MALDI-TOF MS

The mass spectrometric analysis is performed as follows[6–9]: One microliter of chloroform–methanol (1:1, v/v) solution containing about 1–10 pmol of lysoglycosphingolipid or 50–200 pmol of sphingolipid and 1 μl of the matrix solution [10 mg of α-cyano-4-hydroxycinnamic acid (α-CHCA) in 1 ml of a 1:1 mixture of acetonitrile–water containing 0.1% of trifluoroacetic acid, or 10 mg of 2,5-dihydroxybenzoic acid (2,5-DHB) in 1 ml of 8:2 mixture of water–ethanol] are placed in a 1.5-ml Eppendorf tube, shaken vigorously on a vortex mixer, and then spun down in a microcentrifuge (Chibitan, Japan Millipore, Tokyo, Japan).

[9] E. Sugiyama, A. Hara, K. Uemura, and T. Taketomi, *Glycobiology* **7,** 719 (1997).

FIG. 4. (*continued*)

Mass (*m*/*z*)	Ions	LCB	FA	Mass (*m*/*z*)	Ions	LCB	FA
1046	$[M + Na]^+$	d18:1	C16:0	1158	$[M + Na]^+$	d18:1	C24:0
1062	$[M + Na]^+$	d18:1	C16:0h	1174	$[M + Na]^+$	d18:1	C24:0h
1064	$[M + Na]^+$	d18:0	C16:0h		$[M + Na]^+$	d18:0	C24:1h
1074	$[M + Na]^+$	d18:1	C18:0	1186	$[M + Na]^+$	d18:1	C25:1h
1078	$[M + H]^+$	d18:1	C20:1	1188	$[M + Na]^+$	d18:1	C25:0
1130	$[M + Na]^+$	d18:1	C22:0		$[M + Na]^+$	d18:0	C25:1h
1146	$[M + H]^+$	d18:1	C22:0h	1202	$[M + Na]^+$	d18:0	C26:1h
1158	$[M + Na]^+$	d18:0	C24:1		$[M + Na]^+$	d18:1	C26:0h

One microliter of the supernatant is loaded onto the sample plate, then allowed to dry and crystallize for about 20 min at room temperature. The sample plate is loaded into the load position of Voyager Elite XL (6.6-m flight length in the reflector mode) Biospectrometry Workstation (PerSeptive Biosystem, Framingham, MA). A nitrogen laser (337 nm) is used for the desorption ionization. In the positive ion mode with 2,5-DHB or α-CHCA used as the matrix, potassium ($[M]^+$: 38.96) and angiotensin I ($[M + Na]^+$: 1,297.50) are used to calibrate the instrument. Neutral glycosphingolipids and sphingomyelin are usually measured in the positive ion mode. In the negative ion mode with 2,5-DHB or α-CHCA, lysosulfatide (d18:1, $[M - H]^-$: 540.28) and GM1 (d18:1, C_{18}, $[M - H]^-$: 1544.87) are used to calibrate the instrument. Acidic glycosphingolipids are measured in the negative ion made. An adequate N_2 laser step (1400–2800) was freely selected for a good analytical result. The resolution of the ion peak (full-width at half-maximum intensity) is measured by the resolution calculator in the GRAM/386 software. Five-point Savitsky–Golay smoothing is usually applied to the spectra.

Gala-Series Glycosphingolipids

A lyso compound is prepared from galactosylceramide (cerebroside) isolated from monkey brain and it, as well as the intact cerebroside, is separately subjected to DE MALDI-TOF MS analysis.[7] The mass spectrum in Fig. 1A shows molecular weight-related ions of $[M - OH]^+$ at m/z 444.59, $[M + H]^+$ at m/z 462.21, and $[M + Na]^+$ at m/z 484.20, confirming psychosine (lysogalactosylceramide) containing d18:1 mainly. Also, the presence of psychosine containing d18:0 was detected by the following procedure.

The counts of the isotopically resolved ion peaks at m/z 462.21, 463.22, and 464.23 in the enlarged high-resolution spectrum in Fig. 1B accounted for 65.9, 22.5, and 11.6%. The probabilities of the corresponding peaks indicating no ^{13}C atoms (monoisotopic molecule), one ^{13}C atom, and two ^{13}C atoms of psychosine (d18:1) were theoretically calculated by the natural abundance of stable isotope as 76.7, 20.6, and 2.7%.

Because the difference between the found and theoretical values indicated the presence of psychosine (d18:0), the long-chain base composition of monkey brain cerebroside was estimated to be d18:1 (91.1%) and d18:0 (8.9%). Thus, the molecular weight-related ion peaks corresponding to individual molecular species of monkey brain cerebroside in the mass spectrum were identified as described in the legend to Fig. 1C. Although this analysis of the mass spectrum provides a quantitative analysis of individual

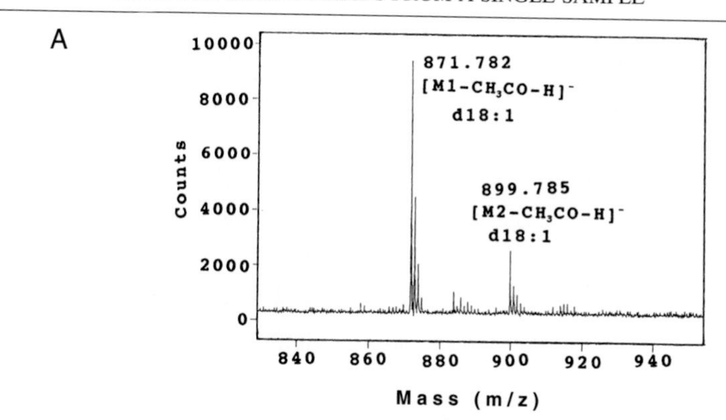

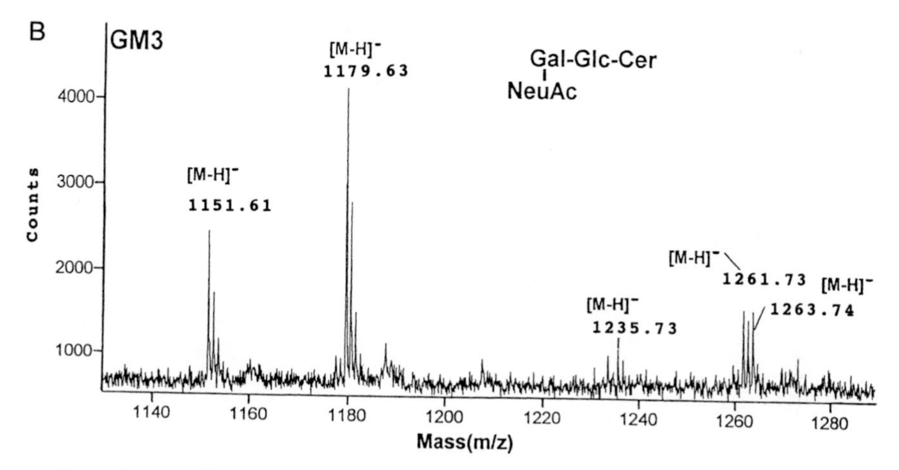

FIG. 5. (A) DE MALDI-TOF mass spectrum of the lyso compound deacylated from GM3 (human brain tumor) in the negative ion mode. Deprotonated deacetylated lysoGM3 (d18:1) and (d20:1), $[M_1\text{-}CH_3CO\text{-}H]^-$ at m/z 871.78 and $[M_2\text{-}CH_3CO\text{-}H]^-$ at m/z 899.79. (B) DE MALDI-TOF mass spectrum of intact GM3 (human brain tumor) in the negative ion mode.

Mass (m/z)	Ions	LCB	FA	Mass (m/z)	Ions	LCB	FA
1151	$[M - H]^-$	d18:1	C16:0	1261	$[M - H]^-$	d18:1	C24:1
1179	$[M - H]^-$	d18:1	C18:0		$[M - H]^-$	d20:1	C22:1
1235	$[M - H]^-$	d18:1	C20:0	1263	$[M - H]^-$	d18:1	C24:0
	$[M - H]^-$	d20:1	C18:0		$[M - H]^-$	d20:1	C22:0

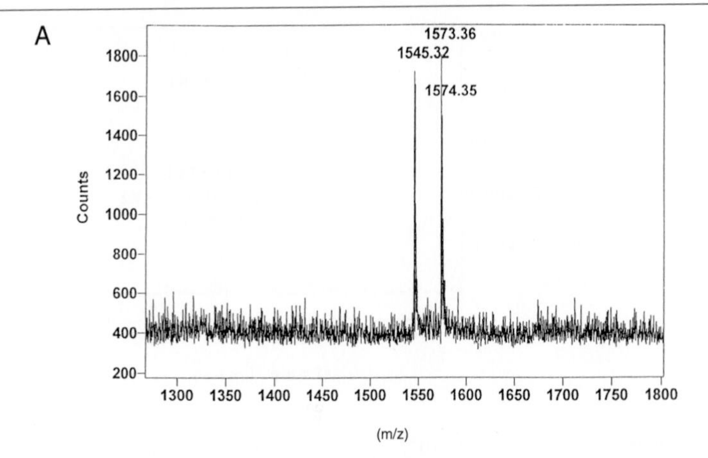
A
1573.36
1545.32
1574.35
Counts
(m/z)

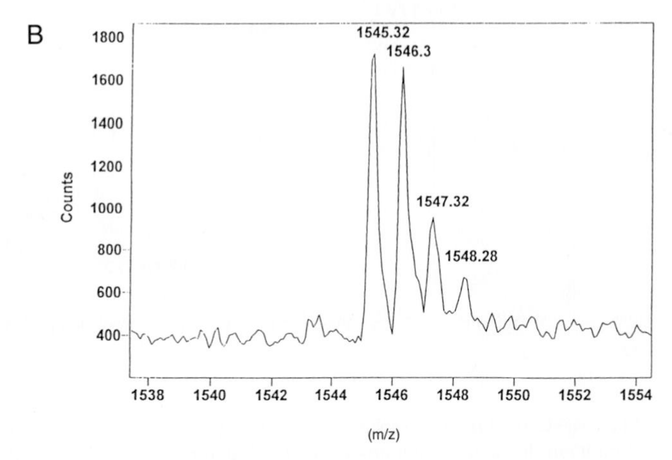
B
1545.32
1546.3
1547.32
1548.28
Counts
(m/z)

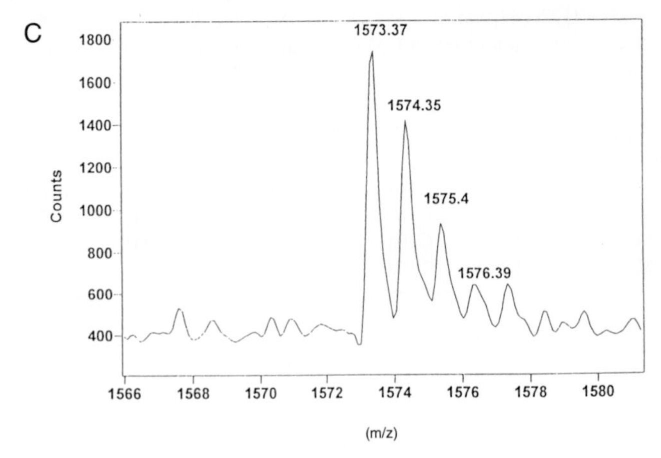
C
1573.37
1574.35
1575.4
1576.39
Counts
(m/z)

molecular species, the ability of this procedure to give exact quantities awaits.

Figure 2A shows the mass spectrum of the lyso compound obtained from another class of sphingolipid, sulfatide prepared from monkey brain provided a molecular weight-related ion of $[M - H]^-$ at m/z 540.42, confirming lysosulfatide (d18:1). The presence of lysosulfatide (d18:0) was also confirmed by the difference between the found and theoretical values of the isotopically resolved peaks in the enlarged high-resolution spectrum in Fig. 2B, thus, the long-chain base composition of monkey brain sulfatide was estimated to be d18:1 (70.6%) and d18:0 (29.4%). The molecular weight-related ions corresponding to individual molecular species of monkey brain sulfatide were identified as shown in the legend to Fig. 2C.

Globo-Series Glycosphingolipids

Globo-series lysoglycosphingolipids[7] are analyzed by this same procedure. Lysolactosylceramide was prepared from porcine erythrocytes and gave the molecular weight-related ions $[M + H]^+$ at m/z 624.29 and $[M + Na]^+$ at m/z 646.26 in the mass spectrum in Fig. 3A, and the ratio of d18:1 to d18:0 in the latter ion peak was estimated to be 95.8 and 4.2%. Thus, the molecular weight-related ion peaks corresponding to individual molecular species of lactosylceramide were decided as shown in the legend to Fig. 3B. Lysoglobotriaosylceramide from the same source gave the molecular weight-related ions $[M + H]^+$ at m/z 786.41 and $[M + Na]^+$ at m/z 808.38 in the mass spectrum in Fig. 4A, and the ratio of d18:1 to d18:0 in the latter peak was estimated to be 97.8 and 2.2%. Thus, the molecular weight-related ion peaks corresponding to individual molecular species of globotriaosylceramide were determined as shown in the legend to Fig. 4B. The same approach can be applied to lysoglobotetraosylceramide and lysoglobopentaosylceramide.[6]

Ganglio-Series Glycosphingolipids: Gangliosides and Asialogangliosides

Gangliosides[5,9] are highly complex oligoglycosylceramides, which contain one or more sialic acid groups, in addition to glucose, galactose, and

FIG. 6. (A) DE MALDI-TOF mass spectrum of intact GM1 (adult human brain) in the negative ion mode. Molecular weight-related ions: $[M_1 - H]^-$ at m/z 1545.32 and $[M_2 - H]^-$ at m/z 1573.36 corresponding to GM1 (d18:1C18:0) and GM1 (d20:1C18:0). (B) Enlarged high-resolution spectrum of the isotopically resolved peaks of deprotonated GM1 (d18:1C18:0) for the determination of GM1 (d18:0C18:0). (C) Enlarged high-resolution spectrum of the isotopically resolved peaks of deprotonated GM1 (d20:1C18:0) for the determination of GM1 (d20:0C18:0).

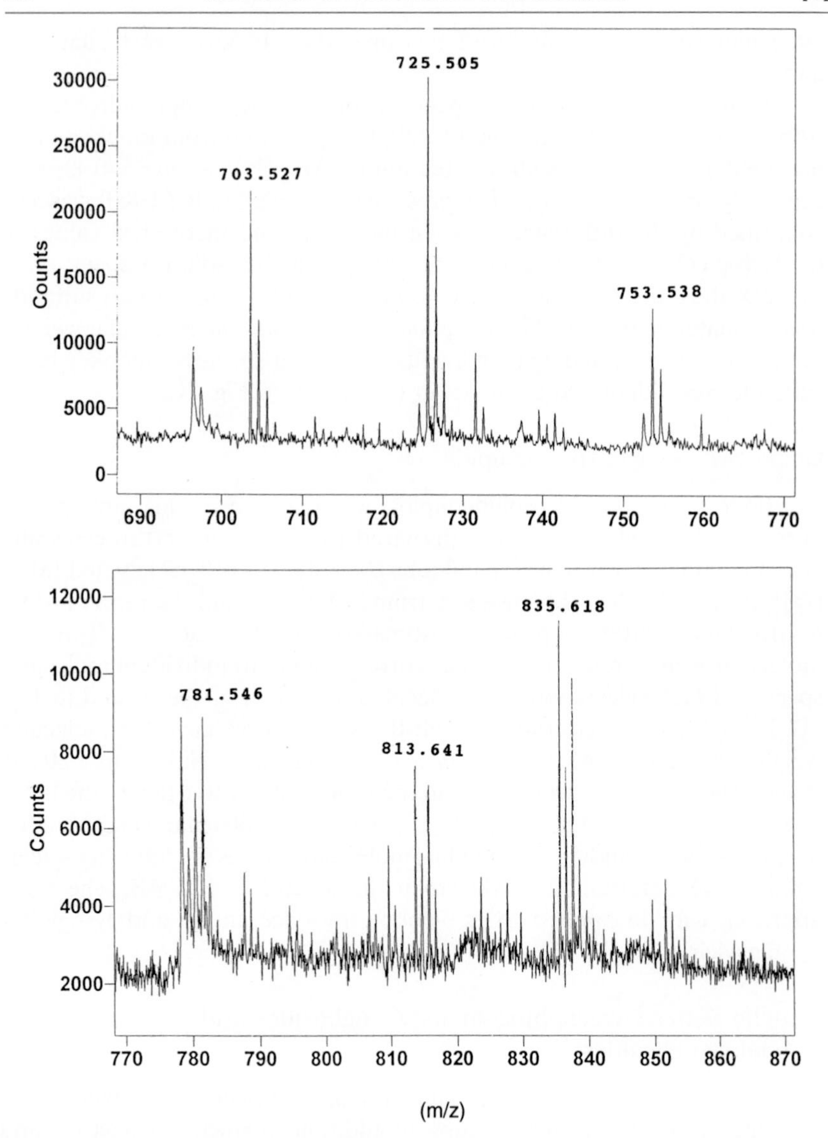

FIG. 7. DE MALDI-TOF mass spectrum of intact sphingomyelin (caprine erythrocytes) in the positive ion mode.

N-acetylgalactosamine. However, most gangliosides contain a major long-chain base of d18:1 and d20:1 and major stearic acid and thus consist of relatively simple individual molecular species.

Regarding GM3 isolated from human brain tumor tissue,[9] the lyso compound prepared from GM3 gave two molecular weight-related ion peaks corresponding to lysoGM3 (d18:1), $[M_1\text{-}CH_3CO\text{-}H]^-$ at m/z 871.78 and lysoGM3 (d20:1), $[M_2\text{-}CH_3CO\text{-}H]^-$ at m/z 899.79 in the mass spectrum in negative ion mode in Fig. 5A. Also, the presence of lysoGM3 (d18:0) and (d20:0) was detected by the difference between the found and theoretical values of three isotopically resolved peaks of each lysoGM3 (d18:1) and (d20:1) in both the enlarged high-resolution spectra (data not shown) and thus the ratios of d18:1 to d18:0 and d20:1 to d20:0 were estimated to be 97.6 and 2.4%, and 99.1 and 0.9%. Together with the percentage of counts of two molecular weight-related ions of lysoGM3 (d18:1) and (d20:1), the long-chain base composition of GM3 was estimated to be d18:1 (77.4%), d18:0 (1.9%), d20:1 (20.5%), and d20:0 (0.2%). Thus, the molecular weight-related ion speaks corresponding to individual molecular species of GM3 of human brain tumor tissue were decided as shown in the legend to Fig. 5B. Analysis of GM2 can be made by this same method.

Finally, regarding GM1 from the brain tissue,[10] the preparation of lyso compound is not always necessary for the determination of the long-chain

[10] T. Taketomi, A. Hara, K. Uemura, and E. Sugiyama, *Acta Biochim. Polon.* **45,** 987 (1998).

FIG. 7. (*continued*)

Mass (m/z)	Ions	LCB	FA	Mass (m/z)	Ions	LCB	FA
696	?	?	?	781	$[M + Na]^+$	d18:1	C20:0
697	$[M + Na]^+$	d18:1	C14:0	806	?	?	?
703	$[M + H]^+$	d18:1	C16:0	809	$[M + Na]^+$	d18:1	C22:0
711	$[M + Na]^+$	d18:1	C15:0	813	$[M + H]^+$	d18:1	C24:1
725	$[M + Na]^+$	d18:1	C16:0	815	$[M + H]^+$	d18:1	C24:0
*	$[M + Na]^+$	d18:0	C16:1	823	$[M + Na]^+$	d18:1	C23:0
731	$[M + H]^+$	d18:1	C18:0	827	$[M + H]^+$	d18:1	C25:1
739	$[M + K]^+$	d18:1	C16:0	835	$[M + Na]^+$	d18:1	C24:1
741	$[M + K]^+$	d18:0	C16:0	837	$[M + Na]^+$	d18:1	C24:0
753	$[M + Na]^+$	d18:1	C18:0	**	$[M + Na]^+$	d18:0	C24:1
759	$[M + H]^+$	d18:1	C20:0	849	$[M + Na]^+$	d18:1	C25:1
778	?	?	?	851	$[M + K]^+$	d18:1	C24:1

* and ** suggested by the isotopically resolved peaks of sphingomyelin $[M + Na]^+$ (d18:1C16:0) and (d18:1C24:0), respectively.

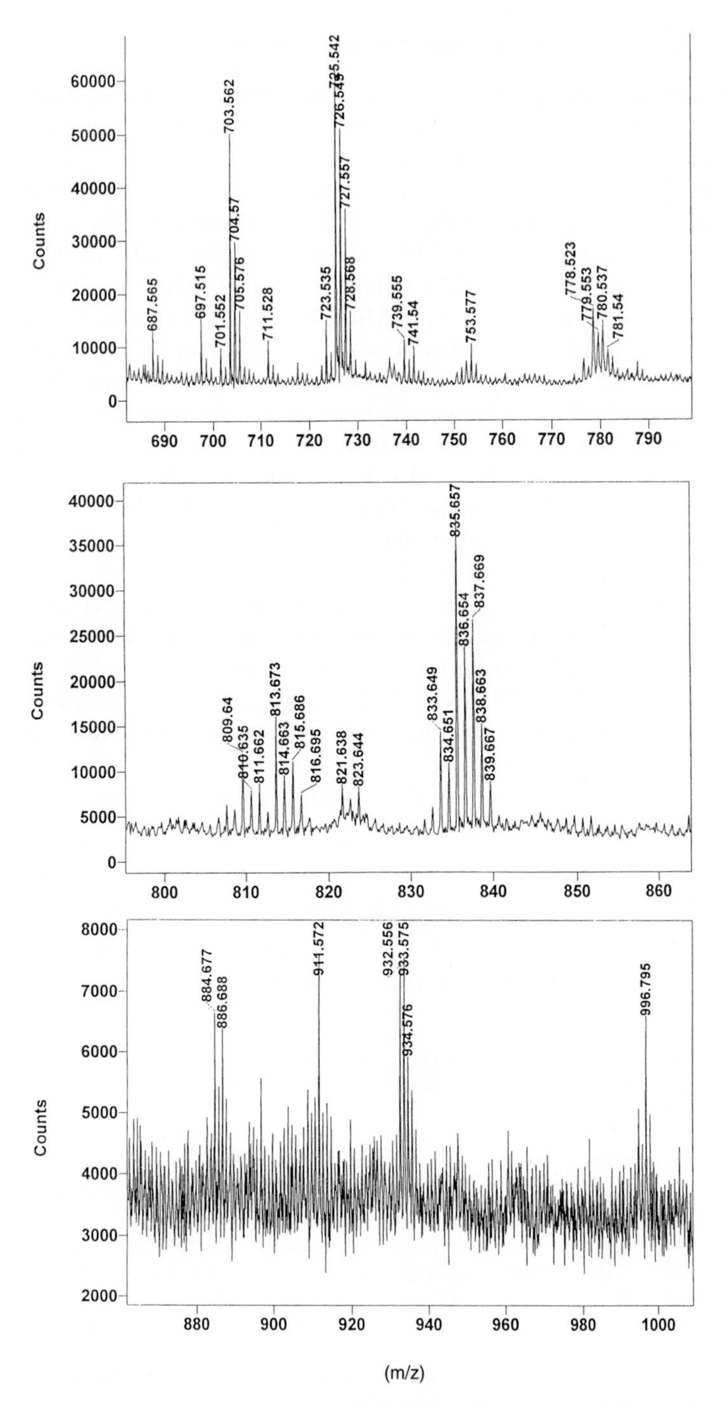

FIG. 8. DE MALDI-TOF mass spectrum of multiple sphingolipids of the alkali-stable lipid fraction from human platelets in the positive ion mode.

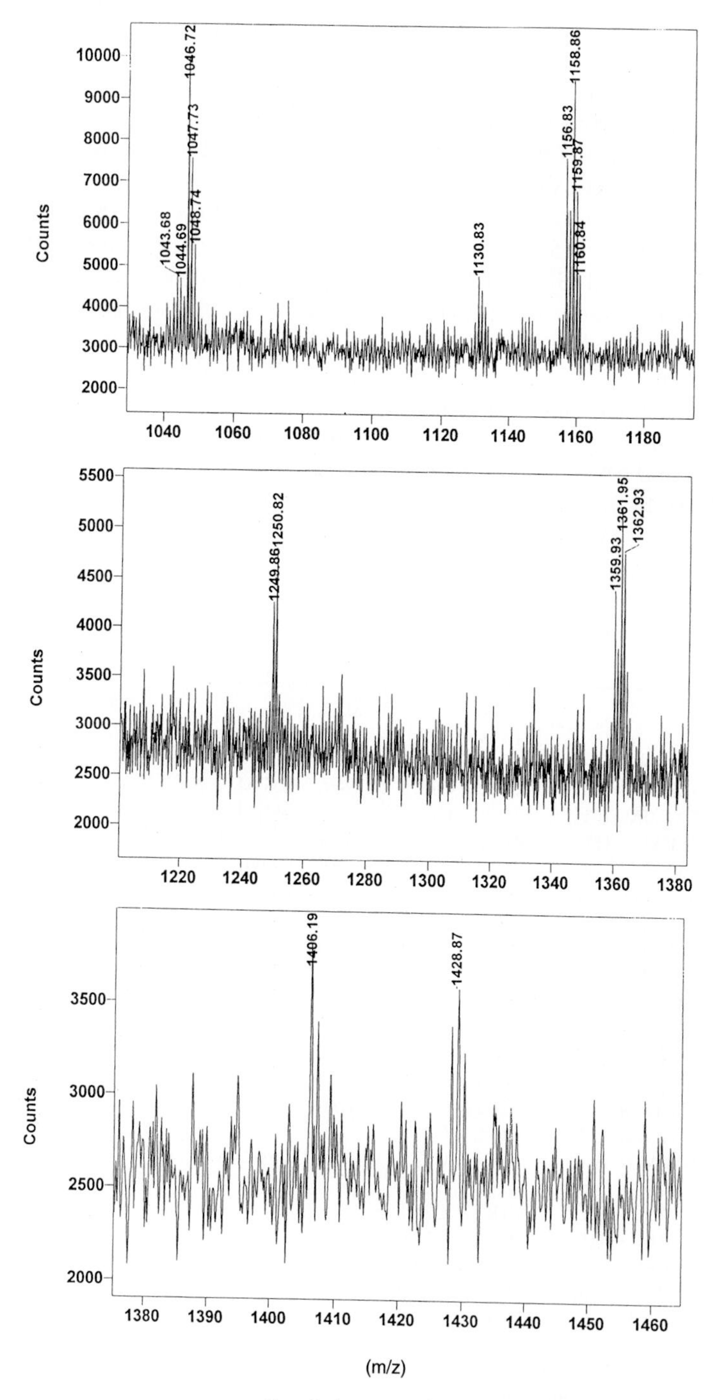

FIG. 8. (*continued*)

base composition, because there is only one fatty acid, stearic acid. Intact GM1 isolated from the brain tissue of an adult patient with sudanophilic leukodystrophy was directly analyzed by DE MALDI-TOF MS. As shown in Fig. 6A, it was exactly identified by two molecular weight-related ions: $[M_1 - H]^-$ at *m/z* 1545.42 corresponding to GM1 (d18:1C18:0) and $[M - H]^-$ at *m/z* 1573.46 corresponding to GM1 (d20:1C18:0). The GM1 (d18:0C18:0) and GM1 (d20:0C18:0) were also detected by the difference between the found and theoretical values of three isotopically resolved peaks of each GM1 (d18:1C18:0) and GM1 (d20:1C18:0) in both the enlarged high-resolution spectra in Figs. 6B and 6C in the same way as already described above, and the ratios of GM1 (d18:1C18) to (d18:0C18:0) and GM1 (d20:1C18:0) to (d20:0C18:0) were estimated to be 95.1 and 4.9%, and 93.4 and 6.6%. Together with the ratios of counts of GM1 (d18:1C18:0) to (d20:1C18:0), the long-chain base composition of GM1 was calculated to be d18:1 (44.2%), d18:0 (2.3%), d20:1 (49.9%), and d20:0 (3.6%). The

FIG. 8. (*continued*)

Mass (*m/z*)	Ions	LCB	FA	Mass (*m/z*)	Ions	LCB	FA
687	SM$[M + H]^+$	d18:1	C15:1	836	SM$[M + Na]^+$	d18:1	C24:1
697	CMH$[M + H]^+$	d18:1	C16:1	838	SM$[M + Na]^+$	d18:1	C24:0
703	SM$[M + H]^+$	d18:1	C16:0	884	CDH$[M + Na]^+$	d18:1	C16:0
	SM$[M + H]^+$	d18:0	C16:1	886	CDH$[M + Na]^+$	d18:0	C16:0
711	SM$[M + Na]^+$	d18:0	C15:0	911	CDH$[M + Na]^+$	d18:1	C18:0
723	SM$[M + Na]^+$	d18:1	C16:1	932	CDH$[M + H]^+$	d18:1	C21:0
725	SM$[M + Na]^+$	d18:1	C16:0	996	CDH$[M + Na]^+$	d18:1	C24:0
	SM$[M + Na]^+$	d18:0	C16:1	1046	CTH$[M + Na]^+$	d18:1	C16:0
727	SM$[M + Na]^+$	d18:0	C16:0	1131	CTH$[M + Na]^+$	d18:1	C22:0
739	SM$[M + K]^+$	d18:1	C16:1	1157	CTH$[M + Na]^+$	d18:1	C24:1
741	SM$[M + K]^+$	d18:1	C16:0	1159	CTH$[M + Na]^+$	d18:1	C24:0
753	SM$[M + Na]^+$	d18:1	C18:0	1250	Gb$[M + Na]^+$	d18:1	C16:0
778	CMH$[M + Na]^+$	d18:1	C20:0	1360	Gb$[M + Na]^+$	d18:1	C24:1
780	CMH$[M + Na]^+$	d18:0	C20:0	1362	Gb$[M + Na]^+$	d18:1	C24:0
809	SM$[M + Na]^+$	d18:1	C22:0	1406	GM2$[M + Na]^+$	d18:1	C18:0
811	SM$[M + Na]^+$	d18:0	C22:0	1428	GM2$[M + 2Na - H]^+$	d18:1	C18:0
813	SM$[M + H]^+$	d18:1	C24:1				
815	SM$[M + H]^+$	d18:1	C24:0				
821	SM$[M + Na]^+$	d18:1	C23:1				
823	SM$[M + Na]^+$	d18:1	C23:0				
834	CMH$[M + Na]^+$	d18:1	C24:0				

SM, sphingomyelin; CMH, glucosylceramide; CDH, lactosylceramide; CTH, globotriaosylceramide; Gb, globotetraosylceramide; GM2, ganglioside GM2.

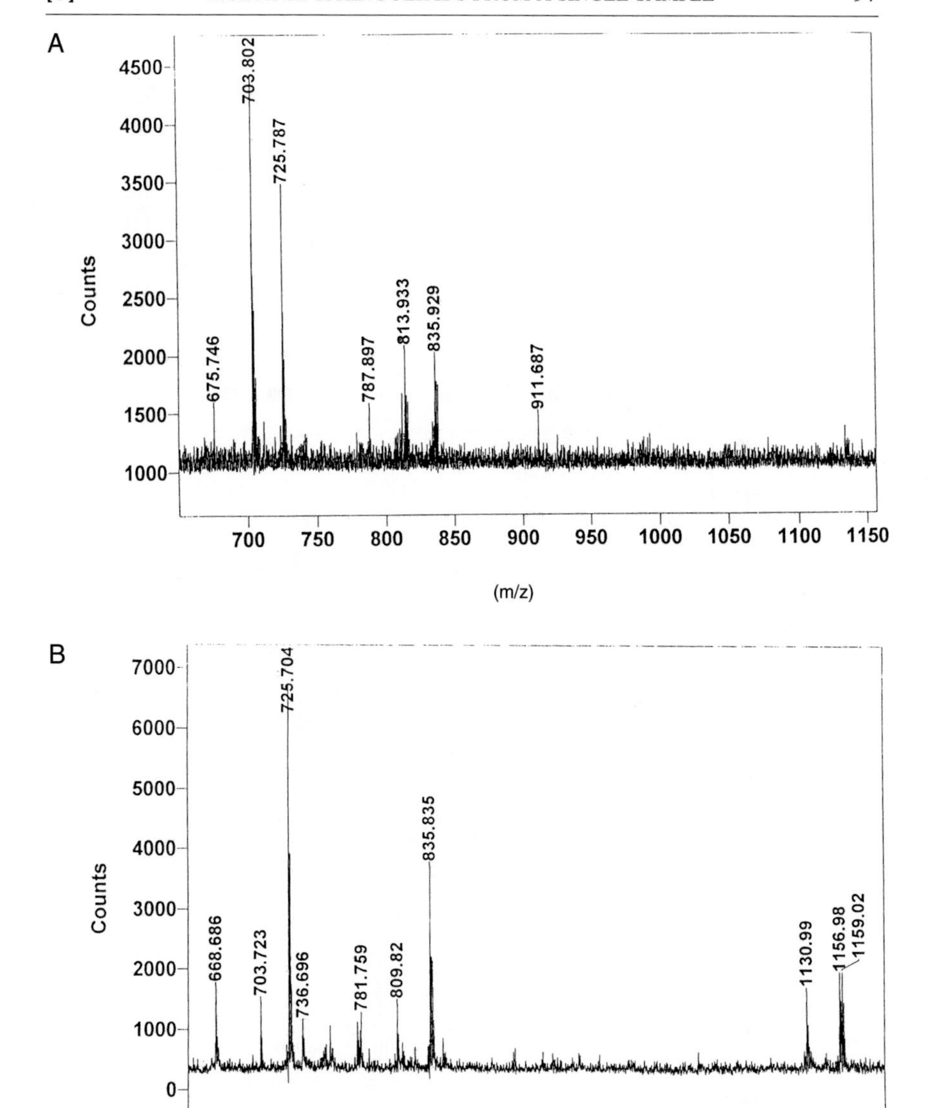

FIG. 9. (A) DE MALDI-TOF mass spectrum of multiple sphingolipids of the alkali-stable lipid fraction from the necrotic tissue of head of femur of a patient as a control in the positive ion mode. (B) DE MALDI-TOF mass spectrum of multiple sphingolipids of the alkali-stable lipid fraction from the necrotic tissue of head of femur of a patient with Fabry disease.

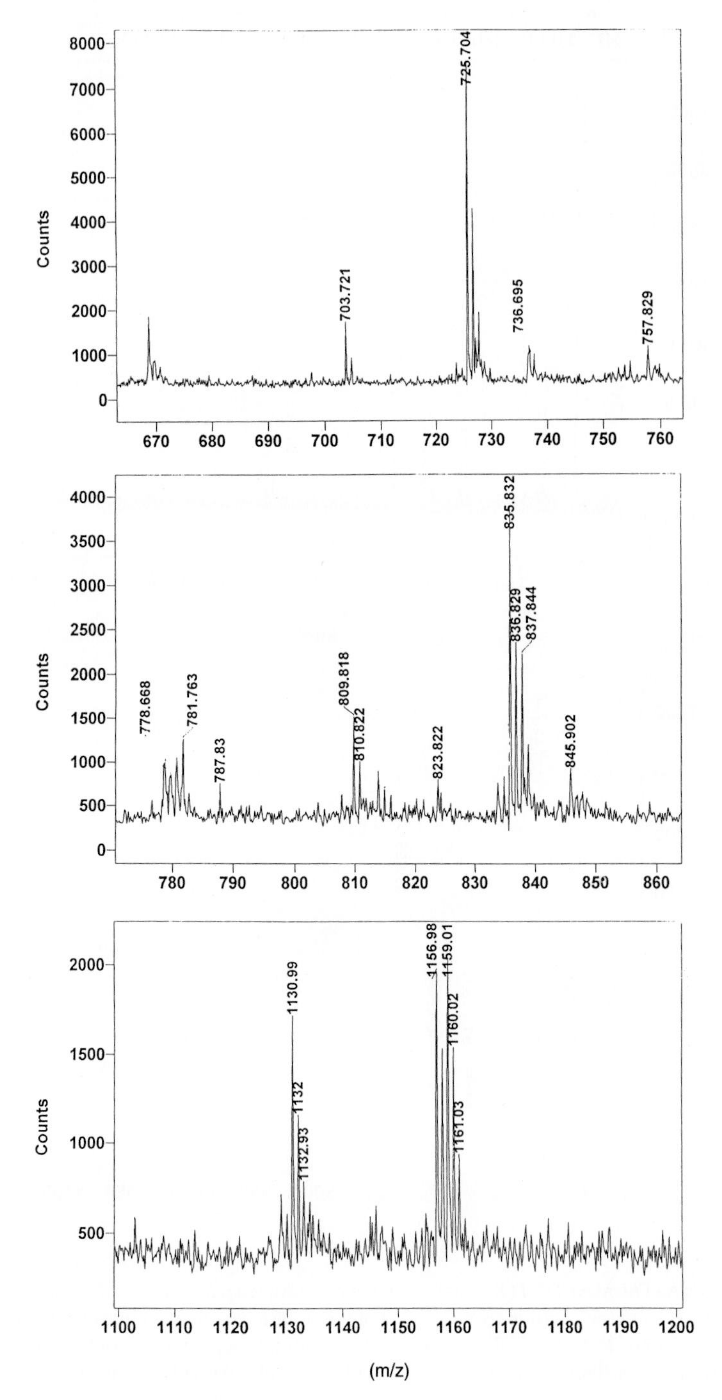

FIG. 10. Enlarged DE MALDI-TOF mass spectrum of Fig. 9B.

long-chain base compositions of other gangliosides containing major stearic acid can be determined in the same way by DE MALDI-TOF MS (data not shown).

Sphingomyelin

Because sphingomyelin in animal tissues has in general long-chain bases of d18:1 and d18:0 and unbranched and non-hydroxy fatty acids, its individual molecular species are relatively simple. When the intact sphingomyelin isolated from caprine erythrocytes was subjected to DE MALDI-TOF MS by the same procedure as described in the section on glycosphingolipids, the mass spectrum was found to indicate the molecular weight-related ions corresponding to the individual molecular species as shown in Fig. 7.

Extraction and DE MALDI-TOF MS Analysis of Multiple Sphingolipids from a Single Sample

DE MALDI-TOF MS analysis of sphingolipids is also possible with the alkali-stable lipid fraction obtained from a biologic sample. We have conducted the analysis with human platelets, human fibroblasts, sciatic nerve, and brain tissues of UDP-galactose ceramide galactosyltransferase (CGT) null mutant mouse (ctg$^{-/-}$) and wild-type mouse, and necrotic tissues from the head of a femur of a patient with Fabry disease and a control subject. A few examples are described in this section.

Fig. 10. (*continued*)

Mass (*m/z*)	Ions	LCB	FA	Mass (*m/z*)	Ions	LCB	FA
703	SM$[M + H]^+$	d18:1	C16:0	836	SM$[M + Na]^+$	d18:1	C24:1
725	SM$[M + Na]^+$	d18:1	C16:0	838	SM$[M + Na]^+$	d18:1	C24:0
737				846	SM$[M + Na]^+$	d18:1	C26:0
758	SM$[M + Na]^+$	d18:1	C20:1	1131	CTH$[M + Na]^+$	d18:1	C22:0
778	CMH$[M + Na]^+$	d18:1	C20:0	1157	CTH$[M + Na]^+$	d18:1	C24:1
788	SM$[M + H]^+$	d18:0	C22:0	1159	CTH$[M + Na]^+$	d18:1	C24:0
810	SM$[M + Na]^+$	d18:1	C22:0				
824	SM$[M + Na]^+$	d18:1	C23:0				
834	CMH$[M + Na]^+$	d18:1	C24:0				

SM, sphingomyelin; CMH, glucosylceramide; CTH, globotriaosylceramide.

Human Platelets

About 50 mg of packed platelets washed with saline are shaken vigorously with 10 ml of chloroform–methanol (2:1, v/v) using a vortex mixer. After centrifugation, the supernatant is decanted and the precipitate is extracted with 5 ml of the same solution. The lipid extracts are combined and evaporated under N_2. The residue is treated with about 1 ml of acetone to remove cholesterol and simple lipids. After decanting, the residue is again extracted with about 1 ml of diethyl ether to remove glycerophospholipids. After decanting again, the residue is treated with 1 ml of 0.5 *N* NaOH in methanol at 50° for 1 hr.

After the saponification, the reaction mixture is acidified with one drop of 6 *N* HCl together with one drop of distilled water and shaken vigorously with 1 ml of hexane using a vortex mixer. After the upper phase containing free fatty acids is removed, the lower phase is again treated with hexane. After this procedure is repeated, the lower phase is evaporated under N_2 and the residue is shaken vigorously with 1 ml of chloroform–methanol–water (86:14:1, v/v; theoretical lower phase) and 1 ml of chloroform–methanol–water (3:48:47, v/v; theoretical upper phase) to remove salts. The upper phase and the lower phase are separated by centrifugation at 3000 rpm for 10 min. After the upper phase is carefully removed with a micropipette, the lower phase is evaporated under N_2. This residue is rextracted to remove salts, and the final residue is dissolved with 20 μl of chloroform–methanol (1:1, v/v) with a microsyringe.

One μl of the solution is mixed vigorously with 1 μl of the matrix solution of 2,5-DHB in an Eppendorf tube and centrifuged. One microliter of the supernatant is applied to the sample plate. As described above, it is subjected to DE MALDI-TOF MS analysis by the same procedure. The mass spectrum displayed molecular weight-related ions corresponding to individual molecular species of sphingomyelin, glucosylceramide, lactosylceramide, globotriaosylceramide, globotetraosylceramide, and GM2 at the wide mass range in the positive ion mode, as identified in Fig. 8. Judging by the counts in the mass spectrum, sphingomyelin was found to be a major sphingolipid and other minor multiple glycosphingolipids were definitely confirmed.

Necrotic Tissues from Head of Femur of a Patient with Fabry Disease and a Control Subject

About 100 mg of wet necrotic tissue from the head of a femur of two patients with or without Fabry disease separately was homogenized vigorously with 10 ml of chloroform–methanol (2:1, v/v) with a pestle in a mortar. After the homogenate was filtered, the total lipid extract (8 ml) was mixed vigorously with 2 ml of distilled water and partitioned into an upper phase and a lower phase by centrifugation. After the upper phase was removed carefully, the lower phase was evaporated under N_2. The residue was treated to obtain the alkali-stable lipid fraction as described above. The solution was subjected to the DE MALDI-TOF MS analysis by the same procedure.

As shown in Fig. 9, the mass spectrum of the patient with Fabry disease certainly suggested the molecular weight-related ions corresponding to accumulated globotriaosylceramides. All molecular weight-related ions of the globotriaosylceramide and other sphingolipids in the alkali-stable lipid fraction of the patient were identified as shown in Fig. 10.

These examples illustrate the utility of this method with impure as well as purified sphingolipids. It is necessary to confirm the identities of the peaks by another method (such as MS/MS); however, once this has been done, DE MALDI-TOF MS is a simple and convenient analytical procedure.

[9] Purification of Sphingolipid Classes by Solid-Phase Extraction with Aminopropyl and Weak Cation Exchanger Cartridges

By JACQUES BODENNEC, GÉRARD BRICHON, GEORGES ZWINGELSTEIN, and JACQUES PORTOUKALIAN

Introduction

Isolation and purification of molecules from complex lipid mixtures is generally done by preparative thin-layer chromatography (TLC),[1] liquid chromatography on different silica-based sorbents,[2] and high-performance liquid chromatography (HPLC).[3] However, TLC can be cumbersome and time consuming. Silicic acid column chromatography often requires considerable amounts of solvents and larger amounts of samples. Isolation of

[1] B. Fried, *in* "Thin-Layer Chromatography" (B. Fried and J. Sherma, eds.), Vol. 81, p. 277. Wiley, New York, 1984.

[2] J. B. Davenport, *in* "Biochemistry and Methodology of Lipids" (A. R. Johnson and J. B. Davenport, eds.), p. 151. Wiley-Interscience, New York, 1971.

[3] W. W. Christie, *in* "Height-Performance Liquid Chromatography and Lipids. A Practical Guide," p. 87. Pergamon Press, Oxford, 1987.

lipids by HPLC is obtained with high resolution but it requires relatively expensive equipment and it is also often time consuming. Solid-phase extraction (SPE) is an interesting tool that has been used for the preparative isolation of defined lipid compounds (reviewed in Refs. 4 and 5) with a high degree of purity and often with a better recovery when compared to TLC.[4,6] The ease is handling such techniques and the possibility of adopting a chromatographic mode sequencing by changing the solvent environment around the solid phase or by changing the solid phase itself enable compounds with subtle differences in chemical nature to be efficiently separated.[7]

The work detailed in this report describes procedures for isolation of sphingolipid classes from complex lipid mixtures by using simultaneously aminopropyl-bonded silica gel (LC-NH_2) and weak cation exchanger (LC-WCX) cartridges. As also shown, these SPE cartridges can be efficiently used to isolate sphingoid bases obtained after hydrolysis of sphingolipids from other products of hydrolysis such as fatty acids and unhydrolyzed parent compounds.

Materials

Solvents used in this method are of analytical grade and can be purchased from SDS (Peypin, France), Carlo Erba (Milan, Italy), Riedel-de Haën (Seelze, Germany), or other suppliers. Standard lipids are available from Sigma (Saint Quentin Fallavier, France); free ceramides containing both phytosphingosine and normal fatty acids are from Coletica (Lyon, France). Neutral glycosphingolipid standards are isolated from melanoma tumors according to Saito and Hakomori[8] and purified by HPLC.[9] SPE cartridges are from Supelco (Saint Quentin Fallavier, France); LC-NH_2 aminopropyl-bonded silica gel (100 mg and 500 mg); LC-WCX, propanoic acid-bonded silica gel (100 mg sodium counterion). All HPTLC plates are glass-backed silica gel 60 plates from Merck (Darmstadt, Germany).

[4] S. E. Ebeler and T. Shibamoto, *in* "Lipid Chromatographic Analysis" (T. Shibamoto, ed.), p. 1. Marcel Dekker, New York, 1994.

[5] S. E. Ebeler and J. D. Ebeler, *Inform* **7,** 1094 (1996).

[6] M. A. Kaluzny, L. A. Duncan, M. V. Meritt, and D. E. Epps, *J. Lipid Res.* **26,** 135 (1985).

[7] E. M. Thurman and M. S. Mills, "Solid-Phase Extraction. Principles and Practice." Wiley, New York, 1998.

[8] S. Saïto and S. Hakomori, *J. Lipid Res.* **12,** 257 (1971).

[9] R. Kannagi, K. Watanabe, and S. Hakomori, *Methods Enzymol.* **138,** 3 (1987).

Chromatography Conditions and Procedures

The general procedure[10,11] for elution of sphingolipid classes is shown in Fig. 1. A 100-mg aminopropyl column is positioned onto a Visiprep DL vacuum apparatus (Supelco) equipped with diposable Teflon liners. The column is preconditioned with 2 ml of hexane. Then the lipid mixture resuspended in 200 μl of chloroform is loaded onto the column. No pressure should be applied during this step in order to optimize the loading of lipids by gravity. As soon as the top of the solvent reaches the column upper frit, the flow is stopped and a 12-ml glass tube is positioned to collect the eluate.

This first elution is made by passing 1.4 ml of solvent A (see Table I) under negative pressure of 10 kPa and by limiting the flow rate to 0.3 ml/min with the help of an individual column valve. The flow is stopped as soon as the top of the solvent reaches the upper frit. Solvent system A allows the recovery of most neutral lipids (see Table II) except free ceramides, monoglycerides, and free fatty acids (FA).

Then the LC-NH_2 column is positioned on a preconditioned 100-mg LC-WCX column by using an adapter valve. The 100-mg LC-WCX cartridge is preconditioned by successive elution with 1 ml hexane, 2 ml of methanol–0.5 *N* acetic acid, 4 ml of methanol, and 4 ml of hexane.[11] Free ceramides are eluted through the two piggybacked cartridges (see Fig. 1) with 1.6 ml of solvent B in a second glass tube. This second elution ensures the recovery of monoglycerides and of the 15% cholesterol remaining on the LC-NH_2 column after passing solvent A. Part of the free sphingoid bases also elute from the aminopropyl column, but these bases are strongly retained on the weak cation exchanger matrix while ceramides pass through.

Normal free fatty acids and α-hydroxy fatty acids are eluted in a third fraction by passing 1.8 ml of solvent C through the two piggybacked columns. Most of the free sphingoid bases are still retained on the LC-WCX matrix. Solvent system C allows for good recovery of fatty acids while neutral glycosphingolipids and phospholipids still bind to the LC-NH_2 matrix. Then, the neutral glycosphingolipids and all remaining free sphingoid bases (sphingosine, dihydrosphingosine, and phytosphingosine) are released with 2 ml of solvent D.

During these eluting steps of the procedure using piggybacked cartridges (solvents B, C, and D), positive pressure is applied on the LC-NH_2 tube using a small pipette rubber bulb rather than applying negative pressure with a vacuum apparatus. This ensures a steady solvent flow through the

[10] J. Bodennec, O. Koul, I. Aguado, G. Brichon, G. Zwingelstein, and J. Portoukalian, *J. Lipid Res.,* in press.

[11] J. Bodennec, C. Famy, G. Brichon, G. Zwingelstein, and J. Portoukalian, *Anal. Biochem.* **279,** 245 (2000).

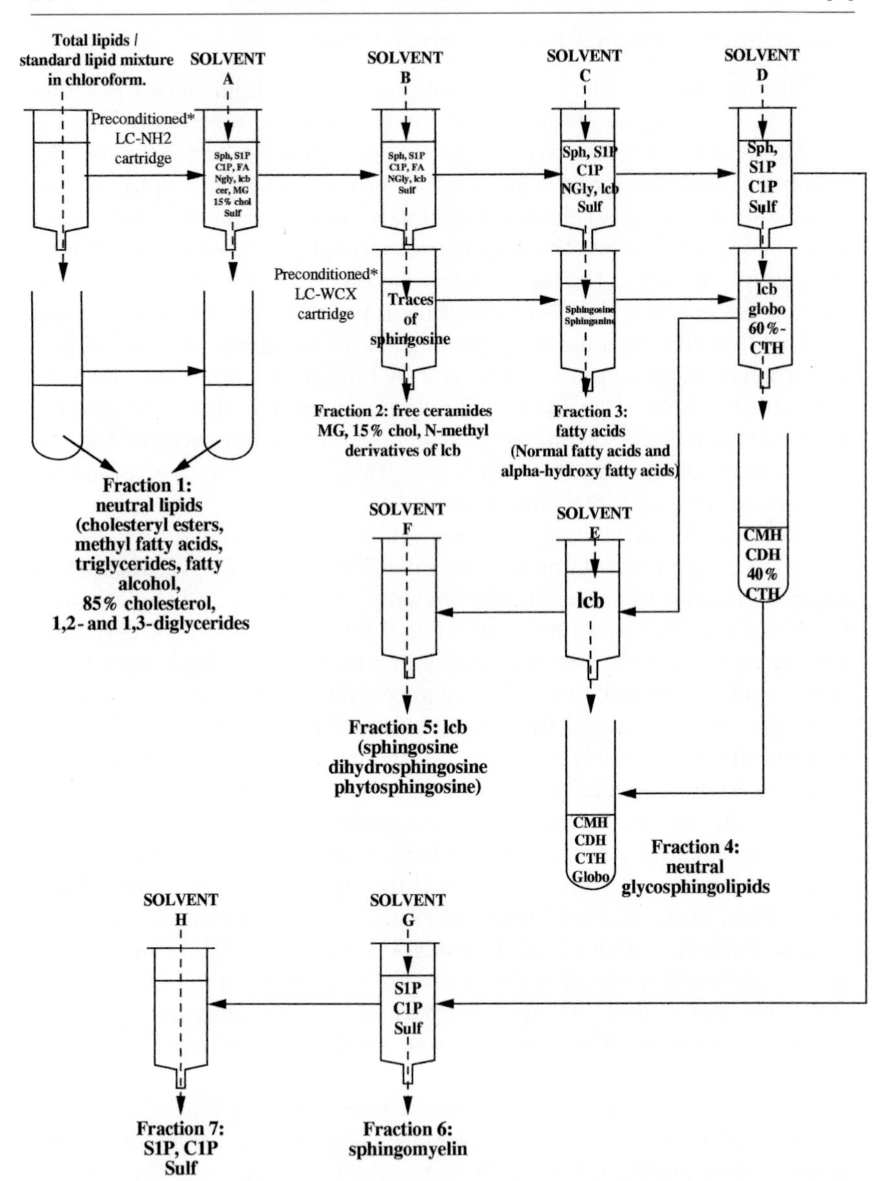

FIG. 1. Scheme of elution of sphingolipid classes from SPE cartridges. This procedure is simultaneously using an aminopropyl (LC-NH_2) and a weak cation exchanger (LC-WCX) column (see the text). Composition of solvents is given in Table I. (*) LC-NH_2 and LC-WCX cartridges are preconditioned as mentioned in the text.

TABLE I
SOLVENTS USED FOR ISOLATION OF SPHINGOLIPID CLASSES ON SPE CARTRIDGES[a]

Solvent	Solvent composition (v/v)	Volume of solvent (matrix weight)	Type of cartridge
A	Hexane–ethyl acetate (85:15)	1.4 ml (100 mg) 2 ml (500 (mg)	NH_2
B	Chloroform–methanol (23:1)	1.6 ml (100 mg) 3 ml (500 mg)	Piggybacked NH_2/WCX
C	Diisopropyl ether–acetic acid (98:5)	1.8 (100 mg) 2 ml (500 mg)	Piggybacked NH_2/WCX
D	Acetone–methanol (9:1.35)	2 ml (100 mg) 4 ml (500 mg)	Piggybacked NH_2/WCX
E	Chloroform–methanol (9:3)	2 ml (100 mg)[a]	WCX
F	1 *N* Acetic acid in methanol	2 ml (100 mg)[a]	WCX
G	Chloroform–methanol (2:1)	2 ml (100 mg) 3 ml (500 mg)	NH_2
H	Chloroform–methanol–3.6 *M* aqueous ammonium acetate (30:60:8, v/v/v)	2 ml (100 mg) 4 ml (500 mg)	NH_2

[a] Only 100-mg LC-WCX cartridges were used. Both 100-mg and 500-mg LC-NH_2 cartridges were used.

two piggybacked columns as soon as the elution process begins. If negative vacuum pressure is used during these steps, the matrix of the lower piggybacked tube (LC-WCX) will dry before the solvent from the upper aminopropyl tube can reach it.

With solvent system D, the different phospholipids remain on the

TABLE II
ELUTION PATTERN OF STANDARD SPHINGOLIPIDS AND OTHER LIPID SPECIES IN SOLVENTS A TO H

Solvent	Eluted lipids
A	Cholesterol (85%), cholesteryl esters, triglycerides, diglycerides, fatty alcohols, fatty acid methyl esters
B	Cholesterol (15%), monoglycerides, free ceramides, *N*-methyl derivatives of sphingoid bases
C	Free fatty acids (normal and α-hydroxy)
D	CMH, CDH, CTH (40%)
E	CTH (60%), globoside
F	Sphingosine, dihydrosphingosine, phytosphingosine
G	Sphingomyelin, lecithin, lysolecithin, phosphatidylethanolamine
H	Sphingosine 1-phosphate, ceramide 1-phosphate, sulfatides, acidic phospholipids

LC-NH_2 column allowing further fractionation, if needed. Two milliliters of solvent D coming from LC-NH_2 and containing neutral glycosphingolipids as well as the remaining free sphingoid bases elutes ceramide monohexosides (CMH), ceramide dihexosides (CDH) and 40% of ceramide trihexosides (CTH) when passing through the piggybacked LC-WCX tube, while globoside, the remaining CTH, and free sphingoid bases bind to the cation exchanger column.

Next, the two cartridges are taken apart and further processed separately. The LC-WCX cartridge is eluted with 2 ml of solvent E to recover the remaining neutral glycosphingolipids (i.e., 60% of CTH and globoside) into the same tube that contains CMH and CDH eluted with solvent D. Free sphingoid bases that are still binding to the LC-WCX column are eluted with 2 ml of solvent F.

The LC-NH_2 column is eluted with solvent G to remove sphingomyelin. This solvent system allows for the recovery of phosphatidylcholine, lysophosphatidylcholine, and a portion of the phosphatidylethanolamine, whereas sphingosine 1-phosphate (S1P), ceramide 1-phosphate (C1P), sulfatides (Sulf), and acidic glycerophospholipids (such as phosphatidylinositol and phosphatidylserine) remain bound to the column. These lipids are eluted with 2 ml of solvent H.

The 100-mg aminopropyl cartridges are particularly useful when small amounts of material are fractionated (see comments). If a larger quantity of lipids is to be separated, 500-mg cartridges are employed in the same way as for small cartridges, but the volumes of solvents should be changed (see Table I).

Comments

This chromatographic procedure allows for the separation of all major sphingolipids contained in the lower phase obtained after solvent partition of a total lipid extract.[12] Hence, free ceramides are recovered in a separate fraction nearly free of other neutral lipids since only monoglycerides and a small part of the cholesterol are eluted in this fraction. If needed, this ceramide fraction can be further purified from most of the remaining cholesterol by passing the fraction onto a second LC-NH_2 cartridge preconditioned as described and eluted with the same solvents used for the first column, i.e., solvents A and B.

Ceramides can also be easily purified from monoglycerides that contaminate fraction 2 by submitting this fraction to alkaline methanolysis. This can also be performed on the whole lipid mixture before separation on

[12] J. Folch, M. Lees, and G. H. S. Stanley, *J. Biol. Chem.* **226,** 497 (1957).

SPE cartridges. The alkaline treatment results in a concentration of the different sphingolipids before separation, hence allowing the purification of larger amounts of sphingolipid products per SPE cartridge. However, methanolysis can also concentrate any alkyl monoglycerides in fraction 2 arising from methanolysis of glycerolipids because the alkyl linkage is resistant to mild alkaline treatment. Such a concentration of alkyl monoglycerides can be troublesome when further separation of free ceramides is realized because these different molecules migrate very close to each other on HPTLC plates in conventional solvent systems.[13,14] Misinterpretations of such chromatograms can result after charring or visualizing the plate with iodine when the different spots overlap. Hence, alkaline treatment may be carried out either before or after the separation of sphingolipids on SPE columns.

When alkaline treatment is done before SPE separation, it also results in fatty acid accumulation from hydrolyzed lipids. The SPE procedure described herein allows for good resolution of free ceramides from the different fatty acids, including α-hydroxy species, which tend to tail and migrate close by in usual solvent systems for monodimensional TLC[13,14] (see also Fig. 2). The resolution is made possible by the stronger binding of fatty acids to the aminopropyl-bonded groups of the LC-NH_2 column.

Part of the free sphingoid bases elutes along with free ceramides when solvent B is passed through the LC-NH_2 column. Similarly, neutral glycosphingolipids coelute with the free sphingoid bases remaining on the LC-NH_2 matrix. Further separation of free LCB from free ceramides and neutral glycosphingolipids is effected with the LC-WCX column, which strongly retains sphingoid bases on appropriate conditioning.[11] Activation of the propanoic acid functional group of the matrix is performed with 0.5 *N* acetic acid in methanol, which removes any sodium ions.

This purification of free sphingoid bases is possible after piggybacking the LC-NH_2 column onto the preconditioned LC-WCX. The latter column is further eluted with 2 ml of chloroform–methanol (9:3, v/v) allowing the recovery of CTH and globoside, which remain on the column when piggybacking with LC-NH_2 and passing the acetone–methanol (9:1.35, v/v) solvent through the two cartridges (see Fig. 3a). Special care should be taken at this step of separation regarding the quality of the solvent to be used, and this is particularly true for acetone. If freshly purified acetone is not used, a large proportion of phytosphingosine may be degraded into an

[13] J. Bodennec, G. Brichon, O. Koul, M. El Babili, and G. Zwingelstein, *J. Lipid Res.* **38,** 1702 (1997).

[14] J. Bodennec, G. Brichon, J. Portoukalian, and G. Zwingelstein, *J. Liq. Chrom. Rel. Technol.* **22,** 1493 (1999).

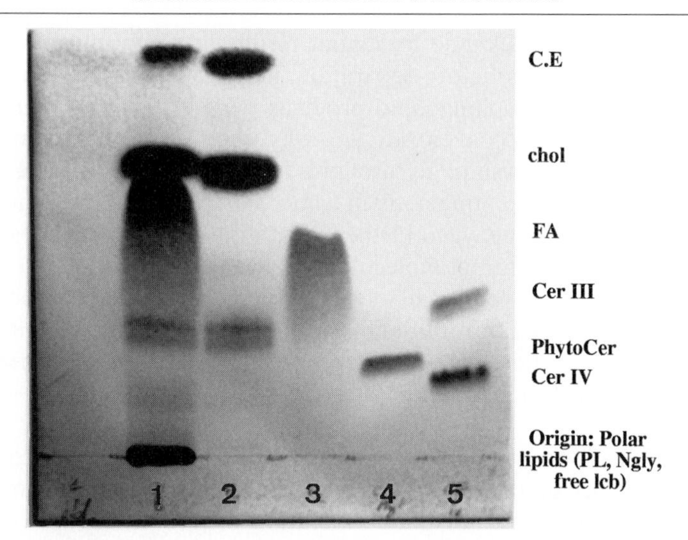

FIG. 2. Solid-phase extraction (SPE) of free ceramides from fresh human melanomas. Total lipids are extracted according to Folch *et al.*[12] and then submitted to alkaline methanolysis to hydrolyze ester-containing lipids for 1 hr at room temperature in the presence of 4 ml of chloroform–methanol–0.6 *N* NaOH (1:1, v/v). The mixture is further acidified with 230 μl of concentrated HCl and hydrolysis is carried out for 1 hr. After hydrolysis, the mixture is washed with 1.6 ml of distilled water, shaken, and the chloroform lower phase is saved. An aliquot of this chloroform phase is spotted on a TLC plate (corresponding to lane 1) and a second aliquot is loaded onto a 100-mg LC-NH_2 cartridge preconditioned with 2 ml of hexane. Free ceramides are directly eluted with 1.6 ml of solvent B (see Table I). This eluate corresponds to fraction 2 of Fig. 1, and the free fatty acids released during alkaline methanolysis are recovered from the column with 1.8 ml of solvent C. These two fractions are dried and applied on the TLC plate, which is developed in chloroform–methanol (50:2, v/v). Lanes 2 and 3 correspond, respectively, to solvents B and C eluted from the LC-NH_2 tube. Lane 4 is a standard of ceramide containing phytosphingosine linked to nonhydroxy fatty acid. Lane 5 is a mixture of ceramide standard type III (upper) and IV (lower) from Sigma (St. Louis, MO). C.E., cholesteryl esters; chol, cholesterol; FA, free fatty acids; Cer, free ceramides; PhytoCer, free ceramide containing phytosphingosine as sphingoid backbone; PL, phospholipids; Ngly, neutral glycosphingolipids; lcb, long-chain bases (free sphingoid bases).

unknown compound migrating faster than sphingosine on the TLC plates. This degradation of phytosphingosine is selective since neither sphingosine nor dihydrosphingosine is unaffected (not shown) indicating that the fourth hydroxyl group is needed to observe this phenomenom. That is prevented when freshly purified acetone is used.

As for all separation and preparative chromatographic techniques, the present method can be successfully applied to different lipid mixtures from different origins. In our hands, this SPE method was used to purify different sphingolipid classes such as free ceramides from lipid mixtures extracted

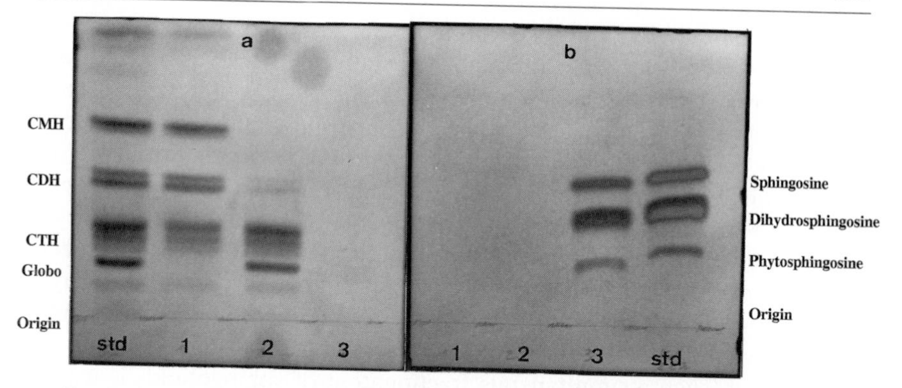

FIG. 3. A complex lipid mixture enriched in neutral glycosphingolipid standards and free sphingoid base standards is separated according to the SPE procedure described in Fig. 1. The two chromatograms show the products recovered after elution of the piggybacked LC-NH_2 and LC-WCX cartridges with solvent D and further elution of the LC-WCX column with solvent E then solvent F (see Fig. 1). The different eluates are dried, resuspended in a small volume of chloroform–methanol (2:1, v/v) and each half respectively applied onto two different HPTLC plates. (a) One of the plates is developed in chloroform–methanol–water (65:25:4, v/v/v) to resolve neutral glycolipids which are visualized with orcinol reagent. (b) The other HPTLC plate is developed in chloroform–methanol–2 *N* ammonia (40:10:1, v/v/v) and sprayed with ninhydrin reagent to detect free sphingoid bases. std, neutral glycosphingolipid standards in (a) and free sphingoid bases standards in (b); 1, HPTLC of the products eluted through the piggybacked LC-NH_2 and LC-WCX column by solvent D; 2, products eluted from LC-WCX cartridge by solvent E; 3, products eluted from LC-WCX cartridge by solvent F; Globo, globoside.

from fish gills,[13,14] fish leukocytes,[15,16] HL 60 cells,[17] and fresh human melanomas.[11]

It is important to keep in mind that, in the experimental conditions described herein, the LC-NH_2 columns show a maximum capacity of 4 mg of total lipids per 100 mg of matrix.[10] However, loading up to this capacity should be avoided because of a possible breakthrough of compounds between the different fractions. We routinely purify sphingolipid fractions with a maximum loading of 2 mg of total lipids per 100 mg of matrix. Under these conditions, a very good recovery (up to 95%) of the various sphingolipid classes is obtained.[10]

When larger amount of lipids are to be separated, 500-mg LC-NH_2

[15] J. Bodennec, G. Zwingelstein, O. Koul, G. Brichon, and J. Portoukalian, *Biochem. Biophys. Res. Commun.* **250,** 88 (1998).

[16] J. Bodennec, G. Brichon, O. Koul, J. Portoukalian, and G. Zwingelstein, *Comp. Biochem. Physiol. B* **25,** 523 (2000).

[17] J. Bodennec, D. Ardail, I. Popa, G. Brichon, G. Zwingelstein, and J. Portoukalian, in preparation.

columns are used (see Table I). The latter are very useful to isolate small amounts of molecules from radiolabeled samples. The procedure we describe allows simultaneous fractionation of all major sphingolipid classes obtained after Folch's partition. This is particularly interesting since it avoids imprecisions inherent to procedures using separate processing of samples when studying different sphingolipids. This point is important since the different sphingolipids are structurally related due to the precursor and/or catabolic relationships and since the intracellular balance between the different sphingolipids may be finely regulated by specific enzyme activities such as sphingomyelinases, ceramidases, sphingoid base kinases, and phosphatases.[18]

However, such a simultaneous fractionation of sphingolipids is useful only if a good recovery of molecules is achieved during the initial extraction and purification steps. This is the case for sphingosine 1-phosphate, which can be partitioned differently between the lower and upper phases according to the pH of the mixture.[19] The solid-phase extraction procedure described in this article can be selectively used according to a specific need to purify a single class of sphingolipids avoiding multifractionation of samples for all the sphingolipid classes. An example of specific use is described in the following section.

Isolation and Purification of Sphingoid Bases from Hydrolysis Mixtures of Parent Sphingolipids

Choice of Conditions of Hydrolysis

The previous SPE procedure on a LC-WCX column can be used to purify sphingoid bases released during hydrolysis of parent sphingolipids such as ceramides.[11]

Hydrolysis of ceramides is an essential step when analyzing the molecular species or localizing radioactivity in the molecule. A variety of methods exists to hydrolyze ceramides,[20] but the most commonly used is the method described by Gaver and Sweeley,[21] and solid-phase extraction can be used to purify the long-chain bases released during the latter procedure. Following hydrolysis, the mixture is washed several times with a nonpolar solvent such as hexane to remove fatty acids and their methyl esters. Then the lower phase mixture, containing the released sphingoid bases, is made

[18] D. K. Perry and Y. A. Hannun, *Biochim. Biophys. Acta* **1436,** 233 (1998).
[19] L. C. Edsall and S. Spiegel, *Anal. Biochem.* **272,** 80 (1999).
[20] K. Karlsson, *Lipids* **5,** 878 (1970).
[21] R. C. Gaver and C. C. Sweeley, *J. Am. Oil Chem. Soc.* **42,** 294 (1965).

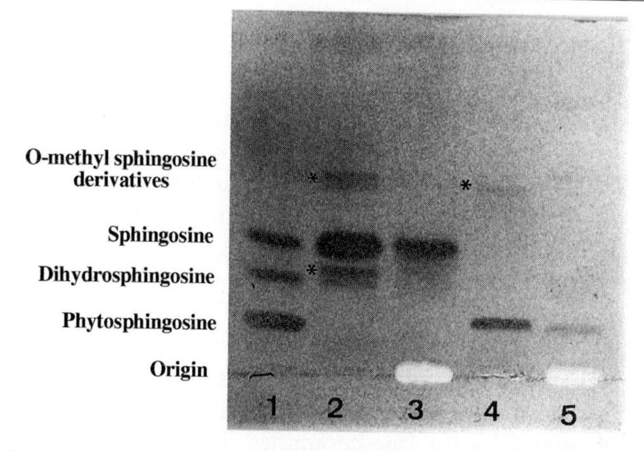

FIG. 4. Solid-phase extraction of long-chain bases from hydrolysis mixtures of free ceramides. Ceramide type III from Sigma and ceramide containing phytosphingosine linked to nonhydroxy fatty acids are hydrolyzed under two different conditions. At the end of hydrolysis, the 2 ml of hydrolysis mixture (see also text for explanation) is extracted with 6 ml of chloroform–methanol (2:1, v/v). The lower phase is directly applied onto a preconditioned 100-mg LC-WCX column and the sphingoid bases are washed and extracted (see chromatography procedure and conditions). The column eluate containing the sphingoid bases is dried under nitrogen, resuspended in a small volume of chloroform–methanol (2:1, v/v) and an aliquot is spotted on an HPTLC plate, which is developed in chloroform–methanol–2 *N* ammonia (40:10:1, v/v/v). Lane 1: standards of sphingoid bases. Lane 2: sphingoid bases obtained after acid hydrolysis of ceramide type III according to Gaver and Sweeley,[21] i.e., 18 hr at 70° in 1 ml of methanol–water–concentrated HCl (100:9.4:8.6, v/v/v) further alkalinized with 1 ml of 2 *N* KOH. Lane 3: sphingoid bases obtained after hydrolysis of ceramide type III onto silica gel in 2 ml of 10 *N* aqueous KOH during 18 hr at 80°. Lanes 4 and 5: sphingoid bases obtained, respectively, after hydrolysis of ceramide containing phytosphingosine according to Gaver and in 10 *N* aqueous KOH. The asterisks (*) indicate the by-products formed during acid hydrolysis such as *O*-methyl derivatives of sphingoid bases (upper *, lanes 2 and 4) and the *threo* isomer of sphingosine mixed with small amount of sphinganine present in ceramide type III (lower *, lane 2). These derivatives are not formed during alkaline hydrolysis.

alkaline and extracted several times with diethyl ether to recover these sphingoid bases.[22]

As shown in Fig. 4, sphingoid bases can be directly recovered onto the SPE column from the chloroform phase obtained when the hydrolysis mixture is extracted with chloroform–methanol. For this purpose, 500 μg of each ceramide standard (ceramide type III and ceramide containing phytosphingosine linked to normal fatty acids) is hydrolyzed separately in 1 ml of methanol–water–concentrated HCl (100:9.4:8.6, v/v/v) for 18 hr

[22] T. Akino, *Sapporo Med. J.* **34,** 255 (1968).

at 70°. After cooling to room temperature, the mixture is alkalinized with 1 ml of 2 *N* KOH. Then, 6 ml of chloroform–methanol (2:1, v/v) is added and the mixture is thoroughly shaken.

After centrifugation (3000 rpm, 5 minutes at 6°), the lower (chloroform) phase (approximately 4 ml) is directly applied to a 100-mg LC-WCX column preconditioned as described in the SPE procedure. The sphingoid bases released during acid hydrolysis bind to the matrix allowing their extraction and further elution. The column is washed with 2 ml of chloroform–methanol (9:3, v/v), which allows the recovery of free fatty acids and of the remaining unhydrolyzed free ceramides. Then the sphingoid bases are eluted from the column with 2 ml of 1 *N* acetic acid in methanol.

If hydrolysis is performed on glycosphingolipids such as cerebrosides, the sphingoid bases generated are also easily washed from their unhydrolyzed compounds as for free ceramides. This step is critical since the extent of hydrolysis can be high or low according to the procedure. Hence, when using the method of Gaver and Sweeley, the hydrolysis of gangliosides is not complete,[23] making the purification of sphingoid bases from parent compounds an important analytical step.

Moreover, when hydrolysis is performed in acidic conditions, some degradation or modification of the sphingoid bases often occurs as a result of the formation of a carbonium ion. This ion can give rise to double bond migration, dehydration products, diastereoisomers at carbon 3 or 5, and *O*-methyl ethers if hydrolysis is performed in methanol-containing mixtures.[20,24,25] In the method of Gaver and Sweeley, formation of these *O*-methyl ether by-products is lower than in other methods, but only if the given conditions of hydrolysis and temperature[26] are strictly respected. Phytosphingosine long-chain bases are also sensitive to these acidic hydrolysis conditions. This is an obvious problem because this long-chain base is often present in low amounts in mammalian tissues as compared to the more abundant sphingosine and dihydrosphingosine species. However, formation of most of the degradation products can be circumvented by performing hydrolysis under alkaline conditions in 10 *N* aqueous KOH–butanol (1:9, v/v) instead of acidic conditions.[27,28]

Hydrolysis of free ceramides can be performed directly on silica gel

[23] H. Kadowaki, E. G. Bremer, J. E. Evans, F. B. Jungalwala, and R. H. McCluer, *J. Lipid Res.* **24,** 1389 (1983).

[24] A. Kisic, M. Tsuda, R. J. Kulmacz, W. K. Wilson, and G. J. Schroepfer, Jr., *J. Lipid Res.* **36,** 787 (1995).

[25] H. E. Carter, O. Nalbandov, and P. A. Tavormina, *J. Biol. Chem.* **192,** 197 (1951).

[26] A. H. Merrill, Jr., and E. Wang, *J. Biol. Chem.* **261,** 3764 (1986).

[27] T. Taketomi and N. Kawamura, *J. Biochem.* (*Tokyo*) **68,** 475 (1970).

[28] L. Riboni, R. Bassi, S. Sonnino, and G. Tettamanti, *FEBS Lett.* **300,** 188 (1992).

after separation of group species by TLC. This hydrolysis on silica gel can be performed in acidic[11] or alkaline conditions. For direct hydrolysis of ceramides on silica gel in alkaline conditions, aliquots of the different ceramide species (ceramide type III and ceramide containing phytosphingosine as backbone) are spotted on a silica gel 60 HPTLC plate. The silica gel areas containing the ceramide species are scraped into screw-capped tubes and 2 ml of 10 *N* aqueous KOH is added. The mixture is thoroughly shaken and hydrolysis is performed for 18 hr at 80° (during which the silica gel is totally dissolved). At that time, hydrolysis is complete and 6 ml of chloroform–methanol (2:1) is added to the tubes. (Extraction with chloroform is preferred to ether because it produces better recovery of sphingoid bases and because ether is more unstable.)

The mixture is shaken and after centrifugation (3000 rpm, 5 minutes at 4°) the chloroform lower phase containing the sphingoid bases is applied directly onto a 100-mg preconditioned LC-WCX column (see above). The washing steps and purification of the sphingoid bases retained on the column are the same as for purification of bases by SPE after hydrolysis under the conditions described by Gaver and Sweeley (see above).

Figure 4 compares the long-chain bases obtained by solid-phase extraction after hydrolysis under the different conditions of the various free ceramides. As can be seen in Fig. 4, no *O*-methyl ether is formed when hydrolysis is performed after adsorption on silica gel as compared with acid hydrolysis. The solid-phase hydrolysis of ceramides containing sphingosine clearly shows a single spot migrating in the same position as the sphingosine standard, and no other ninhydrin-positive spot can be observed (see lane 3, Fig. 4) indicating that sphingosine integrity is preserved under these hydrolysis conditions.

In the case of ceramides containing phytosphingosine, no degradation of this sphingoid base can be observed when hydrolysis is performed by alkaline treatment on silica gel (compare lanes 4 and 5, Fig. 4). The recovery of sphingoid bases is optimal since the ether extract of fraction 1 eluted from LC-WCX and the ether extracts of the upper phase obtained from the hydrolysis mixture are ninhydrin negative (not shown). Hence, these hydrolysis conditions on silica gel should be used to preserve long-chain bases from degradation or allylic rearrangements that may occur during hydrolysis. Moreover, this procedure has the advantage that it can be carried out directly after isolation of ceramides by TLC, thus avoiding elution from the silica gel and the inherent loss of material.[6] The isolation of sphingoid bases produced during hydrolysis by solid-phase extraction of the chloroform extract avoids the classical partition step followed by extraction with ether, a step that often needs to be repeated several times to ensure maximal recovery of the sphingoid bases.

The solid-phase procedure allows the separation of released sphingoid bases from unhydrolyzed parent sphingolipids simultaneously and also removes fatty acid esters that can interfere with gas chromatography.[22] This procedure can be particularly useful when radiolabeled compounds have to be hydrolyzed in order to determine their radioactivity.

Isolation of Free Sphingoid Bases from Serum and Urine

Determination of the sphinganine-to-sphingosine ratio in serum and urine has been used as a potential biomarker to fumonisin exposure.[29,30] These free sphingoid bases are extracted from urine with ethyl acetate[31] or chloroform–methanol mixtures[30] in alkaline conditions. Silica gel minicolumns have been used to clean the chloroformic extract before HPLC analysis of the *o*-phthaldialdehyde derivatives of the sphingoid bases.[30,32]

The solid-phase extraction procedure described here using a 100-mg LC-WCX cartridge can also be used to extract the sphingoid bases from the ethyl acetate or the chloroform extract. This is done by directly passing these solvent fractions onto a LC-WCX column after preconditioning as described in this method. The extracted sphingoid bases are washed with 2 ml of chloroform–methanol (9 : 3, v/v) to remove neutral lipids, fatty acids, and neutral glycosphingolipds. Then, free sphinganine and sphingosine are recovered with 2 ml of 1 *N* acetic acid in methanol.

This procedure of extraction and cleaning is simpler than previously published ones[30–32] because it can be performed on commercially available, ready-made SPE cartridges. Moreover, the results are highly reproducible as long as the cartridges are purchased from the same supplier. The matrix of SPE cartridges provided by other companies may have slightly different physical properties that can require changes in the scheme of elution described in the method reported here (for explanations, see Ref. 7).

[29] R. T. Riley, E. Wang, and A. H. Merrill, Jr., *J. AOCS Int.* **77,** 533 (1994).
[30] M. Solfrizzo, G. Avantaggiato, and A. Visconti, *J. Chromatogr. B* **692,** 87 (1997).
[31] M. Castegnaro, L. Garren, I. Gaucher, and C. P. Wild, *Natural Toxins* **4,** 284 (1996).
[32] G. S. Shephard and L. van der Westhuizen, *J. Chromatogr. B* **710,** 219 (1998).

[10] Ganglioside Analysis by High-Performance Thin-Layer Chromatography

By ROBERT K. YU and TOSHIO ARIGA

Introduction

Gangliosides are a family of glycosphingolipids (GSLs) that contain one or more sialic acid residues (Fig. 1). They occur in practically all vertebrate cells and body fluids and are particularly abundant in the nervous system.[1] Gangliosides, as are GSLs in general, are located primarily, although not exclusively, on the outer surface of cell plasma membranes.[1] Interest in their structural and functional roles in cells has greatly stimulated the development of new techniques for their analysis. In particular, high-performance thin-layer chromatography (HPTLC) has become a method of choice for analyzing GSLs in a complex mixture because it is convenient and affords superior resolving power. HPTLC also allows the simultaneous identification and quantitation of GSLs in small amounts of samples and, in some instances, without extensive sample purification prior to analysis.

The development of the two-dimensional HPTLC technique has further enhanced the resolution of complex samples, which is very useful for analysis of overlapping ganglioside molecular species.[2] Furthermore, the combination of affinity detection on a TLC plate (the HPTLC-overlay technique) has greatly extended the versatility of this method for identification and quantitation of biologically active compounds using specific ligands or antibodies that are capable of binding specifically to glycosphingolipids on the plate.* Many highly specific reagents have been used for TLC-overlay, including specific anti-GSL antibodies and toxins (see reviews[5–11]). The

* The term Eastern blot evolved from laboratory jargon, reflecting the fact that it was first developed by Magnani's laboratory in Bethesda, MD,[3] and further elaborated by our laboratory[4] at Yale University, in New Haven, CT, both on the East Coast.

[1] R. K. Yu and M. Saito, *in* "Neurobiology and Glycoconjugates" (R. U. Margolis and R. K. Margolis, eds.), p. l. Plenum, New York, 1989.

[2] M. Ohashi, *Lipids* **14,** 52 (1979).

[3] J. L. Magnani, D. F. Smith, and V. Ginsburg, *Anal. Biochem.* **109,** 399 (1980).

[4] N. Kasai, M. Naiki, and R. K. Yu, *J. Biochem.* (*Tokyo*) **96,** 261 (1984).

[5] R. K. Yu and T. Ariga, *Ann. N.Y. Acad. Sci.* **845,** 285 (1998).

[6] M. Kotani, H. Ozawa, I. Kawashima, S. Ando, and T. Tai, *Biochim. Biophys. Acta* **1117,** 97 (1992).

[7] H. Ozawa, I. Kawashima, and T. Tai, *Arch. Biochem. Biophys.* **294,** 427 (1992).

0076-6879/00 $30.00

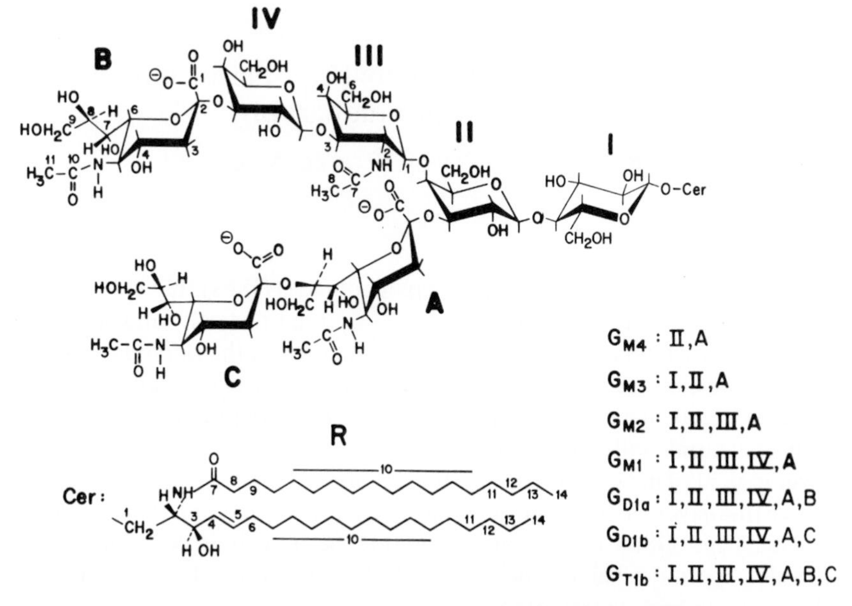

FIG. 1. Structures of some common brain gangliosides.

ability to carry out enzymatic and chemical reactions on the plate to modify the GSL structure further extends the range of this technique.[12,13] Combining HPTLC and soft-ionization mass spectrometry has paved the way for structural elucidation of a small amount of GSLs without purification.[14–16] In this article, some of the recent advances in HPTLC are described. For earlier developments, the readers should consult several related articles for the separation of gangliosides by HPTLC.[17–21]

---

[8] H. Ozawa, M. Kotani, I. Kawashima, and T. Tai, *Biochim. Biophys. Acta* **1123,** 184 (1992).
[9] T. Kanda, H. Yoshino, T. Ariga, M. Yamawaki, and R. K. Yu, *J. Cell Biol.* **126,** 235 (1994).
[10] S. L. Arnsmeier and A. S. Paller, *J. Lipid Res.* **36,** 911 (1995).
[11] G. S. Wu and R. W. Ledeen, *Anal. Biochem.* **173,** 368 (1988).
[12] M. Saito, N. Kasai, and R. K. Yu, *Anal. Biochem.* **148,** 54 (1985).
[13] X. B. Gu, T. J. Gu, and R. K. Yu, *Anal. Biochem.* **185**, 151 (1990).
[14] Y. Kushi, C. Rokukawa, and S. Handa, *Anal. Biochem.* **173**, 167 (1988).
[15] T. Taki, D. Ishikawa, S. Handa, and T. Kasama, *Anal. Biochem.* **225,** 24 (1995).
[16] T. Taki and D. Ishikawa, *Anal. Biochem* **241,** 135 (1997).
[17] R. W. Ledeen and R. K. Yu, *Methods Enzymol.* **83,** 139 (1982).
[18] S. Ando and M. Saito, *J. Chromatogr.* **37,** 266 (1987).
[19] J. Muthing and F. Unland, *Glycoconj. J.* **11,** 486 (1994).
[20] M. Saito and R. K. Yu, *in* "Planar Chromatography in the Life Sciences" (J. C. Touchstone, ed.), p. 59. John Wiley & Sons, New York, 1990.
[21] R. L. Schnaar and L. K. Needham, *Methods Enzymol.* **230,** 371 (1994).

Sample Preparation: Extraction and Isolation of Glycosphingolipids

To achieve best results, it is advisable to begin analysis with a relatively pure sample free of interfering contaminants. The following procedure has been used in the authors' laboratories for many years, consistently yielding excellent results.[17,22–24] Total lipids are first extracted from cells or tissues (less than 0.5 g of wet weight) with 10 volumes each of chloroform : methanol (2 : 1, 1 : 1, v/v), successively. The combined extracts are adjusted to a final solvent ratio of chloroform : methanol : water (30 : 60 : 8, v/v; solvent A) and then applied to a DEAE-Sephadex A-25 column (acetate form, 3-ml bed volume; Pharmacia, Upsala, Sweden). The column is further eluted with 15 ml of solvent A to recover neutral GSLs. After recovery of the neutral GSLs from the column, the acidic GSL fraction, which contains gangliosides and sulfatides, is eluted with 15 ml of chloroform : methanol : 0.8 *M* sodium acetate (30 : 60 : 8, v/v; solvent B). The acidic lipid fraction is evaporated to dryness, and the residue redissolved in 0.5 ml of Solvent A and desalted by passing through a Sephadex LH-20 column (Pharmacia, 26-ml bed volume; 0.7-cm i.d. × 48 cm). The column is then eluted with solvent A.[22]

After the first 7 ml of the effluent is discarded, the second 10 ml of the effluent, which contains the ganglioside fraction, is collected. The ganglioside fraction is evaporated to dryness under a gentle stream of nitrogen, and the residue is redissolved in a small volume of chloroform : methanol (2 : 1 or 1 : 1, v/v). An aliquot of the ganglioside fraction can be examined by HPTLC. In most cases, the sample is sufficiently pure to be analyzed at this stage, although it may still contain many other acidic lipids that may interfere with the ganglioside analysis. If the analysis demands a highly purified sample, the crude ganglioside fraction can be further purified by silicic acid column chromatography as described by Ledeen and Yu.[17] It is important that TLC analysis be performed with at least two different solvent systems for positive identification of the GSLs to be analyzed, as described later.

In certain cases when resolution of ganglioside species becomes a problem, it is advisable to separate the total gangliosides into mono-, di-, and polysialoganglioside fractions and analyze the individual fractions. The ganglioside fraction, dissolved in methanol, is applied to a minicolumn of DEAE-Sephadex A-25 and eluted with an increasing concentration of ammonium acetate (20, 40, and 100 m*M*) in methanol,[23] which elutes the mono-, di-, and polysialoganglioside species, respectively. In addition to

[22] T. Ariga, M. Sekine, R. K. Yu, and T. Miyatake, *J. Biol. Chem.* **257,** 2230 (1982).
[23] T. Ariga, G. M. Blaine, H. Yoshino, G. Dawson, T. Kanda, G. C. Zeng, T. Kasama, Y. Kushi, and R. K. Yu, *Biochemistry* **34,** 11500 (1995).
[24] T. Tanaka, Y. Arai, and Y. Kishimoto, *J. Neurochem.* **52,** 1931 (1989).

DEAE-Sephadex A-25, other ion-exchange media such as DEAE-Fractogel,[24] Q-Sepharose,[25] TMAE-Fractogel,[19] DEAE-Toyopearl 650(M),[18,26] and DEAE-Sepharose[27] have also been used for this purpose.

Desalting of the ganglioside fraction is necessary prior to analysis by HPTLC. This process can be achieved by a small column (bed volume: 9 ml) of Toyopearl C-40[28] or LH-20.[29] In the desalting step, the first 3 ml of effluent (solvent A) is discarded and the next 3 ml of the effluent containing gangliosides collected. Alternatively, the ganglioside fraction is evaporated to dryness and resolved in a small volume of methanol, and then the solution is passed through a SepPak C_{18} cartridge with methanol, followed by chloroform : methanol (1 : 2, v/v) for desalting.[26,30]

A number of ganglioside-extraction procedures that are not based on the use of chloroform : methanol have been developed. For example, Nakamura *et al.*[31] extracted muscle gangliosides with tetrahydrofuran (THF) : water (4 : 1, v/v), and the yield was reported to be comparable to that of the chloroform : methanol : water system (30 : 60 : 6, v/v). Heitmann *et al.*[32] recommended two solvent systems for ganglioside extraction: *n*-propanol : water (40 : 10, v/v) and methyl isobutyl ketone : methanol : water (40 : 80 : 30, v/v), with a yield ranging from 98 to 116% as compared with that based on the chloroform : methanol system. Schwarz *et al.*[33] also described a rapid, simple, and efficient chloroform–methanol-free method of isolating gangliosides using a nonionic hexaethylene glycol mono-*n*-tetradecyl ether and water system. The above procedures are rapid, simple, and applicable for isolating gangliosides from biologic samples.

Solvent Systems for Ganglioside Separation by HPTLC

Commercial aluminum-, plastic-, or glass-backed plates precoated with silica gel are most commonly used for the separation of glycosphingolipids

[25] Y. Hirabayashi, T. Nakao, M. Matsumoto, K. Obata, and S. Ando, *J. Chromatogr.* **445,** 377 (1988).

[26] T. Ariga, N. Kasai, I. Miyoshi, M. Yamawaki, J. N. Scarsdale, R. K. Yu, T. Kasama, and T. Taki, *Biochim. Biophys. Acta* **1254,** 257 (1995).

[27] L. Svennerholm, K. Bostrom, P. Fredman, J.-E. Mansson, B. Rosengren, and B. M. Rynmark, *Biochim. Biophys. Acta* **1005,** 109 (1989).

[28] H. Yoshino, T. Ariga, N. Latov, T. Miyatake, Y. Kushi, T. Kasama, S. Handa, and R. K. Yu, *J. Neurochem.* **61,** 658 (1993).

[29] R. K. Yu, H. Yoshino, M. Yamawaki, J. E. Yoshino, and T. Ariga, *J. Biomed. Sci.* **1,** 167 (1994).

[30] M. A. Williams and R. H. McCluer, *J. Neurochem.* **35,** 266 (1980).

[31] K. Nakamura, T. Ariga, T. Yahagi, T. Miyatake, A. Suzuki, and T. Yamakawa, *J. Biochem. (Tokyo)* **94,** 1359 (1983).

[32] D. Heitmann, M. Lissel, R. Kempken, and J. Muthing, *Biomed. Chromatogr.* **10,** 245 (1996).

[33] A. Schwarz, G. C. Tertappen, and A. H. Futerman, *Anal. Biochem.* **254,** 221 (1997).

by TLC, particularly the HPTLC plates (silica gel 60) from E. Merck Co. (Darmstadt, Germany). Before use, the plates are usually activated for 30 min at 110° and stored over heat-activated silica gel in a desiccator. To achieve consistent results, it is important to minimize deactivation by moisture. Thus, when the sample is applied, it is advisable to cover the silica gel layer with a clean glass plate, leaving only the bottom part open for sample application.

One-Dimensional HPTLC

The mobility and compactness of ganglioside bands are remarkably influenced by the presence of salts in the developing solvent system,[17,18,34] especially for HPTLC plates that use an organic material as a binder. The most frequently used HPTLC developing solvents for gangliosides are mixtures of chloroform : methanol : 0.2% $CaCl_2 \cdot 2H_2O$ (50 : 45 : 10 or 55 : 45 : 10, v/v)[17,22] and chloroform : methanol : 0.4% $CaCl_2 \cdot 2H_2O$: 2.5 *N* ammonium hydroxide (60 : 40 : 4 : 5 or 50 : 45 : 5 : 5, v/v).[22,23] For the separation of polysialogangliosides, the mixture of *n*-propanol : water containing 0.1% $CaCl_2 \cdot 2H_2O$ (80 : 20, v/v) or *n*-propanol : 2.5 *N* ammonium hydroxide : 0.4% $CaCl_2 \cdot 2H_2O$ (80 : 15 : 5, v/v) provides excellent separation.[17,35] A typical HPTLC profile for ganglioside separation is shown in Fig. 2.

Another useful solvent system is that devised by Ando *et al.*,[36] which includes acetonitrile : 2-propanol : water in the presence of salts, amines, or ammonia.[36] This solvent system is excellent for separating gangliosides, especially polysialoganglioside species, with high resolution. Figure 3 shows a typical example of separation using the acetonitrile : 2-propanol : 50 m*M* aqueous potassium chloride system.

Two-Dimensional HPTLC

Two-dimensional HPTLC systems have been devised for improved separation of complex GSL mixtures[18,34] and are more informative in identifying GSLs from various biologic samples. For example, the ganglioside distributions of various fat tissues from human, rabbit, rat, mouse, chicken, and frog have been compared with pig adipose gangliosides by two-dimensional TLC[2,18] as shown in Fig. 4. In this case, the first dimension is developed using the solvent system of chloroform : methanol : 0.2% aqueous $CaCl_2 \cdot 2H_2O$ (60 : 35 : 8, v/v), and the second dimension by *n*-propa-

[34] S. Ando, H. Waki, and K. Kon, *J. Chromatogr.* **405,** 125 (1987).
[35] T. Ariga, M. Sekine, R. K. Yu, and T. Miyatake, *J. Lipid Res.* **24,** 737 (1983).
[36] S. Ando, H. Waki, K. Kon, and Y. Kishimoto, *NATO ASI Series* **H7,** 167 (1987).

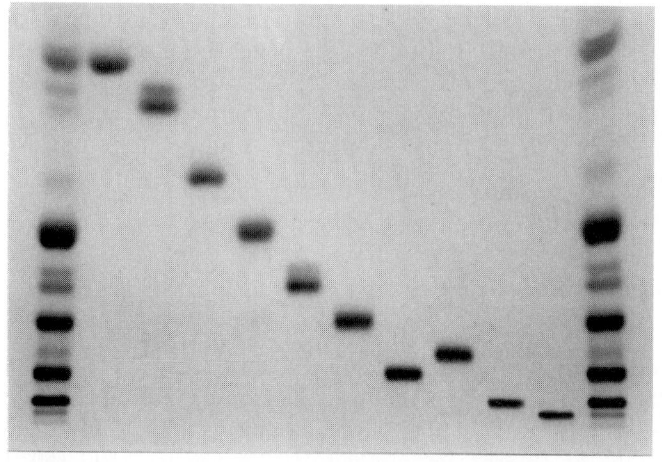

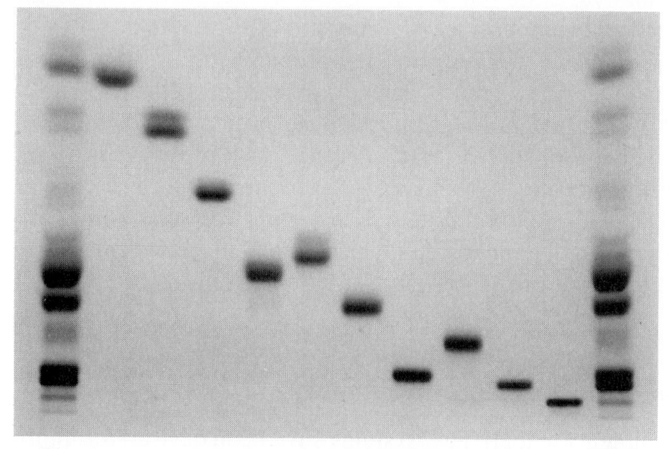

FIG. 2. HPTLC of normal human brain white matter gangliosides and the individual components. *Top:* Plate was developed with chloroform : methanol : 0.2% aqueous $CaCl_2 \cdot 2H_2O$ (60 : 40 : 9, v/v). *Bottom:* Plate was developed with chloroform : methanol : 2.5 N ammonium hydroxide (65 : 35 : 8, v/v). Each white matter ganglioside mixture contains about 5 μg lipid-bound sialic acid, and each individual ganglioside lane contains about 0.5 μg lipid-bound sialic acid. Gangliosides were visualized by spraying with the resorcinol–hydrochloric acid reagent followed by heating the covered plate at 110° for 15 min. Note the reversal of the position of GD3 in the ammoniacal system.

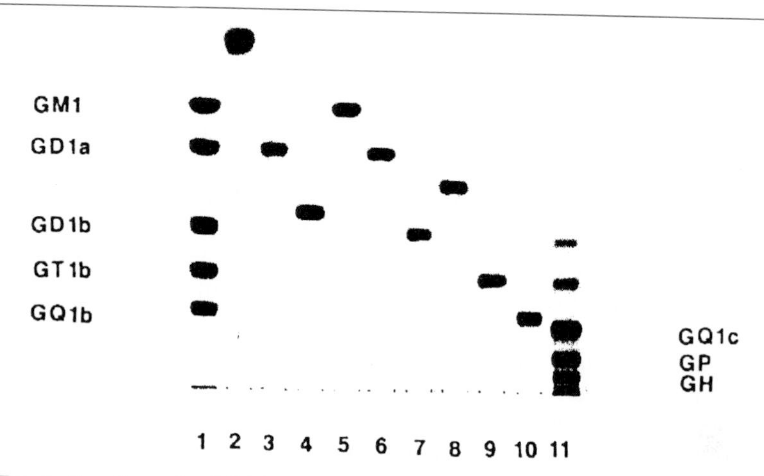

FIG. 3. One-dimensional HPTLC of gangliosides using acetonitrile : 2-propanol : 50 m*M* aqueous KCl (10 : 67 : 23, v/v) as the developing solvent system. Lane 1, a mixture of GM1, GD1a, GD1b, GT1b, and GQ1b; lane 2, GD3; lane 3, GD1a; lane 4, GD2; lane 5, GM1; lane 6, GD1a; lane 7, GD1b; lane 8, GT1a; lane 9, GT1b; lane 10, GQ1b; and lane 11, cod fish brain total gangliosides. (Reproduced with permission from Ref. 36.)

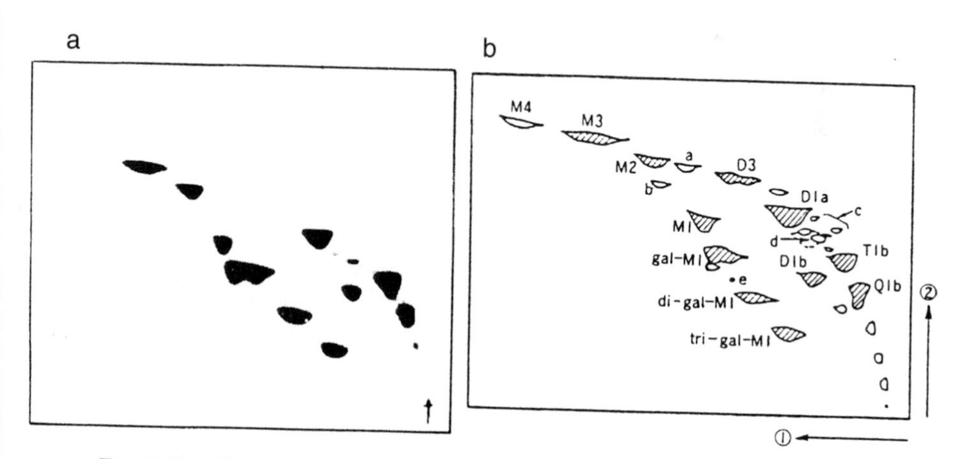

FIG. 4. Two-dimensional HPTLC of gangliosides. (a) Two-dimensional HPTLC of rat brain gangliosides plus authentic gangliosides as developed with the solvent system of chloroform : methanol : 0.2% aqueous CaC1 (60 : 35 : 8, v/v; 1st development) and *n*-propanol : water : 28% ammonium hydroxide (75 : 25 : 5, v/v; 2nd development). (b) Schematic presentation of major and minor gangliosides in (a) M4, GM4; M3, GM3; M2, GM2; M1, GM1; D3, GD3; D1a, GD1a; D1b, GD1b; T1b, GT1b; Q1b, GQ1b; gal-M1, Gal-GM1; di-gal-M1, di-Gal-GM1; tri-gal-M1, tri-Gal-GM1; a, sialosylparagloboside (LM1); b, GM2(NeuAc); c, unidentified gangliosides; d, GD1; e, Fuc,Gal-GM1 (B-GM1). (Reproduced with permission from M. Ohashi.)

nol : water : 28% ammonium hydroxide (75 : 25 : 5, v/v). Adequate drying of the plate between runs is necessary to achieve good resolution. In our laboratory, the plate is first dried with the help of a hair dryer on low heat, followed by placing the plate in a vacuum desiccator for at least 1 hr before the second run.

To identify alkali-labile gangliosides, such as *O*-acetylated gangliosides and ganglioside lactones in brain tissues, Riboni *et al.*[37] devised an elegant two-dimensional HPTLC procedure in which the ganglioside mixture to be analyzed is first developed using chloroform : methanol : aqueous $CaCl_2$ (50 : 40 : 10, v/v). After separation, the plate is placed over concentrated ammonium hydroxide in a solvent tank for 5 hr to hydrolyze the *O*-acetylated or lactonized residue. Following the removal of ammonia, the plate is turned 90 degrees and developed with the same solvent system (Fig. 5).

Another useful approach for carrying out two-dimensional TLC analysis is using *n*-propanol : 0.1% aqueous $CaCl_2$ (80 : 20, v/v) in the first dimension and *n*-butanol : pyridine : 0.1% aqueous $CaCl_2$ (65 : 45 : 15, v/v) in the second dimension.[31] More than 15 different minor gangliosides have been identified in this manner in bovine and hog muscle owing to the presence of small amounts of the *N*-glycolylneuraminic acid (NeuGc)-containing species in addition to the usual NeuAc-containing species.

Recently, an HPTLC-mapping method for the analysis of GSL molecular species has been developed using two-dimensional HPTLC.[38] This is accomplished by use of two types of HPTLC plates. An NH_2-silica gel HPTLC plate developed with chloroform : methanol : 1% diethylamine (50 : 47 : 15, v/v) is used for the first dimension. Acidic GSLs on the plate are then transferred to a normal silica gel HPTLC plate by developing the plate with chloroform : methanol : ammonia after the NH_2 plate is placed on top of the silica gel HPTLC plate. Then, the acidic GSLs on the HPTLC plate are developed with chloroform : methanol : 0.2% $CaCl_2$ (60 : 40 : 10, v/v) for the second dimension (Fig. 6). The HPTLC mapping method is designed to be of general applicability for the determination of ganglioside molecular species.

Chemical Detection of GSLs on TLC Plates

The detection of sialic acid and neutral sugars in gangliosides after HPTLC separation is frequently achieved by using specific colorimetric

[37] L. Riboni, A. Malesci, S. M. Gaini, S. Sonnino, and G. Tettamanti, *J. Biochem.* (*Tokyo*) **96,** 1943 (1984).

[38] K. Watanabe and M. Nishiyama, *Anal. Biochem.* **227,** 195 (1995).

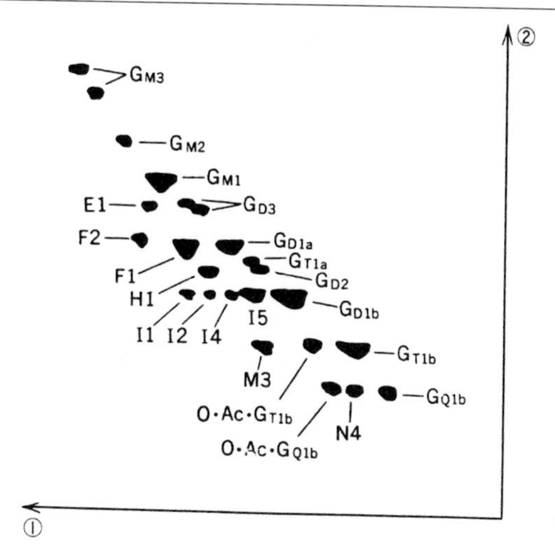

FIG. 5. Two-dimensional HPTLC of *O*-acetylated gangliosides from human brain. The gangliosides were developed with chloroform : methanol : aqueous $CaCl_2$ (50 : 40 : 10, v/v). After the plate was dried, it was placed in a chamber filled with ammonia vapor (generated from concentrated ammonium hydroxide placed at the bottom of the tank) for 5 hr. After evaporation of the ammonia vapor, the plate was developed again with the same solvent system for the second development. Alkali-labile gangliosides, such as *O*-acetylated (O-Ac) gangliosides, were hydrolyzed by the ammonia vapor, therefore they had the same mobilities as the corresponding gangliosides in the second dimension. E1, O-Ac-FucGM1 or O-Ac-GD3; F1 and F2, O-Ac(one or two)-GD1a; H1, O-Ac-GD2 or O-Ac-GT1a; I1–I5, O-Ac-GD1b that contains one or two O-Ac residues; M3, O-Ac(two?)-GT1b; N4, O-Ac-GQ1b(isomer). (Reproduced from Ref. 37.)

reagents such as resorcinol–hydrochloric acid reagent (0.2% resorcinol in 80% concentrated HCl containing 4 mg of $CuSO_4$)[39] and orcinol–sulfuric acid reagent (0.2% orcinol in 12.5% sulfuric acid),[40] respectively. Individual lipid classes can also be quantitated by dipping the TLC plates using a general lipid-detecting reagent such as 3% (w/v) cupric acetate/8% (v/v) phosphoric acid charring reagent,[41] followed by charring at 180° for 15 min in a hot air oven or on a hot plate. Sulfoglycolipids, including sulfatides, mono- and disulfogangliotetraosylceramide, and seminolipid, are stained with the azure A reagent.[42] Sulfated gangliosides have been reported as minor components in oligodendrocyte progenitor cells in the central ner-

[39] L. Svennerholm, *Biochim. Biophys. Acta* **604** (1957).
[40] L. Svennerholm, *J. Neurochem.* **1,** 42 (1956).
[41] L. J. Macala, R. K. Yu, and S. Ando, *J. Lipid Res.* **24,** 1243 (1983).
[42] K. Tadano-Aritomi and I. Ishizuka, *J. Lipid Res.* **24,** 1368 (1983).

A step 1 **1st dimensional TLC**

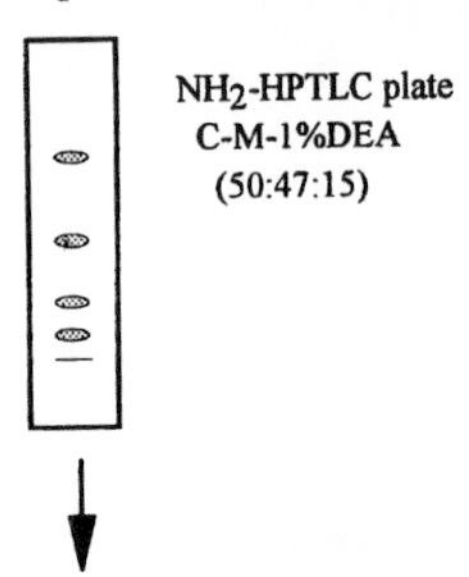

step 2 **standard GSL spot**

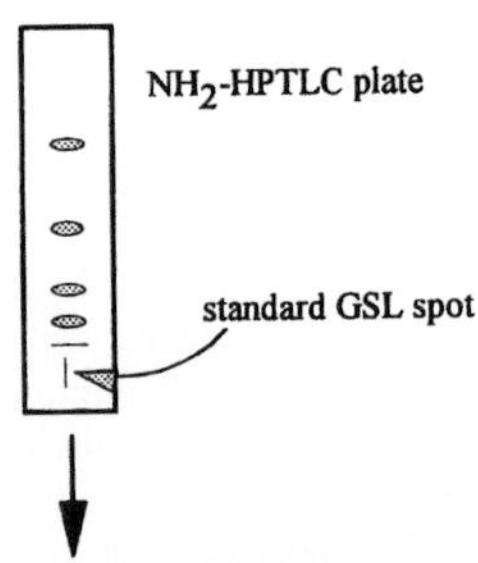

step 3 **GSL transfer**

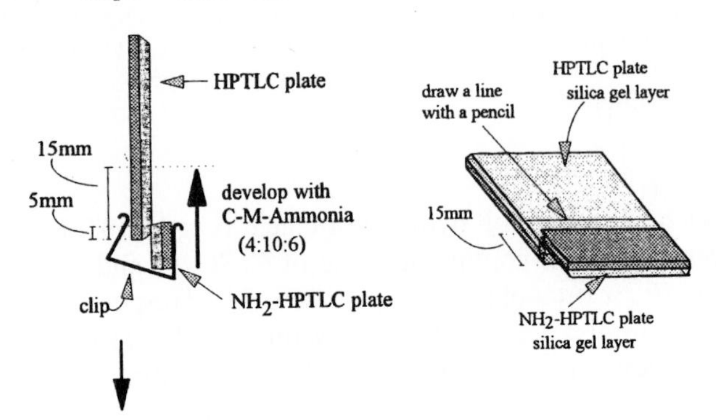

step 4 **2nd dimensional TLC**

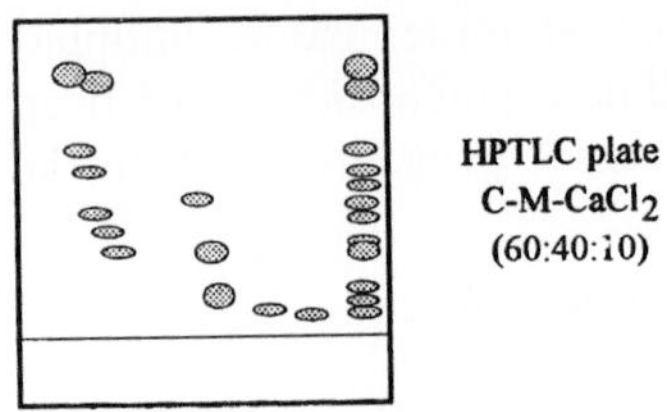

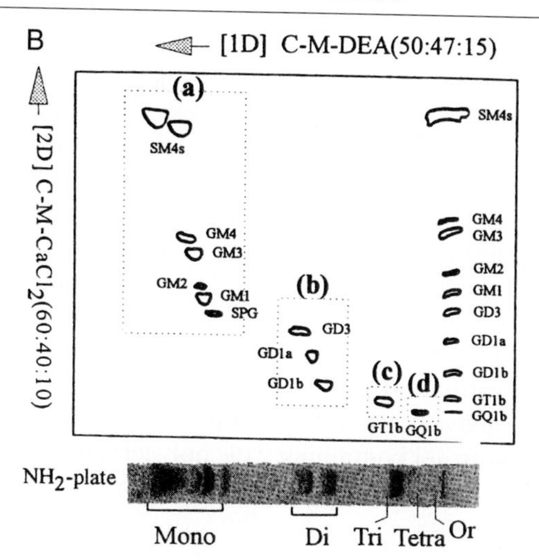

FIG. 6. (A) HPTLC mapping ganglioside molecular species. (B) Two-dimensional HPTLC of standard gangliosides, such as SM2s, GM4, GM3, GM2, GM1, LM1, GD3, GD1a, GD1b, GT1b, and GQ1b. The samples were spotted on an NH_2 plate and developed with chloroform : methanol : 1% diethylamine (50 : 47 : 15, v/v). After the gangliosides were transferred from the NH_2 plate to the silica gel HPTLC plate, the latter was developed with chloroform : methanol : 0.2% aqueous $CaCl_2$ (60 : 40 : 10, v/v). (Reproduced from Ref. 38.)

vous system using two-dimensionnal HPTLC in this manner.[43] A microdetection system for GSL analysis has been developed using 5-hydroxy-1-tetralone as a fluorescence-labeling reagent.[44] Muthing and Kemminer[45] reported the use of nondestructive detection of neutral GSLs with lipophilic anionic fluorochromes. This method offers an easy preparative HPTLC method for obtaining pure neutral GSLs in microgram quantities. For example, gangliosides can be visualized by UV light on silica gel-precoated HPTLC plates following staining with an uncharged lipophilic fluorochrome, 4-(*N,N*-dihexadecyl)amino-7-nitrobenz-2-oxa-1,3-diazole.[46,47]

The oligosaccharides released from GSLs by endoglycoceramidase can be analyzed by a high-pH anionic chromatographic (HPAEC) analysis using thin-layer and column chromatography.[48] The fluorophore-assisted carbohydrate electrophoresis (FACE) analysis of oligosaccharides released

[43] R. G. Farrer and R. H. Quarles, *J. Neurochem.* **68,** 878 (1997).
[44] K. Watanabe and M. Mizuta, *J. Lipid Res.* **38,** 1848 (1995).
[45] J. Muthing and S. E. Kemminer, *Anal. Biochem.* **238,** 195 (1996).
[46] J. Muthing and F. Unland, *Biomed. Chromatogr.* **6,** 227 (1992).
[47] J. Muthing and D. Heitmann, *Anal. Biochem.* **208,** 121 (1993).
[48] S. Ruan and K. O. Lloyd, *Glycoconj. J.* **11,** 249 (1994).

from purified brain glycosphingolipids has been reported.[49] The oligosaccharides are released from 10 μg of GSLs using 50 mU ceramide glycanase in 1.8 m*M* sodium cholate and 50 m*M* phosphate buffer at pH 5.0. The reaction mixture is incubated overnight at 37°. Following incubation, the reaction mixture is extracted with 250 μl of chloroform–methanol (2:1, v/v), centrifuged, and the aqueous phase is recovered and dried in a centrifugal vacuum evaporator. The released oligosaccharides are labeled by the addition of 5 μl of the fluorescence dye, 0.15 *M* potassium 7-aminonaphthalene 1,3-disulfonate in 15% acetic acid, and 5 μl of 1 *M* $NaBH_3CN$ in dimethyl sulfoxide (DMSO). The labeled oligosaccharides are dried, resuspended in 20 μl water, and 2 μl is combined with 2 μl of 12.5% glycerol containing a Thorin dye marker. Separation of oligosaccharides is achieved on a precast profiling gel containing 21% polyacrylamide (Glyko, Novato, CA) at a constant current of 20 mA at 5° for 1 hr.

HPTLC-Overlay of Glycosphingolipids

TLC-overlay is a method for detecting GSLs separated on TLC plates using specific anti-glycolipid antibodies, lectins, or carbohydrate-binding proteins. This procedure was originally developed by Magnani *et al.*[3] who used ^{125}I-labeled cholera toxin as the ligand to detect ganglioside GM1. Thereafter, the TLC-overlay technique has proved to be applicable to detection of essentially all kinds of GSLs by using specific ligands as well as microorganisms, and this technique has become widely used in glycosphingolipid research. The procedure consists of two essential steps: separation of GSLs on a TLC plate, and *in situ* detection of the glycolipids with specific ligands or antibodies. Since HPTLC can achieve high resolution for GSLs, minor components that are difficult to purify can be separated and characterized using this medium. The use of specific anti-GSL antibodies makes it possible to detect glycolipids with strict structural specificity. The sensitivity of this method is generally very high, nanogram or less quantities of GSLs being sufficient. The original technique reported by Magnani *et al.*[3] has been modified in many respects to improve stability and reproducibility. Glass-backed silica gel plates, which generally provide finer separation of glycolipids but tend to lose the silica gel layer more easily during incubation than do aluminum- or plastic-backed plates, have been made more reliable by treatment with an additional coating of an organic polymer, poly(isobutyl methacrylate), at higher concentrations.[4,12]

[49] C. M. Starr, R. I. Masada, C. Hague, E. Skop, and J. C. Klock, *J. Chromatogr.* **720,** 295 (1996).

Identification and Characterization of GSLs and anti-GSL Antibodies

A typical HPTLC-overlay procedure used extensively in our laboratory[12] is as follows: After chromatographic separation of samples by HPTLC, the plate is coated with 0.2% to 0.4% poly(isobutyl methacrylate) (Aldrich Chemical Co., Milwaukee, WI) in *n*-hexane. The higher concentration of polymer is achieved by diluting a 2.5% chloroform solution with *n*-hexane. A serum or monoclonal antibody against GSLs, diluted 1:300 in a dilution buffer (0.3% gelatin in phosphate-buffered saline, pH 7.4) is applied to the plate and incubated at room temperature for 1 hr. The antibody dilution factor may vary depending on the titer of the specific antibody. After washing 3 times with phosphate-buffered saline (PBS), the plate is incubated with peroxidase-conjugated goat anti-human or mouse immunoglobulin G (IgG) or IgM (Cappel, Westchester, PA), diluted 1:1000 in dilution buffer at room temperature for 1 hr. After washing the plate 3 times with PBS, the plate is overlaid with cyclic diacyl hydrazine (chemiluminescence) solution [ECL (enhanced chemiluminescence) detection reagent, Amersham, Arlington Heights, IL] and exposed for a few minutes to an X-ray film.[9,10,50,51] This method reliably detects 100 pg of sulfoglucuronosyl paragloboside. The use of the ECL reagent has the advantage of increased sensitivity and simplicity, and the data can be saved easily for producing a permanent record.

To avoid the use of radioactive materials, such as radiolabeled secondary antibodies, a number of immunoenzymatic methods, such as the immunoperoxidase technique[52,53] and the avidin–biotin enzyme system,[54] have been developed without significant reduction in sensitivity. In addition to cholera toxin (specific to GM1) and anti-GSL antibodies, other kinds of ligands, such as lectins, have also been used.[55] Such ligands, having less strict specificity compared to antibodies, would be especially useful for obtaining information about glycolipids of unknown structures.

Quantitation of GSLs by HPTLC-Overlay

In binding reactions between ligands and glycolipids on TLC plates, the amount of a ligand bound to a glycolipid is assumed to correlate with

[50] M. Yamawaki, T. Ariga, J. W. Bigbee, H. Ozawa, I. Kawashima, T. Tai, T. Kanda, and R. K. Yu, *J. Neurosci. Res.* **44,** 586 (1996).

[51] S. Ladisch, F. Chang, R. Li, P. Cogen, and D. Johnson, *Cancer Lett.* **120,** 71 (1997).

[52] H. Higashi, K. Ikuta, S. Ueda, S. Kato, Y. Hirabayashi, M. Matsumoto, and M. Naiki, *J. Biochem.* (*Tokyo*) **95,** 785 (1984).

[53] T. Ariga, L. J. Macala, M. Saito, R. K. Margolis, L. A. Greene, R. U. Margolis, and R. K. Yu, *Biochemistry* **27,** 52 (1988).

[54] J. Buehler and B. A. Macher, *Anal. Biochem.* **158,** 283 (1985).

[55] T. Momoi, T. Tokunaga, and Y. Nagai, *FEBS Lett.* **141,** 6 (1982).

that of the glycolipid applied to the plate. This assumption has led to the development of quantitative analytical methods for glycolipids using the TLC-overlay technique. Quantitation of lipids is usually achieved by scanning autoradiograms of plates or chromatograms stained with the immunoenzymatic technique. Kohriyama *et al.*[56] have devised a quantitative immunostaining method by scanning autoradiogram, and then analyzed the subcellular distribution of sulfated glucuronic acid-containing glycolipids, such as sulfoglucuronosylparagloboside (SGPG) and sulfoglucuronosyllactosaminylparagloboside (SGLPG). In this procedure, the acidic lipid samples and different amounts of the standard glycolipids are developed with HPTLC. The separated SGPG and SGLPG on the plate are then detected with the serum of a patient with neuropathy with IgM M proteins. The plate is treated with anti-human goat IgG and ^{125}I-labeled staphylococcal protein A, successively, and the resulting autoradiogram is scanned by a densitometer equipped with a data processor. The amount of the glycolipids in each sample is computed based on the standard curves, which are obtained with known amounts of the standard glycolipids developed on the same plate with the samples. Under the conditions used, the dose–response curves are essentially linear up to 50 ng for SGPG and up to 15 ng for SGLPG. The densitometry of chromatograms stained using immunoenzymatic techniques has also been successfully applied to the quantitation of many other GSLs, including the Hangenutzui–Deicher antigens.[52]

Analysis of in Situ Reaction Products of GSLs

HPTLC-overlay indicates that GSLs developed on TLC plates can react with ligands such as antibodies and suggests that they can also be subjected to chemical and enzymatic reactions other than binding with the ligands. This possibility has been confirmed by several authors carrying out various types of reactions involving GSLs on TLC plates. We examined the effect of neuraminidase on gangliosides developed on silica gel plates, and found that, in the absence of detergent, the hydrolytic reaction of gangliosides proceeded more easily on plates than it did in solution.[12] Based on this finding, we developed a new analytical method for determining the carbohydrate structures of gangliosides.

In principle, one can modify unknown glycolipid structures by enzymatic or chemical means after separation of TLC plates and identify the structures of the GSL remnants by specific antibodies. Once this is done, it is possible to deduce the structures of the parent GSLs based on the specificity of the chemical or enzymatic reactions. In this method, gangliosides are modified

[56] T. Kohriyama, S. Kusunuki, T. Ariga, J. E. Yoshino, G. H. DeVries, N. Latov, and R. K. Yu, *J. Neurochem.* **48,** 1516 (1987).

by *Arthrobacter ureafaciens* neuraminidase on the plate, and the generated asialogangliosides are treated with an anti-asialoGMI (anti-Gg_4) antibody. If a ganglioside of unknown structure can be detected with the antibody after treatment with the enzyme, it means that the ganglioside has the same carbohydrate core structure as Gg_4.

Details of the method are as follows: Human brain gangliosides (1.25 μg as sialic acid) are applied to an HPTLC glass-backed plate (Silica gel 60, E. Merck) in 5-mm streaks and developed to 7 cm from the bottom edge of the plate with chloroform : methanol : 0.22% aqueous $CaCl_2 \cdot 2H_2O$ (50 : 45 : 10). After drying *in vacuo* for at least 2 hr, the plate is dipped in a 0.4% poly(isobutyl methacrylate) solution for 1 min. Each lane of the plate is overlayed with an *A. ureafaciens* neuraminidase solution (10, 20, and 40 mU/ml of 0.1 *M* sodium acetate, pH 4.8, about 0.5 ml per lane) and incubated at room temperature for 2 hr in a humid plastic box. The plate is washed with PBS and dried briefly in air. Thereafter, the plate is treated for 1 hr with an anti-Gg_4 rabbit serum diluted with 0.3% gelatin–PBS, and then with ^{125}I-labeled staphylococcal protein A (about 5×10^5 cpm/ml of 0.3% gelatin–PBS) for another hour. Finally, the plate is thoroughly washed with 0.1% Triton X-100 in PBS and exposed to an X-ray film overnight. As shown in Fig. 7, the anti-Gg_4 antibody reacted exclusively with Gg_4 (lane 5) but not with any intact gangliosides (lane 4). After treatment with *A. ureafaciens* neuraminidase, gangliosides such as GM1, GD1a, GT1a, GT1b, and GQ1b became reactive with the antibody, indicating that these gangliosides had the same carbohydrate core structure as Gg_4 (lanes 1 to 3, Fig. 7). The intensity of each positive band increases in proportion to the concentration of the enzyme used. This method has been applied to the analysis of gangliosides of cultured PC12 pheochromocytoma cells.[53]

Various combinations of enzymes and ligands can be used in this method, providing valuable structural information about the parent GSLs. For example, when GD1b was treated with jack bean β-galactosidase (1.875 U/ml of 0.1 *M* citrate–0.2% taurodeoxycholate, pH 4.5) on TLC plates, it was converted to GD2, which was detected with a monoclonal antibody, 3F8, directed toward cultured neuroblastoma cells.[20]

Biosynthetic reactions of glycolipids can also be carried out on plates. For example, we developed a procedure for the assay of GSL glycosyltransferase activities using an HPTLC plate.[18] In this procedure, acceptor GSLs, 4–10 nmol for each component, are spotted on an HPTLC plate and developed with a suitable solvent system. The plate is dried under vacuum for 12–15 hr, then soaked in 0.4% poly(isobutyl methacrylate) in *n*-hexane for 1 min, and completely dried under air. Several HPTLC plates can be prepared at one time and stored at −20° for several months for future use.

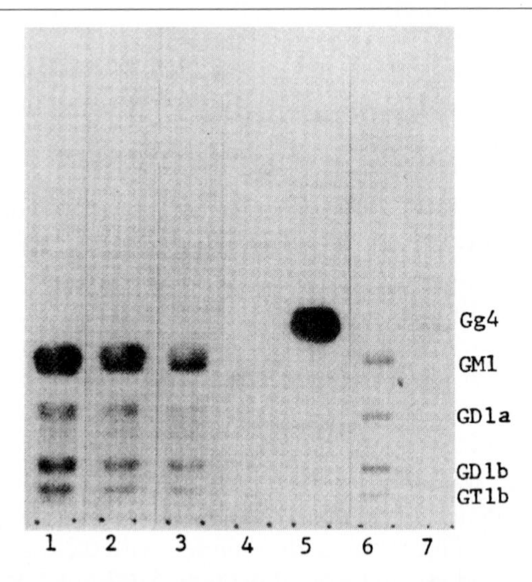

FIG. 7. Detection of gangliosides having the same carbohydrate core structure as Gg_4 in human brain gangliosides. The ganglioside mixture (lanes 1 to 4, lane 6) and Gg_4 (lanes 5 and 7) were developed on an HPTLC plate. Lanes 1 to 5 were incubated with 40 mU/ml (lane 1), 20 mU/ml (lane 2) and 10 mU/ml (lane 3) of an *A. ureafaciens* sialidase solution, or with the buffer (0.1 *M* sodium acetate, pH 4.8, lanes 4 and 5), followed by incubation with anti-Gg_4 antibody and ^{125}I-labeled SPA, successively. Lanes 6 and 7 were visualized by the α-naphthol–sulfuric acid reagent. (Reproduced from Ref. 8.)

An enzyme mixture containing 240 nmol (1×10^6 cmp) of CMP-[^{14}C]NeuAc, 0.1–2 mg of protein, 25 m*M* sodium cacodylate buffer (pH 6.5), 10 m*M* $MnCl_2$, and 0.1% Triton X-100 in a final volume of 1 ml is pipetted onto the plate. After incubation at 37° for 30–60 min, the plate is carefully washed with PBS containing 2% Tween 80. After the positions of gangliosides are revealed by staining with iodine, the radiolabeled products are detected by exposing the plate to an X-ray film. For quantitation, the product on the plate can be scraped into a scintillation vial for counting. This method is sensitive, fast, and reliable, and capable of assaying simultaneously the activities of glycosyltransferases with multiple acceptor specificity.

In addition to the above, Samuelsson[57] has also reported that blood group glycolipids can be fucosylated on TLC plates. Total neutral GSLs prepared from plasma of individual blood donors are first separated on an HPTLC plate. Each lane is overlaid with an incubation medium, which consists of 650 μl of an enzyme preparation, about 1.3×10^6 cpm of GDP-

[57] B. E. Samuelsson, *FEBS Lett.* **167,** 47 (1984).

L-[^{14}C]fucose, 25 μmol Tris-HCl (pH 7.5), 10 μmol adenosine triphosphate (ATP), 7.5 μmol $MgCl_2$, and 0.1% Triton X-100. For the enzyme preparation, a microsomal fraction obtained from pig intestine mucosa is used. Autoradiography of the plate shows that some glycolipids, including Le^b hexaglycosylceramide, are fucosylated.

The glycolipid-overlay technique has also been used to analyze the receptor function of glycolipids for viruses and bacteria. Bock *et al.*[58] examined the interaction of *Escherichia coli 36692* with 34 different glycolipids and reported that only the glycolipids having the structure of Gal(α1-4) Gal at the internal or terminal position interacted with the bacteria on plates. After chromatography and sialidase hydrolysis, a highly sensitive method has been reported for detection and quantitation of GM1 ganglioside using cholera toxin B subunit.[11,59,60] α-Galactose-containing GSLs, particularly Gb_3, are good candidates for receptors for piliated *Burkholderia cepacia* from the sputa of cystic fibrosis patients.[61] The sperm glycolipid that has been tentatively identified as galactosylalkylacylglycolipid was the receptor for gp120 in HIV.[62] The epidemic form of the hemolytic uremic syndrome (HUS) has been associated with a verocytotoxic producing *E. coli* infection. The verocytotoxin-1 (VT-1) binds to a globotriaosylceramide (Gb_3).[63–67] *Lactobacillus* in the intestinal tract has been found to bind to some specific GSLs, asialo-GM1 (Gg_4) and trihexosylceramide.[68] The cell

[58] K. Bock, M. E. Breimer, A. Brignole, G. C. Hansson, K.-A. Karlsson, G. Larson, H. Leffler, B. E. Samuelsson, N. Stromberg, C. S. Eden, and J. Thurin, *J. Biol. Chem.* **260,** 8545 (1985).

[59] P. Davidson, P. Fredman, J.-E. Masson, and L. Svennerholm, *Clin. Chim. Acta* **197,** 105 (1991).

[60] N. Miyatani, M. Saito, T. Ariga, H. Yoshino, and R. K. Yu, *Mol. Chem. Neuropathol.* **13,** 205 (1990).

[61] F. A. Sylvester, U. S. Sajjan, and J. F. Forstner, *Infect. Immun.* **64,** 1420 (1996).

[62] A. Brogi, R. Presentini, D. Solazzo, P. Piomboni, and E. Costantino-Ceccarini, *AIDS Res.* **12,** 483 (1996).

[63] B. Boyd, G. Tyrrell, M. Maloney, C. Gyles, J. Brunton, and C. A. Lingwood, *J. Exp. Med.* **177,** 1745 (1993).

[64] C. A. Lingwood, *in* "Advances in Lipid Research" (R. Bell, Y. A. Hannun, and A. Merrill, Jr., eds.), p. 189. Academic Press, San Diego, 1993.

[65] P. A. Van Setten, L. A. Monnens, R. G. Verstraten, L. P. van den Heuvel, and V. W. van Hinsbergh, *Blood* **88,** 174 (1996).

[66] P. A. Van Setten, V. W. van Hinsberg, T. J. van der Velden, N. C. van de Kar, M. Vermeer, J. D. Mahan, K. J. Assmann, L. P. van den Heuvel, and L. A. Monnens, *Kidney Int.* **51,** 1245 (1997).

[67] S. C. K. Yiu and C. A. Lingwood, *Anal. Biochem.* **202,** 188 (1982).

[68] K. Yamamoto, K. Miwa, H. Taniguchi, T. Nagano, K. Shimamura, T. Tanaka, and H. Kumgai, *Biochem. Biophys. Res. Commun.* **228,** 148 (1996).

surface hsp-70-related heat shock proteins can mediate *Haemophilus influenzae* attached to sulfatoxygalactosylglycerol following heat shock.[69]

The differences in receptor specificity of influenza virus A/PR/8/34(H1N1), A/X-31 (H3N2), and parainfluenza Sendai virus (HNF1, Z-strain) for binding to neolacto-series gangliosides were determined by the HPTLC-overlay technique.[19] A ganglioside that comigrated with GM1 from human respiratory epithelial cells has been suggested as a potential receptor for *H. influenzae*, nontype-able *H. influenzae* (NTHI) 1479.[70] The ganglioside serves as a receptor for Sendai virus.[71] The receptor for parvovirus B19 was reported to be globoside (Gb_4).[72] Using HPTLC-immunostaining, sulfatides and GM3 ganglioside, two major acidic GSLs preferentially bind to *Helicobacter pylori.*[73] Galactosylceramide and sulfatide have been shown to bind to gp160 of human immunodeficiency virus (HIV) on CD4- cell lines, but these GSLs do not apparently act as HIV coreceptors nor are they involved in HIV infection of those cells.[74]

Combined HPTLC and Mass Spectrometry (HPTLC-MS)

Mass spectrometry has been used extensively for elucidating the structures of GSLs with respect to their saccharide sequence, fatty acid and long-chain base compositions, and molecular weights. Mass spectrometry employing soft-ionization methods, such as fast atom bombardment mass spectrometry (FAB-MS), secondary ion mass spectrometry (SIMS), atmospheric pressure ionization mass spectrometry (API-MS), and time-of-flight mass spectrometry, are particularly suitable for GSLs that are nonvolatile and thermolabile. The combination of HPTLC and mass spectrometry is especially attractive and useful because it simultaneously provides chromatographic and structural information for GSLs.[14,75] Kushi *et al.*[76] first reported the use of high-performance liquid chromatography/atmospheric pressure ionization mass spectrometry (HPLC/API-MS)[76] and HPTLC/

[69] E. Hartmann and C. A. Lingwood, *Infect. Immun.* **65,** 1729 (1997).

[70] M. G. Fakih, T. F. Murphy, M. A. Pattoli, and C. S. Berenson, *Infect. Immun.* **65,** 1695 (1997).

[71] R. M. Epand, S. Nir, M. Parolin, and T. D. Flanagan, *Biochemistry* **34,** 1084 (1995).

[72] L. L. Cooling, T. A. W. Koerner, and S. J. Naides, *J. Infect. Dis.* **172,** 1198 (1995).

[73] S. Kamisago, M. Iwamori, T. Tai, K. Mitamura, Y. Yazaki, and K. Sugano, *Infect. Immun.* **64,** 624 (1996).

[74] N. Seddiki, A. Ben Younes-Chennoufi, A. Benjouad, L. Saffa, N. Baumann, J. C. Gluckman, and L. Gattegno, *AIDS Res.* **12,** 695 (1996).

[75] S. Handa and Y. Kushi, *in* "New Trends in Ganglioside Research" (R. W. Ledeen, E. L. Hogan, G. Tettamanti, A. J. Yates, and R. K. Yu, eds.), Fidia Research Series Vol. 14, p. 37. Liviana Press, Padova, 1988.

[76] Y. Kushi, C. Rokukawa, Y. Numajiri, Y. Kato, and S. Handa, *Anal. Biochem.* **182,** 405 (1989).

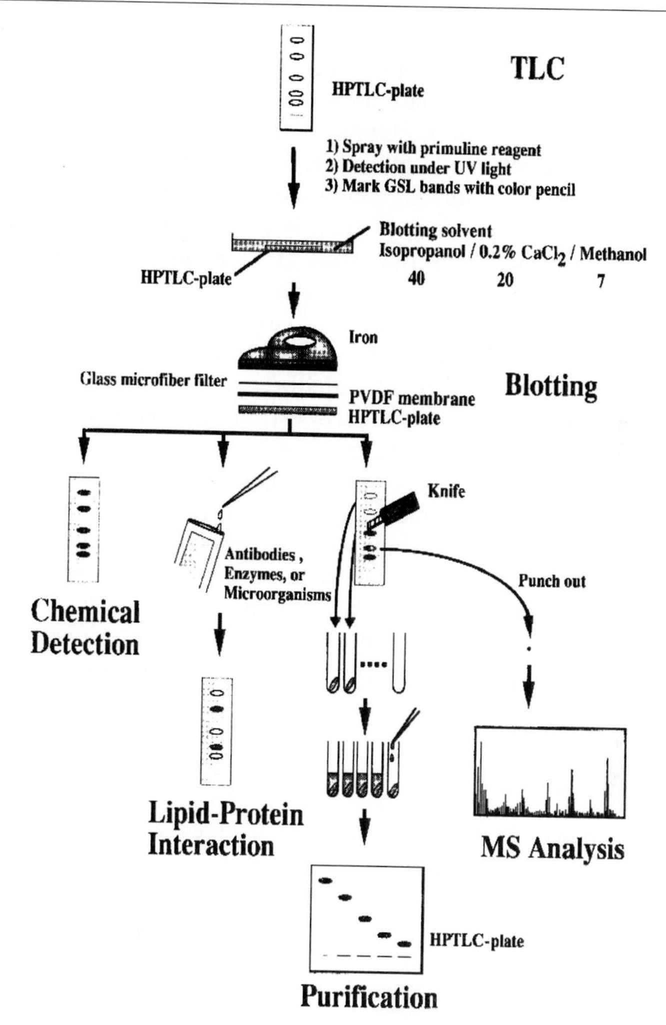

FIG. 8. Applications of Far-Eastern blot/mass spectrometry. (Reproduced from Ref. 77.)

matrix-assisted secondary ion mass spectrometry (HPTLC/SIMS)[77] for the separation and structural characterization of as little as 60 ng of intact GSLs from biologic samples without purification.[76]

A simple and rapid method for the analysis of GSLs using a combination of HPTLC-immunostaining and secondary ion mass spectrometry (Far-

[77] D. Ishikawa, T. Kato, S. Handa, and T. Taki, *Anal. Biochem.* **231,** 13 (1995).

Eastern blot/MS*) has been reported.[15,16,78] The procedure, as outlined in Fig. 8, is as follows: After developing GSLs on an HPTLC plate, a primuline reagent is sprayed over the plate until it is visibly wet. GSLs on the plate are viewed under UV light at 360 nm. Each GSL band is marked with a colored drawing pencil while still under illumination. The plate is then immersed for 20 sec in a solvent composed of 2-propanol : methanol : 0.2% aqueous $CaCl_2$ (40 : 20 : 7, v/v). It then is placed on another glass plate, after which a polyvinylidene difluoride (PVDF) membrane sheet and a glass microfilter sheet are placed over the plate. The "sandwich" is then pressed with a household iron (about 180°) for 30 sec. The PVDF membrane is separated from the HPTLC plate and washed with water to remove the primuline reagent and then dried and stored at −20° until use. The GSL band on the PVDF membrane is excised (2 mm in diameter) and placed on a mass spectrometer probe tip, and a few microliters of triethanolamine added as the matrix. The sample on the membrane is bombarded with a Cs^+ beam at 20 kV. The ion multiplier is kept at 1.5 kV and the conversion dynode at 20 kV.[15] The advantage of this method is that only 1 μg of GSL can be analyzed structurally without purification by repeated column chromatography. The method has been applied to the analysis of gangliosides in tumor cells.[79] More recently, the glycolipids in prions, the agents of transmissible spongiform encephalopathies have been identified using TLC/MS.[80]

Acknowledgment

The work performed in the authors' laboratory has been supported by USPHS grant NS11853.

* The term Eastern blot evolved from laboratory jargon, reflecting the fact that it was first developed by Magnani's laboratory in Bethesda, MD,[3] and further elaborated by our laboratory[4] at Yale University, in New Haven, CT, both on the East Coast.

[78] Y. Kushi, K. Ogura, C. Rokukawa, and S. Handa, *J. Biochem.* (*Tokyo*) **107,** 685 (1990).

[79] T. Kasama, Y. Hisano, M. Nakajima, S. Handa, and T. Taki, *Glycoconj. J.* **13,** 461 (1996).

[80] T. Klein, D. Kirsch, R. Kaufmann, and D. Riesner, *J. Biol. Chem.* **379,** 655 (1998).

[11] Purification and Analysis of Gangliosides

By STEPHAN LADISCH and RUIXIANG LI

Introduction

One problem in the purification of gangliosides from biologic samples is that the ganglioside concentration is frequently low or the samples available are small. One example is human cerebrospinal fluid, which normally contains a very low concentration of total gangliosides ($3–5 \times 10^{-7}$ *M*). The second example is gangliosides shed by tumor cells in the conditioned culture medium (1×10^{-7} *M*). Quantification of these molecules therefore requires both good recovery and a sensitive detection method. In this chapter, we first review the three-step purification method for microscale isolation and purification of gangliosides from the total lipid extract of cultured cells, plasma, cerebrospinal fluid (CSF), and tissues. We then describe the metabolic radiolabeling method that is used for the study of ganglioside synthesis and shedding. Finally, we describe the high-performance liquid chromatography (HPLC) isolation method to obtain highly purified gangliosides required for structural analysis and for assessment of their biologic activity.

Three-Step Purification Method

The three-step method[1,2] consists of (1) preparation of a dry total lipid extract, (2) partition of the total lipid extract in diisopropyl ether (DIPE)/1-butanol/aqueous NaCl (DIPE/butanol partition), and (3) Sephadex G-50 gel filtration. This total lipid extraction reduces the high concentration of proteins and salt relative to gangliosides. The solvent partition step separates gangliosides from most other lipids. Finally, the lower aqueous (ganglioside-containing) phase from this solvent partition is purified by gel filtration to remove low molecular weight contaminants (salts and peptides). The final total ganglioside preparation is lyophilized and stored in the dry state.

Procedures

Total Lipid Extraction. Total lipid extracts of cultured cells, homogenized tissues, plasma, or cerebrospinal fluid are lyophilized in a 50- or 15-ml

[1] S. Ladisch and B. Gillard, *Anal. Biochem.* **146,** 220 (1985).
[2] S. Ladisch and B. Gillard, *Methods Enzymol.* **138,** 300 (1987).

0076-6879/00 $30.00

glass centrifuge tube (Corning, Corning, NY) or a round-bottom flask of appropriate size (i.e., 30 × tissue or plasma volume). The lyophilized samples are pulverized with a glass rod or a spatula and resuspended in chloroform : methanol (1 : 1, v/v; 20 ml/g wet weight tissue or 10^9 cells; 10 ml/ml plasma or serum; 2 ml/ml cerebrospinal fluid) in a Branson bath sonicator (Danbury, CT) to disperse the solid material. The samples are extracted for 18 hr at 4° with magnetic stirring, and then centrifuged at 2000 rpm (750*g*) for 10 min. The clear supernatant is transferred to a 50-ml glass centrifuge tube (the first extract contains more than 90% of total gangliosides). The residue is then reextracted for 4 hr with an additional original volume of fresh chloroform : methanol (1 : 1, v/v) The second extract, which contains ~10% of total gangliosides, is likewise clarified by centrifugation, and the two extracts are combined. The combined extracts are reduced to about one-quarter of the original total volume by evaporation under a stream of nitrogen or by a Rotavapor, and then cooled overnight at −20°. This results in further precipitation of salts and other molecules (e.g., glycoproteins) that are marginally soluble in chloroform and methanol, without loss of gangliosides. After centrifugation, the clear ganglioside-containing supernatant is transferred to a 50- or 15-ml glass centrifuge tube for partition. Finally, the supernatant is completely dried under a stream of nitrogen. Any residual trace of solvents is removed by oil pump vacuum.

DIPE/Butanol Partition. Organic phase (DIPE/1-butanol, 6 : 4, v/v) is added to the dried total lipid extract. The sample is vortexed and sonicated in a Branson bath sonicator to achieve fine suspension of the lipid extract. Then, the aqueous phase is added as follows: 10 ml 0.1% saline/10^9 cells (minimum volume of 1 ml); 2.0 ml distilled water/ml plasma or serum; 10 ml 0.3% saline/g wet weight tissue. The ratio of organic phase to aqueous phase is 2 : 1 (v/v). After the addition of the aqueous phase, the sample is alternately vortexed and sonicated for 2 min, and then centrifuged at 2000 rpm (750*g*) for 10 min. The upper organic phase, which contains neutral lipids and phospholipids and which can be further purified if desired, is carefully removed using a Pasteur pipette, and the lower aqueous (ganglioside-containing) phase is repartitioned with the same volume of fresh DIPE/1-butanol (6 : 4, v/v). The sample is alternately vortexed and sonicated again for 2 min and centrifuged as above. The residual organic solvents are removed by a stream of nitrogen gas from the final aqueous phase, and are then lyophilized to concentrate the sample prior to gel filtration. The recovery rate for the partition step is 93–100%, depending on the sample type and the ganglioside species.[1]

Sephadex G-50 Gel Filtration. The lyophilized samples are redissolved with bath sonication in distilled water to ensure micelle formation and then loaded onto a Sephadex G-50 (Sigma, St. Louis, MO) column equilibrated in distilled water. Specifications of the typical column conditions are as follows: (1) bed volume, 10–15 ml; inside diameter, 8 mm; flow rate, 0.25 ml/min; optimal sample loading volume, 0.3 ml; loaded ganglioside, 5–20 nmol; recovery, 80–100%; (2) bed volume, 40 ml; inside diameter, 12 mm; flow rate, 0.6 ml/min; optimal sample loading volume, 0.5 ml; loaded ganglioside, 20–50 nmol; recovery, 100%; and (3) bed volume, 90 ml; inside diameter, 22 mm; flow rate, 1.2 ml/min; optimal sample loading volume, 1.5 ml; loaded ganglioside, 70–5400 nmol; recovery, 93–100%. The samples are eluted with distilled water, and the eluate is monitored using an LKB Uvicord UV detection monitor (Uppsala, Sweden) at 206 nm. The gangliosides are eluted in the void volume peak (peak 1), collected in a 15-ml centrifuge tube (Corning), and lyophilized with the Teflon-lined cap loosely capped. The gangliosides are redissolved in a small volume of chloroform–methanol (1 : 1,v/v; for example, 0.3–0.4 ml for 1 ml plasma), and centrifuged to remove any insoluble material.

Applications

As an example of the application of the three-step purification method, we have chosen the analysis of cellular and plasma gangliosides from patients with acute myelogenous leukemia (AML). This example demonstrates the applicability of the methodology to purification of cellular gangliosides, as well as the more difficult purification of gangliosides from plasma or serum (1–5 ml), which have a higher protein-to-ganglioside ratio. Normal human plasma contains GM3 as a major ganglioside and a low concentration of SPG (0.2–0.4 nmol/ml; 2–5% of total plasma gangliosides).[3,4] In contrast, GM3 and SPG are present in both plasma and leukocytes of AML patients as two major gangliosides, each migrating as a doublet on the HPTLC plate (Fig. 1). SPG in the plasma of AML patients is significantly elevated (~17% of total gangliosides), and is therefore likely derived from the circulating malignant AML leukocytes (Fig. 1).

This method is also applicable to small CSF samples for immunostaining of brain tumor-associated ganglioside GD3 in cerebrospinal fluid.[6] CSF samples (1–5 ml) are lyophilized after addition of 10 nmol unrelated gangliosides (e.g., GT1b) as carrier molecules. Since CSF did not contain an abundance of lipids potentially interfering with ganglioside detection by

[3] S. K. Kundu, I. Diego, S. Osovitz, and D. M. Marcus, *Arch. Biochem. Biophys.* **238,** 388 (1985).

[4] H. J. Senn, M. Orth, E. Fitzke, H. Wieland, and W. Gerok, *Eur. J. Biochem.* **181,** 657 (1989).

[5] L. Svennerholm, *Biochim. Biophys. Acta* **24,** 604 (1957).

[6] S. Ladisch, F. Chang, R. Li, P. Cogen, and D. Johnson, *Cancer Lett.* **120,** 71 (1997).

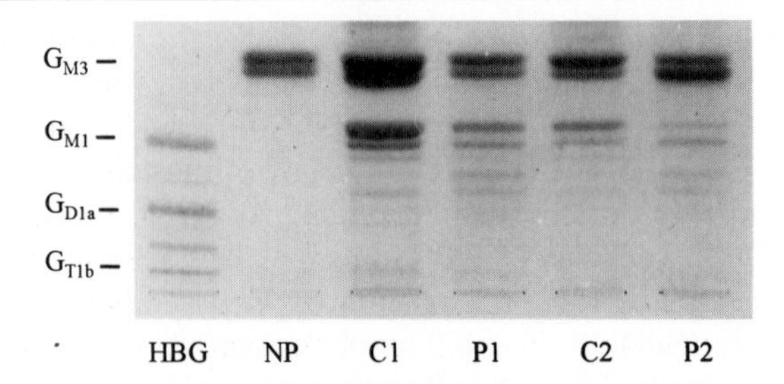

FIG. 1. HPTLC analysis of plasma and leukocyte gangliosides in patients with AML. HBG, Standard human brain gangliosides; NP, normal plasma ganglioside GM3; C1, P1, leukocyte and plasma gangliosides from patient 1; C2, P2, leukocyte and plasma gangliosides from patient 2. Each lane contains 3–6 nmol gangliosides. The HPTLC plate was developed in chloroform/methanol/0.25% aqueous $CaCl_2 \cdot 2H_2O$ (60 : 40 : 9, v/v/v) to separate gangliosides, which were stained by resorcinol–hydrochloric acid.[5]

immunostaining, the solvent partition step was eliminated from the three-step method. Purified total CSF gangliosides and standard GD3 (Matreya, Inc., Pleasant Gap, PA) were applied to aluminum-backed HPTLC plates (Merck, Darmstadt, Germany), chromatographed, immunostained, and revealed by enhanced chemiluminescence. GD3 was quantified by densitometric scanning at 620 nm. The detection limit was 5 pmol GD3.[6]

The combination of the methods described here provides a simple, rapid, and sensitive analysis for ganglioside GD3 in CSF. HPTLC analysis of normal CSF gangliosides by resorcinol staining revealed that the major species were the relatively more the complex gangliosides GT1b, GD1a, and GQ1b (Fig. 2). GD3, prominent in medulloblastomas and astrocytomas, was only a minor component in normal CSF and present in a similar concentration in ventricular and lumbar fluid (mean of 16.0 and 18.2 pmol/ml). The GD3 levels in CSF of patients (mean $\pm$ S.D. 44.7 $\pm$ 8.4, median 36.5 pmol/ml) with medulloblastoma ($n = 9$) and astrocytoma ($n = 10$) were significantly higher than those in CSF of 20 normal controls (mean 18.2 $\pm$ 1.9, median 16.4 pmol/ml; Wilcoxon rank sum test, $p < 0.0002$), clearly showing that these brain tumors shed GD3 into the CSF.[6] It is notable that GD3 migrated as a doublet on the HPTLC plates due to its ceramide heterogeneity. In CSF from medulloblastoma patients, the more slowly migrating lower band (short-chain fatty acyl-containing ceramide subspecies) of GD3 was more prominent, as detected both by immunostaining and resorcinol staining (Fig. 2).

Metabolic Radiolabeling and Analysis of Cellular and Shed Gangliosides

Metabolic radiolabeling is a sensitive approach to the detection of newly synthesized gangliosides. This approach can be applied to the detection of

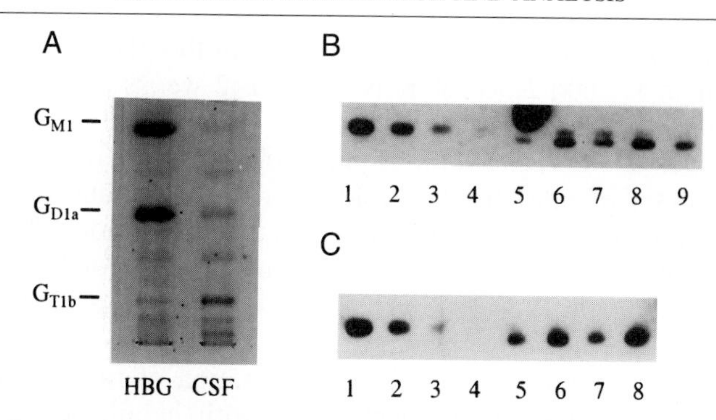

FIG. 2. Detection of ganglioside GD3 in cerebrospinal fluid. (A) Resorcinol staining of total gangliosides from normal human CSF. HBG, normal human brain gangliosides; CSF, gangliosides isolated from 25 ml ventricular CSF. (B and C) Immunostaining detection of GD3 in CSF of patients with brain tumors. Total CSF gangliosides were isolated and separated by HPTLC. GD3 was detected with the monoclonal antibody R24 and revealed by the enhanced chemiluminescence method. Lanes 1–4 in both (B) and (C) are standard buttermilk GD3, 80, 40, 20, and 10 pmol, respectively. (B) Lanes 5–9, GD3 isolated from 1 ml CSF of patients with medulloblastoma (lanes 5–7), or ependymoma (lanes 8–9). (C) Lanes 5–8, GD3 isolated from 1 ml CSF of patients with medulloblastoma.

both cellular gangliosides and gangliosides shed *in vitro* in experimental systems.

Procedures

Metabolic Radiolabeling of Cellular Gangliosides. Cells are cultured in the appropriate medium, e.g., RPMI-1640 medium supplemented with 2 mM L-glutamine and 10% fetal bovine serum (Hyclone), or HB104 serum-free medium (Hana Biologics). The cells are seeded at a density of 3–5 $\times$ 10^5 cells/ml in 10 ml medium in 75-cm^2 flasks.[7,8] After overnight incubation, the cells are metabolically radiolabeled for 24 hr with 1–5 μCi each of D-[1-^{14}C]glucosamine hydrochloride (54.2 mCi/mmol) and D-[1-^{14}C]galactose (56.5 mCi/mmol, New England Nuclear, Boston, MA) per milliliter of culture medium, to maximize the specific activity of the radiolabeled gangliosides. Alternatively, [^{14}C]galactose (52 mCi/mmol) and [^{14}C]glucosamine hydrochloride (50 Ci/mmol; New England Nuclear) are used.[9]

HPTLC Autoradiography. Purified radiolabeled gangliosides (1000–5000 dpm) are spotted on 10- $\times$ 20-cm precoated silica gel 60 HPTLC plates (Merck). The plates are developed in chloroform/methanol/0.25% aqueous $CaCl_2 \cdot 2H_2O$ (60 : 40 : 9, v/v/v) to separate gangliosides. The radio-

[7] R. Li, D. Gage, and S. Ladisch, *Biochim. Biophys. Acta* **1170,** 283 (1993).
[8] Y. Kong, R. Li, and S. Ladisch, *Biochim. Biophys. Acta* **1394,** 43 (1998).
[9] R. Li and S. Ladisch, *J. Biol. Chem.* **272,** 1349 (1997).

labeled gangliosides are revealed by exposure of the plates to XRP X-ray film (Eastman Kodak, Rochester, NY) for 1–2 weeks.[8] ^{14}C-Labeled rat brain gangliosides are used as standard.

Preparation of ^{14}C-Labeled Rat Brain Gangliosides. To obtain a valuable and relevant set of marker gangliosides for the metabolic radiolabeling studies, we prepare ^{14}C-radiolabeled rat brain gangliosides. The specific radioactivity of gangliosides is ~300 dpm/nmol. The radiolabeled gangliosides are stable for at least 10 years when stored in chloroform/methanol (1:1, v/v) at 4° or at room temperature after being dried. In this work, one 3-week-old albino rat is used for preparation of radiolabeled brain gangliosides. After the rat is anesthetized with halothane (2-bromo-2-chloro-1,1,1-trifluoroethane), 15 μCi of D-[1-^{14}C]-*N*-acetylmannosamine (53.8 mCi/mmol; New England Nuclear) in 10 μl saline is injected intracerebrally at 2 μl/min into each of the cerebral hemispheres with a Hamilton syringe that has a stop marker to control the depth of injection (4.5 mm). The needle is inserted through the skull slightly to the left or right of the midline and 1–2 mm posterior to the coronal structure. The rat is decapitated one day later and the brain tissue removed, weighed (~1.34 g), and rinsed 3 times with phosphate-buffered saline (PBS). Finally, it is homogenized in distilled water and lyophilized. The radiolabeled gangliosides are then isolated by the three-step purification method and quantified as nanomoles of lipid-bound sialic acid by the modified resorcinol method.[10,11]

Applications

YAC-1 murine lymphoma cells and DAOY human medulloblastoma cells are used for measurement of ganglioside shedding (Fig. 3). After radiolabeling, cells are washed with culture medium to remove unincorporated radiolabeled sugars. The cells are then cultured in 10–20 ml fresh medium for an additional 24–48 hr to detect radiolabeled gangliosides shed from the cells into the conditioned culture medium. The cells are harvested and the culture supernatant collected. The cell suspension is centrifuged at 300*g* for 10 min, and washed once with PBS. The culture supernatant is centrifuged at 1000*g* for 10 min to remove any further cell fragments.[8]

Gangliosides from the lyophilized cell pellets and culture supernatant are purified by the three-step purification method. Total lipids are extracted twice with chloroform : methanol (1 : 1, v/v; 20 ml/10^9 cells; 1 ml/ml culture

[10] P. Fredman, O. Nilsson, J. L. Tayot, and L. Svennerholm, *Biochim. Biophys. Acta* **618,** 42 (1980).

[11] R. W. Ledeen and R. K. Yu, *Methods Enzymol.* **83,** 139 (1982).

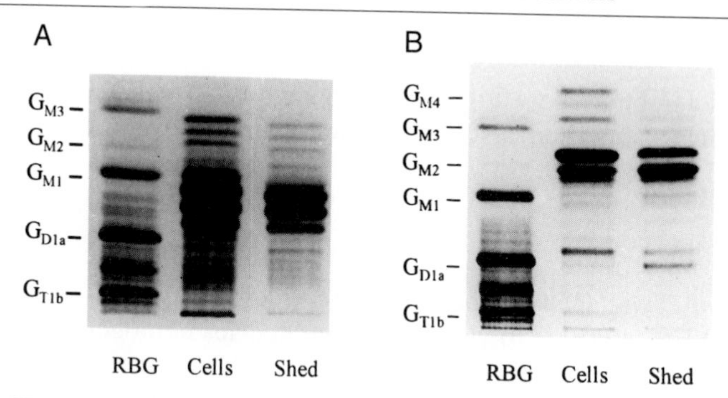

FIG. 3. HPTLC analysis of gangliosides shed by tumor cells. YAC-1 cells (A) or DAOY cells (B) were metabolically radiolabeled for 24 hr, washed, and cultured in fresh medium for 24 hr (A) or 48 hr (B). Radiolabeled cellular gangliosides (Cells) and shed supernatant gangliosides (Shed) were analyzed by HPTLC autoradiography. The HPTLC plate was developed in chloroform/methanol/0.25% aqueous $CaCl_2 \cdot 2H_2O$ (60:40:9, v/v/v). RBG, ^{14}C-labeled rat brain gangliosides. Each lane contained ~1000 dpm ^{14}C-labeled gangliosides. The plate was exposed to X-ray film for 2 weeks.

supernatant). In the case of the culture supernatant, 15 nmol of purified human brain gangliosides is added before lyophilization as carrier molecules to optimize the recovery of the shed radiolabeled gangliosides. For DIPE/butanol partition, the aqueous phase is as follows: 10 ml 0.1% saline/10^9 cells; 0.25 ml 0.1% saline/ml culture supernatant. Gangliosides, in the lower aqueous phase, are further purified by gel filtration.

YAC-1 cells synthesize GM1b and GalNAcGM1b as their major gangliosides (18.6 nmol/10^8 when cultured in 10% fetal calf serum (FCS) medium and 25.8 nmol/10^8 cells when cultured in serum free medium). These cells rapidly shed gangliosides when cultured in 10% fetal bovine serum (240 pmol/10^8 cells/hr, or ~1% of total gangliosides/hr) or serum-free HB 104 medium (300 pmol/10^8 cells/hr), or when passaged in mouse ascites fluid (12.8 nmol/ml when ascites fluid was recovered 7 days later).[7] DAOY cells contain a high concentration of gangliosides (141 ± 13 nmol LBSA/10^8 cells), and the major gangliosides have been characterized as GM2, GM3, and GD1a. DAOY cells shed 169 pmol gangliosides/10^8 cells/hr, which represents 0.12% of total cellular gangliosides per hour.[12] Shed gangliosides in culture supernatant were observed to exist, in their natural states, in three forms—membrane vesicles, micelles, and monomers.[8]

[12] F. Change, R. Li, and S. Ladisch, *Exp. Cell Res.* **234,** 341 (1997).

Purification of Gangliosides by HPLC

HPLC can be used to further purify gangliosides isolated by the three-step purification method described earlier for use in structural analysis and functional studies.

Procedures

Normal-Phase HPLC Isolation. Total gangliosides are first separated by normal-phase HPLC to yield molecular species homogeneous in carbohydrate structure.[13,14] This is accomplished by dissolving 600 nmol of total gangliosides in 100 μl water, and by chromatographing the solution on a LiChrosorb-NH_2 column (250-mm length, 10-mm i.d., Merck) using a Perkin-Elmer (Norwalk, CT) HPLC system. The eluting solvent system is composed of acetonitrile/5 m*M* Sorensen's phosphate buffer (83:17, pH 5.6, solvent A) and acetonitrile/20 m*M* Sorensen's phosphate buffer (1:1, pH 5.6, solvent B).

Gangliosides are eluted using a gradient from solvent A to solvent B designed according to the ganglioside composition of the samples. For example, the following program can be used to isolate the AML leukocyte and plasma gangliosides described earlier: solvent A is first maintained for 25 min, followed by a linear gradient from 100% solvent A to solvent A/solvent B (66:34) over 20 min, to solvent A/solvent B (36:64) over 5 min, and finally to 100% B over 10 min, at a flow rate of 6.25 ml/min. The elution profile is monitored by flow-through detection at 215 nm.

Reversed-Phase HPLC Isolation. The individual gangliosides, which are homogeneous in carbohydrate structure, are further separated by reversed-phase HPLC to yield ganglioside species also homogenous in ceramide structure.[7,14,15] For example, gangliosides (10–50 nmol GM3 or SPG) in 50 μl are chromatographed on a LiChrosorb RP-8 column (250-mm length, 4-mm i.d., Merck). The solvent system consists of acetonitrile and 5 m*M* sodium phosphate buffer (pH 7.0), which are maintained first in a ratio of 55:45 for 10 min for GM3 or for 20 min for SPG, and then increased linearly to 70:30 over 40 min at a flow rate of 0.52 ml/min. The elution profile is monitored by flow-through detection at 195 nm.

Following purification, the carbohydrate structure of gangliosides is characterized by negative-ion fast atom bombardment mass spectrometry (FAB MS), and the ceramide structure is elucidated by negative ion fast atom bombardment collisionally activated dissociation tandem mass spec-

[13] G. Gazzotti, S. Sonnino, and R. Ghidoni, *J. Chromatogr.* **348,** 371 (1985).

[14] S. Ladisch, C. C. Sweeley, H. Becker, and D. Gage, *J. Biol. Chem.* **264,** 12097 (1989).

[15] G. Gazzotti, S. Sonnino, and R. Ghidoni, *J. Chromatogr.* **315,** 395 (1984).

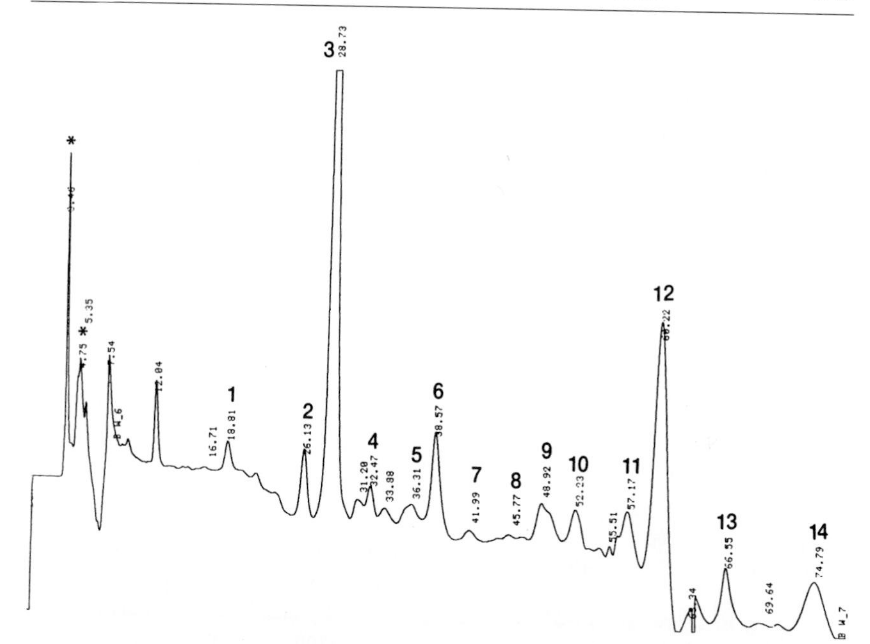

FIG. 4. Isolation of AML plasma GM3 subspecies by reversed-phase HPLC. Ganglioside GM3 from AML plasma was separated by reversed-phase HPLC to yield individual ganglioside ceramide subspecies. Peak 1, d18:1-C14:0; 2, d18:1-hC16:0; 3, d18:1-C16:0; 4, d18:2-C18:0; 5, d18:1-hC18:0; 6, d18:1-C18:0; 7, d18:2-C20:0; 8, d18:1-hC20:0; 9, d18:1-C20:0; 10, d18:2-C22:0; 11, d18:1-hC22:0/d18:1-hC24:1; 12, d18:1-C22:0/d18:1-C24:1; 13, d18:1-C23:0; 14, d18:1-C24:0. Solvent-related peaks near the origin of the HPLC tracings are marked by an asterisk (*).

trometry (CAD MS/MS).[14,16] These methods do not require prior derivatization of the gangliosides.

Applications

One example is the HPLC purification and detection of cellular and plasma gangliosides from patients with AML. GM3 and SPG from plasma and leukocytes of patients with AML and from normal human plasma are isolated by normal-phase HPLC. The reversed-phased HPLC profiles revealed the ceramide heterogeneity of gangliosides (Fig. 4). This method can be used for isolation of ceramide subspecies of a given ganglioside, using as little as 10 nmol of ganglioside. It can also be used to prepare ganglioside ceramide subspecies for biologic assays.[17]

[16] B. Domon and C. E. Costello, *Biochemistry* **27,** 1534 (1988).
[17] S. Ladisch, R. Li, and E. Olson, *Proc. Natl. Acad. Sci. U.S.A.* **91,** 1974 (1994).

The carbohydrate structure of GM3 and SPG, isolated by normal-phase HPLC, is characterized by negative-ion FAB MS. The ceramide subspecies of GM3 and SPG, isolated by reversed-phase HPLC, are characterized by negative-ion FAB CAD MS/MS as well as by FAB MS. HPLC isolation and mass spectrometric analysis provide a rapid and reliable approach for ganglioside structural characterization.[14] For example, SPG is first isolated from the total gangliosides by normal-phase HPLC and characterized by

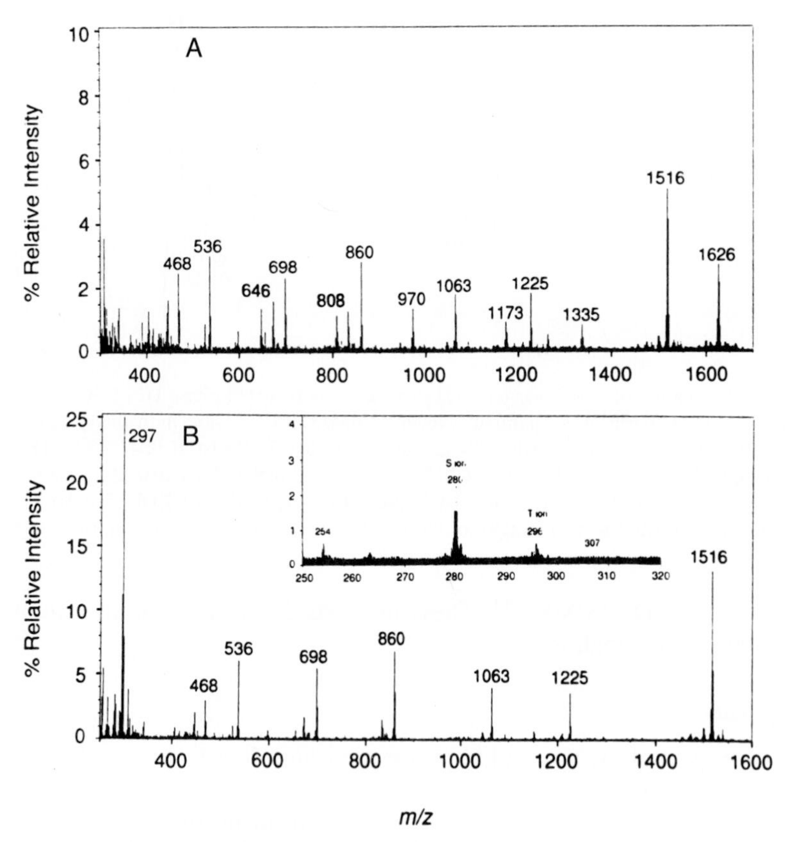

FIG. 5. Negative-ion FAB MS of SPG isolated by normal phase (A) and reversed-phase (B) HPLC. (A) Two predominant peaks at *m/z* 1626, 1516, representing the two major molecular ions of SPG. Two corresponding sets of characteristic fragment ions at *m/z* 1335, 1173, 970, 808, 646 and at *m/z* 1225, 1063, 860, 698, 536 resulted from the loss of *N*-acetylneuraminic acid, galactose, *N*-acetylglucosamine, galactose, and glucose. (B) Mass spectrum of one major ceramide subspecies of SPG isolated by reversed-phase HPLC. Only one set of fragment ions corresponds to the molecular ion at *m/z* 1516. *Inset:* FAB CAD MS/MS of the precursor ion $[CerO]^-$ at *m/z* 536 is shown. The product ions, S (*m/z* 280) and T (*m/z* 296), indicate the ceramide structure is d18:1-C16:0.

negative-ion FAB MS (Fig. 5A). The peaks at m/z 1626, 1516 represent the two main molecular ions of SPG (NeuAc 2-3Gal 1-4GlcNAc 1-3Gal 1-4Glc 1-1Cer). These two major ceramide subspecies of SPG as well as the minor ones are isolated by reversed-phase HPLC. The existence of only one set of ions in the mass spectrum of negative-ion FAB MS indicates the homogeneity of both carbohydrate and ceramide structure of the ganglioside molecule (Fig. 5B). The ceramide structure is elucidated by negative-ion FAB CAD MS/MS. The product S ion at m/z 280 and T ion at m/z 296 indicate that the long-chain base is sphingosine (d18:1), the fatty acid is C16:0, and the ceramide structure is d18:1-C16:0 (Fig. 5B). This combination of the three-step purification method with further purification by HPLC has particular value in isolating gangliosides for structural analysis and for assessment of their biologic activity.[17]

Acknowledgments

This work was supported by grant CA 42361 from the National Cancer Institute. The authors thank Ms. Yu Kong for assistance in purification and analysis of the AML plasma samples, and Dr. Douglas Gage, Department of Biochemistry and the NIH-MSU Mass Spectrometry Facility, Michigan State University, for performing the mass spectrometric analysis.

[12] Thin-Layer Chromatography Blotting Using Polyvinylidene Difluoride Membrane (Far-Eastern Blotting) and Its Applications

By DAI ISHIKAWA and TAKAO TAKI

Introduction

Nucleic acid research and protein research have been advanced by the simplification of assay systems; for example, the automatization of DNA and peptide sequences and their synthesis, and the commercial preparation of many restriction enzymes and proteases. Many new technologies, such as polymerase chain reaction (PCR),[1] two-hybrid system,[2] Southern, Northern, and Western blotting, have also played a role.[3–5]

[1] R. K. Saiki, S. Scharf, F. Faloona, K. B. Mullis, G. T. Horn, H. A. Erlich, and N. Arnheim, *Science* **230,** 1350 (1985).

[2] C. T. Chien, P. L. Bartel, R. Sternglanz, and S. Fields, *Proc. Natl. Acad. Sci. U.S.A.* **88,** 9578 (1991).

[3] E. M. Southern, *J. Mol. Biol.* **98,** 503 (1975).

0076-6879/00 $30.00

In lipid research, a new technology that transfers glycosphingolipids separated on a high-performance thin-layer chromatography (HPTLC) plate to a plastic [polyvinylidene difluoride (PVDF)] membrane has been developed and the various applications of the technology have been demonstrated. Glycosphingolipid biochemistry has advanced to a new level due to this technology.

Thin-layer chromatography (TLC) is one of the most widely used methods in the research of glycosphingolipids and has the following advantages: (1) glycosphingolipid compositions can be analyzed in a short time with simple equipment; (2) many samples can be compared simultaneously on one plate; (3) structural information on samples can be obtained from their mobilities and chemical reaction of the TLC plate; (4) the amounts of glycosphingolipids can be determined by a densitometer after chemical detection; and (5) enzyme reactions and binding experiments to glycosphingolipids can also be done on the TLC plate. The TLC overlay procedure using monoclonal antibodies (MAb) has contributed to the development of the structure analysis of glycosphingolipids.[6,7]

In our research we have used a plastic membrane, which is easier to handle than the silica-coated glass plate. The transfer method using a nitrocellulose membrane was reported by Towbin *et al.* previously.[8] However, nitrocellulose membrane is not adequate for chemical detection because the membrane is unstable toward acid solutions when heated, and organic solvents such as methanol and chloroform. By employing a PVDF membrane, the transfer method and the usefulness are dramatically improved because of the high capacity of the membrane to bind amphiphilic molecules by presumably hydrophobic interaction and their resistance to organic solvents, heating, and acid solutions.[9–11] This method can be used not only for glycosphingolipids but also for any hydrophobic molecules such as phospholipids, cholesterol, drugs, and their metabolites, which can be separated on an HPTLC plate. We proposed the name Far-Eastern blotting for

---

[4] J. C. Alwine, D. J. Kemp, and G. R. Stark, *Proc. Natl. Acad. Sci. U.S.A.* **74,** 5350 (1977).

[5] H. Towbin, T. Staehelin, and J. Gordon, *Proc. Natl. Acad. Sci. U.S.A.* **76,** 4350 (1979).

[6] M. Saito, N. Kasai, and R. K. Yu, *Anal. Biochem.* **148,** 54 (1985).

[7] T. Taki, M. Takamatsu, A. Myoga, K. Tanaka, S. Ando, and M. Matsumoto, *J. Biochem.* **103,** 998 (1988).

[8] H. Towbin, C. Schoenenberger, R. Ball, D. G. Braun, and G. Rosenfelder, *J. Immunol. Methods* **72,** 471 (1984).

[9] F. Chabraoui, E. A. Derrington, F. Mallie-Didier, C. Confavreux, C. Quincy, and C. Caudie, *J. Immunol. Methods* **165,** 225 (1993).

[10] T. Taki, S. Handa, and D. Ishikawa, *Anal. Biochem.* **221,** 312 (1994).

[11] T. Taki and D. Ishikawa, *Anal. Biochem.* **251,** 135 (1997).

this transfer procedure. Figure 1 shows the scheme for Far-Eastern blotting. In this chapter, we describe the Far-Eastern blotting procedure together with its applications.

Far-Eastern blotting

Reagents and Other Materials

1. Thin-layer chromatography plate, high-performance thin-layer chromatography plate (HPTLC plate, silica gel 60, E. Merck, Darmstadt, Germany)
2. Solvent system for TLC (solvent I, chloroform/methanol/0.2% aqueous $CaCl_2$, 60/35/8, by volume, and solvent II, chloroform/methanol/0.2% aqueous $CaCl_2$, 55/45/10, by volume)
3. Polyvinylidene difluoride (PVDF) membrane (membrane-P, ATTO Co., Ltd., Tokyo)
4. Glass fiber filter sheets (GF/A, Whatman International Ltd., Maidstone, England)
5. Teflon membrane (PTFE membrane, ATTO Co., Ltd., Tokyo)
6. Iron or TLC thermal blotter (AC-5970, ATTO Co., Ltd., Tokyo)
7. Blotting solvent (2-propanol/0.2% aqueous $CaCl_2$/methanol, 40/20/7, by volume)
8. Primuline reagent.[12] Primuline (100 mg) is dissolved in 100 ml of distilled water (stock solution). The stock solution (1 ml) is added to 100 ml of a solvent mixture of acetone/distilled water (4/1, by volume).
9. Orcinol–H_2SO_4 reagent.[13] Orcinol (200 mg) in 11.4 ml of sulfuric acid is added to 88.6 ml of distilled water.
10. Resorcinol–HCl reagent.[14] Resorcinol (200 mg) in 20 ml of distilled water is mixed with 80 ml of concentrated HCl, 0.25 ml of 0.1 *M* $CuSO_4$.
11. Drawing pencil
12. UV light

Procedure

Glycosphingolipids are separated on an HPTLC plate. After separation of glycosphingolipids, the HPTLC plate is sprayed with primuline reagent until the plate becomes wet, then is air-dried thoroughly. The glycosphingo-

[12] V. P. Skipski, *Methods Enzymol.* **35,** 396 (1975).
[13] L. Svennerholm, *J. Neurochem.* **1,** 42 (1956).
[14] L. Svennerholm, *Biochim. Biophys. Acta* **24,** 604 (1957).

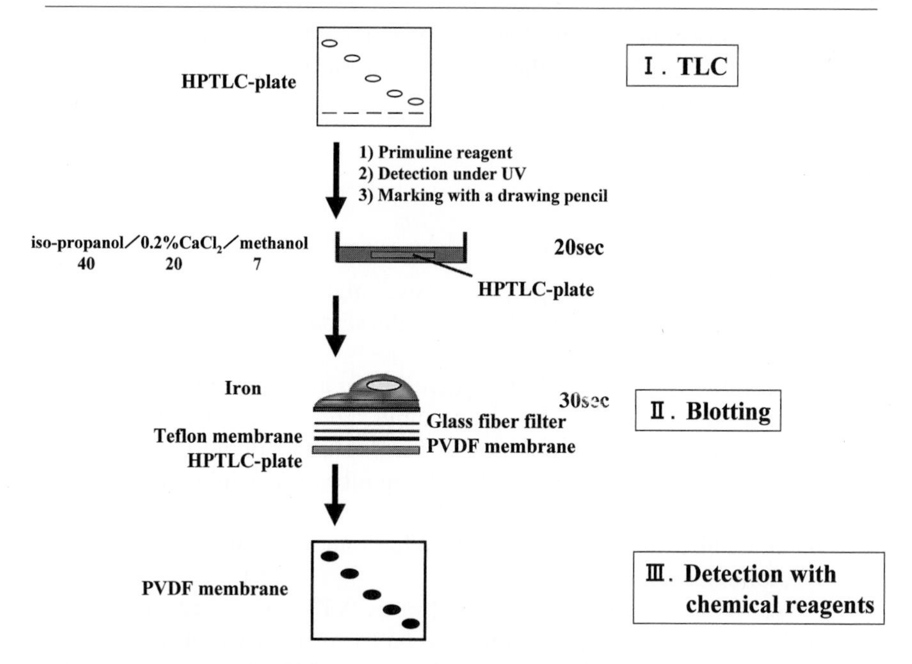

FIG. 1. Scheme of Far-Eastern blotting.

lipids are visualized under UV light at 365 nm, and then the bands are marked with a soft drawing pencil. Other contaminants can be made visible if they are present in the samples. The HPTLC plate is dipped in the blotting solvent for 20 sec, after which the HPTLC plate is immediately placed on a flat glass plate. First, a PVDF membrane, a Teflon membrane, and then a glass fiber filter sheet are placed over the plate. This assembly is pressed evenly for 30 sec with a 180° iron or put in a TLC thermal blotter. The PVDF membrane, removed from the HPTLC plate, is then air-dried. The glycosphingolipids are transferred to the PVDF membrane with pencil marks. The pencil marks are transferred to the membrane facing the HPTLC plate, but the glycosphingolipids are exclusively located on the reverse side.

Chemical Detection

All the reagents used for the detection of glycosphingolipids transferred on a PVDF membrane are prepared as usual, and then diluted with methanol to 50% to impregnate the membrane with the reagents. Neutral glycosphingolipids are made visible with orcinol–H_2SO_4 reagent diluted 50% (v/v) with methanol. The membrane then is placed in an oven or on a hot

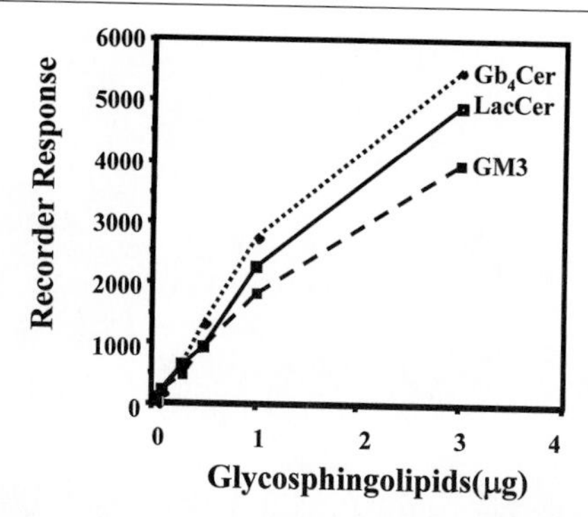

FIG. 2. Dose-dependent blotting of glycosphingolipids by Far-Eastern blotting. Various amounts of lactosylceramide (LacCer), globoside (Gb_4Cer), and GM3 were transferred to a PVDF membrane by Far-Eastern blotting, stained with orcinol–H_2SO_4 reagents, and each color intensity was measured with a densitometer.

plate (100°) for 5 min. For the detection of gangliosides, the membrane is immersed in resorcinol–HCl reagent diluted 50% with methanol (v/v) and then positioned between two pieces of glass plate and placed on a hot plate (100°) for 5 min.

Most glycosphingolipids are present on the reverse side of the membrane rather than on the side attaching to the HPTLC plate. Glycosphingolipids are assumed to have passed through the membrane by capillary action during heating. After being stained with the chemical reagents, the membrane is washed with distilled water to remove all traces of sulfuric acid or hydrochloric acid. By washing the PVDF membrane with distilled water, the color remains stable for more than a year, indicating that further degradation of the visible compound has stopped by the removal of the sulfuric and hydrochloric acids. Efficiency of blotting assessed by a chromatoscanner[15] after staining of the glycosphingolipids on the PVDF membrane is more than 80%. Linearity of the blotting is observed for 0.1–3.0 μg of each glycosphingolipid that has been tested (Fig. 2).

Far-Eastern blotting has the following advantages: (1) the entire procedure can be done within 1 min; (2) it gives high blotting efficiency; (3) detection of the transferred glycosphingolipids is more sensitive than that on an HPTLC plate; (4) the detection reagents for TLC can be used; and (5) the colors on the membrane are stable. A Far-Eastern blotting apparatus (called TLC thermal blotter) has been developed (Fig. 3).

[15] S. Ando, N. C. Chang, and R. K. Yu, *Anal. Biochem.* **89,** 437 (1978).

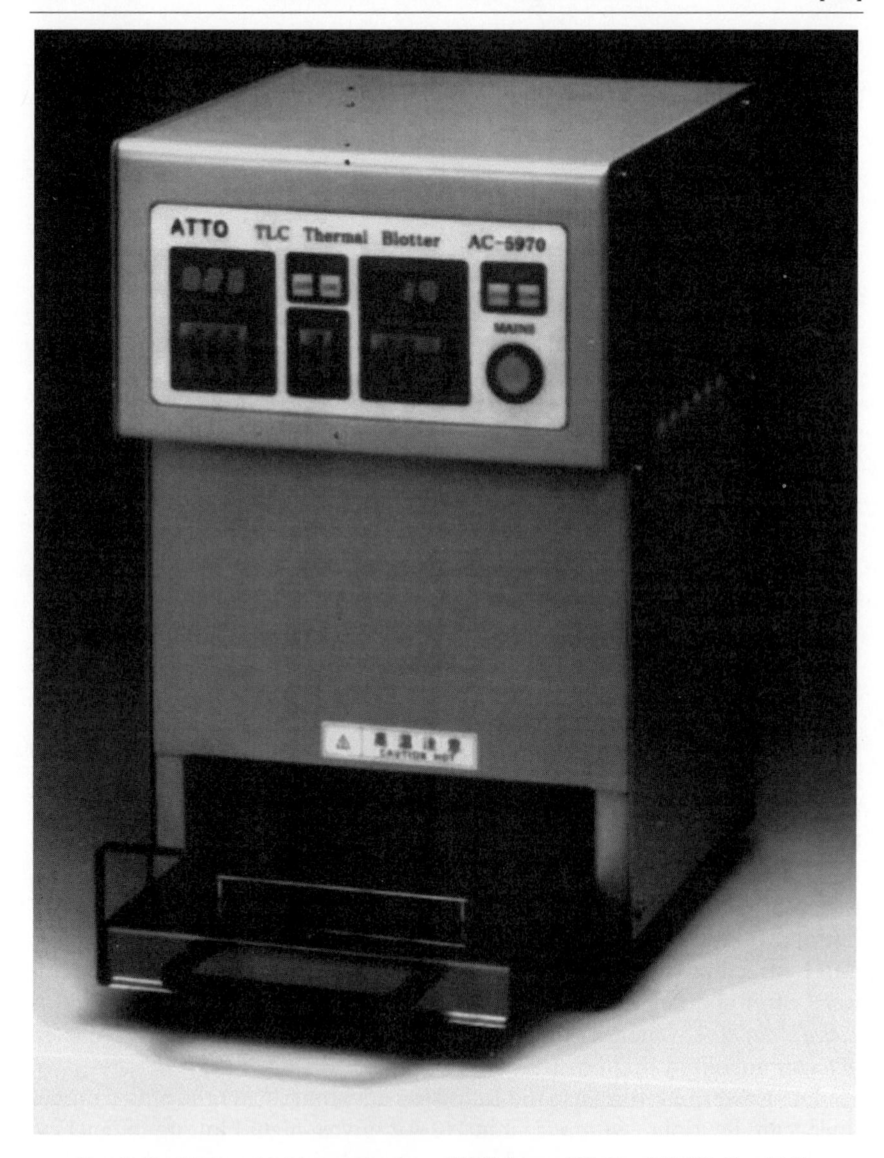

FIG. 3. Far-Eastern blotting apparatus, a TLC thermal blotter (ATTO Co., Ltd.)

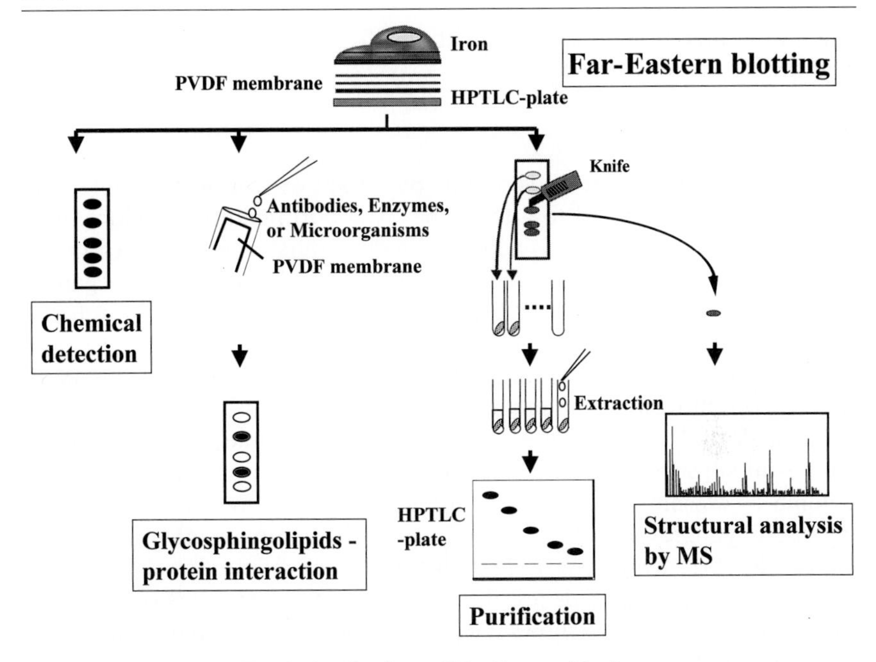

FIG. 4. Applications of Far-Eastern blotting.

Rapid Purification of Glycosphingolipids by Far-Eastern Blotting

Various applications of Far-Eastern blotting have been developed and are shown in Fig. 4. First, rapid and quantitative purification of glycosphingolipids is described.[16]

Reagents and Materials

1. Knife
2. Silica beads column (6RS-8060, Iatron Laboratories Co., Tokyo, Japan)
3. Some items from previous list

Procedure

For purification of glycosphingolipids from a minute sample, the glycosphingolipids are first separated by two-dimensional TLC. After being developed, the glycosphingolipid bands are made visible by spraying the HPTLC plates with primuline reagent, marked with a soft drawing pencil, and transferred to a PVDF membrane by Far-Eastern blotting. Then

[16] T. Taki, T. Kasama, S. Handa, and D. Ishikawa, *Anal. Biochem.* **223,** 232 (1994).

the PVDF membrane is washed with distilled water to remove primuline and dried.

Extraction

The areas marked for glycosphingolipid are excised with a knife, and each piece is transferred to a test tube. A 500-μl mixture of chloroform/methanol (2/1, by volume) is added to the test tubes to extract each glycosphingolipid. This extraction step is repeated twice. For extraction of gangliosides, methanol is used for the first extraction and a mixture of chloroform/methanol (2/1, by volume) for the second extraction.

The extracts are evaporated, but the residues are sometimes contaminated. To remove contaminants, the extracted glycosphingolipids are dissolved in 500 μl of chloroform/methanol (9/1, by volume) and then applied to a 1-ml silica bead column. The contaminants are passed through the column by washing with chloroform/methanol (9/1, by volume), and the glycosphingolipids are eluted with 2 ml of chloroform/methanol/distilled water (20/80/8, by volume). The eluates are evaporated and individual purified glycosphingolipids are dissolved in a few microliters of adequate organic solvent. Figure 5 shows the purification of acidic glycosphingolipids from bovine brain.

Far-Eastern Blotting/Mass Spectrometry (MS)

When mass spectrometry (MS) is used for structural analysis of glycosphingolipids with a minute sample, the glycosphingolipids usually have to be purified to homogeneity with no contaminants. By applying Far-Eastern blotting, glycosphingolipids on a PVDF membrane can be analyzed by mass spectrometry directly without other purification steps.[17,18] This method avoids the difficulties found in glycosphingolipid analyses and makes it possible to study the glycosphingolipids in small amounts of samples such as those obtained from clinical laboratories.

Procedure

Glycosphingolipids developed on the HPTLC plate are made visible with primuline reagent under UV light. Each glycosphingolipid area is marked with a drawing pencil, after which the glycosphingolipids are transferred to a PVDF membrane by Far-Eastern blotting. Each marked area

[17] T. Taki, D. Ishikawa, S. Handa, and T. Kasama, *Anal. Biochem.* **225,** 24 (1995).

[18] S. C. Li, Y. Y. Wu, E. Sugiyama, T. Taki, T. Kasama, R. Casellato, S. Sonnino, and Y. T. Li, *J. Biol. Chem.* **270,** 24246 (1995).

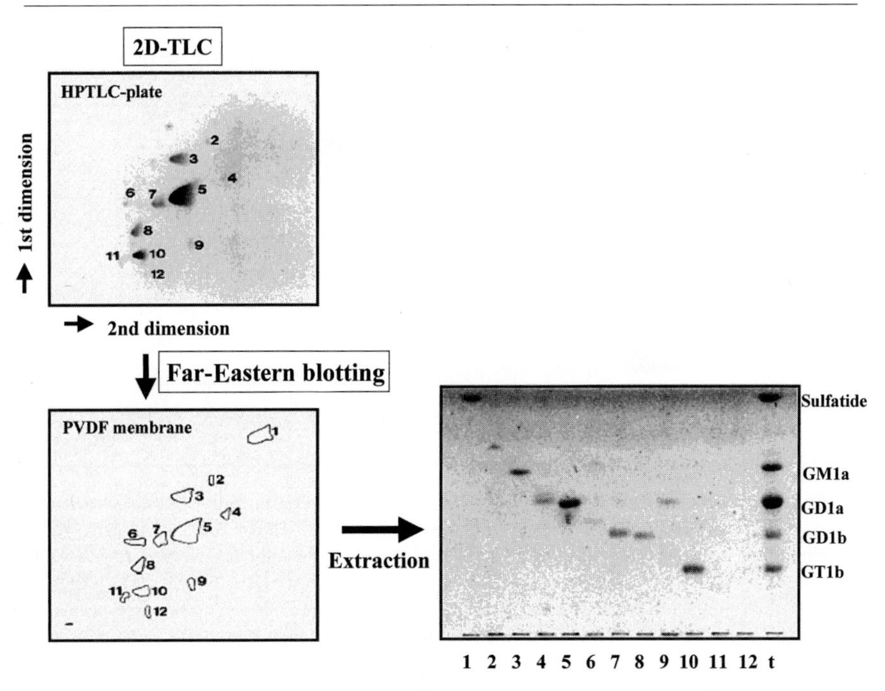

FIG. 5. Rapid purification of acidic glycosphingolipids from bovine brain.[16] Acidic gangliosides were extracted from bovine brain, separated on two HPTLC plates by two-dimensional TLC. Chloroform/methanol/0.2% $CaCl_2$ (55/45/10, by volume) was used for first dimension and a solvent mixture of propanol/distilled water/25% NH_4OH (75/25/5, by volume) was used for second dimension, after which, gangliosides on one HPTLC plate were detected with resorcinol–HCl reagent and those on the other plate were made visible with primuline reagent. Eleven and 12 bands were detectable by resorcinol–HCl reagent and primuline reagent, respectively. After detection with primuline reagent, the acidic glycosphingolipids were subjected to purification using Far-Eastern blotting. After extraction of glycosphingolipids from the PVDF membrane, each glycosphingolipid was developed on an HPTLC plate and made visible with orcinol–H_2SO_4 reagent. Lane numbers on the right correspond to bands shown in the two-dimensional TLC picture (left).

on the PVDF membrane is excised and trimmed to form a circle 2 mm in a diameter that fits the probe tip of the MS equipment. A few microliters of triethanolamine as a matrix is placed on the PVDF membrane. After heating the probe tip, MS spectra are obtained using a mass spectrometer.

For negative secondary ion mass spectrometry (SIMS), glycosphingolipids on the PVDF membrane are bombarded with a Cs^+ beam at 20 kV. The ion multiplier is 1.5 kV and the conversion dynode is 20 kV. Mass spectra of monosialoganglioside GM1a by Far-Eastern blotting/MS are

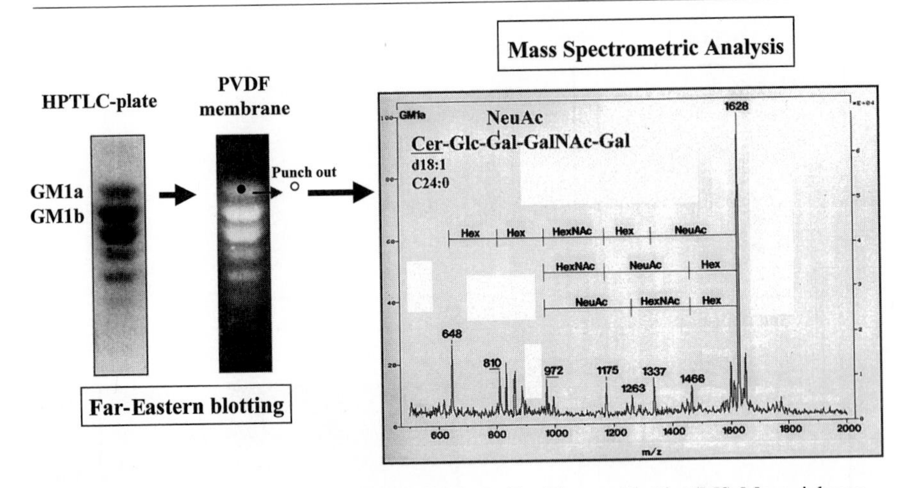

FIG. 6. Spectra of monosialoganglioside GM1a by Far-Eastern blotting/MS. Monosialogangliosides extracted from murine lymphosarcoma cell (RAW117-H10 cell,[20] 5×10^6 cells) were developed on an HPTLC plate, made visible with primuline, and transferred to a PVDF membrane by Far-Eastern blotting, after which the area corresponding to GM1a was punched out and then subjected to mass spectrometric analysis. Molecular ion 1628 corresponds to GM1a and a peak at m/z 648 is ceramide with a d18:1 long-chain base and C24:0 fatty acid.

shown in Fig. 6. Far-Eastern blotting/MS has some advantages: (1) analyses can be done with very low noise; (2) the glycosphingolipids on the PVDF membrane can be dissolved rapidly in a matrix; (3) the sample to be analyzed can be selected easily, because all separated glycosphingolipids are visible; (4) sample loss is negligible; and (5) a minute sample of glycosphingolipid (1 μg) extracted from $0.5–1 \times 10^7$ cells can be analyzed.[19]

Binding Study on PVDF Membrane

Glycosphingolipids function as ligands of bacterial toxin such as cholera toxin and verotoxin, and certain viruses.[21,22] *N*-Glycolylneuraminic acid-containing glycosphingolipids on the cell surface are reported to be the ligands in *Escherichia coli* K99.[23] Overlay binding assays on an HPTLC plate[23a] have been used to determine ligand glycosphingolipids for bacteria.

[19] T. Kasama, Y. Hisano, M. Nakajima, S. Handa, and T. Taki, *Glycoconj. J.* **13,** 461 (1996).
[20] T. Taki, D. Ishikawa, M. Ogura, M. Nakajima, and S. Handa, *Cancer Res.* **57,** 1882 (1997).
[21] J. Pereira, B. Boyd, J. Newbigging, C. Lingwood, and P. M. Strasberg, *J. Mol. Neurosci.* **5,** 121 (1994).
[22] W. V. Williams and D. B. Weiner, *AIDS Res. Hum. Retroviruses* **9,** 175 (1993).
[23] S. Teneberg, P. Willemsen, F. K. De Graaf, and K. A. Karlsson, *FEBS Lett.* **263,** 10 (1990).
[23a] D. Ishikawa and T. Taki, *Methods Enzymol.* **312,** [13], (2000) (this volume).

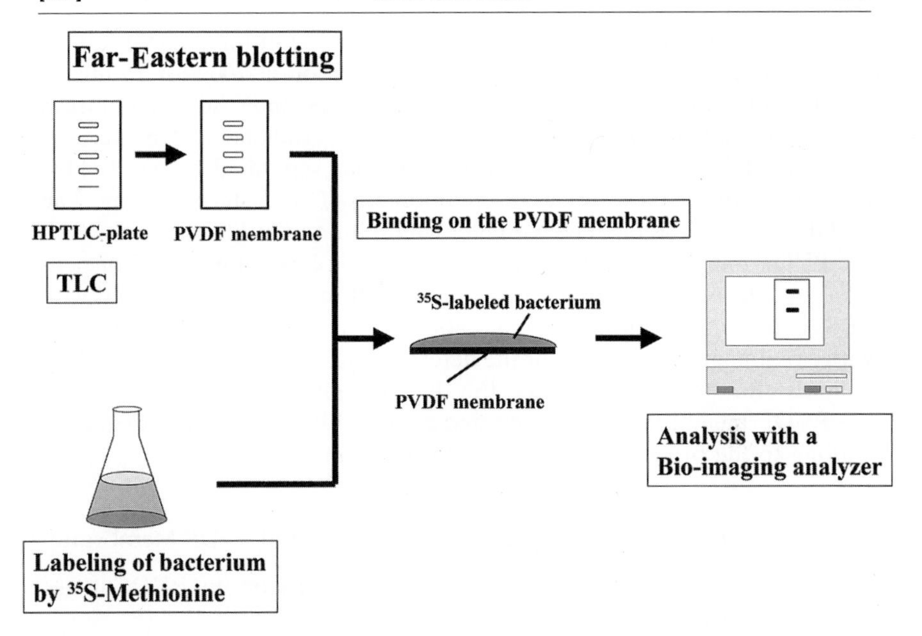

FIG. 7. Scheme of overlay binding assay on the PVDF membrane.

During incubation and washing of the HPTLC plate, silica often sloughs off the plates. To improve TLC-overlay assays, a combination technique with Far-Eastern blotting was developed (Fig. 7). Here, we describe the binding assay of *E. coli* K99 strains to *N*-glycolylneuraminic acid-containing glycosphingolipids transferred to a PVDF membrane.[24]

Reagents

1. *E. coli* strains H1915 (O8:K25:K99$^+$)
2. Medium (standard "minimal" medium)
3. Blocking solution [4% casein containing 0.1% methionine in phosphate-buffered saline (PBS)]
4. [^{35}S]Methionine
5. Bioimaging analyzer (BAStation, Fuji Photo Film, Co., Ltd., Tokyo)

Procedure

Escherichia coli strains H1915 (O8:K25:K99$^+$) isolated from an infected calf[25] are grown in minimal medium and metabolically labeled with

[24] T. Isobe, M. Naiki, S. Handa, and T. Taki, *Anal. Biochem.* **236,** 35 (1996).
[25] E. Ono, K. Abe, M. Nakazawa, and M. Naiki, *Infect. Immun.* **57,** 907 (1989).

[^{35}S]methionine (0.25 μCi/ml) at 37° during the logarithmic phase (16–18 hr). The bacteria are precipitated by centrifugation at 700*g* for 15 min at room temperature, washed with PBS several times, and suspended in fresh minimum medium. The specific activity of the radiolabeled bacteria is approximately 10^4 cpm/cell. Glycosphingolipids separated on an HPTLC plate are transferred to a PVDF membrane by Far-Eastern blotting. The PVDF membrane is blocked with blocking solution for 30 min, after which the PVDF membrane is incubated with the radiolabeled bacterial suspension for 2 hr at room temperature. The PVDF membrane is washed with PBS, and the location of the radiolabeled *E. coli* K99 strain is observed with a bioimaging analyzer. For the determination of the ligand glycosphingolipids after the binding assay, the detected glycosphingolipid areas on the PVDF membrane are excised and analyzed by mass spectrometry according to the procedure for Far-Eastern blotting/MS.

Enzyme Reaction on Glycosphingolipid-Transferred PVDF Membrane

As an example, the detection of sialidase activity on the PVDF membrane is described.[26]

Reagents

1. Glycosphingolipids, sialyl(α2-3)paragloboside (IV^{3}NeuAcαnLc$_4$Cer, NeuAcα2-3Galβ1-4GlcNAcβ1-3Galβ1-4Glcβ1-1Cer), paragloboside (nLc$_4$Cer, Galβ1-4GlcNAcβ1-3Galβ1-4Glcβ1-1Cer)
2. *Clostridium perfringens* sialidase (Sigma)
3. Reaction buffer [50 m*M* acetate buffer, pH 5.5, containing 0.005% Triton CF-54 and 0.5% bovine serum albumin (BSA)]
4. Nylon bag
5. Anti-paragloboside monoclonal antibody H11 (mouse IgM)[27]
6. Second antibody, horseradish peroxidase-conjugated anti-mouse IgM
7. 4-Chloro-1-naphthol[23a]

Assay

After Far-Eastern blotting, sialyl(α2-3)paragloboside as a substrate glycosphingolipid on the PVDF membrane is incubated with 1.0 mU of *C. perfringens* sialidase/ml of the reaction buffer in a nylon bag at 37° for

[26] D. Ishikawa, T. Kato, S. Handa, and T. Taki, *Anal. Biochem.* **231,** 13 (1995).

[27] A. Myoga, T. Taki, K. Arai, K. Sekiguchi, I. Ikeda, K. Kurata, and M. Matsumoto, *Cancer Res.* **48,** 1512 (1988).

1 hr. After incubation, the PVDF membrane is washed with PBS several times, and the amount of paragloboside, the product glycosphingolipid, is determined by immunostaining with anti-paragloboside antibody H11. This method is based on the HPTLC-overlay assays.[28,29] This method is advantageous when searching for enzymes that cleave the carbohydrate structure and for finding novel glycosphingolipid substrates against certain glycosidases or glycosyltransferases.

[28] S. L. Spitalnik, J. F. Scwartz, J. L. Magnani, D. D. Roberts, P. F. Spitalnik, C. I. Civin, and V. Ginsburg, *Blood* **66,** 319 (1985).
[29] B. E. Samuelsson, *FEBS Lett.* **167,** 47 (1984).

[13] Thin-Layer Chromatography Immunostaining

By DAI ISHIKAWA and TAKAO TAKI

Introduction

A number of monoclonal antibodies directed to glycosphingolipids have been produced and they are useful in analyzing the structure of novel glycosphingolipids, their biologic functions, and tissue distribution.[1,2] For example, patients suffering from type I diabetes and certain autoimmune diseases produce anti-glycosphingolipid antibodies.[3–5] Overlay binding assays on high-performance thin-layer chromatography (HPTLC) plates have been used to provide structural information about unknown epitopes of disease-associated glycosphingolipids for the antibodies.[6–9] These assays take advantage of the fact that the HPTLC plate has a high-resolution

[1] T. Taki, C. Rokukawa, T. Kasama, K. Kon, S. Ando, T. Abe, and S. Handa, *J. Biol. Chem.* **267,** 11811 (1992).
[2] I. Kawashima, I. Nagata, and T. Tai, *Brain Res.* **732,** 75 (1996).
[3] B. K. Gillard, J. W. Thomas, L. J. Nell, and D. M. Marcus, *J. Immunol.* **142,** 3826 (1989).
[4] N. Yuki, T. Taki, F. Inagaki, T. Kasama, M. Takahashi, K. Saito, S. Handa, and T. Miyatake, *J. Exp. Med.* **178,** 1771 (1993).
[5] N. Baumann, M. L. Harpin, Y. Marie, K. Lemerle, B. Chassande, P. Bouche, V. Meininger, R. K. Yu, and J. M. Leger, *Ann. N.Y. Acad. Sci.* **845,** 322 (1998).
[6] M. Brockhaus, J. L. Magnani, M. Herlyn, Z. Steplewski, H. Koprowski, and V. Ginsburg, *Arch. Biochem. Biophys.* **217,** 647 (1982).
[7] N. Kasai, M. Naiki, and R. K. Yu, *J. Biochem.* **96,** 261 (1984).
[8] M. Saito, N. Kasai, and R. K. Yu, *Anal. Biochem.* **148,** 54 (1985).
[9] T. Taki, M. Takamatsu, A. Myoga, K. Tanaka, S. Ando, and M. Matsumoto, *J. Biochem.* **103,** 998 (1988).

METHODS IN ENZYMOLOGY, VOL. 312
0076-6879/00 $30.00

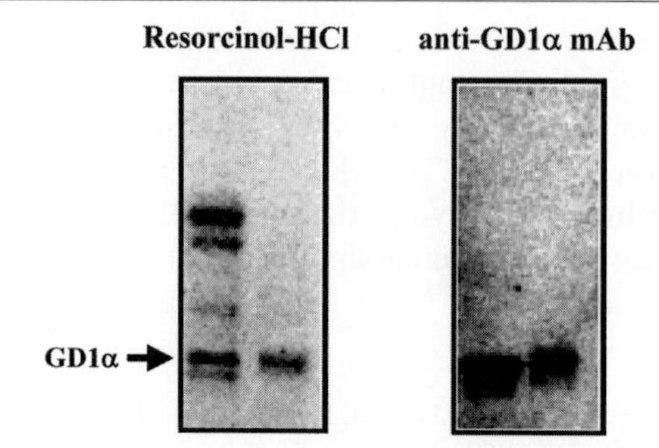

FIG. 1. TLC immunostaining of disialoganglioside GD1α. GD1α containing ganglioside fraction was extracted from murine metastatic lymphoma cell line RAW117-H10 cells.[10] Standard GD1α and the ganglioside fraction were separated on two HPTLC plates, in which one HPTLC plate was used for chemical detection with resorcinol-HC1 reagent, and the other for TLC immunostaining with anti-GD1α monoclonal antibody (mouse IgM). Solvent system for developing is chloroform/methanol/0.2% aqueous $CaCl_2$ (55/45/10, by volume). Horseradish peroxidase-conjugated anti-mouse IgM was used as second antibody. Only GD1α was detected by anti-GD1α monoclonal antibody in each lane. Ganglioside fraction and standard GD1α, respectively, correspond to the lane on the left and right in each picture.

property for glycosphingolipid separation from mixed samples. In this chapter, we describe the immunologic detection of glycosphingolipids with monoclonal antibodies on an HPTLC plate.

Reagents

1. Thin-layer chromatography plate, high-performance thin-layer chromatography-plate (HPTLC plate, silica gel 60, E. Merck, Darmstadt, Germany)
2. Solvent systems for TLC (chloroform/methanol/0.2% aqueous $CaCl_2$, 60/35/8, by volume, and 55/45/10, by volume, for neutral glycosphingolipids and gangliosides, respectively)
3. 0.4% Polyisobutylmethacrylate (PIM solution, 2.5% polyisobutylmethacrylate in chloroform is diluted to 0.4% with hexane)
4. 0.1% Bovine serum albumin (BSA) and 1% reconstituted powdered milk in phosphate-buffered saline (PBS), pH 7.4
5. Monoclonal antibodies against glycosphingolipids
6. Second antibody, horseradish peroxidase-conjugated anti-mouse im-

[10] T. Taki, D. Ishikawa, M. Ogura, M. Nakajima, and S. Handa, *Cancer Res.* **57,** 1882 (1997).

munoglobins (Cappel, West Chester, PA), 500–1000 fold dilution with 0.1% BSA and 1% reconstituted powdered milk in PBS

7. 4-Chloro-1-naphthol solution (100 μl of 3% 4-chloro-1-naphthol in ethanol in 10 ml Tris, pH 7.5, plus 10 μl of 30% H_2O_2; prepare before use)

Procedure

The antigen glycosphingolipids (few micrograms) in the sample are separated on an HPTLC plate. After TLC, the HPTLC plate is dried thoroughly with a hair dryer to remove the organic solvent. The HPTLC plate is immersed in 0.4% PIM solution for 30–60 sec and dried. Then the antibody solution (or antisera) is diluted with 0.1% BSA and 1% reconstituted powdered milk containing PBS and is overlaid on the glycosphingolipid-developed area of the HPTLC plate. A Parafilm sheet is placed over this area and the HPTLC plate is put in a humidified box to avoid evaporation and incubated overnight at 4°. The next day, the HPTLC plate is washed with PBS to remove nonbound antibodies, and then horseradish peroxidase-conjugated antibody as a second antibody is mounted on the HPTLC plate and left for 2 hr at room temperature. After washing the HPTLC plate with PBS several times, the bound antimonoclonal antibody to the antigen glycosphingolipid is detected by the addition of freshly prepared 4-chloro-1-naphthol as a peroxidase substrate. Alkaline phosphatase-conjugated or fluorescein-conjugated antibodies can also be used as the second antibody instead of peroxidase (Fig. 1).

Lectins, bacterial toxins, and adhesion molecules such as laminin can also be used for the TLC-overlay assay instead of antibodies.[11,12] TLC immunostaining is a simple and rapid method for the detection of antigen glycosphingolipids. However, handling of the HPTLC plate should be done very carefully so as not to disturb the surface of the plate, when the glycosphingolipids on the HPTLC-plates are incubated with antibodies in buffer.

[11] G. C. Hansson, K. A. Karlsson, G. Larrson, N. Stromberg, and J. Thurin, *Anal. Biochem.* **146,** 158 (1985).

[12] D. D. Roberts, C. N. Rao, J. L. Magnani, S. L. Spitalnik, L. A. Liotta, and V. Ginsberg, *Proc. Natl. Acad. Sci. U.S.A.* **82,** 1306 (1985).

[14] Monoclonal Anti-Glycosphingolipid Antibodies

By REIJI KANNAGI

Introduction

Monoclonal antibodies directed to glycolipid determinants have been widely utilized for the study of expression, distribution, and function of carbohydrate determinants in a variety of cells and tissues. Microanalysis of glycolipids and glycoproteins is greatly facilitated by the application of monoclonal antibodies whose specificity for specific carbohydrate sequences and structures is well established. Some antibodies directed to carcinoma-associated carbohydrate determinants are clinically applied for diagnosis of cancers. Anti-glycolipid antibodies are applied for analysis of carbohydrate-mediated cell adhesion and for cloning of glycosyltransferases and related genes. This chapter describes the methods for preparation of monoclonal antibodies against carbohydrate antigens and for determination of the specificity of established monoclonal antibodies.

Strategy for Generation of Anti-Glycolipid Antibodies

Immunization with Whole Cells and/or Tissues

In early 1980s, many monoclonal antibodies (MAb) were generated that detect specific types of tumor cells or embryonic cells, and later not a few of these antibodies turned out to recognize carbohydrate determinants. Such monoclonal antibodies include anti-SSEA-1, anti-SSEA-3, anti-melanoma, anti-Burkitt lymphoma, and anti-adenocarcinoma antibodies. Most of the earlier monoclonal anti-glycolipid antibodies were obtained historically in this way. This indicated that the immunization with whole cells or tissues is sufficiently effective in obtaining anti-glycolipid antibodies. On the other hand, the conventional immunization protocol for carbohydrate antigens, which had been used for generation of polyclonal antibodies in goats or rabbits, did not prove very effective in generating monoclonal antibodies in mice and rats.

These findings imply that the immunization of mice with whole cells or tissues is more effective than immunization with pure glycolipids for obtaining a good antibody response against carbohydrate antigens. For the intentional application of this immunization method to obtain antibodies directed to a specific glycolipid determinant, first cells or tissues strongly expressing the target glycolipid should be chosen, and the appropriate

0076-6879/00 $30.00

hybridoma clones are then selected according to their reactivity to the target glycolipid antigen. The enzyme-linked immunosorbent assay (ELISA) using immobilized pure glycolipids prepared from those cells and tissues is frequently used in the screening procedure.

Immunization with Pure Carbohydrate Antigens

It is not easy to obtain anti-carbohydrate antibody using glycoproteins as immunization antigen. When animals are immunized with xenogeneic or allogeneic glycoproteins, they produce antibodies specific to the protein portion, but the carbohydrate moiety is not strongly immunogenic. When animals are immunized with syngeneic glycoproteins, they usually do not produce antibodies to either the protein portion or carbohydrate determinant. A few exceptions include carbohydrate determinants that are very short and linked directly to the polypeptide portion without any linker portion (e.g., such as Tn, sialyl Tn, and T determinants).

In contrast, a significant antibody response has long been reported to occur against the carbohydrate portion of glycolipids when animals are immunized with a glycolipid together with a foreign protein or included in a lipid bilayer liposome or cell membrane. In particular, glycolipids noncovalently adsorbed onto acid-treated *Salmonella minnesota* provided effective immunization in mice against a number of glycolipids. So far, the use of salmonella is the best method for obtaining the anti-glycolipid antibodies. Immunization with liposomes containing lipid A can be used as an alternative for acid-treated *S. minnesota.*

Many carbohydrate epitopes, such as ABO, Ii, Le^a, and Le^x, are commonly carried by both glycolipids and glycoproteins. Once the specific antibodies directed to those determinants are obtained by immunizing glycolipid antigens, the antibodies usually detect the same determinant carried by glycoproteins as well, and are useful for the analyses of both glycolipids and glycoproteins in cells and tissues.

Use of Chemically Synthesized Glycolipid Antigens

More recently, chemically synthesized glycolipids have been applied for immunization and screening of anti-glycolipid antibodies. The antibodies directed to a minor lactoganglio-series glycolipid in murine leukemia cells were successfully generated by using the synthetic glycolipid antigen as immunogen,[1] which is an early application of synthetic glycolipid for immunization. The advantage of the synthetic antigen is that it can be obtained

[1] K. Shigeta, Y. Kirihata, Y. Ito, T. Ogawa, S. Hakomori, and R. Kannagi, *J. Biol. Chem.* **262,** 1358 (1987).

in larger amounts by organochemical synthesis than by purifying it from a natural source. At least 100 μg of glycolipid antigen is required for immunization of a mouse; this amount is sometimes difficult to obtain from a natural source when the glycolipid is an extremely minor component.

Another advantage of synthetic glycolipid is that it is possible to intentionally produce an artificially designed glycolipid by chemical synthesis, which is not known to be present in a natural source. For instance, the sequence Galβ1$\rightarrow$4GlcNACβ1$\rightarrow$6GalNAcα is specifically found on *O*-linked carbohydrate side chains of glycoproteins but not in glycolipids. The antibody directed to this determinant was obtained by immunizing a chemically synthesized artificial glycolipid, Galβ1$\rightarrow$4GlcNACβ1$\rightarrow$6GalNAcα1$\rightarrow$Cer. The antibody was useful in detecting *O*-linked carbohydrate side chains in gastric cancer tissues.[2] Similarly, sialyl-Lewis X glycolipid having internal 6-sulfated GlcNAc has never been described in a natural source, but was chemically synthesized. Using this artificial glycolipid as immunogen, monoclonal antibodies directed to sialyl-6-sulfo-Lewis X determinant was successfully generated, and they turned out to detect glycoproteins that carry the sialyl-6-sulfo-Lewis X determinant on their carbohydrate side chains.[3] As stated earlier, carbohydrate side chains of glycoproteins are poor immunogens compared to glycolipids, and this indicates that the antibodies directed to the glycoprotein-specific carbohydrate determinant, which is not carried by glycolipids, is hard to obtain. This problem can be overcome by the use of synthetic glycolipid carrying a glycoprotein-specific oligosaccharide structure.

Instead of organochemical synthesis of the entire glycolipid moiety, oligosaccharides obtained from glycoproteins and other natural sources can be covalently attached to lipid moiety such as phospholipids or appropriately derivatized cholesterol. These "neoglycolipids"[4] are mainly used for the characterization of established monoclonal antibodies, although the reducing terminus of oligosaccharides is not authentic in materials derivatized in this manner.

Selection of Mouse Strains for Immunization and Other Aspects

There is a significant strain difference among mice in the immune response against glycolipids. Use of a high responder strain facilitated the

[2] Y. Yamashita, Y. S. Chung, K. Murayama, R. Kannagi, and M. Sowa, *J. Natl. Cancer Inst.* **87,** 441 (1995).

[3] C. Mitsuoka, M. Sawada-Kasugai, K. Ando-Furui, M. Izawa, H. Nakanishi, S. Nakamura, H. Ishida, M. Kiso, and R. Kannagi, *J. Biol. Chem.* **273,** 11225 (1998).

[4] T. Feizi, M. S. Stoll, C. T. Yuen, W. Chai, and A. M. Lawson, *Methods Enzymol.* **230,** 484 (1994).

generation of ganglio-series glycolipids. Some researchers successfully established specific anti-glycolipid antibodies using autoimmune mice such as NZB mice or tight skin (TSK) mice.[5–7] Use of glycosyltransferase knockout mice would also be a good strategy to obtain certain anti-glycolipid antibodies.

Certain anti-glycolipid antibodies are known to be encoded by a limited set of germline V_H and V_L genes.[8] Taking advantage of this, high affinity antibodies directed to the Le^x determinant were successfully generated by genetic manipulation using a phage display library.[9]

Methods for Generation of Anti-Glycolipid Monoclonal Antibodies

Acid-Treated Salmonella minnesota Strain R595

Salmonella minnesota strain R595 (ATCC, Rockville, MD, Number: 49284) is usually kept frozen at −85° under anaerobic conditions as an agar stick culture at room temperature in a dark and dry place. To obtain single colonies, the strain is transferred with a sterile platinum loop onto 10-cm agar dishes. One isolated colony is picked up with a platinum loop and transferred to a 250-ml Erlenmeyer flask containing 70–100 ml of bouillon. The composition of bouillon is 15.0 g Difco casamino acids (0230-02-0), 5.0 g Difco yeast extract (0127-02-6), 3.0 g NaCl, 2.0 g $Na_2HPO_4 \cdot 12H_2O$, 0.2 g $MgSO_4 \cdot 7H_2O$, and 10.0 g D-glucose per liter of medium. The Erlenmeyer flask is sealed with a cotton stopper and incubated overnight at 37° in a shaking water bath. This preculture is transferred to a 4-liter Erlenmeyer flask with 2 liters of bouillon and again incubated for 16 hr until a highly turbid suspension is obtained. The bacteria are spun down by centrifugation (5000*g*, 20 min at 4°), washed twice with 1% acetic acid, then suspended in 1% acetic acid (1 g/50 ml) and heated at 100° for 2 hr. The acid-treated bacteria are washed with distilled water and lyophilized.

Immunization Protocol Utilizing Glycolipid-Coated Salmonella minnesota

Glycolipids are adsorbed to the bacteria by the procedure described by Galanos *et al.*[10] Distilled water (2 ml) is heated to 50° in a tube. To this,

[5] M. Matic, S. Shibata, and H. M. Fillit, *Immunol. Invest.* **26,** 371 (1997).

[6] Y. Sanai, M. Yamasaki, and Y. Nagai, *Biochim. Biophys. Acta* **958,** 368 (1988).

[7] S. Furuya, T. Hashikawa, F. Irie, A. Hasegawa, T. Nakao, and Y. Hirabayashi, *Neurosci. Res.* **22,** 411 (1995).

[8] J. G. Snyder, Q. Dinh, S. L. Morrison, E. A. Padlan, M. Mitchell, L.-Y. Yu-Lee, and D. M. Marcus, *J. Immunol.* **153,** 1161 (1994).

[9] Q. Dinh, N. P. Weng, M. Kiso, H. Ishida, A. Hasegawa, and D. M. Marcus, *J. Immunol.* **157,** 732 (1996).

[10] C. Galanos, O. Luderitz, and O. Westphal, *Eur. J. Biochem.* **24,** 116 (1971).

30 μg glycolipid dissolved in ethanol is injected with a Hamilton syringe, and the solution is mixed with a vortex mixer. The concentration of ethanol in water should not exceed 5%. Dry powder of acid-treated *S. minnesota* (120 μg) is added to the tube and again mixed with a vortex mixer. The ratio of glycolipid : bacteria is usually 1 : 4, w/w. The mixture is incubated for 10 min at 50°.

The mixture is aliquotted to the desired amount, lyophilized, and stored in a freezer (−20°). The aliquots of lyophilized glycolipid–bacteria complex are resuspended in PBS and injected into mice intravenously or intraperitoneally. The standard immunization schedule is 5 μg of glycolipid on day 0, 10 μg on day 4, 15 μg on day 8, 20 μg on day 12, and 25 μg on day 16. After a rest of at least 2 weeks, the mice receive a final booster injection with 25 μg glycolipid. Three days after the final injection, the spleen cells are harvested and fused with HAT-sensitive mouse myeloma cells.

Screening Hybridoma Clones with Glycolipid Antigens by ELISA

A glycolipid solution is prepared by dissolving 1 μg of glycolipid antigen in 2 ml of ethanol. Twenty microliters of the ethanol solution is added to each well of a 96-well plate, and ethanol is allowed to evaporate at room temperature. Gangliosides and sulfatides are more efficiently adsorbed and form a stable lipid film on a plastic surface when a mixture of glycolipid–egg yolk phosphatidylcholine–cholesterol (1 : 10 : 5, w/w/w) in an ethanol solution is added to a plastic well and the ethanol is evaporated. A control plate with immobilized phosphatidylcholine and cholesterol but no glycolipids is necessary when phosphatidylcholine and cholesterol are used, because phosphatidylcholine reacts with antibodies secreted by some hybridomas that do not have anti-glycolipid activity, especially when mice were immunized with strong adjuvant.

Nonspecific binding of antibodies to the plate is blocked by incubation with 100 μl of 5% bovine serum albumin–phosphate-buffered saline (BSA-PBS) per well at 37° for 2 hr. After washing with 0.5% BSA-PBS, 50–100 μl of culture fluid of each hybridoma clone is incubated in each of the glycolipid-coated wells for 1 hr at 37°. Nonbound immunoglobulin is then removed from the wells by washing three times with 0.5% BSA-PBS. Wells are subsequently incubated with a 1 : 1000 dilution of peroxidase-labeled goat anti-mouse immunoglobulin at 37° for 1 hr, and washed three times with 0.5% BSA-PBS. To this, substrate solution [50 ml of 0.05 *M* citrate–0.1 *M* Na_2HPO_3 buffer, pH 5.0, containing 20 mg *O*-phenylenediamine (Wako Pure Chemicals Co. Ltd., Osaka, Japan) and 10 μl of 30% H_2O_2] is added, and the reaction is stopped by the addition of 3 *N* H_2SO_4. Absorbance is measured at 492 nm with an appropriate plate spectrophotometer. Clones that correspond to the positive reaction on the plate are recloned.

When the antigenic glycolipid preparation used in the assay is contaminated with other glycolipids having longer carbohydrate chains than the antigen glycolipid, the reactivity of the antigen is sometimes masked and false-negative results can be obtained. Even if the contaminants are glycolipids having a similar carbohydrate chain as the antigenic glycolipid, significant masking of antigenicity is observed when the amount of contaminant is not negligible.[11]

Immunostaining of Glycolipid Antigens on TLC Plates

This method was first described by Magnani *et al.*[12] A mixture of glycolipids is first separated by TLC. The TLC plates are subsequently reacted with antibody solution and ^{125}I-labeled protein A solution. Bands of glycolipids that react with the antibody are detected by autoradiography. The method has been modified by using HPTLC plates, BSA as a blocking reagent, and peroxidase-labeled second antibody instead of using radioactive materials.

A sample glycolipid is first chromatographed on a HPTLC plate (SiHPF, Baker Chemical Co., Phillipsburg, NJ) with an appropriate solvent. After drying at room temperature, the TLC plate is soaked in 5% BSA-PBS in a tissue culture dish of appropriate size for 2 hr at room temperature to block nonspecific absorption of antibodies. The TLC plate is soaked and gently washed in 0.5% BSA-PBS with two changes of buffer. The TLC plate is then soaked in the first antibody solution and incubated for 2 hr at 4°. The plate is then gently washed with 0.5% BSA-PBS with three changes of the buffer. The plate is soaked in the peroxidase-labeled second antibody solution, incubated at 4° for 1 hr, and washed with 0.5% BSA-PBS in the same way. Positive reaction is visualized using ECL (enhanced chemiluminescence) Western blotting detection reagents (Amersham Life Science Ltd., Buckinghamshire, UK).

This is an excellent method for qualitative detection of a glycolipid antigen with high sensitivity. Less than 10 ng of glycolipid antigen can be detected under the best conditions. Besides information on the antigenicity of the glycolipid, information on the structure of the antigen can be obtained from its mobility on TLC.

[11] R. Kannagi, R. Stroup, N. A. Cochran, D. L. Urdal, W. W. Young, Jr., and S. Hakomori, *Cancer Res.* **43,** 4997 (1983).

[12] J. L. Magnani, M. Brockhaus, D. F. Smith, and V. Ginsburg, *Methods Enzymol.* **83,** 235 (1982).

Properties of Monoclonal Antibodies Directed to Glycolipid Antigens

The properties of monoclonal antibodies directed to carbohydrate antigens are described in this section. Established monoclonal anti-carbohydrate antibodies and their properties are listed in Table I.

Antibodies Directed to Ganglio-Series Glycolipids

A monoclonal antibody directed to asialo-GM2 (ganglio-*N*-triaosylceramide) has been successfully prepared in earlier days, but ganglio-series glycolipids appear to be poorly immunogenic in mice in general, and it has been hard to generate specific monoclonal antibodies. An anti-GM3 antibody was found to be secreted by a hybridoma raised against melanoma cells, but the immunization with lactonized GM3 was necessary to obtain an antibody with similar specificity using pure glycolipid as immunogen. Immunogenicity seems to partly depend on the strain of mice, and a set of antiganglio-series antibodies was obtained by immunizing C3H/HeN mice instead of BALB/c.[13] Nowadays antibodies covering almost all members of ganglio-series glycolipids are obtained, including antibodies that discriminate differences between *N*-acetyl and *N*-glycolyl sialic acid, and those specific to their *O*-acetylated derivatives or extremely minor lactoganglio-series glycolipids. Anti-GM2 antibody 10-11 was successfully used for the expression cloning of UDP-GalNAc: GM3/GD3 $\beta 1 \rightarrow 4N$-acetylgalactosaminyltransferase,[14] and antibodies directed to *O*-acetylated derivatives, D1.1 and 493D4, were used for the expression cloning of an *O*-acetylation-related gene.[15]

Antibodies Directed to Globo-Series Glycolipids

A glycolipid antigen defined by the rat monoclonal IgM antibody directed to Burkitt lymphoma (Daudi) turned out to be CTH (Gb3). Monoclonal antibodies directed to Forssman were successfully prepared. Anti-SSEA-3 and -4 antibodies, which specifically react to the stage-specifically expressed embryonic antigens, were found to recognize novel glycolipids belonging to this series yet at that time not known. This led to the discovery of so-called "extended globo-series glycolipids." Not only P-blood group active epitopes, but also blood group H- (globo-H) or A-active (globo-A) epitopes are carried by this series of glycolipids and are specifically detected

[13] I. Kawashima, O. Nakamura, and T. Tai, *Mol. Immunol.* **29,** 625 (1992).

[14] Y. Nagata, S. Yamashiro, J. Yodoi, K. O. Lloyd, H. Shiku, and K. Furukawa, *J. Biol. Chem.* **267,** 12082 (1992).

[15] A. Kanamori, J. Nakayama, M. N. Fukuda, W. B. Stallcup, K. Sasaki, M. Fukuda, and Y. Hirabayashi, *Proc. Natl. Acad. Sci. U.S.A.* **94,** 2897 (1997).

TABLE I

MONOCLONAL ANTIBODIES DIRECTED TO GLYCOLIPIDS AND RELATED CARBOHYDRATE DETERMINANTS

Recognized glycolipids	Immunogens	Clone name	References[a]
A. Ganglio-series glycolipids			
Asialo-GM2	Pure lipid	2D4 (IgM)[b]	Young *et al.* (A-1)
Asialo-GM1	Pure lipid	SH-34 (IgM)	Solomon *et al.* (A-2)
	Pure lipid	103HT30 (IgM)	Jacquemart *et al.* (A-3)
	Pure lipid	MW-1, -2, -3, -4, -5 (IgM)	Shimada *et al.* (A-4)
	Pure lipid	H2G10 (IgM)	Harrison *et al.* (A-5)
GM4	Pure lipid	AMR10 (IgM)	Ozawa *et al.* (A-6)
GM3	Mouse melanoma	M2590 (IgM)	Hirabayashi *et al.* (A-7)
	Melanoma patient	FCM1 (HIgM)	Yamaguchi *et al.* (A-8ab)
	GM3 lactone	DH2 (IgG_3)	Dohi *et al.* (A-9)
GM2	Melanoma	5-3, 10-11 (IgM)	Natoli *et al.* (A-10)
	Pure lipid	MK1-16[c], 1-17, 1-8 (IgM)	Miyake *et al.* (A-11)
	Melanoma patient	OFAI-1 (L55, HIgM)	Cahan *et al.* (A-12)
	Pure lipid	GMB28 (IgM)	Kotani *et al.* (A-13)
GM2 (NeuGc)	Melanoma patient	2-39M (HIgM)	Furukawa *et al.* (A-14)
	Rabbit thymocyte	YHD-07 (IgM)	Sanai *et al.* (A-15)
	Pure lipid	MK2-34 (IgM)[c]	Miyake *et al.* (A-11)
GM2 and GD2	Melanoma patient	3-207 (HIgM)	Yamaguchi *et al.* (A-16)
GM1	Pure lipid	GMB16 (IgM)	Kotani *et al.* (A-13)
	Pure lipid	AGM-1, -2, -3 (IgM)	Watarai *et al.* (A-17)
GM1b	Pure lipid	NA-6 (IgM)	Furuya *et al.* (A-18)
GalNAcGD1a	Liver cancer	2A3D2, 2D11E2 (IgM)[c]	Hiraiwa *et al.* (A-19)
GD3	Human melanoma	R24 (IgG_3)[b]	Pukel *et al.* (A-20)
	Human melanoma	4.2. (IgM)	Nudelman *et al.* (A-21)
	Human melanoma	LeoMel3 (IgM κ)	Werkmeister *et al.* (A-22)
	Melanoma patient	HJM1 (HIgM)	Yamaguchi *et al.* (A-8ab)
GD3 $(NeuGc)_2$	Melanoma patient	32-27M (HIgM)	Furukawa *et al.* (A-14)
GD2	Neuroblastoma	14.18 (IgG_3), 126 (IgM)[b]	Mujoo *et al.* (A-23)
	Neuroblastoma	2F7, 3A7, 3G6 (IgM), 3F8 (IgG_3)	Cheung *et al.* (A-24)
	Human glioma	BW625 (IgG_3), BW704 (IgG_3)	Bosslet *et al.* (A-25)
	Melanoma patient	OFA-I-2 (L72, HIgM)	Tai *et al.* (A-26)
	Pure lipid	GMR7 (IgM)	Ozawa *et al.* (A-27)
GD3 and GD2	Human melanoma	ME361 (IgG_3)[b]	Thurin *et al.* (A-28)

(*continued*)

TABLE I (*continued*)

Recognized glycolipids	Immunogens	Clone name	References[a]
9-*O*-acetyl-GD3	Neuroepithelial tumor (B49)	antiD1·1 (IgM)	Levine *et al.* (A-29a)
			Cheresh *et al.* (A-29b)
	Retinal epithelia	JONES (IgM)	Bonafede *et al.* (A-30)
	Rat embryonic brain cells	RB13-2 (IgM)	Reinhardt-Maelicke *et al.* (A-31)
	Human melanoma	ME311 (IgG_3)	Thurin *et al.* (A-32)
O-Acetyldisialoganglioside	Fetal rat brain	3G5 (IgM)[b]	Nayak *et al.* (A-33ab)
GD1a	Pure lipid	anti-G_{D1a} (IgM)	Nishinaka *et al.* (A-34)
GD1b	Pure lipid	anti-G_{D1b} (IgM)	Fredman *et al.* (A-35)
	Pure lipid	GGR12 (IgG_{2b})	Ozawa *et al.* (A-27)
	Pure lipid	MAb-5G6 (IgM)	Maehara *et al.* (A-36)
GD1c	Pure lipid	YK-3 (IgM)	Nakamura *et al.* (A-37)
GT3	Chick embryo retinal cells	18B8 (IgM), A_2B_5 (IgM)	Dubois *et al.* (A-38ab)
	Neural tissue of 3-day chick embryos	M6703 (IgG_3)	Hirabayashi *et al.* (A-39)
9-*O*-Acetyl-GT3	Optic nerve of 10-day chick embryos	493D4 (IgM)	Zhang *et al.* (A-40)
IV^2FucGM1	Pure lipid	F3,F12,F14 (IgG)	Nilsson *et al.* (A-41)
		F-1, F-2, F4 (IgM)	
GT1a	Pure lipid	GMR11 (IgM)	Kotani *et al.* (A-13)
GT1b	Pure lipid	GMR5 (IgM)	Ozawa *et al.* (A-27)
GQ1b	Pure lipid	4F10 (IgG_3), 7F5 (IgG_{2a}), 4E7 (IgM)	Yamamoto *et al.* (A-42)
	Pure lipid	GMR13 (IgM)	Ozawa *et al.* (A-27)
GQ1Bα, GT1bα	Pure lipid	GGR41 (IgG_{2a})	Kusunoki *et al.* (A-43)
GT1c, GQ1c, GP1c	Chick brain	Q211 (IgM)	Rosner *et al.* (A-44ab)
B. Lacto- and neolacto-series glycolipids			
LacCer	ANLL	T5A7 (IgM)	Symington *et al.* (B-1)
Lc_3	Syngenic testis	J-1 (IgM)	Fenderson *et al.* (B-2ab)
Lc_4 (Le^c)	Teratocarcinoma	Fc-10.2 (IgM)	Gooi *et al.* (B-3)
	Embryonal carcinoma cells (PA-1)	K21 (IgM)	Rettig *et al.* (B-4ab)
	Uterine cancer	HMST-1 (HIgM)	Nozawa *et al.*et al. (B-5)
	A375 cells	11-50 (HIgM)	Yago *et al.* (B-6)
nLc_4	Pure lipid	1B2 (IgM)	Young *et al.* (B-7)
	Granulocytes	My28 (IgM)	Spitalnik *et al.* (B-8)
	Germ cell tumor	2G10 (IgM)	Fujimoto *et al.* (B-9)
	Lipid mixture	αPGF1H11 (IgM)	Myoga *et al.* (B-10)
nLc_5	Pure lipid	TE-4, TE-5, TE-7 (IgM), TE-6 (IgG_3)	Holmes *et al.* (B-11)

nLc_6	Lung cancer	MH21-134 ($HIgG_3$)	Miyake *et al.* (B-12)
nLc_6	Lung cancer	NCC-1004 (HIgM)	Hirohashi *et al.* (B-13)
nLc_6	EBV	GL-1, GL-2 (HIgM)	Nagatsuka *et al.* (B-14)
$isonLc_8$	Mouse spermatogenic cells	C6 (IgM)	Fenderson *et al.* (B-15ab)
IV^2-$FucLc_4$ (Le^d)	Epidermoid cancer	101 (IgM)	Fredman *et al.* (B-16)
IV^2-$FucnLc_4$	Pure lipid	BE2 (IgM)	Young *et al.* (B-7)
	Hepatocellular carcinoma cells	S1 (IgG2a), S3 (IgG_3)	Imai *et al.* (B-17)
	Endometrial carcinoma cells	C12 (IgM)	Tsuji *et al.* (B-18)
III^4FucLc_4(Le^a)	Lipid mixture	CA37-4 (IgG_1), CF4-C4 ($IgG_{2}a$)[b] BC9-ES (IgG_3), DG4-1 (IgG_1)	Young *et al.* (B-19)
	Pure lipid	3E8 (IgM)	Shigeta *et al.* (B-20)
$III^4V^4Fuc_2Lc_6$ (extended Le^a, Le^a/Le^a)	Gastric cancer	NCC-ST-421 (IgG_3)	Stroud *et al.* (B-21)
$III^3FucnLc_4$ (Le^x)	Gastric cancer	WGHS29 (IgM)	Brockhaus *et al.* (B-22)
	Colon cancer	ZWG13,14,III (IgM)	Huang *et al.* (B-23)
	Lung cancer	J525,34,38 (IgM)	Cuttita *et al.* (B-24)
	Leukemia cell	My-1 (IgM)	Huang *et al.* (B-25)
	Leukemia cell	VEP8,9 (IgM)	Gooi *et al.* (B-26)
	Leukemia cell	1G10 (IgM)	Urdal *et al.* (B-27)
	Pure lipid	FH-2 (IgM)	Fukushi *et al.* (B-28)
$III^3FucnLc_6$ (Le^x-i)	Teratocarcinoma	αSSEA-1 (IgM)	Gooi *et al.* (B29a)
			Kannagi *et al.* (B-29bc)
$III^3V^3Fuc_2nLc_6$ (extended Le^x)	Pure lipid	SH-2 (IgG)	Singhal *et al.* (B-30)
$III^3V^3VII^3Fuc_3nLc_{10}$ (internal Le^x)	Gastric cancer	ACFH-18 (IgM)	Nudelman *et al.* (B-31)
III^3 $IV^2Fuc_2Lc_4$ (Le^b)	Colon cancer	NS-10 (IgM)	Brockhous *et al.* (B-32)
$III^3IV^2Fuc_2nLc_4$ (Le^y)	Colon adenoma	C14/1/16/10 (IgM)	Brown *et al.* (B-33)
	Lung cancer	F_3 (IgM)	Lloyd *et al.* (B-34)
	Teratocarcinoma	75.12 (IgM)	Blaineau *et al.* (B-35)
	Gastric cancer	AH-6 (IgM)	Abe *et al.* (B-36)
	Lung cancer	NCC-ST-433 (IgM)	Hirohashi *et al.* (B-37)
	Lung cancer	Luca 6 (IgM)	Kyoizumi *et al.* (B-37)
$III^3V^3VI^2Fuc_3nLc_6$ (extended Le^y)	Pure lipid	KH-1 (IgM)	Kaizu *et al.* (B-38)
	Colon cancer	CC-1 (IgM), CC-2 (IgM)	Sun *et al.* (B-39)
$IV^3NeuAcIII^4FucLc_4$ (sialyl Le^a)	Colon cancer	N-19-9 (IgG)	Koprowski *et al.* (B-40ab)
	Colon cancer	CC3C195 (IgM)	Fukuta *et al.* (B-41)
	Pancreas cancer	SPan1 (IgM, κ)	Chung *et al.* (B-42)
	Colon cancer	KM01 (IgG_1)	Kano *et al.* (B-43)
	Colon cancer	MSW113 (IgG_3)	Kitagawa *et al.* (B-44)
	Pure lipid	2D3[c], 4E2 (IgM), 1H4[c] (IgG_3)	Shigeta *et al.* (B-45)
	Colon cancer	C-50 (IgM)	Lindholm *et al.* (B-46)

(*continued*)

TABLE I (*continued*)

Recognized glycolipids	Immunogens	Clone name	References[a]
IV^{3}NeuAcLc$_4$	Pancreas cancer	DU-PAN-2 (IgM)	Metzgar *et al.* (B-47ab)
	Embryonal carcinoma cells (PA-1)	K4 (IgM)	Rettig *et al.* (B-4ab)
	Unknown	SL-50 (IgM)	Svennerholm *et al.* (B-48)
IV^{3}NeuAcnLc$_4$	Pure lipid	NS24 (IgM)	Suzuki (B-49)
IV^{6}NeuAcnLc$_4$	Pure lipid	1B9 (IgG$_3$)	Hakomori *et al.* (B-50)
	Pure lipid	LM-4 (IgG$_{2a}$)	Nilsson (B-51)
	Meconium glycolipid mixture	MSG15 (IgM)	Taki *et al.* (B-52)
	Lipid mixture	1-15B (IgG$_3$)	Harada *et al.* (B-20)
IV^{3}NeuAcIII^{3}FucnLc$_4$ (sialyl Le^x)	Colon cancer	CSLEX-1 (IgM)[b,e]	Terasaki *et al.* (B-53)
	Pure lipid	SNH-3 (1-12F, IgM)	Singhal *et al.* (B-54a)
	Pure lipid	SNH-4 (IgG$_3$)	Polley *et al.* (B-54b)
	Lung cancer	KM-93 (IgM)[c]	Shitara *et al.* (B-55)
	Lipid mixture	2H5 (IgM)[d]	Sawada *et al.* (B-56)
	Pure lipid	2F3 (IgM)[d]	Ohmori *et al.* (B-57)
IV^{3}NeuAc III$^{3/4}$Fuc(n)Lc$_4$	Tonsil stroma	HECA-452 (RIgM)[b]	Berg *et al.* (B-58)
VI^{3}NeuAcV^3 III^{3}Fuc$_2$nLc$_6$, VI^{3}NeuAcV^{3}FucnLc$_6$ (sialyl Le^x-i)	Pure lipid	FH-6 (IgM)[b]	Fukushi *et al.* (B-59)
		3-5E (IgG$_3$)	Singhal *et al.* (B-54a)
IV^3 III^{6}NeuAc$_2$ III^{4}FucLc$_4$	Lipid mixture	FH-7 (IgG$_3$)	Nudelman *et al.* (B-60)
IV^3 III^{6}NeuAc$_2$Lc$_4$	Lipid mixture	FH-9 (IgG$_2$a)	Fukushi *et al.* (B-61)
IV^{3}GalαnLc$_4$	Rabbit erythrocyte	Gal-13 (IgG$_1$)	Galili *et al.* (B-62)
III^{3}V^{4}Fuc$_2$Lc$_2$nLc$_4$	Milk-fat membrane	115C2, 115G3 (IgG$_1$)	Hilkens *et al.* (B-63ab)
	Lung cancer cells	43-9F (IgM)	Martensson *et al.* (B-64ab)
IV^{3}VIII^{3}NeuAc$_2$isonLc$_8$	Human placenta	NUH-2 (IgM)[b]	Nudelman *et al.* (B-65)
VI^{3}NeuGcnLc$_6$	Pure lipid	SHS-1 (IgM)	Watarai *et al.* (B-66)
C. Globo- and extended globo-series glycolipids			
Gb$_3$(CTH)	Burkitt lymphoma	38-13 (RIgM)	Nudelman *et al.* (C-1)
	Teratocarcinoma	1A4 (IgM)	Ishigami *et al.* (C-2)
	Lipid mixture	H-8 (IgM)	Harada *et al.* (C-3)
III^{3}GalαGb$_3$	Pure lipid	BGR47 (IgG$_3$)	Kotani *et al.* (C-4)

Gb_4	PC12 cells	CC1 (IgM)	Schwarting *et al.* (C-5)
	Tonsillar lymphocytes	HJ6 (IgM)	Madassery *et al.* (C-6)
	Pure lipid (*N*-glycolyl G_{M1})	12D4 (IgM)	Madassery *et al.* (C-6)
	Pure lipid	BMR26 (IgM)	Kotani *et al.* (C-4)
Iso Gb_4 and an isoglobo-lacto series glycolipid	Pure lipid	14.2, 14.10, 14.3 (IgM)	Brodin *et al.* (C-7)
Gb_5	Early mouse embryo	αSSEA-3 (MC631,RIgM)	Kannagi *et al.* (C-8)
	Pure glycolipid (GLY)	5A3 (IgM)	Marcus *et al.* (C-9)
	Lung cancer patient	J309, D579 (HIgM)	Schrump *et al.* (C-10)
$V^3NeuAcGb_5$	Teratocarcinoma	αSSEA-4 (IgM)	Kannagi *et al.* (C-11)
$V^3IV^6NeuAc_2Gb_5$	Renal cancer	RM2 (IgM)	Satoh *et al.* (C-12ab)
$IV^3GalNAcGb_4$ (Forssman)	Mouse spleen cells	M1/22.25 (RIgM)	Stern *et al.* (C-13ab)
	Influenza patient	H1-C4 (HIgM)	Nowinski *et al.* (C-14)
	Mouse mammary tumor	33B12 (RIgG_1 or $_2$c), 117C9 (RIgM)	Sonnenberg *et al.* (C-15)
	Pure lipid (*in vitro* immunization)	4C3 (RIgM)	Gathuru *et al.* (C-16)
V^2FucGb_5	Breast cancer cell	MBr-1 (IgM)	Bremer *et al.* (C-17a)
			Kannagi *et al.* (C-17b)
D. Lactoganglio-series glycolipids			
$II^3GlcNAcGg_3$ ($II^4GalNAcLc_3$)	Synthetic lipid	YI328-18 (IgG_3), YI328-51 (IgG_3)	Shigeta *et al.* (D-1)
Le^x-GM_1	Frog skin extract	188C1 (IgM)	Hirabayashi *et al.* (D-2)
E. Sulfated glycolipids and carbohydrate determinants			
SM_4g (seminolipid)	Pure lipid (*in vitro* immunization)	CA10 (IgM)	Eddy *et al.* (E-1)
SM_4s (sulfatide)	Pure lipid	AIC3IA2 (IgG_3)	Hofstetter *et al.*(E-2)
	PWM	M14-376 (HIgM)	Miyake *et al.* (E-3)
GlcUA-PG-sulfate	Human lymphocytes	HNK-1 (IgM)	Chou *et al.* (E-4abc)
	Glycolipid mixture	L9 (IgM)	Borroni *et al.* (E-5)
SB_1a	Lipid mixture	2H11E5 (IgM)[c]	Hiraiwa *et al.* (E-6)
	Liver cancer cells	4A9E10 (IgG_3)[c]	Hiraiwa *et al.* (E-6)
SM_3	Liver cancer cells	49-D6 (IgM),[c] 7-E10 (IgM)[c]	Hiraiwa *et al.* (E-7)
$III^6SO_4III^3FucnLc_4$ (6-sulfo Le^x)	Synthetic glycolipid	AG223, AG107 (IgM)	Uchimura *et al.* (E-8)
$IV^3NeuAcIII^6SO_4nLc_4$ (sialyl 6-sulfo paragloboside)	Synthetic glycolipid	G72 (IgM)	Mitsuoka *et al.* (E-9)
$IV^3NeuAcIII^3FucIII^6SO_4nLc_4$ (sialyl 6-sulfo Le^x)	Synthetic glycolipid	G152 (IgM)	Mitsuoka *et al.* (E-9)
$IV^3NeuAcIII^3FucIV^6III^6(SO_4)_2nLc_4$ (sialyl 6,6′-disulfo Le^x)	Synthetic glycolipid	G2706, G27011, G27037, G27039 (IgM)	Mitsuoka *et al.* (E-9)

(*continued*)

TABLE I (*continued*)

Recognized glycolipids	Immunogens	Clone name	References[a]
F. Mucin-type carbohydrate determinants			
T (Galβ1→3GalNAcα1→Ser/Thr)	Desialylated erythrocytes	49H-24 (IgM)	Rahman *et al.* (F-1)
		49H-8 (IgM)	Longenecker *et al.* (F-2)
Tn (GalNAcα1→Ser/Thr)	Synthetic antigen	164H.1	Longenecker *et al.* (F-3)
	Lung cancer	NCC-LU-35, NCC-LU-81 (IgG)	Hirohashi *et al.* (F-4)
	Tn erythrocytes	FBT3 (IgM)	Roxby *et al.* (F-5)
	Tn erythrocytes	BRIC111 (IgG_1)	King *et al.* (F-6)
	Deglycosylated bovine submaxillary mucin	CU-1 (IgG_3)	Takahashi *et al.* (F-7)
	Deglycosylated bovine submaxillary mucin	5F4 (IgM)	Thurnher *et al.* (F-8)
	Colon cancer (LS180 cells)	MLS128 (IgG_3)	Numata *et al.* (F-9)
	Deglycosylated colon cancer	91S8 (IgM), 10F4 (IgG_3)	Huang *et al.* (F-10)
	Cells (LS174T cells)	15D3a (IgM)	
Sialyl Tn (NeuAcα2→6GalNAcα1→Ser/Thr)	Colon cancer	MLS102 (IgM)	Kurosaka *et al.* (F-11)
	Mucin	TKH-1 (IgM), TKH-2 (IgG)	Kjeldsen *et al.* (F12)
	Breast cancer	B72.3 (IgG_1)	Springer *et al.* (F-13ab)
T-Sialyl Tn (Galβ1→3[NeuAcα2→6]GalNAcα1→Ser/Thr)	Glycoprotein	CC49(IgM)	Hanisch *et al.* (F-14)
Disialyl T (NeuAcα2→3Galβ1→3[NeuAcα2→6]-GalNAcα1→Ser/Thr)	Glycophorin A	QSH2	Qiu *et al.* (F-15)
Core 2/4(Galβ1→4GlcNAcβ1→6GalNAcα1→Ser/Thr)	Synthetic glycolipid	F1α79 (IgM)	Yamashita *et al.* (F-16)
Sialyl Le^x on core 2/4 (NeuAcα2→3Galβ1→4[Fucα2→3]GlcNAcβ1→6GalNAcα1→Ser/Thr)	Gastric cancer	NCC-ST-439 (IgM)	Kumamoto *et al.* (F-17)
O-GlcNAc	Streptococcal group A carbohydrate	HGAC86, HGAC39 (IgG_3)	Turner *et al.* (F-18)

[a] *Key to references:* Capital letter preceding a reference number refers to that section in Table I.

A. Ganglio-series glycolipids:

1. W. W. Young, Jr., E. M. MacDonald, R. C. Nowinski, and S. I. Hakomori, *J. Exp. Med.* **150,** 1008 (1979).
2. F. R. Solomon and T. J. Higgins, *Mol. Immunol.* **24,** 57 (1987).
3. F. Jacquemart, G. Millot, C. Goujet-Zalc, G. Mahouy, and B. Zalc, *Hybridoma* **7,** 323 (1988).

4. S. Shimada and D. Iwata, *Microbiol. Immunol.* **31,** 923 (1987).
5. B. A. Harrison, R. MacKenzie, T. Hirama, K. K. Lee, and E. Altman, *J. Immunol. Methods* **212,** 29 (1998).
6. H. Ozawa, M. Kotani, I. Kawashima, M. Numata, T. Ogawa, T. Terashima, and T. Tai, *J. Biochem. (Tokyo)* **114,** 5 (1993).
7. Y. Hirabayashi, A. Hamaoka, M. Matsumoto, T. Matsubara, M. Tagawa, S. Wakabayashi, and M. Taniguchi, *J. Biol. Chem.* **260,** 13328 (1985).
8a. H. Yamaguchi, K. Furukawa, S. R. Fortunato, P. O. Livingston, K. O. Lloyd, H. F. Oettgen, and L. J. Old, *Proc. Natl. Acad. Sci. U.S.A.* **84,** 2416 (1987).
8b. K. Furukawa, H. Yamaguchi, H. F. Oettgen, L. J. Old, and K. O. Lloyd, *Cancer Res.* **49,** 191 (1989).
10. E. J. Natoli, Jr., P. O. Livingston, C. S. Pukel, K. O. Lloyd, H. Wiegandt, J. Szalay, H. F. Oettgen, and L. J. Old, *Cancer Res.* **46,** 4116 (1986).
11. M. Miyake, M. Ito, S. Hitomi, S. Ikeda, T. Taki, M. Kurata, A. Hino, N. Miyake, and R. Kannagi, *Cancer Res.* **48,** 6154 (1988).
12. L. D. Cahan, R. F. Irie, R. Singh, A. Cassidenti, and J. C. Paulson, *Proc. Natl. Acad. Sci. U.S.A.* **79,** 7629 (1982).
13. M. Kotani, H. Ozawa, I. Kawashima, S. Ando, and T. Tai, *Biochim. Biophys. Acta Gen. Subj.* **1117,** 97 (1992).
14. K. Furukawa, H. Yamaguchi, H. F. Oettgen, L. J. Old, and K. O. Lloyd, *J. Biol. Chem.* **263,** 18507 (1988).
15. Y. Sanai, M. Yamasaki, and Y. Nagai, *Biochim. Biophys. Acta* **958,** 368 (1988).
16. H. Yamaguchi, K. Furukawa, S. R. Fortunato, P. O. Livingston, K. O. Lloyd, H. F. Oettgen, and L. J. Old, *Proc. Natl. Acad. Sci. U.S.A.* **87,** 3333 (1990).
17. S. Watarai, K. Kiura, R. Shigeto, T. Shibayama, I. Kimura, and T. Yasuda, *J. Biochem. (Tokyo)* **116,** 948 (1994).
18. S. Furuya, T. Hashikawa, F. Irie, A. Hasegawa, T. Nakao, and Y. Hirabayashi, *Neurosci. Res.* **22,** 411 (1995).
19. N. Hiraiwa, K. Tsuyuoka, Y. T. Li, M. Tanaka, T. Seno, Y. Okubo, Y. Fukuda, H. Imura, and R. Kannagi, *Cancer Res.* **50,** 5497 (1990).
20. C. S. Pukel, K. O. Lloyd, L. R. Travassos, W. G. Dippold, H. F. Oettgen, and L. J. Old, *J. Exp. Med.* **155,** 1133 (1982).
21. E. Nudelman, S. Hakomori, R. Kannagi, S. B. Levery, M. Y. Yeh, K. E. Hellström, and I. Hellström, *J. Biol. Chem.* **257,** 12752 (1982).
22. J. A. Werkmeister, T. Triglia, I. R. Mackay, J. P. Dowling, G. A. Varigos, G. Morstyn, and G. F. Burns, *Cancer Res.* **47,** 225 (1987).
23. K. Mujoo, D. A. Cheresh, H. M. Yang, and R. A. Reisfeld, *Cancer Res.* **47,** 1098 (1987).
24. N. K. Cheung, U. M. Saarinen, J. E. Neely, B. Landmeier, D. Donovan, and P. F. Coccia, *Cancer Res.* **45,** 2642 (1985).
25. K. Bosslet, H. D. Mennel, F. Rodden, B. L. Bauer, F. Wagner, A. Altmannsberger, H. H. Sedlacek, and H. Wiegandt, *Cancer Immunol. Immunother.* **29,** 171 (1989).
26. T. Tai, J. C. Paulson, L. D. Cahan, and R. F. Irie, *Proc. Natl. Acad. Sci. U.S.A.* **80,** 5392 (1983).
27. H. Ozawa, K. Kotani, I. Kawashima, and T. Tai, *Biochim. Biophys. Acta Lipids Lipid Metab.* **1123,** 184 (1992).
28. J. Thurin, M. Thurin, M. Herlyn, D. E. Elder, Z. Steplewski, W. H. Clark, Jr., and H. Koprowski, *FEBS Lett.* **208,** 17 (1986).
29a. J. M. Levine, L. Beasley, and W. B. Stallcup, *J. Neurosci.* **4,** 820 (1984).
29b. D. A. Cheresh, R. A. Reisfeld, and A. P. Varki, *Science* **225,** 844 (1984).
30. D. M. Bonafede, L. J. Macala, M. Constantine-Paton, and R. K. Yu, *Lipids* **24,** 680 (1989).
31. S. Reinhardt-Maelicke, V. Cleeves, A. Kindler-Rohrborn, and M. F. Rajewsky, *Dev. Brain Res.* **51,** 279 (1990).
32. J. Thurin, M. Herlyn, O. Hindsgaul, N. Stromberg, K. A. Karlsson, D. Elder, Z. Steplewski, and H. Koprowski, *J. Biol. Chem.* **260,** 14556 (1985).
33a. R. C. Nayak, A. B. Berman, K. L. George, G. S. Eisenbarth, and G. L. King, *J. Exp. Med.* **167,** 1003 (1988).
33b. D. K. Chou, S. Flores, and F. B. Jungalwala, *J. Neurochem.* **54,** 1598 (1990).
34. T. Nishinaka, D. Iwata, S. Shimada, K. Kosaka, and Y. Suzuki, *Neurosci. Res.* **17,** 171 (1993).
35. P. Fredman, S. Jeansson, E. Lycke, and L. Svennerholm, *FEBS Lett.* **189,** 23 (1985).
36. T. Maehara, K. Ono, K. Tsutsui, S. Watarai, T. Yasuda, H. Inoue, and A. Tokunaga, *Neurosci. Res.* **29,** 9 (1997).
37. K. Nakamura, H. Suzuki, Y. Hirabayashi, and A. Suzuki, *J. Biol. Chem.* **270,** 3876 (1995).
38a. C. Dubois, J. C. Manuguerra, B. Hauttecoeur, and J. Maze, *J. Biol. Chem.* **265,** 2797 (1990).
38b. C. Dubois, J. L. Magnani, G. B. Grunwald, S. L. Spitalnik, G. D. Trisler, M. Nirenberg, and V. Ginsburg, *J. Biol. Chem.* **261,** 3826 (1986).

(continued)

TABLE I (*continued*)

39. Y. Hirabayashi, M. Hirota, M. Matsumoto, H. Tanaka, K. Obata, and S. Ando, *J. Biochem.* (*Tokyo*) **104,** 973 (1988).
40. G. Zhang, L. Ji, S. Kurono, S. C. Fujita, S. Furuya, and Y. Hirabayashi, *Glycoconj. J.* **14,** 847 (1997).
41. Nilsson, F. T. Brezicka, J. Holmgren, S. Sorenson, L. Svennerholm, F. Yngvason, and L. Lindholm, *Cancer Res.* **46,** 1403 (1986).
42. H. Yamamoto, S. Tsuji, and Y. Nagai, *J. Neurochem.* **54,** 513 (1990).
43. S. Kusunoki, A. Chiba, Y. Hirabayashi, F. Irie, M. Kotani, I. Kawashima, T. Tai, and Y. Nagai, *Brain Res.* **623,** 83 (1993).
44a. C. Greis and H. Rösner, *Brain Res.* **517,** 105 (1990).
44b. H. Rösner, C. Greis, and S. Henke-Fahle, *Brain Res.* **470,** 161 (1988).

B. Lacto- and neolacto-series glycolipids:

1. F. W. Symington, I. D. Bernstein, and S. Hakomori, *J. Biol. Chem.* **259,** 6008 (1984).
2a. B. A. Fenderson, D. A. O'Brien, C. F. Millette, and E. M. Eddy, *Dev. Biol.* **103,** 117 (1984).
2b. F. W. Symington, B. A. Fenderson, and S. Hakomori, *Mol. Immunol.* **21,** 877 (1984).
3. H. C. Gooi, L. K. Williams, K. Uemura, E. F. Hounsell, R. A. McIlhinney, and T. Feizi, *Mol. Immunol.* **20,** 607 (1983).
4a. P. P. Wilkins, K. L. Moore, R. P. McEver, and R. D. Cummings, *J. Biol. Chem.* **270,** 22677 (1995).
4b. M. N. Fukuda, B. Bothner, K. O. Lloyd, W. J. Rettig, P. R. Tiller, and A. Dell, *J. Biol. Chem.* **261,** 5145 (1986).
5. S. Nozawa, S. Narisawa, K. Kojima, M. Sakayori, R. Iizuka, H. Mochizuki, T. Yamauchi, M. Iwamori, and Y. Nagai, *Cancer Res.* **49,** 6401 (1989).
6. K. Yago, K. Zenita, I. Ohwaki, R. Harada, S. Nozawa, K. Tsukazaki, M. Iwamori, N. Endo, N. Yasuda, M. Okuma, and R. Kannagi, *Mol. Immunol.* **30,** 1481 (1993).
7. W. W. Young, Jr., J. Portoukalian, and S. Hakomori, *J. Biol. Chem.* **256,** 10967 (1981).
8. S. L. Spitalnik, J. F. Schwartz, J. L. Magnani, D. D. Roberts, P. F. Spitalnik, C. I. Civin, and V. Ginsburg, *Blood* **66,** 319 (1985).
9. J. Fujimoto, J. Hata, E. Ishii, S. Tanaka, R. Kannagi, Y. Ueyama, and N. Tamaoki, *Lab. Invest.* **57,** 350 (1987).
10. A. Myoga, T. Taki, K. Arai, K. Sekiguchi, I. Ikeda, K. Kurata, and M. Matsumoto, *Cancer Res.* **48,** 1512 (1988).
11. E. H. Holmes and T. G. Greene, *Arch. Biochem. Biophys.* **288,** 87 (1991).
12. M. Miyake, N. Koohno, E. D. Nudelman, and S. Hakomori, *Cancer Res.* **49,** 5689 (1989).
13. S. Hirohashi, H. Clausen, E. Nudelman, H. Inoue, Y. Shimosato, and S. Hakomori, *J. Immunol.* **136,** 4163 (1986).
14. Y. Nagatsuka, S. Watarai, T. Yasuda, H. Higashi, T. Yamagata, and Y. Ono, *Immunol. Lett.* **46,** 93 (1995).
15a. B. A. Fenderson, E. J. Nichols, H. Clausen, and S. I. Hakomori, *Mol. Immunol.* **23,** 747 (1986).
15b. E. J. Nichols, B. A. Fenderson, W. G. Carter, and S. Hakomori, *J. Biol. Chem.* **261,** 11295 (1986).
16. P. Fredman, N. D. Richert, J. L. Magnani, M. C. Willingham, I. Pastan, and V. Ginsburg, *J. Biol. Chem.* **258,** 11206 (1983).
17. K. Imai, T. Sasanami, T. Nakanishi, T. Noguchi, and A. Yachi, *Tumour. Biol.* **6,** 257 (1985).
18. Y. Tsuji, M. Yoshioka, T. Ogasawara, T. Takemura, and S. Isojima, *Cancer Res.* **47,** 3543 (1987).
19. W. W. Young, Jr., H. S. Johnson, Y. Tamura, K. A. Karlsson, G. Larson, J. M. Parker, D. P. Khare, U. Spohr, D. A. Baker, O. Hindsgaul, and R. U. Lemieux, *J. Biol. Chem.* **258,** 4890 (1983).
20. K. Zenita, K. Hirashima, K. Shigeta, N. Hiraiwa, A. Takada, E. Fujimoto, K. Hashimoto, S. Hakomori, and R. Kannagi, *J. Immunol.* **144,** 4442 (1990).
21. M. R. Stroud, S. B. Levery, E. D. Nudelman, M. E. Salyan, J. A. Towell, C. E. Roberts, M. Watanabe, and S. Hakomori, *J. Biol. Chem.* **266,** 8439 (1991).
22. M. Brockhaus, J. L. Magnani, M. Herlyn, M. Blaszczyk, Z. Steplewski, H. Koprowski, and V. Ginsburg, *Arch. Biochem. Biophys.* **217,** 647 (1982).
23. L. C. Huang, M. Brockhaus, J. L. Magnani, F. Cuttitta, S. Rosen, J. D. Minna, and V. Ginsburg, *Arch. Biochem. Biophys.* **220,** 318 (1983).
24. F. Cuttitta, S. Rosen, A. F. Gazdar, and J. D. Minna, *Proc. Natl. Acad. Sci. U.S.A.* **78,** 4591 (1981).
25. L. C. Huang, C. I. Civin, J. L. Magnani, J. H. Shaper, and V. Ginsburg, *Blood* **61,** 1020 (1983).
26. H. C. Gooi, S. J. Thorpe, E. F. Hounsell, H. Rumpold, D. Kraft, O. Forster, and T. Feizi, *Eur. J. Immunol.* **13,** 306 (1983).

27. D. L. Urdal, T. A. Brentnall, I. D. Bernstein, and S. I. Hakomori, *Blood* **62,** 1022 (1983).
28. Y. Fukushi, S. Hakomori, E. Nudelman, and N. Cochran, *J. Biol. Chem.* **259,** 4681 (1984).
29a. H. C. Gooi, T. Feizi, A. Kapadia, B. B. Knowles, D. Solter, and M. J. Evans, *Nature* **292,** 156 (1981).
29b. R. Kannagi, E. Nudelman, S. B. Levery, and S. Hakomori, *J. Biol. Chem.* **257,** 14865 (1982).
29c. R. Kannagi, E. Nudelman, and S. Hakomori, *Proc. Natl. Acad. Sci. U.S.A.* **79,** 3470 (1982).
30. A. K. Singhal, T. F. Orntoft, E. Nudelman, S. Nance, L. Schibig, M. R. Stroud, H. Clausen, and S. Hakomori, *Cancer Res.* **50,** 1375 (1990).
31. E. Nudelman, S. B. Levery, M. R. Stroud, M. E. Salyan, K. Abe, and S. Hakomori, *J. Biol. Chem.* **263,** 13942 (1988).
32. M. Brockhaus, J. L. Magnani, M. Blaszczyk, Z. Steplewski, H. Koprowski, K. A. Karlsson, G. Larson, and V. Ginsburg, *J. Biol. Chem.* **256,** 13223 (1981).
33. A. Brown, T. Feizi, H. C. Gooi, M. J. Embleton, J. K. Picard, and R. W. Baldwin, *Biosci. Rep.* **3,** 163 (1983).
34. K. O. Lloyd, G. Larson, N. Stromberg, J. Thurin, and K. A. Karlsson, *Immunogenetics* **17,** 537 (1983).
35. C. Blaineau, J. Le Pendu, D. Arnaud,F. Connan, and P. Avner, *EMBO J.* **2,** 2217 (1983).
36. K. Abe, J. M. McKibbin, and S. Hakomori, *J. Biol. Chem.* **258,** 11793 (1983).
37. K. Hirashima, K. Zenita, A. Kitahara, G. Ishihara, R. Harada, K. Ohmori, S. Hirohashi, S. Akiyama, and R. Kannagi, *J. Immunol.* **145,** 224 (1990).
38. T. Kaizu, S. B. Levery, E. Nudelman, R. E. Stenkamp, and S. Hakomori, *J. Biol. Chem.* **261,** 11254 (1986).
39. Q. Sun, B. Siddiqui, E. Nudelman, S. Hakomori, J. J. L. Ho, and Y. S. Kim, *Cancer J.* **1,** 213 (1987).
40a. H. Koproswki, M. Herlyn, Z. Steplewski, and H. F. Sears, *Science* **212,** 53 (1981).
40b. J. L. Magnani, B. Nilsson, M. Brockhaus, D. Zopf, Z. Steplewski, H. Koprowski, and V. Ginsburg, *J. Biol. Chem.* **257,** 14365 (1982).
41. S. Fukuta, J. L. Magnani, P. K. Gaur, and V. Ginsburg, *Arch. Biochem. Biophys.* **255,** 214 (1987).
42. Y. S. Chung, J. J. Ho, Y. S. Kim, H. Tanaka, B. Nakata, A. Hiura, H. Motoyoshi, K. Satake, and K. Umeyama, *Cancer* **60,** 1636 (1987).
43. Y. Kano, T. Taniguchi, Y. Uemura, K. Yokoyama, K. Uesaka, M. Yamamoto, H. Ohyanagi, and Y. Saitoh, *Hybridoma* **9,** 363 (1990).
44. H. Kitagawa, H. Nakada, Y. Numata, A. Kurosaka, S. Fukui, I. Funakoshi, T. Kawasaki, and I. Yamashina, *J. Biochem.* (*Tokyo*) **104,** 817 (1988).
45. A. Takada, K. Ohmori, N. Takahashi, K. Tsuyuoka, K. Yago, K. Zenita, A. Hasegawa, and R. Kannagi, *Biochem. Biophys. Res. Commun.* **179,** 713 (1991).
46. O. Nilsson, J-E. Månsson, L. Lindholm, J. Holmgren, and L. Svennerholm, *FEBS Lett.* **182,** 398 (1985).
47a. Y. Hamanaka, S. Hamanaka, Y. Shinagawa, T. Suzuki, F. Inagaki, M. Suzuki, and A. Suzuki, *FEBS Lett.* **353,** 48 (1994).
47b. S. Kawa, M. Tokoo, H. Oguchi, S. Furuta, T. Homma, Y. Hasegawa, H. Ogata, and K. Sakata, *Pancreas* **9,** 692 (194).
48. L. Svennerholm, K. Bostrom, P. Fredman, J. E. Mansson, B. Rosengren, and B. M. Rynmark, *Biochim. Biophys. Acta* **1005,** 109 (1989).
49. Y. Suzuki, H. Nishi, K. Hidari, Y. Hirabayashi, M. Matsumoto, T. Kobayashi, S. Watarai, T. Yasuda, J. Nakayama, and H. Maeda, *J. Biochem.* (*Tokyo*) **109,** 354 (1991).
50. S. Hakomori, C. M. Patterson, E. Nudelman, and K. Sekiguchi, *J. Biol. Chem.* **258,** 11819 (1983).
51. O. Nilsson, L. Lindholm, J. Holmgren, and L. Svennerholm, *Biochim. Biophys. Acta* **835,** 577 (1985).
52. T. Taki, K. Yamamoto, M. Takamatsu, K. Ishii, A. Myoga, K. Sekiguchi, I. Ikeda, K. Kurata, J. Nakayama, and S. Handa, *Cancer Res.* **50,** 1284 (1990).
53. K. Fukushima, M. Hirota, P. I. Terasaki, A. Wakisaka, H. Togashi, D. Chia, N. Suyama, Y. Fukushi, E. Nudelman, and S. Hakomori, *Cancer Res.* **44,** 5279 (1984).
54a. M. L. Phillips, E. Nudelman, F. C. A. Gaeta, M. Perez, A. K. Singhal, S. Hakomori, and J. C. Paulson, *Science* **250,** 1130 (1990).
54b. M. J. Polley, M. L. Phillips, E. Wayner, E. NUdelman, A. K. Singhal, S. Hakomori, and J. C. Paulson, *Proc. Natl. Acad. Sci. U.S.A.* **88,** 6224 (1991).
55. K. Shitara, N. Hanai, and H. Yoshida, *Cancer Res.* **47,** 1267 (1987).
56. M. Sawada, A. Takada, I. Ohwaki, N. Takahashi, H. Tateno, J. Sakamoto, and R. Kannagi, *Biochem. Biophys. Res. Commun.* **193,** 337 (1993).
57. K. Ohmori, A. Takada, I. Ohwaki, N. Takahashi, Y. Furukawa, M. Maeda, M. Kiso, A. Hasegawa, M. Kannagi, and R. Kannagi, *Blood* **82,** 2797 (1993).
58. E. L. Berg, M. K. Robinson, O. Mansson, E. C. Butcher, and J. L. Magnani, *J. Biol. Chem.* **266,** 14869 (1991).

(*continued*)

TABLE I (*continued*)

59. Y. Fukushi, R. Kannagi, S. Hakomori, T. Shepard, B. G. Kulander, and J. W. Singer, *Cancer Res.* **45,** 3711 (1985).
60. E. Nudelman, Y. Fukushi, S. B. Levery, T. Higuchi, and S. Hakomori, *J. Biol. Chem.* **261,** 5487 (1986).
61. Y. Fukushi, E. Nudelman, S. B. Levery, T. Higuchi, and S. Hakomori, *Biochemistry* **25,** 2859 (1986).
62. U. Galili, C. B. Basbaum, S. B. Shohet, J. Buehler, and B. A. Macher, *J. Biol. Chem.* **262,** 4683 (1987).
63a. J. Hilkens, F. Buijs, J. Hilgers, P. Hageman, J. Calafat, A. Sonnenberg, and M. van der Valk, *Int. J. Cancer* **34,** 197 (1984).
63b. H. C. Gooi, N. J. Jones, E. F. Hounsell, P. Scudder, J. Hilkens, J. Hilgers, and T. Feizi, *Biochem. Biophys. Res. Commun.* **131,** 543 (1985).
64a. D. E. Pettijohn, P. L. Stranhan, C. Due, E. Ronne, H. R. Sorensen, and L. Olsson, *Cancer Res.* **47,** 1161 (1987).
64b. S. Martensson, C. Due, P. Pahlsson, B. Nilsson, H. Eriksson, D. Zopf, L. Olsson, and A. Lundblad, *Cancer Res.* **48,** 2125 (1988).
65. E. D. Nudelman, U. Mandel, S. B. Levery, T. Kaizu, and S. Hakomori, *J. Biol. Chem.* **264,** 18719 (1989).
66. S. Watarai, Y. Kushi, R. Shigeto, N. Misawa, Y. Eishi, S. Handa, and T. Yasuda, *J. Biochem.* (*Tokyo*) **117,** 1062 (1995).

C. Globo- and extended globo-series glycolipids:

1. E. Nudelman, R. Kannagi, S. Hakomori, M. Parsons, M. Lipinski, J. Wiels, M. Fellous, and T. Tursz, *Science* **220,** 509 (1983).
2. H. Kojima, S. Tsuchiya, K. Sekiguchi, R. Gelinas, and S. Hakomori, *Biochem. Biophys. Res. Commun.* **143,** 716 (1987).
3. S. Toda, S. Y. Yokobori, Q. Sen, R. Kannagi, T. Taki, K. Kasama, S. Tanaka, K. Takata, and H. Kimura, Globotriaosylceramide (Gb3) in bull sperm, testis and gonadal fluids," *in* XVIth Int. Symp. Glycoconjugates," Zurich, Switzerland, September 7–12, 1997.
4. M. Kotani, I. Kawashima, H. Ozawa, K. Ogura, T. Ariga, and T. Tai, *Arch. Biochem. Biophys.* **310,** 89 (1994).
5. G. A. Schwarting and J. E. Crandall, *Brain Res.* **547,** 239 (1991).
6. J. V. Madassery, B. Gillard, D. M. Marcus, and M. H. Nahm, *J. Immunol.* **147,** 823 (1991).
7. N. T. Brodin, J. Thurin, K. A. Karlsson, S. Martensson, and H. O. Sjogren, *Int. J. Cancer* **43,** 317 (1989).
8. R. Kannagi, S. B. Levery, S. Ishigami, S. Hakomori, L. H. Shevinsky, B. B. Knowles, and D. Solter, *J. Biol. Chem.* **258,** 8934 (1983).
9. D. M. Marcus, S. Gilbert, M. Sekine, and A. Suzuki, *Arch. Biochem. Biophys.* **262,** 620 (1988).
10. D. S. Schrump, K. Furukawaw, H. Yamaguchi, K. O. Lloyd, and L. J. Old, *Proc. Natl. Acad. Sci. U.S.A.* **85,** 4441 (1988).
11. R. Kannagi, N. A. Cochran, F. Ishigami, S. Hakomori, P. W. Andrews, B. B. Knowles, and D. Solter, *Eur. Mol. Biol. Organ. J.* **2,** 2355 (1983).
12a. S. Saito, S. B. Levery, M. E. K. Salyan, R. I. Goldberg, and S. Hakomori, *J. Biol. Chem.* **269,** 5644 (1994).
12b. M. Satoh, K. Handa, S. Saito, S. Tokuyama, A. Ito, N. Miyao, S. Orikasa, and S. Hakomori, *Cancer Res.* **56,** 1932 (1996).
13a. P. L. Stern, K. R. Willison, E. Lennox, G. Galfre, C. Milstein, D. Secher, and A. Ziegler, *Cell* **14,** 775 (1978).
13b. K. R. Willison and P. L. Stern, *Cell* **14,** 785 (1978).
14. R. Nowinski, C. Berglund, J. Lane, M. Lostrom, I. Bernstein, W. W. Young, Jr., S. I. Hakomori, L. Hill, and M. Cooney, *Science* **210,** 537 (1980).
15. A. Sonnenberg, P. van Balen, T. Hengeveld, G. J. Kolvenbag, R. P. Van Hoeven, and J. Hilgers, *J. Immunol.* **137,** 1264 (1986).
16. J. K. Gathuru, I. Miyoshi, and M. Naiki, *J. Immunol. Methods* **137,** 95 (1991).
17a. E. G. Bremer, S. B. Levery, S. Sonnio, R. Ghidoni, S. Canevari, R. Kannagi, and S. Hakomori, *J. Biol. Chem.* **259,** 14773 (1984).
17b. R. Kannagi, S. B. Levery, and S. Hakomori, *FEBS Lett.* **175,** 397 (1984).

D. Lactoganglio-series glycolipids:

1. K. Shigeta, Y. Kirihata, Y. Ito, T. Ogawa, S. Hakomori, and R. Kannagi, *J. Biol. Chem.* **262,** 1358 (1987).
2. Y. Hirabayashi, S. C. Fujita, K. Kon, and S. Ando, *J. Biol. Chem.* **266,** 10268 (1991).

E. Sulfated glycolipids and carbohydrate determinants:

1. E. M. Eddy, C. H. Muller, and C. A. Lingwood, *J. Immunol. Methods* **81,** 137 (1985).

2. W. Hofstetter, L. Bologa, A. Wetterwald, A. Z'graggen, K. Blaser, and N. Herschkowitz, *J. Neurosci. Res.* **11,** 341 (1984).
3. M. Miyake, T. Taki, R. Kannagi, and S. Hitomi, *Cancer Res.* **52,** 2292 (1992).
4a. K. H. Chou, A. A. Ilyas, J. E. Evans, R. H. Quarles, and F. B. Jungalwala, *Biochem. Biophys. Res. Commun.* **128,** 383 (1985).
4b. V. E. Shashoua, P. F. Daniel, M. E. Moore, and F. B. Jungalwala, *Biochem. Biophys. Res. Commun.* **138,** 902 (1986).
4c. D. K. Chou, A. A. Ilyas, J. E. Evans, C. Costello, R. H. Quarles, and F. B. Jungalwala, *J. Biol. Chem.* **261,** 11717 (1986).
5. E. Borroni, E. A. Derrington, and V. P. Whittaker, *Cell Tissue Res.* **256,** 373 (1989).
6. N. Hiraiwa, N. Iida, I. Ishizuka, S. Itai, K. Shigeta, Y. Fukuda, R. Kannagi, and H. Imura, *Cancer Res.* **48,** 6769 (1988).
7. N. Hiraiwa, Y. Fukuda, H. Imura, K. Tadano-Aritomi, K. Nagai, I. Ishizuka, and R. Kannagi, *Cancer Res.* **50,** 2917 (1990).
8. K. Uchimura, H. Muramatsu, K. Kadomatsu, Q. W. Fan, N. Kurosawa, C. Mitsuoka, R. Kannagi, O. Habuchi, and T. Muramatsu, *J. Biol. Chem.* **273,** 22577 (1998).
9. C. Mitsuoka, M. Sawada-Kasugai, K. Ando-Furui, M. Izawa, H. Nakanishi, S. Nakamura, H. Ishida, M. Kiso, and R. Kannagi, *J. Biol. Chem.* **273,** 11225 (1998).

F. Mucin-type carbohydrate determinants:

1. A. F. Rahman and B. M. Longenecker, *J. Immunol.* **129,** 2021 (1982).
2. B. M. Longenecker, A. F. Rahman, J. B. Leigh, R. A. Purser, A. H. Greenberg, D. J. Willans, O. Keller, P. K. Petrik, T. Y. Thay, and M. R. Suresh, *Int. J. Cancer* **33,** 123 (1984).
3. B. M. Longenecker, D. J. Willans, G. D. MacLean, S. Selvaraj, M. R. Suresh, and A. A. Noujaim, *J. Natl. Cancer Inst.* **78,** 489 (1987).
4. S. Hirohashi, H. Clausen, T. Yamada, Y. Shimosato and S. Hakomori, *Proc. Natl. Acad. Sci. U.S.A.* **82,** 7039 (1985).
5. D. J. Roxby, A. A. Morley, and M. Burpee, *Br. J. Haematol.* **67,** 153 (1987).
6. M. J. King, S. F. Parsons, A. M. Wu, and N. Jones, *Transfusions* **31,** 142 (1991).
7. H. K. Takahashi, R. Metoki, and S. Hakomori, *Cancer Res.* **48,** 4361 (1988).
8. M. Thurnher, H. Clausen, N. Sharon, and E. G. Berger, *Immunol. Lett.* **36,** 239 (1993).
9. Y. Numata, H. Nakada, S. Fukui, H. Kitagawa, K. Ozaki, M. Inoue, T. Kawasaki, I. Funakoshi, and I. Yamashina, *Biochem. Biophys. Res. Commun.* **170,** 981 (1990).
10. J. Huang, J. C. Byrd, B. Siddiki, M. Yuan, E. Lau, and Y. S. Kim, *Dis. Markers* **10,** 81 (1992).
11. A. Kurosaka, S. Fukui, H. Kitagawa, H. Nakada, Y. Numata, I. Funakoshi, T. Kawasaki, and I. Yamashina, *FEBS Lett.* **215,** 137 (1987).
12. T. Kjeldsen, H. Clausen, S. Hirohashi, T. Ogawa, H. Iijima, and S. Hakomori, *Cancer Res.* **48,** 2214 (1988).
13a. M. Nuti, Y. A. Teramoto, R. Mariani-Costantini, P. H. Hand, D. Colcher, and J. Schlom, *Int. J. Cancer* **29,** 539 (1982).
13b. G. F. Springer, P. R. Desai, M. K. Robinson, H. Tegtmeyer, and E. F. Scanlon, *Prog. Clin. Biol. Res.* **204,** 47 (1986).
14. F. G. Hanisch, G. Uhlenbruck, H. Egge, and J. Peter Katalinic, *Biol. Chem. Hoppe Seyler* **370,** 21 (1989).
15. S. Saito, S. B. Levery, M. E. K. Salyan, R. I. Goldberg, and S. Hakomori, *J. Biol. Chem.* **269,** 5644 (1994).
16. Y. Yamashita, Y.S. Chung, K. Murayama, R. Kannagi, and M. Sowa, *J. Natl. Cancer Inst.* **87,** 441 (1995).
17. K. Kumamoto, C. Mitsuoka, M. Izawa, N. Kimura, N. Otsubo, H. Ishida, M. Kiso, T. Yamada, S. Hirohashi, and R. Kannagi, *Biochem. Biophys. Res. Commun.* **247,** 514 (1998).
18. J. R. Turner, A. M. Tartakoff, and N. S. Greenspan, *Proc. Natl. Acad. Sci. U.S.A.* **87,** 5608 (1990).

[b] Listed in ATCC catalog.
[c] Listed in Seikagaku Kogyo catalog.
[d] Listed in Pharmingen catalog.
[e] Listed in Becton-Dickinson catalog.

by monoclonal antibodies. Recently, a disialylated glycolipid of this series was known to have a cell adhesive activity, and the antibody specific to this structure was successfully applied in the cell adhesion experiments.[16]

Antibodies Directed to Lacto- and Neolacto-Series Glycolipids

Lacto- and neolacto-series glycolipids seem to be more immunogenic to mice, and many monoclonal antibodies directed to this series of glycolipids have been derived from mice immunized with tumor cells or pure glycolipids. These antibodies include those directed to fucose-containing lacto-series glycolipids. The anti-SSEA-1 antibody that was raised against F9 cells turned out to recognize Le^x-i hapten,[17] while many other antibodies directed to leukemic cells or lung cancer cells turned out to recognize simply Le^x hapten.[18] Human blood group A antigen carried by a lacto-series carbohydrate structure elicits a good antibody response in mice, but blood group B antigen carried by a similar carbohydrate core seems to be rather poorly immunogenic.

Antigenic epitopes carried by lacto- and neolacto-series glycolipids are expressed also on glycoproteins, and these antibodies are useful for characterization of glycoprotein carbohydrate side chains as well. Certain lacto- and neolacto-series epitopes are expressed strongly on cancer cells of the digestive and respiratory organs, and these antibodies have been used for characterization of cancer-associated carbohydrate determinants.[19] Many antibodies raised against cancer cells turned out to recognize sialyl Le^x and sialyl Le^a, which are known to serve as ligands for the cell adhesion molecules of the selectin family. Anti-sialyl Le^x and anti-sialyl Le^a antibodies having more refined specificity were obtained by immunizing pure and/or synthetic glycolipids. These antibodies are very useful for functional dissection of selectin-mediated cell adhesion.[20–24]

[16] M. Satoh, K. Handa, S. Saito, S. Tokuyama, A. Ito, N. Miyao, S. Orikasa, and S. Hakomori, *Cancer Res.* **56,** 1932 (1996).

[17] R. Kannagi, E. Nudelman, S. B. Levery, and S. Hakomori, *J. Biol. Chem.* **257,** 14865 (1982).

[18] J. L. Magnani, E. D. Ball, M. W. Fanger, S. I. Hakomori, and V. Ginsburg, *Arch. Biochem. Biophys.* **233,** 501 (1984).

[19] S. Hakomori, *Cancer Res.* **56,** 5309 (1996).

[20] J. B. Lowe, L. M. Stoolman, R. P. Nair, R. D. Larsen, T. L. Berhend, and R. M. Marks, *Cell* **63,** 475 (1990).

[21] M. L. Phillips, E. Nudelman, F. C. A. Gaeta, M. Perez, A. K. Singhal, S. Hakomori, and J. C. Paulson, *Science* **250,** 1130 (1990).

[22] G. Walz, A. Aruffo, W. Kolanus, M. Bevilacqua, and B. Seed, *Science* **250,** 1132 (1990).

[23] E. L. Berg, M. K. Robinson, O. Mansson, E. C. Butcher, and J. L. Magnani, *J. Biol. Chem.* **266,** 14869 (1991).

[24] A. Takada, K. Ohmori, N. Takahashi, K. Tsuyuoka, K. Yago, K. Zenita, A. Hasegawa, and R. Kannagi, *Biochem. Biophys. Res. Commun.* **179,** 713 (1991).

Antibodies Directed to Sulfated Glycolipids and Sulfated Carbohydrate Determinants

Antibodies directed to sulfated carbohydrate determinants include those against GlcUA-PG-sulfate, an important determinant expressed on peripheral nerves as well as on NK cells. More recently, antibodies against sulfated Le^x and sulfated sialyl Le^x were raised using synthetic glycolipids as immunogen. The carbohydrate ligands for L-selectin on high endothelial venules of human lymph nodes are identified as sialyl 6-sulfo Le^x using these antibodies.[3]

Antibodies Directed to Mucin-Type Carbohydrate Determinants

These groups contain antibodies directed to important carcinoma-associated antigens such as T, Tn, and sialyl Tn determinants. Recently, antibodies directed to more complex mucin-type determinants were described, such as sialyl Le^x linked to mucin core 2 or core 4 structure.[25]

[25] K. Kumamoto, C. Mitsuoka, M. Izawa, N. Kimura, N. Otsubo, H. Ishida, M. Kiso, T. Yamada, S. Hirohashi, and R. Kannagi, *Biochem. Biophys. Res. Commun.* **247,** 514 (1998).

[15] Immunolocalization of Gangliosides by Light Microscopy Using Anti-Ganglioside Antibodies

By ANDREAS SCHWARZ and ANTHONY H. FUTERMAN

Introduction

The sialic acid-containing glycosphingolipids (GSLs), the gangliosides, play important roles in both neuronal[1] and nonneuronal[2,3] tissues. Most of the proposed functions of gangliosides in neurons are related to their localization at the plasma membrane, where biochemical analyses suggest that they are highly enriched.[4] However, biochemical analysis of the subcellular localization of gangliosides is fraught with problems and depends, among other things, on the biochemical purity of the particular fraction.

[1] R. W. Ledeen and G. Wu, *Trends Glycosci. Glycotech.* **4,** 174 (1992).
[2] S. Hakomori, *Chem. Phys. Lipids* **42,** 209 (1986).
[3] C. B. Zeller and R. B. Marchase, *Am. J. Physiol.* **262,** C1341 (1992).
[4] R. K. Yu and M. Saito, *in* "Neurobiology of Glycoconjugates" (R. U. Margolis and R. K. Margolis, eds.), p. 1. Plenum Press, New York, 1989.

0076-6879/00 $30.00

TABLE I
COMMERCIALLY AVAILABLE ANTI-GANGLIOSIDE ANTIBODIES

Ganglioside or GSL	Commercial source
Monoclonal antibodies	
GD2	Matreya[a]
A2B5	Boehringer-Mannheim
GD3 (clone R24)	Matreya
GalCer	Matreya, Sigma, Boehringer-Mannheim
9-*O*-Acetyl-GD3 (clone JONES)	Sigma
Polyclonal antibodies	
GM1	Matreya
GM2	Matreya
Asialo-GM1	Matreya
Asialo-GM2	Matreya
GL4	Matreya

[a] Matreya, Inc., 500 Tressler Street, Pleasant Gap, PA 16823; fax: (814)359-5062.

Another issue that complicates the interpretation of subcellular fractionation studies from neuronal tissues is the variety of cell types found in the brain (i.e., many different types of both neurons and glia), rendering difficult the determination of the precise association between a particular ganglioside and a specific cell type.

An alternative approach used to localize gangliosides is to use antibodies. However, use of anti-ganglioside antibodies also poses a number of problems, and the current chapter discusses some of the lessons that we have learned during the past few years using these antibodies. A comprehensive review summarizing the extensive but often confusing literature on this subject has recently been published,[5] as have some of our experimental findings.[6]

Factors That Affect Binding of Anti-Ganglioside Antibodies to Gangliosides in Biologic Membranes

Unlike most glycerolipids and sphingolipids (SLs), gangliosides contain antigenic carbohydrate moieties, which has facilitated the production of a large number of polyclonal and monoclonal antibodies against all of the major and most of the minor gangliosides.[5] Some anti-ganglioside antibodies can be purchased from commercial sources, and the antibodies with which we have had the most success are listed in Table I. In addition, many

[5] A. Schwarz and A. H. Futerman, *Biochim. Biophys. Acta* **1286,** 247 (1996).

[6] A. Schwarz and A. H. Futerman, *J. Histochem. Cytochem.* **45,** 611 (1997).

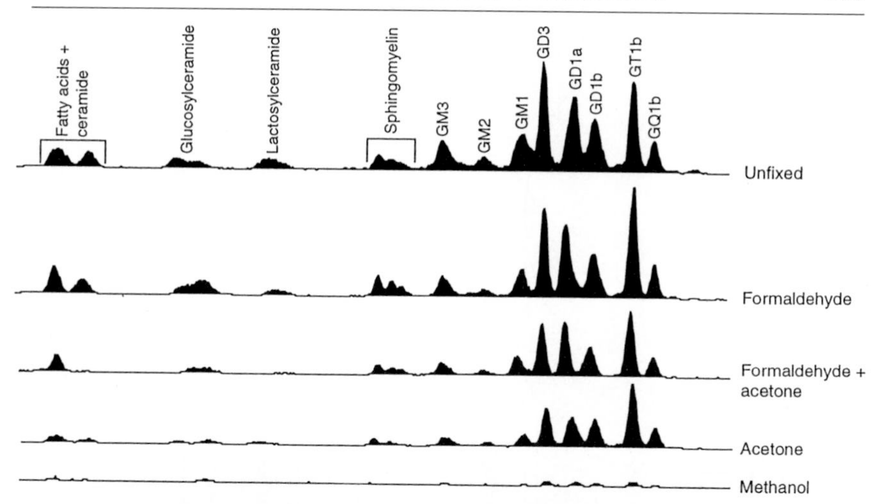

FIG. 1. TLC analysis of cell-associated [^{3}H]sphingolipids and [^{3}H]gangliosides after various fixation methods. The origin of the TLC plate is on the right-hand side. Note that fixation using various fixatives commonly used for immunolocalization results in loss of different gangliosides depending on the fixative, and in some cases, in total loss of gangliosides. [Reproduced with permission from A. Schwarz and A. H. Futerman, Determination of the localization of anti-ganglioside antibodies: comparison of fixation methods. *J. Histochem. Cytochem.* **45,** 611 (1997).]

laboratories have produced polyclonal and monoclonal anti-antibodies, and these are listed in recent reviews.[5,6a]

Two major problems have to be taken into consideration when using anti-ganglioside antibodies. The first, and most crucial, is the choice of fixative. Organic solvents, such as acetone and methanol, are particularly problematic as fixatives because both solubilize gangliosides to various extents. Methanol completely solubilizes gangliosides, at least from cultured neurons, and acetone solubilizes gangliosides by >50% (Fig. 1). Other common fixatives, such as formaldehyde, may extract some gangliosides from membranes more readily than others (Fig. 1). Thus, it is often best to incubate with antibodies prior to fixation (see below).[6]

The second problem concerns ganglioside accessibility in the membrane. Antibodies may be prevented by steric hindrance, due to membrane proteins or glycoproteins, from binding to the antigenic sugar moieties of gangliosides (Fig. 2), and protease treatment has been shown to expose "cryptic" gangliosides in a number of studies. Thus, no labeling with an

[6a] R. Kannagi, *Methods Enzymol.* **312,** [14], (2000) (this volume).

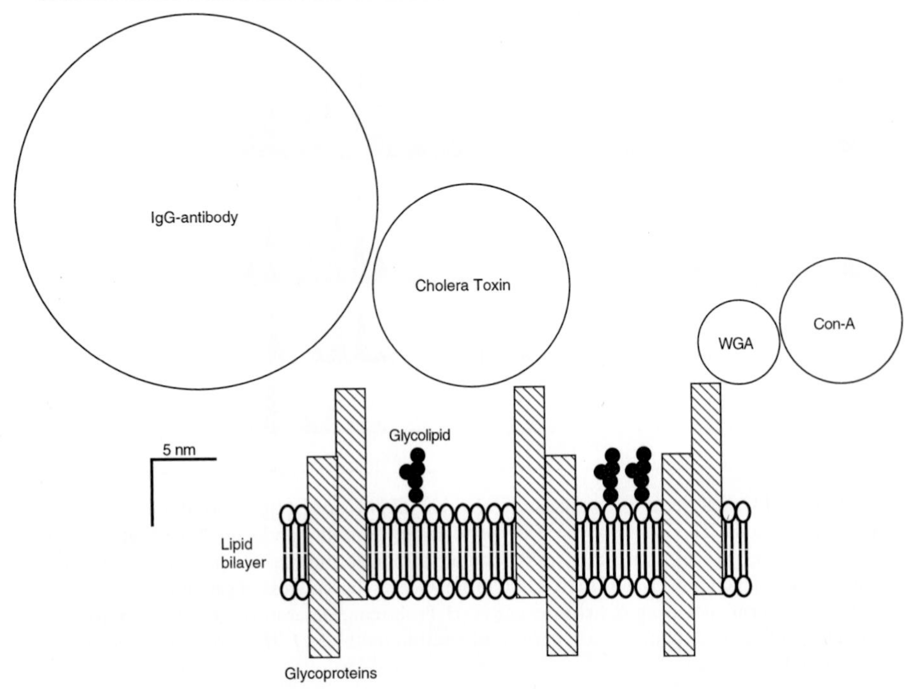

FIG. 2. A comparison of the size of antibodies and other ganglioside-binding molecules. This figure, and the size of the antibodies and other ganglioside-binding molecules, was adapted from Schrével *et al.*[21] The approximate molecular mass of wheat germ agglutinin (WGA) is 36 kDa, of Con-A is 52 kDa, the holoenzyme of *Cholera* toxin is 87 kDa, and IgG is 150 kDa. Glycoproteins and other membrane molecules can hinder the access of anti-ganglioside antibodies to gangliosides, whereas other smaller ganglioside-binding molecules might have more access.

antibody could be detected before treatment with pronase or trypsin,[7,8] and trypsinization of rat retina increased the extent of labeling with an anti-GD3 antibody.[9] The accessibility of gangliosides can also be influenced by other cell-surface glycoconjugates or other SLs.[10]

Other issues also affect antibody binding, although these are less well

[7] G. Uhlenbruck, H. Otten, U. Rehfeldt, U. Reifenger, and O. Prokop, *Allergie Klin. Immunol.* **134,** 476 (1968).

[8] S. Hakomori, C. Teather, and H. Andrews, *Biochem. Biophys. Res. Commun.* **33,** 563 (1968).

[9] N. A. Gregson, A. R. Johnson, D. F. Wheeler, and M. J. Voaden, *Biochem. Soc. Trans.* **17,** 222 (1989).

[10] R. Kannagi, R. Stroup, N. A. Cochran, D. L. Urdal, W. W. Young, and S. Hakomori, *Cancer Res.* **43,** 4997 (1983).

defined. For instance, the reason that three monoclonal antibodies to GM1 bind to totally different areas of the central nervous system is not known.[11] To summarize, great care must be taken when assigning gangliosides to specific cell populations or to intracellular locations unless extensive and careful controls are performed, as discussed below.

Experimental Protocols for Using Anti-Ganglioside Antibodies

We have found that incubation with anti-ganglioside antibodies prior to fixation (prefixation) yields reproducible patterns of immunofluorescence using all of the antibodies that we have tested, whereas incubation with the antibodies after fixation (postfixation) often leads to ambiguities about the localization of a particular ganglioside. We describe below our method for prefixation using formaldehyde as a fixative, and also show some examples of cells labeled using various postfixation techniques, resulting in "artifactual" labeling, i.e., labeling that cannot possibly be ascribed to binding of the antibody to a ganglioside or gangliosides.

It should be stated at the outset that prefixation can normally only be used successfully for monolayers of cultured cells, where fixation and permeabilization are not required for the antibody to gain access to cells that, for instance, are buried deep within a tissue slice. Likewise, prefixation cannot usually be used for electron microscopy. Thus prefixation is essentially limited to light microscopic studies of dissociated cells in culture, and would need to be modified for tissue slices and other biologic specimens. However, some workers have been able to lable tissue slices using prefixation protocols. For instance, Stainier and Gilbert recommend overnight incubation with a primary antibody for unfixed tissue sections, using a slice thickness of about 100–120 μm.[12] Using a postfixation protocol on tissue slices,[13] labeling was observed using an antibody to GD3 on tissue that had been prefixed with formaldehyde, but subsequent addition of either methanol, acetone, ethanol/acetic acid, or Triton X-100 resulted in total loss of labeling.

Preparation of Formaldehyde

In a fume hood, paraformaldehyde (4 g) is added to 60–70 ml of distilled water that has been heated to ~60°. A few drops of 1 *N* NaOH are added, while stirring, to hydrolyze the paraformaldehyde to formaldehyde, until the milky solution turns clear. Subsequently, 10 ml of a 10× phosphate-buffered

[11] H. Laev and S. P. Mahadik, *Neurosci. Lett.* **102,** 7 (1989).

[12] D. Y. Stainier and W. Gilbert, *J. Neurosci.* **9,** 2468 (1989).

[13] R. Reynolds and G. P. Wilkin, *Development* **102,** 409 (1988).

solution (PBS) (82 g NaCl; 4.3 g $NaH_2PO_4 \cdot 2H_2O$; 17.4 g Na_2HPO_4, pH 7.3; in a final volume of 1 liter) and 4 g sucrose are added, and H_2O is added to give a final volume of 100 ml. Any remaining cloudiness can be removed by filtration. The formaldehyde solution can be used fresh or from frozen aliquots. Frozen aliquots should be used only once.[14]

Incubation with Primary Antibodies

Coverslips are washed in the buffer of choice [i.e., Hanks' balanced salt solution (HBSS)] that contains 3% (w/v) bovine serum albumin (BSA) for 5 min, and subsequently incubated directly with primary antibodies (37°, 30 min), diluted in buffer (i.e., HBSS) containing BSA (3%) or normal goat serum (1%).[6,15] The time of incubation can be varied depending on the dilution of the primary antibody; however, we do not recommend less than 10 min or longer than 1 hr, because antibodies may be internalized by endocytosis during long incubations. Endocytosis can be avoided by lowering the temperature to below 15°[16]; however, many cells do not survive such low temperatures. Times and temperatures of incubation with primary antibodies are empirical and need to be determined for each antibody and cell type.

Subsequent to incubation with the primary antibody, cells are fixed with 4% formaldehyde for 30 min at 37°. As for the primary antibodies, incubation conditions with the secondary antibody used for detection are empirical, but since the cells are already fixed, fewer problems are usually encountered. We normally incubate at 37° for 30 min. We recommend secondary antibodies obtained from Jackson Immunoresearch Labs, Inc. (Westgrove, PA) or Boehringer-Mannheim (Germany).

Examples of Immunolocalization of Gangliosides.

Figures 3A and B show an example of a cultured hippocampal neuron labeled using the technique described above, with an antibody to ganglioside GD1b. Labeling can be detected over all of the cell surface, the membrane on each neuronal process can be resolved, and filipodia are labeled along the length of the processes. Similar examples can be found in published studies.[6,15] Figures 3C and D show the same cells, under the same fixation conditions, but first treated with 50 μM PDMP (an inhibitor

[14] K. Goslin and G. Banker, *in* "Culturing Nerve Cells" (G. Banker and K. Goslin, eds.), p. 251. The MIT Press, Cambridge, Massachusetts, 1991.

[15] R. Harel and A. H. Futerman, *J. Biol. Chem.* **268,** 14476 (1993).

[16] A. Sofer and A. H. Futerman, *J. Biol. Chem.* **270,** 12117 (1995).

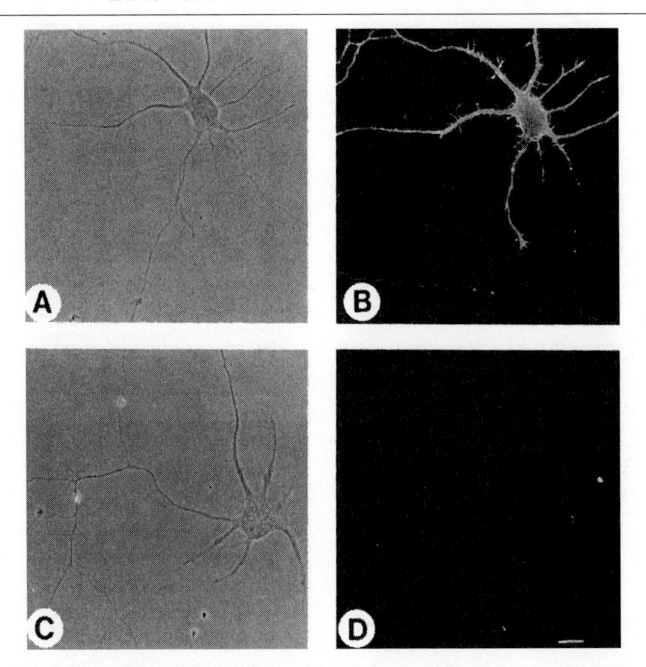

FIG. 3. Effect of depleting ganglioside levels on the binding of an anti-ganglioside antibody to cultured hippocampal neurons. Control neurons (A, B) or neurons that had been treated with 50 μM PDMP[17] for 3 days (C, D) were incubated with an anti-GD1b antibody before fixation (prefixation) on day 3 in culture. (A, C) Phase contrast micrographs; (B, D) fluorescence micrographs. Bar: 10 μm. The anti-GD1b antibody (B17) was obtained in a study examining the antigenic epitopes on mast cells.[22] [Reproduced with permission from A. Schwarz and A. H. Futerman, Determination of the localization of anti-ganglioside antibodies: comparison of fixation methods. *J. Histochem. Cytochem.* **45,** 611 (1997).]

of GSL synthesis)[17] for 3 days prior to immunolabeling. No surface labeling of GD1b is observed,[6] consistent with biochemical data showing that GD1b levels are significantly depleted after this treatment.[18,19] Thus, we have found that depletion of GSLs or SLs by chemical inhibitors is an important control for verifying whether labeling observed with a particular antibody is indeed due to the presence of the ganglioside in the membrane.[6,15] Similar approaches might involve modifying sialic acid residues with, for instance, *Vibrio cholerae* neuraminidase, which converts all higher gangliosides to GM1.

[17] J. A. Shayman and A. Abe, *Methods Enzymol.* **311,** 42–49, (2000).

[18] A. Schwarz and A. H. Futerman, *J. Neurosci,* **17,** 2929 (1997).

[19] A. Schwarz, E. Rapaport, K. Hirschberg, and A. H. Futerman, *J. Biol. Chem.* **270,** 10990 (1995).

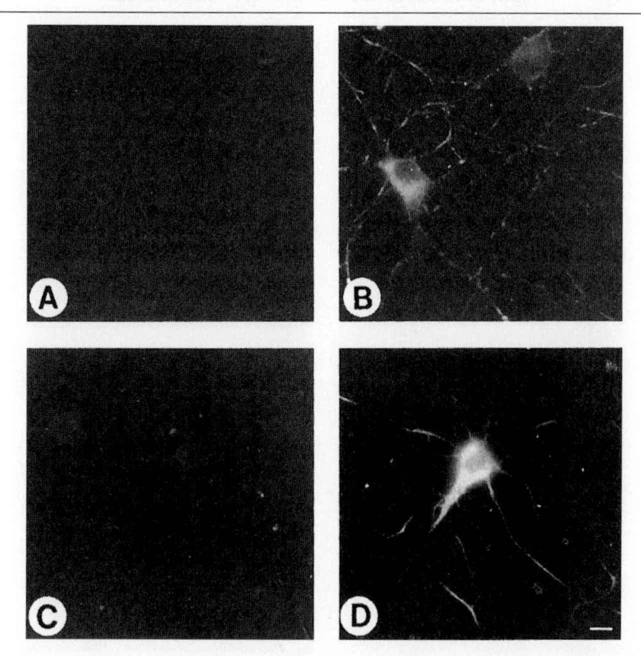

FIG. 4. Effect of fixation using formaldehyde on antibody binding to cultured hippocampal neurons. Neurons were treated with formaldehyde either prefixation (A, C) or postfixation (B, D), and then incubated with an anti-GM4 polyclonal antibody (A, B) or a monoclonal antibody to GD2 (C, D). Bar: 10 μm. Note that neither GM4 nor GD2 is synthesized by hippocampal neurons. [Reproduced with permission from A. Schwarz and A. H. Futerman, Determination of the localization of anti-ganglioside antibodies: comparison of fixation methods. *J. Histochem. Cytochem.* **45,** 611 (1997).]

An example of the problems encountered when incubating with antibodies postfixation can be seen in Fig. 4, using antibodies to GM4 and GD2. Neither of these gangliosides is synthesized in hippocampal neurons,[20] and, indeed, when neurons were incubated with antibodies prefixation, no labeling was observed (Figs. 4A and 4C).[6] However, on incubation with the antibody postfixation, a striped-labeling pattern was observed (Figs. 4B and 4D). Although it could be argued that fixation exposed a cryptic epitope, this cannot be the explanation for this case, since hippocampal neurons do not contain gangliosides GM4 and GD2.

In conclusion, we recommend that optimal fixation techniques be estab-

[20] K. Hirschberg, R. Zisling, G. van Echten-Deckert, and A. H. Futerman *J. Biol. Chem.* **271,** 14876 (1996).

[21] J. Schrével, D. Gros, and M. Monsigny, *Progr. Histochem. Cytochem.* **14,** 1 (1981).

[22] E. Ortega, A. Licht, Y. Biener, and I. Pecht, *Mol. Immunol.* **27,** 1269 (1990).

lished for each anti-ganglioside antibody, and suggest that biochemical analysis (or other nonimmunohistochemical analyses) should be used, where possible, to confirm that a particular ganglioside is present in a particular tissue.

[16] Cloud-Point Extraction of Gangliosides using Nonionic Detergent C14EO6

By GEORG C. TERSTAPPEN, ANTHONY H. FUTERMAN, and ANDREAS SCHWARZ

Introduction

Gangliosides are glycosphingolipids (GSLs) that are highly enriched in the mammalian nervous system.[1,2] These lipids consist of a long-chain base (normally sphingosine or sphinganine) to which a fatty acid is attached via the amino group at carbon-2, and to which a variety of oligosaccharides are attached via an ester bond at carbon-1. The characteristic feature of gangliosides is that at least one (and often more) of the oligosaccharide residues is sialic acid (*N*-acetyl or *N*-glycoloylneuraminic acid).

A number of different methods have been described for ganglioside extraction and isolation.[3–7] However, many of these procedures require time-consuming extraction steps, and in some cases, some gangliosides are lost during extraction. For instance, the method of Folch *et al.*[3] requires long extraction and filtration steps, and less-polar gangliosides such as GM3 and GM4 are difficult to extract quantitatively.[8] Similar problems have also been reported for ganglioside isolation using a three-component solvent

[1] A. Schwarz and A. H. Futerman, *Biochim. Biophys. Acta* **1286,** 247 (1996).

[2] R. K. Yu and M. Saito, *in* "Neurobiology of Glycoconjugates" (R. U. Margolis and R. K. Margolis, eds.), pp. 1–42. Plenum Press, New York, 1989.

[3] J. Folch, M. B. Lees, and G. H. Sloane Stanley, *J. Biol. Chem.* **226,** 497 (1957).

[4] S. Ladisch and B. Gillard, *Anal. Biochem.* **146,** 220 (1985).

[5] G. Tettamanti, F. Bonali, S. Marchesini, and V. Zambotti, *Biochim. Biophys. Acta* **296,** 160 (1973).

[6] G. van Echten-Deckert, *Methods Enzymol.* **312,** [7], (2000) (this volume).

[7] D. Heitmann, M. Lissel, R. Kempken, and J. Müthing, *Biomed. Chromatogr.* **10,** 245 (1996).

[8] M. C. Byrne, M. Sbaschnig-Agler, D. A. Aquino, J. R. Sclafani, and R. W. Ledeen *Anal. Biochem.* **148,** 163 (1985).

0076-6879/00 $30.00

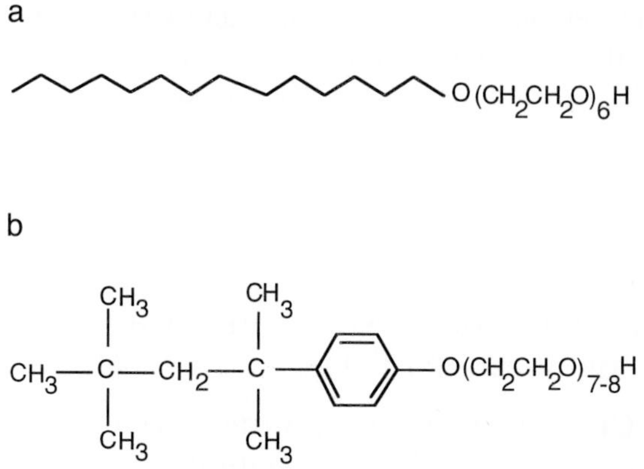

a

$O(CH_2CH_2O)_6H$

b

$CH_3-C(CH_3)_2-CH_2-C(CH_3)_2-C_6H_4-O(CH_2CH_2O)_{7\text{-}8}H$

FIG. 1. Structures of the polyoxyethylene detergents. (a) C14EO6 and (b) Triton X-114.

system comprised of diisopropyl ether/1-butanol/aqueous 50 m*M* NaCl.[4,9] An alternative approach using tetrahydrofuran[5] is more efficient than the method of Folch, but ganglioside fractions are contaminated by high levels of phospholipids.[10] One of the most efficient methods for extracting gangliosides uses $CHCl_3/CH_3OH/H_2O$/pyridine, but extraction requires incubating for 2 days at 50°.[6,11]

In this chapter, we describe a novel detergent-based method for ganglioside isolation, based on "cloud-point" extraction, that is much less time consuming than the techniques described above; extraction, purification, and TLC analysis can be performed within 1 day.[12] Moreover, the procedure does not require the use of toxic solvents for extraction. Before describing a detailed protocol, we briefly discuss the general principles of cloud-point extraction and the characteristics of the polyoxyethylene detergent, C14EO6, used in this procedure.

Principles of Cloud-Point Extraction

Nonionic polyoxyethylene detergents are comprised of polar ethylene oxide moieties that render them relatively water soluble. The hydrophobic moiety usually consists of an alkyl chain (as is the case for C14EO6, Fig. 1a), which may also contain an aromatic residue (as is the case for the

[9] H.-J. Senn, M. Orth, E. Fitzke, H. Wieland, and W. Gerok, *Eur. J. Biochem.* **181,** 657 (1989).

[10] R. W. Ledeen and R. K. Yu, *Methods Enzymol.* **83,** 139 (1982).

[11] G. van Echten, H. Iber, H. Stotz, A. Takatsuki, and K. Sandhoff, *Eur. J. Cell Biol.* **51,** 135 (1990).

[12] A. Schwarz, G. C. Terstappen, and A. H. Futerman, *Anal. Biochem.* **254,** 221 (1997).

Triton series of detergents, Fig. 1b). These detergents generally form clear micellar solutions in water above the critical micellar concentration (CMC), but on increasing the temperature, the solution becomes turbid at the "cloud point," followed by separation into two distinct phases. One of the phases, the "coacervate" phase, is smaller in volume than the aqueous phase, and contains most of the detergent, whereas the aqueous phase is largely detergent depleted.[13] The reason for this behavior has been ascribed to the reversible dehydration of the polar ethylene oxide headgroups of the detergent on increasing temperature.[14] The appearance of turbidity and subsequent phase separation are probably due to the formation of large detergent aggregates, such as micelles and lamellar structures.[15,16] Any component that binds to the micellar aggregates is extracted from the original solution and becomes concentrated in the coacervate phase. Cloud-point extraction has been used for isolating a wide variety of different materials including metal chelates, biomaterials such as hydrophobic proteins and vitamins, as well as organic compounds. Their use has recently been reviewed.[17]

Use of Nonionic Polyoxyethylene Detergent, C14EO6, for Cloud-Point Extraction

C14EO6 (hexaethylene glycol mono-*n*-tetradecyl ether) has been shown to be the most effective nonionic polyoxyethylene detergent, including the widely used Triton X-114, for cloud-point extraction of lipophilic proteins.[18] C14EO6 contains 14 carbon atoms in its linear alkyl chain, and an average of 6 polar ethylene oxide units (Fig. 1a). The cloud point of C14EO6 is 35° [1% (w/v) in water], which is compatible with the extraction of thermally labile materials such as proteins. The CMC is 0.0005% (w/v), and it exhibits a good solubility in water [$>$10% (w/v)]. The coacervate phase is less dense (0.990 g/cm^3) than the detergent-depleted phase (0.992 g/cm^3),[19] which greatly facilitates its recovery and prohibits potential cross-contamination. This is in contrast to the Triton series of detergents, in which the coacervate phase is the lower phase in the two-phase system. Moreover, in the presence

[13] T. Nakagawa, *in* "Nonionic Surfactants" (M. J. Schick, ed.), p. 558. M. Dekker, New York, 1966.

[14] K. Shinoda, T. Nakagawa, B.-I. Tamamushi, and T. Isemura, "Colloidal Surfactants: Some Physicochemical Properties." Academic Press, New York, 1963.

[15] J. M. Di Meglio, L. Paz, M. Dvolaitzky, and C. Taupin, *J. Phys. Chem.* **88,** 6036 (1984).

[16] F. Kopp and R. Heusch, *Prog. Colloid Polymer Sci.* **73,** 146 (1987).

[17] W. L. Hinze and E. Pramauro, *Crit. Rev. Anal. Chem.* **24,** 133 (1993).

[18] G. C. Terstappen, R. A. Ramelmeier, and M.-R. Kula, *J. Biotech.* **28,** 263 (1993).

[19] G. C. Terstappen, A. J. Geerts, and M.-R. Kula, *Biotechnol. Appl. Biochem.* **16,** 228 (1992).

of salts, the density difference between the phases becomes even more pronounced, facilitating phase separation (which is driven by the density difference between the phases, according to Stokes law). All of these features render C14EO6 well suited for cloud-point extraction of hydrophobic materials.

C14EO6 can be obtained in different grades of chemical homogeneity. Chemical synthesis usually results in formation of mixtures with respect to the ethylene oxide (EO) moieties, and the given number of EO units normally represents the average number of these polar groups in a particular batch of the detergent. Subsequent purification using fractionated distillation or chromatography is necessary to obtain a completely homogenous product. The C14EO6 used in the experiments described in this chapter was obtained from Henkel KGaA, COF Specialities, D-40191 Duesseldorf, Germany; phone: (49)-211-797-2441, fax: (49)-211-798-3604 (the trade name is Agrimul NRE1406). It exhibits a narrow range distribution of EO units with an average of 6, and is homogeneous with respect to alkyl chain length (C14).

Detergents like C14EO6 should be stored at 4° in the dark to avoid radical formation and subsequent degradation. A working solution of 4% (w/v) is prepared in distilled water and stirred for about 2 hr at room temperature to achieve complete solubilization. This solution can be stored at room temperature for several weeks. The integrity of the solution is determined by analyzing the cloud point, by heating an aliquot of a 1% (w/v) detergent solution in a water bath until the solution becomes turbid. The solution is then allowed to cool down until it becomes clear. The two respective temperatures are measured (usually they differ by no more than 0.5°) and the average is the cloud point.

The cloud point of a detergent solution can be manipulated in either direction by various substances, such as salts.[20] The effect of salt usually follows the lyotropic or Hofmeister series,[21] e.g., phosphates and sulfates lower the cloud point, whereas thiocyanates and iodides increase it. For instance, the addition of 600 m*M* NH_4SCN to a solution of 1% Triton X-114 leads to an increase in the cloud point of 21°, whereas the addition of 400 m*M* $(NH_4)_2SO_4$ results in a decrease of about 23°. In the case of C14EO6, 400 m*M* $(NH_4)_2SO_4$ decreases the cloud point by 12°, whereas ethanol can be used to increase the cloud point (Fig. 2). Thus, the cloud point and the temperature needed to achieve phase separation can be easily manipulated.

[20] A. R. Ramelmeier, G. C. Terstappen, and M.-R. Kula, *Bioseparation* **2,** 315 (1991).

[21] P. H. Hippel, *in* "Structure and Stability of Biological Macromolecules" (S. N. Timasheft and G. D. Fasman, eds.), p. 417. M. Dekker, New York, 1969.

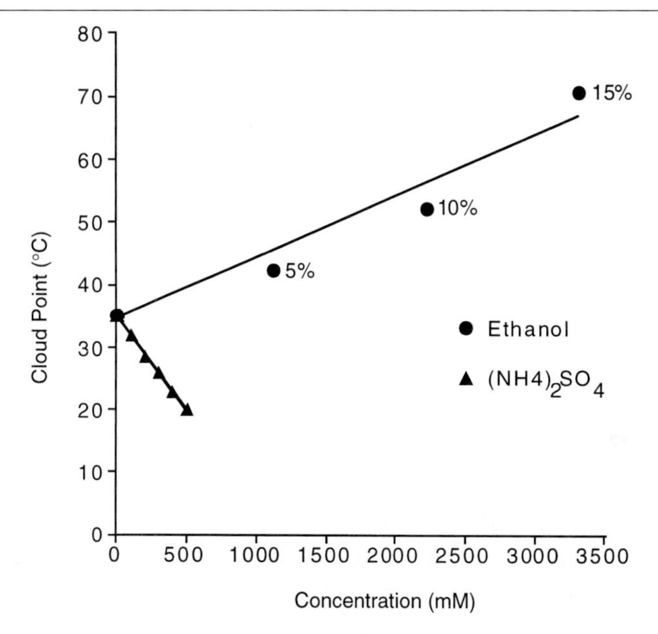

FIG. 2. Manipulation of the cloud point of C14EO6 by two different additives. The concentration of the detergent was 1% (w/v, in water). For ethanol, the concentration is also indicated as a percentage (v/v).

Protocol for Cloud-Point Extraction of Gangliosides from Cultured Cells and Tissue Samples

The protocol described below has, to date, been successfully used for ganglioside isolation from primary cultures of hippocampal neurons, and from rat cerebellar tissue.[12] A schematic outline of the procedure is shown in Fig. 3.

Cloud-Point Extraction from Cultured Cells

When small amounts of material are used, it is sometimes necessary to lyophilize cell material to avoid loss. Using primary cultures of neurons, we routinely remove cells from coverslips by scraping with a rubber policeman, centrifuge, and lyophilize the cell pellet; however, cloud-point extraction can also be performed on cells harvested directly from culture dishes.[22] The dry material is then resuspended in 0.5 ml of 4% C14EO6 and sonicated for 2 min prior to addition of 1.5 ml Tris-HCl (10 m*M*, pH 7.0) containing 270 m*M* $(NH_4)_2SO_4$. This mixture is briefly vortexed and sonicated again

[22] T. Minuth, J. Thoemmes, and M.-R. Kula, *Biotechnol. Appl. Biochem.* **23,** 107 (1996).

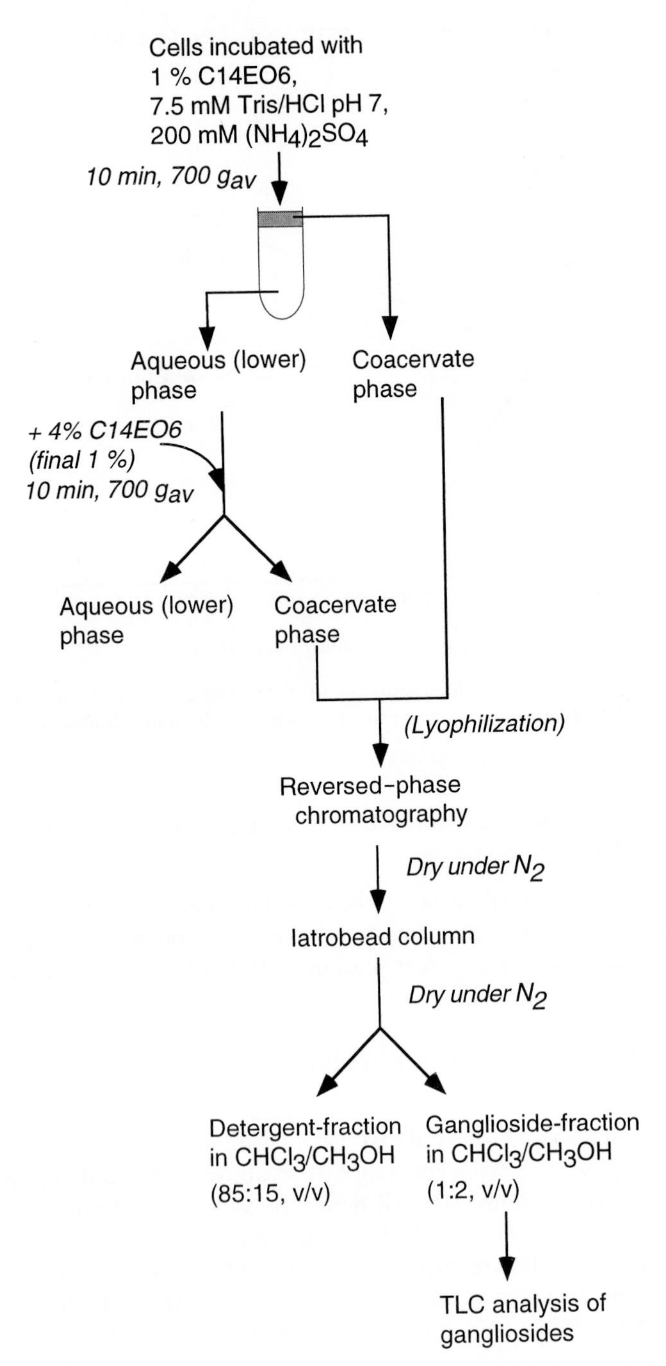

FIG. 3. Schematic diagram of a typical procedure used for ganglioside extraction from cell material. For more details, see text.

for 1 min. To achieve phase separation, the solution is briefly warmed at 37° in a water bath until clouding occurs (about 1 min) and immediately centrifuged for 10 min at room temperature ($700g_{av}$). The upper (coacervate) phase (about 0.3 ml volume), which contains the gangliosides, is carefully collected and stored at 4°. The lower phase is reextracted with detergent to increase the overall yield of gangliosides. For this purpose, C14EO6 is added to the lower phase to give a final concentration of 1%. After brief vortexing, phase separation is performed as described above. The upper phases are combined and stored at 4°.

Notes. (1) The concentration of $(NH_4)_2SO_4$ was determined empirically, and the working concentration may have to be adjusted in order to permit clouding of the solution at a convenient temperature. Note, however, that increasing the amount of $(NH_4)_2SO_4$ results in a decrease in the volume of the coacervate phase. For instance, addition of 200 m*M* $(NH_4)_2SO_4$ to a solution of 1% C14EO6 decreases the relative volume of the coacervate phase from 0.24 to 0.15, i.e., from 10 ml of starting detergent solution, 2.4 ml of coacervate phase is obtained at 37°, and 1.4 ml in the presence of the salt. This is important when small sample volumes are extracted because a very small coacervate phase may be difficult to collect. Increasing the temperature above the cloud point to induce phase separation has a similar effect. (2) Although phase separation is normally accelerated by brief centrifugation, it also occurs simply by letting the solution stand at 37° in a water bath for about 1–2 hr.

Cloud-Point Extraction of Tissue Samples

Tissue (~1 g of material) is directly homogenized in 1 ml C14EO6 (4%, w/v). Subsequently, 13 ml C14EO6 (10%, w/v, in 10 m*M* Tris-HCl, pH 7.0) containing 200 m*M* $(NH_4)_2SO_4$ is added to the homogenate. Tissue debris is removed by passing the homogenate through a fine nylon mesh, and the solution is then vortexed for 1 min. Phase separation is achieved by warming the solution at 37° followed by centrifugation, as described above. The upper coacervate phase is collected (volume about 1.5 ml), and the lower phase reextracted one or two times with C14EO6 to increase ganglioside yield. The combined coacervate phases may be lyophilized in order to reduce the volume.

Notes. The remarks made above are also valid here. Moreover, depending on the type and amount of tissue used, the volume and concentration of detergent required for homogenization and cloud-point extraction can be adjusted, i.e., more tissue needs more detergent. Nevertheless, this method is most useful for small amounts of material, because the volume required for extraction increases significantly as the amount of tissue increases.

Removal of $(NH_4)_2SO_4$ and Detergent by Chromatography on RP-18 and Iatrobead RS6-8060 Columns

Because detergent and salts disturb subsequent ganglioside analysis by TLC, the ganglioside-containing coacervate phase must be further purified. We have found that the most convenient and rapid way to remove salts and detergent is to use RP-18 and Iatrobead RS6-8060 columns, respectively. These techniques have been described in detail by others[6] and are described briefly below.

Desalting of samples is achieved by reversed-phase chromatography on RP-18 columns (Merck, Germany). Samples (e.g., 0.3 ml) are resuspended in 1 ml methanol and sonicated for 1 min. Subsequently, 1 ml of 300 m*M* CH_3COONH_4 is added and sonication repeated to reduce the amount of nondissolved salt. Samples are then applied to an RP-18 column; quantitative recovery of gangliosides is achieved by rinsing the tube with 0.5 ml methanol/300 m*M* CH_3COONH_4 (1:1, v/v) prior to application to the column. Salts are eluted with 6 ml distilled water and lipids subsequently eluted with 1 ml methanol followed by 8 ml $CHCl_3$/methanol (1:1, v/v). The lipid-containing fraction is dried under N_2 and dissolved in 3 ml $CHCl_3$/methanol (85:15, v/v).

Detergent is removed by using an Iatrobead RS6-8060 column (Iatron Lan., Inc., Tokyo, Japan), which is prepared by pouring the bead material to a height of about 2 cm in a Pasteur pipette. The column is equilibrated with about 5 ml $CHCl_3$/methanol (1:1, v/v) followed by 5 ml $CHCl_3$/methanol (85:15, v/v). Samples are applied to the column and detergent eluted with ~2 ml $CHCl_3$/methanol (85:15, v/v). Gangliosides are eluted with 2 × 10 ml $CHCl_3$/methanol (1:2, v/v) and concentrated by drying with N_2. If necessary, detergent concentration can be determined as described elsewhere.[23]

Notes. (1) The various steps of column chromatography usually do not lead to a significant loss of gangliosides. Volumes may have to be adjusted depending on the sample volume used. (2) If necessary, phospholipids and 9-*O*-acetylated gangliosides can be degraded by dissolving the extracts in a suitable volume of 100 m*M* methanolic NaOH (e.g., 5 volumes per volume of coacervate phase) and incubated for 2 hr at 37° under constant stirring.[6]

TLC Analysis of Gangliosides

Gangliosides are separated by TLC as described.[6] Generally, we use silica gel plates (e.g., Merck, Germany) and $CHCl_3$/methanol/9.8 m*M* $CaCl_2$

[23] G. C. Terstappen and M.-R. Kula, *Anal. Lett.* **23,** 2175 (1990).

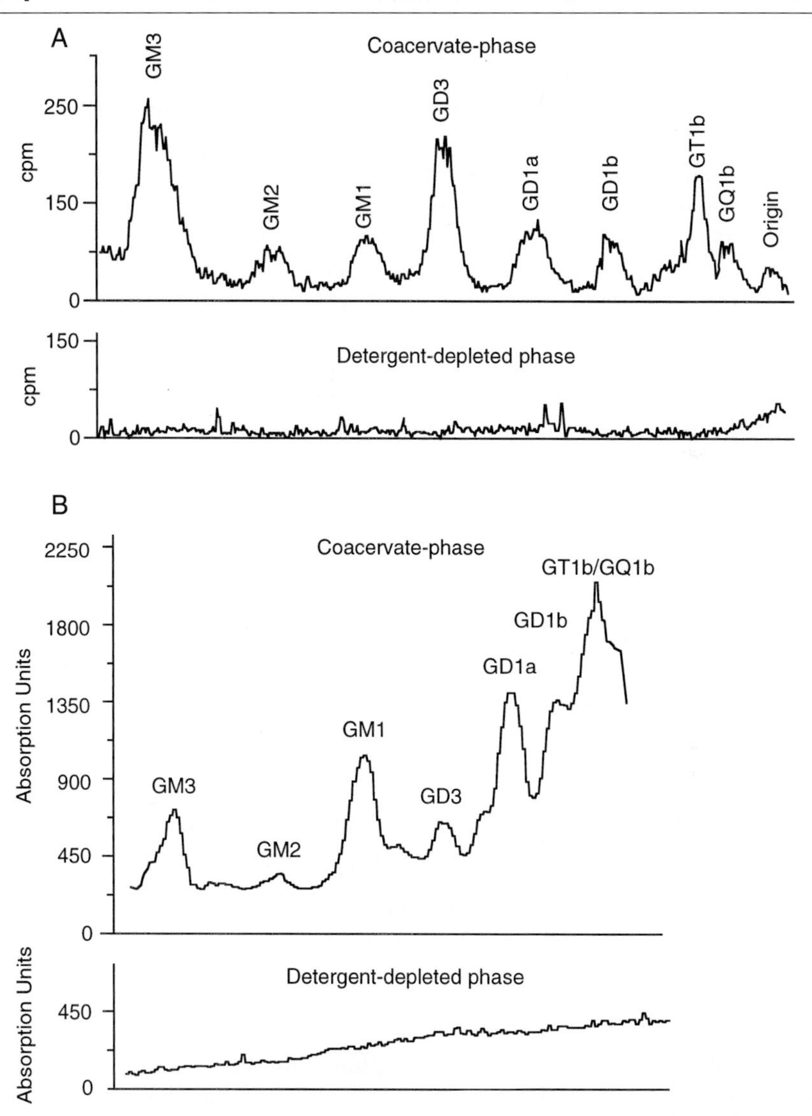

FIG. 4. (A) [^{3}H]ganglioside profile of cultured hippocampal neurons obtained after cloud-point extraction with C14EO6. *Top:* TLC analysis of gangliosides recovered in the coacervate phase. *Bottom:* Gangliosides recovered in the detergent-depleted phase. (B) Densitometric scan of resorcinol-stained gangliosides obtained after cloud-point extraction of rat cerebellum. *Top:* Gangliosides in the coacervate phase. *Bottom:* Gangliosides in the detergent-depleted phase.

(60 : 35 : 8, v/v/v) as the developing solvent.[24] Gangliosides can be visualized by staining with resorcinol, i.e., spraying the plates with 1 ml resorcinol (2%)/8 ml HCl (37%)/1 ml H_2O/0.05 ml 0.1 *M* $CuSO_4$, followed by heating at 100° for 10 min.

An example of cloud-point extraction of gangliosides from cultured hippocampal neurons and from rat cerebellar tissue, using the methods described in this chapter, are shown in Figs. 4A and 4B, respectively. As can be seen, this method is extremely efficient for ganglioside extraction because recovery in the coacervate phase is essentially quantitative. Importantly, less-polar gangliosides such as GM3 are also efficiently recovered. Thus, this fast and easy, solvent-free extraction method is a suitable alternative to other methods of ganglioside extraction that are much more time consuming and require toxic solvents.

Acknowledgment

The authors would like to thank Henkel KGaA, Duesseldorf, Germany, for supplying C14EO6.

[24] K. Hirschberg, R. Zisling, G. van Echten-Deckert, and A. H. Futerman, *J. Biol. Chem.* **271,** 14876 (1996).

[17] Analyses of Glycosphingolipids Using Clam, *Mercenaria mercenaria,* Ceramide Glycanase

By SARA DASTGHEIB, SHIB S. BASU, ZHIXIONG LI, MANJU BASU, and SUBHASH BASU

Introduction

Carbohydrate components of glycosphingolipids (GSLs) play many important biologic roles.[1,2] To investigate the structure–function relationships in biologically active GSLs, it is frequently necessary to release intact carbohydrate moieties from GSLs. The GSL degrading enzyme, called endoglycoceramidase (EGCase) or ceramide glycanase (CGase), is a new endo-type glycosidase that catalyzes the hydrolysis of the glycosidic linkage between oligosaccharides and ceramides of a variety of GSLs. CGases have

[1] R. Dwek, D. R. Wing, A. C. Willis, and A. N. Barclay, *Glycobiology* **3,** 339 (1993).
[2] R. L. Schnaar, *Glycobiology* **1,** 477 (1991).

0076-6879/00 $30.00

so far been discovered from the following sources: bacteria,[3,4] leeches,[5,6] earthworm,[7] oysters,[8] rat,[7–10] and rabbit mammary tissues.[11,12] Our recent report on CGase from edible clam, *Mercenaria mercinaria,*[13,14] provides for an inexpensive available source of enzyme to be purified and used for GSL structural investigation. The present article includes both purification and application of clam CGase using the flurorophore-assisted carbohydrate electrophoresis (FACE) technique.

Preparation of Purified Clam Ceramide Glycanase

All steps for enzyme preparation are conducted at 4°. In a typical preparation, livers from 3–5 dozen cherrystone clams are rinsed with distilled water, and homogenized by Polytron in 100–150 ml in 50 m*M* citrate–phosphate buffer (McIlvaine's buffer) at pH 5.5. The homogenate is dialyzed against homogenizing buffer with two changes. An acid-precipitable material obtained during dialysis is removed by centrifugation at 100,000*g* for 1 hr and the supernatant was heat-treated at 60° for 5 minutes and the heat-treated supernatant was obtained after centrifuging at 100,000*g* for 1 hour. This heat-treated supernatant was subjected to 70% ammonium sulfate saturation. The precipitates were collected and dissolved in 20 m*M* sodium phosphate buffer (pH 7.0) and subjected to overnight dialysis against 20 m*M* sodium phosphate buffer (pH 7.0). This supernatant is referred to hereafter as clam acid and heat-treated supernatant (CAHS).[14]

Purification of Clam CGase by Ion Exchange, Hydrophobic, and Dye Matrices

The purification scheme of the clam CGase protein is shown in Table I. Acid and heat-treated clam CGase is bound to Q-Sepharose (Amer-

[3] M. Ito and T. Yamagata, *J. Biol. Chem.* **261,** 4278 (1986).
[4] M. Ito and T. Yamagata, *J. Biol. Chem.* **264,** 9510 (1989).
[5] S.-C. Li, R. DeGasperi, J. E. Muldrey, and Y. T. Li, *Biochem. Biophys. Res. Commun.* **141,** 346 (1986).
[6] Z. Bing, S. C. Li, R. D. Lain, R. T. C. Huang, and Y.-T. Li, *J. Biol. Chem.* **264,** 12272 (1989).
[7] M. Ito and T. Yamagata, *Biochem. Biophys. Res. Commun.* **162,** 1093 (1989).
[8] N. V. Pavlova, K. Noda, S.-C. Li, and Y.-T. Li, *Glycoconjugate J.* **10** (1994).
[9] M. Basu, M. Girzadas, S. Dastgheib, J. Baker, R. Federica, N. Radin, and S. Basu, *IJBB* **34,** 142 (1996).
[10] M. Basu, S. Dastgheib, J. Baker, and S. Basu, *Glycoconjugate J.* **12,** 439 (1995).
[11] C. Westervelt, J. W. Hawes, K. K. Das, M. Basu, M. J. Beuter, A. Shukla, and S. Basu, *Glycoconjugate J.* **6,** 386 (1989).
[12] M. Basu, S. Dastgheib, M. A. Girzadas, P. H. O'Donnell, C. W. Westervelt, Z. Li, J.-I. Inokuchi, and S. Basu, *Acta Biochim. Pol.* **45,** 327 (1998).
[13] I. Concha-Slebe, K. Presper, T. De, and S. Basu, *Carbohyd. Res.* **155,** 73 (1986).
[14] S. Ghosh, S. Lee, T. Brown, M. Basu, J. Hawes, D. Davidson, and S. Basu, *Anal. Biochem.* **196,** 252 (1991).

TABLE I
PURIFICATION OF CERAMIDE GLYCANASE FROM CLAM, *Mercenaria Mercenaria*, HEPATOPANCREAS

Fraction	Volume (ml)	Total protein (mg)	TA (nmol/hr)	SA (nmol/mg/hr)	Purity (fold)	Yield (%)
Clam supernatant (CSDAH)	50	2390	32.2	0.014	1	100
$(NH_4)_2SO_4$ precipitation	4.5	2103	29.2	0.016	1.1	118
Sephacryl S-100	16	53	20.6	0.39	28	64
Green 19	6	5	14.9	3.11	222	46
Q-Sepharose	4	0.96	9.8	10.2	729	30
Octyl-Sepharose	3	0.12	1.5	12.4	885	5

sham Pharmacia Biotech; Piscataway, NJ) column at neutral pH and is eluted in a batchwise manner using a gradient of increasing NaCl concentrations (0.1–1.0 *M*). The majority of the contaminating proteins did not bind to the anionic resin. Most of the CGase activity was eluted at a salt concentration of 0.5 *M*. CGase was very weakly retained by the S-Sepharose column (Amersham Pharmacia Biotech; Piscataway, NJ) at pH 4.0 and eluted in 0.1 *M* NaCl fraction.[14,15]

In addition to phenyl-Sepharose, octyl-Sepharose (Amersham Pharmacia Biotech; Piscataway, NJ) matrices are quite effective for hydrophobic interaction with clam CGase. The enzyme is bound on the hydrophobic column in the presence of high salt concentration (1 *M*); clam CGase is more efficiently bound to octyl-Sepharose in the presence of 1 *M* phosphate ions, than in the presence of acetate and chloride ions.[15] The average recovery of clam CGase activity from the octyl-Sepharose column is about 50%.[15] Among six different dyes tested, the matrix bound to Green 19 dye (Sigma, St. Louis, MO) is most effective as an affinity column for further purification of clam CGase. The majority of the contaminating proteins are eluted in high ionic strength buffers (0.6–1 *M* NaCl), whereas purified CGase is eluted with 0.4–0.5 *M* NaCl. The final overall purification of clam CGase is 855-fold with a recovery of 5%. SDS–PAGE analysis[15] of the purified CGase revealed two closely moving bands at 64 and 57 kDa.

Enzyme Assay

The complete incubation mixture contains the following components in the final volume of 0.05 ml: 10 μl of 1.0 *M* sodium acetate buffer,

[15] S. Dastgheib, Ph.D. Thesis, University of Notre Dame (1997).

(pH 4.0), glycosphingolipid substrate (nLcOse5[^{3}H]ceramide), 1 nmol (23,000 cpm); sodium taurodeoxycholate (100 μg), and an appropriate amount (100 μg) of CGase to maintain a 1:1 detergent-to-protein ratio. After incubation at 37° for 4 hr, the reaction is terminated by freezing at −18°. Then 50 μl of heptane and 50 μl 2-propanol are added and the reaction mixture is vortexed and then centrifuged at 1500*g* for 5 min to obtain a distinct phase separation. The organic phase (upper layer) contains the released [^{3}H]ceramide or Glc-Cer, and the aqueous phase (lower layer) contains nLcOse5[^{3}H]Cer and oligosaccharides (di to higher) from other glycolipids. The organic layer (40 μl) is spotted directly on SG-81 paper and subjected to descending paper chromatography (DPC)[16] using $CHCl_3/CH_3OH$ (9:1, v/v) as the developing solvent.[7,15] The [^{3}H]ceramide obtained from CGase-catalyzed reaction migrated at R_f of 0.7–0.8 and can be identified by comparison with standards. The paper is cut into pieces and the [^{3}H]ceramide area is quantitated by a toluene-based liquid scintillation system.

Properties of Purified Clam CGase

Among the various detergents tested, the anionic detergent taurodeoxycholate (TDC), at a protein-to-detergent ratio of 1:1, yielded the highest activity. Note that the anionic detergent, sodium cholate, gives optimum activity with earthworm[17] and leech[6,18] CGase; however, only 10–20% activity was obtained with sodium cholate and clam CGase.

Clam CGase activity is not inhibited by EDTA (16 m*M*), indicating that CGase enzyme activity was not dependent on the presence of a metal ion. Among an array of divalent metals analyzed (Ca^{2+}, Co^{2+}, Zn^{2+}, Ni^{2+}, Mg^{2+}, and Mn^{2+})[18], clam CGase activity was inhibited (60–80%) in the presence of 1–5 m*M* Hg^{2+}, Cu^{2+}, Ag^{2+}, or Fe^{2+} similar to rat[9] or rabbit[12] CGase activities. This implies a role for a thiol group at the enzymatic active site. *p*-Hydroxymercuribenzoate and thimerosal completely abolished CGase activity. NaCl also inhibits at high concentrations, and removal of salt (NaCl) after column elution is necessary for measuring accurate CGase activity (70% inhibition is noted in the presence of 0.4 *M* NaCl).

An optimum pH of 4.25 was observed with the purified clam CGase fraction obtained from Green 19 affinity chromatography. The same frac-

[16] M. Basu, T. De, K. K. Das, J. W. Kyle, H. C. Chon, R. J. Schaeper, and S. Basu, *Meth. Enzymol.* **138,** 575 (1987).

[17] B.-Z. Carter, S.-C. Li, and Y.-T. Li, *J. Biol. Chem.* **285,** 619 (1992).

[18] B.-Z. Carter, S.-C. Li, Y.-T. Li, R. A. Laine, and R. T. C. Huang, *J. Biol. Chem.* **264,** 12272 (1989).

tion showed an isoelectric point (p*I*) value of 4.8 similar to the CGases from the other sources.[17,18]

Clam CGase is a heat-stable enzyme: almost 90% of the activity is retained when it is heated at 100° for 10 min. This unique characteristic of clam CGase was used as an initial step in its purification scheme to precipitate and eliminate contaminating proteins. The most effective stabilizer for clam CGase activity at 4° and −20° is the addition of sucrose (0.25 *M*). The second most effective preservative of activity is maintaining CGase at a pH of 3.5–5.0 (0.2 *M* citrate–phosphate buffer). Other conventional protein stabilizers such as polyethylene glycol (PEG) 3000 and glycerol did not seem to be effective. Freezing and thawing of purified clam CGase protein without cryoprotectants results in a rapid loss of activity.[15]

Glycosphingolipid Substrate Specificity of CGase from Clam

The kinetic parameter is examined using a variety of glycosphingolipids (globo-, lacto-, and gangliosides), as shown in Table II with purified CGase from clam as the enzyme source. In the globo series of GSLs, GbOse4[^{3}H]Cer shows a higher K_m (132 μM) and a slightly higher V_{max} than the rates of GbOse5[^{3}H]Cer. The substrates examined in the lacto series of GSLs (nLcOse4[^{3}H]Cer and nLcOse5[^{3}H]Cer) appear to have

TABLE II
GLYCOSPHINGOLIPID SUBSTRATE SPECIFICITY STUDIES WITH PURIFIED CGASE FROM CLAM[a]

Glycosphingolipid	K_m (app) (μM)	V_{max} (μmol/ml/hr)
GgOse4[^{3}H]Cer	31	0.95
GbOse4[^{3}H]Cer	132	0.57
GbOse5[^{3}H]Cer	13.7	0.20
nLcOse4[^{3}H]Cer	32	0.78
nLcOse5[^{3}H]Cer	34	3.8
GM1[^{3}H]Cer	62	4.0
GM2[^{3}H]Cer	62	3.5
GM3[^{3}H]Cer	5.5	0.29

[a] An octyl-sepharose eluent fraction containing purified CGase activity from clam was used as the enzyme source. Specific activities of the substrates were maintained between 10–20 × 10^6 cpm/mmol. CGase activities were assayed as described in the text. The kinetic parameters of each substrate were determined from their Lineweaver–Burke plots using EZ-FIT program, version 2.02, by Perrella Scientific Inc., 1989.

fairly similar K_m values (32 and 34 μM); however, the V_{max} is 5 times higher for mLcOse5-Cer. The K_m values of the ganglio series of GSL are similar for both GM1 and GM2 ($\approx$ 62 μM), and lower (5.5 μM) for GM3. Overall, there is an increasing trend in the V_{max} values as the oligosaccharide chain length is increased. The kinetic parameters of each substrate were determined from their Lineweaver–Burke plots.[15]

Isolation and Visualization of Oligosaccharides after Hydrolysis by Clam CGase

Glycosphingolipids (30 nmol) are incubated with clam CGase in an appropriate ratio of buffer (1 *M* sodium acetate, pH 4.0) and detergent. The free oligosaccharides are then separated from the ceramide moiety by adding 5 volumes of chloroform–methanol (2:1) to the reaction mixtures (1.0 ml). The upper layer containing the oligosaccharide portion of the hydrolyzed GSL is applied onto a Sep-pak C_{18} column (1 ml), preequilibrated with water. The column is washed with 2 column volumes (2 ml) of water. This water-based eluent is lyophilized and resuspended in a minimum volume of distilled water (0.0.5 ml), which is spotted on a TLC plate and developed in an *n*-butanol/acetic acid/water (2:1:1) solvent system. The cleaved oligosaccharides are visualized with the diphenylamine-developing agent (Fig. 1).

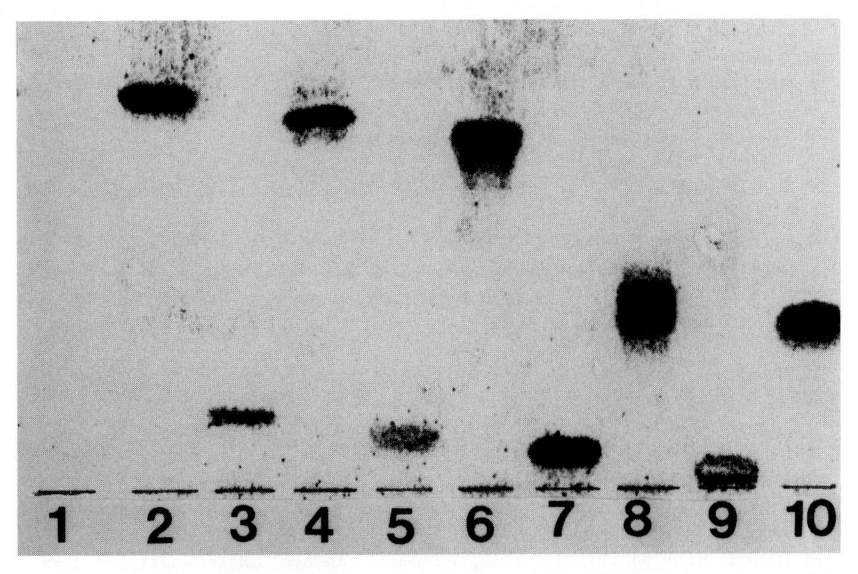

FIG. 1. Thin-layer chromatography profiling the hydrolysis of oligosaccharides from GSLs by clam CGase. Lane 1, clam CGase only; lane 2, GM3 only; lane 3, GM3 plus CGase; lane 4, GM2 only; lane 5, GM2 plus CGase; lane 6, GM1 only; lane 7, GM1 plus CGase; lane 8, GD1a only; lane 9, GD1a plus CGase; lane 10, sialic acid.

Analysis of GSLs in Conjunction with Fluorophore-Assisted Carbohydrate Electrophoresis (FACE)

The method for analysis of GSL oligosaccharides[14,19] by fluorophore-assisted carbohydrate electrophoresis (FACE)[20,21] was developed to provide a means for precise structural characterization of GSLs used as substrates of glycosyltransferases.[16,22–35] It is a simple, rapid, inexpensive, and highly sensitive method, capable of sequencing nanomole quantities of GSL oligosaccharides. This technique also provides valuable information when used in conjunction with linkage-specific exo- and endoglycosidases. Hence, this method (Fig. 2) is an alternative to other analytical methods presently in use for structural determination. Because this technique allows analysis of multiple samples in parallel, it can conceivably be applied in monitoring expression patterns of GSLs in normal developing and differentiating tissues or neoplastic cells.[36] It also can be applied as a diagnostic tool for pathogenesis and neurologic disorders.[37]

The FACE N-linked oligosaccharide profiling kit (Glyko-Kit, Inc., Novato, CA) and FACE monosaccharide composition kit (Glyko-Kit, Inc.,

[19] S. S. Basu, S. Dastgheib-Hosseini, G. Hoover, Z. Li, and S. Basu, *Anal. Biochem.* **222,** 270 (1994).
[20] P. Jackson, *Anal. Biochem.* **216,** 243 (1994).
[21] P. Jackson, *Methods Enzymol.* **230,** 250 (1994).
[22] M. Basu and S. Basu, *J. Biol. Chem.* **247,** 1489 (1972).
[23] M. Basu and S. Basu, *J. Biol. Chem.* **248,** 1700 (1973).
[24] S. Basu and M. Basu, *Glycoconjugates* **3,** 265 (1982).
[25] M. Basu, S. Basu, A. Staffyn, and P. Staffyn, *J. Biol. Chem.* **257,** 12765 (1982).
[26] M. Basu and S. Basu, *J. Biol. Chem.* **259,** 12557 (1984).
[27] H. Higashi, M. Basu, and S. Basu, *J. Biol. Chem.* **260,** 824 (1985).
[28] M. Basu, J. W. Hawes, Z. Li, S. Ghosh, F. A. Khan, B. J. Zhang, and S. Basu, *Glycobiology* **1,** 527 (1991).
[29] K. K. Das, M. Basu, S. Basu, D. H. Chou, and F. Jungalwala, *J. Biol. Chem.* **266,** 5238 (1991).
[30] S. Ghosh, J. W. Kyle, S. Dastgheib, F. Daussin, Z. Li, and S. Basu, *J. Glycoconjugate* **12,** 838 (1995).
[31] S. Basu, M. Basu, and S. S. Basu, *in* "Biology of the Sialic Acids" (Abraham Rosenberg, ed.), pp. 69–94. Plenum Press, New York, 1995.
[32] M. Basu, S.-A. Weng, H. Tang, F. Khan, B.-J. Zhang, and S. Basu, *Glycoconjugate J.* **13,** 423 (1996).
[33] S. S. Basu, M. Basu, Z. Li, and S. Basu, *Biochemistry,* **35,** 5166 (1996).
[34] R. Gornati, S. Basu, G. Bernardinin, M. Rizzo, F. Rossi, and B. Berra, *Mol. Cellular Biochem.* **166,** 117 (1997).
[35] S. Basu, M. Basu, S. Dastgheib, and J. W. Hawes, *in* "Comp. Natural Product Chem.-Carbohydrate" (B. M. Pinto, ed.), pp 107–128. Elsevier, New York, 1999.
[36] M. Basu, P. Kelly, M. Girzadas, Z, Li, and S. Basu, *Methods Enzymol.* **311,** 287 (1999).
[37] R. K. Yu and T. Ariga, *in* "Sphingolipid as Signaling Modulators in the Nervous System" (R. W. Ledeen, S.-I. Hakomori, A. J. Yates, J. S. Schneider, and R. K. Yu, eds.), pp. 285–306. The New York Academy of Sciences, New York, 1998.

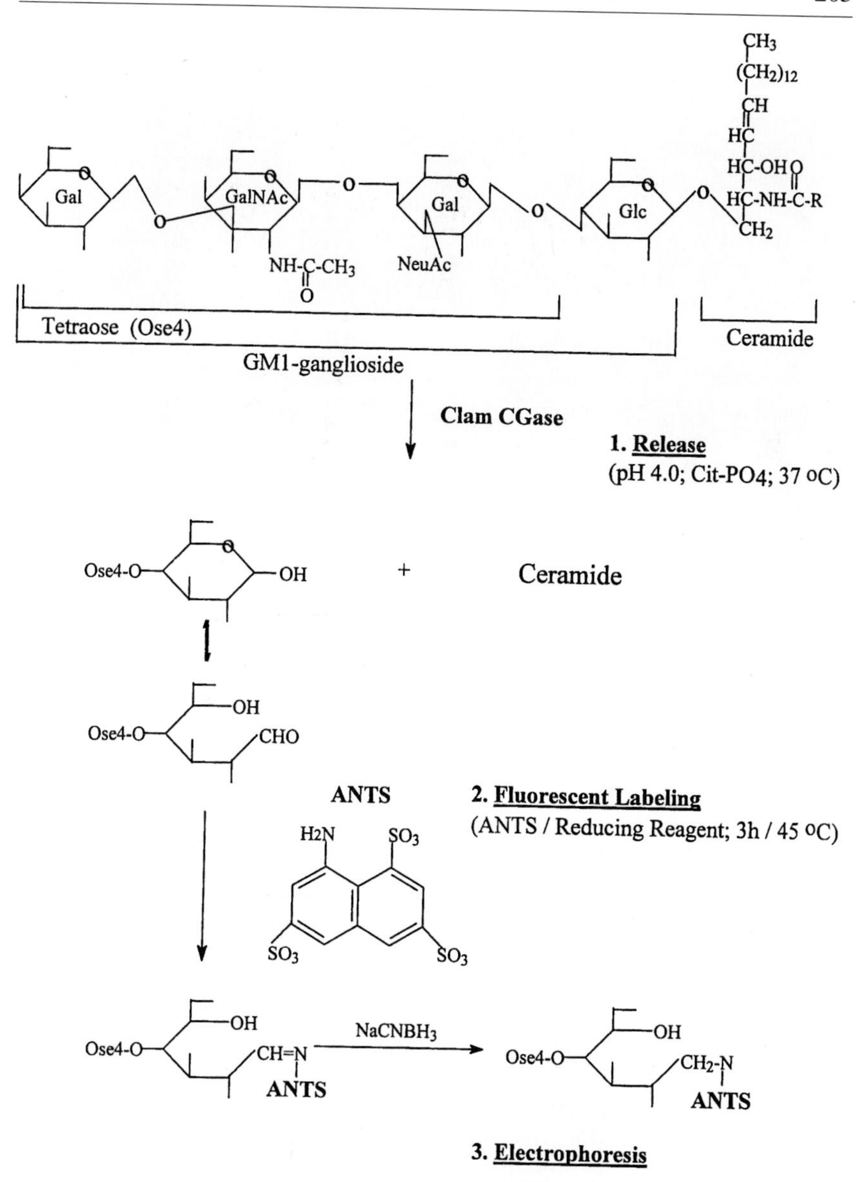

FIG. 2. Schematic representation of release of oligosaccharide from GSLs and their reaction with ANTS for electrophoresis imaging.

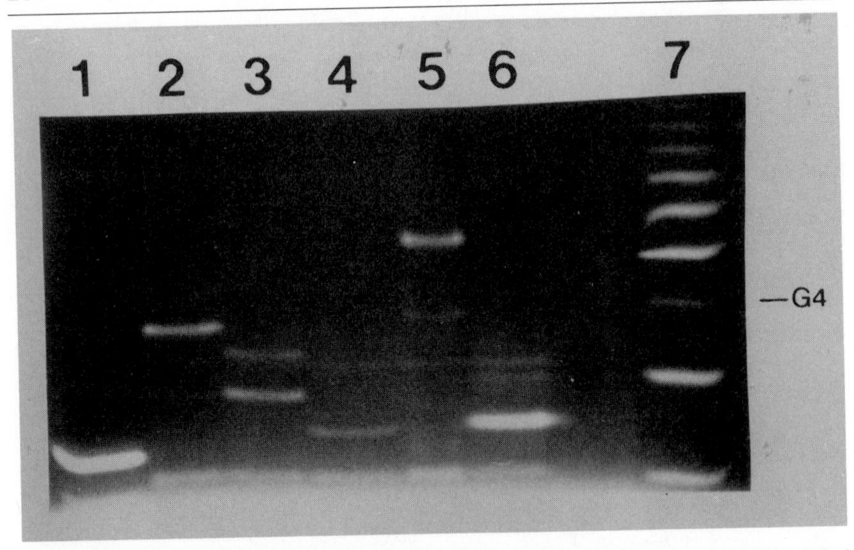

FIG. 3. Image of an electrofluorogram showing oligosaccharide profiles of ANTS labeled glycans released by clam CGase. Lane 1, NeuAcα(2-3)lactosamine (NeuAcα2,3Ga1β1,4Glc NAc); lane 2, GM1 oligosaccharide (Galβ1,3GalNAcβ1,4Galβ1,4Glc); lane 3, GM2 oligosaccharide (GalNAcβ1,4Galβ1,4Glc); lane 4, GD3 oligosaccharide (NeuAcα2,8NeuAcα2,3 Galβ1,4Glc); lane 5, nLcOse 5 oligosaccharide (Galα1,3Galβ1,4GlcNAcβ1,3Galβ1,4Glc); lane 6, GbOse 3 oligosaccharide; (Galα1,4Galβ1,4Glc), standard ladder of glucose polymer with maltotetraose.

TABLE III

R_f VALUES OF THE ANTS-LABELED OLIGOSACCHARIDES RELEASED FROM GLYCOSPHINGOLIPIDS[a]

GSL	Oligosaccharide (structure)	R_f value
GM1	[Gal(β1-3)GalNAc(β1-4)(NeuAc(α2-3))Gal(β1-4)Glc]	1.12
GM2	[GalNAc(β1-4)(NeuAc(α2-3))Gal(β1-4)Glc]	1.41
GD3	[NeuAc(α2-8)NeuAc(α2-3)Gal(β1-4)Glc]	1.59
Gg_4	[Gal(β1-3)GalNAc(β1-4)Gal(β1-4)Glc]	1.04
Gg_3	[GalNAc(β1-4)Gal(β1-4)Glc]	1.36
Lc_3	[GlcNAc(β1-3)Gal(β1-4)Glc]	1.31
nLc_4	[Gal(β1-4)GlcNAc(β1-3)Gal(β1-4)Glc]	1.02
nLc_5	[Gal(α1-3)Gal(β1-4)GlcNAc(β1-3)Gal(β1-4)Glc]	0.72
LM1	[NeuAc(α2-3)Gal(β1-4)GlcNAc(β1-3)Gal(β1-4)Glc]	1.02
Gb_3	[Ga(α1-4)Gal(β1-4)Glc]	1.54
NeuAc-Lactosamine	[NeuAc(α2-3)Gal(β1-4)GlcNAc]	1.62

[a] The R_f values were calculated from the mobility of the GSL oligosaccharides through the polyacrylamide gel (Fig. 3) relative to glucose polymer standard. R_f values of 0.0 and 1.0 are assigned to glucose-12 and glucose-4, respectively.

Novato, CA) are used for the analysis of the glycans from GSLs. Oligosaccharide hydrolysis and fluorophore labeling of the saccharides are described in the instruction manuals accompanying the kits. A fraction (15 μl) of the glycan released from GSL by clam CGase is either directly labeled with 8-aminonaphthalene-1,3,6-trisulfonic acid (ANTS) or hydrolyzed for monosaccharide analysis under different conditions (using specific glycosidases). To analyze the sialic acid content in the oligosaccharide, hydrolysis is carried on for 30 min at 80° in 0.1 *N* HCl. For determination of neutral monosaccharide (glucose, galactose, and fucose), the samples are treated with 2 *N* trifluoroacetic acid (TFA), at 100° for 1 hr. Determination of amino sugar (*N*-acetylgalactosamine and *N*-acetylglucosamine) content is done by hydrolysis using 4 *N* HCl at 100° for 45 min, followed by reacetylation with acetic anhydride at pH 7.8. The product of hydrolysis is labeled with ANTS[20,21] according to the procedure provided by Glyco-Kit. An aliquot (1–3 μl) of fluorophore-labeled saccharide is loaded per lane on FACE gel and electrophoresed. The image of an electrofluorogram is shown in Fig. 3 (R_f values in Table III).

[18] Quantitative Analyses of Binding Affinity and Specificity for Glycolipid Receptors by Surface Plasmon Resonance

By C. ROGER MACKENZIE and TOMOKO HIRAMA

Introduction

The involvement of cell surface oligosaccharides, presented on glycolipids and glycoproteins, in a wide range of human disease states is well recognized.[1,2] Standard assay methods for these interactions have used microtiter plates or thin-layer chromatography (TLC) overlays. The rigid immobilization of glycoconjugates in these assays gives oligosaccharide presentation geometries that may not permit maximum binding valencies. Protein–carbohydrate interactions are typically multivalent to compensate for low intrinsic affinities. Also, conventional assays do not allow for the investigation of the effects of the composition of the lipid bilayer microenvironment on binding. This is important because it is recognized that factors

[1] R. A. Dwek, *Chem. Rev.* **96,** 683 (1996).

[2] I. Brockhausen and W. Kuhns, "Glycoproteins and Human Disease." Chapman and Hall, London, 1997.

0076-6879/00 $30.00

such as fatty acid content and ceramide hydroxylation can modulate carbohydrate epitope presentation.[3] The development of assays that approximate the membrane surface venue of protein–carbohydrate interactions is central to the advancement of our understanding of the biologic events mediated by glycolipids.

In this chapter we describe surface plasmon resonance (SPR) methods for the assay of protein binding to glycolipids contained in artificial membranes. Liposomes incorporating glycolipids are captured on[4] or fused to[5] sensor chip surfaces, and the binding of analytes such as bacterial toxins or antibodies to oligosaccharide ligands presented on the membrane surfaces can be characterized. Glycolipid presentation in a membrane environment overcomes the problems associated with conventional assays. In addition, monitoring interactions by surface plasmon resonance avoids requirements for labeled ligand or analyte and provides data from which rate and affinity constants can be derived.[6]

Optical Principle of Surface Plasmon Resonance Detection

Evanescent wave biosensors are becoming increasingly popular as tools for the study of interacting macromolecules. The first commercial instrument was introduced by Pharmacia Biosensor (now Biacore AB, Uppsala, Sweden) in 1990. The BIACORE instruments available from Biacore AB are still the most widely used evanescent wave biosensors. These instruments use SPR as the optical detection principle. A removable sensor chip with a gold film forms part of the optical system. Polarized light is reflected from the gold film and is detected on a diode array. SPR results in a decrease in the intensity of reflected light at a specific angle of incidence and this angle is sensitive to refractive index changes in the vicinity of the gold film. One reactant is covalently coupled to the sensor chip surface and the second is pumped continuously over the surface. An autoinjector and miniaturized fluidics cartridge are used to transport the mobile reactant to the surface. Binding is recorded in resonance units (RU) with 1 RU corresponding to an immobilized protein concentration of approximately 1 pg/mm^2.[7]

[3] C. A. Lingwood, *Glycoconj. J.* **13,** 495 (1996).

[4] C. R. MacKenzie, T. Hirama, K. K. Lee, E. Altman, and N. M. Young, *J. Biol. Chem.* **272,** 5533 (1997).

[5] A. L. Plant, M. Brigham-Burke, E. C. Petrella, and D. J. O'Shannessy, *Anal. Biochem.* **226,** 342 (1995).

[6] U. Jönsson, L. Fägerstam, B. Ivarsson, B. Johnsson, R. Karlsson, K. Lundh, S. Löfås, B. Persson, H. Roos, I. Rönnberg, S. Sjölander, E. Stenberg, R. Ståhlberg, C. Urbaniczky, H. Östlin, and M. Malmqvist, *BioTechniques* **11,** 620 (1991).

[7] E. Stenberg, B. Persson, H. Roos, and C. Urbaniczky, *J. Coll. Interface Sci.* **143,** 513 (1991).

The IAsys instruments (Affinity Sensors, Cambridge, UK) are similar to the SPR-based BIACORE instruments. They are evanescent wave biosensors that detect refractive index changes but are based on resonant mirror principles,[8] not surface plasmon resonance.

Glycolipid Immobilization on Sensor Chip Surfaces

Somewhat atypical strategies are required for the immobilization of glycolipids on sensor chip surfaces. With the BIACORE instruments, ligands are generally attached covalently to a carboxylated dextran layer on the gold film of CM5 sensor chips (Biacore AB) by amine, aldehyde, or thiol coupling strategies.[9] Alternatively, streptavidin and chelating chips are also available for the capture of biotinylated or oligohistidine-tagged ligands, respectively.[9] None of these options is suitable for the immobilization of glycolipids on sensor chip surfaces nor are such approaches desirable since they would not present the ligands in a membrane setting. Incorporation of glycolipids into liposomes provides a means of presenting them in well-defined lipid bilayer environments. Two methods have been described for generation of glycolipid surfaces on BIACORE sensor chips by using liposomes. One approach employs liposome capture using an antigen–antibody pair with the glycolipid antigen being incorporated into the liposomes and the antibody covalently coupled to CM5 sensor chips.[4] In the second approach, liposomes are fused to an alkane thiol monolayer on the gold film to give a hybrid bilayer in which the glycolipid is contained in the outer leaflet.[5] These two approaches give curved and flat bilayer surfaces, respectively.

Liposome Preparation

Small unilamellar vesicles are suitable for the preparation of glycolipid surfaces and are easily prepared by extrusion of hydrated lipid mixtures (Fig. 1). Dimyristoylphosphatidylcholine (DMPC) is often the phospholipid of choice[4] but others can be used.[10] Cholesterol may also be added as a third membrane lipid. In the procedure outlined in Fig. 1, *Salmonella* serogroup B lipopolysaccharide (LPS) is added at a level of 1% (w/w) of the total lipid. The LPS serves as the capture molecule in the liposome capture approach to glycolipid immobilization.[4] The extrusion step is conveniently performed using the Liposofast apparatus available from Avestin Inc. (Ottawa, On-

[8] R. Cush, J. M. Cronin, W. J. Steward, C. H. Maule, J. Molley, and N. J. Goddard, *Biosens. Bioelectron.* **8,** 347 (1993).

[9] P. Schuck, *Annu Rev. Biophys. Biomol. Struct.* **26,** 541 (1997).

[10] G. M. Kuziemko, M. Stroh, and R. C. Stevens, *Biochemistry* **35,** 6375 (1996).

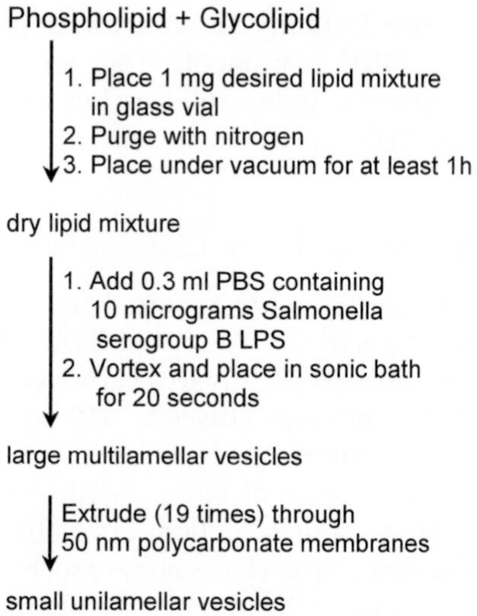

Phospholipid + Glycolipid

1. Place 1 mg desired lipid mixture in glass vial
2. Purge with nitrogen
3. Place under vacuum for at least 1h

dry lipid mixture

1. Add 0.3 ml PBS containing 10 micrograms Salmonella serogroup B LPS
2. Vortex and place in sonic bath for 20 seconds

large multilamellar vesicles

Extrude (19 times) through 50 nm polycarbonate membranes

small unilamellar vesicles

FIG. 1. Preparation of glycoliposomes containing *Salmonella* capture antigen.

tario, Canada). An odd number of extrusions are performed so that particulate matter is retained by the membrane and, therefore, not present in the final liposome preparations. Unincorporated LPS can be removed by passing the liposome preparation through a 1-ml Sepharose CL-4B (Pharmacia Biotech, Piscataway, NJ) column but this is not usually necessary at an LPS concentration of 1%. The size distribution of liposomes can be determined using a particle sizer such as the Nicomp Model 370 (Nicomp, Santa Barbara, CA).

Liposome Capture on Sensor Chips

Liposome capture via *Salmonella* serogroup B LPS is highly efficient.[4] To capture liposomes presenting the oligosaccharide component of this molecule on their surfaces, a monoclonal antibody, Se155-4, which is specific for this antigen is immobilized in the carboxylated dextran layer on CM5 sensor chips by amine coupling using reagents obtained from Biacore AB.[4] Typically, as much immunoglobulin G (IgG) as possible (approximately 12,000 RU) is coupled to the chip surface. The polysaccharide portion of the *Salmonella* LPS consists of multiple repeats of the tetrasaccharide {→ 2)[αD-Abe(1 → 3)]αD-Man(1 → 4)αL-Rha(1 → 3)αD-Gal(1 →}. The monoclonal antibody, an $IgG_{1\lambda}$, recognizes a trisaccharide epitope on the

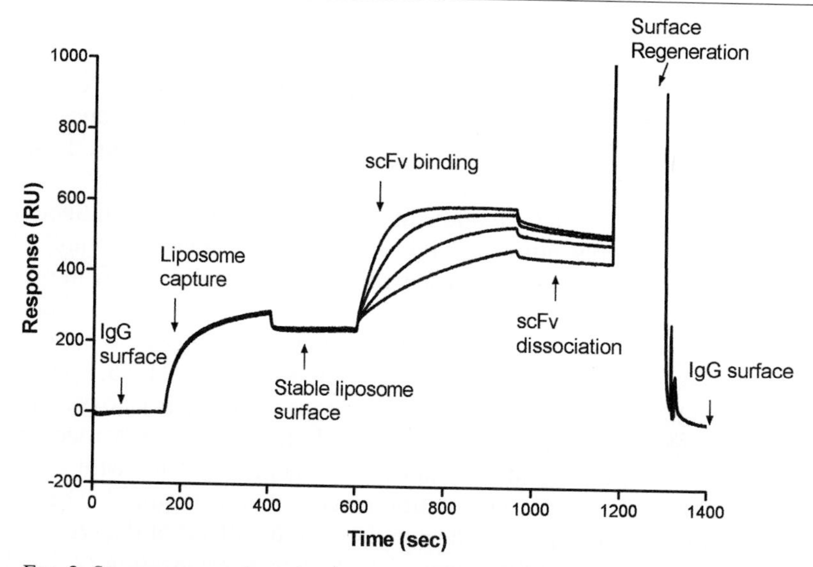

FIG. 2. Sensorgram overlays showing cycles of liposome capture, analyte binding, and surface regeneration. The analyte in this instance was a scFv version of the capture antibody Se155-4 and was injected at concentrations of 80, 120, 160, and 200 n*M*.

repeating unit and dissociates from the antigen at pH 4.5.[11] This permits surface regeneration under relatively mild conditions.

A typical binding cycle involves liposome capture followed by analyte binding, partial dissociation of the analyte, and a regeneration step to restore the original IgG surface (Fig. 2). In the example shown in Fig. 2, liposomes containing 99% DMPC and 1% serogroup B LPS are captured by the IgG surface. Measurements are carried out in 10 m*M* HEPES, pH 7.4, 150 m*M* NaCl, 3.4 m*M* EDTA. The multivalent nature of the attachment gives a stable surface with little or no loss of immobilized liposomes during continuous buffer flow. Various concentrations of a bivalent single-chain Fv (scFv) version of Se 155-4, designated B5-1,[12] are injected over the liposome surfaces and bound to the LPS antigen. Analyte binding occurs over the course of the scFv injections. Immediately following the injections, analyte dissociation from the liposomes is observed. With the BIACORE instruments, the injection volumes can be between 5 and 750 μl, and can be varied depending on factors such as the rate constants

[11] D. R. Bundle, E. Eichler, M. A. J. Gidney, M. Meldal, A. Ragauskas, B. W. Sigurskjold, B. Sinnott, D. C. Watson, M. Yaguchi, and N. M. Young, *Biochemistry* **33,** 5172 (1994).

[12] S.-j. Deng, C. R. MacKenzie, T. Hirama, R. Brousseau, T. L. Lowary, N. M. Young, D. R. Bundle, and S. A. Narang, *Proc. Natl. Acad. Sci. U.S.A.* **92,** 4992 (1995).

for the interaction, the need to achieve equilibrium binding, and sample availability. After sufficient analyte has dissociated (at least 10% of the total bound), the IgG surface is regenerated with 0.1% Triton X-100 in 10 m*M* sodium acetate, pH 4.5 (Fig. 2). This regeneration procedure does not significantly damage the IgG surface as demonstrated by reproducible capture of liposomes over four binding–regeneration cycles.

Although Se155-4 IgG readily dissociates from its LPS antigen at pH 4.5, liposomes presenting the LPS polysaccharide cannot be completely removed from the IgG surfaces by lowering the pH to this level. Detergent is also required. Triton X-100 is preferred because the IgG surface is tolerant to this detergent. However, when certain analytes are bound to captured liposomes, 0.1% Triton X-100 in 10 m*M* sodium acetate, pH 4.5, does not give complete regeneration. In such instances, sodium taurodeoxycholate in combination with 10 m*M* sodium acetate, pH 4.5, is an effective regeneration reagent. This detergent can damage the IgG surface and care should be taken not to use more regeneration reagent than is required. The IgG surfaces are apparently protected from the harmful effects of the regeneration solution by residual liposomes. A detergent concentration of 1 m*M* should be tried initially and if this does not give full regeneration, the concentration can be increased to 4 m*M*, keeping the contact times as short as possible.

Liposome Fusion to Alkane Thiol Monolayers

Liposomes can be fused to sensor chips with an alkane thiol layer covalently linked to the gold film.[5] The HPA chips available from Biacore AB are constructed in this way. Prior to use, the HPA surface should be cleaned by injection of octyl glucoside, a nonionic detergent.[13] Liposomes of the desired composition are repeatedly injected over the HPA surface at a flow rate of 5 μl/min until a maximum response is obtained. The liposomes are thought to spontaneously fuse to the HPA surface by a mechanism similar to that proposed by Kalb *et al.*[14] for the fusion of phospholipid vesicles with hydrophobic surfaces. The result is a hybrid bilayer in which the outer half is of liposomal origin and available for analyte binding. The time course of hybrid bilayer formation varies depending on liposome size and composition. Following the liposome loading, 10 m*M* NaOH is injected at high flow rate to remove any multilamellar structures and give a stable baseline.[13] Prolonged buffer washing has also been used to

[13] M. A. Cooper, D. H. Williams, and Y. R. Cho, *Chem. Commun.* **1997,** 1625 (1997).
[14] E. Kalb, S. Frey, and L. Tamm, *Biochim. Biophys. Acta* **1103,** 307 (1992).

provide baseline stabilization.[15] Finally a solution of bovine serum albumin, which binds strongly to the self-assembled monolayer alone, is injected to confirm complete coverage of the chip with lipid.[13] The fusion process typically increases the baseline response by 1500–2000 RU. Following analyte binding, it is desirable to regenerate the hybrid bilayer surface. Regenerating to the self-assembled monolayer considerably lengthens the analysis time and is also undesirable because there is a limit to the number of times that the alkane thiol monolayer can be reloaded with liposomes.

At this early stage in the use of membrane-based SPR methods, it is difficult to know which approaches to glycolipid immobilization will find general application. A comparison of the liposome capture and liposome fusion methods for characterization of a monoclonal IgM antibody specific for the glycolipid asialo-GM1 (Galβ1-3GalNAcβ1-4Galβ1-4Glcβ1-1-ceramide) showed that fused liposome surfaces had a higher capacity for antibody binding than captured liposome surfaces.[16] However, it is important to realize that the alkane thiol surface of the HPA chip is extremely sticky and to ensure that none of this surface is exposed to analyte. The captured liposome approach has the advantages of using the widely used CM5 sensor chips and providing a fresh membrane surface for each analyte injection since both bound analyte and captured liposomes are removed during the regeneration step. In addition, the captured liposome surfaces are more defined since it is difficult to know the exact topography of the hybrid bilayers on HPA chips.

Binding Specificities

The artificial membrane approaches described here offer a convenient and accurate means of determining the specificities of protein–glycolipid interactions. Compared to conventional assays, the liposome capture approach showed more restricted oligosaccharide specificities for bacterial toxin binding.[4] For the cholera toxin B pentamer (CTB) and heat-labile enterotoxin (LT), the binding specificities were in excellent agreement with the binding site features observed in crystal structures. The GM1 (Galβ1-3GalNAcβ1 (NeuAcα2-3)4Galβ1-4Glc-ceramide) binding site in CTB has been described as a two-fingered grip with the Galβ1-3GalNAc moiety representing the "forefinger" and sialic acid representing the

[15] M. A. Cooper, J. Carroll, E. R. Travis, D. H. Williams, and D. J. Ellar, *Biochem. J.* **333,** 677 (1998).

[16] B. A. Harrison, R. MacKenzie, T. Hirama, K. K. Lee, and E. Altman, *J. Immunol. Methods* **212,** 29 (1998).

"thumb."[17] The strongest interactions involve the terminal galactose that inserts into a deep pocket in the binding site and the sialic acid, which occupies a shallower depression on the toxin surface. This is consistent with the lack of binding of CTB to LacCer (Galβ1-4Glc-ceramide), asialo-GM1 (Galβ1-3GalNAcβ1-4Galβ1-4Glc-ceramide), GM2 (GalNAcβ1(NeuAcα2-3)4Galβ1-4Glc-ceramide), GD1a (NeuAcα2-3Gal-β1-3GalNAcβ1(NeuAcα2-3)4Galβ1-4Glc-ceramide), GT1b (NeuAcα2-3Galβ1-3GalNAcβ1(NeuAcα2-8NeuAcα2-3)4Galβ1-4Glc-ceramide), and GQ1b (NeuAcα2-8NeuAcα2-3Galβ1-3GalNAcβ1(NeuAcα2-8NeuAcα2-3)4Gal-β1-4Glc-ceramide). It is also consistent with some tolerance to the addition of a second internal sialic acid as in GD1b (Galβ1-3GalNAcβ1(NeuAcα2-8NeuAcα2-3)4Galβ1-4Glc-ceramide) (Fig. 3A). The structure of LT, which is very similar to that of CTB, complexed with lactose (Galβ1-4Glc) revealed a subtle difference in ligand binding compared to CTB. Whereas the His-13 residue in CTB donates a hydrogen bond from its backbone amide to sialic acid with no side-chain contribution to binding, Arg-13, found at this position in most variants of LT, can also contribute side-chain binding to the GalNAcβ1-4Galβ1-4Glc portion of GM1 and other gangliosides.[18,19] These structural features are in agreement with the BIACORE profiles observed for toxin binding to liposomes, which showed that, unlike CTB, LT does not display an absolute requirement for the internal sialic acid as shown by the weak binding of LT to asialo-GM1.[4]

The SPR-liposome capture has also been used to determine the fine specificity of a monoclonal IgM that showed specificity for asialo-GM1 in microtiter plate assays.[16] The antibody showed strong binding to liposomes containing asialo-GM1 with little or no binding to liposomes containing other glycolipids (Fig. 3B). The lack of binding to asialo-GM2 was indicative of a key role for the terminal galactose in antigen binding.

In performing these experiments, it is extremely important to monitor analyte binding to reference surfaces. For toxin and antibody binding to captured liposomes, the control surfaces were captured liposomes that contained only DMPC and the capture antigen. This provided a means of quantitating refractive index differences between the sample and running buffers and of monitoring nonspecific binding of the analyte to surfaces. Refractive index differences between the sample and running buffers result in rapid positive or negative responses at the beginning and end of sample

[17] E. A. Merritt, S. Sarfaty, F. van den Akker, C. L. L.'Hoir, J. A. Martial, and W. G. J. Hol, *Protein Sci.* **3,** 166 (1994).

[18] K.-A. Karlsson, *Curr. Opin. Struct. Biol.* **5,** 622 (1995).

[19] E. A. Merritt, T. K. Sixma, K. H. Kalk, B. A. M. van Zanten, and W. G. J. Hol, *Mol. Microbiol.* **13,** 745 (1994).

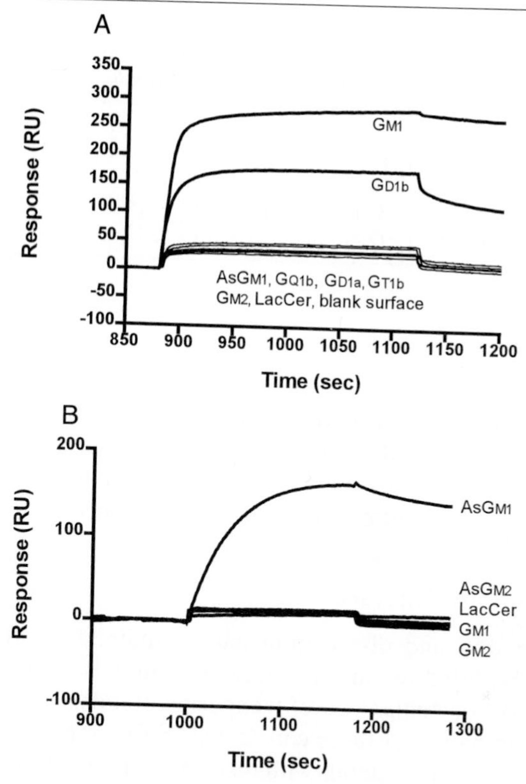

FIG. 3. Sensorgram overlays showing the binding of cholera toxin B-pentamer[4] (A) and H2G10 IgM[16] (B) to captured liposomes containing different glycolipids. The toxin fragment was injected at a concentration of 100 n*M* and a flow rate of 10 μl/min over 1000 RU of liposomes containing 2% glycolipid. The antibody was injected at a concentration of 250 n*M* and a flow rate of 10 μl/min over 1000 RU of liposomes containing 10% glycolipid.

injections. Nonspecific binding can either be rapid or can occur over the course of the injection. In the examples given here, there could be nonspecific binding to either the dextran matrix of the sensor chip or the phospholipid headgroups on the liposomes but this was not observed. To obtain a binding profile devoid of buffer and nonspecific binding effects, sensorgram data from active and reference surfaces can be subtracted using the BIAevaluation software available from Biacore AB.

Hybrid bilayer surfaces formed by the fusion of liposomes containing glycolipids to HPA sensor chips have also been used for the determination of cholera toxin specificity for glycolipid receptors.[10] This approach indicated binding to several glycolipids with a binding strength of GM1 > GM2 > GD1a > GM3 > GT1b > GD1b > asialo-GM1. This binding

TABLE I
AFFINITIES AND RATE CONSTANTS FOR CHOLERA TOXIN B-PENTAMER BINDING TO LIPOSOMES CONTAINING GLYCOLIPIDS THAT EXHIBIT BINDING BY SPR[a]

Ganglioside	k_a (M^{-1} s^{-1})	k_d (s^{-1})	K_D[b] (M)
GM1	6.2×10^5 (±5.5)[c]	4.5×10^{-4} (±18.3)	7.3×10^{-10} (±19.1)
GD1b	4.1×10^5 (±12.4)	3.1×10^{-3} (±17.4)	8.0×10^{-9} (±21.4)

[a] From Ref. 4.
[b] k_a/k_d.
[c] Numbers in parentheses are the S.E., expressed as a percentage.

profile is difficult to rationalize on the basis of the CTB-GM1 pentasaccharide crystal structure and contradicts titration microcalorimetry data.[20] Nonspecific binding is a possible explanation although binding to reference surfaces consisting only of phospholipid was not observed.[10]

Kinetic and Affinity Analyses

The association and dissociation data contained in BIACORE sensorgrams can be fitted to various interaction models for the derivation of rate and affinity constants.[21] Careful experimental design is crucial for the accurate determination of these constants.[17,22] The injected component or analyte should be homogeneous with respect to binding valency and ideally should be monovalent to allow for data fitting to a simple one-to-one interaction model.[23] However, the intrinsic, or per binding site, affinities for protein–carbohydrate interactions are typically very weak, and monovalent binding, which can have a K_D as low as 2 mM,[24] may be too weak to measure by SPR. Also, many carbohydrate binding proteins, such as the AB_5 group of bacterial toxins, have multiple copies of the carbohydrate binding subunits that are not amenable to isolation as monovalent molecules.[18] Nonetheless, under conditions that permit full valency binding to glycolipid surfaces, meaningful values can be obtained. This requires using low analyte concentrations to minimize competition for available ligand. Linearity of the transformed dissociation data is indicative of homogeneity

[20] M. Masserini, E. Freire, P. Palestini, E. Calappi, and G. Tettamanti, *Biochemistry* **31,** 2422 (1992).
[21] D. G. Myszka, *Curr. Opin. Biotechnol.* **8,** 50 (1997).
[22] R. Karlsson and A. Fält, *J. Immunol. Methods* **200,** 121 (1997).
[23] C. R. MacKenzie, T. Hirama, S.-j. Deng, D. R. Bundle, S. A. Narang, and N. M. Young, *J. Biol. Chem.* **271,** 1527 (1996).
[24] P. M. St. Hilaire, M. K. Boyd, and E. J. Toone, *Biochemistry* **33,** 14452 (1994).

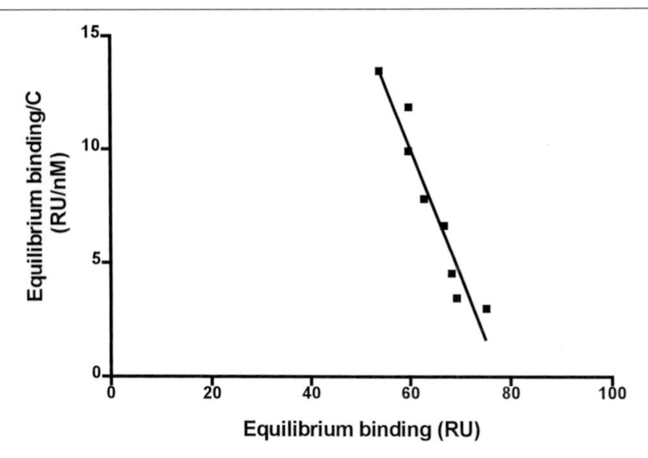

FIG. 4. Scatchard plot for the determination of the affinity of cholera toxin B-pentamer for GM1.[4] The toxin fragment, at concentrations of 5–30 n*M*, was injected over 1000 RU of liposomes containing 2% GM1.

with respect to binding valency.[4] The association and dissociation rate constants obtained under these conditions are a measure of the functional affinities of these multivalent interactions under conditions of optimum binding. For the cholera toxin B-pentamer, optimum binding conditions gave similar association rate constants for binding to liposomes containing GM1 and GD1b (Table I).[4] However, the dissociation rate constant for binding to GD1b was approximately 10 times faster than that for binding to GM1, resulting in a 10-fold lower affinity with the latter ligand. The faster dissociation rate for GD1b binding is clearly seen in Fig. 3A.

Affinity constants can also be derived from Scatchard plots (bound/free versus bound) of steady-state binding data. Injection volumes and flow rates can be adjusted to permit binding to reach equilibrium at all analyte concentrations. With BIACORE data, the concentration of free analyte is the same as the injected concentration since the continuous flow system continuously replenishes analyte. In some instances, only steady-state data can be obtained because of low affinities and rapid kinetics. For example, monovalent antibody fragments of anti-carbohydrate antibodies exhibit very rapid dissociation rates and at analyte concentrations required for adequate response, equilibrium binding is reached too quickly for collection of association data.[23,25] Under these conditions, "square" sensorgrams are obtained. Obviously, affinity constants derived from rate constants and

[25] S. I. Patenaude, C. R. MacKenzie, D. Bilous, R. J. To, S. E. Ryan, N. M. Young, and S. V. Evans, *Acta Cryst*, **D54**, 1456 (1998).

steady-state binding should be in good agreement and any discrepancy is an indication of faults in experimental design. Scatchard analysis of steady-state binding of CTB to liposomes containing 2% GM1 gave a K_D of 1.7 nM (Fig. 4). This is in good agreement with the value of 0.7 nM derived from the association and dissociation rate constants (Table I).

It was observed that the kinetics and affinity of CTB binding to liposomes containing GM1 was influenced by the glycolipid content of the liposomes.[4] More rapid association kinetics were observed with liposomes containing lower values (1–2%) compared to those containing 3–4%. This effect was confirmed by Scatchard analysis of equilibrium binding data. The K_D of the GM1–CTB interaction was determined to be 1.7 nM using 2% GM1 liposomes and 6.8 nM for 4% liposomes.

Summary

In recent years, rapid expansion has been seen of the application of SPR to a wide range of biomolecular interactions. For protein–carbohydrate interactions, SPR techniques offer the means of presenting glycolipids, and potentially glycoproteins, in an environment that closely resembles their *in vivo* situation. It is already clear that the technology can provide much additional insight into these interactions because it monitors the interactions under conditions that approach physiologic ones and can allow for the investigation of parameters that are not accessible by conventional approaches. Strategies for the immobilization of glycolipids on BIACORE sensor chips are not limited to the two approaches described here. For example, liposomes containing biotinylated phospholipid could be captured by avidin or streptavidin surfaces; this strategy has been used for the capture of natural membrane vesicles.[26] Also, Biacore AB has announced plans to introduce a version of the CM5 chip that has been derivatized with an undisclosed group that mediates the capture of large amounts of liposomes.

Acknowledgments

We thank Dr. N. M. Young for critical review of the manuscript. This is National Research Council of Canada publication 42429.

[26] L. Masson, A. Mazza, and R. Brousseau, *Anal. Biochem.* **218,** 405 (1994).

[19] Use of Circular Dichroism for Assigning Stereochemistry of Sphingosine and Other Long-Chain Bases

By AKIRA KAWAMURA, KOJI NAKANISHI, and NINA BEROVA

Introduction

Sphingolipid signal transduction pathways are mediated by a wide variety of enzyme–substrate interactions. The key factor that ensures the fidelity of this pathway is the specificity between each enzyme and substrate. The specificity can be understood in terms of the molecular-level interactions, in which the asymmetric protein binding site accommodates its chiral substrate stereospecifically. In other words, stereochemistry of a substrate determines the binding affinity to its receptor enzyme—hence the subsequent course of the physiologic processes. In fact, each stereoisomer of sphingosine and dihydrosphingosine has been found to possess its own biologic properties,[1–11] which underscores the significance of stereochemical assignment for both synthetic and natural sphingolipids.

This paper describes the application of an exciton chirality method to the stereochemical assignment of sphingosines and dihydrosphingosines,

[1] A. H. Merrill, S. Nimkar, D. Menaldino, Y. A. Hannun, C. Loomis, R. M. Bell, S. R. Tyagi, J. D. Lambeth, V. L. Stevens, R. Hunter, and D. C. Liotta, *Biochemistry* **28,** 3138 (1989).

[2] Y. Igarashi, S. Hakomori, T. Toyokuni, B. Dean, S. Fujita, M. Sugimoto, T. Ogawa, K. El-Ghaendy, and E. Racker, *Biochemistry* **28,** 6796 (1989).

[3] T. J. Mullmann, M. I. Siegel, R. W. Egan, and M. M. Billah, *J. Biol. Chem.* **266,** 2013 (1991).

[4] A. Bielawska, C. M. Linardic, and Y. A. Hannun, *J. Biol. Chem.* **267,** 18493 (1992).

[5] R. Chao, W. Khan, and Y. A. Hannun, *J. Biol. Chem.* **267,** 23459 (1992).

[6] M. Y. Pushkareva, W. A. Khan, A. V. Alessenko, N. Sahyoun, and Y. A. Hannun, *J. Biol. Chem.* **267,** 15246 (1992).

[7] B. M. Buehrer and R. M. Bell, *J. Biol. Chem.* **267,** 3154 (1992).

[8] A. Bielawska, H. M. Crane, D. Liotta, L. M. Obeid, and Y. A. Hannun, *J. Biol. Chem.* **268,** 26226 (1993).

[9] A. Olivera, H. Zhang, R. O. Carlson, M. E. Mattie, R. R. Schmidt, and S. Spiegel, *J. Biol. Chem.* **269,** 17924 (1994).

[10] R. A. Wolff, R. T. Dobrowsky, A. Bielawska, L. M. Obeid, and Y. A. Hannun, *J. Biol. Chem.* **269** (1994).

[11] N. Divecha and R. F. Irvine, *Cell* **80,** 269 (1995).

0076-6879/00 $30.00

the structural core of sphingolipids.[12,13] Exciton chirality is a very simple and highly sensitive analytical protocol utilizing circular dichroic (CD) spectroscopy. It requires only a low microgram quantity of the original sample material and allows stereochemical assignment of all possible configurational isomers of sphingosine and dihydrosphingosine. The method has already been employed for the stereochemical assignment of several sphingolipids from marine sources.[14,15] Note that all of the chemical reactions and the purification methods are simple and straightforward, and the readers with limited chemical background should soon be able to obtain reproducible results after some practice.

Circular Dichroic Exciton Chirality Method

Circular dichroism is defined as the difference in absorption for the left and the right circularly polarized light ($\Delta\varepsilon = \varepsilon_L - \varepsilon_R$, where ε_L and ε_R represent the absorptions of the left and the right helical rays, respectively). Because all chiral substances exhibit circular dichroism, CD spectroscopy has been used as a powerful tool for stereochemical analysis of many organic and inorganic compounds. Most of the commercial CD spectrometers are made for the detection of circular dichroism in the UV/VIS wavelength region. CD spectroscopy, therefore, is widely used for the conformational study of biopolymers, such as proteins and nucleic acids, both of which contain intrinsic UV absorbing chromophores. CD has also been applied to the stereochemical analysis of small organic molecules, one of the most successful applications being the exciton chirality method.[16,17] This method is based on the through-space interaction between two or more chromophores, which may be preexisting within the substrate or introduced by derivatization of hydroxyl groups, amino groups, etc.; thus the original molecule of interest could be devoid of UV absorbing moieties. The CD of the substrate or derivatized substrate measures the through-space inter-

[12] V. Dirsch, J. Frederico, N. Zhao, G. Cai, Y. Chen, S. Vunnam, J. Odingo, H. Pu, K. Nakanishi, N. Berova, D. Liotta, A. Bielawska, and Y. Hannun, *Tetrahedron Lett.* **36,** 4959 (1995).

[13] A. Kawamura, N. Berova, V. Dirsch, A. Mangoni, K. Nakanishi, G. Schwartz, A. Bielawska, Y. Hannun, and I. Kitagawa, *Bioorg. Med. Chem.* **4,** 1035 (1996).

[14] M. Ojika, G. Yoshino, and Y. Sakagami, *Tetrahedron Lett.* **38,** 4235 (1997).

[15] V. Costantino, E. Fattorusso, A. Mangoni, M. Di Rosa, and A. Ianaro, *J. Am. Chem. Soc.* **119,** 12465 (1997).

[16] N. Harada and K. Nakanishi, "Circular Dichroic Spectroscopy—Exciton Coupling in Organic Stereochemistry." University Science Books, Mill Valley, California, 1983.

[17] N. Berova and K. Nakanishi, *in* "Circular Dichroism" (N. Berova, K. Nakanishi, and R. W. Woody, eds.), p. 337, Wiley-VCH, New York, 2000.

actions among the chromophores. Because the extent of interaction is proportional to the square of the chromophoric extinction coefficient, usage of strongly absorbing chromophores enables the analytical scale to be lowered to levels of just a few micrograms or less. In the case of conformationally rigid molecules, the exciton chirality method leads to a nonempirical determination of absolute configuration as shown in the following simple example of (1*S*,2*S*)-cyclohexanediol dibenzoate (**1**, Fig. 1).

The absolute sense of twist between the two hydroxyl groups is represented by the twist between the two benzoate electric transition moments since they are parallel to the two hydroxylic C–O bonds. The interaction splits the "excited states" of the chromophores into two energy levels, which is the so-called exciton coupling (Fig. 2). One of the two electric transitions, therefore, gives rise to a red-shifted absorption band (longer wavelength), while the other affords a blue-shifted band (shorter wavelength), and the two bands exhibit circular dichroism (or Cotton effects) of different signs to each other. When the absolute sense of twist between the two chromophores is clockwise (positive chirality), as is the case for **1** (Fig. 1), the red-shifted band absorbs the left circularly polarized light more than the right counterpart ($\Delta\varepsilon > 0$), while the blue-shifted band absorbs

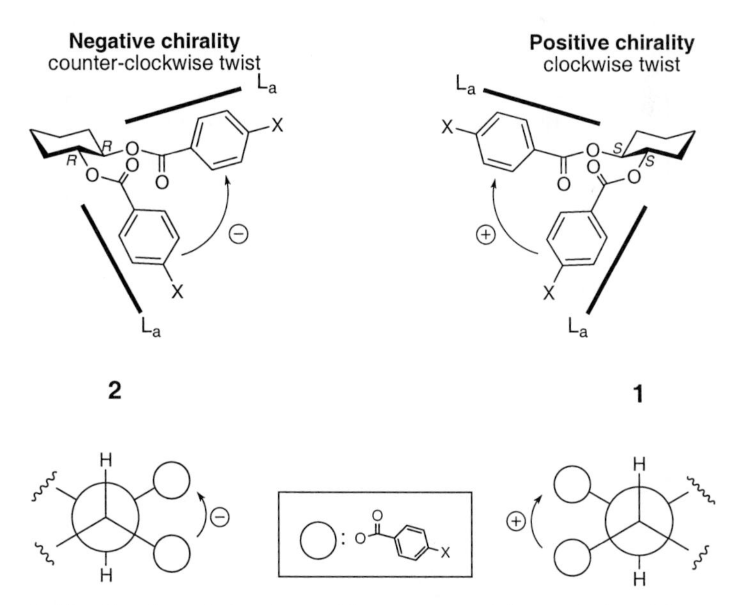

FIG. 1. The directions of electric transition moments of the *p*-substituted benzoate, the chromophoric L_a transitions, are represented by the thick lines. Positive chirality is defined by a clockwise twist between the transition moments as in **1**.

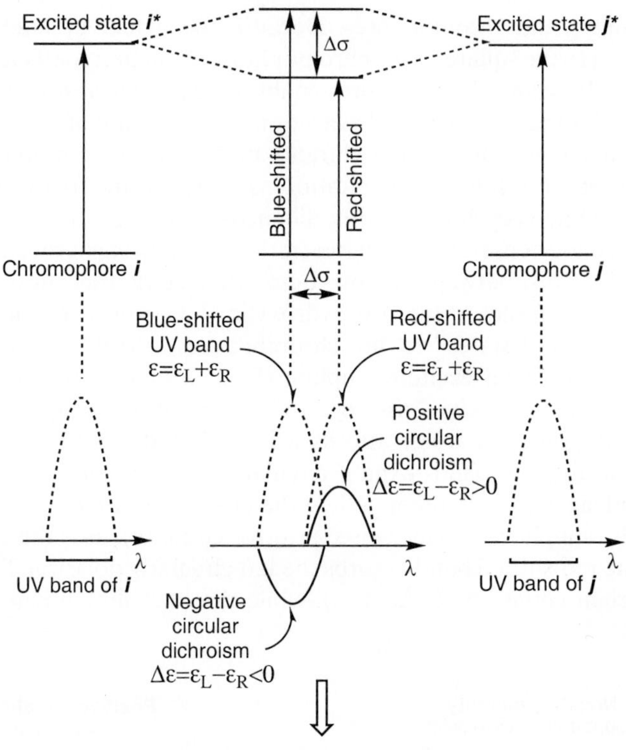
Exciton coupling
Excited state i*
Excited state j*
Δσ
Blue-shifted
Red-shifted
Chromophore i
Chromophore j
Δσ
Blue-shifted
UV band
ε=εL+εR
Red-shifted
UV band
ε=εL+εR
Positive
circular
dichroism
Δε=εL−εR>0
λ
λ
λ
UV band of i
UV band of j
Negative
circular
dichroism
Δε=εL−εR<0

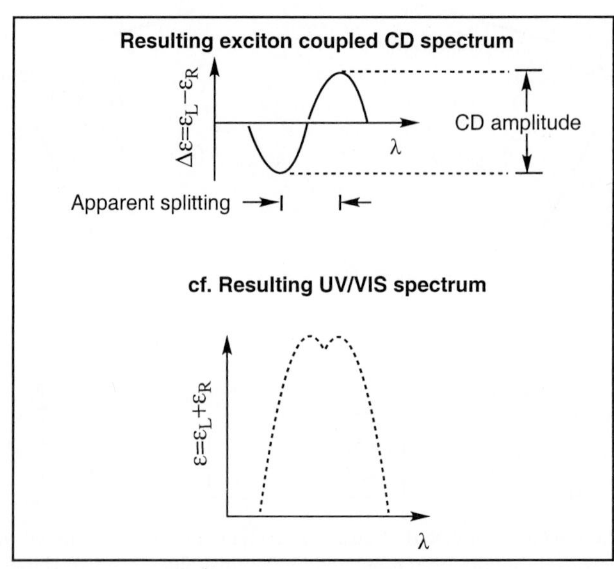
Resulting exciton coupled CD spectrum
Δε=εL−εR
CD amplitude
λ
Apparent splitting
cf. Resulting UV/VIS spectrum
ε=εL+εR
λ

the right circularly polarized light more than the left one ($\Delta\varepsilon < 0$). Therefore, the resulting CD curve, which is the summation of the two absorption bands with opposite signs, consists of a positive component at the longer wavelength side (first Cotton effect) and a negative component at the shorter wavelength side (second Cotton effect). Such a curve is defined as a "positive" exciton split CD or a positive CD couplet. On the other hand, a negative chirality, as in (1*R*,2*R*)-cyclohexanediol dibenzoate **2** (Fig. 1), leads to a negative CD couplet, i.e., negative first Cotton and positive second Cotton effects. Without exception, positive and negative chiralities afford positive and negative CD couplets, respectively. The absolute stereochemistry of conformationally rigid molecules, therefore, can be determined nonempirically by this method. The theoretical background of exciton chirality method[16] and its applications to actual stereochemical problems of natural products[17] have been described elsewhere.

With respect to conformationally flexible molecules, such as sphingosines, however, the analysis becomes more complicated. Because a chromophoric derivative of a flexible molecule gives a CD curve that is arising from a mixture of many conformers,[18] it is not possible to unequivocally assign the stereochemistry by the sign of the CD couplet. However, for compounds with significant biologic importance, it is worthwhile to prepare a set of reference spectra for all possible stereoisomers, and use this library for the stereochemical assignment of the unknown. This alternative approach is still highly sensitive and does not require any authentic sample as reference. In addition to sphingosines and dihydrosphingosines,[12,13] which are discussed below, CD reference curves for other classes of acyclic compounds, such as bacteriohopanoids,[19] brassinosteroids,[20] and heptopyranosides/heptofuranosides,[21] have also been prepared for such purposes.

[18] N. Zhao, P. Zhou, N. Berova, and K. Nakanishi, *Chirality* **7,** 636 (1995).

[19] N. Zhao, N. Berova, K. Nakanishi, M. Rohmer, P. Mougenot, and U. J. Jurgens, *Tetrahedron* **52,** 2777 (1996).

[20] A. Kawamura, N. Berova, K. Nakanishi, B. Voigt, and G. Adam, *Tetrahedron* **53,** 11961 (1997).

[21] I. Akritopoulou-Zanze, K. Nakanishi, H. Stepowska, H. Grzeszczyk, B. Zamojski, and N. Berova, *Chirality* **9,** 699 (1997).

FIG. 2. Exciton coupling of two identical chromophores with positive chirality, as in **1** of Fig. 1. The red-shifted and blue-shifted bands show positive and negative circular dichroism, respectively. Summation of the two bands with opposite signs gives rise to the exciton coupled CD spectrum shown, which is defined as a positive CD couplet.

Application of Exciton Chirality Method to Sphingosine and Dihydrosphingosine

Sphingosine and dihydrosphingosine contain two stereogenic centers at the sites of the 2-amino and 3-hydroxyl groups, thus giving rise to a total of eight isomers: D-*erythro* (2*S*,3*R*), L-*threo* (2*S*,3*S*), L-*erythro* (2*R*,3*S*), and D-*threo* (2*R*,3*R*) of sphingosines and dihydrosphingosines (Fig. 3). All eight isomers can be differentiated by CD spectroscopy after appropriate chromophoric derivatization, i.e., 1,3-bisnaphthoate-2-*N*-naphthimido derivative, which can be prepared by a simple two-step reaction sequence (Fig. 4). Reference CD spectra for all isomers have already been prepared (Fig. 5).[13] Therefore, the configuration of any sphingosine sample, natural or synthetic, can be checked by chemical derivatization, followed by CD

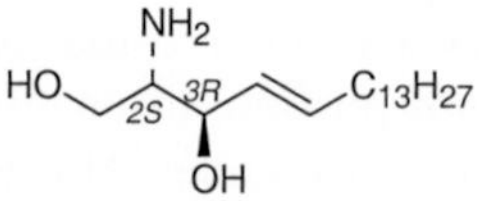

D-*erythro*-(2*S*,3*R*)-sphingosine **3A**

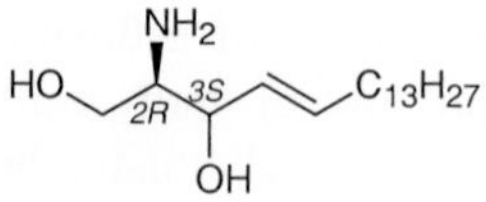

L-*erythro*-(2*R*,3*S*)-sphingosine **5A**

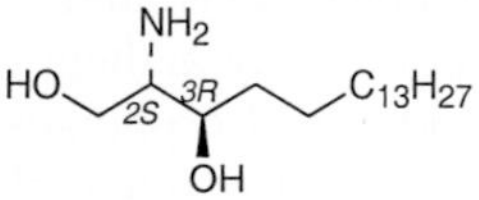

D-*erythro*-(2*S*,3*R*)-dihydrosphingosine **3a**

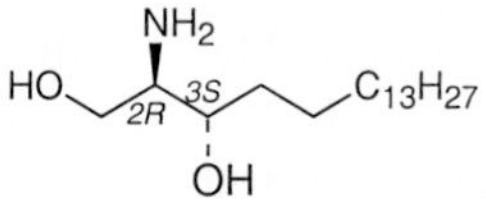

L-*erythro*-(2*R*,3*S*)-dihydrosphingosine **5a**

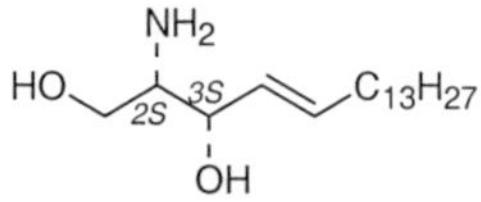

L-*threo*-(2*S*,3*S*)-sphingosine **4A**

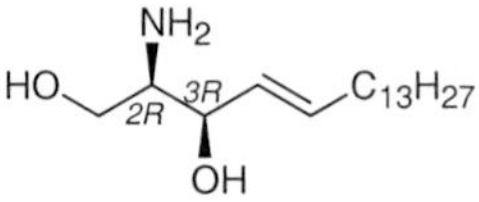

D-*threo*-(2*R*,3*R*)-sphingosine **6A**

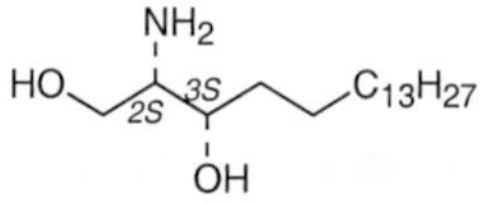

L-*threo*-(2*S*,3*S*)-dihydrosphingosine **4a**

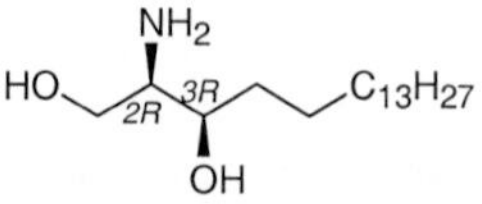

D-*threo*-(2*R*,3*R*)-dihydrosphingosine **6a**

FIG. 3. Structures of sphingosines and dihydrosphingosines.

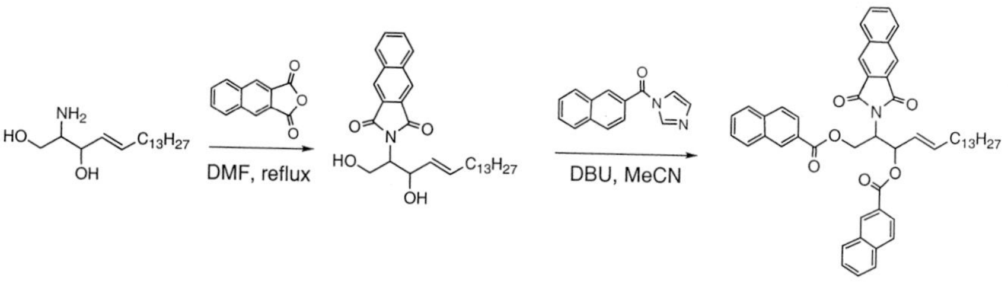

FIG. 4. Two-step chromophoric derivatization.

measurements. Note that the CD spectra of sphingosines and dihydrosphingosines are not the same, reflecting the difference in conformation of the unsaturated and saturated series (Fig. 5). The CD spectra of L-*erythro* and D-*threo* isomers are the mirror images of D-*erythro* and L-*threo* isomers, respectively.[13] Because conformational distribution of the derivative depends on the solvent, each isomer was measured in two different solvents, acetonitrile (polar) and methylcyclohexane (nonpolar). The CD measurements in two different conditions ensure the stereochemical identity of unknown species in this analysis.

Selection of chromophores was the key to the successful differentiation of these structurally close isomers (Fig. 4).[12] Monochromophoric derivatization, such as perbenzoylation or pernaphthoylation, cannot clearly differentiate the isomers since all stereochemical information is condensed within a narrow UV range of the single chromophore. Differentiation becomes easier if the derivative contains several different chromophores so that stereochemical information is spread out over a wider absorption range. The naphthimide group was selected for the derivatization of the primary amino group at C-2 because of its intense UV band (ε 64,000, λ_{max} 258 nm, acetonitrile) and fluorescence, both positive attributes for microscale manipulation. On the other hand, the 1,3-hydroxyl groups were derivatized to 2-naphthoates,[22] also an intensely absorbing chromophore (ε 58,000, λ_{max} 234 nm, acetonitrile) with an absorption maximum well separated from that of the naphthimide. The combination of the two different chromophores and the resulting complex chromophoric interactions made it possible to obtain characteristic CD curves covering the range of 200–320 nm for each isomer.

Two-Step Chromophoric Derivatization

Before the derivatization of sphingosine sample, one reagent, 2,3-naphthalenedicarboxylic acid anhydride, has to be prepared from the corre-

[22] N. Ikemoto, L.-C. Lo, and K. Nakanishi, *Angew. Chem. Int. Ed. Engl.* **31,** 890 (1992).

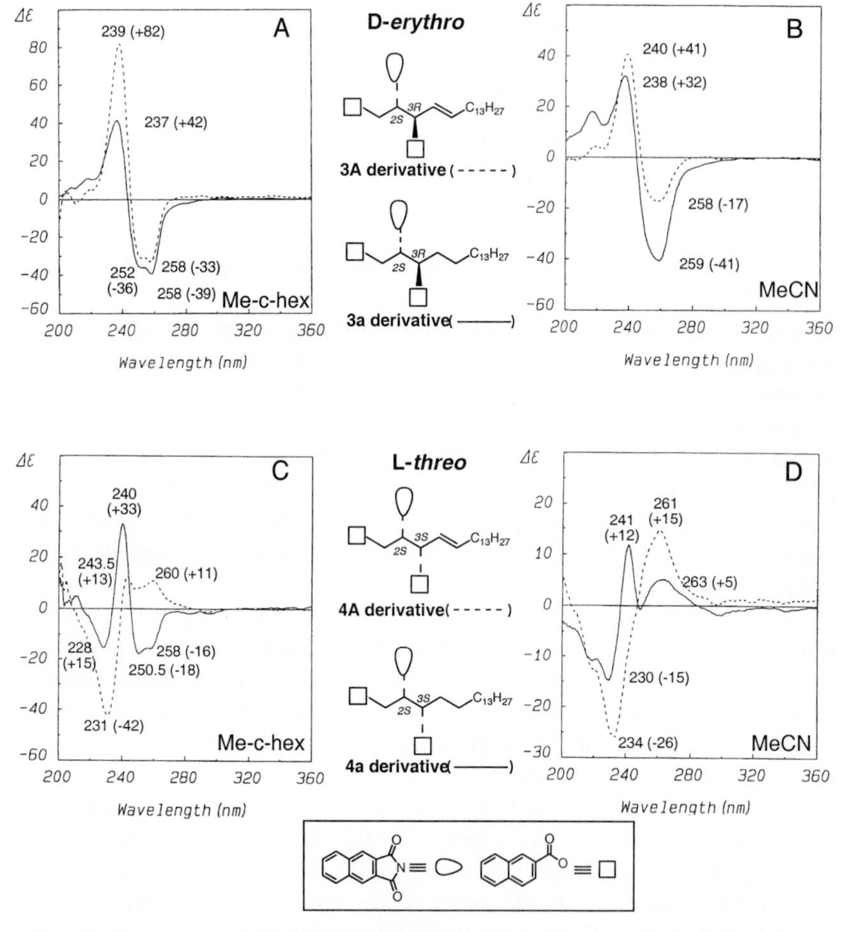

FIG. 5. CD spectra of 1,3-bisnaphthoate-2-*N*-naphthimide derivatives. (A) D-*erythro*-sphingosine (dotted line) and D-*erythro*-dihydrosphingosine (solid line) derivatives in methylcyclohexane. (B) D-*erythro*-sphingosine (dotted line) and D-*erythro*-dihydrosphingosine (solid line) derivatives in acetonitrile. (C) L-*threo*-sphingosine (dotted line) and L-*threo*-dihydrosphingosine (solid line) derivatives in methylcyclohexane. (D) L-*threo*-sphingosine (dotted line) and L-*threo*-dihydrosphingosine (solid line) derivatives in acetonitrile.

sponding diacid,[23] which is commercially available; all the other chemicals used in the derivatization can be obtained from commercial sources and used without further purification. 2,3-Naphthalenedicarboxylic acid (500 mg, 2.3 mmol) is placed in a dried 10-ml round-bottomed flask. Acetic

[23] B. H. Nicolet, J. A. Bender, F. C. Whitmore, and W. F. Singleton, *in* "Organic Synthesis" 2nd ed. (H. Gilman and A. H. Blatt, eds.), Coll. Vol. 1, p. 410. John Wiley & Sons, New York, 1941.

anhydride (2 ml, 21 mmol) is added, and the mixture is refluxed for 3 hr under inert gas (argon, nitrogen, etc.). The dark brown solution is cooled to room temperature. The crude product, which should be solidified by then, is collected by filtration with a Büchner funnel and washed with acetic acid and ether. The material is then dried at 100° overnight. The brownish anhydride is then purified by sublimation and stored in a desiccator.

The two-step derivatization is carried out as follows (Fig. 4). The first derivatization step, which uses the reagent prepared above, converts the 2-amino group into the naphthimide as follows. Pyridine was used as the solvent for this step in the previous report,[13] but a subsequent study found that dimethylformamide (DMF) gives more reproducible results, possibly by preventing by-product formation. The sample sphingosine (50 μg, 0.16 μmol) and 2,3-naphthalenedicarboxylic acid anhydride (36 μg, 0.18 μmol, 1.1 equivalent) are placed in a dried 10-ml round-bottomed flask. The mixture is dissolved in anhydrous DMF (200 μl) and refluxed under inert gas for 12 hr.

After cooling to room temperature, the sample is dried under vacuum pumping, and purified by TLC (5 $\times$ 20 cm, E. Merck silica gel 60 F-254, 250 μm). [*Note*: Wash the TLC plate before use with ethyl acetate since the purified product is extracted to ethyl acetate. The washing can be done by developing a new TLC plate (20 $\times$ 20 cm) with ethyl acetate, which is then dried and cut into the appropriate size.] When hexane–ethyl acetate (1:1) is used as the mobile phase, R_f values of the *erythro* isomers are around 0.26, while that of the *threo* isomers are around 0.38. The product spot on the developed TLC plate, which is seen under the UV lamp, is scraped and extracted to ethyl acetate. The yield can be checked with the UV absorption (*N*-naphthimide, ε 64,000, λ_{max} 258 nm, acetonitrile). At least 50% yield (44 μg of the product) should be attainable after some practice.

The purified product, *N*-naphthimide, is then submitted to the second step, in which the 2-naphthoate group is attached to the 1,3-hydroxyl groups.[22] *N*-Naphthimide (44 μg, 0.08 μmol) is placed in a dried 10-ml round-bottomed flask, dissolved in anhydrous acetonitrile (200 μl), 1-(2-naphthoyl)imidazole (200 μg, excess) is added, and a catalytic amount of DBU (1 drop of 10% DBU solution in anhydrous acetonitrile) is introduced. The mixture is stirred at room temperature for 1 hr. The solvent is evaporated and the final product is purified by TLC (5 $\times$ 20 cm, E. Merck silica gel 60 F-254, 250 μm, prewashed with ethyl acetate). Hexane–ethyl acetate (3:1) gives R_f values of ca. 0.36 and 0.39 for the *erythro* and *threo* isomers, respectively. The product spot is scraped and extracted with ethyl acetate. The yield of this step should easily exceed 70% after some practice. The purified 1,3-bisnaphthoate-2-*N*-naphthimido derivative is then submitted to the CD measurements in acetonitrile (polar) and methylcyclohexane (nonpolar).

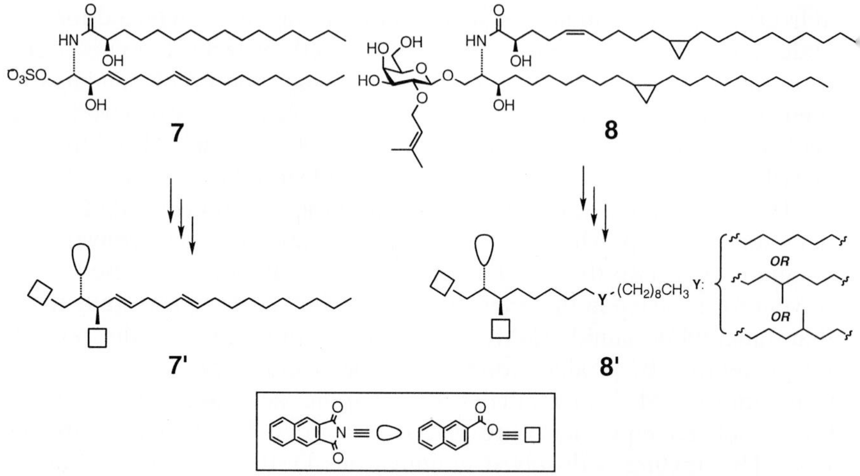

FIG. 6. Applications to unknown sphingolipids; a ceramide 1-sulfate **7** and a glycosphingolipid **8**. The sphingosine moieties of **7** and **8** were derivatized to the 1,3-bisnaphthoate-2-*N*-naphthimides, **7′** and **8′**, which gave CD curves close to those of D-*erythro*-sphingosine and D-*erythro*-dihydrosphingosine, respectively.

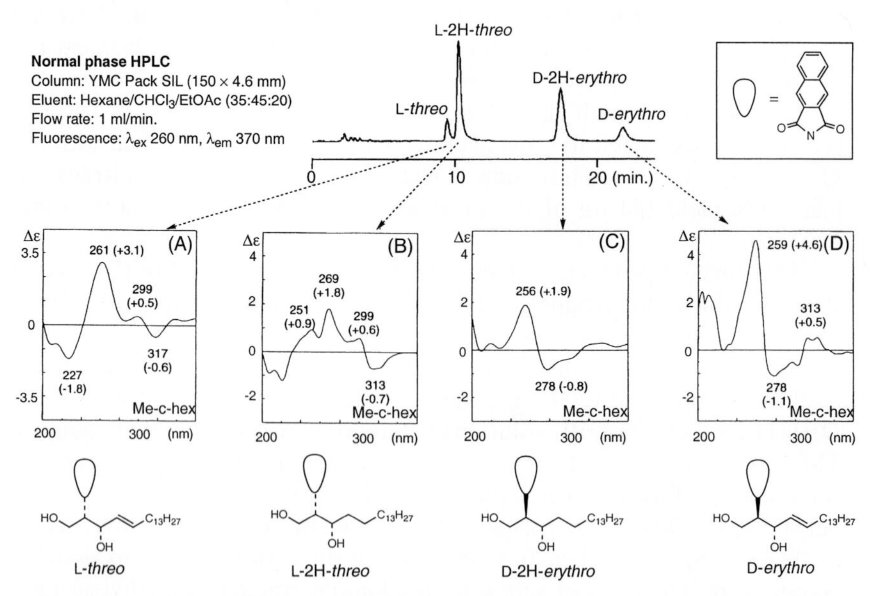

FIG. 7. Normal-phase HPLC separation of *N*-naphthimide derivatives, and their CD spectra in methylcyclohexane. (A) L-*threo*-Sphingosine derivative; (B) L-*threo*-dihydrosphingosine derivative; (C) D-*erythro*-dihydrosphingosine derivative; and (D) D-*erythro*-sphingosine derivative.

This CD method has been employed for the configurational assignment of several sphingolipids from marine sources (Fig. 6).[14,15] The sphingosine moiety from a ceramide 1-sulfate **7** was converted into the 1,3-bisnaphthoate-2-*N*-naphthimido derivative **7′**, which gave a negative CD curve closely resembling that of the D-*erythro*-sphingosine derivative.[14] Another sphingosine from a glycosphingolipid **8** was also derivatized to **8′**, which afforded a negative CD close to D-*erythro*-dihydrosphingosine derivative.[15] It seems that the double bond at C-4 affects the conformational profiles around the chromophore attachment sites, which is reflected to the CD shapes of the sphingosine derivative and **7′**. On the other hand, the double bond at C-8 in **7′** and cyclopropyl group at C-11 in **8′** do not appear to affect the CD outcome.

Attempts to Lower the Scale of Stereochemical Analysis

In addition to the CD method, a high-performance liquid chromotography (HPLC) protocol has been developed for the stereochemical analysis of sphingosine and dihydrosphingosine.[13] The sample material is derivatized to its *N*-naphthimide in this method, and analyzed by the sequential normal and chiral phase HPLC, which can differentiate all stereoisomers. The advantages of this HPLC protocol are that the method requires only one-step derivatization and the fluorescence detection of *N*-naphthimido derivative makes it possible to lower the analysis scale to the low nanogram level: Because each *N*-naphthimido derivative shows a weak but distinct CD curve, stereochemical assignment could also be done by CD spectroscopy provided the purified derivative is enough for the detection of the weak CD arising from the chiral environment surrounding the naphthimido group (Fig. 7). The drawbacks of the sequential HPLC protocol are that the method requires authentic derivatives for HPLC calibration; and the nanogram-level derivatization and purification requires experience handling small quantities of these materials.

There is an approach to improve the sensitivity of the exciton chirality method, in which fluorescence of chromophoric derivative is measured to obtain the CD curve, so-called fluorescence detected CD (FDCD).[24] Although FDCD has not been applied to sphingolipid analysis, it would theoretically lower the CD detection level to the nanogram scale. The limiting factor here, however, is not the sensitivity of FDCD, but the two-step derivatization, which is not easily scaled down to the nanogram level.

[24] J.-G. Dong, A. Wada, T. Takakuwa, K. Nakanishi, and N. Berova, *J. Am. Chem. Soc.* **119,** 12024 (1997).

[20] Infrared Determination of Conformational Order and Phase Behavior in Ceramides and Stratum Corneum Models

By RICHARD MENDELSOHN and DAVID J. MOORE

Introduction

Infrared (IR) spectroscopy offers many advantages for the study of lipid structure and interactions in macromolecular assemblies. The technique provides a submolecular probe of diverse organizational elements including acyl chain conformation and packing, chain segment orientation, interfacial region hydrogen bonding, and headgroup ionic interactions. The technology is versatile in that many biologically relevant states, including vesicle dispersions, multilayer arrays, monolayers *in situ* at air/water interfaces, crystalline solids, aqueous solutions, etc., can be directly sampled. In addition, structural data arise directly from the molecules under investigation without requiring the use of spectroscopic probes. This eliminates ambiguities that arise with probe-based techniques such as fluorescence or ESR spectroscopies in which data interpretation requires extrapolation of results from the labeled to the native molecule. Occasionally, deuterium substitution in the hydrocarbon chains under investigation or the substitution of D_2O for H_2O as solvent is useful for IR studies. This substitution is innocuous in that it produces negligible perturbations of molecular structural parameters. Finally, as a practical matter, small amounts of material (0.1–1 mg for solids, 50 μl of aqueous suspensions at 1–5% by weight) suffice for routine measurements. Much smaller amounts suffice for samples examined by microscopy or monolayer approaches. As will be discussed below, the ordered phases generally adopted by ceramides and stratum corneum models under physiologic conditions are particularly well suited for examination by IR spectroscopic methods.

This chapter outlines some sampling techniques and summarizes the type of information available from the IR spectra of lipid molecules, with emphasis on ceramides. We illustrate the power of the technology with appropriate measurements from ceramides or related systems. Finally, we briefly describe two recently developed novel experiments with obvious potential for studying ceramide-related systems, namely IR reflection–absorption spectroscopy (IRRAS) of monolayer films at the air/water interface and IR microscopic imaging with a focal plane array detector.

There are at least seven major ceramide species in the stratum corneum,

0076-6879/00 $30.00

the outer layer of the skin, which are routinely labeled according to their polarity as ceramides 1 through 7. For simplicity and clarity all data shown in this chapter come from our studies of ceramide 2 alone or mixed with hexadecanoic acid and cholesterol (a standard model of the stratum corneum lipid barrier). Ceramide 2 consists of nonhydroxy fatty acids, amide linked to sphingosine, and is one of the most abundant stratum corneum ceramides. The nomenclature "ceramide 2" is used throughout this chapter.

Experimental Techniques

The widespread and increasing application of IR spectroscopy to biologic investigations, including membrane biophysics, is directly attributable to the development of affordable bench-top Fourier transform IR instruments in the last 25 years. Therefore, before outlining some of the experimental techniques employed in such studies, including sample preparation, data collection, and data manipulation, a brief description of a Fourier transform infrared (FT-IR) spectrometer will be given. For a more detailed description, the reader is referred to one of the standard works in the field.[1]

Components of IR Spectrometer

A typical FT-IR spectrometer consists of an infrared source, a Michelson interferometer, an infrared detector, a laser, and a computer. The interferometer consists of a beam splitter, a fixed mirror, and a moving mirror. Radiation from the source illuminates the beam splitter at an angle of 45° at which point (ideally) 50% of the light is transmitted and 50% is reflected. When the light from the two arms of the interferometer is recombined at the beam splitter, interference occurs between the reflected and transmitted radiation. The interferogram recorded at the detector measures the intensity of the light that has been recombined at the beam splitter (containing all infrared frequencies) as a function of the position of the moving mirror. The light intensity $I'(x)$ at a single frequency (ν) as a function of retardation (x, defined as twice the difference in path length between the two arms of the interferometer), where $I(\nu)$ is the light intensity of the source, is given by

$$I'(x) = 0.5I(\nu)[1 + \cos(2\pi\nu x)]$$

Thus the intensity of light at the detector consists of two components, a constant part $[0.5I(\nu)]$ and a varying component $0.5I(\nu)\cos(2\pi\nu x)$. The

[1] P. R. Griffiths and J. A. de Haseth, "Fourier Transform Infrared Spectrometry." John Wiley & Sons, New York, 1986.

varying component is referred to as the interferogram and is of greatest interest for spectroscopy.

A variety of instrumental factors affect the interferogram intensity. However, for a given instrument configuration, the only parameter that varies between measurements is $I(\nu)$. The interferogram is usually modified with a frequency-dependent correction factor combined with $I(\nu)$ and defined as $B(\nu)$, the single-beam spectral intensity. $B(\nu)$ gives the intensity of the source at a wave number (ν) as modified by the instrument. Therefore, the interferogram can be written as:

$$I(x) = B(\nu)\cos(2\pi\nu x)$$

For a polychromatic source the interferogram can be expressed as the integral:

$$I(x) = \int B(\nu)\cos(2\pi\nu x)\,d\nu$$

This is one-half of a cosine Fourier transform pair, the other half being:

$$B(\nu) = \int I(x)\cos(2\pi\nu x)\,dx$$

Thus, by collecting an interferogram (i.e., a recording of intensity versus mirror position) and performing a Fourier transform, a spectrum is generated of the light intensity as a function of frequency. The helium–neon laser present in the spectrometer provides an extremely precise signal that initiates data acquisition by the detector.

The detector most widely used in research-grade FT-IR spectrometers is liquid N_2-cooled mercury cadmium telluride (usually abbreviated as either HgCdTe or MCT). Within these photoconductive elements, the energy from an IR photon promotes a bound electron to a free state (conduction band), thereby increasing the electrical conductivity of the detector. Normal FT-IR spectra, as seen in most of the figures in this chapter, are acquired by placing a sample in the path of the recombined IR radiation coming from the beam splitter. The sample vibrational frequencies are absorbed (i.e., removed) from the light before light hits the detector. The Fourier transform of the sample interferogram is ratioed to the Fourier transform of a background interferogram (collected without a sample present) to produce the final absorbance spectrum characteristic of the sample. The final IR spectrum is usually a plot of sample absorbance [$-\log_{10}$(transmitted intensity/incident intensity)] as a function of wave number.

There are three major advantages in the above design of spectrometer that have enabled IR spectroscopy to become a powerful technique for studying biologic samples. First, the Connes advantage arises from the very high accuracy with which data are encoded as a function of retardation. This factor permits precise spectral subtractions. Second, the "multiplexing" or Fellgett advantage arises because an FT-IR spectrometer detects all frequencies simultaneously. For experiments (such as FT-IR) where the detector noise is independent of the signal, the signal/noise (*S/N*) ratio for a constant time of data acquisition, improves as $E^{1/2}$, where E is the number of resolution elements constituting the spectrum. Thus for the 3600-cm^{-1} spectral range (4000–400 cm^{-1}) that characterizes the mid-IR spectral region, if the resolution desired is 4 cm^{-1}, then Fellgett's advantage provides a gain of a factor of 30 in *S/N*, i.e., $(3600/4)^{1/2}$. This reduces by a factor of $(30)^2$ the time required for acquisition of a spectrum with a given *S/N* ratio compared with a grating instrument. The final advantage of FT-IR instrumentation results from the high intensity of IR radiation within the spectrometer. This is known as the Jacquinot advantage and produces a frequency-dependent enhancement of the *S/N* ratio.

Preparing Lipid Samples for IR Spectroscopy

Transmission spectroscopy is the technique of choice for studies of lipid vesicles. In these experiments, lipids are initially dried from organic solvents using N_2 flow or a rotary evaporator. The lipid film is then placed under vacuum for several hours to remove traces of solvent. Buffer is added and the sample is taken through sufficient heating and cooling cycles to ensure full hydration. Physical agitation of the sample such as mixing by vortex action or by sonication helps speed the hydration process. To prepare vesicles of a given size, samples may be extruded through filters and size fractionated with gel permeation chromatography. To prepare lipid vesicles for IR spectroscopic examination, a sample is sandwiched between two IR disks (usually water-insoluble crystals of CaF_2 or BaF_2) separated by a thin Teflon spacer and placed within a temperature-controlled cell. The holder is placed in the sample compartment of the spectrometer. For studies of the thermotropic characteristic of lipid suspensions, spectra are acquired as a function of temperature. The sample volume required for these experiments is ~20 μl, depending on the thickness of the spacer. However, note that for aqueous samples, such as most biologic samples, path length is limited to less than ~50 μm, due to the intense absorption of liquid water.

The horizontal attenuated total reflection (HATR) technique offers some specific advantages for studying lipid samples because the sample is physically accessible throughout the experiment. In this approach, the IR radiation is internally reflected through a horizontally mounted long flat crystal that is covered with a sample. A small amount of radiation (the evanescent wave) penetrates into the sample when the crystal is covered with an absorbing material. Detailed descriptions of the physics of

ATR-IR spectroscopy are available.[2] In many commercially available HATR units, the ATR crystal is mounted in a trough. Samples are prepared by casting a lipid film on the crystal via evaporation from organic solvent. Buffer is then added to cover the sample. This arrangement permits access to the buffer throughout the experiment. Variables such as pH, ionic strength, and ionic concentration may be altered and their effect on the lipid bilayers directly monitored. Temperature-controlled HATR units are available that augment the versatility of this technique for investigations of ceramides. An added advantage of the ATR approach is that samples may be oriented on the crystal surface. Experiments with polarized radiation permit determination of the orientation of transition dipoles on the surface.[3]

When studying lipid dispersions, either with transmission or HATR methods, spectra of a given resolution are generated after ratioing background and sample interferograms. To avoid interfering absorption from water vapor, spectra are normally acquired with the spectrometer under N_2 or dry air purge. Spectra of high signal-to-noise ratios ($>5000:1$) may be generated with 0.5–5 min of scanning.

The success of biologic FTIR depends to a large extent on the experimenter's ability to subtract one spectrum from another with great precision. This permits, for example, a buffer spectrum to be subtracted from a spectrum of hydrated lipids, leaving only the spectrum of the lipid molecules. Other useful techniques for enhanced visualization of spectra include derivative spectroscopy in which a second derivative spectrum is mathematically generated from an original absorbance spectrum. The technique enhances weak (but relatively narrow) bands and thereby permits their frequency to be determined with a precision of better than 0.1 cm^{-1}.

IR Spectral Regions Sensitive to Ceramide Structure and Organization

For illustrative purposes, a mid-IR spectrum of a ceramide 2 film is shown in Fig. 1. The vibrational modes are labeled. As is evident from Fig. 1, the information content of the spectrum is quite high; vibrations arising from both the ceramide chains and the headgroups are readily observed. The vibrations of the hydrocarbon chains are most sensitive to ceramide conformational order and packing, whereas the amide bond vibrations of the polar head regions are sensitive to hydrogen bonding interactions and hydrogen–deuterium ($H \rightarrow D$) exchange.

[2] N. J. Harrick, "Internal Reflection Spectroscopy." Harrick Scientific Corporation, Ossining, New York, 1967.

[3] L. K. Tamm and S. A. Tatulian, *Quart. Rev. Biophys.* **30,** 365 (1997).

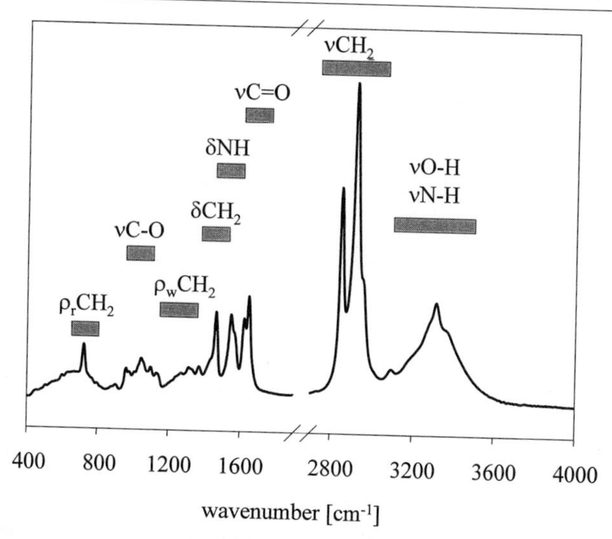

FIG. 1. A representative IR spectrum of ceramide 2. The positions of some of the major headgroup and chain vibrational modes discussed in this chapter are labeled.

An important advantage for the application of IR to investigate the structure of ceramides is the availability of detailed spectral assignments and spectra–structure correlations, especially those arising from the hydrocarbon chains. These assignments and correlations have been derived to a large extent from the pioneering, seminal studies of alkanes in the solid, liquid, and rotator phases by R. G. Snyder and associates.[4–9] The spectral assignments and frequency ranges for a variety of modes used to probe lipid structure, are summarized in Table I.

Chain Vibrations

Methylene Stretching Modes. The most intense chain modes in the IR spectra of lipids arise from the methylene CH_2 (CD_2 in deuterated chains) stretching vibrations. The symmetric ($\nu_{sym}CH_2$) and asymmetric ($\nu_{asym}CH_2$) modes occur at ~2850 and 2920 cm^{-1} in proteated chains and ~2090 ($\nu_{sym}CD_2$) and ~2195 ($\nu_{asym}CD_2$) cm^{-1} in perdeuterated chains. The variation of these frequencies with temperature or with sample composition

[4] R. G. Snyder, *J. Mol. Spec.* **4,** 411 (1960).
[5] R. G. Snyder, *J. Mol. Spec.* **7,** 116 (1961).
[6] R. G. Snyder and J. H. Schachtschneider, *Spectrochim. Acta* **19,** 85 (1963).
[7] R. G. Snyder, S. L. Hsu, and S. Krimm, *Spectrochim. Acta* **34A,** 395 (1978).
[8] R. G. Snyder, *J. Chem. Phys.* **47,** 1316 (1967).
[9] Y. Kim, H. L. Strauss, and R. G. Snyder, *J. Phys. Chem.* **93,** 7520 (1989).

TABLE I
IR SPECTRAL REGIONS THAT MONITOR LIPID CONFORMATION

Frequency (cm^{-1})	Vibrational mode	Sensitivity
2846–2855	CH_2 symmetric stretch	Frequency increases when chains disorder
2916–2924	CH_2 asymmetric stretch	Frequency increases when chains disorder
2090–2100	CD_2 symmetric stretch	Frequency increases when chains disorder
2180–2195	CD_2 asymmetric stretch	Frequency increases when chains disorder
1720–1740	Ester C=O stretch	Frequency is lowered on H-bond formation
	CH_2 scissoring modes	
1462, 1473		Orthorhombic phase doublet
1468		Hexagonal phase
1466		Disordered phase
	CH_2 wagging (disordered)	
1368		*gtg* or kink (*gtg*′) conformer sequence
1353		*gg* sequence
1341		End gauche conformer
1180–1350	CH_2 wagging (ordered)	In ordered phases, these modes split in a fashion characteristic of the particular chain length
1250	PO^-_2 asymmetric stretch	Ion binding lowers this frequency
1086–1094	CD_2 scissoring mode	Mode splits into a doublet in orthorhombic phases
720	CH_2 rocking mode CD_2 rocking mode (isolated CD_2 group)	Mode splits into a doublet in orthorhombic phases
~680		*gg* conformational sequence
650		*tg* conformation in isolated *g* states
620		*tt* conformational sequence

provides a qualitative indication of alterations in chain packing or conformational order, and has been utilized in studies of ceramides.[10,11] While the sensitivity of the stretching modes to conformational change has been widely recognized and extensively applied to characterize conformational order changes during gel → liquid crystal phase transitions in phospholipids and the effects of the interaction of membrane proteins with phospholipids, the sensitivity of these vibrations to altered chain packing is not as widely recognized.[12] Yet for the ordered phases that occur in ceramides, chain

[10] D. J. Moore, M. E. Rerek, and R. Mendelsohn, *Biochem. Biophy. Res. Comm.* **231,** 797 (1997).

[11] D. J. Moore, M. E. Rerek, and R. Mendelsohn, *J. Phys. Chem. B.* **101,** 8933 (1997).

[12] R. A. MacPhail, H. L. Strauss, R. G. Snyder, and C. A. Ellinger, *J. Phys. Chem.* **88,** 334 (1984).

packing alterations are quite likely to occur under physiologically relevant conditions.

As an example of the use of the CH_2 stretching frequencies to monitor structural alterations in ceramides, the temperature dependence of ν_{sym} CH_2 in a sample of ceramide 2 is presented in Fig. 2. Two transitions are evident from the melting profile. The higher temperature transition, centered at ~80°, is characterized by a frequency increase from ~2849.5 to ~2853 cm^{-1}. (Note that these frequencies can be easily measured with a precision of 0.05–0.1 cm^{-1}, so that although a 4- or 5-cm^{-1} shift may seem relatively small compared to the range of mid-IR frequencies, structural transitions characterized by shifts of this magnitude can be readily detected.) This range of frequency shift over a fairly narrow temperature interval is characteristic of an order → disorder transition in the hydrocarbon chains. At temperatures below the cooperative transition, the chains are fully extended and conformationally ordered (all-*trans*), while at temperatures >80°, a substantial amount of conformational disorder has been introduced through the cooperative introduction of *gauche* conformers into the chains.

In contrast to the high-temperature transition, which arises from an order → disorder transition, the low-temperature transition seen in Fig. 2, characterized by a midpoint of ~60° and a frequency increase from 2848 to 2849 cm^{-1}, arises from packing changes. In general, transitions detected

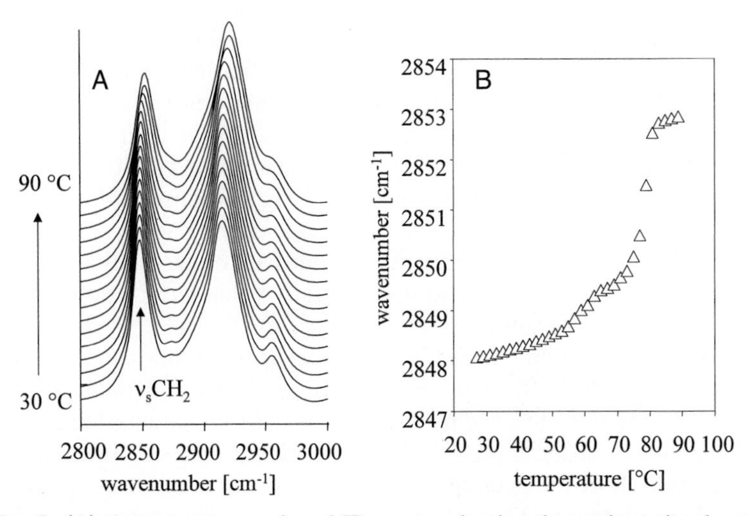

FIG. 2. (A) A temperature series of IR spectra showing the conformational-sensitive methylene stretching region (2800–3000 cm^{-1}) modes of hydrated ceramide 2. (B) The frequency of the symmetric CH_2 stretching mode plotted as a function of temperature showing transitions at 60° and 80°.

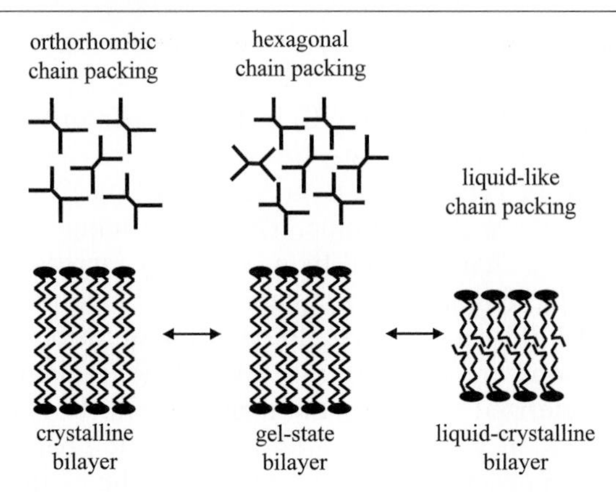

FIG. 3. Schematic of inter- and intramolecular hydrocarbon chain packing observed in lamellar lipid bilayers.

from cooperative changes in methylene stretching frequencies are assumed to arise from packing alterations if the final frequency position is $<2850\ cm^{-1}$, and from the introduction of conformational disorder if the final frequency is $>2850\ cm^{-1}$. Although this apparently sharp division of positions may seem a bit arbitrary, a comparison of X-ray diffraction studies[13] with IR results confirms the suggested delineation.

Although the methylene stretching vibrations provide a convenient means to detect structural and packing alterations in ceramide chains, the information from these modes is essentially qualitative. Thus, it has proven difficult to correlate the magnitude of the increase in CH_2 stretching frequency during gel → liquid crystal phase transitions with the extent of *gauche* rotamer formation either at a particular position in the acyl chains or averaged over all chain positions. Similarly, the observation of a CH_2 stretching frequency increase below $2850\ cm^{-1}$ during solid → solid phase transitions cannot be used to identify the particular phases involved. Other spectral parameters provide more quantitative characterizations of structural alterations in ceramides.

Methylene Scissoring Modes. The CH_2 scissoring modes, $\delta(CH_2)$, which span the range 1460–1475 cm^{-1} for proteated chains, and 1085–1095 cm^{-1} for perdeuterated chains, are particularly useful for characterizing packing arrangements in ordered phases. The scissoring modes on adjacent chains interact via a short-range coupling that occurs in a direction perpendicular to the chain direction. This interaction produces a broadening or splitting

[13] W.-S. Sun, S. Tristram-Nagle, R. M. Suter, and J. F. Nagle, *Biochim. Biophys. Acta* **1279,** 17 (1996).

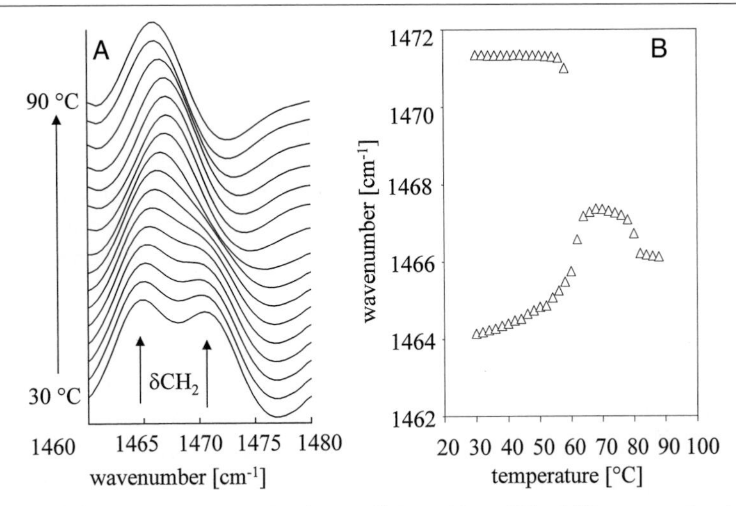

FIG. 4. (A) A temperature series (every 4° from 30 to 90°) of IR spectra showing the methylene bending mode region (~1468 cm^{-1}) of hydrated ceramide 2. Two peaks can be clearly observed in the spectra below 60°. (B) The frequencies of the symmetric CH_2 bending modes plotted as a function of temperature. The collapse of the splitting near 60° indicates a solid–solid phase transition from an orthorhombic perpendicular to a hexagonal packing motif.

of the scissoring contour only when neighboring chains are conformationally ordered and, in addition, are packed in orthorhombic perpendicular fashion (various common chain packing arrangements are depicted in Fig. 3). The appearance of broadening or splitting in the scissoring mode contour thus directly identifies the presence of this solid-phase structure. The utility of the $\delta(CH_2)$ modes for identifying an orthorhombic perpendicular packing in ceramide 2 is shown in Fig. 4. The observation of splitting immediately identifies the orthorhombic perpendicular chain packing arrangement below 60°, while the temperature-dependent collapse of the splitting at 60° (see Fig. 4) to a single peak at 1467 cm^{-1} identifies a transition from an orthorhombic to a hexagonal packing of the chains (see Table I). The order → disorder transition at 80° to a conformationally disordered phase produces an additional 1-cm^{-1} decrease in the scissoring frequency.

Quantitative evaluation of the $\delta(CH_2)$ contour by Snyder and collaborators has shown that the magnitude of the splitting or broadening monitors the extent of microaggregation (lateral domain formation) in orthorhombic phases.[14,15] The utility of the scissoring modes for evaluation of the domain size distribution function is based on the fact that the scissoring vibrational

[14] R. G. Snyder, G. Conti, H. L. Strauss, and D. L. Dorset, *J. Phys. Chem.* **97,** 7342 (1993).
[15] R. G. Snyder, H. L. Strauss, and D. A. Cates, *J. Phys. Chem.* **99,** 8432 (1995).

frequencies of proteated chains (1467 cm^{-1}) versus deuterated chains (1090 cm^{-1}) are sufficiently different so as to eliminate vibrational coupling between them. In lipid mixtures containing proteated ceramide and deuterated fatty acid chains, the CH_2 scissoring modes of the ceramide domains will not interact with the CD_2 scissoring modes of the fatty acid domains.[10] From a quantitative point of view, the magnitude of the scissoring band splitting is a function of the domain size. The latter can therefore be determined from the spectrum providing that the relationship between scissoring splitting and size is elucidated. Snyder *et al.*[15] have discussed this point and determined this relationship. The method is most sensitive for domain sizes below 100 chains,[15,16] and will prove useful for studying mixtures of proteated ceramides with chain perdeuterated fatty acids, where various types of solid phases and miscibilities may occur. For example, the components of the mixture may be ideally miscible in the solid state. Alternatively, they may be phase separated into domains with a wide possible distribution of lateral dimensions.

Another type of experiment that becomes feasible, thanks to the sensitivity of scissoring contour to chain packing, is the examination of the kinetics of microdomain formation.[17] The time evolution of orthorhombic phase formation in a sample of ceramide 2 is shown in Fig. 5. The sample was initially heated to 90°, then cooled rapidly to room temperature where spectra were acquired at the desired period of time after quenching. The time evolution of the splitting of the scissoring modes is evident from the series of difference spectra generated by subtracting the spectrum of the time 0 spectrum from spectra acquired at succeeding times.

Methylene Wagging Modes. The CH_2 wagging modes, $\rho_w(CH_2)$, although substantially weaker in intensity than either the stretching or scissoring vibrations, provide a unique means to characterize chain conformations in both ordered[4] and disordered[8] hydrocarbon chain phases. The nature of the vibrational interactions between the wagging modes of adjacent methylenes within a chain differs dramatically from the scissoring modes. Whereas the former are sensitive to interchain interactions, the latter couple strongly within a single chain. In addition, the coupling of the wagging modes between adjacent CH_2 groups within a single chain changes substantially in going from conformationally ordered to disordered states. This fact has been used to probe conformational states of phospholipids.[18,19] In conforma-

[16] R. Mendelsohn, G. L. Liang, H. L. Strauss, and R. G. Snyder, *Biophys. J.* **69,** 1987 (1996).

[17] R. G. Snyder, M. C. Goh, V. J. P. Srivatsavoy, H. L. Strauss, and D. L. Dorset, *J. Phys. Chem.* **96,** 10008 (1992).

[18] L. Senak, D. J. Moore, and R. Mendelsohn, *J. Phys. Chem.* **96,** 2749 (1992).

[19] L. Senak, M. A. Davies, and R. Mendelsohn, *J. Phys. Chem.* **95,** 2565 (1991).

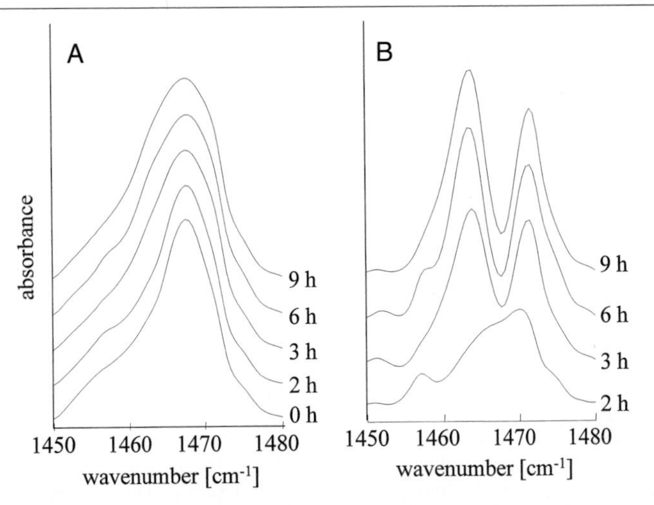

FIG. 5. (A) A series of spectra taken over 9 hr showing the CH_2 bending mode region of ceramide 2 in a model stratum corneum system also containing cholesterol and perdeuterated hexadecanoic acid. The broadening indicates the formation of orthorhombic perpendicular domains of ceramide 2 chains. (B) The corresponding difference spectra generated by subtraction of the 0-hr spectrum from spectra at later times. The time-dependent formation of orthorhombic perpendicular ceramide 2 domains is clearly revealed in these spectra.

tionally ordered phases, the wagging modes behave as a system of coupled harmonic oscillators. The eigenvalues for the vibrational secular equation are

$$4\pi^2\nu^2 = H_0 + 2\Sigma H_m \cos(m\Phi_k) \qquad (1)$$

In Eq. (1), the Φ values are phase differences between adjacent oscillators, i.e., $\Phi_k = k\pi/(n + 1)$, where $k = 1, 2, 3, \ldots, n$ and n is the number of CH_2 groups, and H_m defines the coupling strength. The progression is strongest between 1180 and 1300 cm^{-1}, although peaks may extend to 1360 cm^{-1}. Two important points regarding the use of these coupled modes for the study of ordered phases are, first, that the appearance of the progression depends mostly on the presence of an all-*trans* chain and, second, that the actual number and position of bands depends very strongly on the chain length of the all-*trans* segment. Thus, ordered chains can be readily identified in a complex mixture, even in the technically difficult case of intact cells.[20,21] This spectral region is demonstrated in Fig. 6. The wagging modes of ceramide 2 are plotted at temperatures above and below the

[20] D. J. Moore, M. Wyrwa, C. P. Reboulleau, and R. Mendelsohn, *Biochem.* **32,** 6281 (1993).
[21] D. J. Moore and R. Mendelsohn, *Biochem.* **33,** 4080 (1994).

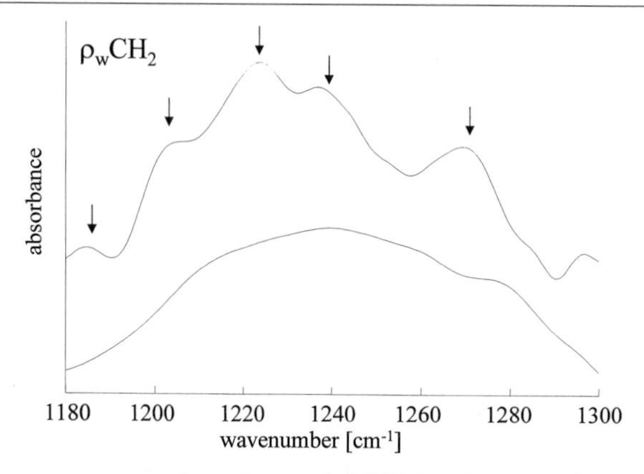

FIG. 6. The upper spectrum shows the coupled CH_2 wagging modes (marked with arrows) of hydrated ceramide 2 at physiologic temperature (30°). These wagging modes are extremely sensitive to chain conformation and are characteristic of all-*trans* chains. The lower spectrum was taken at a higher temperature (80°) and the wagging modes are no longer present in the conformationally disordered chains.

chain melting transition. The observed progression vanishes at the melting temperature.

As noted above, in disordered phases the coupling between adjacent oscillators changes completely. The presence of conformational disorder greatly diminishes the vibrational coupling along the chain, thus the wagging modes become much more localized. Spectral features arise from particular, fairly localized, two- or three-bond conformational states. The progression of bands that characterizes the all-*trans* conformation is replaced by three features when the chains are disordered as follows (Table I): a band at 1341 cm^{-1} arising from end *gauche* conformers (rotations about the penultimate C–C bond), a band at 1353 cm^{-1} arising from double *gauche* states, and a band at 1368 cm^{-1} arising from the sum of *gauche–trans–gauche* and *gauche–trans–gauche′* states.[19] The utility of this spectral region for characterization of conformational order in disordered chains is shown in Fig. 7. Spectra of hydrated ceramide 2 at a temperature (85°) above the gel–liquid crystal phase transition are shown along with spectra at 73° (between the solid–solid and gel–liquid crystal transitions) and 30° (below the solid–solid phase transition). A spectrum of the molecule in $CHCl_3$ solution is included for comparison. The localized conformations giving rise to spectral features are marked in Fig. 7. In $CHCl_3$ solution (where ceramide 2 is either monomeric or in an inverse micellar state) the 1341-, 1353-, and 1368-cm^{-1} bands are all visible with substantial intensity. The spectrum of vesicles at 85° resembles that of the $CHCl_3$ solution. This

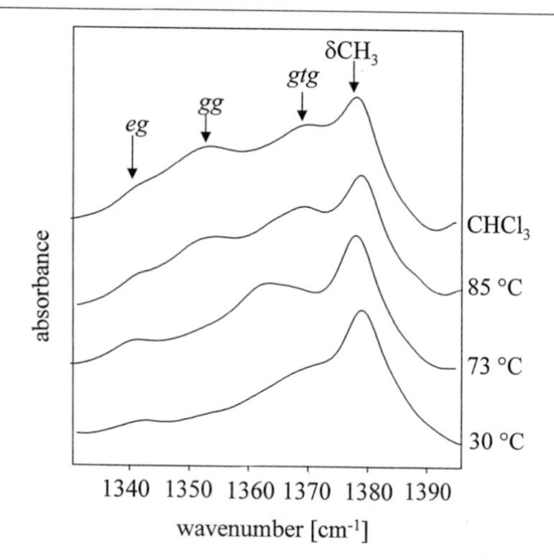

FIG. 7. The localized CH_2 wagging modes of conformationally disordered hydrated ceramide 2 in $CHCl_3$ solution, and at 80, 73, and 30°, respectively. The particular 2- and 3-bond conformations that give rise to these modes are indicated. The bottom spectrum was acquired at the physiologic temperature (30°) for skin where the intensities of these modes are substantially reduced.

immediately delineates the liquid-like, highly disordered nature of the chains. In contrast, in spectra acquired at 30°, the *eg* and *gtg* intensities are substantially reduced in intensity compared to their high-temperature counterparts, whereas the *gg* peak intensity is nearly eliminated. The elimination of *gg* intensity is a manifestation of the packing disruption induced by that conformation in the ordered phases.

The wagging modes offer a further advantage in that they present the possibility of quantitatively determining the extent of a particular form of conformational disorder. The approach, outlined in Ref. 19, uses IR spectra of liquid alkanes as reference standards to calibrate the IR intensities of the bands arising from particular chain states. The occurrence of particular conformations in alkanes can be calculated accurately theoretically. Comparison of the experimental IR spectra of liquid alkanes with the theoretical calculation of conformational populations permits the determination of extinction coefficients for each of the aforementioned localized wagging modes. Comparisons of lipid spectra with alkane spectra then lead to quantitative estimates of the occurrence of a particular type of disorder in lipid phases.

CD_2 Rocking Modes. A final spectral region that provides a quantitative estimate of conformational disorder at particular chain positions arises from CD_2 rocking modes (610–690 cm^{-1}) in (synthesized) lipid derivatives

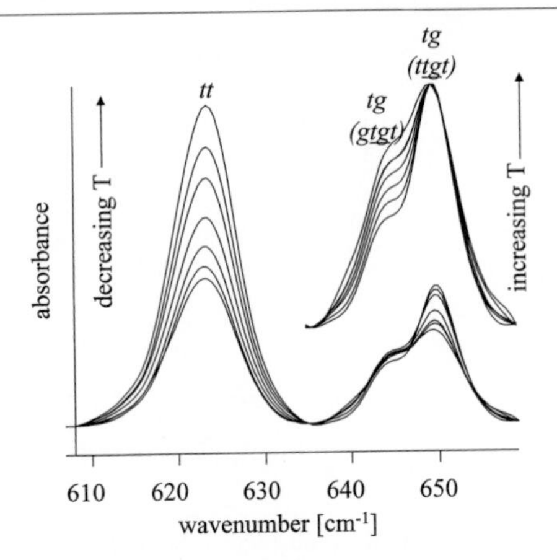

FIG. 8. Temperature dependence of the CD_2 rocking modes in liquid heptane -4-d_2. Six spectra are plotted in 10-degree intervals ranging from −70° to −20°. The bands arising from the *tt* conformation around the central C–C–C skeleton absorb at 622 cm^{-1} and diminish in intensity with increasing temperature. Concomitantly, two marker bands from disordered states containing isolated *gauche* bonds (644, 649 cm^{-1}), increase in intensity as the temperature is raised. The lower frequency band increases faster in intensity as the temperature is raised.

possessing isolated CD_2 groups in the chains. The spectroscopic approach is based on the observation of Snyder and Poore[22] that the rocking frequency of an isolated CD_2 group along a chain is dependent on the conformation of the C–C bonds on either side of the group.[23,24] The rocking frequencies associated with *trans–trans, trans–gauche*, and *gauche–gauche* conformations of the three-carbon fragment containing the central CD_2 appear at ~622, 650, and 680 cm^{-1}, respectively. The 650-cm^{-1} band is further slightly split by conformational effects further along the chain (see Fig. 8). The relative extinction coefficients of the *trans–trans* and *trans–gauche* marker bands have been determined both from theoretical models and from experiments with cycloalkanes. This provides the pathway to quantitative determination of conformational order as a function of acyl chain position in phospholipids.[25] From an experimental viewpoint, detection of the rocking modes in hydrated lipid assemblies is technically difficult, due to solvent (both H_2O and D_2O) absorption in this spectral region, which renders the accurate background compensation needed for quantitative intensity

[22] R. G. Synder and M. W. Poore, *Macromolecules* **6,** 708 (1973).
[23] M. Maroncelli, H. L. Strauss, and R. G. Snyder, *J. Phys. Chem.* **89,** 4390 (1985).
[24] R. G. Snyder, M. Maroncelli, and H. L. Strauss, *J. Am. Chem. Soc.* **105,** 133 (1983).
[25] R. Mendelsohn, M. A. Davies, J. W. Brauner, H. F. Schuster, and R. A. Dluhy, *Biochem.* **28,** 8934 (1989).

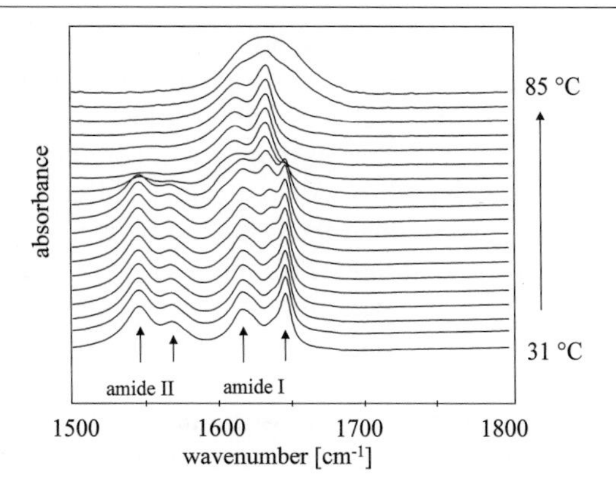

FIG. 9. The amide I and II spectral region of ceramide 2 hydrated in D_2O. At lower temperatures both the amide I and II modes are split. At the orthorhombic → hexagonal transition the amide II mode disappears due to H → D exchange and the amide I modes shift but remain split until 80°.

measurements of these very weak spectral features extremely challenging. Nevertheless, this mode is probably the single most powerful IR spectral feature for the characterization of chain conformational disorder. This spectral region is shown in Fig. 8 for a sample of heptane-4-d_2 in the liquid phase over a wide range of temperatures. The increased relative intensity of the *trans–gauche* marker bands, present as a doublet near 650 cm^{-1}, compared to the 622 cm^{-1} *trans–trans* rocking mode as the temperature is raised, is evident.

Polar Region Vibrations

The amide bonds of the ceramides provide substantial insight into the structure and accessibility to water in the polar regions of the molecule.[11] The amide I (~1650 cm^{-1}) and II (1550 cm^{-1}) modes are widely used in protein secondary structure determination.[26] The former arises from (mostly) C=O stretch while the latter arises from a mixture of N–H in-plane bending and C–N stretching internal coordinates. Spectra of the amide I and II regions of ceramide 2 in D_2O (Fig. 9) show remarkable alterations as the temperature is increased. At low temperatures, both modes are split, a consequence of interactions between amide groups in the structure. As the temperature is raised, the penetration of D_2O and its exchange for the amide hydrogen is evident from two spectral alterations. First, when H → D exchange is complete, each component of the amide I

[26] T. Miyazawa and E. R. Blout, *J. Am. Chem. Soc.* **83,** 712 (1961).

mode doublet is shifted to lower frequency. The frequency shifts because the amide I mode is not a pure C=O stretching mode but contains minor contributions from other internal coordinates that are sensitive to H → D exchange. At temperatures where exchange is incomplete, four peaks are evident in the contour. Second, H → D exchange shifts the amide II mode from ~1550 to ~1450 cm^{-1}, since the coupling between N–H stretch and C–N stretch that defines the amide II frequency is greatly altered by deuterium incorporation. Thus the penetration of water into the polar regions of the ceramide can easily be monitored by disappearance of the original amide II mode (Fig. 9). It is interesting to note that H → D exchange is complete at the temperature of the orthorhombic → hexagonal solid–solid phase transition, whereas the splitting of the amide I contour persists until the higher temperature order → disorder transition.

An additional mode in this spectral region arising from the acid C=O stretch appears in mixtures containing fatty acids. For example, in a three-component ceramide 2/hexadecanoic acid-d_{31}/cholesterol model for the stratum corneum lipid barrier, the C=O stretching mode frequency increases in a highly cooperative manner from 1698 cm^{-1} (characteristic of a strong H bond) to 1712 cm^{-1} (characteristic of a weaker H bond) at ~45°. This shift (which is paralleled in the amide II mode of the ceramide component) implies a large structural reorganization of the headgroup region.

New Methods in IR Spectroscopy of Biologic Molecules

Infrared Reflection–Absorption Spectroscopy (IRRAS) of Monomolecular Films in Situ at the Air/Water Interface

Monolayers at the air/water interface provide a useful experimental paradigm for studying interactions between membrane components. The availability of molecular-level structural information from monolayers took a significant step forward in the mid-1980s when Dluhy and co-workers demonstrated[27,28] the feasibility of acquiring IR spectra of a monolayer film using an external reflectance technique known as IR reflection–absorption spectroscopy (IRRAS). The theory and extension of the method to proteins has been summarized recently.[29]

[27] R. A. Dluhy and D. G. Cornell, *J. Phys. Chem.* **89,** 3195 (1985).
[28] R. A. Dluhy, *J. Phys. Chem.* **90,** 1373 (1986).
[29] R. Mendelsohn, J. W. Brauner, and A. Gericke, *Ann. Rev. Phys. Chem.* **46,** 305 (1995).

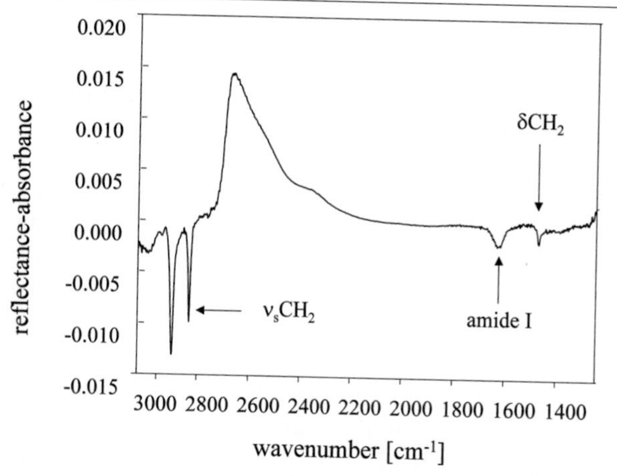

FIG. 10. An infrared reflection–absorbance (IRRAS) spectrum of a monolayer of ceramide 2 at the air/water (D_2O) interface.

The quality of the spectra obtainable are shown in Fig. 10, where an IRRAS spectrum of ceramide 2 is presented. The CH_2 stretching modes are clearly evident, along with the amide I and CH_2 scissoring vibrations. The ability to examine monolayers *in situ* at the air/water interface will permit the distinction between structural elements perpendicular to the normal to the monolayer from those parallel to the normal.[30]

Infrared Imaging with Array Detector

Recently, an IR experiment has been developed that permits microscopic imaging of IR parameters in domains spatially resolved at the diffraction limit (5–10 μm).[31,32] This is an important advance that will provide a direct measure of the molecular structures and interactions in well-defined spatial regions of intact tissues. The instrumentation is based on IR detectors in chips formed from arrays of MCT elements. These arrays are optically mapped onto the focal plane of an IR microscope. During each pass of the interferometer, spectra are generated from each element in the array. Therefore, any measure of lipid structure (e.g., conformational order from the methylene stretching frequency) can be mapped throughout the sample.

[30] C. R. Flach, A. Gericke, and R. Mendelsohn, *J. Phys. Chem.* **101,** 58 (1997).
[31] E. N. Lewis, P. J. Treado, R. C. Reeder, G. M. Story, A. E. Dowrey, C. Marcott, and I. W. Levin, *Anal. Chem.* **67,** 3377 (1995).
[32] C. Marcott, R. C. Reeder, E. P. Paschalis, D. N. Tatakis, A. L. Boskey, and R. Mendelsohn, *Cell. Mol. Biol.* **44,** 109 (1998).

The overall effect of this technology is to increase data collection rates by a factor of $>10^3$ compared with conventional IR spectrometers, with only a relatively small loss in sensitivity for each individual spectrum. IR images of the spatial distribution of water, hexadecanoic acid-d_{31}, and ceramide 2 in a stratum corneum model are shown as an application of this technology in Fig. 11. These were acquired and plotted in ~8 min with a 64 × 64 array of MCT detectors imaged in the transmission mode onto a 400- × 400-μm sample spot of an IR microscope. The spectral resolution was 8 cm^{-1} while the spatial resolution varied between the greater of the diffraction limit and the spot size limit imposed by instrument geometry (~7 μm). The spatial variation in water concentration is imaged from the O–H stretching intensity near 3600 cm^{-1}, whereas the palmitic acid levels are imaged from the CD_2 stretching contour intensity, and the ceramide levels from the CH_2 stretching contour intensity. The heterogeneous nature of the sample is evident, as revealed by the presence of domains much larger than that examined by a single detector element (7 × 7 μm). The potential of this approach for the study of tissues is evident.[32]

Summary

The temperature-induced variation of several spectral features such as those shown above illustrates the power of IR spectroscopy for the study of lipid structure and organization in ceramides and models for the stratum corneum. Because the various spectral parameters monitor diverse aspects

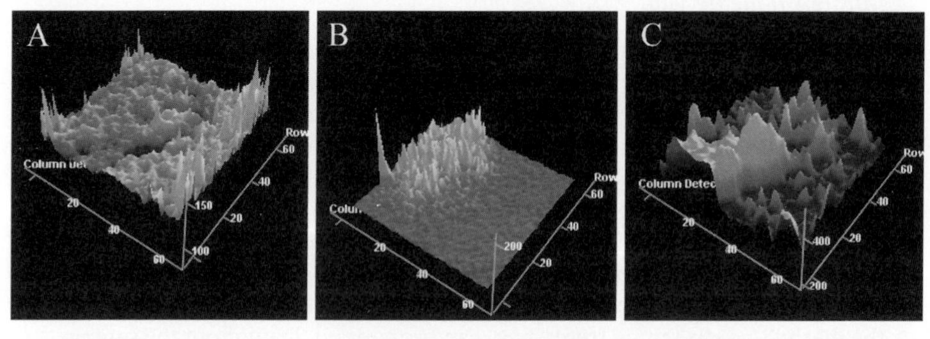

FIG. 11. IR images of the following spectral parameters in a ceramide 2/hexadecanoic acid-d_{31} cholesterol (1 : 1 : 1) mixture. (A) The spatial distribution of ceramide 2 as monitored from the integrated CH_2 stretching intensities; (B) the spatial distribution of the hexadecanoic acid-d_{31} as monitored from the integrated CD_2 stretching intensities; and (C) the spatial distribution of the water present as monitored from the O–H stretching intensities. To acquire the image, a 64 × 64 array of MCT detectors was imaged onto the focal plane of an IR microscope. The horizontal and vertical axis each represent a 400-μm length.

of molecular structure, a rather detailed picture of the structural organization that exists in both the polar and nonpolar regions of the molecules is acquired. In future applications, it is evident that the ability to monitor these changes in a detailed way will be extended to more complex experimental paradigms such as the normal and pathologic states of the stratum corneum.

Acknowledgments

We thank Dr. Carol R. Flach for acquiring the IRRAS spectrum of ceramide 2, which is displayed in Fig. 10 and Bio-Rad, Inc. (Cambridge, MA) for permitting us to acquire the IR imaging data (Fig. 11) of the stratum corneum on their Sting-Ray system. Finally, we thank Dr. Mark Rerek for continuing discussions and encouragement.

[21] Use of Nuclear Magnetic Resonance Spectroscopy in Evaluation of Ganglioside Structure, Conformation, and Dynamics

By DOMENICO ACQUOTTI and SANDRO SONNINO

Introduction

Evaluation of the structure, conformation, and dynamics of sphingolipids can provide substantial help in better understanding sphingolipid–ligand interaction mechanisms. The oligosaccharide structure of an intact glycosphingolipid can be established *ab initio* by nuclear magnetic resonance (NMR) spectroscopic analysis (Table I) without the necessity of resorting to any other chemical or spectroscopic methods; important information on the structural features of the ceramide moiety can also be derived from the same analyses. Moreover, NMR provides information on secondary structure and molecular dynamics of the molecule in a solution state. The main advantage of NMR spectroscopy is that the experiments are nondestructive. On the other hand, the major weakness of the method is its poor sensitivity, even though this can be overcome, in part, by new techniques and pulse sequences.

In this chapter we do not report on the general NMR technical procedures that have already been discussed in previous issues of this series and in a number of books and reviews,[1–9] but we do discuss several aspects of

[1] *Methods Enzymol.* **176** and **177** (1989)

[2] R. R. Ernst, G. Bodenhausen, and A. Wokaun, "Principle of Nuclear Magnetic Resonance in One and Two Dimensions." Oxford University Press, London, 1987.

0076-6879/00 $30.00

TABLE I
STRUCTURAL INFORMATION, NMR INFORMATION, AND NMR METHODS

Structural information	NMR information	NMR methods
Number of sugar residues	^{1}H and ^{13}C chemical shifts Relative abundance	Integrated 1D ^{1}H NMR spectrum 1D ^{13}C NMR spectrum 2D ^{1}H–^{1}H correlation spectroscopy (COSY, TOCSY) 2D ^{1}H–^{13}C correlation spectroscopy (HMQC, HSQC)
Monosaccharide residues	^{1}H and ^{13}C chemical shifts ^{1}H vicinal coupling constants	1D ^{1}H NMR spectrum 1D ^{13}C NMR spectrum 1D selective ^{1}H–^{1}H correlation spectroscopy (TOCSY) 2D ^{1}H–^{1}H correlation spectroscopy (COSY, TOCSY) 2D ^{1}H–^{13}C correlation spectroscopy (HMQC, HSQC)
Anomeric configuration	^{1}H and ^{13}C chemical shifts ^{1}H vicinal coupling constants ^{1}H–^{13}C coupling constants Interresidue NOE	1D ^{1}H NMR spectrum 1D ^{13}C NMR spectrum 1D selective ^{1}H–^{1}H correlation spectroscopy (TOCSY) 1D ^{1}H–^{1}H selective dipolar spectroscopy (NOESY, ROESY) 2D ^{1}H–^{1}H dipolar spectroscopy (NOESY, ROESY)
Linkage site and sequence	^{1}H and ^{13}C chemical shifts Interresidue NOE Long-range homo- and heteronuclear correlation	1D ^{1}H NMR spectrum 1D ^{13}C NMR spectrum 1D ^{1}H–^{1}H selective dipolar spectroscopy (NOESY, ROESY) 2D ^{1}H–^{1}H dipolar spectroscopy (NOESY, ROESY) 2D ^{1}H–^{1}H correlation spectroscopy (COSY, TOCSY) 2D ^{1}H–^{13}C correlation spectroscopy (HMQC, HSQC)
Linkage position	^{1}H and ^{13}C chemical shifts Interresidue NOE Long-range homo- and heteronuclear correlation	1D ^{1}H NMR spectrum 1D ^{13}C NMR spectrum 1D ^{1}H–^{1}H selective dipolar spectroscopy (NOESY, ROESY) 2D ^{1}H–^{1}H dipolar spectroscopy (NOESY, ROESY) 2D ^{1}H–^{1}H correlation spectroscopy (COSY, TOCSY) 2D ^{1}H–^{13}C correlation spectroscopy (HMQC, HSQC)

the NMR spectroscopy of gangliosides largely related to their chemical and physicochemical features.

Gangliosides form aggregates of large molecular weight and hydrodynamic radius in water varying from 60 to 250 Å (Table II), showing such a low mobility so as to prevent obtaining high-resolution NMR spectra.[10] Thus three different approaches are possible for NMR analysis: (1) to solubilize samples into organic solvents where they are present as monomers; (2) to investigate them as mixed micelles with perdeuterated detergents such as dodecylphosphocholine in aqueous solutions[11]; and (3) to investigate gangliosides as lipid-modified ganglioside small spherical micelles in aqueous solutions.

Samples of ganglioside monomers are useful for ganglioside structural characterization allowing NMR of the highest resolution. Mixed micelles of gangliosides and dodecylphosphocholine mimic the membrane situation; these kind of micelles show almost spherical size and are much smaller in comparison to the pure ganglioside aggregates (12–16 versus 300–18,000 kDa) and consequently show a much higher rotational mobility, which allows high-resolution NMR spectra.[11–13] Micelles of gangliosides are considered a cluster-like lateral distribution of gangliosides.[14]

Ganglioside Purity and Preparation of Samples

NMR spectroscopy of gangliosides requires samples to be as pure as possible. NMR is a powerful tool that reveals contaminants, including those that often cannot be displayed by colorimetric procedures after chromatography. Signals of the contaminants make the interpretation of the ganglioside NMR spectra, which are themselves quite crowded with signals, diffi-

[3] H. Kessler, M. Gehrke, and C. Griesinger, *Angew. Chem. Int. Ed.* **27,** 490 (1988).
[4] J. Dabrowski, *Methods Enzymol.* **179,** 122 (1989).
[5] S. Homans, *Prog. NMR Spectrosc.* **22,** 55 (1990).
[6] H. van Halbeek, *Methods Enzymol.* **230,** 132 (1994).
[7] H. van Halbeek, *Curr. Opin. Struct. Biol.* **4,** 697 (1994).
[8] E. F. Hounsell, *Prog. NMR Spectrosc.* **27,** 445 (1995).
[9] P. K. Agrawal, *Phytochemistry* **31,** 3307 (1992).
[10] S. Sonnino, L. Cantù, M. Corti, D. Acquotti, and B. Venerando, *Chem. Phys. Lipids* **71,** 21 (1994).
[11] H. L. Eaton, and S. I. Hakomori, presented at 195th ACS Natl. Mfg., 91-CARB (Abstr.) (1988).
[12] D. Acquotti, L. Poppe, J. Dabrowski, C. W. von der Lieth, S. Sonnino, and G. Tettamanti, *J. Am. Chem. Soc.* **112,** 7772 (1990).
[13] L. Poppe, H. van Halbeek, D. Acquotti, and S. Sonnino, *Biophys. J.* **66,** 1642 (1994).
[14] P. Brocca, P. Berthault, and S. Sonnino, *Biophys. J.* **74,** 309 (1998).

TABLE II

HYDRODYNAMIC RADIUS R_h, AGGREGATION NUMBER N, MOLECULAR MASS M, AND AXIAL RATIO R_a/R_b OF GANGLIOSIDE AGGREGATES IN AQUEOUS SOLUTIONS[a]

Gangliosides	Aggregate shape	R_h (Å)	N	M (kDa)	R_a/R_b
Natural[b]					
GM4	Interdigitated vesicle	~270	~18,000	~18,200	—
GM3	Interdigitated vesicle	~250	~14,000	~16,700	—
GM2	Ellipsoidal micelle	66.0	529	740	3.1
GM1	Ellipsoidal micelle	58.7	301	470	2.3
GD1a	Ellipsoidal micelle	58.0	226	418	2.0
GD1b	Ellipsoidal micelle	52.0	169	311	1.8
GalNAc-GD1a	Ellipsoidal micelle	60.0	246	509	2.1
GD1b	Ellipsoidal micelle	53.2	176	378	1.8
Modified					
GM1(acetyl)	Spherical micelle	34.0	76	102	1.1
GD1a(acetyl)	Spherical micelle	33.0	59	96	1.0
Mixed with dodecylphosphocholine, 1:40 by mol	Spherical micelle			12–16	

[a] The ceramide moiety of natural gangliosides contains mainly the stearoyl chain (over 90% of the total acyl content), that of modified gangliosides, the acetyl group.

[b] GM4, I^3-Neu5AcGalCer, α-Neu5Ac-(2-3)-β-Gal-(1-1)-Cer.

GM3, II^3-Neu5AcLacCer, α-Neu5Ac-(2-3)-β-Gal-(1-4)-β-Glc-(1-1)-Cer.

GM2, II^3-Neu5AcGgOse$_3$Cer, β-GalNAc-(1-4)-[α-Neu5Ac-(2-3)]-β-Gal-(1-4)-β-Glc-(1-1)-Cer.

GM1, II^3-Neu5AcGgOse$_4$Cer, β-Gal-(1-3)-β-GalNAc-(1-4)-[α-Neu5Ac-(2-3)]-β-Gal-(1-4)-β-Glc-(1-1)-Cer.

GD1a, IV^3-Neu5AcII^3Neu5AcGgOse$_4$Cer, α-Neu5Ac-(2-3)-β-Gal-(1-3)-β-GalNAc-(1-4)-[α-Neu5Ac-(2-3)]-β-Gal-(1-4)-β-Glc-(1-1)-Cer.

GalNAc-GD1a, IV^4-GalNAcIV^3Neu5AcII^3Neu5AcGgOse$_4$Cer, β-GalNAc-(1-4)-[α-Neu5Ac-(2-3)]-β-Gal-(1-3)-β-GalNAc-(1-4)-[α-Neu5Ac-(2-3)]-β-Gal-(1-4)-β-Glc-(1-1)-Cer.

GD1b, II^3-(Neu5Ac)$_2$GgOse$_4$Cer, β-Gal-(1-3)-β-GalNAc-(1-4)-[α-Neu5Ac-(2-8)-α-Neu5Ac-(2-3)]-β-Gal-(1-4)-β-Glc-(1-1)-Cer.

cult. In particular, gangliosides must not be contaminated by other glycosphingolipids, amphiphilic compounds, or lipids, which can lead to very large mixed aggregates not suitable for NMR analyses in aqueous samples. The use of an oil vacuum pump needs a liquid nitrogen trap to condense the oil vapors, otherwise a high signal could be recorded around 0 ppm of the NMR spectrum. The sample solution must be free of bivalent cations that strongly bind to the ganglioside oligosaccharide chain and lead to signal broadening and slow proton deuterium exchange.[15] To do this, gangliosides are treated as follows.

Ganglioside (5–8 mg) solubilized in 1 ml of water is dialyzed against 2 m*M* EDTA followed by water. The ganglioside solution is then passed through a 5 × 0.5-cm, pH 7, Chelex 100 column and collected in a small pear-shaped flask. The solution is lyophilized and dried under high vacuum in the same flask, maintained under vacuum through a tap cap, and left in a nitrogen-fluxed dry box.

Preparation of Samples Dissolved in Dimethyl Sulfoxide (DMSO)

The small conical flask containing 5–8 mg of highly homogenous ganglioside, prepared as reported above, is opened inside the dry box and solubilized in 0.4–0.5 ml of DMSO-d_6 or in 0.4–0.5 ml of DMSO-D_6 : ^2H_2O, 98:2 by volume, and transferred into a 5-mm NMR tube. The sample must be kept under vacuum and must be dissolved and transferred to the NMR tube under nitrogen flux to avoid the inclusion of the H_2O from the atmosphere. The water signal can pose a major problem because it can obscure different methine signals. Although the solvent signal can be eliminated by a number of Fourier transform (FT) techniques, the less solvent, the better the spectrum.[16] For the same reason very high-quality deuterated solvent must be chosen.

Preparation of Mixed Micelles of Ganglioside and Perdeuterated Dodecylphosphocholine (DPC)

Synthetic DPC mixed with a small amount of ganglioside forms small spherical micelles of 12–16 kDa in aqueous solutions, as determined by NMR data.[13] Ganglioside (5–8 mg) is mixed with DPC at a molar ratio of 1:40, and dissolved in deuterated potassium phosphate buffer (50 m*M*, 0.4 ml, p^{2}H 6). The mixture is exchanged twice with ^2H_2O, with intermediate lyophilization, dried under high vacuum, and finally inserted into the dry box under a nitrogen flux, dissolved in ^2H_2O (0.4 ml), and transferred into a 5-mm NMR tube.

Preparation of Lipid-Modified Ganglioside Micelles

Alkaline hydrolysis of gangliosides in butanol–aqueous tetramethylammonium hydroxide yields a mixture of the derivatives lacking the acetyl group(s) and acyl chain.[17] Acetylation of these compounds followed by

[15] P. Brocca, D. Acquotti, and S. Sonnino, *Glycoconj. J.* **10,** 441 (1993).

[16] M. Gueron, P. Plateau, and M. Decorps, *Prog. NMR Spectrosc.* **23,** 135 (1991).

[17] S. Sonnino, G. Kirschner, R. Ghidoni, D. Acquotti, and G. Tettamanti, *J. Lipid Res.* **26,** 24 (1985).

purification yields a single-lipid-chain ganglioside derivative, which, in aqueous solution, and due to the reduced hydrophobic volume, forms spherical micelles of low molecular mass that are suitable for NMR measurements (Table II).[15,18,19]

Thirteen μmol of (GM1) or (GD1a) are dissolved in 2 ml of 1-butanol at 100°, and mixed with 0.2 ml of aqueous 1 *M* tetramethylammonium hydroxide. The reaction mixture is refluxed at 100° under continuous stirring. After 13 hr the pH is adjusted to 6–7 by dropwise addition of acetic acid and the solution is dried in a rotary evaporator at 50°. The residue is dissolved in a few milliliters of water and applied to 5 × 2-cm RP-18 column. The column is then washed with 100 ml of water to remove salt, and with 200 ml of chloroform–methanol, 2:1 by vol, to recover the formed deacetyl- and deacetyldeacyl-ganglioside derivatives. The chloroform–methanol solution is dried, the residue dissolved in 2 ml of magnesium dehydrated methanol, and mixed with 10 μl of acetic anhydride. After stirring for 30 min at 50°, the solution is dried and purified by chromatography on a 50 × 1-cm silica gel column equilibrated and eluted with chloroform–methanol–water, 50:42:11 by vol, as the solvent system. Column elution is monitored by TLC (silica gel HPTLC plates developed in chloroform–methanol–0.3% aqueous $CaCl_2$, 50:42:11 by vol, and revealed by colorimetric detection with anisaldehyde spray reagent). The fractions containing the single-chain ganglioside are collected, pooled, and evaporated until completely dry. Samples are then treated to remove divalent cations and to prepare the NMR tubes as described above.

Nuclear Magnetic Resonance Experimental Section

One-dimensional homonuclear Hartman–Hahn spectroscopy (1D HOHAHA) experiments are performed by selective excitation with a DANTE pulse sequence, followed by a MLEV-17 sequence for spin-lock and a *z*-filter for purging phase or multiplet distortions.[20–22] The mixing time varies from 50 to 160 ms in 1-ms steps to obtain various degrees of magnetization transfer; 800–1600 scans are accumulated per spectrum. 2D-HOHAHA spectra are run in time-proportional phase incrementation (TPPI) mode with the MLEV-17 pulse sequence for spin-lock.[23] The mixing

[18] S. Sonnino, L. Cantù, M. Corti, D. Acquotti, G. Kirschner, and G. Tettamanti, *Chem. Phys. Lipids* **56,** 49 (1990).

[19] P. Brocca, L. Cantù, and S. Sonnino, *Chem. Phys. Lipids* **77,** 41 (1995).

[20] G. A. Morris and R. J. Freeman, *J. Magn. Reson.* **29,** 433 (1978).

[21] S. Subramanian and A. Bax, *J. Magn. Reson.* **71,** 325 (1987).

[22] O. Sorensen, M. Rance, and R. R. Ernst, *J. Magn. Reson.* **56,** 527 (1984).

[23] A. Bax and D. G. Davis, *J. Magn. Reson.* **65,** 355 (1985).

time is 170 ms and the spectral width 3 kHz. Transient 1D-NOE and -ROE experiments are performed using a composite 90° selective excitation pulse for selective irradiation.[24] Two-dimensional (2D) ROESY spectra are measured using the pulse sequence devised by Rance.[25] To avoid any offset dependence of the spin-lock and any Hartmann–Hahn interference, the radio-frequency (RF) carrier is placed at a distance of 1–1.5 kHz from the center of the spectrum during the spin-lock time, and at the center during evolution and acquisition; the spin-lock is sandwiched between two 170°× pulses whose amplitudes are twice the locking RF power (2.6 kHz) to improve the magnetization transfer efficiency.[26–28] Baseline distortions are minimized by applying a Hahn-echo sequence immediately before the acquisition and adjusting the receiver phase to pure absorption mode.[29,30] The spectra are recorded at 300–324 K with 512 t_1 increments and 64 scans collected for each t_1. The mixing time is in the range of 75–300 ms; the spectral width is 4–5 kHz. After zero filling, the time-domain spectra are transformed to give a 4K × 2K real points matrix with the resolution of 3.91 and 1.95 Hz/point in the F1 and F2 dimension, respectively. The 2D-heteronuclear ^{1}H-detected experiments are carried out using TPPI for quadrature detection in the F1 dimension.[31] The 2D HSQC (32 scans per t_1 increment, 0.25-sec mixing time) spectra are recorded with t_2 and t_1 acquisition times of 340 and 28.2 ms, respectively, and a relaxation delay of 1.1 sec. Pulse trains for ^{13}C relaxation, T_1 and $T_{1\rho}$, based on the INEPT or DEPT-45°, and heteronuclear NOE measurements are as previously described.[13]

Nuclear Magnetic Resonance Spectroscopy of Gangliosides: Choice of Solvent and Temperature

Glycosphingolipids, and in particular gangliosides with the general structure shown in Fig. 1, are soluble in DMSO-d_6 where they are present in a monomeric form. Thus, high-resolution spectra suitable for primary and secondary structural characterization can be recorded (Table III). In this solvent hydroxyl and amido groups do not exchange at room temperature (Table IV); this provides a supplementary source of structural information

[24] V. Sklenar and J. Feigon, *J. Am. Chem. Soc.* **113,** 3236 (1990).
[25] M. Rance, *J. Magn. Reson.* **74,** 557 (1987).
[26] Z. Dezheng, T. Fujiwara, and K. Nagajama, *J. Magn. Reson.* **81,** 628 (1989).
[27] T. Farmer II and L. R. Brown, *J. Magn. Reson.* **72,** 197 (1987).
[28] A. Bax, *J. Magn. Reson.* **77,** 134 (1988).
[29] G. Davis, *J. Magn. Reson.* **79,** 603 (1988).
[30] D. Marion and A. Bax, *J. Magn. Reson.* **81,** 352 (1989).
[31] D. Marion and K. Wüthrich, *Biochem. Biophys. Res. Comm.* **113,** 967 (1983).

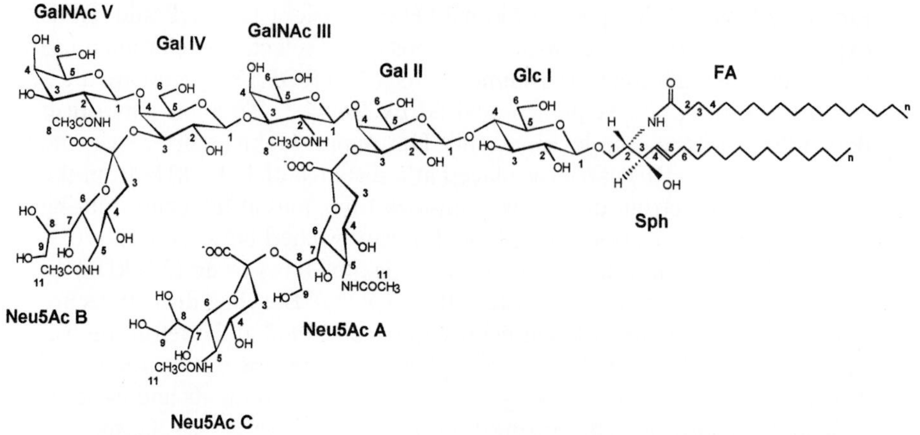

FIG. 1. Representation of the carbon numbering and of the code for the identification of glycosphingolipid residues. The represented ganglioside structure has not yet been isolated. Gangliosides quoted in this chapter derive from this general structure by detaching one or more sugar residues.

(see below) but, on the other hand, makes spectra more complicated. If determination of chemical shifts of hydroxyl and amido groups is not the aim of the work, they can be exchanged by dissolving samples in DMSO-d_6:^2H_2O (98:2, v:v). In this solution, the exchangeable protons of the sample are substituted by deuterium so that their signals disappear. Such exchange is not necessary to measure ^{13}C spectra. Ganglioside micelles are formed in ^2H_2O, where exchangeable protons disappear, or in H_2O where OH groups can be maintained at a low temperature (Table V).

The effect of temperature on chemical shifts is usually small for nonlabile protons but hydroxyl and amido protons are temperature dependent. Their temperature coefficient gives information on long-distance interactions such as hydrogen bonds. Moreover, by adjusting the temperature, the residual HDO signal frequency will shift to a less crowded region of the spectrum.

Chemical shifts are referenced indirectly by setting the signal of residual DMSO-d_5 at 2.49 ppm and at 39.5 ppm for ^1H and ^{13}C spectra, respectively. In water solution ^1H and ^{13}C chemical shifts are respectively referenced to separated resonances such as the sphingosine olefinic H-4 resonance at 5.45 ppm and to the anomeric resonance of Gal (IV) at 104.8 ppm, respectively. In DMSO, some chemical shifts may be sensitive to the water content.

Primary Structure

Monosaccharide Composition and Sugar Configuration

All the experiments for ganglioside structural characterization are carried out in DMSO-d_6 and DMSO-d_6 : ^2H_2O where gangliosides are in monomeric form and the NMR resolution is at its highest.

1D ^1H NMR spectrum of a glycosphingolipid shows only a few well-resolved signals in the range of 4.1–6.1 ppm, such as the anomeric and hydroxyl protons, amide protons over 6.5 ppm, methylene and acetamido groups at 1.6–2.5 ppm. These, in conjunction with vicinal ^1H–^1H coupling constants ($^3J_{HH}$) analysis and with the relative intensities of isolated signals allow the identification of the number and kind of monosaccharide residues. In pyranosides, the six-membered rings form a chair of fixed configuration with protons arranged in axial or equatorial position. In brain mammalian glycosphingolipids monosaccharides are predominantly hexopyranoses, having a $^4C_1$ configuration, the same as the D-sugars. Nevertheless, fucose and sialic acid are L-sugars. The coupling pattern is characteristic for the ring stereochemistry and allows us to fix the anomeric configuration. $^3J_{HH}$ is directly correlated to the dihedral angle by the Karplus relation, so if a $^3J_{1,2}$ of 2–4 Hz is observed for an anomeric proton, then a *gauche* configuration (dihedral angle of ca. 60°) of protons 1 and 2 leads to the α-configuration, where glycosidic linkage is oriented below the plane of the ring, as in the case of fucose residue of Fuc-GM1. On the other hand, as in the majority of sugar residues, a larger $^3J_{1,2}$ (6–9 Hz) indicates a *trans* diaxial relationship (dihedral angle of ca. 180°) due to a β-configuration, where glycosidic linkage is oriented above the plane of the ring.

Most of the sugar protons appear within a very narrow frequency range (3.0–4.0 ppm) and have consequent severe overlap problems: these derive from the bulk of nonanomeric methine and methylene protons of the different sugar rings that have very similar chemical shifts. This leads to a strongly coupled spectrum where many vicinal and geminal coupling constants are almost the same, causing a number of assignment problems. These difficulties can be overcome by the use of a number of high-field NMR techniques (Table I), such as the homonuclear mono- (1D) and two-dimensional (2D) shift-correlated spectroscopy experiments, to obtain a complete through-bond connectivity analysis and settle all the different spin systems corresponding to the individual sugar rings. The relative intensities of isolated resonances can be used to establish the homogeneity of the compound.

A conventional COSY spectrum contains information on spin-coupling networks within each residue of the oligosaccharide chain through the

TABLE III

^{1}H NMR Chemical Shifts of Nonlabile Protons of Neu5Ac and Stearic Acid Containing GM3, GM2, GM1, GD1a, GalNAc-GD1a and GD1b Ganglioside Species[a]

Residue	Ganglioside[b]	Chemical shift (ppm) of proton at position									
		1	2	3	4	5	6	7	8	9 [sugar] or 4 ⟷ (n − 1) [FA] or 7 ⟷ (n − 1) [Sph]	11 [sugar] or -CH_3 [Sph and FA]
Glc I	GM3	4.15	3.04	3.33	3.29	3.26	3.62 3.75				
	GM3(αHFA)	*4.20*		*3.35*	*3.32*	*3.30*					
	GM2	4.14	3.04	3.33	3.27	3.32	3.62 3.73				
	GM1	4.14	3.03	3.32	3.27	3.39	3.63 3.73				
	GM1(Neu5Gc)			*3.30*		*3.34*					
	GD1a	4.15	3.04	3.33	3.28	3.28	3.64 3.74				
	GalNAc-GD1a	4.14	3.03	3.33	3.28	3.28	3.65 3.72				
	GalNAc-GD1a(Neu5Ac B, Neu5Gc A)				*3.30*	*3.26*	*3.63 3.71*				
	GalNAc-GD1a(Neu5Gc B, Neu5Ac A)				*3.30*	*3.26*	*3.63 3.71*				
	GD1b	4.14	3.04	3.34	3.32	3.30	3.63 3.73				
	GD1b-lactone				*3.27*	*3.27*	*3.61 3.74*				
Gal II	GM3	4.19	3.33	3.93	3.75	3.32	3.46 3.51				
	GM2	4.26	3.18	3.75	3.95	3.46	3.50 3.62				
	GM1	4.26	3.13	3.70	3.94	3.47	3.47 3.64				
	GM1(Neu5Gc)			*3.74*		*3.74*	*3.62 3.63*				
	GM1-lactone	*4.62*	*4.54*	*4.10*							
	GD1a	4.25	3.18	3.76	3.94	3.44	3.45 3.64				
	GalNAc-GD1a	4.26	3.13	3.70	3.95	3.45	3.62 3.62				
	GD1b	4.29	3.24	3.78	3.99	3.46	3.42 3.52				
	GD1b-lactone	*4.21*	*3.21*								
GalNAc III	GM2	4.83	3.74	3.36	3.53	3.63			1.76		
	GM1	4.89	3.91	3.43	3.69	3.64	3.45 3.47		1.73		
	GM1(Neu5Gc)			*3.50*	*3.73*						
	GM1-lactone	*4.81*		*3.48*					*1.87*		
	GD1a	4.80	3.96	3.62	3.74	3.62	3.48 3.48		1.75		
	GalNAc-GD1a	4.87	3.92	3.38	3.65	3.62	3.43 3.45		1.76		
	GalNAc-GD1a(Neu5Ac B, Neu5Gc A)			*3.46*		*3.59*	*3.47 3.47*				
	GalNAc-GD1a(Neu5Gc B, Neu5Ac A)			*3.46*		*3.59*	*3.47 3.47*				
	GD1b	4.93	3.94	3.83	3.85	4.03	3.30 3.49		1.73		
	GD1b-lactone	*4.89*	*3.91*	*3.42*	*3.66*	*3.23*	*3.42 3.52*		*1.75*		

Gal IV	GM1	4.18	3.32	3.28	3.59	3.36	3.49 3.53				
	GD1a	4.22	3.33	3.95	3.76	3.25	3.53 3.53				
	GalNAc-GD1a	4.18	3.18	3.72	3.91	3.37	3.64 3.64				
	GalNAc-GD1a(Neu5Ac B, Neu5Gc A)						*3.63 3.36*				
	GalNAc-GD1a(Neu5Gc B, Neu5Ac A)						*3.63 3.36*				
	GD1b	4.32	3.32	3.27	3.57	3.54	3.58 3.58				
	GD1b-lactone	*4.26*			*3.65*	*3.38*					
GalNAcV	GalNAc-GD1a	4.82	3.72	3.37	3.51	3.64	3.47 3.48		1.75		
	GalNAc-GD1a(Neu5Ac B, Neu5Gc A)					*3.60*	*3.44 3.44*				
	GalNAc-GD1a(Neu5Gc B, Neu5Ac A)					*3.60*	*3.44 3.44*				
Neu5x A	GM3			2.74e 1.37a	3.61	3.40	3.61	3.21	3.34	3.41 3.61	1.88
	GM2			2.54e 1.60a	3.71	3.35	3.13	3.16	3.50	3.32 3.61	1.86
	GM1			2.53e 1.67a	3.71	3.34	3.05	3.16	3.47	3.34 3.63	1.87
	GM1(Neu5Gc)			*2.56e 1.64a*	*3.84*	*3.42*	*3.22*				*3.86 3.86*
	GM1-lactone			*2.37e 1.59a*	*4.11*	*3.75*					
	GD1a			2.59e 1.61a	3.72	3.38	3.17	3.21	3.52	3.34 3.62	1.88
	GalNAc-GD1a			2.52e 1.62a	3.76	3.39	3.09	3.16	3.48	3.35 3.60	1.87
	GalNAcGD1a(Neu5Ac B, Neu5Gc A)				*3.86*	*3.43*	*3.18*			*3.32 3.59*	*3.85 3.85*
	GD1b			2.04e 1.74a	4.41	3.52	3.79	3.28	3.83	3.46 3.54	1.87
	GD1b-lactone			*2.28e 1.59a*	*4.28*	*3.40*	*3.73*	*3.51*	*4.10*	*4.70 4.47*	
Neu5x B	GD1a			2.72e 1.38a	3.59	3.41	3.40	3.14	3.43	3.33 3.61	1.88
	GalNAc-GD1a			2.50e 1.64a	3.68	3.39	3.07	3.13	3.49	3.37 3.59	1.87
	GalNAc-GD1a(Neu5Gc B, Neu5Ac A)				*3.78*	*3.45*	*3.19*			*3.32 3.60*	*3.85 3.85*
Neu5x C	GD1b			2.91e 1.48a	3.47	3.30	3.06	3.14	3.53	3.32 3.53	1.87
	GD1b-lactone			*2.26e 1.53a*	*4.12*	*3.58*	*3.32*	*3.26*	*3.38*	*3.42 3.58*	
Sph	GM3	3.95 3.44	3.79	3.91	5.36	5.54	1.93			1.23	0.85
	GM3(αHFA)	*3.92 3.52*									
	GM2	3.95 3.44	3.70	3.89	5.35	5.52	1.93				
	GM1	3.97 3.43	3.77	3.89	5.35	5.53	1.93				
	GM1(αHFA)	*3.52 3.26*	*3.80*	*3.99*							
	GD1a	3.96 3.45	3.77	3.82	5.38	5.55	1.93				
	GalNAc-GD1a	3.98 3.40	3.76	3.87	5.34	5.52	1.92				
	GD1b	3.97 3.43	3.78	3.90	5.35	5.53	1.92				
FA	GM3		2.03	1.45						1.23	0.85
	GM1(αHFA)		*3.80*	*1.42a 1.56b*							

[a] In DMSO-d_6 solution at 300–324 K. Characteristic values for gangliosides containing Neu5Gc or α-hydroxy fatty acids or lactonic rings are reported as right-shifted *italic* data.

[b] GD1b-lactone, II^3-(α-Neu5Ac-(2-8,1-9)-α-Neu5Ac)GgOse$_4$Cer, β-Gal-(1-3)-β-GalNAc-(1-4)-[α-Neu5Ac-(2-8,1-9)-α-Neu5Ac-(2-3)]-β-Gal-(1-4)-β-Glc-(1-1)-Cer; Ganglioside(Neu5Gc), ganglioside containing Neu5Gc; Ganglioside(αHFA), ganglioside containing α-hydroxy fatty acids; Neu5Ac, *N*-acetylneuraminic acid; Neu5Gc, *N*-glycolylneuraminic acid.

TABLE IV
^{1}H NMR CHEMICAL SHIFTS OF LABILE PROTONS OF GM3, GM2, GM1, GD1a, GalNAc-GD1a AND GD1b GANGLIOSIDE SPECIES[a]

Residue	Ganglioside[b]	Chemical shift (ppm) of proton at position							
		2	3	4	5	6	7	8	9
Glc I	GM3	4.98				4.33			
	GM2	5.06	4.51			4.48			
	GM1	5.13	4.55			4.59			
	GD1a	4.98	4.51			4.43			
	GalNAc-GD1a	5.12	4.55			4.55			
	GD1b	5.06	4.61			4.55			
Gal II	GM3	4.39		4.27		4.50			
	GM2	4.87				4.23			
	GM1	5.03				4.30			
	GD1a	4.72				4.19			
	GalNAc-GD1a	4.99				4.38			
	GD1b	4.94				4.15			
GalNAc III	GM2	7.35	4.48	4.24		4.47			
	GM1	7.68		4.34		4.58			
	GD1a	7.18		4.04		4.43			
	GalNAc-GD1a	7.61		4.65		4.52			
	GD1b	9.79		4.16		5.01			
Gal IV	GM1	3.68	4.78	4.40		4.86			
	GD1a	3.87		4.22		5.03			
	GalNAc-GD1a	4.46				4.45			
	GD1b		4.71	4.30		5.26			
GalNAc V	GalNAc-GD1a	7.43	4.30	4.30		4.49			
Neu5Ac A	GM3			4.77	8.05		4.55	6.15	3.94
	GM2			4.67	7.98		4.63	6.1	4.54
	GM1			4.72	8.03		4.71	5.97	4.89
	GD1a			4.74	8.08		4.65	5.98	4.59
	GalNAc-GD1a			4.75	8.02		4.69	5.96	4.91
	GD1b			4.51	7.97		4.18		3.65
Neu5Ac B	GD1a			4.69	7.90		4.49	6.10	4.36
	GalNAc-GD1a			4.67	8.01		4.75	6.07	4.72
Neu5Ac C	GD1b			6.09	7.97		4.92	6.19	3.97
Sph	GM3	7.37	4.74						
	GM2	7.41	4.80						
	GM1	7.42	4.80						
	GD1a	7.38	4.73						
	GalNAc-GD1a	7.47	4.85						
	GD1b	7.43							

[a] In DMSO-d_6 solution at different temperatures. GM3 at 323 K. GM2 at 310 K (II-6, A-8, and A-9 at 315 K). GM1 at 297 K, GD1a at 324 K, GalNAc-GD1a at 300 K, GD1b at 313 K.

TABLE V

^{1}H NMR Chemical Shifts of Nonlabile Protons of GM1, GD1a, and GalNAc-GD1a Mixed with DPC, and GM1(acetyl) and GD1a(acetyl), in Water Solutions

Residue	Ganglioside[a]	Chemical shift (ppm) of proton at position 1	2	3	4	5	6	7	8	9
Glc I	GM1	4.53	3.44	3.72	3.72	3.64	4.28 3.87			
	GM1-acetyl	4.49	3.39	3.68	3.66	3.60	3.86 4.00			
	GD1a	4.48	3.35	3.65	3.63	3.61	3.82 4.00			
	GD1a-acetyl	4.50	3.38	3.69	3.67	3.61	3.83 4.01			
	GalNAc-GD1a	4.51	3.38	3.66	3.65	3.63	3.86 4.01			
Gal II	GM1	4.59	3.40	4.22	4.18					
	GM1.acetyl	4.56	3.38	4.18	4.17	3.78	3.82 3.82			
	GD1a	4.55	3.38	3.62	4.15	3.78	3.83 3.83			
	GD1a-acetyl	4.55	3.41	4.18	4.15	3.77	3.85 3.85			
	GalNAc-GD1a	4.56	3.41	4.18	4.15	3.79	3.86 3.86			
GalNAc III	GM1	4.86	4.11	3.86	4.22	3.77				
	GM1-acetyl	4.85	4.09	3.83	4.19	3.77	3.80 3.80			
	GD1a	4.82	4.07	3.83	4.19	3.75	3.79 3.79			
	GD1a-acetyl	4.83	4.09	3.84	4.20	3.76	3.81 3.81			
	GalNAc-GD1a	4.84	4.10	3.83	4.16	3.75	3.82 3.82			
Gal IV	GM1	4.60	3.58	3.69	3.98	3.72				
	GM1-acetyl	4.58	3.56	3.67	3.94	3.72	3.79 3.79			
	GD1a	4.64	3.56	4.11	3.97	3.69	3.75 3.75			
	GD1a-acetyl	4.64	3.57	4.11	3.98	3.70	3.76 3.76			
	GalNAc-GD1a	4.65	3.40	4.16	4.13	3.79	3.78 3.78			
GalNAc V	GalNAc-GD1a	4.83	3.97	3.71	3.98	3.79	3.82 3.82			
Neu5Ac A	GM1			2.72eq 1.98ax	3.59	3.86	3.55	3.64	3.79	3.68 3.93
	GM1-acetyl			2.70eq 1.95ax	3.85	3.83	3.53	3.63	3.79	3.67 3.91
	GD1a			2.72eq 1.91ax	3.82	3.83	3.53	3.61	3.78	3.66 3.89
	GD1a-acetyl			2.70eq 1.94ax	3.83	3.83	3.53	3.63	3.80	3.68 3.93
	GalNAc-GD1a			2.74eq 1.95ax	3.85	3.85	3.55 or 3.56	3.62	3.73	3.70 3.92
Neu5Ac B	GD1a			278eq 1.80ax	3.72	3.86	3.65	3.61	3.91	3.67 3.90
	GD1a-acetyl			2.77eq 1.83ax	3.72	3.87	3.65	3.62	3.92	3.68 3.93
	GalNAc-GD1a			2.72ax 1.95eq	3.85	3.85	3.55 or 3.56	3.62	3.73	3.70 3.92
Sph	GM1		4.10	4.13	5.45	5.83				
	GM1-acetyl	4.12 3.82	3.99	4.12	5.45	5.76				
	GD1a	4.18 3.79	3.95	4.10	5.45	5.75				
	GD1a-acetyl	4.11 3.82	4.00	4.12	5.45	5.77				
	GalNAc-GD1a	3.79 4.19	3.96	4.12	5.46	5.76				

[a] Ganglioside-acetyl, ganglioside containing an acetyl group as acyl moiety.

observation of cross-peaks. Assignment of this spectrum requires a starting signal for the identification of each individual spin system. Usually the anomeric protons are free enough to be a convenient starting point. Within a typical aldohexopyranosyl ring, H-1 couples to H-2, H-2 to H-1 and H-3, H-3 to H-2 and H-4, and so on. If the coupling constant $^3J_{H,H}$ is large enough (5–8 Hz) an intense cross-peak arises, but if no or small couplings (2–3 Hz) are present between vicinally related protons, no or very small cross-peaks arise. For example, for a galactopyranosyl residue where the $J_{4,5}$ is 2–3 Hz, it is necessary to optimize the magnetization transfer for this J to reach protons 5 and 6s (RELAY-COSY). A pure absorption phase sensitive COSY gives cross-peaks containing the entire coupling information concerning the involved protons. However, a single COSY spectrum is not sufficient for an unequivocal assignment of all the resonance of an oligosaccharide chain because of the overlap of many proton signals. A number of experiments can be run to overcome this problem. The *double-quantum filtered COSY* (DQF-COSY) suppresses the detection of the isolated protons, such as those arising from solvents or isolated methyl and acetyl groups. Moreover, this experiment provides pure absorption cross- and diagonal-peaks showing a thinner diagonal and allowing a better resolution of cross-peaks close to the diagonal. The *triple-quantum filtered COSY* (TQF-COSY) suppresses all the spin systems that contain less than three or more mutually coupled spins; such a system in hexopyranosides is H-5, H-6, and H-6′, which are often difficult to assign. The *total correlation spectroscopy* (TOCSY) or *homonuclear Hartmann–Hahn spectroscopy* (HOHAHA) is a very useful experiment in which the net magnetization is transferred along the chain of spins under spin-locking conditions. In these spectra a complete spin system can be identified if there is at least one well-isolated resonance to be used as a starting point and reasonably large couplings between all the spins of the lattice. In this way, a slice along the diagonal leads to a ^{1}H subspectrum containing all scalar-coupled protons within a sugar ring. 2D-TOCSY gives poor information about J couplings and even in these spectra, resonance overlapping often does not permit the unambiguous identification of all the protons of the oligosaccharide chain. A very potent tool to definitively circumvent these problems is the selective 1D-TOCSY experiment that allows us to obtain high-resolution subspectra from which coupling constants are safely measured. This is the case of the galactosyl residues, where the $J_{4,5}$ is small and the magnetization transfer is recognized up to H-5, not only to H-4 as in a 2D-TOCSY spectrum.

The hydroxy and amido protons usually show well-separated signals so that their identification is very useful both for the assignment of resonances and for spatial connectivity investigation. They can be used as the starting points in selective 1D-TOCSY experiments to assign, for example, the sialic acid side chains or the methylene protons of the galactosyl residues. The net magnetization can be transferred from OH-6 resonance to H-6, H-6′, and H-5: conjoining these spectra with those deriving from the selective excitation of the anomeric protons, which transfer magnetization from H-1 to H-5, it is possible to assign all the galactose ring resonances. When H-5 is separated enough to be selectively excited the full galactosyl residue subspectrum is obtained.

Sometimes proton resonances are too overlapped to be disentangled by the above-mentioned homonuclear techniques. ^{13}C NMR spectroscopy is potentially a very powerful tool for glycosphingolipid analysis because of its larger chemical shift range, which minimizes overlapping and lack of complexities due to spin–spin coupling.[32] In practice, 1D spectra require very long acquisition time or very large amount of samples because of the natural low abundance of ^{13}C. Alternatively, inverse-detection techniques provide very useful heteronuclear correlation maps that relate a carbon atom to the resonance of its directly bonded proton that is split by a large $^{1}J_{CH}$ coupling. The *heteronuclear single quantum coherence* (HSQC) experiment shows the cross-peaks obtained at carbon frequency F1 (vertical axis) and proton frequency F2 (horizontal axis), anti-phased split by the $^{1}J_{CH}$ along the F2 axis; this simplifies the complete assignment of both the ^{1}H and ^{13}C resonances.[31]

The last step is the location of the appended groups, if present. The substitution of hydroxyl and amido groups with an acyl, a glycolyl, a glycolic, or a glycoside group leads to chemical shift changes at the substitution site and some other nearby resonance in both ^{1}H and ^{13}C spectra. The position of these groups can be verified with nuclear Overhauser effect (NOE) experiments. The chemical shifts of the neuraminic glycerol side chain protons have been assigned observing the magnetization transfer from OH-8 to OH-9, OH-7, H-7, H-8, H-9*R*, and H-9*S* in a 1D-HOHAHA spectrum. The side chain proton set can be assigned to the proper sialic acid ring by NOE contact determination between OH-8 and some protons of the closer Gal residue.

Linkage Position and Sugar Sequence

Once each sugar residue has been completely characterized using a number of the above-described methods, it is necessary to identify the glycosidic linkages and their sequence to complete the structure determination. The protons across the glycosidic linkages are four bonds apart, so

[32] P. Brocca, D. Acquotti, and S. Sonnino, *Glycoconj. J.* **13,** 57 (1996).

that they do not show any scalar coupling and, consequentially, no cross-correlation can be observed using COSY and TOCSY experiments. Two different approaches can be used: the *glycosylation-induced shift* method that is quick and easy if the oligosaccharide can be related to any other structure previously examined by NMR, and the *ab initio* method that does not require any prior knowledge. The first method is based on the fact that substitution by another sugar residue at a sugar ring induces chemical shift changes of both protons and carbons of the glycosylated unit, particularly those at the linkage site and those vicinal to it, whereas the other protons are much less affected. This method can be applied to oligosaccharides for which ^{1}H assignments are already unambiguously available. Glycosylation shifts in ^{13}C NMR are relatively regular: the glycosylated carbon shifts to lower fields by 4–10 ppm and the resonances of the adjacent carbon atoms shift upfield by a small amount, whereas the other carbon resonances remain virtually unaffected. These effects can be correlated with the configuration at the anomeric center of both pyranose units because they depend on the spatial disposition of protons that cause polarization of the C–H bonds.[33,34] The correlation between ^{13}C chemical shifts for both glycone and aglycone carbons with one of the torsional angles (ψ) can be used for the determination of the conformation of the glycosidic linkage.

The sequence of an oligosaccharide chain can be assigned without prior knowledge of the type of the structure using homo- and heteronuclear nuclear Overhauser effect (NOE) data. NOE depends on the spin–spin dipolar interaction and, consequently, on the distance between protons so that the NOE connectivity between the anomeric and the aglycone protons are usually very effective (Table VI). The identification of the aglyconic proton at the linkage site is unequivocal, except when there is an equatorial proton vicinal to it, e.g., H-4 in 3-glycosylated galactose residues, which exhibits NOEs of approximately the same magnitude. In these cases the problem is circumvented with the chemical shift analysis: substitution by a second sugar unit at a sugar ring induces a larger chemical shift change for the proton at the linkage site than to the vicinal one. For sialic acids, the diagnostic contacts are between the H-3ax of the sialic ring and some protons of the adjacent sugar residue, i.e., H-4 of the inner Gal residue of many gangliosides.

Measurements of the average spin lattice relaxation time for sugar carbon, which increases with the growth of distance in glycosides from the

[33] H. Baumann, P. E. Jansson, and L. Kenne, *J. Chem. Soc. Perkin Trans. I* 2299 (1991).
[34] K. Bock, A. Brignole, and B. W. Sigurskjold, *J. Chem. Soc. Perkin Trans. II* 1711 (1986).

TABLE VI

INTERRESIDUE ^{1}H–^{1}H NOE INTERACTIONS FOR GANGLIOSIDE SACCHARIDE SEQUENCES, AND PART OF THEM, THAT CAN BE FOUND FOR THE BRANCHED OLIGOSACCHARIDE STRUCTURE GalNAc-[Neu5Ac-]Gal-GalNAc-[Neu5Ac-Neu5Ac-]Gal-Glc-Cer SHOWN IN FIG. 1

α-Neu5Ac-(2-3)-β-Gal-(1-3)-β-GalNAc- *NOE interactions*		
Neu5Ac-3ax : Gal-3	Gal-1 : GalNAc-3	Neu5Ac-8 : GalNAc-CH_3
Neu5Ac-OH8 : Gal-3	Gal-1 : GalNAc-4	
Neu5Ac-8 : Gal-3	Gal-1 : GalNAc-2	
Neu5Ac-3eq : Gal-OH2	Gal-1 : GalNAc-NH	

β-GalNAc-(1-4)-[α-Neu5Ac-(2-3)]-β-Gal- *NOE interactions*		
GalNAc-1 : Gal-4	Neu5Ac-3ax : Gal-3	Neu5Ac-8 : GalNAc-1
GalNAc-NH : Gal-2	Neu5Ac-3ax : Gal-OH2	Neu5Ac-OH8 : GalNAc-1
GalNAc-CH_3 : Gal-OH2	Neu5Ac-3eq : Gal-OH2	Neu5Ac-OH8 : GalNAc-5
GalNAc-CH_3 : Gal-2	Neu5Ac-OH7 : Gal-4	Neu5Ac-OH8 : GalNAc-NH
GalNAc-OH6 : Gal-OH6	Neu5Ac-OH8 : Gal-4	Neu5Ac-OH9 : GalNAc-OH6
		Neu5Ac-OH9 : GalNa-5
		Neu5Ac-9R : GalNAc-OH6

β-*Gal*-(1-3)-β-GalNAc-(1-4)-[α-*Neu5Ac*-(2-8)-α-Neu5Ac-(2-3)]-β-Gal- *NOE interactions*			
GalNAc-1 : Gal-4	Neu5Ac-3ax : Gal-3	*Neu5Ac-3eq* : GalNAc-1	*Neu5Ac-3ax* : Neu5Ac-6
	Neu5Ac-8 : Gal-4	*Neu5Ac-3eq* : GalNAc-5	*Neu5Ac-6* : Neu5Ac-9a
			Neu5Ac-8 : Neu5Ac-9a
			Neu5Ac-3eq : Neu5Ac-8
Neu5Ac-NH : *Gal-OH6*	Neu5Ac-8 : GalNAc-1		
	Neu5Ac-8 : GalNAc-5		

β-Gal-(1-4)-β-Glc- *NOE interactions*	β-Glc-(1-1)-Cer *NOE interactions*
Gal-1 : Glc-4	Glc-1 : Cer-1′
Gal-1 : Glc-6	Glc-1 : Cer-2
Gal-1 : Glc-OH3	Glc-1 : Cer-3
Gal-OH2 : Glc-6	Glc-OH2 : Cer-NH
Gal-OH2 : Glc-6′	
Gal-OH2 : Glc-OH6	
Gal-1 : Glc-OH6	
Gal-6 : Glc-OH3	

aglycone, together with the total assignment of the OH resonances can confirm the NOE results.[35]

Characterization of Ceramide Moiety

Some proton resonances of ceramide are easily recognizable in the spectra: the triplet of methyl groups, the alkyl, allylic, and α-carbonyl methylene resonances are assigned at 0.85, 1.23, 1.92, and 2.03 ppm, respectively. The olefinic proton resonances of sphingosine, which are assigned between 5.3 and 5.6 ppm, show vicinal coupling constant $J_{4,5}$ of 15 Hz, clearly indicating a *trans* configuration. The NH and the OH of sphingosine are approximately 7.5 and 4.8 ppm, respectively, in the temperature range 300–323 K. The presence of α-hydroxylated fatty acids is indicated by the appearance of a new doublet at 5.5 ppm; moreover the α-carbonyl methylene protons shift to low field at 3.8 ppm. Sometimes the presence of this α-OH can influence the glucose anomeric resonance, as in GM3. Unambiguous assignment of the sphingosine and the fatty acid moieties can be obtained with a 1D-HOHAHA spectrum by selective excitation of the two different OH groups and/or of the olefinic resonances.[36]

The content of sphingosine on the total long-chain base content, sphingosine and sphinganine, can be determined by comparison of the olefinic and anomeric resonance intensities. The evaluation of the length of the lipid chains, as the average total number of carbons, requires a very accurate integration of ^{13}C spectrum, while the precise knowledge of the long-chain base and fatty acid distributions requires the use of other analytical methods.

Secondary Structure

Detailed information on the conformational properties of gangliosides is obtained on gangliosides dissolved in dimethyl sulfoxide, the structures being determined for an isolated molecule.[12,36–43] In view of extending the

[35] K. Bock and C. Pedersen, *J. Chem. Soc. Perking Trans. II* 293 (1974).

[36] S. Sonnino, D. Acquotti, L. Cantù, V. Chigorno, M. Valsecchi, R. Casellato, M. Masserini, M. Corti, P. Allevi, and G. Tettamanti, *Chem. Phys. Lipids* **69,** 95 (1994).

[37] S. Sabesan, K. Bock, and R. U. Lemieux, *Can. J. Chem.* **62,** 1034 (1984).

[38] L. Poppe, J. Dabrowski, C. W. von der Lieth, M. Numata, and T. Ogawa, *Eur. J. Biochem.* **180,** 337 (1989).

[39] J. N. Scarsdale, J. H. Prestegard, and R. Y. Yu, *Biochemistry* **29,** 9843 (1990).

[40] D. Acquotti, G. Fronza, E. Ragg, and S. Sonnino, *Chem. Phys. Lipids* **59,** 107 (1991).

[41] S. B. Levery, *Glycoconj. J.* **8,** 484 (1991).

[42] H. C. Siebert, G. Reuter, R. Schauer, C. W. von der Lieth, and J. Dabrowski, *Biochemistry* **31,** 6962 (1992).

[43] D. Acquotti, L. Cantù, E. Ragg, and S. Sonnino, *Eur. J. Biochem.* **225,** 271 (1994).

results to the physiologic membrane, the need to study aqueous solution systems becomes significant. Pure ganglioside aggregates present in water media are generally too big (see Table II) for high-resolution NMR and, moreover, do not allow us to consider any conformational changes reflecting possible ganglioside–phospholipid interactions. Thus, both the effects of the aqueous solvent and of a phospholipid surface on the ganglioside secondary structure are investigated using small micelles of dodecylphosphocholine containing one ganglioside molecule per micelle.[11–14,32,38,42–44] The model system consisting of the small spherical micelles formed by semisynthetic gangliosides bearing a single tail lipid moiety, like GD1a(acetyl) and GM1(acetyl), also gives information on the conformational, dynamic, and interaction properties of the ganglioside oligosaccharide polar head in an environment enriched in gangliosides, thus mimicking a ganglioside-clustered membrane microdomain.[14]

All together, the NMR results show that only a few, but very strong and cooperative, intramolecular interresidual interactions primarily determine, independently from the environment and the particular composition of the lipidic portion, the conformational properties of ganglioside oligosaccharide.

Conformation

NMR studies of solution conformation of biomolecules of any class are mainly based on the proton–proton distance information gained from the quantitative evaluation of NOE data.[45] NOEs depend on spectrometer angular frequency ω_0 and molecular correlation time τ_c that is itself dependent on molecular size. Thus, oligosaccharides may show positive or negative NOE levels, as a direct result of their molecular size, so that NOEs become undetectable when $\omega_0\tau_c$ is approximately equal to unity. On the contrary, in the rotating frame all molecules behave as if they were in the positive NOE regime and NOE increases from 38.5 to 67.5%, and does not vanish for any $\omega_0\tau_c$ value. Moreover, ROESY experiments are less prone to spin diffusion and allow one to discriminate real NOE from proton exchange cross-peaks.[46] Therefore for tri- to decasaccharides NOE experiments in the rotating frame (ROE) are often preferred to the conventional NOE experiments.[47]

In contrast to proteins, oligopeptides, and nucleic acids, which usually

[44] L. Poppe, C. W. von der Lieth, and J. Dabrowski, *J. Am. Chem. Soc.* **112,** 7762 (1990).
[45] D. Neuhaus and M. P. Williamson, "The Nuclear Overhauser Effect." VCH, New York, 1989.
[46] J. Dabrowski and L. Poppe, *J. Am. Chem. Soc.* **111,** 1510 (1989).
[47] A. A. Bothner-By, R. L. Stephens, and J. Lee, *J. Am. Chem. Soc.* **106,** 811 (1984).

display a sufficient number of NOE connectivities for the modeling of well-founded three-dimensional (3D) structure, oligosaccharides show very few structurally relevant NOE contacts, very often only one contact per glycosidic linkage.[48,49] This situation can be improved if hydroxy and amido groups are used as long-range sensors to provide additional contact information.[47] Since NOE can be correlated to the internuclear distance and since these groups protrude approximately twice as far as C-linked protons, a number of additional interatomic contacts can be observed for each disaccharide fragment of the oligosaccharide chain, and even for not directly linked rings. In this way, the amount of experimental data available for a verification of theoretically calculated conformations can be greatly extended. Moreover, the analysis of the coupling constants and of the temperature coefficients for amido and hydroxy protons allows the identification of intra- and/or interresidual hydrogen bonds that give further conformational information.

An oligosaccharide structure can be seen as a linear or branched chain of quite rigid sugar residues that have a well-defined 3D structure. Usually, the connected groups also assume a preferred conformation. In particular, the sialic acid side chain, of gangliosides, which in principle should be very flexible, adopts a well-defined conformation up to atom C-8, where H6-C6-C7-H7 (θ_1) and H7-C7-C8-H8 (θ_2) dihedral angles can be derived from vicinal $^3J_{6,7}$ and $^3J_{7,8}$, less than 2 and 8.5 Hz, respectively. Although the Karplus equation gives two alternative solutions for each ^3J value, the correct one can be chosen by "NOE labeling," i.e., by locating the vicinal protons firmly localized in the molecule. In this way dihedral angles $\theta_1 = -60°$ and $\theta_2 = -160°$ give the best agreement for the observed NOE contacts between ring and side chain protons and vicinal coupling constants where small $J_{6,7}$ and large $J_{7,8}$ may be directly compared to $J_{4,5}$ in galactose ($J < 1.5$ Hz, H5-C5-C4-H4 $= -60°$) and glucose ($J = 10$ Hz, H5-C5-C4-H4 $= 180°$) rings of a ganglioside molecule. Thus the small $^3J_{8,OH8}$ coupling constant (<2 Hz versus 6 Hz for a freely rotating OH group), the small temperature shift coefficient of the OH-8, and the strong OH-8/H-6 NOE contact all point to a hydrogen bond between OH-8 and either the carboxylic O-1 oxygen or the O-6 oxygen. The former of these two hydrogen bonds seems to be more probable in view of the exceptionally large low-field shift of OH-8 (ca. 6 ppm), as compared with the chemical shift of Glc I-OH3 (ca. 4, 6 ppm) hydrogen bonded to Gal II-O5 ring oxygen.

[48] K. Wüthrich, "NMR of Proteins and Nucleic Acids." John Wiley & Sons, New York, 1986.

[49] H. Pepermans, D. Toerwe, G. van Binst, R. Boelens, R. M. Scheek, W. F. van Gunsteren, and R. Kaptein, *Biopolymers* **27,** 323 (1988).

The situation does not change after deletion of the *N*-acetyl group of the neuraminic acid. Indeed, the vicinal coupling constants $J_{6,7}$ and $J_{7,8}$ remain unchanged in deacetyl-GM1, and proton OH-8 also shows the same behavior. Moreover, it appears that proton OH-7 is hydrogen bonded with carbonyl oxygen. Actually, on going from deac-GM1 to GM1, this proton has a significantly smaller coupling constant (8.2 versus 3.7 Hz) and chemical shift temperature coefficient (6.12 versus 3.27×10^{-3} ppm), and the NOE distance to proton H-5 changes from 0.29 to 0.23 nm. Additional support for the existence of the aforementioned hydrogen bonding is the favorable conformation of the acetamido group, characterized by a single torsional angle H-N-C5-H5 close to 150°, a value that is in excellent agreement with measured NOE contacts and observed vicinal coupling constant. The vicinal coupling constants for prochiral methylene protons OH-9*R* and OH-9*S* have significantly different coupling constants and can be readily assigned by so-called "NOE labeling" to proton H-7.[38] Assuming that the hydroxymethyl group can occupy three staggered conformations, the population ratio for *gauche–gauche, gauche–trans*, and *trans–gauche* conformers obtained from extended Karplus equation is equal to 0.5 : 0.5 : 0.[50] Moreover, the conformation of the sialic acid side chain remains almost the same when gangliosides are solubilized in organic solvents or anchored in DPC micelles in aqueous solution, as inferred from the coupling constants and NOE interaction with the GalNAc residue. Therefore, if every ring has a fixed rigid conformation, then the 3D structure of an oligosaccharide chain can be described by a set of torsional angles H1′-C1′-O1-Cx (ϕ) and C1′-O1′-Cx-Hx (ψ) connecting two sugar units through the glycosidic bond. For sialic acids, these are defined by C1′-C2′-O2′-Cx (ϕ) and C2′-O2′-Cx-Hx (ψ).[51]

As already seen, interresidue NOEs can serve as a basis for the determination of glycosylation sites and sequences, and can be converted into interatomic distances. The cross-peak volumes obtained from ROESY spectra, measured in the linear buildup region of NOE, have been used to calculate reliable distance values. Several proton pairs for the reference distance could be found in the saccharide rings; for instance, all of the H1–H5, H1–H3 and H3–H5 interactions arise from protons separated by 0.25 nm from each other. The interatomic distances are therefore calculated safely by applying the two-spin approximation with equation $r_{ij} = r_{kl}(V_{kl}/V_{ij})^{1/6}$, where V_{ij} and V_{kl} are the cross peak volumes for unknown (r_{ij}) and calibration (r_{kl}) distances, respectively. This equation relies on the assumption that all dipole–dipole interactions are modulated by a

[50] G. Haasnot, F. A. A. M. de Leeuw, and C. Altona, *Tetrahedron* **36,** 2783 (1980).
[51] W. Klyne and V. Prelog, *Experientia* **16,** 521 (1960).

TABLE VII
GLYCOSIDIC TORSIONAL ANGLES (ϕ, ψ) PAIRS FOR VARIOUS GANGLIOSIDE SPECIES

	ϕ, ψ (°)						
Compound	V-IV	B-IV	C-A	IV-III	III-II	A-II	II-I
GM1				25, 30 30, −40	30, 25	−162, −29	5 → 55, −50 → 0
GD1b			89, 10	53, 10	41, 9	175, 5	
GD1a		−76, 2 −159, −18		36, 8 −26, −21	31, 18	−162, −29	
GalNAc-GD1a	36, 14	−163, −27		54, 21 −32, −23	31, 19	−162, −28	

[a] From Refs. 12, 14, 38, 40, 42, 43, and 52. Accuracy is ±(10–15)°.

single correlation time, which has been checked by calculating, with the same reference volume and interatomic distance, several r_{kl} concerning proton pairs at a known distance and located in different sugar moieties. All derived distances were in agreement with the known values, to within ±0.02 nm, on average. This also gave an indication of the quality of the measured cross-peak volumes. The derived interatomic distances can be directly used to determine the torsional angles ϕ, ψ (Table VII) with a simple geometrical iterative calculation, such as the distance-mapping procedure. Results can be confirmed using unrestrained molecular mechanics (MM) and dynamics (MD) calculations. An unconstrained Monte Carlo/energy (MC/EM) minimization conformational search performed on GM1 ganglioside led to results that are fully compatible with the experimental NMR data.[52]

Application of ^{1}H NMR to large micelles of a single ganglioside component leads to a detailed study of a surface system with a high carbohydrate density, which might mimic the presence of phase-separated surface domains on cell membranes. High-resolution ^{1}H NMR studies of the micelles show few interresidual dipole–dipole interactions for the CH protons of the oligosaccharide chain; thus, studies are also needed for hydroxyl proton interactions (Table VIII). On the basis of NOESY and ROESY results, the data interpretation was simplified by the finding that the detected interactions were in complete agreement with those expected for a ganglioside headgroup that maintained the same conformational features already observed for a monomer in DMSO and for a single ganglioside molecule per DPC micelle in water.[12,13,19] It can be seen that, qualitatively, there are

[52] Bernardi, and L. Raimondi, *J. Org. Chem.* **60,** 3370 (1995).

TABLE VIII
^{1}H NMR Chemical Shifts of Labile Protons of GM1(acetyl) Micelles and DPC/GD1a Mixed micelles[a]

		Chemical shift (ppm) of proton at position							
Residue	Ganglioside	2	3	4	5	6	7	8	9
Glc I	GM1(acetyl)	n.d.	5.84			n.d.			
	GD1a	6.95	5.85			5.41			
Gal II	GM1(acetyl)	6.51				n.d.			
	GD1a	6.60				n.d.			
GalNAc III	GM1(acetyl)	7.53		n.d.		n.d.			
	GD1a	7.60		5.85		n.d.			
Gal IV	GM1(acetyl)	n.d.	n.d.	n.d.		n.d.			
	GD1a	5.62		5.95		n.d.			
Neu5Ac A	GM1(acetyl)			6.34	8.22		5.86	6.24	n.d.
	GD1a			6.45	8.25		5.75	6.25	3.66
Neu5Ac B	GD1a			6.41	8.25		5.75	6.22	n.d.
Sph	GM1(acetyl)	8.15							
	GD1a	8.22	6.20						

[a] In water solution at 285 and 305 K, respectively; n.d., not determined.

no new contacts and no missing contacts at the level of the saccharidic portion of the GM1(acetyl) micelles with respect to the other mentioned systems. Thus, it appears unnecessary to assume any significant effect due to intermonomer interactions, allowing us to regard all of the contacts as belonging to a single monomer. Even if NMR-invisible contact(s) cannot be excluded, the results suggest that the carbohydrate-enriched surface does not influence the main conformational features of the single ganglioside oligosaccharide.

Possible contacts between monomers inserted into a surface can be investigated in ganglioside mixed micelles in D_2O formed by Neu5Ac and Neu5Gc containing compounds. The spectral parameters of the sialic acid ring protons for GM1 (Neu5Ac containing GM1) and GM1(Neu5Gc) are different, the protons concerned being those at positions 3, 4, 5, and 6 and those of the acyl group[32]; these differences make it possible to check for any NOE originating from interresidual interactions between two sialic acid residues of distinct monomers. Note that the sialic acid belongs to the branched region of the oligosaccharide chain, possibly a favorable location for intermonomer interactions, but no Neu5Ac/Neu5Gc NOE contacts are detected.

A further investigation of ganglioside conformation is carried out by analyzing the water–ganglioside exchange process. By applying off-resonance ROESY and setting the angle θ between the effective field and the external static field equal to 35.3°, the dipolar cross-relaxation is made to vanish so that only chemical exchange is effective in determining magnetization transfer from water protons to ganglioside OH and NH.[56,57] Thus, it is possible to establish that at 285-276 K and pH 7.4, the water–ganglioside interaction is completely dominated by exchange in the low mixing time range. The low-temperature, long correlation time, and aggregation state of the sample lead to a water–ganglioside exchange in the time scale of 10 ms, an NMR-accessible time range. Hydroxyl proton exchange is then followed in the evolution time range of 1–120 ms. The exchange rates are accurately calculated only for Gal II-OH2, Neu5Ac A-OH4, and Neu5Ac A-OH8, their resonances being isolated in the spectrum. By comparing the exchange times of Gal II-OH2, Neu5Ac A-OH4, and Neu5Ac A-OH8, similar behavior can be seen for Gal II-OH2 and Neu5Ac A-OH4, whereas the Neu5Ac A-OH8 exchange time and its temperature dependence appear quite peculiar. In comparing the exchange times of Neu5Ac A-OH4 and Gal II-OH2, the latter's constant larger exchange time can be attributed to a reduced solvent accessibility, according to the position depth of the group within the hydrophilic shell of the micelle. On the contrary, despite its relatively external position, Neu5Ac A-OH8 exchanges less easily than Neu5Ac A-OH4 and Gal II-OH2 and its exchange time shows a brusque increment at the lowest temperature studied. The fact that the activation energy for the Neu5Ac A-OH8 proton exchange process is higher than that for the Gal II-OH2 and Neu5Ac A-OH4 confirms that Neu5Ac A-OH8 is involved in intramolecular interactions. This is in agreement with the observation that it belongs to the interaction region between the sialic acid lateral chain and GalNAc, which is probably less accessible in water. Moreover, Neu5Ac A-OH8 has been shown to have a hydrogen bond with the carboxylic group (note the strong Neu5Ac A-OH8: Neu5Ac A-H6 NOE contact), thus making the sialic acid structure quite rigid.[12,13]

Dynamics

A critical analysis of the cross-peak contacts, and consequently of the derived distances is necessary. Sometimes two contacts are absolutely incompatible with a single conformer, i.e., the terminal fragment Gal-GalNAc is described by the Gal IV-1 : GalNAc III-2, Gal IV-1 : GalNAc III-4 and Gal IV-1 : GalNAc III-NH interactions (see Table IV) that cannot be satisfied simultaneously by a single rigid structure.[43] Thus a more accurate model has to take into account the possibility of a flexible linkage with the presence of at least two different conformations. Another interesting example is referred to as the terminal sialic acid, i.e., in GD1a, that behaves

TABLE IX
RELAXATION PARAMETERS[a] FOR ^{13}C NUCLEI IN GD1a/DPC MICELLES[b]

Residue	Parameter	C-1	C-2	C-3	C-4	C-5	C-6	C-8	C-9
Glc I	T_{1p}	0.08	0.08	0.08	0.08	0.08	0.09		
	NOE	1.2	1.2	1.2	1.3	1.2	1.4		
Gal II	T_{1p}	0.07	0.08	0.08	0.08	0.07	0.16		
	NOE	1.1	1.1	1.0	1.0	1.1	1.7		
GalNAc III	T_{1p}	0.07	0.08	0.07	0.06	0.08	0.16		
	NOE	1.1	1.2	1.1	1.2	1.2	1.9		
Gal IV	T_{1p}	0.14	0.15	0.15	0.07	0.14	0.22		
	NOE	1.5	1.7	1.6	1.2	1.5	2.05		
Neu5Ac A	T_{1p}			n.d.	0.08	0.08	0.08	0.08	0.13
	NOE			1.1	1.1	1.2	1.3	1.1	1.5
Neu5Ac B	T_{1p}			0.14	0.13	0.14	0.13	0.15	0.17
	NOE			1.6	1.6	1.6	1.6	1.7	1.6
Sph	T_{1p}		0.07	0.08					
	NOE		1.2	1.4					

[a] nT_{1p} [s], NOE.
[b] In D_2O solution at 305 K. The estimated average standard error is 10% for T_{1p} and 20% for NOE values. Data for the C-7 atoms of internal NeuAc and terminal NeuAc residues are not included in the analysis because of spectral overlap. The relaxation parameters for the methylene carbons should be treated with caution since they might contain systematic errors due to cross-correlation effects.[53,54,55]

differently from the internal one. In GD1a, the NOE contact Neu5Ac B-OH8 : Gal IV-3 is observed, and the Gal IV-3 : Neu5Ac B-3ax distances from 0.36 to 0.23 nm. A molecular dynamics simulation revealed the existence of four conformers where the ϕ, $\psi_{B\text{-}IV}$ flips between two distinct angles, −74, 1 and −159, −18, and the two distances Neu5Ac B-OH8 : Gal IV-3 and Gal IV-3 : Neu5Ac B-3ax changed from 0.42 to 0.23 nm indicating that the experimental values might correspond to an average distance. These results are compatible with those other authors have found in com-

TABLE X
AVERAGE ^{13}C RELAXATION RATES AND DYNAMICS PARAMETERS[a]

Residue	R_1 [s^{-1}]	R_1 [s^{-1}]	NOE	τ_m [ns]	τ_i [ns]	s^2
Neu5Ac B, Gal IV	2.3 ± 0.2	7.1 ± 1.0	1.6 ± 0.2	2.8 ± 0.1	0.34 ± 0.1	0.55 ± 0.08
GalNAc III, Neu5Ac A, Gal II, Glc I	2.2 ± 0.2	12.5 ± 2.0	1.2 ± 0.2	2.8 ± 0.1	0	1

[a] By nonlinear least-squares optimization, for GD1a/DPC micelles in aqueous solution. The relaxation data for the methylene carbons and Gal IV-4 were not included in the analysis.

pounds carrying the same terminal fragment, among them GM4, GM3, and GD1a.[38,42,53]

The conformation and the conformational dynamics of GD1a ganglioside anchored into a DPC micelle were investigated by means of multinuclear magnetic resonance spectroscopy.[13] The observed $^1H/^1H$ NOE interactions, once again, revealed conformational averaging of the terminal α-Neu5Ac-(2-3)-β-Gal and β-Gal-(1-3)-β-GalNAc glycosidic linkages. The pronounced flexibility of this terminal trisaccharide moiety was substantiated by probing $^1H/^{13}C$ dipolar interactions through two-dimensional proton-detected ^{13}C, T_{1p}, and NOE measurements. With the exception of the Gal IV-4 spin, the data in Table IX show significantly different T_{1p} and NOE values for the terminal Gal IV and Neu5Ac B residues as compared to the rest of the molecule. The relaxation data are converted into motional parameters assuming two correlation times for the terminal fragment NeuAc-Gal and a single correlation time for the GalNAc[Neu5Ac-]-Gal-Glc (Table X). The presence of the fast internal motions of the Gal IV and Neu5Ac B rings, reflected in the τ_i and S^2 parameters, agrees with the proton NOE data and those previously reported.[38,42,53]

Acknowledgment

This work was supported by the "Cofinanziamento del MURST 1997. Glicobiologia: struttura fine dei glicoconiugati come base del loro coinvolgimento in processi biologici e patologici."

[53] J. Breg, L. M. J. Kroon-Batemburg, G. Strecker, J. Montreuil, and J. F. G. Vliegenthart, *Eur. J. Biochem.* **178,** 727 (1989).
[54] L. G. Werbelow and D. M. Grant, *Adv. Magn. Reson.* **9,** 189 (1977).
[55] L. E. Kay and T. E. Bull, *J. Magn. Reson.* **99,** 615 (1992).
[56] H. Desvaux, P. Berthault, N. Birlirakis, and M. Goldman, *J. Mag. Reson. A.* **108,** 219 (1994).
[57] H. Desvaux, P. Berthault, N. Birlirakis, M. Goldman, and M. Piotto, *J. Mag. Reson. A.* **113,** 47 (1995).

[22] Fluorescence Quenching Assay of Sphingolipid/Phospholipid Phase Separation in Model Membranes

By Erwin London, Deborah A. Brown, and Xiaolian Xu

Introduction

Recently, considerable experimental support has been obtained for a complex model of plasma membrane structure in eukaryotic cells. This

0076-6879/00 $30.00

model proposes that a lipid phase separation results in the formation of regions enriched in sphingolipids and cholesterol (lipid domains or "rafts") that coexist with regions enriched in unsaturated phospholipids.[1–3] Early speculation on whether lipid phase separation occurred in natural membranes centered on the possibility that membranes might contain domains of gel (solid-like) phase lipid coexisting with fluid lipid. However, studies on model and natural membranes suggest that domains enriched in unsaturated-chain phospholipids in the fluid liquid–crystalline state (also called liquid-disordered state) coexist with sphingolipid and cholesterol-rich lipid domains that are in a liquid-ordered like state *in vivo*.[4–9] The liquid-ordered state, which requires cholesterol to form, has properties intermediate between those of the fully fluid liquid–crystalline and gel states.[2]

Progress in understanding lipid domain structure has been limited by the difficulty of detecting lipid phase separation in both model and cellular membranes. This difficulty may be due to domains being of submicroscopic size, and/or from their having physical properties that make them hard to distinguish from one another. In this report, we describe a method that allows facile detection of lipid phase separation in model membrane systems.

Principles of Using Fluorescence Quenching to Detect Phase Separation: Behavior of Simple Binary Lipid Mixtures

Fluorescence quenching (the decrease in intensity induced by a quencher molecule) offers a powerful approach to detection of sphingolipid phase separation.[4] The basic quenching experiment involves the use of mixed lipid bilayers in which one lipid carries a fluorescence quenching group. Phase separation results in the formation of one set of domains enriched in the quencher-carrying lipid, and a second set of domains depleted in the quencher-carrying lipid. When a fluorescent probe is incorporated into bilayers exhibiting phase separation, probe fluorescence intensity is very different from that measured in bilayers containing only a single lipid phase (see below).

[1] D. A. Brown and J. K. Rose, *Cell* **68,** 533 (1992).
[2] D. A. Brown and E. London, *J. Membr. Biol.* **164,** 103 (1998).
[3] A. Reitvald and K. Simons, *Biochim. Biophys. Acta* **1376,** 467 (1998).
[4] S. N. Ahmed, D. A. Brown, and E. London, *Biochemistry* **36,** 10944 (1997).
[5] R. Schroeder, E. London, and D. A. Brown, *Proc. Natl. Acad. Sci. U.S.A.* **91,** 12130 (1994).
[6] R. Schroeder, S. N. Ahmed, Y. Zhu, E. London, and D. A. Brown, *J. Biol. Chem.* **273,** 1150 (1998).
[7] R. Varma and S. Mayor, *Nature* **394,** 798 (1998).
[8] T. Kurzchalia and T. V. Kurzchalia, *Nature* **394,** 802 (1998).
[9] T. L. Thomas, D. Holowka, B. Baird, and W. W. Webb, *J. Cell. Biol.* **125,** 795 (1994).

Quenching is uniquely suited to phase separation studies because it can detect relatively small domains. In addition, unlike other spectroscopic assays, it does not require that the two phases have different physical properties. Only a difference in lipid composition in each phase is necessary to detect phase separation. We have found that quenching is useful in detecting both liquid–crystalline/gel and liquid–crystalline/liquid-ordered phase separation. The gel and liquid-ordered phases have in common a high degree of acyl chain order, which distinguishes these phases from the liquid–crystalline phase. In discussing phase separation in general terms, then, we will use the term *ordered* to refer either to a gel or liquid-ordered phase, and the term *fluid* to refer to the liquid–crystalline phase.

To understand the consequences of phase separation for fluorescence quenching, the phase behavior of binary lipid mixtures must be considered. Lipid phase separation depends on both lipid composition and temperature, as can be illustrated by a schematic phase diagram for such a mixture (Fig. 1A). The two rows of schematic membrane bilayers (Figs. 1B and 1C) crudely illustrate lipid organization as the fraction of two lipids in a bilayer is varied under different conditions. Note that the behavior of bilayers under the conditions labeled a–d and a′–d′ in the phase diagram (Fig. 1A) is depicted in the corresponding schematic diagrams in Figs. 1B and 1C. Consider the case in which a quencher-carrying lipid (designated by black polar headgroups in Fig. 1) having a low T_m (the temperature at which the transition between an ordered/solid and fluid phase occurs) is mixed with an unlabeled lipid (designated by white polar headgroups) having a high T_m. When the lipid composition is varied at a temperature above the T_m of both lipids (e.g., upper dashed line, Fig. 1A) the lateral distribution of the quencher is random at all compositions (as illustrated in Fig. 1B). In other words, the two lipids are fully miscible.

Lipid organization is quite different under conditions where the lipids exhibit only partial miscibility. For example, consider lipid behavior at a temperature below the T_m of the unlabeled lipid but above T_m of the quencher-labeled lipid (e.g., along the lower dashed line in Fig. 1A). At low quencher concentration, the quencher-carrying lipid distributes randomly in a single ordered phase, and its lateral distribution is similar to that at higher temperature (illustrations a and a′ in Fig. 1C). However, when the concentration of quencher-labeled lipid is increased, the ordered phase rich in high T_m lipids becomes saturated with quencher-carrying lipid, and a separate fluid phase appears. This fluid phase is enriched in quencher-carrying lipid (illustration b′ in Fig. 1C). As the concentration of quencher-carrying lipid is increased further, the proportion of the bilayer that is fluid increases (illustration c′ in Fig. 1C). Notice that the composition of each phase

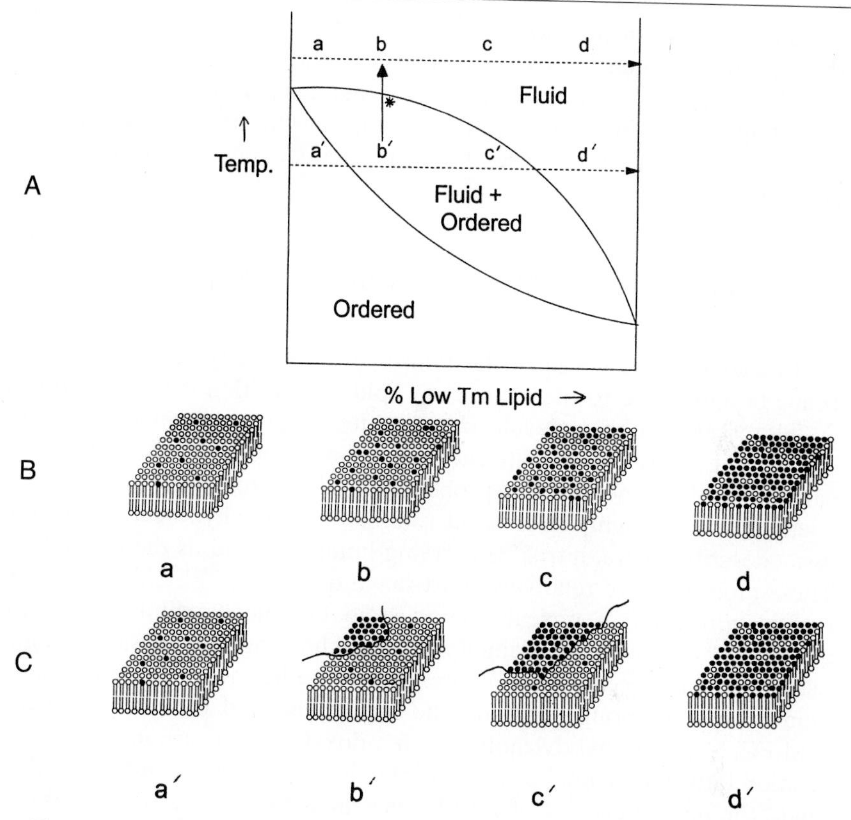

FIG. 1. Phase behavior for mixtures of two lipids with different melting temperatures (T_m). (A) Hypothetical phase diagram for mixtures of a low and high T_m lipid. The "ordered" phase can be a solid-like phase such as the gel phase or the liquid ordered phase. Notice that the fraction of low T_m lipid increases along the x axis. The dashed horizontal arrows correspond to conditions explored in experiments in which composition is varied. The vertical arrow represents conditions explored in experiments in which temperature is varied. The * denotes T_{mix}. (B and C) Schematic illustrations of lateral distributions of the high T_m (white polar headgroup) and low T_m (filled polar headgroup) lipids. The headgroup composition is only shown for the upper leaflet of the bilayer. Illustrations a–d and a′–d′ denote compositions with increasing amounts of the lipid with the low melting temperature. (B) The lateral distribution under conditions of miscibility (i.e., a single phase present at all compositions). (C) Behavior under conditions of partial miscibility (i.e., there are two phases present at some compositions).

remains constant in this two-phase region of the phase diagram. Finally, there is enough quencher-carrying lipid to dissolve all of the high T_m lipid into the fluid phase. At this or higher quencher concentrations, the distribution of lipids is again random, as it is in samples with the identical composition at higher temperature (i.e., Figs. 1B and 1C; compare illustrations d and d′).

Effect of Composition on Fluorescence in Fully Miscible Lipid Mixtures

Consider the response of the fluorescence intensity of a small membrane-bound probe to the changes in lipid composition described above. A fluorescent probe molecule incorporated into the membrane bilayer will exhibit a fluorescence intensity that depends on how many quencher-carrying lipids are nearby. The probe used in these studies is diphenylhexatriene (DPH), a highly fluorescent rod-like aromatic hydrocarbon. In the assay described here, nitroxide-carrying lipids are used as the quenchers. These molecules are relatively short-range quenchers,[10,11] and if the fluorescent probe has a typical excited state lifetime it will be primarily quenched by nitroxide-carrying lipids in neighboring sites. Basically, complete quenching occurs if the fluorescent probe has at least one nitroxide lipid in a neighboring position. The quenching lipid used in the assay, 12SLPC, is a phosphatidylcholine with a doxyl ring on carbon-12 of the 2 position fatty acyl chain. Because this ring inhibits close packing with other lipids, the phase behavior of 12SLPC is similar to that of a highly unsaturated lipid.[4]

Consider a case in which a mixture of quencher and unlabeled lipids forms a single fluid phase at all compositions. Assume that, even in the fluid state, molecules pack closely in the bilayer. Also assume that the fluorophore has dimensions close to those of a lipid, and that each fluorophore is located within one of the leaflets of the bilayer (as opposed to being at the exact bilayer center). Under these conditions, the fluorescent molecule will have about six neighboring lipids. Because quenching will occur if any of these neighboring lipids carries a nitroxide, it will show an approximately sixth-power dependence on the concentration of nitroxide lipid, as illustrated by the solid line in Fig. 2.[12] (More sophisticated analyses

[10] A. Chattopadhyay and E. London, *Biochemistry* **26,** 39 (1987).
[11] S. A. Green, D. J. Simpson, G. Zhou, P. S. Ho, and N. V. Blough, *J. Am Chem. Soc.* **112,** 7337 (1990).
[12] E. London and G. W. Feigenson, *Biochemistry* **20,** 1932 (1981).

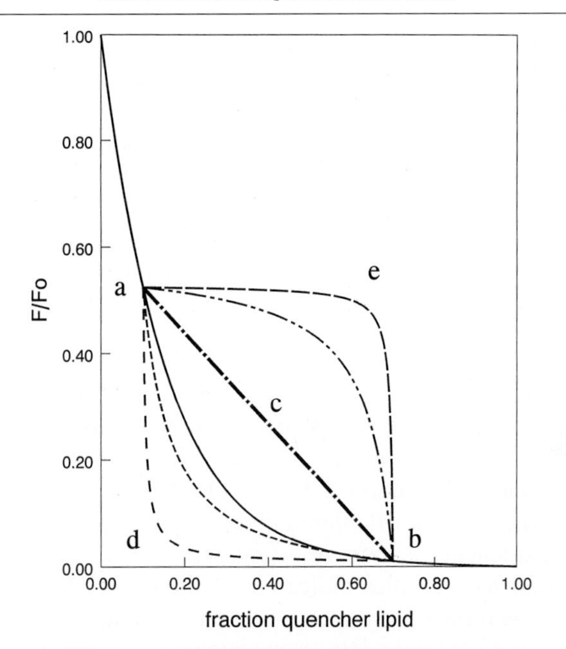

FIG. 2. Theoretical quenching patterns showing the effect of probe partition coefficient on quenching curves. The fraction of quencher lipid in a binary mixture with a nonquenching lipid is shown on the x axis. F/F_0 is the ratio of fluorescence in the presence of quencher to that in its absence, and is shown for: (___) a mixture forming a uniform phase at all compositions; or mixtures forming two phases between 10% quencher lipid (fraction 0.1) and 70% quencher (fraction 0.7) lipid with the fluorescent probe having a partition coefficient (K_p) of (– – –) 100, (----) 10, (–·–·–) 1, (–··–··–) 0.1 and (———) 0.01. [Adapted with permission from Ahmed *et al., Biochemistry* **36,** 10944 (1997). Copyright American Chemical Society.]

can be used to predict a more precise concentration dependence, but this has no effect on the phenomena discussed here.[10,13,14])

Effect of Composition on Fluorescence in Partially Immiscible Lipid Mixtures

The dependence of fluorescence on lipid concentration will be more complex at temperatures low enough for phase separation to occur. In bilayers in which only one phase is present, quenching will be similar to that described above. However, in all lipid mixtures that contain two phases (i.e., between points a and b in Fig. 2), fluorescence intensity will depend

[13] M. D. Yeager and G. W. Feigenson, *Biochemistry* **29,** 4380 (1990).

[14] A. S. Ladokhin and P. W. Holloway, *Biophys. J.* **69,** 509 (1995).

critically on the amount of fluorescent molecule in each phase. At these points, discontinuities in the slopes of the dependence of fluorescence intensity on the fraction of quencher-carrying lipid curves reveal the boundaries between compositions at which one phase is present, and those in which there are two coexisting phases (points a and b, Fig. 2). The shape of the quenching curve in the two-phase region depends on how the fluorescent probe partitions between the two phases. The theoretical concentration dependence of fluorescence in a two-phase region is illustrated quantitatively for a number of different partition coefficients (K_p) in Fig. 2. If the fluorescent molecule shows no preference for one phase over the other, fluorescence will decrease linearly as the concentration of quencher-carrying lipid is increased (curve c, Fig. 2). However, if the probe partitions favorably into the quencher-rich phase, it will be quenched more strongly as the quencher concentration is increased than if it partitioned equally well into both phases (e.g., curve d, Fig. 2). On the other hand, if it partitions favorably into the quencher-depleted phase, its quenching will remain relatively weak as quencher is increased (e.g., curve e, Fig. 2).[4,15]

In practice, discontinuities indicative of the onset of phase separation can sometimes be difficult to discern. For this reason, it is best to compare the quenching curve of lipid mixtures in which phase separation is suspected to occur to a control curve, showing quenching in mixtures known to be fully miscible throughout the concentration range. For instance, in Fig. 3A the curve showing quenching of DPH in mixtures of fully miscible 12SLPC and dioleoylphosphatidylcholine (DOPC) (circles) is contrasted with the quenching curve of DPH in mixtures of 12SLPC and dipalmitoylphosphatidylcholine (DPPC) (triangles). The two curves are coincident at both ends of the concentration range, under conditions in which the DPPC/12SLPC mixtures are in either a single gel phase (high DPPC), or a single liquid–crystalline phase (low DPPC). However, discontinuities in the DPPC/12SLPC curve indicative of phase separation are seen at about 20 and 80% DPPC. In the two-phase region between these points, fluorescence is quenched less strongly than in the control single-phase DOPC/12SLPC mixture. Such decreased quenching relative to uniform mixtures containing the same amount of quencher lipid is observed when fluorescent probes with some ability to partition into the quencher-depleted phase (i.e., K_p equal to or less than 1) are used.

Note that these experiments can also be used to measure the partition coefficient of a fluorescent probe between two phases (Fig. 2). This parameter can potentially be used to obtain the relative affinity of fluorescent-labeled molecules for sphingolipid-rich and sphingolipid-depleted domains.

[15] E. London and G. W. Feigenson, *Biochim. Biophys. Acta* **649,** 89 (1991).

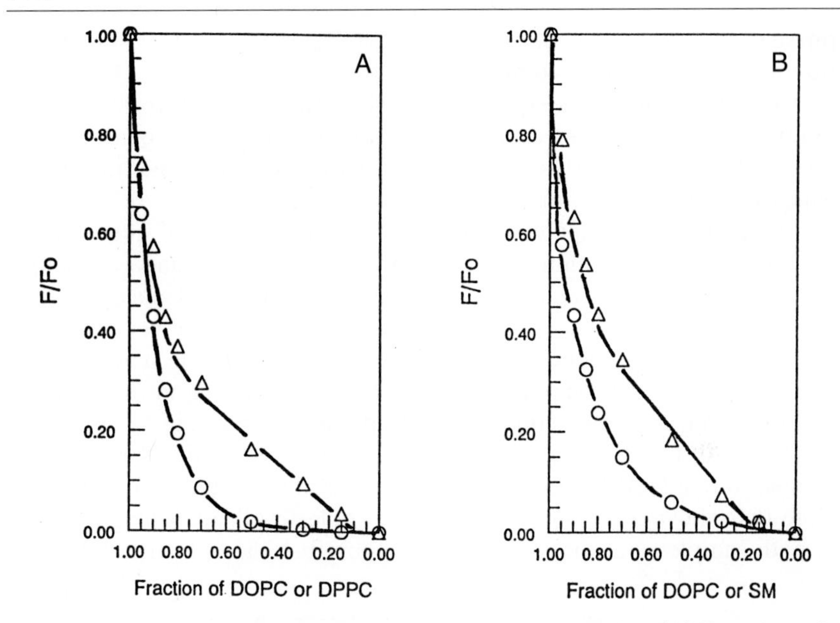

FIG. 3. Comparison of quenching curves for various lipid mixtures. (A) Data shown for (circles) DOPC/12SLPC mixtures and (triangles) DPPC/12SLPC mixtures at 23°. (B) Data shown for (circles) DOPC/12SLPC/cholesterol mixtures and (triangles) sphingomyelin/12SLPC/cholesterol mixtures with 33 mol% cholesterol at 37°. The x axis shows the concentration of DOPC, DPPC, or sphingomyelin as a mole fraction of 12SLPC + DOPC, 12SLPC + DPPC, or 12SLPC + sphingomyelin, respectively. F/F_0 values have been normalized as described in the text. [Adapted with permission from Ahmed *et al., Biochemistry* **36,** 10944 (1997). Copyright American Chemical Society.]

The probe used in our experiments, diphenylhexatriene (DPH), has a partition coefficient close to 1 in most cases.

Effect of Temperature on Fluorescence

It is often more desirable to measure how lipid phase behavior is affected by increasing temperature at a fixed lipid composition (vertical arrow, Fig. 1), than it is to measure the effect of varying lipid composition at a fixed temperature.

Varying temperature is particularly useful when comparing the effect of changing lipid structure on phase separation (e.g., examining sphingolipids having different acyl chain or polar headgroup structures). The stronger the tendency of a particular mixture to phase separate, the higher the temperature at which phase separation should disappear, i.e., the higher the "mixing" temperature (T_{mix}; indicated in Fig. 1A). (Note that if one phase is ordered and the other is fluid, then T_{mix} is also the lowest temperature at which melting is complete.)

Consider a quenching experiment in which the temperature dependence of DPH fluorescence is measured in a sample containing a nitroxide-labeled lipid and sphingolipid. There are two ways to determine conditions under which two phases are present in this sample, and the value of T_{mix}. The first method is to compare the temperature dependence of quenching in the sphingolipid-containing "test" sample to that in a "standard" sample that contains the same amount of quencher lipid, but exists in a single fluid phase at all temperatures. If two phases are present in the test sample, it will exhibit weaker quenching (and higher F/F_0) than in the standard sample. T_{mix} will be the temperature above which quenching in the standard and test sample become equal. An example is shown in Fig. 4, in which a DPPC/12SLPC mixture that can undergo phase separation (triangles) is compared to a DOPC/12SLPC mixture that is always fluid (open squares). Note that above T_{mix} (about 35° without cholesterol) quenching is strong in both samples, but that below this temperature the two curves diverge,

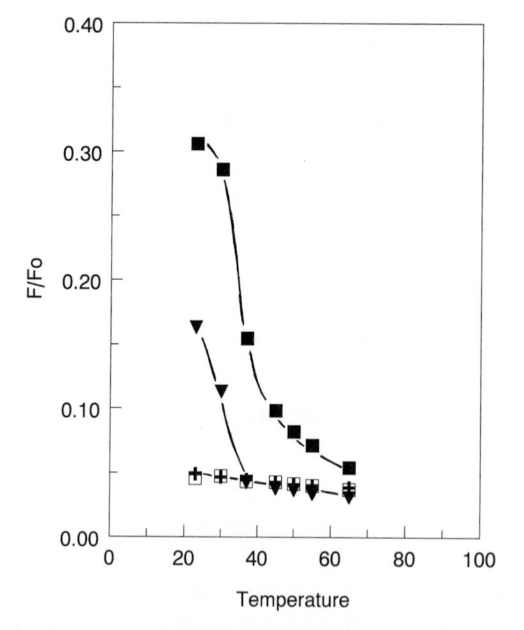

FIG. 4. Comparison of the temperature dependence of quenching in lipid mixtures with and without cholesterol. Thermal dependence of F/F_0 values is shown for: (triangles) 1:1 DPPC/12SLPC; (open squares) 1:1 DOPC/12SLPC; (closed squares) 1:1 DPPC/12SLPC with 15 mol% cholesterol; (crosses) 1:1 DOPC/12SLPC with 15 mol% cholesterol. The behavior of sphingomyelin-containing mixtures is similar to that of mixtures containing DPPC (not shown).

and quenching in the DPPC/12SLPC sample is weaker. The second method of analysis does not require comparison to a control curve, but is based on the fact that while quenching should be relatively temperature independent in a bilayer containing a single fluid phase, it can be strongly temperature dependent in a bilayer containing two phases. In this case, T_{mix} can be detected by the abrupt decrease in the temperature dependence of quenching when the second phase disappears (i.e., a large decrease in the slope of quenching versus temperature) as occurs at 35° in the DPPC/12SLPC mixtures without cholesterol and close to 40–45° in DPPC/12SLPC mixtures with cholesterol (Fig. 4).

One advantage of studying phase separation by varying temperature instead of composition is that many fewer samples are required (see below). Another advantage is that the reversibility of thermally-induced phase mixing is easily established using cooling experiments. If a phase separation exists, and is abolished by heating, but reforms on cooling, it effectively demonstrates that the phase-separated state is thermodynamically stable at lower temperatures and unstable at high temperatures. Such observations rule out artifacts in which quenching is weaker than expected in a single phase simply due to an inhomogeneous lipid distribution during vesicle formation resulting in formation of populations of vesicles with different lipid compositions.

Relationship between Lipid Structure, Lipid Composition, and Domain Formation

It is worthwhile to consider what questions can be examined by the analysis of phase separation. One central question is: What is the relationship between lipid structure and phase separation? There are many different sphingolipids, with different polar headgroup structures.[16] Although virtually all sphingolipids have long saturated hydrocarbon chains and high melting temperatures, their actual T_m values can vary considerably, depending on the structure of the polar headgroup. The same is true for glycerophospholipids. In addition, each class of glycerophospholipids can show a significant variability in the degree of unsaturation in their acyl chains. These differences could have profound effects on the ability to undergo phase separation. Finally, the biologic membranes thought to undergo phase separation are sterol rich.[2] There are subtle structural differences between cholesterol and the sterols of plants and fungi. Does sterol chemical structure affect domain formation?

Just as important as the role of structure in phase separation is the

[16] W. Curatolo, *Biochim. Biophys. Acta* **906,** 111 (1987).

question of how it is affected by the organization and concentration of lipids and proteins in a membrane. For example, protein clustering seems to influence domain structure.[9] In addition, there are indications that cholesterol can both strongly promote, or inhibit phase separation, depending on its concentration in the membrane.[17] Because the fluorescence quenching assay is well suited to studying the effect of cholesterol on phase separation, and its behavior is likely to have important biologic consequences,[2] we will spend some time discussing inclusion of cholesterol in the assay system.

Thus, to gain a full understanding of phase separation in natural membranes, the phase behavior of many lipid combinations, and how they are affected by protein will have to be explored. Applying the fluorescence quenching method to model membranes, where lipid composition can be controlled, should allow for systematic investigation of the effects of lipid structure and composition on phase behavior.

Sample Composition in Experiments in Which Lipid Composition Is Varied

As mentioned above, the fluorescence quenching assay can be performed by varying either lipid composition or temperature. We will discuss sample composition for the two protocols in turn.

To determine phase behavior as a function of lipid composition, the desired mixtures of quencher-carrying phospholipid, sphingolipid, and DPH with or without cholesterol, must be prepared. The total amount of lipid used is close to 50 nmol (enough to make 1 ml of a 50 μM lipid sample). The quencher-carrying phospholipid is 12SLPC. Samples with a [12SLPC]/([sphingolipid + 12SLPC]) ratio of 0, 0.1, 0.2, 0.3, 0.4, 0.5, 0.6, 0.7, 0.8, 0.85, 0.9, 0.95, and 1.0 are prepared. The larger number of samples with low sphingolipid (i.e., high 12SLPC) concentrations allows easier detection of the lowest sphingolipid composition at which phase separation occurs. A low (e.g., 0.5 mol%) DPH concentration is used. The concentration of cholesterol can be varied (at least up to the saturation point[18]). For example, if 33 mol% cholesterol is desired, the samples would contain 67% (12SLPC + sphingolipid) and 33% cholesterol, with all the various fractions of 12SLPC/(12SLPC + sphingolipid) shown above. (The small amount of DPH can be ignored in these calculations.)

As noted earlier, a second set of "standard" samples in which a single phase is present at all quencher concentrations is also needed. The lipids 12SLPC and DOPC are fully miscible, and their mixtures are used as the

[17] J. R. Silvius, D. del Guidice, and M. Lafleur, *Biochemistry* **35,** 15198 (1996).
[18] J. Huang, J. T. Buboltz, and G. W. Feigenson, *Biochim. Biophys. Acta* **1417,** 89 (1999).

standard set. Samples with a [12SLPC]/([DOPC + 12SLPC]) ratio of 0, 0.1, 0.2, 0.3, 0.4, 0.5, 0.6, 0.7, 0.8, 0.85, 0.9, 0.95 and 1.0 are prepared. The samples also contain DPH, and cholesterol (if the test sample contains cholesterol).

The above samples should be prepared in duplicate, with the exception of the samples lacking 12SLPC (F_0 samples). These should be prepared in triplicate.

Finally, background fluorescence samples are needed. These contain lipid, but lack DPH. In general, background fluorescence intensities are low for the mixtures described above, and often their intensity is only weakly dependent on lipid composition. Furthermore, it is undesirable to waste the lipid 12SLPC to make such samples. Therefore, only three background samples, one with 12SLPC, one with sphingolipid, and one with DOPC (all with or without cholesterol as appropriate), are prepared. The background values for samples containing mixtures of these lipids can be approximated from linear combinations of the values obtained with these three samples by assuming a linear dependence of background on lipid composition.

Sample Composition in Experiments in Which Temperature Is Varied

A set of eight samples is required for experiments in which temperature is varied. The first two samples are reserved for measuring the fluorescence in the presence of quencher. The test sample contains 12SLPC, sphingolipid, and DPH with or without cholesterol, and a second (single phase) "standard" sample contains 12SLPC, DOPC and DPH with or without cholesterol. Care should be taken in choosing the relative amounts of 12SLPC and sphingolipid in the test sample. These concentrations should be chosen to maximize the difference in fluorescence intensity between the test and control sample when phase separation occurs (i.e., a lipid composition at which the two curves in Fig. 3A would be farthest apart). We have found for sphingomyelin/12SLPC mixtures that equimolar sphingomyelin and 12SLPC is ideal, although for some applications it would be desirable to vary this ratio. Samples with a DPH concentration of 0.5 μM and total lipid concentration as low as 50 μM can be used.

Two analogous samples lacking 12SLPC should also be prepared. These would be used to determine the fluorescence in the absence of quenching (F_0 values). In these samples, DOPC is substituted for the 12SLPC used in the first two samples. This substitution is acceptable because DOPC and 12SLPC have similar phase behavior.[4]

The remaining four samples are used for fluorescence background measurements and have the same lipid composition as the four samples described above, except that DPH is omitted.

Sample Preparation: Lipid Mixing

Lipids (purchased from Avanti Polar Lipids, Alabaster, AL) and DPH (Aldrich Chemical, Milwaukee, WI) are dissolved in ethanol and stored at −20° until use. The appropriate amounts of the lipids and DPH are transferred to a glass tube. The amounts of each component to be used, determined as described in the preceding section, are such that the final total lipid concentration (including cholesterol if present) is 50 μM. This low concentration minimizes light scattering, and conserves the expensive 12SLPC lipid. (However, higher total lipid concentrations may be needed when a lipid does not bind tightly to vesicles. We have experienced this difficulty in preliminary experiments with ganglioside GM1 (L. Wang, E. London, and D. A. Brown, 1999, unpublished observations).)

To obtain precise lipid compositions in each sample, accurate pipetting is important. Air displacement pipettors and plastic tips are not recommended because of (1) the possibility of leaching fluorescent contaminants from plastic tips and (2) the tendency of air displacement pipettors to drip organic solvents. To avoid these problems, pipetting can be performed with Drummond positive displacement pipettors. These employ disposable glass capillaries, and have continuously variable volume settings. They are available in different models for pipetting of volumes up to 100 μl.

After the lipid mixture is prepared, the sample is dried under a flow of N_2, redissolved in $CHCl_3$, redried under N_2, and further dried under high vacuum for 1 hr. The redissolving step is included to help ensure that all of the components are fully mixed. One concern is that, during drying, one of the components will reach its solubility limit and precipitate out of solution before the other components. This could result in inhomogeneous preparations in which different vesicles have a different lipid composition. Inhomogeneous incorporation of cholesterol is of particular concern.[18]

An other concern that should be mentioned is that the fraction of 12SLPC molecules actually carrying the fluorescence quenching nitroxide group can be variable. We have found that the fraction of molecules carrying an active nitroxide group in different preparations fall in the range 60–90%.[10,19,20] Although influencing the absolute level of quenching, this does not pose a serious problem in the data analysis procedure because the quenching values are normalized to a percent of maximal quenching (see

[19] F. S. Abrams and E. London, *Biochemistry* **31,** 5312 (1992).
[20] R. D. Kaiser and E. London, *Biochemistry* **37,** 8180 (1998).

below). Nevertheless, normalization does not totally eliminate the effect of a difference in the number of active quencher molecules, and it is important to use the same preparation of 12SLPC, or at least preparations that give equally strong quenching, for all of the experiments in a particular set of studies.

Dispersing Lipid Samples in Buffer at Ambient Temperatures

To obtain samples for the quenching experiments, multilamellar vesicles (MLVs) are prepared from the dried mixtures described above. This avoids problems of residual solvent, detergent, and curvature effects that can result from other vesicle preparation procedures. At the low concentrations used, light scattering is not a serious limitation. Interbilayer quenching is also not of concern because both the quencher (12SLPC nitroxide) and DPH locate deeply within the bilayer.[13,20]

At room temperature (23°), MLVs are prepared by vigorously mixing (vortexing) the dried samples in phosphate-buffered saline (PBS, 10 m*M* sodium phosphate, 150 m*M* NaCl, pH 7), for 60 sec. To ensure that all the dried lipid is dispersed into the buffer, it is important to make certain the buffer is in contact with the lipid during the vortexing process. Dispersal of the lipid may also be aided by a few cycles of intermittent vortexing, and by varying the angle at which the tube is placed onto the vortex head. When possible, samples should be visually inspected to ensure that dried lipid does not remain on the tube wall. After vortexing, the samples are incubated in the dark for 30 min to 1 hr.

Dispersing Lipid Samples in Buffer at Elevated Temperatures

In many cases it is desirable to disperse the lipid at high temperature. Tubes containing the lipid mixtures are hydrated with (PBS) warmed to the desired temperature in a water bath. Each tube is placed in a small beaker (50 ml) partly filled with water also warmed to the desired temperature. Then the tube and beaker are vortexed together for 15 sec. This can be done by holding the tube at an angle so that it touches the bottom of the beaker at the angle between the beaker wall and beaker bottom, and then placing the tube and beaker so that the tube is as close as possible to the head of the vortex mixer. The beaker and tube should be covered with Parafilm to prevent splashing. The purpose of this procedure is to disperse the lipid by vortexing, while avoiding the appreciable cooling that would occur without the warmed solution in the beaker. After this step, the samples are incubated for 5 min at the desired elevated temperature (in a water bath), and revortexed for 30 sec without the beaker (to allow more vigorous mixing). Finally, the samples are incubated at the elevated temperature in the dark for 30 min to 1 hr prior to fluorescence measurements. This elaborate procedure is necessary because dried lipids tend to disperse poorly into buffer at temperatures below their T_m. In addition, hydration at too low a temperature increases the chances that an inhomogenous preparation, in which different vesicles have different lipid composition, will be obtained.

As a consequence, hydration should be performed at temperatures above T_m of the lipid with the highest melting temperature. This is usually more than sufficient as in mixtures containing low T_m lipids, melting of the high T_m lipid is complete at temperatures below its T_m in the pure state (Fig. 1). We have prepared samples of sphingomyelin/12SLPC at 65°; cerebrosides/sphingomyelin/12SLPC at 81°; dipalmitoylphosphatidylcholine(DPPC)/12SLPC and DPPC/DOPC/12SLPC at 50°; and DOPC/12SLPC at 23°.

It is not yet clear whether samples that contain high T_m lipid and cholesterol need to be prepared at elevated temperatures. With 33 mol% cholesterol, we found no consistent differences between experiments in which samples were prepared at room temperature, and those in which they were prepared at 81°.[4] We have also had very reproducible results with samples prepared at room temperature when they contain 1:1 mixtures of DPPC and DOPC with 15 mol% of a variety of sterols. Such samples show phase mixing at high temperature that is reversible on cooling (data not shown). Nevertheless, we suggest that the preparation conditions for samples containing cholesterol be determined on a case-by-case basis.

Fluorescence Measurements

Fluorescence measurements can be made on a standard fluorimeter such as the Spex 212 Fluorolog spectrophotometer (Jobin-Yvon, Inc., Edison, NJ). This instrument has double excitation and emission monochrometers to minimize interference from light scattering. Fluorescence can be measured in semimicro fluorescence cuvettes (1-cm excitation, 4-mm emission path length).

Care should be taken to minimize the exposure of DPH to light. Samples should be kept in the dark until transfer to a cuvette. The time and extent of exposure to light from the fluorimeter lamp should also be minimized. To minimize the latter, relatively small excitation slit settings should be used. (On the Spex instruments we have used 1.25-mm excitation slits.) After placing the cuvette with the desired sample in the fluorimeter sample chamber, samples are incubated in the dark for 1 min at the appropriate temperature, and then the excitation shutter is opened and fluorescence is measured over a 3-sec interval. The excitation and emission wavelengths used are 358 and 430 nm, respectively. The temperature in the cuvette is regulated by a temperature bath connected to a cuvette holder through which the bath solution circulates. The temperature within the cuvette is verified with an electronic thermometer placed in a cuvette containing water. Similar instrumental settings are used for samples when the temperature dependence of fluorescence is measured. In the first step, intensity measurements on all samples, including background controls, are made at room temperature (23°). The temperature dependence of fluorescence is then measured in the two samples containing quencher and DPH. Temperature settings on the external bath are increased in 5° to 10° steps (measuring the actual cuvette temperature in control experiments). Samples are allowed to reach the new temperature, which takes a few minutes, and then incubated for an additional 5 min prior to reading fluorescence. At the highest temperature, the fluorescence in all eight samples is measured. Controls show that the intensity in background samples lacking DPH has a negligible dependence on temperature, and that samples lacking quencher exhibit an almost linear decrease of intensity as a function of temperature. Thus, the fluorescence at intermediate temperatures can be calculated by extrapolation both for the background samples and those lacking quencher. The linear extrapolation for samples that lack quencher is not always valid, and can lead to an error in F/F_0 values. However, we find the maximum overestimate of F/F_0 due to this approximation is about 0.01, which is not serious enough to affect any conclusions from the data. Furthermore, this procedure greatly simplifies the experiments, since the sample compartment of the fluorescence instrument can accommodate only a limited number of cuvettes.

Data Analysis

The first step in data analysis, after subtraction of background values, is the calculation of the ratio of fluorescene in the presence of the quencher 12SLPC (F) to that in its absence (F_0). F/F_0 values are also normalized to a value of 0 at 100% 12SLPC (not counting cholesterol) using the formula: $F/F_{0\,\text{normalized}} = (F/F_0 - F/F_{0\,100\%\,12\text{SLPC}})/(1 - F/F_{0\,100\%\,12\text{SLPC}})$. Normalization makes it easier to compare results using different unlabeled lipids, because the fluorescence intensity of DPH in the absence of quencher (F_0) can vary depending on the unlabeled lipid used. Normalization has only a small effect on curve shape as residual fluorescence at 100% 12SLPC is only 2–6% of F_0 in our experiments.

For the protocol in which lipid composition is varied, the F/F_0 values for samples containing sphingolipid are then graphed versus [12SLPC]/([12SLPC] + [sphingolipid]) or [sphingolipid]/([12SLPC] + [sphingolipid]).

The values for samples containing DOPC are graphed similarly, substituting DOPC for sphingolipid in these formulas. As noted earlier, the range of sphingolipid mole fractions in which a phase separation occurs can be discerned using either of two methods. One is by identifying the range of sphingolipid concentrations bounded by the discontinuities in the F/F_0 graph for 12SLPC/sphingolipid mixtures. The second is the range of concentrations in which F/F_0 values for the 12SLPC/sphingolipid mixture exceed that in 12SLPC/DOPC. In practice, both methods work well for detecting the lowest sphingolipid concentration at which phase separation occurs (e.g., Fig. 3B, low sphingomyelin concentration). However, although our method assumes that quenching is identical in any uniform phase, this may not always be precisely true (see below). Thus, because F/F_0 values in a fluid phase may not be identical to those in an ordered phase, there can be a difference in F/F_0 values even though in only one phase in each sample. An example of this behavior is shown in Fig. 3B. A discontinuity in the sphingomyelin-containing curve (triangles) occurs at about 80% sphingomyelin, suggesting that mixtures with higher sphingomyelin concentrations are in a uniform liquid-ordered phase. However, F/F_0 for these mixtures is higher than that of control DOPC-containing mixtures in a uniform liquid–crystalline phase that contain the same amount of quencher (circles). Therefore, in this case identifying a discontinuity in the quenching curve appears to be a more reliable method for detecting the onset of phase separation.

The analysis is basically similar when analyzing the data in experiments in which temperature is varied. After subtracting background fluorescence, F/F_0 values are calculated, dividing the fluorescene in samples containing 12SLPC by the intensity in the samples in which DOPC has been substituted for 12SLPC. Unlike the analysis above, no additional normalization of F/F_0 is performed. F/F_0 in both the sphingolipid-containing and the (standard) fluid-phase control samples are then plotted as a function of temperature. The curves will diverge at temperatures in which two phases are present in a sphingolipid/12SLPC sample; in this region F/F_0 values in the sphingolipid/12SLPC sample will be higher than in the control. Thus, at T_{mix} for the sphingolipid/12SLPC sample, there can be both a "discontinuity" in the slope of the F/F_0 curve, and a (near) coincidence of it and the control curve. However, in some cases only the points of discontinuity can be clearly discerned (Fig. 4).

Factors Limiting Sensitivity and Interpretation of Quenching Experiments

Fluorescence quenching experiments have several limitations. One is that the extent of the decrease in quenching due to phase separation depends on the precise composition of the two phases. Note that the ordered and fluid phases in a two-phase mixture do not contain pure order- and fluid-preferring lipids, respectively. Instead, each phase contains small amounts of the lipid that prefers the other phase. Thus, a certain amount of fluid-preferring quencher lipid will always be present in the ordered phase in our assay. The larger the difference between the quencher concentration in the two phases, the larger will be the decrease in quenching resulting from phase separation. As a consequence, in order to detect phase separation the two phases must have very different lipid compositions.[4] For facile detection, it is best that the lipid in the quencher-enriched phase be composed of over 50% quencher, while the lipid in the quencher-depleted phase has 20% or less quencher.

For the same reason, it can be difficult to interpret a change in the level of quenching on changing conditions in a single sample containing a mixture in which two phases are present. For example, if an increase in quenching in such a sample were observed upon increasing temperature, it might mean that an increased fraction of total lipid was in the fluid phase. Alternatively, however, it might mean that the composition of the two phases became more similar, although the fraction of total lipid in the fluid phase stayed constant.

A related problem arises from the fact that in designing this assay, we have assumed that quenching will be the same in any single-phase bilayer, and will not depend on the properties of that phase. However, quenchers may actually behave differently in ordered and fluid bilayers, as discussed above for the example shown in Fig. 3B. This is true because quenching is affected by properties that are different in fluid and ordered phases (i.e., motion and the depth of the quenching and fluorescent groups[19]). As a consequence, it is not certain whether the small difference in quenching often observed in samples containing less than 20% 12SLPC (e.g., Fig. 3B) represents a difference in the number of phases present, or simply an intrinsic difference in the strength of quenching in an ordered and fluid phase. Nevertheless, if there is an easily discerned "discontinuity" in the slope of the quenching curve, we generally take this as evidence of a change in the number of phases present. A different type of limitation is that microclusters of a few lipid molecules cannot be detected by quenching. The quenching method depends on the size of a cluster (domain) being sufficient so that on the average a fluorescent molecule is not close to the edge of the cluster, where it would be accessible to quenching by quencher lipid in both phases. Approximate calculations suggest that cluster sizes of less than 50–100 molecules cannot be easily detected by quenching (calculation not shown).

The properties of the fluorescent probe can also limit sensitivity. DPH, which in most situations partitions nearly equally between ordered and fully fluid phases, is suitable for most experiments. However, a probe that partitions very strongly into the more ordered, quencher-depleted phase generally gives the most readily apparent discontinuities at the phase separation compositional boundaries, and greatest difference in quenching from that in a single phase sample (Fig. 2).[21] Unfortunately, few probes that strongly partition into ordered phases are available.

The need to use a quencher-labeled phospholipid as the predominant fluid (low T_m) lipid is another problem in terms of biologic relevance. This requirement makes it difficult to examine the phase behavior of mixtures of sphingolipids with natural low-T_m phospholipids.

A final complication arises from the fact that many experiments examine the phase behavior of ternary lipid mixtures. The effects of including cholesterol in ternary mixtures are of particular interest, because it appears that an important function of cholesterol in cell membranes is to promote phase separation between sphingolipids and phospholipids. However, certain assumptions are valid in binary, but not in ternary, lipid mixtures. For example, in a binary mixture the composition of each phase is constant at all compositions in which two phases are present. Only the amount of each phase varies. However, in a ternary mixture exhibiting a phase separation the composition of each phase is not necessarily constant. In addition, ternary mixtures can from three coexisting phases over a considerable range of concentrations.[17] If there are three phases, but two have similar quencher concentrations, quenching will not detect all three phases. In other words, quenching only reveals whether quencher-rich and quencher-depleted phases coexist, not the total number of phases present.

Note also that under some conditions metastable phase behavior may exist. Hints of such behavior can be seen in quenching studies of cerebroside phase separation.[21] In this case, there is a range of compositions in the one phase region at low quencher concentrations within which quenching increases beyond the value found at a higher quencher concentration, where two phases are present.[17] This anomalously strong quenching probably results from the presence in some lipid mixtures of a single ordered phase that is supersaturated with quencher.

[21] J. R. Silvius, *Biochemistry* **31,** 3398 (1992).

Section II

Methods for Analyzing Aspects of Sphingolipid Metabolism in Intact Cells

[23] Synthesis of Fluorescent Substrates and Their Application to Study of Sphingolipid Metabolism *in Vitro* and in Intact Cells

By ARIE DAGAN, VERED AGMON, SHIMON GATT, and TAMA DINUR

Introduction

In the course of our studies, the following enzymes have been isolated and partially purified: ceramidase[1–3]; two sphingomyelinases, the lysosomal, with an acidic pH optimum,[4] and a neutral, magnesium-dependent[5–7] enzyme; β-glucosidase and α- or β-galactosidase[8–11]; *N*-acetylhexosaminidase[12–14]; neuraminidase[15]; phospholipase A_1[16,17]; lysophospholipase[18–20]; and brain lipase.[21–23] For assaying these enzymes we have used labeled substrates or colored and fluorescent derivatives. In *Methods in Enzymology,* Volume 72,[24–32] we presented a comprehensive description of assay

[1] S. Gatt, *J. Biol. Chem.* **238,** 3131 (1963).
[2] S. Gatt, *J. Biol. Chem.* **241,** 3724 (1966).
[3] Y. Yavin and S. Gatt, *Biochemistry* **8,** 1692 (1969).
[4] Y. Barenholz, A. Roitman, and S. Gatt, *J. Biol. Chem.* **241,** 3731 (1966).
[5] S. Gatt, *Biochem. Biophys. Res. Comm.* **68,** 235 (1976).
[6] S. Gatt, T. Dinur, and J. Kopolovic, *J. Neurochem.* **31,** 547 (1978).
[7] S. Gatt, T. Dinur, and Z. Leibovitz-Ben Gershon, *Biochim. Biophys. Acta* **531,** 206 (1978).
[8] S. Gatt and M. M. Rapport, *Biochim. Biophys. Acta* **113,** 567 (1966).
[9] S. Gatt and M. M. Rapport, *Biochem. J.* **101,** 680 (1966).
[10] S. Gatt, *Biochem. J.* **101,** 687 (1966).
[11] S. Gatt and E. A. Baker, *Biochim. Biophys. Acta* **206,** 125 (1970).
[12] Y. Z. Frohwein and S. Gatt, *Biochim. Biophys. Acta* **128,** 216 (1966).
[13] Y. Z. Frohwein and S. Gatt, *Biochemistry* **6,** 2775 (1967).
[14] Y. Z. Frohwein and S. Gatt, *Biochemistry* **6,** 2783 (1967).
[15] Z. Leibovitz and S. Gatt, *Biochim. Biophys. Acta* **152,** 136 (1968).
[16] S. Gatt, Y. Barenholz, and A. Roitman, *Biochem. Biophys. Res. Comm.* **24,** 169 (1966).
[17] S. Gatt, *Biochim. Biophys. Acta* **159,** 304 (1968).
[18] Z. Leibovitz and S. Gatt, *Biochim. Biophys. Acta* **164,** 439 (1968).
[19] Z. Leibovitz-Ben Gershon, I. Kobiler, and S. Gatt, *J. Biol. Chem.* **247,** 6840 (1972).
[20] Z. Leibovitz-Ben Gershon and S. Gatt, *J. Biol. Chem.* **249,** 1525 (1974).
[21] M. C. Cabot and S. Gatt, *Biochim. Biophys. Acta* **431,** 105 (1976).
[22] M. C. Cabot and S. Gatt, *Biochemistry* **16,** 2330 (1977).
[23] M. C. Cabot and S. Gatt, *Biochim. Biophys. Acta* **530,** 508 (1978).
[24] S. Gatt, Y. Barenholz, R. Goldberg, T. Dinur, G. Besley, Z. Leibovitz-Ben Gershon, J. Rosenthal, R. J. Desnick, E. A. Devine, B. Shafit-Zagardo, and F. Tsuruki, *Methods Enzymol.* **72,** 351 (1981).
[25] R. Goldberg, Y. Barenholz, and S. Gatt, *Methods Enzymol.* **72,** 351 (1981).

0076-6879/00 $30.00

procedures using sphingolipids linked with trinitrophenol as a colored (yellow) marker or the fluorescent probes anthracene, NBD, pyrene, carbazole, and dansyl. More recently, we replaced these with the more polar fluorescent probe, Lissamine rhodamine (LR) or with BODIPY (Molecular Probes, Eugene, OR) linked via a 12-carbon spacer (LR12 and BOD12) for use in assaying lipid hydrolases *in vitro* (e.g., Refs. 33–39) and in intact cells in culture (e.g., Refs. 40–47). This latter approach provided a fluorescence-based evaluation of the intracellular metabolism of the respective sphingolipids and permitted diagnosis of the lipidoses in the intact, living cells and, in some cases, in the culture medium.[48] It also led to the development of a novel procedure for selecting lipidotic cells that have been corrected by transduction with a retroviral vector containing the cDNA encodingthe normal gene.[46,49–51] The latter approach was developed

---

[26] S. Gatt, T. Dinur, and Y. Barenholz, *Methods Enzymol.* **72,** 356 (1981).
[27] G. Besley and S. Gatt, *Methods Enzymol.* **72,** 360 (1981).
[28] G. Besley and S. Gatt, *Methods Enzymol.* **72,** 362 (1981).
[29] Z. Leibovitz-Ben Gershon, T. Dinur, Y. Barenholz, and S. Gatt, *Methods Enzymol.* **72,** 364 (1981).
[30] S. Gatt, T. Dinur, Y. Barenholz, Z. Leibovitz-Ben Gershon, J. Rosenthal, E. A. Devine, B. Shafit-Zagardo, and R. J. Desnick, *Methods Enzymol.* **72,** 367 (1981).
[31] S. Gatt and F. Tsuruki, *Methods Enzymol.* **72,** 372 (1981).
[32] S. Gatt and F. Tsururki, *Methods Enzymol.* **72,** 373 (1981).
[33] T. Dinur, G. A. Grabowski, J. R. Desnick, and S. Gatt, *Anal. Biochem.* **136,** 223 (1984).
[34] R. Salvayre and S. Gatt, *Enzyme* **33,** 175 (1985).
[35] G. Bach, A. Dagan, B. Herz, and S. Gatt, *Clin. Genet.* **31,** 211 (1987).
[36] R. Klar, T. Levade, and S. Gatt, *Clin. Chim. Acta* **176,** 259 (1988).
[37] S. Marchesini, P. Viani, B. Cestaro, and S. Gatt, *Biochim. Biophys. Acta* **1002,** 14 (1989).
[38] A. Negre, A. Dagan, and S. Gatt, *Enzyme* **42,** 110 (1989).
[39] A. Negre, R. Salvayre, A. Dagan, and S. Gatt, *Biochim. Biophys. Acta* **1006,** 84 (1989).
[40] T. Levade, S. Gatt, and R. Salvayre, *Biochem. J.* **275,** 211 (1991).
[41] T. Levade, S. Gatt, A. Maret, and R. Salvayre, *J. Biol. Chem.* **266,** 13519 (1991).
[42] V. Agmon, T. Dinur, S. Cherbu, A. Dagan, and S. Gatt, *Exp. Cell Res.* **196,** 151 (1991).
[43] A. Negre-Salvayre, A. Dagan, S. Gatt, and R. Salvayre, *Biochem. J.* **294,** 885 (1993).
[44] T. Levade, F. Vidal, S. Gatt, S. Vermeersch, N. Andieu, and R. Salvayre, *Biochim. Biophys. Acta* **1258,** 277 (1995).
[45] V. Agmon, R. Khosravi, S. Marchesini, T. Dinur, A. Dagan, S. Gatt, and R. Navon, *Clin. Chim. Acta* **247,** 105 (1996).
[46] S. Erlich, S. R. P. Miranda, J. W. M. Visser, A. Dagan, S. Gatt, and E. H. Schuchman, *Blood* **93,** 80 (1999).
[47] M. Chatelut, M. Leruth, K. Harzer, A. Dagan, S. Marchesini, S. Gatt, R. Salvayre, P. Courtoy, and T. Levade, *FEBS Lett.* **426,** 102 (1998).
[48] V. Agmon, E. Monti, A. Dagan, A. Preti, S. Marchesini, and S. Gatt, *Clinica Chimica Acta* **218,** 139 (1993).
[49] M. Suchi, T. Dinur, R. J. Desnick, S. Gatt, L. Pereira, E. Gilboa, and H. Schuchman, *Proc. Natl. Acad. Sci. U.S.A.* **89,** 3227 (1992).

in collaboration with Dr. E. H. Schuchman and an article is presented in this volume.[51a]

Synthesis of Fluorescent Sphingolipids

Lissamine-Rhodamine Dodecanoic Acid (LR12)

Two hundred milligrams of 12-aminododecanoic acid (Aldrich, Milwaukee, WI) is dissolved in 200 ml of 2-propanol:H_2O, 4:1, containing 400 mg K_2CO_3 by heating at 60° for 10 min; 100 mg of Lissamine-rhodamine sulfonyl chloride (Molecular Probes) is added; and, the solution is stirred for 24 hr in a 500-ml Erlenmeyer flask equipped with a magnetic stirrer and protected from light. The mixture is then transferred to a 1-liter separatory funnel and 300 ml dichloromethane and 150 ml water are added. Following shaking, the phases are separated, the lower dichloromethane phase is collected, and the upper aqueous–alcoholic phase is shaken with 100 ml of a mixture of dichloromethane:methanol, 3:1 (by volume). The lower phase is combined with the original one, washed 2 times with 100 ml each of water, dried by adding $MgSO_4$, filtered and evaporated to dryness in a rotary evaporator under vacuum. The residue is dissolved in dichloromethane:methanol, 4:1, loaded onto a silica gel column and the product, i.e., Lissamine-rhodamine dodecanoic acid (LR12) eluted by increasing ratios of methanol and dichloromethane. Samples of the eluent are spotted on silica thin-layer chromatography (TLC) plates that are developed in chloroform:methanol:2*N* HCl, 80:20:2 (by volume), and the product identified by a long-wavelength ultraviolet lamp (UVA). Suitable eluents are combined and the solvent evaporated under vacuum. The product is comprised of two main isomers: one (defined as normal) (LR12-N) fluoresced in the entire pH region; the second (defined as pH sensitive) (LR12-pHs) fluoresced in the acidic pH region, but lost its fluorescence in the neutral or alkaline pH range. For separating these two respective isomers the product is dissolved in a small volume of methanol, spotted onto a preparative TLC silica plate, and developed in the above solvent mixture. Two fluorescent bands are observed. To identify LR12-pHs, a drop of concentrated ammonia is applied (or the plate inverted and held over ammonia

[50] T. Dinur, E. H. Schuchman, E. Fibach, A. Dagan, M. Suchi, R. J. Desnick, and S. Gatt, *Human Gene Therapy* **3,** 633 (1992).

[51] P. L. Yeyati, V. Agmon, T. Dinur, C. Fillat, A. Dagan, R. J. Desnick, S. Gatt, and E. H. Schuchman, *Human Gene Ther.* **6,** 975 (1995).

[51a] E. H. Schuchman, S. Erlich, S. R. P. Miranda, T. Dinur, A. Dagan, and S. Gatt, *Methods Enzymol.* **312,** [26], (2000) (this volume).

vapor for several seconds); the band which lost fluorescence is defined as "pH sensitive." The two respective bands are scraped, applied to a glass column, and the LR12 is eluted with methanol.

In the procedures describing the syntheses of the respective LR–sphingolipids, the starting materials are either a normal or pH-sensitive fatty acid.

Lissamine-Rhodamine Dodecanoic Acid–Ceramide (LR12–Cer)

Five milligrams each of D-*erythro*-sphingosine (Sigma) and Lissamine-rhodamine dodecanoic acid are dissolved in 15 ml of water : acetonitrile (8 : 2, v/v) in a 50-ml Erlenmeyer flask. Two milligrams of 1-hydroxybenzotriazole (HOBT) is added, and the solution is stirred in the dark with a magnetic stirrer for 10 min. Eight milligrams of 1-ethyl-3-(3-dimethylaminopropyl)carbodiimide (EDC) is added and the solution is stirred for an additional 24 hr in the dark.

The reaction mixture is transferred to a small separatory funnel and shaken with 5 ml of dichloromethane. The lower phase is collected and the upper phase reshaken with 5 ml of dichloromethane : methanol, 3 : 1. The lower phases (containing the ceramide) are combined and evaporated to dryness. The residue is dissolved in a minimal amount of dichloromethane : methanol, 2 : 1, applied to a TLC plate and developed in chloroform : methanol, 9 : 1. The fluorescent ceramide band, visualized under a UVA lamp, is scraped and applied to a glass column, and the ceramide is eluted with dichloromethane : methanol, 1 : 1.

Lissamine-Rhodamine Dodecanoic Acid–Glucocerebroside (LR12-Glc-Cer)

Five milligrams of Lissamine-rhodamine dodecanoic acid and 8 mg of glucopsychosine (glucosylsphingosine, Sigma) are dissolved in 10 ml of water : acetonitrile (8 : 2) in an Erlenmeyer flask. Twenty milligrams of EDC is added to the magnetically stirred solution, and the solution is stirred for 24 hr in the dark. The extraction and purification are the same as described for LR12–ceramide, with the TLC plate developed in chloroform : methanol : water, 80 : 20 : 2 by volume.

Lissamine-Rhodamine Dodecanoic Acid–Galactocerebroside (LR12–Gal-Cer)

Lissamine-rhodamine dodecanoic acid and galactopsychosine (galactosylsphingosine, Sigma) are reacted and purified as described for Lissamine-rhodamine–glucocerebroside.

Lissamine-Rhodamine Dodecanoic Acid–Sphingomyelin (LR12–SPM)

Ten milligrams of sphingosylphosphorylcholine (SPC, Sigma) are dissolved in 10 ml of water in a 25-ml Erlenmeyer flask; 8 mg of Lissamine-rhodamine dodecanoic acid is added and the mixture is stirred for 15 min; 2 mg of HOBT is added, followed by 30 mg EDC, and the mixture is stirred for 24 hr in the dark. The extraction is conducted as described for LR12–ceramide, except that chloroform : methanol, 2 : 1, is used in the first step. The TLC plate is developed in chloroform : methanol : water, 65 : 35 : 5, and the LR12–SPM eluted with chloroform : methanol : water, 1 : 2 : 1.

Lissamine-Rhodamine Dodecanoic Acid–Ceramide Dihexoside (LR12–DHC) and Lissamine-Rhodamine Dodecanoic Acid–Ceramide Trihexoside (LR12–THC)

Two milligrams of lysoceramide dihexoside (or trihexoside) (prepared by hydrazinolysis as per Ref. 52) are dissolved in 10 ml of water. To this solution is added 4 mg of Lissamine-rhodamine dodecanoic acid and the solution is stirred in an Erlenmeyer flask protected from light. Two milligrams of HOBT is added, and the temperature is raised to 60° for 10 min, cooled to room temperature, and 10 mg of EDC is added. The solution is stirred for 24 hr in the dark then extracted and purified as described for LR12–sphingomyelin.

Lysosulfatide (Sphingosylgalactosyl Sulfate)

Thirty milligrams of sulfatide (ceramide galactosyl sulfate) is dried under vacuum for 24 hr, then transferred to a screw-capped test tube, and 2 ml of anhydrous hydrazine containing 2% hydrazine sulfate is added. The test tube is sealed and heated at 160° for 24 hr in an oil bath. The hydrazine is evaporated to dryness under a stream of N_2 in a hood. The residue is dissolved in dichloromethane : methanol (1 : 1) and applied to a preparative thin layer chromatography plate, which is developed in chloroform : methanol : water (65 : 35 : 5). The lysosulfatide is scraped and eluted from the silica with chloroform : methanol : water, 1 : 2 : 1.

Lissamine-Rhodamine Dodecanoic Acid–Sulfatide (LR12–CS)

Four milligrams of lysosulfatide (sphingosylgalactosyl sulfate) is dissolved in 5 ml of 2-propanol in a 10-ml Erlenmeyer flask. Lissamine-rhodamine dodecanoic acid is added, followed by 2 mg of HOBT, and the mixture stirred for 15 min. Twenty milligrams of EDC is added, the mixture stirred

[52] Y. Suzuki, Y. Hirabayashi, and M. Matsumoto, *J. Biochem.* **95,** 1219 (1984).

for 24 hr protected from light, and the solvent evaporated under a stream of nitrogen. Chloroform : methanol (1 : 1) is added and the reaction mixture applied to a TLC plate that is developed in chloroform : methanol : water (75 : 25 : 4), and LR12–CS is eluted from the silica with chloroform : methanol : water (1 : 2 : 1).

BODIPY Dodecanoic Acid–Ceramide (BOD12–Cer)

Five milligrams of D-*erythro*-sphingosine (Sigma) is added to 3 mg of 4,4-difluoro-5,7-dimethyl-4-bora-3a,4a-diaza-9-indacene-3-dodecanoic acid (BODIPY12, Molecular Probes) in 3 ml of dichloromethane : methanol (1 : 1). The solution is stirred for 5 min in a 25-ml Erlenmeyer flask, evaporated to dryness under a nitrogen stream, and 10 ml of water is added, followed by 4 mg of HOBT. The solution is stirred for 15 min, 10 mg of EDC is added, and the solution is further stirred for 24 hr protected from light. The product is extracted and purified as described for LR12–sphingomyelin.

BODIPY Dodecanoic Acid–Sphingomyelin (BOD12–SPM)

Three milligrams of BODIPY12 is dissolved in dichloromethane : methanol (1 : 1) and transferred to a 25-ml Erlenmeyer flask. The solvent is evaporated to dryness under a stream of nitrogen, and 5 mg of sphingosylphosphorylcholine (SPC) is added. Ten milliliters of water is added and the solution stirred for 15 min. Twenty milligrams of EDC is added and the solution is stirred overnight in the dark. Ten milligrams of EDC is again added, and the solution is stirred overnight. Extraction and purification of the product are as described for LR12–sphingomyelin.

BODIPY Dodecanoic Acid–Glucocerebroside (BOD12–Glc–Cer)

Ten milligrams of glucopsychosine (Sigma) is dissolved in 2 ml of dimethylformamide in a 25-ml Erlenmeyer flask, and 5 mg of BODIPY12, predissolved in 3 ml of dimethylformamide, is added. The solution is stirred for 10 min and 30 mg of EDC is added. The mixture is stirred for 24 hr protected from light, then 5 ml of dichloromethane and 5 ml of water added, the mixture is stirred, and the lower phase is collected. The upper phase is reextracted with dichloromethane, and again with 3 ml of dichloromethane : methanol (3 : 1). The lower phases are combined, washed 4 times with 3 ml each of water, and evaporated, followed by purification by TLC, as described for LR12–glucocerebroside.

BODIPY Dodecanoic Acid–Dihexosyl- or Trihexosylceramide (BOD12 DHC or THC)

To 1 mg of lyso-DHC or THC (prepared by hydrazinolysis, per Ref. 52) is added a solution of 1 mg of BODIPY dodecanoic acid (BODIPY12) in 5 ml of chloroform : methanol (2 : 1), in a 25-ml Erlenmeyer flask, followed by 1 mg of HOBT, and the solution stirred for 10 min. Ten milligrams of EDC is added and the reaction mixture stirred for 24 hr protected from light. The reaction mixture is evaporated under a stream of nitrogen, the residue is dissolved in 6 ml dichloromethane : methanol (2 : 1), and 3 ml of water is added. The lower phase is washed twice with 2 ml water, evaporated, and the residue is dissolved in a small volume and applied to a TLC plate developed with chloroform : methanol : water (65 : 35 : 5). The BODIPY dodecanoic acid–glycolipids are eluted from the silica using chloroform : methanol : water (1 : 2 : 1).

Fluorescence-Based *in Vitro* Assay Procedure Using LR or BODIPY Sphingolipids

In the assays described in *Methods in Enzymology,* Vol. 72,[24–32] the volume of the reaction mixture was 200 μl and the products of the enzymatic reactions were separated from the undegraded substrate by solvent partition. The current assays using LR12– or BODIPY12–sphingolipids are done in volumes of 10 μl. Following incubation with the respective enzyme, the entire reaction mixture is applied to a silica TLC plate. The plate is developed in a suitable solvent system and the fluorescent product is quantified by an optical reader. Should a plate-reader not be available, the silica spot containing the fluorescent product is located using an ultraviolet (UVA) lamp, then scraped and transferred to a glass tube. The fluorescent lipid is extracted by heating for 10 min at 55° with a suitable solvent mixture and fluorescence is quantified in a spectrofluorometer.

The following assays will be exemplified using a final volume of 10 μl, but the volume can be varied from 5 to 20 μl if needed.

Assay of Lysosomal Acid Sphingomyelinase

A stock solution for 20 assays is prepared by evaporating 20 nmol of LR12–SPM or BOD12–SPM to dryness in an Eppendorf vial under a stream of nitrogen. Then 100 μl of 0.2 *M* sodium acetate, pH 5.0, containing 1% (w/v) of Triton X-100 is added and the vial is stirred. Five microliters of this stock solution are pipetted into an Eppendorf vial and 5 μl of a solution of sphingomyelinase (pure, partially purified, or, alternatively, a cell sonicate) is added. A control tube has 5 μl phosphate-buffered saline (PBS) instead of enzyme. The tubes are incubated at 37° and the entire reaction mixture is applied onto a silica gel sheet (Merck 1.05553) or alternatively onto a lined, silica gel TLC plate with a concentrating zone (Whatman, Clifton, NJ, LK6D silica gel 60A; size 20 × 20 cm; thickness 250 μm).

The sheet or plate is dried and developed in chloroform : methanol : water (90 : 10 : 1, by volume) for BOD12–SPM and 85 : 15 : 1.5 for LR12–SPM. After drying, the plate is scanned in a fluorescence plate scanner (Fluor-S^{IM}-MultiImager, Bio-Rad, Hercules, CA) and the fluorescence of the respective spots quantified. Should a scanner not be available, the plate is illuminated, in a dark room, with a UVA lamp and the spots of the product (i.e., LR12– or BOD12–ceramide) are marked, and scraped into glass test tubes. Two milliliters of a mixture of chloroform : methanol, 1 : 2 (by volume), is added. The test tubes are covered with a thin nylon sheet, heated for 10 min at 55°, centrifuged, the supernatants transferred into a glass cuvette, and the fluorescence recorded as follows: For Lissamine rhodamine: excitation 560 nm, emission 580 nm; for BODIPY: excitation 505 nm, emission 520 nm.

Protein is determined by a Bradford reagent (Sigma) and the fluorescence per milligram of protein calculated.

Assay of Neutral Magnesium-Dependent Sphingomyelinase

For assaying the neutral magnesium-dependent sphingomyelinase, the procedure used for the lysosomal sphingomyelinase is modified as follows: To 5 μl of a stock solution containing 1 nmol of LR12–SPM or BODIPY12–SPM dispersed in 5 μl of Tris-HCl, pH 7.4 (0.2 *M*), containing 1% Triton X-100 and $MgCl_2$ (0.02 m*M*) is added 5 μl of an organ or cell extract containing the sphingomyelinase activity, and the incubation and quantification of the fluorescent product (i.e., LR– or BODIPY–ceramide) is as described for the acidic sphingomyelinase.

Sources that contain both the acidic and neutral magnesium-dependent sphingomyelinases can be assayed for the acidic enzyme in the absence of magnesium ion by adding 40 nmol (i.e., 2 m*M*) of ethylenediaminetetraacetic acid (EDTA) to the reaction mixture. The latter compound binds the endogenous divalent metals and thereby eliminates any activity of the magnesium-dependent, neutral sphingomyelinase.

Assay of Lysosomal β-Glucosidase or β-Galactosidase

For these activities, the procedure described for the assay of the lysosomal, acid sphingomyelinase is followed with two modifications: (1) The substrate is LR12– or BODIPY12–glucosyl or galactosylceramide (also named gluco- or galactocerebroside); and (2) 5 μl of the stock solution will contain 1 nmol of the substrate dispersed in 0.2 *M* citrate–phosphate buffer, pH 5.0, containing 0.5% Triton X-100 and 0.5% sodium taurocholate (Calbiochem, La Jolla, CA).

Assay of Other Glycosphingolipids

The procedure described for the fluorescent glucosyl- or galactosylceramides can also be used for assaying the fluorescent derivatives of ceramide dihexoside (dihexosylceramide, DHC, galactose-β-glucose-β-ceramide); ceramide trihexoside (trihexosylceramide, THC, galactose-α-galactose-β-glucose-β-ceramide); sulfatide (cerebroside sulfate, CS, sulfogalactose-β-cerebroside); and gangliosides, e.g., GM1 (galactose-β-*N*-acetyl-galactosamine-β-galactose-(sialic acid)-β-glucose-β-ceramide) or GM2 (*N*-acetylgalactosamino-β-galactose-(sialic acid)-β-glucose-β-ceramide), but the solvent mixture used for separating the product from the bulk of the unreacted substrate by TLC is adjusted for each substrate. Thus, for separating glucosylceramide from the unreacted DHC, DHC from the unreacted THC, and galactosylceramide from the unreacted sulfatide, mixtures of chloroform : methanol : water in ratios ranging from 80 : 20 : 2 to 75 : 25 : 4 can be used. For separating the products from assays using fluorescent GM1 or GM2, chloroform : methanol : water, 60 : 35 : 8, is used.

Assay of Lysosomal Acid Ceramidase

The procedure described for the lysosomal sphingomyelinase is followed, with the following modifications: (1) 1 nmol of LR12– or BODIPY12–ceramide is dispersed in 5 μl of 0.5 *M* sodium acetate, pH 4.2, containing 0.5% Triton X-100 and 0.5% sodium taurocholate. (2) Following incubation with 5 μl of a ceramidase source (e.g., extract of kidney or brain), the reaction mixture is spotted onto a silica gel plate and the plate is developed with chloroform–methanol–2 *N* HCl, 97 : 3 : 0.3 (by volume).

Administration of Fluorescent Sphingolipids to Cultured Cells and Analysis of Metabolic Products in Cells, Cell Extracts, or Culture Medium

Administration into Cells and Lysosomes

The purpose of *in situ* experiments is to introduce the respective fluorescent sphingolipids into living cells and follow their intracellular metabolism. Since our interest has been mostly in intralysosomal hydrolysis, two approaches have been used. When lipids labeled with a nonpolar fluorescent probe (e.g., pyrene) were used, they were administered in a form that would be taken up by receptor-mediated endocytosis. This liposomal dispersion is prepared as follows: Chloroform–methanol solutions of the respective pyrene lipid and a 9-fold excess of lecithin (phosphatidylcholine, from egg or soybean) are evaporated in a glass tube under a stream of nitrogen. To ascertain the complete evaporation of the solvent, the tubes are further kept under vacuum for 30 min. Saline is added and small unilamellar vesicles (liposomes) are prepared by ultrasonic irradiation with a probe sonicator (Microson, Heat Systems, Ultrasonics, Farmingdale, NY), at a 40% output for 3–5 min. Apolipoprotein E (about 100 μg for a liposomal dispersion containing 50 nmol of the pyrene–lipid) is added and the apoE liposomal dispersion is introduced into the culture medium. This approach utilizes the linking of the apolipoprotein E–liposomes to the LDL (apoE/B) receptor on the cell surface, which will be taken up by endocytosis and trafficked to the lysosomes. For pyrene sphingomyelin, liposomes are prepared in the absence of lecithin, or interacted with apolipoportein and introduced into the culture medium.

When using lipids labeled with Lissamine rhodamine (LR) or sulforhodamine (SR, Texas Red) as well as BODIPY, the above procedure is replaced by a simpler procedure in which the respective lipid, without apoE, is dissolved in dimethyl sulfoxide (DMSO) and the solution added directly to the culture medium. The final volume of the DMSO should not exceed 0.1% of that of the cell culture medium.

Cells

We have used these methods with a large variety of adherent cells, such as skin fibroblasts, macrophages, lymphoblasts, embryonic cells, CHO, COS, and cancer cells (e.g., MCF-7-wt and MCF-7-AdrR) as well as cells grown in suspension, e.g., HL60 and P388 cancer cells. Depending on the specific requirements for the experiment, the cells are grown in six-well dishes or flasks and the volume of the culture medium varied accordingly. The respective fluorescent lipid is added as a liposomal dispersion or DMSO solution and the cells incubated at 37°.

Incubation and Termination

Depending on the respective experiments, two modes of incubation can be used: a "pulse" or a "pulse-chase." In the former, the fluorescent lipid is incubated with the cells for the period required and the reaction terminated. For a "pulse-chase," the fluorescent lipid is incubated with the cells for the period required (pulse); the medium is then removed, the cells washed with sterile saline, fresh medium (devoid of fluorescent lipid) is added, and the incubation is continued for the chase period desired.

For termination of the reaction, cells in suspension are sedimented and washed twice with PBS; for adherent cells, the cells are washed with PBS and removed from the plate by an appropriate method (e.g., trypsin and EDTA). The pelleted cells are suspended in 200–500 μl of PBS, and sonicated in a probe or bath sonicator up to 10 sec. An aliquot of 5–10 μl is removed for protein determination (Bradford procedure, Sigma). The rest is transferred to a glass tube or vial, water is added to a volume of 1 ml, followed by 1.0 ml of methanol containing 2% acetic acid, and the tube is vortexed; 1.0 ml of chloroform is added and, following vortexing, the tubes are centrifuged and two phases are obtained (the protein is at the interface). The lower, organic phase is pipetted into a vial and evaporated to dryness. Then 100 μl of chloroform : methanol, 2 : 1, is added and a portion is applied to a TLC plate. The portion applied depends on the sensitivity of the fluorescent probes. Thus, when using Lissamine rhodamine lipids, a quantity derived from 300–500 μg protein is used; and for BODIPY lipids, 50–100 μg is usually sufficient.

Analysis of Fluorescence

In Intact Cells. The fluorescence of intact cells can be viewed by fluorescence microscopy or by computerized fluorescence microscopy (also entitled imaging or microspectrofluorometry). The latter permits quantifying the fluorescence in a single cell and its intracellular components. For quantifying the fluorescence of intact cells, a suspension of the cells is introduced into a cuvette and fluorescence emission is recorded using the excitation and emission wavelength suitable for the specific probe. When cell populations with varying fluorescence exist, they can be sorted as a suspension on the basis of their respective fluorescence emission in a fluorescence activated cell sorter (FACS). The FACS also permits separation of cell populations with differing fluorescence and collection, under sterile conditions, for regrowing in culture. (For an example, see Schuchman *et al.*[51a])

In Cell Extracts. The lower, organic phase of the extracted, cellular lipids contains the original fluorescent lipid as well as metabolic products. The latter can be hydrolytic products or, alternatively, synthetic ones. Analysis is usually performed by TLC or high-performance liquid chromatography (HPLC). For TLC, an aliquot of the lipid extract is applied to a suitable plate (which, in most cases, is silica, but plates coated with aluminum oxide or a reversed-phase material can also be used). The plate is then developed in a closed container using a suitable solvent system. The composition of the latter depends on the polarity of the lipids to be separated. Nonpolar lipids are separated in mixtures of hexane : diethyl ether : acetic acid (80 : 20 : 1). The more polar lipids are separated using mixtures of chloroform, methanol, and water, e.g., 85 : 15 : 1.5 for cerebrosides; 75 : 25 : 4 for oligohexoside ceramides or sulfatide, 65 : 35 : 3.5 for sphingomyelin, and 50 : 50 : 10 or even 1 : 2 : 1 for highly polar lipids such as gangliosides. The lipids of interest are quantified as described in the preceding section.

In Culture Medium. Some metabolic products are secreted into the culture medium (e.g., LR12–SPM[48]). For quantifying their fluorescence, the culture medium is stirred vigorously with an equal volume of chloroform:methanol, 1:1. Following centrifugation, the upper aqueous methanolic phase is removed and discarded. To remove residual protein particles, 1 ml of methanol is added to the lower chloroform phase, and the solution is filtered through glass wool or white tissue paper in Pasteur pipette. The solvent is evaporated in a steam of nitrogen, the residue is dissolved in 100 μl of chloroform:methanol, 2:1, and a portion is applied to a TLC plate and analyzed as above.

[24] Selection of Mammalian Cell Mutants in Sphingolipid Biosynthesis

By KENTARO HANADA and MASAHIRO NISHIJIMA

Introduction

Sphingolipid biosynthesis is initiated by condensation of L-serine with palmitoyl-CoA to generate 3-ketodihydrosphingosine, which is then converted to dihydroshingosine (Fig. 1).[1] Dihydrosphingosine is *N*-acylated to form dihydroceramide, followed by desaturation to produce ceramide (Cer). These early synthetic steps occur at the cytosolic surface of the endoplasmic reticulum (ER). Ceramide is translocated from the ER to the Golgi apparatus, and then converted either to sphingomyelin (SM) in the lumenal side of the Golgi apparatus or to glucosylceramide (GlcCer) on the cytosolic surface of the Golgi apparatus. After translocation into the Golgi lumen, GlcCer is further converted to lactosylceramide and more complex glycosphingolipids.

A powerful approach toward understanding the metabolic mechanisms and biologic functions of membrane lipids is the biochemical and cell biological characterization of cell mutants with specific defects in lipid metabolism. Several mutants have been isolated but the repertory of sphingolipid-synthetic mutants available thus far is not comprehensive. A temperature-sensitive Chinese hamster ovary (CHO) cell mutant defective in serine palmitoyltransferase (SPT), the enzyme catalyzing the first step of sphingo-

[1] A. H. Merrill, Jr., and C. C. Sweeley, *in* "Biochemistry of Lipids, Lipoproteins and Membranes" (D. E. Vance and J. Vance, eds.), p. 309. Elsevier Science B.V., The Netherlands, 1996.

0076-6879/00 $30.00

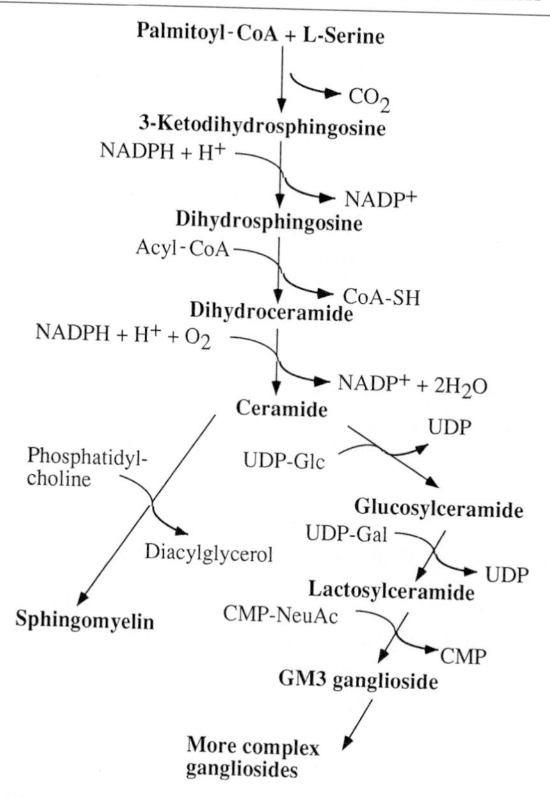

FIG. 1. The main pathway of sphingolipid synthesis in mammalian cells.

lipid synthesis, was isolated after screening colonies of mutagenized cells by an *in situ* assay for SPT activity.[2] This CHO mutant strain (SPB-1) requires externally supplied sphingolipids for proliferation, providing the first genetic evidence that sphingolipids play essential roles in growth of mammalian cells.[3] Studies with SPB-1 cells have also shown that membrane sphingolipids interact with glycosylphosphatidylinositol-anchored proteins,[4] and that sphingolipids and cholesterol are coordinately responsible for formation of detergent-resistant membrane domains.[5] Functional complementation of the SPB-1 strain with the hamster *LCB1* cDNA has shown

[2] K. Hanada, M. Nishijima, and Y. Akamatsu, *J. Biol. Chem.* **265,** 22137 (1990).

[3] K. Hanada, M. Nishijima, M. Kiso, A. Hasegawa, S. Fujita, T. Ogawa, and Y. Akamatsu, *J. Biol. Chem.* **267,** 23527 (1992).

[4] K. Hanada, K. Izawa, M. Nishijima, and Y. Akamatsu, *J. Biol. Chem.* **268,** 13820 (1993).

[5] K. Hanada, M. Nishijima, Y. Akamatsu, and R. E. Pagano, *J. Biol. Chem.* **270,** 6254 (1995).

that this cDNA encodes a component of the SPT enzyme.[6] A mouse B16 melanoma mutant (GM-95) defective in GlcCer synthase was isolated by screening with anti-GM3 antibody,[7] and the cDNA encoding GlcCer synthase was cloned by a functional rescue method with the GM-95 cell line.[8] The finding that the severe growth defect of SPB-1 cells is rescued by exogenous SM but not glucosylceramide and that GM-95 cells normally grow without any glycosphingolipids[3,7] suggests that SM is important for mammalian cell growth in culture.

Recently, after selection of cell variants resistant to SM-directed cytolysin from mutagenized CHO cells, we isolated not only another SPT-defective CHO mutant (LY-B strain) but also a novel type of SM-deficient CHO mutant (LY-A strain), which is defective in intracellular translocation of ceramide from the ER to the Golgi apparatus for SM synthesis.[9,10] Here we describe details of the isolation of these LY-A and LY-B strains.

CHO Cells as Somatic Cell Genetic Tools

CHO cells have been used for isolation of variants affecting various cellular functions including the metabolism of lipids.[11,12] The most important nature of CHO cells as a somatic cell genetic tool is a high frequency of recessive mutant phenotypes. This is probably due to the functional hemizygosity of some bialleic genes and a high frequency of segregation-like events (such as aberrant separation of homologous chromosomes instead of sister chromatids to opposite poles at mitosis, or somatic crossing-over between mutant and wild-type alleles), which unmask otherwise recessive mutations.[13,14] Moreover, CHO cells can be easily transfected with DNA,

[6] K. Hanada, T. Hara, M. Nishijima, O. Kuge, R. C. Dickson, and M. M. Nagiec, *J. Biol. Chem.* **272,** 32108 (1997).

[7] S. Ichikawa, N. Nakajo, H. Sakiyama, and Y. Hirabayashi, *Proc. Natl. Acad. Sci. U.S.A.* **91,** 2703 (1994).

[8] S. Ichikawa, H. Sakiyama, G. Suzuki, K. I.-P. J. Hidari, and Y. Hirabayashi, *Proc. Natl. Acad. Sci. U.S.A.* **93,** 4638 (1996).

[9] K. Hanada, T. Hara, M. Fukasawa, A. Yamaji, M. Umeda, and M. Nishijima, *J. Biol. Chem.* **273,** 33787 (1998).

[10] M. Fukasawa, M. Nishijima, and K. Hanada, *J. Cell Biol.* **144,** 673 (1999).

[11] M. M. Gottesman, *in* "Molecular Cell Genetics" (M. M. Gottesman, ed.), p. 883. John Wiley & Sons, New York, 1985.

[12] M. Nishijima, O. Kuge, and K. Hanada, *Trends Cell Biol.* **7,** 324 (1997).

[13] L. Siminovitch, *Cell* **7,** 1 (1976).

[14] R. G. Worton and S. G. Grant, *in* "Molecular Cell Genetics" (M. M. Gottesman, ed.), p. 831. John Wiley & Sons, New York, 1985.

and there are reliable techniques for replica plating of CHO colonies formed on polyester disks.[15,16]

Mutagenesis of CHO-K1 Cells

To obtain a reasonable frequency of mutants, it is necessary to mutagenize parental cells. We routinely mutagenize CHO cells with ethylmethane sulfonate (EMS). *Caution:* EMS is volatile. Bottles contains EMS should be opened under a chemical fume hood while wearing disposable gloves. Apparatus in contact with EMS (pipettes, tips, etc.) should be soaked well in 5% (w/v) sodium thiosulfate, and EMS-containing medium should be diluted 2-fold with 10% sodium thiosulfate[17] to decompose the EMS. *Never discard the EMS-containing medium before decomposition.*

Materials

CHO-K1 cells (ATCC, Rockville, MD CCL 61)
10% NCS/F12 medium: Ham's F12 medium (Asahi Technoglass, Tokyo) supplemented with 10% newborn calf serum (ICN Pharmaceuticals, Costa Mesa, CA), penicillin G (100 units/ml) and streptomycin sulfate (100 μg/ml)
Ethylmethane sulfonate, EMS (Sigma, St. Louis, MO).
T-25 tissue culture flask (Corning, Corning, NY)

Procedures

1. Seed 7.5×10^4 CHO-K1 cells in each of four T-25 flasks containing 10 ml of 10% NCS/F12 medium. Culture the cells at 33° for 2 days.
2. Dilute 7 μl of EMS into 21 ml of 10% NCS/F12 medium (final concentration 400 μg EMS/ml). Remove the culture medium from the flasks, add 5 ml of the EMS-containing medium to each flask, cap the flasks tightly, and incubate at 33° for 18 hr. Hereafter, each flask should be independently handled to provide separate batches of mutagenized cells.
3. Remove the EMS-containing medium, rinse the cell monolayer four times with 5 ml of 10% NCS/F12 medium, and add 10 ml of 10% NCS/F12 medium to each flask. Culture the cells in a 5% (v/v) CO_2 atmosphere in 100% humidity at 33° for ~4 days to confluence.

[15] J. D. Esko and C. R. H. Raetz, *Proc. Natl. Acad. Sci. U.S.A.* **75,** 1190 (1978).
[16] C. R. H. Raetz, M. M. Wermuth, T. M. McIntyre, J. D. Esko, and D. C. Wing, *Proc. Natl. Acad. Sci. U.S.A.* **79,** 3223 (1982).
[17] F. Winston, *in* "Current Protocols in Molecular Biology" (K. Janssen, ed.), Vol. 2, Unit 13.3.1. John Wiley & Sons, New York, 1994.

4. Harvest the cells by trypsinization (if necessary, plate 200 cells; see below) and prepare 2–4 frozen stocks from each flask. The success of mutagenesis can be preliminarily checked by determination of survival rates as follows: separately plate 200 mutagenized cells and unmutagenized CHO-K1 cells in 100-mm dishes, culture at 37° for ~10 days, and count the number of colonies. The colony number of the mutagenized cells should be 10–20% of the unmutagenized controls.

Isolation of CHO Mutants Resistant to SM-Directed Cytolysin

Lysenin, a 41-kDa protein derived from the coelomic fluid of the earthworm *Eisenia foetida,* was initially discovered as a factor that induced contraction of rat vascular smooth muscle and hemolysis of mammalian red cells.[18,19] It was subsequently shown that lysenin is a cytolysin that binds tightly to SM as a specific membrane receptor[9,20]; therefore, selection of lysenin-resistant variants is a rational method for isolation of cell mutants defective in SM metabolism.

Materials

10% NCS/F12 medium (see above)

Sterile phosphate-buffered saline (PBS): NaCl, 8 g; KCl, 0.2 g; $Na_2HPO_4 \cdot 12H_2O$, 2.9 g; KH_2PO_4, 0.2 g per 1000 ml

Serum-free Ham's F12 medium

250 μM Oleic acid associated with bovine serum albumin (BSA) stock: Add 1 ml of 10 mM oleic acid, sodium salt (Sigma) to 39 ml of 2.5% (w/v) BSA, fatty acid-free bovine serum albumin (Sigma) in PBS. Sterilize by filtration (0.22-μm pore size, Millex GS, Millipore, Bedford, MA), and store in freezer

Nutridoma medium: Ham's F12 medium supplemented with 1% Nutridoma-SP (Boehringer Mannheim) and gentamicin (10 μg/ml)

Nutridoma-BO medium: Nutridoma medium freshly supplemented with 0.1% fetal calf serum and 10 μM oleic acid associated with BSA[21]

Lysenin purified from the coelomic fluid of *E. foetida:* a gift from

[18] Y. Sekizawa, K. Hagiwara, T. Nakajima, and H. Kobayashi, *Biomed. Res.* **17,** 197 (1996).

[19] Y. Sekizawa, T. Kubo, H. Kobayashi, T. Nakajima, and S. Natori, *Gene* **191,** 97 (1997).

[20] A. Yamaji, Y. Sekizawa, K. Emoto, H. Sakuraba, K. Inoue, H. Kobayashi, and M. Umeda, *J. Biol. Chem.* **273,** 5300 (1998).

[21] S. R. Panini, R. J. Lutz, L. Wenger, J. Miyake, S. Leonard, A. Andalibi, A. J. Lusis, and M. Sinensky, *J. Biol. Chem.* **265,** 14118 (1990).

Drs. Yoshiyuki Sekizawa and Hideshi Kobayashi (Zenyaku Kogyo Co., Tokyo)

Tissue culture dishes (Corning: 100-mm, 60-mm, and 35-mm diameters)

Selection of Lysenin-Resistant CHO Cell Variants

1. Seed mutagenized CHO cells (2×10^5 cells) in a 100-mm dish containing 5 ml of 10% NCS/F12 medium, and cultured at 33° for 1 day.
2. Rinse the cell monolayer twice with 5 ml of PBS, add 10 ml of Nutridoma-BO medium to the dish, and culture the cells at 37° or 40° for 2 days. See also Notes 1 and 2 below.
3. Rinse the cell monolayer twice with 5 ml of serum-free F12 medium. Add 5 ml of serum-free F12 medium containing lysenin (100 ng/ml, see Note 3 below) to the dish.
4. Incubate at 37° or 40° for 1 hr (Figs. 2A–2C).

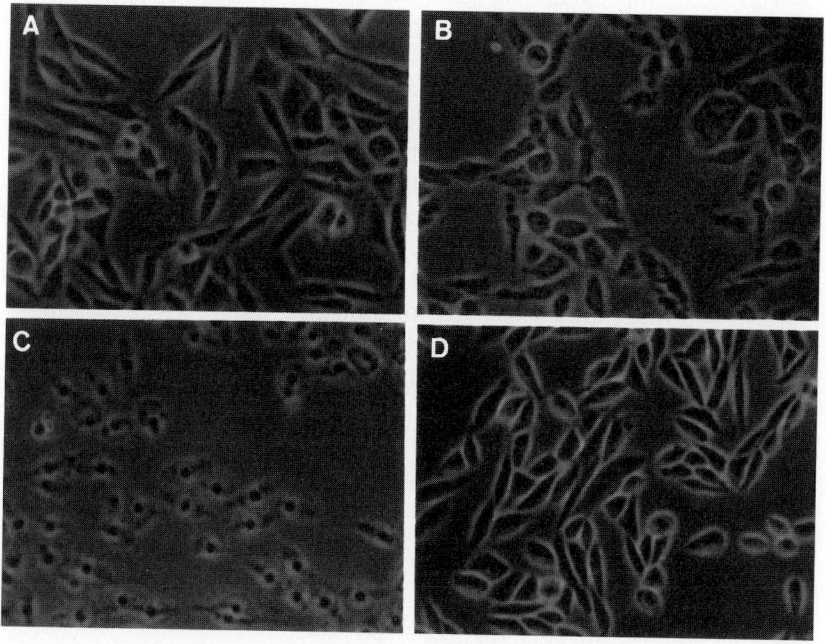

FIG. 2. Effects of lysenin on the morphology of CHO cells. After culture in Nutridoma-BO medium at 40° for 2 days, CHO-K1 (A–C) or LY-B (D) cells were exposed to lysenin (100 ng/ml) at 40° for various times, and viewed with a phase contrast microscope. Compared with the control cells unexposed to lysenin (A), the exposure of CHO-K1 cells to lysenin rendered the cell surface blebby in 5 min (B), and almost completely lysed the cells in 1 hr (C). In contrast, the morphology of LY-B cells was not affected by the exposure to lysenin for 1 hr (D).

5. Rinse the dish four times with 5 ml of 10% NCS/F12 medium, and culture the surviving cells in 10 ml of 10% NCS/F12 medium at 33° for about 7 days.
6. Harvest the propagated cells by trypsinization. For a second cycle of the lysenin treatment, seed the cells in 10% NCS/F12 medium, and culture at 33° for 1 day. The culture scale can be reduced according to the number of harvested cells. Note that the cell monolayer at the lysenin exposure should be well dispersed with less than ca. 70% confluence, because densely packed cells in confluent monolayers (or even single colonies) cause pseudoresistance of cells to lysenin.
7. Repeat steps 2–5 two more times.
8. Purify lysenin-resistant variants by limiting dilution (Note 4, below).
9. Check the lysenin resistance of the purified clones (Fig. 2D). Propagate the purified lysenin-resistant cell clones, and prepare frozen stocks for storage.

Note 1. Serum contains significant amounts of sphingolipids,[3] and CHO cells can use exogenous sphingolipids.[3] Therefore, culture SPT-defective CHO cells (SPB-1 cells) in sphingolipid-deficient medium for 2 days to reduce the SM level in SPB-1 cells enough to display the lysenin-resistant phenotype (K. Hanada, 1996, unpublished observation).

Note 2: During the screening, cells are propagated at 33° and exposed to lysenin at 37 or 40° because lysenin-resistant variants may be temperature sensitive for growth, although none of the lysenin-resistant clones obtained so far have exhibited such stringent temperature sensitivity for growth.

Note 3: The titer of the cytolytic activity of lysenin depends on purification batch and storage. The lysenin concentration for screening should be adjusted to kill about 95% of parental CHO-K1 cells.

Note 4: Seed 0.2 ml of a diluted cell suspension (~5 cells/ml in 10% NCS/F12 medium) into each well of a 96-well plate. Culture at 33° for about 2 weeks. Retrieve the cells from wells in which only one colony forms.

Metabolic Labeling of Lipids with Radioactive Serine

To characterize the purified lysenin-resistant mutant clones, we first examined sphingolipid synthetic rates in intact cells by labeling with [^{14}C]serine. Because serine is the initial precursor in the sphingolipid synthetic pathway (Fig. 1), the pattern of metabolic labeling provides important information concerning the metabolic defects of the mutant cells. For example, the loss of labeling of all sphingolipids implies that there is a defect in the SPT reaction (as with the SPB-1 and LY-B strains), while a striking reduction of the labeling of SM (but not ceramide) suggests a deficiency in the conversion of ceramide to sphingomyelin (as with the LY-A strain) (Fig. 3A). Moreover, the labeling of phosphatidylserine and phosphatidylethanolamine (Fig. 3A) can be used as a control because these two phospholipids incorporate serine by reactions distinct from the sphingolipid synthetic pathway (thus, helping differentiate between cells with sphingolipid synthesis versus other processes, such as serine uptake or acyl-CoA synthesis).

We isolated 27 lysenin-resistant clones after screening a total number of about 2×10^6 mutagenized cells from three independent batches of

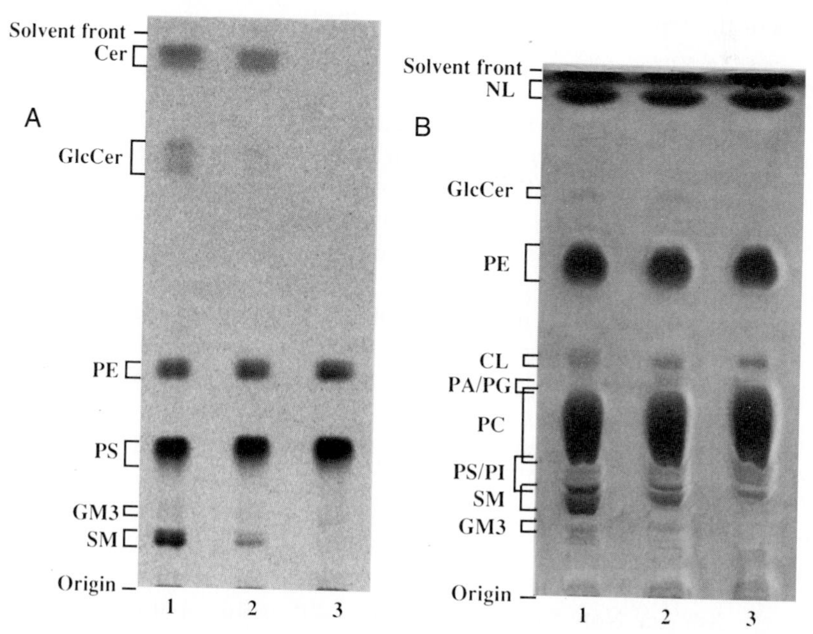

FIG. 3. Analysis of the synthetic rate and chemical content of sphingolipids in CHO cells. (A) Patterns of radioactive lipids metabolically labeled with [^{14}C]serine. CHO cells were incubated with [^{14}C]serine at 40° for 2 hr, and, after extraction of lipids from the cells, the lipids were separated by TLC with TLC solvent I. The patterns of radioactive lipids separated on the TLC plate were detected with an image analyzer. Lane 1, parental CHO-K1 cells; lanes 2 and 3, lysenin-resistant CHO cell mutant LY-A and LY-B strains, respectively. (B) CBB staining of lipids separated on a TLC plate. After culture in Nutridoma-BO medium at 40° for 2 days, CHO cell monolayers were harvested. Lipids were extracted from the cells equivalent to 4 mg of protein, and subjected to TLC. After separation with the TLC solvent II, lipids on the palate were stained with CBB. Lane 1, CHO-K1 cells; lane 2, LY-Acells; lane 3, LY-B cells. PS, Phosphatidylserine; PI, phosphatidylinositol; PC, phosphatidylcholine; PA, phosphatidate; PG, phosphatidylglycerol; CL, cardiolipin; PE, phosphatidylethanolamine; NL, neutral lipids.

mutagenesis. Based on the activity of *de novo* sphingolipid synthesis, we tentatively classified the lysenin-resistant clones into three types. In type I mutants (including 11 clones), the *de novo* synthetic rate of SM was less than 20% of the wild-type level while the synthetic rates of Cer and GlcCer were more than 50% of the wild-type levels. Type II mutants (including 6 clones) did not synthesize SM, GlcCer, or Cer appreciably. In type III mutants (including 10 clones), the synthetic rate of SM was nearly normal (more than 70% of the wild-type level). All of the type I mutant clones were derived from one batch of mutagen treatment, suggesting that the type I clones were siblings of a mutant progenitor. Likewise, the type II mutant clones were probably siblings of another mutant progenitor. We therefore chose two clones, named LY-A and LY-B strains, as representatives of the type I and II mutants, respectively.[9]

Materials

Nutridoma medium; PBS: see above
L-[U-^{14}C]Serine: 5.99 GBq/mmol, 1.85 MBq/ml (Amersham Pharmacia Biotech, Uppsala, Sweden)
0.1% (w/v) Sodium dodecyl sulfate (SDS); chloroform/methanol (1/2, v/v); chloroform; chloroform/methanol (19/1, v/v); 25 m*M* HCl containing 0.15 *M* NaCl
Pyrex glass tubes (13 mm × 100 mm) with screw caps; disposable glass tubes and Pasteur pipettes
Thin-layer chromatography (TLC) plates: Silica gel 60 (Merck, Darmstadt, Germany)
TLC solvent I: methyl acetate/1-propanol/chloroform/methanol/0.25% KCl (25/25/25/10/9, v/v)[22]

Procedures

1. Seed 1 × 10^6 cells in 5 ml of 10% NCS/F-12 medium in a 60-mm dish (see Note 5 below). Culture at 40° for 16–24 hr (see also Note 6 below).
2. Rinse the cell monolayer twice with 1 ml of serum-free F12 medium. Add 1.5 ml of Nutridoma medium to the dish.
3. Add 15 μl of L-[U-^{14}C]serine (5.99 GBq/mmol, 1.85 MBq/ml) to the dish. Culture at 40° for 2 hr (see also Note 6 below).
4. Remove the medium, and rinse the cell monolayer twice with 1 ml of ice-cold PBS.

The remaining steps are performed at room temperature.

[22] F. Vitiello and J.-P. Zanetta, *J. Chromatogr.* **166,** 637 (1978).

5. Add 1000 μl of 0.1% SDS for lysis of cells, and suspend the cell lysate by pipetting. Use 800 μl of the lysate for lipid extraction (steps 7–11),[23] and 50–100 μl of the lysate for determination of protein concentration.[24]
6. Add 800 μl of the cell suspension to a Pyrex glass tube containing 3 ml of chloroform/methanol (1/2, v/v), and mix well.
7. Add 1 ml of chloroform and 1 ml of PBS to the tube (see also Note 7 below), and mix well.
8. Centrifuge (1000*g*, 2 min). Remove the upper phase.
9. Transfer the lower organic solvent phase to a glass tube carefully with a glass Pasteur pipette. Dry the sample under a nitrogen-gas stream.
10. Dissolve the sample in ~50 μl of chloroform/methanol (19/1, v/v). Streak the dissolved sample on a TLC plate with a capillary glass.
11. Develop the sample on the plate with the TLC solvent I. Dry the plate under air.
12. Expose the plate to an Imaging Plate (Fuji Photo Film Co., Tokyo, Japan) for 2 days, or to an X-ray film with an intensified screen at −70° for 2 weeks.
13. Analyze the radioactive image with a BAS2000 Image Analyzer (Fuji Photo Film Co.) (Fig. 3A) or, after scraping the radioactive spots of the TLC plate, determine radioactivity of each sample in 2 ml of a scintillation cocktail (see Note 8 below) by liquid scintillation counting. The rates of the lipid labeling are normalized for protein amount.

Note 5: Standard procedures for wild-type cells are described here. The number to be seeded depends on cell lines and culture temperatures, because growth rates of cells depend on these factors. We routinely use 70–90% confluent cell monolayers for the metabolic labeling.

Note 6: In this protocol, cells are incubated at 40°, because mutant cell lines may have temperature-sensitive defects in sphingolipid synthesis. If necessary, the culture temperature can be changed to 33 or 37°.

Note 7: For better recovery of GM_3 ganglioside into the organic solvent phase, use 25 m*M* HCl containing 0.15 *M* NaCl in place of PBS. For better recovery of sphingoid bases, use 0.5 *M* ammonium.

Note 8: ACS-II (Amersham, Uppsala, Sweden) or 0.4% (w/v) Omnifluor (Dupont, Wilmington, DE) in toluene.

[23] E. G. Bligh and W. J. Dyer, *Can. J. Biochem. Physiol.* **37,** 911 (1959).
[24] O. H. Lowry, N. J. Rosebrough, A. L. Farr, and R. J. Randall, *J. Biol. Chem.* **193,** 265 (1951).

Analyses of the Contents of SM and GM3 in CHO Cells

For characterization of the mutant cell lines, we also determined the contents of major sphingolipids in CHO cells. The fact that not only the synthetic rate but also the chemical level of sphingolipids are affected in a mutant cell line provides compelling evidence that sphingolipid metabolism is affected in the mutant cell line. As mentioned above, since CHO cells can metabolically use exogenous sphingolipids associated with serum, cultivation of cells in Nutridoma-BO medium, a sphingolipid-deficient medium, is necessary to see dramatic reduction of sphingolipid levels in mutant cells defective in sphingolipid synthesis. Thus, we routinely culture CHO cells for 2 days in Nutridoma-BO medium before examination of sphingolipid levels. Prolonged culture in the sphingolipid-deficient medium is toxic to the mutant cells completely defective in *de novo* sphingolipid synthesis.[3,9]

After separation of lipids by TLC, we quantify the chemical amount of each phospholipid type including SM by a lipid phosphorus assay.[25] To determine the amount of GM3 (the sole ganglioside in CHO cells[26,27]), we use a modification of the method by Nakamura and Handa,[28] which stains lipids on TLC plates with Coomassie Brilliant Blue (Fig. 3B). Since the latter method is also useful for semiquantitative comparison of SM levels among cell lines (Fig. 3B), the details are described below.

Materials

- 10% NCS/F-12 medium; Nutridoma-BO medium; PBS, TLC plates: see above
- Chloroform/methanol (1/2, v/v); chloroform; chloroform/methanol (19/1, v/v); 25 m*M* HCl containing 0.15 *M* NaCl
- 150-mm tissue culture dishes; 50-ml plastic conical tubes; a cell scraper; Pyrex glass tubes with screw caps; disposable glass tubes; Pasteur pipettes
- GM3 standard (0.25 mg/ml): Dilute a GM3 solution (0.5 mg/ml; Matreya, Inc., Pleasant Gap, PA) 2-fold with chloroform
- TLC solvent II: Chloroform/methanol/0.2% $CaCl_2$; 80/30/5 (v/v)
- CBB staining solution: 0.1 *M* NaCl-0.05% (w/v) CBB in 25% methanol (NaCl, 2.16 g; Brilliant Blue R (Sigma), 0.2 g; methanol, 100 ml; H_2O, 300 ml). The CBB staining solution is reusable several times
- Destaining solution: 0.1 *M* NaCl in 25% ethanol (NaCl 2.16 g; ethanol, 100 ml; H_2O, 300 ml)

[25] G. Rouser, A. N. Siakotos, and S. Fleischer, *Lipids* **1,** 85 (1966).
[26] E. B. Briles, E. Li, and S. Kornfeld, *J. Biol. Chem.* **252,** 1107 (1977).
[27] G. Yogeeswaren, R. K. Murray, and J. A. Wright, *Biochem. Biophys. Res. Commun.* **56,** 1010 (1974).
[28] K. Nakamura and S. Handa, *Anal. Biochem.* **142,** 406 (1984).

Procedures

1. Seed 3×10^6 CHO-K1, 4×10^6 LY-A cells, or 6×10^6 LY-B cells (see Note 9 below) in a 150-mm dish containing 15 ml of 10% NCS/F12 medium, and culture at 33° for 18–24 hr.
2. Rinse the monolayer twice with 10 ml of PBS, and add 35 ml of Nutridoma-BO medium to the dish. Culture at 40° for 2 days.
3. Rinse the monolayer twice with 10 ml of cold PBS.
4. After addition of 3 ml of cold PBS to the dish, scrape the cell monolayer, and transfer the cell suspension to a 50-ml conical tube. Repeat this step once more for thorough cell harvest.
5. Centrifuge (300*g*, 5 min, 4°). Remove the supernatant.
6. Suspend the cell pellet with 500 μl of cold PBS by pipetting. Store the cell suspension on ice until step 8.
7. Using 5–10 μl of the cell suspension, determine the protein concentration.[24]

The remaining steps are performed at room temperature.

8. Add the cell suspension (equivalent to 1 mg protein for phospholipid analysis, or 2–4 mg for GM3 analysis) to a Pyrex glass tube. The protein amount added to the tube should be equal among the samples to be compared. Adjust the aqueous volume to 800 μl with PBS.
9. Add 3 ml of chloroform/methanol (2/1, vol/vol) to the tube, and mix well.
10. Add 1 ml of chloroform and 1 ml of 25 m*M* HCl containing 0.15 *M* NaCl to the tube, and mix well.
11. Centrifuge (1000*g*, 5 min). Remove the upper phase.
12. Transfer the lower phase to a glass tube carefully with a glass Pasteur pipette. Dry the sample under a nitrogen-gas stream.
13. Dissolve the dried sample in ~50 μl of choloform/methanol (19/1, v/v). Streak the dissolved sample on a TLC plate with a capillary glass. For calibration, streak 0–10 μg of the GM3 standard to the plate.
14. Develop the sample on the plate with the TLC solvent II. Dry the plate under air.
15. Immerse the plate into ~400 ml of CBB staining solution in a tray. Incubate for 10 min without agitation. Remove the staining solution by decantation.
16. Transfer the plate to another tray containing 100–200 ml of destaining solution to the tray. Incubate for 1–2 min. Remove the destaining solution by decantation.
17. Add 100–200 ml of destaining solution to the tray. Incubate for 1–2 min. Remove the destaining solution by decantation. Repeat this step.

18. Dry the plate under air for 1 day.
19. Measure the densities of stained SM or GM3 on the plate with a densitometer (Fig. 3B) (Note 10 below). Determine the amounts of GM3 by comparing the standard calibration curve, or compare relative levels of SM and GM3 among cell lines (Note 11 below).

Note 9: Adjust the number of cells seeded to obtain subconfluent cell monolayers. The growth rate of LY-B cells in Nutridoma-BO medium is half or less of the wild-type level.[9]

Note 10: For the densitometric analysis, we use the NIH Image software after scanning the stained patterns to a desktop computer (EPSON, GT6000).

Note 11: The SM bands partially overlap the phosphatidylserine bands in this TLC system (Fig. 3B). Thus, we use another TLC solvent (chloroform/methanol/acetic acid; 60/25/10, v/v) for quantification of SM levels, although this solvent is not suitable for separating GM3 from other lipid types on a TLC plate.

Identification of Defective Steps in LY-A and LY-B Strains

Determination of SPT activity showed that LY-B cells do not have this activity. LY-B cells lacked expression of the LCB1 protein, a subunit of SPT, and transfection of LY-B cells with the hamster *LCB1* cDNA restored both SPT activity and sphingolipid synthesis to the cells. Collectively, these results indicate that the LY-B strain is a CHO cell mutant specifically defective in SPT activity due to the lack of the LCB1 protein.[9] In addition, we found that the SPT enzyme comprises both the LCB1 and LCB2 proteins, by showing that expression of an affinity peptide-tagged LCB1 protein in LY-B cells causes the endogenous LCB2 protein to adsorb to a tag affinity matrix.[9,29]

Metabolic labeling experiments with [^{14}C]serine, [^{3}H]sphingosine, and [^{14}C]choline indicated that conversion of Cer to SM was impaired in LY-A cells, but that the formation of Cer and phosphatidylcholine, both of which are immediate precursors for SM formation, is normal. However, enzyme assays in cell lysates showed no difference in the activity of SM synthase, neutral sphingomyelinase, or acid sphingomyelinase between LY-A and wild-type cells,[10] raising the possibility that LY-A cells have a defect in translocation of Cer from the ER to the Golgi apparatus where SM

[29] K. Hanada, T. Hara, and M. Nishijima, *J. Biol. Chem.* **275,** 8409 (2000).

synthase exists. Consistent with this possibility, treatment of cells with brefeldin A, which cause fusion of the Golgi apparatus with the ER, restored *de novo* SM synthesis in LY-A cells to the wild-type level.[10] In addition, pulse-chase experiments with a fluorescent Cer analog, *N*-(4,4-difluoro-5,7-dimethyl-4-bora-3a,4a-diaza-*s*-indacene-3-pentanyol)-D-*erythro*-sphingosine (C_5-DMB-Cer), revealed that in wild-type cells C_5-DMB-Cer was redistributed from intracellular membranes to the Golgi apparatus in an intracellular ATP-dependent manner, and that LY-A cells were defective in the energy-dependent redistribution of C_5-DMB-Cer.[10] Under ATP-depleted conditions, conversion of C_5-DMB-Cer to C_5-DMB-SM and of [^{3}H]sphingosine to [^{3}H]SM in wild-type cells decreased to the levels in LY-A cells.[10] ER-to-Golgi apparatus trafficking of glycosylphosphatidylinositol-anchored or membrane-spanning protein in LY-A cells appeared to be normal.[10] From these results, we concluded that the predominant pathway of ER-to-Golgi apparatus trafficking of Cer for *de novo* SM synthesis is ATP dependent and that this pathway is almost completely impaired in LY-A cells.[10] In addition, the specific defect of SM synthesis in LY-A cells suggested different pathways of Cer transport for glycosphingolipids versus SM synthesis.[10]

Acknowledgments

We thank Dr. Masayoshi Fukasawa and Ms. Tomoko Hara of our laboratory for their continued contribution to characterization of the LY-A and LY-B strains. This work was supported in part by Grants-in-Aid from the Ministry of Education, Science and Culture of Japan, CREST of Japan Science and Technology Corporation, and a Special Coordination Fund for Promoting Science and Technology from the Science and Technology Agency of Japan.

[25] Selection of Yeast Mutants in Sphingolipid Metabolism

By TERESA M. DUNN, KEN GABLE, ERIN MONAGHAN, and DAGMAR BACIKOVA

Introduction

The sphingolipids are essential components enriched in the plasma membrane of all eukaryotic cells. The ceramide backbone of sphingolipids is a long-chain base (LCB) joined through an amide linkage with a fatty acid (Fig. 1A). Ceramide is structurally similar to the diacylglycerol moiety

0076-6879/00 $30.00

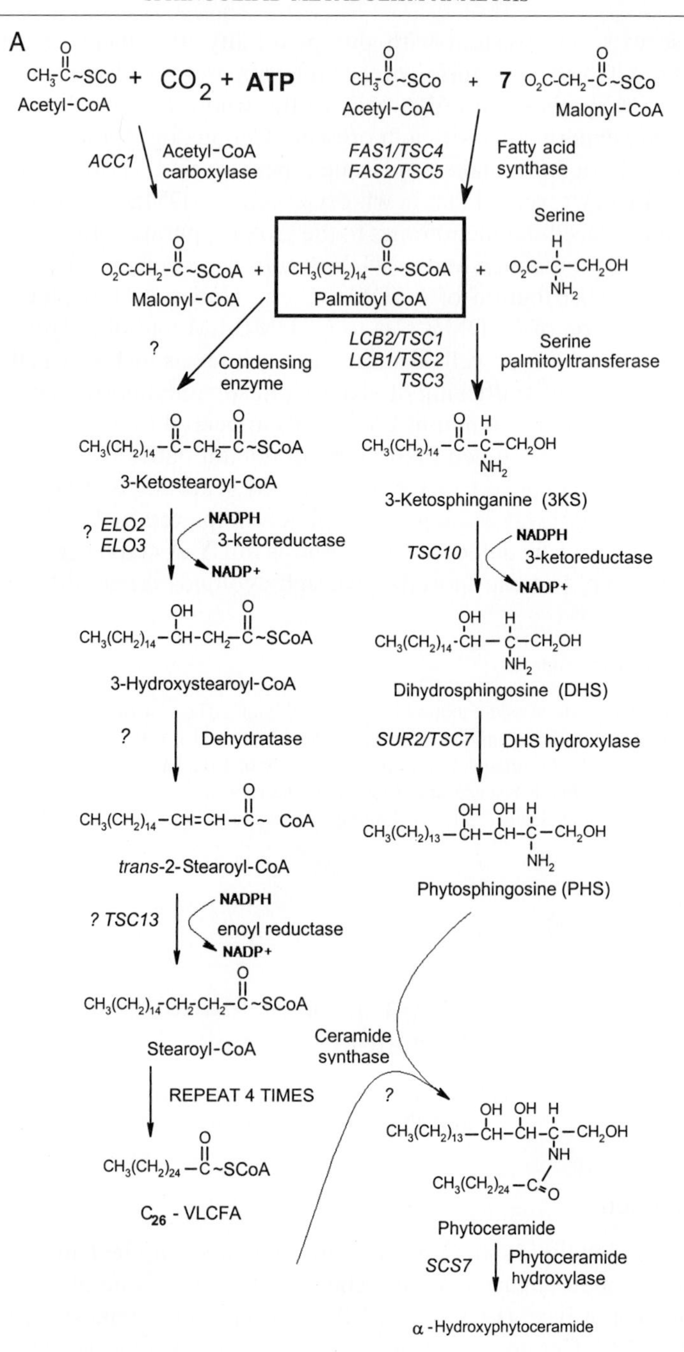

A
Acetyl-CoA
Acetyl-CoA
Malonyl-CoA
ACC1
Acetyl-CoA carboxylase
FAS1/TSC4
FAS2/TSC5
Fatty acid synthase
Serine
Malonyl-CoA
Palmitoyl CoA
?
Condensing enzyme
LCB2/TSC1
LCB1/TSC2
TSC3
Serine palmitoyltransferase
3-Ketostearoyl-CoA
3-Ketosphinganine (3KS)
? ELO2
ELO3
NADPH
3-ketoreductase
NADP+
TSC10
NADPH
3-ketoreductase
NADP+
3-Hydroxystearoyl-CoA
Dihydrosphingosine (DHS)
?
Dehydratase
SUR2/TSC7
DHS hydroxylase
trans-2-Stearoyl-CoA
Phytosphingosine (PHS)
? TSC13
NADPH
enoyl reductase
NADP+
Stearoyl-CoA
Ceramide synthase
REPEAT 4 TIMES
?
C26 - VLCFA
Phytoceramide
SCS7
Phytoceramide hydroxylase
α-Hydroxyphytoceramide

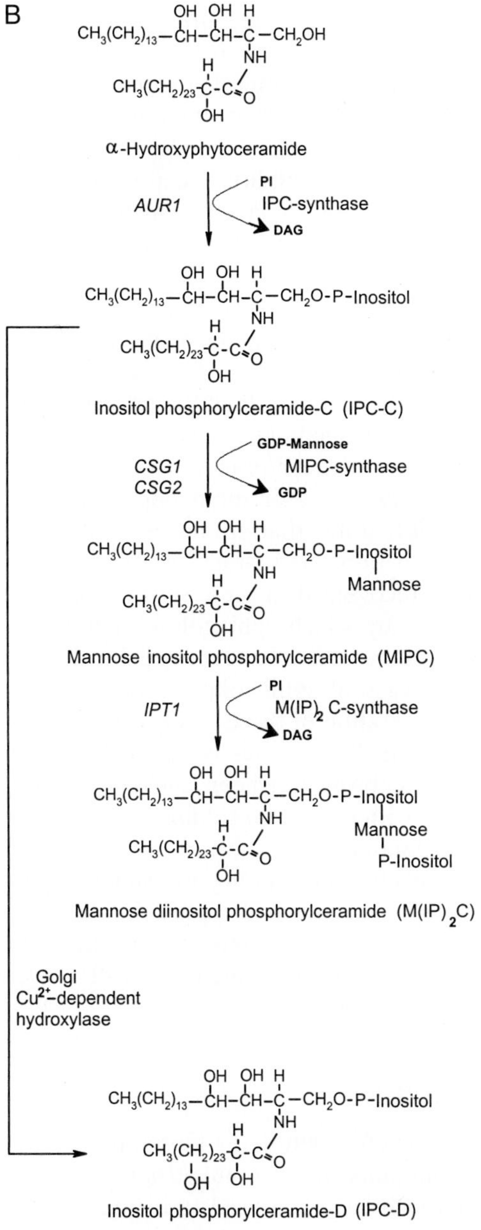

B

α-Hydroxyphytoceramide

AUR1 — IPC-synthase (PI → DAG)

Inositol phosphorylceramide-C (IPC-C)

CSG1, CSG2 — MIPC-synthase (GDP-Mannose → GDP)

Mannose inositol phosphorylceramide (MIPC)

IPT1 — $M(IP)_2C$-synthase (PI → DAG)

Mannose diinositol phosphorylceramide ($M(IP)_2C$)

Golgi Cu^{2+}-dependent hydroxylase

Inositol phosphorylceramide-D (IPC-D)

FIG. 1. (A) Enzymatic reactions and genes of ceramide biosynthesis in yeast. (B) Enzymatic reactions and genes of complex sphingolipid biosynthesis in yeast.

of the glycerophospholipids, but in *Saccharomyces cerevisiae,* ceramide differs from the diacylglycerol base of the glycerophospholipids in three significant ways. First, the two ceramide hydrocarbon chains are saturated whereas most glycerophospholipids contain at least one unsaturated fatty acid. The increased saturation allows the alkyl chains to pack more tightly, thereby decreasing membrane fluidity. Second, the ceramide base contains four hydroxyl groups that are available for hydrogen bond formation. The amide bond also contributes both proton donors and acceptors for hydrogen bonding. Because these constituents reside in a hydrophobic region, the hydrogen bonds that form are stronger than they would be in aqueous medium. These interlipid hydrogen bonds increase lateral cohesion, and therefore also decrease membrane fluidity. Finally, the sphingolipid fatty acid moiety is 8–10 carbons longer than the fatty acids of the glycerophospholipids.[1,2] This very long-chain fatty acid (VLCFA) is expected to affect the membrane structure and the interaction of sphingolipids with membrane proteins. The unique functions of sphingolipids are likely to depend on these three structural features that are not provided by other lipids.

In addition to their presumed structural roles, sphingolipids and their metabolites are now recognized to act as second messengers as well. However, the mechanisms by which sphingolipid metabolites regulate signal transduction processes are not well understood. Furthermore, little is known about how the pathways of sphingolipid synthesis and degradation are regulated to allow generation of these second messengers. The pathway of sphingolipid synthesis in *S. cerevisiae* is outlined in Fig. 1. Purification of the enzymes in this pathway has been challenging, but recently several *S. cerevisiae* mutants with defects in sphingolipid synthesis have been isolated and the corresponding genes have been cloned (Table I). The cloned genes provide the tools for determining how sphingolipid synthesis is regulated and for developing strategies for purifying the encoded proteins. In addition, the mutants that are deficient in the synthesis of the long-chain bases, ceramides, and the mature sphingolipids will be useful for elucidating the essential functions of these lipids.

Principle of Genetic Selection

A classical approach for identifying biosynthetic mutants in a pathway relies on isolating mutants that are auxotrophic for intermediates in the pathway. This approach has been used to identify genes required for the

[1] R. L. Lester and R. C. Dickson, *Adv. Lipid Res.* **26,** 253 (1993).

[2] P. Hechtberger, E. Zinser, R. Saf, K. Hummel, F. Paltauf, and G. Daum, *Eur. J. Biochem.* **225,** 641 (1994).

TABLE I
SPHINGOLIPID SYNTHESIS GENES

Function	Gene	Ref.
3-Ketospinganine synthesis, SPT subunit	LCB1/TSC2/SCS2	*a, b*
3-Ketospinganine synthesis, SPT subunit	LCB2/TSC1/SCS1	*b–d*
3-Ketospinganine synthesis	TSC3	*e*
3-Ketospinganine reductase	TSC10	*b*
Dihydrospingosine hydroxylase	SUR2/TSC7/SYR2	*f, g*
Phytoceramide hydroxylase	SCS7/FAH1	*g–i*
VLCFA synthesis	FEN1/ELO2/GNS1/VBM1	*j–n*
VLCFA synthesis	SUR4/ELO3/APA1/VBM2	*j, k, l, n, o*
VLCFA synthesis	TSC13	*p*
IPC synthesis	AUR1/ABR1	*q*
MIPC synthesis	CSG1/SUR1/BCL21/LPE15	*r, s, t*
MIPC synthesis	CSG2/CLS2	*c, r*
$M(IP)_2C$ synthesis	IPT1/SYR4	*u, v*
Golgi Cu^{2+}-ATPase	CCC2	*w*

[a] R. Buede, C. Rinker-Schaffer, W. J. Pinto, R. L. Lester, and R. C. Dickson, *J. Bacteriol.* **173,** 4325 (1991).
[b] T. Beeler, Bacikova, K. Gable, L. Hopkins, C. Johnson, H. Slife, and T. Dunn, *J. Biol. Chem.* **273,** 30688 (1998).
[c] C. Zhao, T. Beeler, and T. Dunn, *J. Biol. Chem.* **269,** 21480 (1994).
[d] M. M. Nagiec, J. A. Baltisberger, G. B. Wells, R. L. Lester, and R. C. Dickson, *Proc. Natl. Acad. Sci. U.S.A.* **91,** 7899 (1994).
[e] K. Gable, H. Slife, D. Bacikova, E. Monaghan, and T. Dunn, *J. Biol. Chem.* **275,** 7597 (2000).
[f] M. M. Grilley, S. D. Stock, R. C. Dickson, R. L. Lester, and J. Y. Takemoto, *J. Biol. Chem.* **273,** 11062 (1998).
[g] D. Haak, K. Gable, T. Beeler, and T. Dunn, *J. Biol. Chem.* **272,** 29704 (1997).
[h] T. M. Dunn, D. Haak, E. Monaghan, and T. J. Beeler, *Yeast* **14,** 311 (1998).
[i] A. G. Mitchell and C. E. Martin, *J. Biol. Chem.* **272,** 28281 (1997).
[j] S. Silve, P. Leplatois, A. Josse, P. H. Dupuy, C. Lanau, M. Kaghad, C. Dhers, C. Picard, A. Rahier, M. Taton, G. Le Fur, D. Caput, P. Ferrara, and G. Loison, *Mol. Cell Biol.* **16,** 2719 (1996).
[k] E. Revardel, M. Bonneau, P. Durrens, and M. Aigle, *Biochim. Biophys. Acta* **1263,** 261 (1995).
[l] C. S. Oh, D. A. Toke, S. Mandala, and C. E. Martin, *J. Biol. Chem.* **272,** 17376 (1997).
[m] M. el-Sherbeini and J. A. Clemas, *J. Bacteriol.* **177,** 3227 (1995).
[n] D. David, S. Sundarababu, and J. E. Gerst, *J. Cell Biol.* **143,** 1167 (1998).
[o] M. Garcia-Arranz, A. M. Maldonado, M. J. Mazon, and F. Portillo, *J. Biol. Chem.* **269,** 18076 (1994).
[p] K. Gable, D. Bacikova, and T. Dunn, unpublished (1999).
[q] M. M. Nagiec, E. E. Nagiec, J. A. Baltisberger, G. B. Wells, R. L. Lester, and R. C. Dickson, *J. Biol. Chem.* **272,** 9808 (1997).
[r] Y. Takita, Y. Ohya, and Y. Anraku, *Mol. Gen. Genet.* **246,** 269 (1995).
[s] L. Desfarges, P. Durrens, H. Juguelin, C. Cassagne, M. Bonneu, and M. Aigle, *Yeast* **9,** 267 (1993).
[t] T. J. Beeler, D. Fu, J. Rivera, E. Monaghan, K. Gable, and T. M. Dunn, *Mol. Gen. Genet.* **255,** 570 (1997).
[u] R. C. Dickson, E. E. Nagiec, G. B. Wells, M. M. Nagiec, and R. L. Lester, *J. Biol. Chem.* **272,** 29620 (1997).
[v] A. Leber, P. Fischer, R. Schneiter, S. D. Kohlwein, and G. Daum, *FEBS Lett.* **411,** 211 (1997).
[w] D. Fu, T. J. Beeler, and T. M. Dunn, *Yeast* **11,** 283 (1995).

synthesis of the long-chain base (LCB) moiety of sphingolipids in *S. cerevisiae*. The *LCB* mutants were identified in a screen for mutants that require phytosphingosine (PHS, Fig. 1A) for viability.[3] However, the approach is limited to identifying the genes required for LCB synthesis because the solubilities of the VLCFAs, ceramides, and inositol phosphoceramides are too low to allow screening for mutants that require these compounds for viability.

An alternative screen for mutants deficient in sphingolipid synthesis relies on the observation that overaccumulation of an intermediate in sphingolipid biosynthesis, inositol phosphoceramide, IPC-C (Fig. 1B), renders the cells Ca^{2+} sensitive. Two genes, *CSG1* and *CSG2*, are required for conversion of IPC-C to mannosylinositol phosphoceramide (MIPC) (Fig. 1B).[4–6] These genes were identified in a screen for mutants that are sensitive to Ca^{2+} but resistant to Sr^{2+}.[7] The discovery that mutations that decrease the accumulation of IPC-C suppress the Ca^{2+} sensitivity of the *csg1* and *csg2* mutants led to the realization that the *CSG* genes themselves are involved in sphingolipid synthesis.[5,6] Because the *csg* mutants are unable to synthesize mature mannosylated sphingolipids they accumulate high levels of IPC-C. Increased levels of IPC-C lead (for unknown reasons) to the Ca^{2+}-sensitive phenotype. Therefore, analysis of suppressors of the Ca^{2+}-sensitive phenotype has proven to be a powerful method for identifying genes required for IPC-C synthesis.[4,6,8–11]

Suppressors of Ca^{2+}-Sensitive Phenotype of *csg2* Null Mutants

Two screens for suppressors of the Ca^{2+} sensitivity of the *csg2*Δ null mutant have been conducted. In the first screen the *scs* (suppressor of Ca^{2+} sensitivity) suppressors were selected at 37° on YPD (1% yeast extract, 2% bacto-peptone, 2% agar) plates containing 100 m*M* Ca^{2+}. Complementation analysis of the *scs* mutants identified using this protocol revealed that a large number of genes could mutate to reverse Ca^{2+} sensitivity. In some

[3] G. B. Wells and R. L. Lester, *J. Biol. Chem.* **258,** 10200 (1983).
[4] T. M. Dunn, D. Haak, E. Monaghan, and T. J. Beeler, *Yeast* **14,** 311 (1998).
[5] T. J. Beeler, D. Fu, J. Rivera, E. Monaghan, K. Gable, and T. M. Dunn, *Mol. Gen. Genet.* **255,** 570 (1997).
[6] C. Zhao, T. Beeler, and T. Dunn, *J. Biol. Chem.* **269,** 21480 (1994).
[7] T. Beeler, K. Gable, C. Zhao, and T. Dunn, *J. Biol. Chem.* **269,** 7279 (1994).
[8] D. Fu, T. Beeler, and T. Dunn, *Yeast* **10,** 515 (1994).
[9] D. Fu, T. J. Beeler, and T. M. Dunn, *Yeast* **11,** 283 (1995).
[10] T. Beeler, D. Bacikova, K. Gable, L. Hopkins, C. Johnson, H. Slife, and T. Dunn, *J. Biol. Chem.* **273,** 30688 (1998).
[11] D. Haak, K. Gable, T. Beeler, and T. Dunn, *J. Biol. Chem.* **272,** 29704 (1997).

cases the suppressing mutations conferred an inability to grow on medium containing low Ca^{2+}. This restrictive growth phenotype provided a positive selection for cloning the wild-type gene that had mutated to confer suppression.[6] However, for most of the suppressors, the only phenotype was ability to grow on Ca^{2+} despite the presence of the *csg2Δ* mutation, and therefore the recovery of the wild-type suppressing gene depended on the ability of a plasmid carrying the gene to revert the strain to Ca^{2+} sensitivity. Using this strategy, the wild-type *SCS7* gene was cloned.[4] As discussed below, during these studies it became clear that the cloning and analysis of the suppressor genes would be facilitated if an associated conditional lethal phenotype could be identified for the suppressing mutations. Therefore, the suppressor screen was modified.

The observation that bypass suppressors of the Ca^{2+} sensitivity of *csg2Δ* mutants identified reduced-function alleles of genes encoding enzymes required for sphingolipid synthesis suggested that many of the suppressor genes would prove to be essential since sphingolipids are required for viability. Furthermore, it was desirable to identify an associated conditional growth phenotype that resulted from the suppressing mutation. Therefore, a second screen for suppressing mutations that simultaneously conferred temperature-sensitive (ts) lethality was initiated. When suppressors were selected on YPD medium containing 50 m*M* Ca^{2+} at 26°, nearly 8% of the suppressing mutations were found to simultaneously confer temperature-sensitive lethality (inability to grow on YPD medium at 37°). The genes identified in this screen are called the *TSC* genes for Ts suppressors of Ca^{2+} Sensitivity.[12] In many cases the suppressor mutation reduces the function of an essential gene product required for IPC-C synthesis sufficiently at 26° to suppress the Ca^{2+} sensitivity of the *csg2Δ* mutant. However, at 37° the altered gene product does not support sufficient sphingolipid synthesis to sustain viability. The high percentage of suppressing mutations that simultaneously confer ts lethality suggests that the requirement for sphingolipid synthesis is higher at 37° than at 26°. When cells are shifted from 26° to 37°, there is a transient increase in the synthesis of phytosphingosine,[13] as well as increased ceramide accumulation.[14]

The TSC screen has facilitated the analysis of the suppressor genes in several ways. All of the *tsc* mutations are recessive, so the suppressor mutations can be placed in complementation groups based on the ts growth

[12] T. Beeler, D. Bacikova, K. Gable, L. Hopkins, C. Johnson, H. Slife, and T. Dunn, *J. Biol. Chem.* **273,** 30688 (1998).

[13] R. C. Dickson, E. E. Nagiec, M. Skrzypek, P. Tillman, G. B. Wells, and R. L. Lester, *J. Biol. Chem.* **272,** 30196 (1997).

[14] G. B. Wells, R. C. Dickson, and R. L. Lester, *J. Biol. Chem.* **273,** 7235 (1998).

phenotype of diploids formed from pairwise matings of the *tsc* mutants. The wild-type *TSC* genes can be cloned in a positive selection for plasmids carrying genomic inserts that complement the ts phenotype. The collection is enriched in suppressor mutations that reside in essential genes, and the ts alleles facilitate phenotypic characterization of the mutant by providing a restrictive condition under which the mutant phenotype is more penetrant. Finally, as discussed below, these mutants can be used in genetic screens to identify other genes required for the synthesis of sphingolipids and for elucidating the essential functions of sphingolipids.

Isolation and Characterization of the Suppressor Mutants

Selecting Suppressor Mutants

Saccharomyces cerevisiae cells are grown according to standard procedures.[15] The YPD + Ca^{2+} plates are prepared using a 2 *M* $CaCl_2$ stock solution. The long-chain bases are added to the autoclaved growth medium from a 25 m*M* stock solution in ethanol to a final concentration of 10–25 μM. The suppressor mutants have been selected using strains TDY2040 (*Mat***a** *ade 2-101 ura3-52 trp 1* Δ *leu2*Δ *csg2::LEU2*), and TDY2038 (*Mat*α *lys2 ura3-52 trp1*Δ *leu2*Δ *csg2::LEU2*). The suppressor mutations are necessarily extragenic bypass suppressors because they suppress the *csg2::LEU2* null mutant, which is lacking most of the coding sequence of the *CSG2* gene.

Pseudorevertants are isolated by smearing haploid *csg2::LEU2* cells from single large colonies (grown on YPD plates at 26°) using a flat toothpick onto YPD + 50 m*M* Ca^{2+} plates divided into eight sectors. The plates are incubated for 2 or 3 days at 26°, and only one fast growing revertant per colony is selected to ensure that independent suppressor mutations are recovered. The revertant is picked with a toothpick and purified by streaking onto a YPD + Ca^{2+} plate that is incubated at 26° for 2 days to allow single colonies to form. Approximately equal numbers of suppressor mutants are recovered in strains of opposite mating types (TDY2040 and TDY2038) to facilitate the complementation analysis. A single colony from each of the purified suppressors is picked with a toothpick and streaked onto a set of YPD plates with and without $CaCl_2$, which are incubated at 26° and 37° to screen for an associated ts phenotype. Suppressor mutants that are able to grow on YPD + Ca^{2+} at 26°, but that are unable to grow on YPD plates at 37°, are stored as the *tsc* mutants.

[15] F. Sherman, G. R. Fink, and J. B. Hicks, "Methods in Yeast Genetics." Cold Spring Harbor Laboratory, Cold Spring Harbor, 1986.

Placing TSC Suppressor Mutants into Complementation Groups

The suppressor mutants are mated with the *csg2* null mutant and the selected diploids are tested for both the ts and the suppressor of Ca^{2+} sensitivity phenotypes. For all the *tsc* mutants tested thus far (more than 180 isolates), both the ts and suppressor phenotypes are recessive. Complementation analysis is performed to determine how many genes mutate to simultaneously reverse Ca^{2+} sensitivity and confer temperature sensitivity. The *tsc* strains of one mating type are pairwise-mated with those of the other mating type and the diploids are selected on SD medium lacking adenine and lysine. The diploids are streaked on YPD plates with and without $CaCl_2$ and incubated for 2 days at either 26° or 37°. Because the suppressors are recessive for both the temperature-sensitive and the suppressing phenotypes, diploids having mutations in the same gene fail to grow at 37°, and suppress the Ca^{2+} sensitivity at 26°. Several complementation groups have been identified through this analysis.[12]

Establishing That the Same Mutation That Causes Temperature Sensitivity Causes Suppression

The majority of the *tsc* mutants (168/187) fall into 21 complementation groups. Since spontaneously arising suppressor mutants were isolated, it was considered likely that a single mutation would be found to cause both the suppressing and the ts phenotype. However, the possibility that the suppressing mutation provides selective pressure for an unlinked mutation that confers the temperature sensitivity had to be considered. Furthermore, some *tsc* mutants fail to fall into a complementation group, and in that case the suppressing mutation and the ts mutation might be unlinked. Therefore, a representative from each complementation group, and all of the *tsc* mutants that failed to fall into a complementation group, were backcrossed to the *csg2*Δ mutant and the selected diploids were sporulated and dissected. With one exception (TSC6), the same two products of meiosis (POM) that displayed suppression were also ts indicating that a single mutation was responsible for both phenotypes. The single isolate of TSC6 was found to have a non-ts suppressing mutation in the *SCS7* gene,[4] as well as a second mutation in an unlinked gene responsible for the ts phenotype.

Cloning and Disrupting Wild-Type TSC Genes

The wild-type *TSC* genes have been cloned based on the ability of a genomic plasmid carrying the gene to complement the ts lethality conferred by the recessive *tsc* mutation.[16] The *tsc* mutant is grown in YPD medium

[16] M. D. Rose and J. R. Broach, *Methods Enzymol.* **194,** 195 (1991).

at 26°, and competent cells are prepared. A YCp50-based yeast genomic library[17] is transformed into the cells and uracil prototrophs are selected on SD medium at 26°. The transformants are transferred to YPD plates by replica-plating and incubated at 37° for 2 days. Temperature-resistant uracil prototrophs are recovered at frequencies of 10^{-2} to 10^{-4} depending on the *tsc* mutant.

To confirm that the plasmid is responsible for reversing the temperature sensitivity of the *tsc* mutant, transformants that have lost the plasmid are tested for reversion back to the ts phenotype. Transformants are placed on YPD plates at 26°, and allowed to segregate the plasmid. Cells that have lost the plasmid carrying the *URA3* gene are selected by patching onto plates containing 5-fluoroorotic acid (FOA), which kills uracil prototrophs.[18] If the cells that have lost the plasmid revert back to temperature sensitivity, it can be concluded that the plasmid carried a gene responsible for complementing the ts phenotype.

The ends of the genomic fragment that resides in the *Bam*HI site of the YCp50 plasmid are sequenced using primers that anneal on either side of the cloning site. The sequence is compared to the *Saccharomyces* Genome Database (SGD), which provides the entire sequence of the genomic fragment (in the range of 10–20 kilobases) that contains the complementing sequence. A number of strategies are used to establish which ORF on the genomic fragment is responsible for complementation. In some cases, the fragment encodes an open reading frame (ORF) that is a likely candidate for a gene involved in sphingolipid synthesis. Subclones are constructed using defined restriction fragments or PCR-generated restriction fragments and tested to define the minimal complementing ORF.

The complementing ORF may correspond to the wild-type allele of the *tsc* gene or it may represent a dose-dependent suppressor. Genetic linkage analysis is performed to distinguish between these two possibilities. The cloned gene is used to direct the integration of a YIp plasmid with a selectable marker, e. g., the *URA3* gene, at the homologous genomic locus in a wild-type haploid.[19] This marked haploid is mated with the *tsc* mutant, the selected diploid is sporulated, and tetrads are dissected. If the *URA3*-marked locus segregates away from the ts mutation (in every tetrad the two *URA3* prototrophic POM are temperature resistant and the *ura3* auxotrophic POM are ts), the cloned complementing gene is determined to be genetically linked to the wild-type allele of the *tsc* mutant. Otherwise, the cloned gene is identified as a dose-dependent suppressor of the *tsc* mutant.

[17] M. D. Rose, P. Novick, J. H. Thomas, D. Botstein, and G. R. Fink, *Gene* **60,** 237 (1987).
[18] J. D. Boeke, J. Trueheart, G. Natsoulis, and G. R. Fink, *Methods Enzymol.* **154,** 164 (1987).
[19] R. Rothstein, *Methods Enzymol.* **194,** 281 (1991).

Dose-dependent suppressors can also provide clues about the function of the product of the *TSC* gene.

The *TSC* gene is disrupted to determine whether it is essential for viability.[20] The majority of the *tsc* mutants have been found to harbor ts mutations in genes that are essential for viability at all temperatures. However, in some cases, the *TSC* gene (e.g., the *TSC3* gene) is only required for viability at high temperatures.

Characterizing tsc *Mutants for Potential Defects in Sphingolipid Biosynthesis*

Rapid phenotypic screens for determining whether the *tsc* mutants are deficient in sphingolipid synthesis have been developed. The effect of long-chain bases and palmitate on the growth phenotype of the *tsc* mutants is initially determined and has been useful for identifying mutations that result in deficiencies in LCB, palmitate and VLCFA synthesis.

The ts phenotype of *tsc* mutants deficient in LCB synthesis is rescued by addition of LCBs to the growth medium. The ts lethality of mutants in the *TSC1*, *TSC2*, and *TSC3* genes is reversed by 3-ketosphinganine (3-KS), dihydrosphingosine (DHS), and phytosphingosine (PHS) as would be expected if they encoded proteins required for serine palmitoyltransferase (SPT) activity (Fig. 1A). All three genes were in fact demonstrated to be required for SPT activity by directly assaying the SPT activity in microsomes prepared from the mutants. The *TSC1* gene was found to be allelic to *LCB2/SCS1*, and *TSC2* was found to be allelic to *LCB1/SCS2*. The *TSC3* gene encodes a novel protein that, although not absolutely required for SPT activity, enhances the activity severalfold. The ts phenotype of the *tsc10* mutants is rescued by PHS and DHS, but not by 3-KS, suggesting that the mutations reside in 3-KS reductase (Fig. 1A). Indeed, the *TSC10* gene was found to encode 3-KS reductase.[12]

Some of the *tsc* mutants display increased sensitivity to exogenous LCBs. This phenotype has been reported for mutants that are deficient in the degradation of the LCB 1-phosphates.[21–23] Mutations that decrease VLCFA synthesis (*ELO2*, *ELO3*, and *TSC13*) also confer sensitivity to long-chain

[20] R. J. Rothstein, *Methods Enzymol.* **101,** 202 (1983).

[21] J. D. Saba, F. Nara, A. Bielawska, S. Garrett, and Y. A. Hannun, *J. Biol. Chem.* **272,** 26087 (1997).

[22] C. Mao, M. Wadleigh, G. M. Jenkins, Y. A. Hannun, and L. M. Obeid, *J. Biol. Chem.* **272,** 28690 (1997).

[23] L. Qie, M. M. Nagiec, J. A. Baltisberger, R. L. Lester, and R. C. Dickson, *J. Biol. Chem.* **272,** 16110 (1997).

bases. Mutants in two other complementation groups, TSC8 and TSC9, display increased sensitivity to LCB. The *TSC8* gene is allelic to *SIN3,* and the *TSC9* gene is allelic to *RPD3*. Interestingly, the Rpd3p and Sin3p proteins copurify in a complex that is involved in histone deacetylation,[24–26] but if and how this alters sphingolipid metabolism is not yet understood.

The *ts* phenotype of the *tsc4*, *tsc5*, and *tsc21* mutants is rescued by exogenously added myristate or palmitate, suggesting that they might be deficient in fatty acid synthesis. Therefore, genetic linkage studies were done and it was found that the *tsc4* and *tsc5* mutants have mutations in *FAS2*, and the *tsc21* mutant has a mutation in *FAS1*. The *FAS1* and *FAS2* genes encode the two subunits of fatty acid synthase, which are both multifunctional proteins.[27–30] The mutations in the TSC4 group complement those in the TSC5 complementation group, presumably because they have defects in distinct enzymatic activities of the multifunctional Fas2p.

Analysis of the sphingolipids, ceramides, LCBs, and VLCFAs has also been important for establishing how the *tsc* mutants influence sphingolipid metabolism. In some cases the mutations result in decreased levels of the inositol phosphorylated sphingolipids. In other cases the levels are unaltered but the mobility of the sphingolipids during thin-layer chromatography (TLC) changes. For example, increased hydrophobicity of the inositolated sphingolipids, ceramides, and LCBs results from mutations in the enzyme (Sur2p) that hydroxylates C-4 of dihydrosphingosine.[4,11,31] Similarly, the hydrophobicity of sphingolipids and ceramides is increased in the mutants that lack the enzyme (Scs7p) that hydroxylates C-2 of the VLCFA.[4,11,32] Mutations in enzymes required for VLCFA synthesis (Tsc13p, Elo2p, and Elo3p) lead to the accumulation of sphingolipids and ceramides that are relatively hydrophilic because they have fatty acids with chain lengths shorter than C-26. The defects suggested by preliminary TLC analyses of the inositolated sphingolipids and ceramides can be confirmed by more rigorous methods including TLC or high-performance liquid chroma-

[24] D. Kadosh and K. Struhl, *Mol. Cell Biol.* **18,** 5121 (1998).

[25] D. Vannier, D. Balderes, and D. Shore, *Genetics* **144,** 1343 (1996).

[26] M. M. Kasten, S. Dorland, and D. J. Stillman, *Mol. Cell Biol.* **17,** 4852 (1997).

[27] A. Knobling and E. Schweizer, *Eur. J. Biochem.* **59,** 415 (1975).

[28] A. H. Mohamed, S. S. Chirala, N. H. Mody, W. Y. Huang, and S. J. Wakil, *J. Biol. Chem.* **263,** 12315 (1988).

[29] E. Schweizer, K. Werkmeister, and M. K. Jain, *Mol. Cell Biochem.* **21,** 95 (1978).

[30] H. J. Schuller, A. Hahn, F. Troster, A. Schutz, and E. Schweizer, *Embo J.* **11,** 107 (1992).

[31] M. M. Grilley, S. D. Stock, R. C. Dickson, R. L. Lester, and J. Y. Takemoto, *J. Biol. Chem.* **273,** 11062 (1998).

[32] A. G. Mitchell and C. E. Martin, *J. Biol. Chem.* **272,** 28281 (1997).

tography analysis of the purified LCBs[33] and GCMS analysis of the fatty acids.[34]

Conclusions and Future Uses of Suppressor Mutants

Analysis of the suppressors of the Ca^{2+} sensitivity of the *csg* mutants has proven to be a very powerful method for identifying the sphingolipid biosynthetic genes (Fig. 1). In addition, these suppressor mutants have provided the necessary tools for addressing the important question of how sphingolipid synthesis is regulated, and they will permit the development of strategies for purifying the proteins. The value of these mutants is clear when the importance of the ts collections of the *cdc*[35–37] and *sec*[38] mutants is considered. The *scs* and *tsc* mutants can be used in powerful genetic screens designed to identify proteins that interact with the sphingolipid biosynthetic enzymes and proteins that depend on sphingolipids for their function. Examples of two of these screens, briefly described next, illustrate how the mutants will be used in future studies of sphingolipid metabolism.

The ts alleles of the sphingolipid biosynthetic genes can be used for the selection of extragenic suppressors of the temperature-sensitive lethality that will help identify other genes and proteins involved in sphingolipid metabolism and function. The *tsc* mutant collection has multiple different ts alleles of several of the essential sphingolipid biosynthetic genes. This is important for the suppressor studies, because the second-generation suppressors that reverse the temperature sensitivity conferred by one mutant allele can then be tested for ability to suppress the other ts alleles in the same gene. Allele-specific suppression (the extragenic suppressor mutation reverses the ts phenotype associated with some amino acid changes in the protein, but not others) is an indication that the two proteins may interact directly.

Genetic screens designed to identify sphingolipid-dependent proteins will be used for determining what the essential functions of sphingolipids are. For example, many of the *tsc* mutants display reduced synthesis of sphingolipids at 26°; therefore, genes encoding proteins that are especially dependent on sphingolipids for their function may be identified in screens

[33] A. H. Merrill, Jr., E. Wang, R. E. Mullins, W. C. Jamison, S. Nimkar, and D. C. Liotta, *Anal. Biochem.* **171,** 373 (1988).

[34] C. S. Oh, D. A. Toke, S. Mandala, and C. E. Martin, *J. Biol. Chem.* **272,** 17376 (1997).

[35] L. H. Hartwell, J. Culotti, and B. Reid, *Proc. Natl. Acad. Sci. U.S.A.* **66,** 352 (1970).

[36] J. Culotti and L. H. Hartwell, *Exp. Cell. Res.* **67,** 389 (1971).

[37] L. H. Hartwell, J. Culotti, J. R. Pringle, and B. J. Reid, *Science* **183,** 46 (1974).

[38] P. Novick, C. Field, and R. Schekman, *Cell* **21,** 205 (1980).

for mutations that are synthetically lethal in combination with the *tsc* mutations at 26°. Other suppressor mutants synthesize normal levels of sphingolipids that are structurally altered. It is expected that the identification of proteins that require these modifications for optimal activity will elucidate the functions of the modifications. For example, a mutation that reduces the activity of a protein that interacts with the hydroxylated sphingolipids might be identified in a genetic screen designed to identify mutations that are synthetically lethal in combination with the *scs7* mutation that precludes α-hydroxylation of the VLCFA. These examples illustrate the value of the sphingolipid mutants beyond their obvious contribution in delineating the sphingolipid biosynthetic pathway.

[26] Fluorescence-Based Selection of Gene-Corrected Hematopoietic Stem and Progenitor Cells Based on Acid Sphingomyelinase Expression

By EDWARD H. SCHUCHMAN, SHAI ERLICH, SILVIA R. P. MIRANDA, TAMA DINUR, ARIE DAGAN, and SHIMON GATT

Introduction

Although hematopoietic stem cells (HSCs) are important targets for gene therapy,[1–4] until now they have remained refractable to most gene transfer techniques because of their low numbers and lack of proliferation. Therefore, before such cells can be widely used in a clinical setting, new gene transfer vectors must be developed and improved methods of enriching for small numbers of transduced cells must be obtained.

Toward this latter goal, we have developed a method to enrich populations of gene-corrected hematopoietic stem and progenitor cells based on the expression of lysosomal acid sphingomyelinase (ASM; sphingomyelin phosphodiesterase; EC 3.1.4.12) activity. The procedure takes advantage of the fact that cells expressing a functional ASM can metabolize fluorescent sphingomyelin derivatives, unlike those that lack ASM activity (e.g., from

[1] M. K. Brenner, *J. Hematother.* **2,** 7 (1993).

[2] M. K. Brenner, J. M. Cunningham, B. P. Sorrentino, and H. E. Heslop, *Br. Med. Bull.* **51,** 167 (1995).

[3] R. Huss, *Infusions Ther. Transfusionsmed.* **23,** 147 (1996).

[4] E. M. Kaye, *Curr. Opin. Pediatr.* **7,** 650 (1995).

0076-6879/00 $30.00

acid sphingomyelinase-deficient [ASMKO] mice[5] or human patients with the genetic disorder, Niemann-Pick disease [NPD][6]). The product of ASM degradation, fluorescent ceramide, is released from lysosomes and can be used for new sphingolipid synthesis but, ultimately, the fluorescence is lost from the cell. Thus, after labeling with fluorescent sphingomyelin, normal skin fibroblasts expressing ASM and nonexpressing NPD counterparts can be discriminated by fluorescence microscopy or flow cytometry.[7,8] The same is true for NPD cells and NPD cells that have been genetically engineered to express ASM activity. This chapter describes the methods used to isolate gene-corrected hematopoietic stem and progenitor cells from the bone marrow of ASMKO mice based on their ability to metabolize fluorescent sphingomyelin. Similar methods can be easily applied to hematopoietic cells obtained from human NPD patients for future gene therapy.

General Methods

Cell Preparations

To obtain adult nucleated bone marrow cells, the tibia and femurs of 12- to 16-week-old C57BL/SV129 normal and ASMKO mice are flushed with buffered Hanks' solution (10 m*M* HEPES, pH 7.5). Single cell suspensions are obtained by passing the cells through a 0.4-μm mesh (Becton Dickinson Labware, Franklin Lakes, NJ). Low-density bone marrow cells (<1.085 g/cm^3) are isolated by discontinuous density gradient centrifugation using Nycoprep (Nycomed Pharma AS, Oslo, Norway),[9] and washed with buffered Hanks' solution containing 5% heat-inactivated fetal calf serum (HSA).

Lineage Depletion

To obtain an enriched population of hematopoietic stem and progenitor cells, a lineage depletion procedure is used to eliminate differentiated cells.[10]

[5] K. Horinouchi, S. Erlich, D. P. Perl, K. Ferlinz, C. L. Bisgaier, K. Sandhoff, R. J. Desnick, C. L. Stewart, and E. H. Schuchman *Nat. Genet.* **10,** 288 (1995).

[6] E. H. Schuchman and R. J. Desnick, "The Metabolic and Molecular Bases of Inherited Disease," Vol. 2, p. 2601. McGraw-Hill, New York, 1995.

[7] T. Dinur, E. H. Schuchman, E. Fibach, A. Dagan, M. Suchi, R. J. Desnick, and S. Gatt, *Hum. Gene Ther.* **3,** 633 (1992).

[8] P. L. Yeyati, V. Agmon, C. Fillat, T. Dinur, A. Dagan, R. J. Desnick, S. Gatt, and E. H. Schuchman, *Hum. Gene Ther.* **6,** 975 (1995).

[9] I. Bertoncello, S. H. Bartelmez, T. R. Bradley, and G. S. Hodgson, *J. Immunol.* **139,** 1096 (1987).

[10] I. Bertoncello, T. R. Bradley, and S. M. Watt, *Exp. Hematol.* **19,** 95 (1991).

Isolated adult bone marrow cells are incubated for 30 min on ice with a cocktail of the biotinylated, lineage-specific antibodies anti-TER-119, anti-CD45R/B220, and anti-Ly-6G (Pharmingen, San Diego, CA). The concentration of each antibody is 1 μg/10^6 cells. The cells are then washed with HSA, resuspended in buffered Hanks' solution, and incubated with streptavidin-coated magnetic beads (Dynabeads M-280 streptavidin, Dynal, Lake Success, NY) at 4° for 30 min at a 10:1 bead:cell ratio. Magnetic force is then applied for 1 min to precipitate the differentiated cells, and the supernatant (i.e., Lin^- cells enriched for stem and progenitors) collected for further studies.

Anti-c-Kit Labeling

The cell surface receptor (c-Kit) is used as a marker for stem and early progenitor cells. Cultured or freshly collected Lin^- cells are washed once with HSA and counted. The cells are then incubated with phycoerythrin-conjugated anti-CD117 (c-Kit) antibodies (Pharmingen) at a concentration of 1 μg/10^6 cells for 30 min on ice. After labeling, the cells are washed once with HSA and resuspended in buffered Hanks' solution.

Synthesis of Fluorescent Sphingomyelin

Sphingomyelin to which the fluorescent probe BODIPY is covalently linked via a 12-carbon spacer (BODIPY dodecanoylsphingosylphosphocholine; B12SPM) is synthesized as previously described for Lissamine-rhodamine sphingomyelin,[7] except that BODIPY dodecanoic acid (Molecular Probes Inc., Eugene, OR) is condensed with sphingosylphosphocholine.[10a] To incorporate B12SPM into liposomes, B12SPM is mixed with phosphatidylcholine (PC; Sigma, St. Louis, MO) at a molar ratio of 1:4. The solvent is evaporated under a stream of nitrogen and the mixture resuspended in buffered Hanks' solution followed by a 1-min sonication.

"Pulse-Chase" Labeling with B12SPM

B12SPM/PC liposomes (final concentration 0.5–1 nmol B12SPM/ml) are incubated at 37° for 4 hr with Lin^- cells that have been suspended in buffered Hanks' solution. Labeling is terminated by centrifuging the cells (400*g*) and washing the pellets once with HSA. Fresh medium is then added and the cells are further incubated for 48 hr in standard culture media containing Iscove's modified Dulbecco's medium (IMDM; Gibco-BRL,

[10a] A. Dagan, V. Agmon, S. Gatt, and T. Dinur, *Methods Enzymol.* **312,** [23], (2000) (this volume).

Gaithersburg, MD), 10% heat-inactivated fetal calf serum (Gibco-BRL, Gaithersburg, MD) and antibiotics, but no B12SPM/PC liposomes.

Retroviral Transduction

To achieve retroviral transduction, Lin^- cells are cocultured for 48 hr with amphotropic or ecotropic retroviral producing cells containing an ASM/MFG retroviral vector.[8] Control cells are cocultured with the parental producer cells which are not making the ASM/MFG virus. Cocultures are carried out in 0.4-mm Transwell dishes (Corning Costar, Cambridge, MA) containing the producer cells in the upper compartment. SCF (50 ng/ml), interleukin (IL)-3 (20 ng/ml), IL-6 (10 ng/ml) (Genzyme, Cambridge, MA), and Polybrene (8 μg/ml) are added to the media unless indicated otherwise to stimulate the quiescent stem cells to divide and to facilitate retroviral transduction.

FACS Analysis and Expansion of Sorted Cells

Cells are analyzed using a FACScan instrument (Becton Dickinson Immunocytometry Systems, San Jose, CA) and the WinMDI program. Cells are sorted using a FACStar flow cytometer (Becton Dickinson Immunocytometry Systems). Sorted cells are resuspended in bone marrow expansion media (IMDM, Gibco-BRL) containing 10% heat-inactivated fetal calf serum, SCF (50 ng/ml), IL-6 (10 ng/ml), IL-3 (20 ng/ml), and antibiotics, and then grown at 37° for 10 days.

ASM Assays

Fresh or cultured adult bone marrow cells are harvested, washed once with HSA, and incubated on ice for 15 min in 0.2% Triton X-100. Total protein is determined by the method of Stein *et al.*[11] The standard 15 μl ASM assay mixture consists of 10 μl of protein source and 5 μl of a stock solution containing 2 nmol of B12SPM suspended in 0.1 *M* sodium acetate buffer, pH 5.2, containing 0.6% Triton X-100 and 5 m*M* EDTA. After incubating the assay mixture at 37° (up to 3 hr), the samples are loaded onto thin-layer chromatography plates (TLC LK6 D silica gel; Whatman, Clifton, NJ) and resolved using chloroform/methanol (95:5, v/v). After resolution, the band containing the fluorescently labeled ceramide (the product of B12SPM hydrolysis) is scraped from the plates, extracted in chloroform/methanol/water (1:2:1, v/v) for 15 min at 55°, and quantified

[11] S. Stein, P. Bohlen, J. Stone, W. Dairman, and S. Udenfriend, *Arch. Biochem. Biophys.* **155,** 202 (1973).

in a spectrofluorometer (fluorescence spectrophotometer 204-A, Perkin-Elmer, Norwalk, CT). The instrument settings are excitation, 505 nm, and emission, 530 nm.

cfu-S Assays

To confirm that these procedures have produced an enriched population of gene-corrected hematopoietic stem cells, an *in vivo* assay system for stem cells (cfu-S; colony-forming units–spleen) is used. Adult nucleated bone marrow cells are obtained from ASMKO mice that had been pretreated with 5-fluorouracil (FU) (150 mg/kg) 2 days before harvesting, retrovirally transduced with the ecotropic ASM/MFG vector, and "pulse-chase" labeled with B12SPM as described above, and then sorted by FACS. Cells representing the least 25% fluorescent in FL-1 ($B12^{low}$) are collected and 4×10^4 cells injected into the tail veins of lethally irradiated (800 cGy) adult ASMKO mice. For comparison, the same number of nonselected, transduced cells can be injected into another set of animals. After 14 days the mice are sacrificed, and the spleens removed and then fixed in a 70% solution of formalin : acetic acid : ethanol (1 : 1 : 20, v/v/v) for 3 days.

To prepare DNA from the cfu-S colonies, a modification of the method of Frank *et al.*[12] is used. The cfu-S colonies are separated and washed individually 3 times overnight in 1× TE (pH 8.2) at 4°, and then minced and incubated overnight again at 37° in a solution containing 50 m*M* Tris (pH 8.2) and 200 ng/μl proteinase K (Boehringer, Mannheim, Germany). The microcentrifuge tubes containing the digested materials are then immersed in boiling water for 8 min, and the extracted DNA placed on ice. For the ASM-specific polymerase chain reaction (PCR), the DNA solutions are diluted 1 : 100 and 40 μl is used. The PCR can be performed as previously described[8] using 40 cycles. A positive colony is defined as a colony in which the human ASM-specific PCR product is found in at least three independent amplification reactions.

Typical Results from Pulse-Chase Labeling of Normal, ASMKO, and Transduced Cells

Figures 1A and 1B show the B12 fluorescence from two cell populations of Lin^- bone marrow cells (enriched for stem and progenitors) from normal or ASMKO mice that have been incubated for 4 hr with a BODIPY-conjugated sphingomyelin (B12SPM), grown at 37° for 48 hr in standard culture media without B12SPM (i.e., the chase), and then labeled with

[12] T. Frank, S. Svoboda-Newman, and E. D. Hsi, *Diagn. Mol. Pathol.* **5,** 220 (1996).

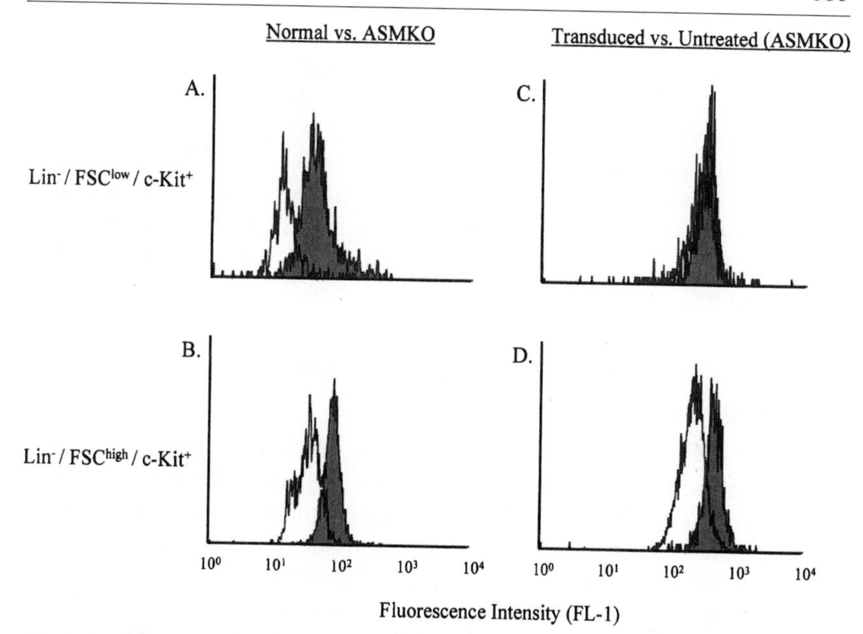

FIG. 1. Analysis of B12SPM labeling of hematopoietic stem- and progenitor-enriched cell populations. (A and C) $Lin^-/FSC^{low}/c\text{-}Kit^+$ stem cell-enriched population; (B and D) $Lin^-/FSC^{high}/c\text{-}Kit^+$ progenitor cell-enriched population. (A and B) Shaded area: ASMKO cells; open area: normal cells. (C and D) Shaded area: nontransduced ASMKO cells; open area: transduced ASMKO cells. Note that FL-1 measures B12SPM. The experiment was repeated at least three times, and the representative datum from one experiment is shown.

a phycoerythrin-conjugated anti-c-Kit antibody. The $Lin^-/FSC^{low}/c\text{-}Kit^+$ population is highly enriched for stem cells, while the $Lin^-/FSC^{high}/c\text{-}Kit^+$ population is highly enriched for progenitor cells. Together, these two cell populations represent ~3% of the total mononuclear cells in the bone marrow.[13]

Comparison of the BODIPY fluorescence (FL-1) of normal and ASMKO bone marrow cells after labeling with B12SPM shows that in both the stem cell-enriched ($Lin^-/FSC^{low}/c\text{-}Kit^+$) and progenitor cell-enriched ($Lin^-/FSC^{high}/c\text{-}Kit^+$) populations, the normal cells are less fluorescent than those from ASMKO animals. These results demonstrate that stem and progenitor cells could internalize B12SPM and that both types of normal cells expressed a functional ASM.

When the same experiment is repeated on ASMKO cells that have

[13] S. Erlich, S. R. P. Miranda, J. W. M. Visser, A. Dagan, S. Gatt, and E. H. Schuchman, *Blood* **93,** 80 (1999).

been retrovirally transduced with an amphotropic MFG retroviral vector expressing human ASM (amphotropic vectors are capable of transducing a wide range of mammalian cell types, including human and murine cells), the results are shown in Figs. 1C and 1D. A significant shift to lower BODIPY fluorescence is observed only in the progenitor-enriched cells, while only a very minor shift is observed in the stem cell population. These results confirm previous studies[14,15] suggesting that murine progenitor cells (but not stem cells) can be efficiently transduced by amphotropic vectors.

Comparison of Retroviral Transduction Efficiencies Using Amphotropic vs. Ecotropic Viruses

The transduction efficiency can be improved using an ecotropic vector, as has been reported for murine hematopoietic stem cells using ecotropic vs. amphotropic retroviral vectors.[13,14] Ecotropic vectors, in contrast to amphotropic vectors, are specific for mouse cells. As shown in Fig. 2, using the B12SPM selection technique, the transduction efficiencies of Lin^-/FSC^{low}/c-Kit^+ stem cells with the amphotropic vs. ecotropic ASM/MFG vectors can be directly compared and the ecotropic vector led to a more than 50-fold increase in transduction in this stem cell-enriched population over that found with the amphotropic vector.

cfu-S Analysis of Sorted Cells

To ensure that these analytical techniques are truly identifying transduced hematopoietic stem cells, Lin^- ASMKO bone marrow cells are transduced with the ecotropic ASM/MFG vector. $B12^{low}$ cells are collected by flow cytometry, and then transplanted into lethally irradiated adult ASMKO mice. Fourteen days later the spleens are harvested and the number of colony-forming units (cfu-S) positive for the retroviral vector are determined by PCR. As shown in Table I, in the absence of selection 32% of the cfu-S colonies were vector positive, while among animals transplanted with B12SPM-selected cells, 76% of the colonies were vector positive. Thus, these results demonstrated that a significant enrichment of transduced repopulating stem cells can be obtained using this procedure.

[14] D. Orlic, L. J. Girard, C. T. Jordan, S. M. Anderson, A. P. Cline, and D. M. Bodine, *Proc. Natl. Acad. Sci. U.S.A.* **93,** 11097 (1996).

[15] D. Orlic, L. J. Girard, S. M. Anderson, B. K. Do, N. E. Seidel, C. T. Hordan, and D. M. Bodine, *Stem Cells* **23,** (1997).

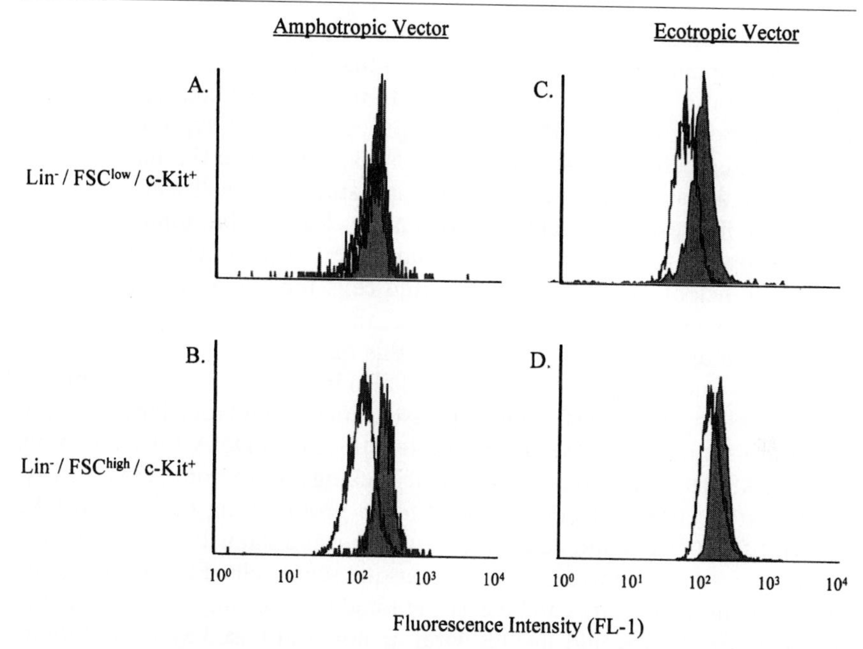

FIG. 2. Comparison of the transduction efficiencies using amphotropic vs. ecotropic retroviral vectors. (A and C) $Lin^-/FSC^{low}/c\text{-}Kit^+$ stem cell-enriched population; (B and D) $Lin^-/FSC^{high}/c\text{-}Kit^+$ progenitor cell-enriched population. The shaded areas represent nontransduced ASMKO cells, and the open areas depict transduced cells. Note that FL-1 measures B12SPM. The experiment was repeated at least three times, and the representative datum from one experiment is shown.

Comments

This is a novel, fluorescence-based method for the analysis of gene transfer into hematopoietic stem and progenitor cells. In our view there are two applications of this technology, one specific for the treatment of NPD, and the other more general. With regards to NPD, using these techniques, enriched populations of retrovirally transduced NPD hemato-

TABLE I
PCR ANALYSIS OF cfu-S COLONIES

Cell type injected	Total colonies analyzed	PCR-positive colonies
B12-selected	37	28 (76%)
Nonselected	28	9 (32%)

poietic stem and progenitor cells can be identified and collected. Such cells can be transplanted into ASMKO mice so that their engraftment potential and clinical usefulness can be compared to transduced hematopoietic cells that have not undergone the selection procedure. It is hypothesized that such selected cells may be clinically advantageous since the likelihood of engrafting a transduced, long-term repopulating stem cell will be greater due to the fluorescent enrichment procedure. If successful, similar methods can be applied to $CD34^+$ cells from human NPD patients. Although obtaining sufficient numbers of such sorted cells for human transplantation may be problematic at the present time, the development of high-speed flow cytometers promises to overcome this limitation.

This simple analytical system should also be useful as a general procedure to assess gene transfer and expression in any cell type. Three components are required: B12SPM, the full-length ASM cDNA for expression, and the ASMKO mice as a source of cells lacking endogenous ASM activity. Because the target cells do not need to be proliferating in order to take up B12SPM (Erlich and Schuchman, unpublished data, 1998), these techniques are amenable to HSCs and other quiescent cells. Furthermore, any gene transfer procedure can be easily studied by inserting the ASM cDNA as a marker. This would include viral or nonviral based systems. Importantly, the system relies on expression of the transferred gene as its final endpoint, and is fluorescence based, making it highly sensitive. Also of note, the corrected cells can be isolated from the noncorrected ones using a FACS, even when they represent 1% or less of the total cell population, and the properties of the two groups can be directly compared *in vitro* or *in vivo*. Finally, as compared to other fluorescence-based gene marking systems, such as those that utilize the green fluorescence protein, in our system the fluorescent compound is removed from the transduced cells, while in those systems the fluorescence is retained. In addition, the marker gene produces a mammalian enzyme, ASM, with no known deleterious effects.

Acknowledgments

This work was supported by research grants from the National Institutes of Health (HD 28607 and HD 32654), March of Dimes Birth Defects Foundation (1-2224), a grant (RR 0071) from the National Center for Research Resources for the Mount Sinai General Clinical Research Center, and a grant (93-00015) from the US–Israel Binational Science Foundation.

[27] Mammalian Ganglioside Sialidases: Preparation and Activity Assays

By ERHARD BIEBERICH, SEAN S. LIOUR, and ROBERT K. YU

Introduction

Sialic acid is the generic name of a family of acidic carbohydrates that consist of either *N*-acetyl- or *N*-glycolylneuraminic acid as the basic backbone. These carbohydrates occur in a variety of glycoconjugates, including glycoproteins, glycolipids, and oligosaccharides and are generally covalently bonded to other carbohydrate units in α2,3-, 2,6-, or 2,8-glycosidic linkages.[1] The enzymatic reaction giving rise to a sialylated glycoconjugate is catalyzed by a specific sialyltransferase using the activated nucleotide sugar CMP-sialic acid as a substrate. The degradation of the sialylated product occurs by hydrolytic cleavage of the sialic acid moiety catalyzed by sialidases, which are widely distributed among vertebrates, fungi, viruses, and bacteria.[2] (*Note:* Although the term *neuraminidase* has been widely used for the same enzyme, the term *sialidase* is preferred.) The generation of sialylated glycoconjugates is often the result of an interdependence between the reactions of sialyltransferases and sialidases that originate in the same or in different subcellular organelles. Glycolipids containing one or more sialic acid residues are defined as gangliosides and are synthesized by the sequential addition of carbohydrate units catalyzed by specific glycosyltransferases starting with *N*-acylated sphingosine (ceramide) as the precursor.

The biologic function of gangliosides is still under intensive investigation.[3,4] Aside from their functions in cell-to-cell recognition and adhesion, as well as in modulating the action of many bioeffectors, gangliosides have long been recognized as important antigens in neurologic disorders. Figure 1 presents the most common gangliosides and their biosynthetic pathways, and Fig. 2 shows the common pathways for their degradation. Although most of the glycosyltransferases participating in the sequential synthesis of gangliosides are well described and their cDNAs cloned, few advances have

[1] R. W. Ledeen and R. K. Yu, *Methods Enzymol.* **83,** 139 (1982).
[2] M. Saito and R. K. Yu, *in* "Biology of the Sialic Acids" (A. Rosenberg, ed.), p. 261. Plenum, New York, 1995.
[3] R. K. Yu and M. Saito, *in* "Neurobiology of Glycoconjugates" (R. U. Margolis and R. K. Margolis, eds.), p. 1. Plenum, New York, 1989.
[4] R. K. Yu, *Prog. Brain. Res.* **101,** 31 (1994).

0076-6879/00 $30.00

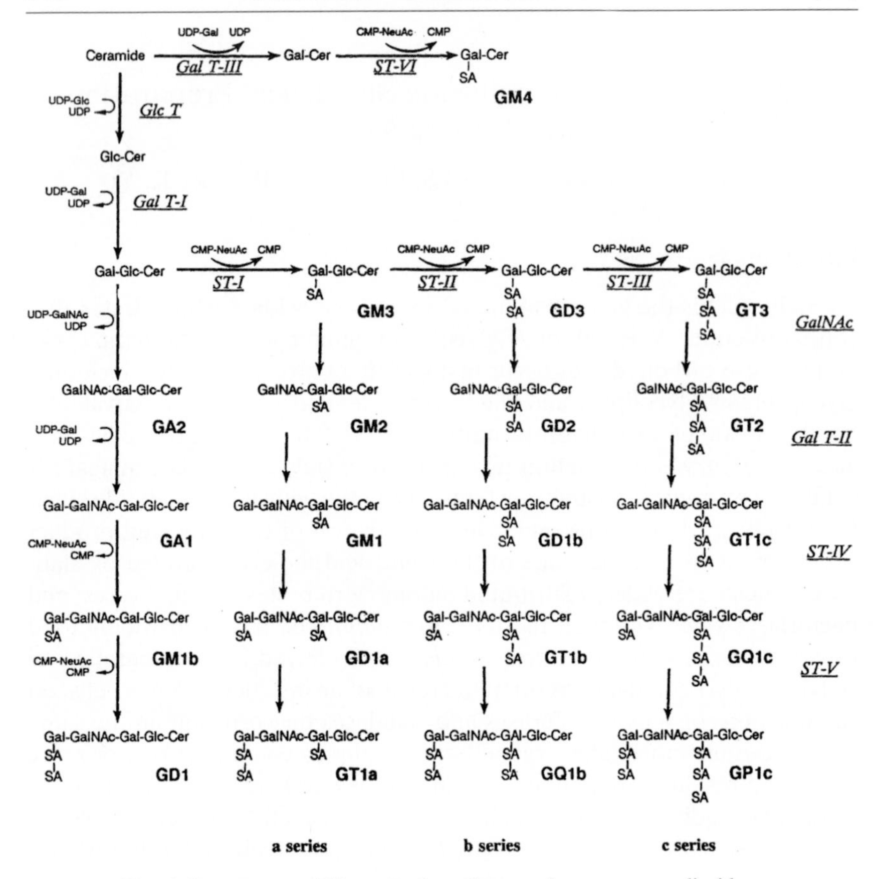

FIG. 1. Structures and biosynthetic pathways of common gangliosides.

been made regarding the sialidases acting on gangliosides in vertebrate species. It is known, however, that mammalian sialidases can be distinguished by their subcellular localization and substrate specificity. Table I summarizes the properties of most of the reported mammalian sialidases. In general, they can be divided into four groups with respect to their subcellular localization: cytosolic, lysosomal, plasma membrane bound, and nuclear. In each group particular species of sialidases can be distinguished by their kinetic properties or their nature as soluble or membrane-bound enzymes. Some sialidases (e.g., membrane-bound sialidase II) can occur in different subcellular compartments (e.g., plasma membranes or lysosomal membranes).

The soluble enzymes are localized in the cytosol or in the lumen of lysosomes and, can act on both gangliosides and sialoglycoproteins. The soluble cytosolic and lysosomal enzymes were the first mammalian sialidases

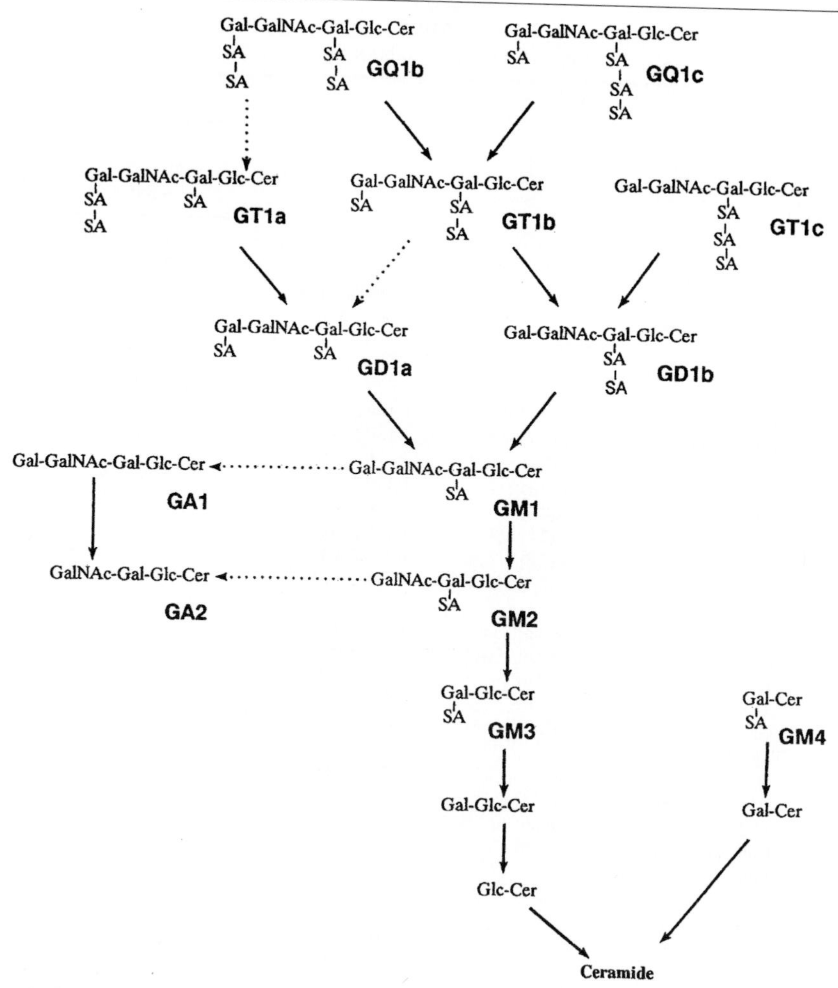

FIG. 2. The degradative pathways of gangliosides. The dotted arrows indicate minor pathways.

that were purified and their cDNAs cloned.[5–7] The membrane-bound enzymes appear to be of narrow specificity and act on only one type of glycoconjugates: gangliosides or sialoglycoproteins. More recently, two plasma membrane-bound enzymes were purified from bovine and human

[5] T. Miyagi, K. Konno, Y. Emori, H. Kawasaki, K. Suzuki, A. Yasui, and A. Tsuik, *J. Biol. Chem.* **268,** 26435 (1993).

[6] J. Ferrari, R. Harris, and T. G. Warner, *Glycobiology* **4,** 367 (1994).

[7] A. V. Pshezhetsky, C. Richard, L. Michaud, S. Igdoura, S. Wang, M. A. Elslinger, J. Qu, D. Leclerc, R. Gravel, L. Dallaire, and M. Potier, *Nature Genet.* **15,** 316 (1997).

TABLE I
PROPERTIES OF MAMMALIAN SIALIDASES[a]

Type			Cytosolic	
Localization:				
Species:			Rat	
Tissue:			Liver	Muscle
Molecular mass:			43 kDa	
Substrate specificity:	Linkage	Standard assay concentration[d]		
Gangliosides				
Mixture			62 (1.73 m*M*)	58
GM3 (N-Ac)	α(2,3)	0.1–0.2 m*M*	180	—
GM3 (N-Gly)	α(2,3)	0.1–0.2 m*M*	—	—
GM2	α(2,3)	0.1–0.2 m*M*	0	—
GM1	α(2,3)	0.1–0.2 m*M*	0	—
GD1a	α(2,3)	0.1–0.2 m*M*	71	—
GD3	α(2,8);α(2,3)	0.1–0.2 m*M*	—	—
GD1b	α(2,8);α(2,3)	0.1–0.2 m*M*	54	—
GT1b	α(2,8);α(2,3)	0.1–0.2 m*M*	—	—
Glycoproteins				
Fetuin	α(2,3) > α(2,6)	0.2–0.5 m*M*	33	50
Acid glycoprotein	α(2,6) > α(2,3)	0.2–0.5 m*M*	48	—
Transferrin	α(2,6)	0.2–0.5 m*M*	25	—
Submaxillary mucin	α(2,6)	0.2–0.5 m*M*	0	—
Glycophorin	α(2,3);α(2,6)	0.2–0.5 m*M*	—	—
Oligo-/Polysaccharides				
Sialyllactose (II^3, II^6)	α(2,3)α(2,6)	0.2–2.0 m*M*	—	100
II^3-Neu5AcLac	α(2,3)	0.2–2.0 m*M*	100 (1.67 m*M*)	
II^6-Neu5AcLac	α(2,6)	0.2–2.0 m*M*	15 (1.47 m*M*)	
Colominic acid	α(2,8)	0.2–2.0 m*M*	18	
Synthetic				
4MU-Neu5Ac	α(2,7)	0.1–0.5 m*M*	190 (670 μ*M*)	190
Detergent effect				
Triton X-100		0.05–0.2% (w/v)	Inh	—
Sodium cholate		0.05–0.2% (w/v)	Act	—
Octylglucoside		0.1–0.5% (w/v)	—	—
CHAPS		0.1–0.5% (w/v)	—	—
pH optimum			6.0–6.5	6.0
Inhibitors				
Neu5Ac-en		0.01–2.0 m*M*	—	—
Heparin			—	—
Heparan sulfate			—	—
Metal ions				
Ca, Mg, Mn		2.0 m*M*	Inh	—
Cu		2.0 m*M*	Inh	—
Reference			16,17	18

[a] Inh, inhibition; Act, activation; n.e., no effect.
[b] Exact intralysosomal localization is not clear.
[c] Membrane-bound sialidase II from rat brain lysosomes and synaptic membranes appear to be identical.
[d] Concentration in millimolar, except % in (w/v).

TABLE I (*continued*)

Lysomal				
Lumenal			Membrane-bound	
Human		Mouse	Human	Rat
Placenta	Fibroblast[b]	Brain[b]	Liver 70 kKa	Brain (II)[c]
—	—	—	271	—
441 (146 μM)	yes	—	476	164
—	—	—	252	97
57	—	—	—	84
59	—	—	33	16
197	14	5 (15 μM)	—	179
—	—	—	—	—
147	—	—	—	—
—	—	—	—	—
20	—	—	310	64
7	—	—	—	21
7	—	—	—	—
0	—	—	90	31
—	—	—	—	—
—	—	—	—	
100	—	—	100	100
36	—	—	33	—
57	—	—	100	—
3,000 (91 μM)	100	100 (42 μM)	—	95
Inh	—	Inh	—	—
Act	Act	—	—	—
—	—	—	—	—
—	—	—	—	—
3.8–5.0	4.0	4.0	4.0–4.5	4.0–4.4
—	—	—	—	—
—	—	—	—	—
—	—	—	—	—
—	—	—	—	—
—	no effect	—	—	—
19–21	12, 14	22	23, 24	25, 26

TABLE I

PROPERTIES OF MAMMALIAN SIALIDASES[a] (*continued*)

Type			Plasma membrane					
Localization:			Membrane-bound				Myelin-associated	
Species:			Rat		Human	Human	Rat	Mouse
Tissue:			Liver	Erythrocyte	Brain	Fibroblast	Brain	Brain
Molecular mass:					60 kDa			
Substrate specificity:	Linkage	Standard assay concentration[d]						
Gangliosides								
Mixture			—	—	—	—	—	—
GM3 (N-Ac)	α(2,3)	0.1–0.2 m*M*	100	100	100 (5.0 μ*M*)	100	100 1.0– (12 μ*M*)	100
GM3 (N-Gly)	α(2,3)	0.1–0.2 m*M*	—	—	—	—	—	—
GM2	α(2,3)	0.1–0.2 m*M*	—	—	<3	—	40	—
GM1	α(2,3)	0.1–0.2 m*M*	11	14	<3	—	4	2
GD1a	α(2,3)	0.1–0.2 m*M*	99	97	50	35	—	141
GD3	α(2,8);α(2,3)	0.1–0.2 m*M*	—	—	40	—	—	—
GD1b	α(2,8);α(2,3)	0.1–0.2 m*M*	—	—	33	—	—	—
GT1b	α(2.8);α(2,3)	0.1–0.2 m*M*	—	—	—	—	—	—
Glycoproteins								
Fetuin	α(2,3) > α(2,6)	0.2–0.5 m*M*	0	2	<3	—	46	—
Acid glycoprotein	α(2,6) > α(2,3)	0.2–0.5 m*M*	3	0	—	—	—	—
Transferrin	α(2,6)	0.2–0.5 m*M*	—	—	—	—	—	—
Submaxillary mucin	α(2,6)	0.2–0.5 m*M*	—	—	—	—	—	—
Glycophorin	α(2,3);α(2,6)	0.2–0.5 m*M*	—	—	—	—	—	—
Oligo-/Polysaccharides								
Sialyllactose (II^3, II^6)	α(2,3)α(2,6)	0.2–2.0 m*M*	3	3	—	—	—	—
II^3-Neu5AcLac	α(2,3)	0.2–2.0 m*M*	—	—	—	—	95	—
II^6-Neu5AcLac	α(2,6)	0.2–2.0 m*M*	—	—	—	—	—	—
Colominic acid	α(2,8)	0.2–2.0 m*M*	—	—	<3	—	—	—
Synthetic								
4MU-Neu5Ac	α(2,7)	0.1–0.5 m*M*	8	4	0	3	—	—
Detergent effect								
Triton X-100		0.05–0.2% (w/v)	—	—	Act	Act	effects <20%	—
Sodium cholate		0.05–0.2% (w/v)	—	—	Act	Act	—	—
Octylglucoside		0.1–0.5% (w/v)	—	—	Act	—	—	—
CHAPS		0.1–0.5% (w/v)	—	—	Act	—	—	—
pH optimum			4.5	4.5	5.0–5.0	4.5	3.6–5.9 (broad)	—
Inhibitors								
Neu5Ac-en		0.01–2.0 m*M*	—	—	10 μ*M*	—	1.3 μ*M*	—
Heparin			—	—	5 μg/ml	—	—	—
Heparan sulfate			—	—	14 μg/ml	—	—	—
Metal ions								
Ca, Mg, Mn		2.0 m*M*	—	—	—	—	n.e.	—
Cu		2.0 m*M*	—	—	—	Inh	Inh <25%	—
Reference			11,35	11	8	12–15	30,31	34

[a] Inh, inhibition; Act, activation; n.e., no effect.

[b] Exact intralysosomal localization is not clear.

[c] Membrane-bound sialidase II from rat brain lysosomes and synaptic membranes appear to be identical.

[d] Concentration in millimolar, except % in (w/v).

TABLE I (*continued*)

Plasma membrane					
Synaptosomal					Nuclear
Rat		Mouse	Bovine		Rat
Brain (I)	Brain (II)[c]	Brain	Brain		
70 kDa	70 kDa		52 kDa		
			With TX-100	With cholate	
—	—	—	70	68	—
100 (125 μM)	100 (22 μM)	100	25 (457 μM)	92 (142 μM)	100
58	59	—	11	43	—
17	51	—	3	4	—
7	10	<1	0	1	—
97	109	234	57	59	314
92	99	—	100 (219 μM)	100 (131 μM)	—
—	—	—	16	53	219
—	—		67	50	—
5	39	—	2	2	—
3	13	—	2	1	—
—	—	—	—	—	—
2	19	—	3	0	—
20	61	—	3	2	—
—	—	—	9	7	—
7	61	—	—	—	4
—	25	—	—	—	—
—	—	—	—	—	—
9	58	—	15 (265 μM)	13 (282 μM)	—
Inh	n.e.	Act	Act	—	—
—	—	—	—	Act	—
—	—	—	—	—	—
—	—	Act	—	—	—
5.0–5.5	4.5–5.0	—	4.6–4.8	4.6–4.8	—
—	—	—	—	—	—
—	—	—	—	—	—
—	—	—	—	—	—
n.e.	n.e.	—	n.e.	n.e.	—
Inh by 80%	Inh <20%	—	Inh	Inh	—
25,26	25,26	34	9	9	36

brain tissues and the cDNA of the bovine enzyme cloned.[8–10] These enzymes reveal a high preference for the gangliosides GD1a, GD3, and GM3 as substrates and show almost no activity toward sialoglycoproteins, sialo-oligosaccharides, or synthetic substrates (Table I). In the nervous system, plasma membrane-bound sialidases are mainly localized in the synaptosomal or myelin membrane, suggesting a potential role in synaptogenesis or myelinogenesis. Plasma membrane-bound sialidases are difficult to purify owing to their presence in very low concentrations and their instability once solubilized with detergents. Accordingly, investigation of their enzymatic activity depends critically on sensitive assay methods. In this article, several methods for the preparation of mammalian sialidases and determination of their enzymatic activity are described.

Methods

Enzyme Preparation and Activity Assays: General Remarks

Assays for sialidase activity can be divided into the use of nonradioactive and radioactive substrates. The classical nonradioactive assay is based on direct colorimetric determination of free sialic acid released from sialylated compounds by the periodate/thiobarbituric acid method. The advantage of this method arises from its versatility allowing use of a vast variety of sialylated compounds as substrates, including sialoglycoproteins, sialo-oligosaccharides, and gangliosides. However, the periodate/thiobarbituric acid method is intrinsically rather insensitive. When sensitivity is crucial, it is advisable to employ the more sensitive fluorometric method, based on the use of 4-methylumbelliferyl-*N*-acetylneuraminic acid (4-MU-SA) as a fluorogenic substrate. A limitation of the fluorometric method, however, is that it may not be feasible for plasma membrane-bound enzymes that require a high substrate specificity for gangliosides (see Table I). With plasma membrane-bound enzymes, the use of natural substrates, such as radioactively labeled gangliosides, is preferable. However, these radioactive substrates are not always commercially available. Thus, we will describe methods for radioactive labeling of gangliosides and sialyllactose. The various assay methods will be described in detail using enzyme samples of 50–100 μl containing 20–500 μg of protein solubilized from various tissues or cells. These methods have been tested in the authors' laboratory, and

[8] J. Kopitz, K. Sinz, R. Brossmer, and M. Cantz, *Eur. J. Biochem.* **248,** 527 (1997).

[9] K. Hata, T. Wada, A. Hasegawa, M. Kiso, and T. Miyagi, *J. Biochem.* **123,** 899 (1998).

[10] T. Miyagi, T. Wada, A. Iwamatsu, K. Hata, Y. Yoshikawa, S. Tokuyama, and M. J. Sawada, *J. Biol. Chem.* **274,** 5004 (1999).

several can be carried out with enzymes electrophoretically separated on polyacrylamide gels.

Preparation of Enzymes for Activity Assays

Enzyme Preparation from Cultivated Cells

Frozen or freshly prepared cells containing 2–4 mg of protein are first resuspended and homogenized in 1.0 ml of 1–5 m*M* potassium phosphate buffer, pH 6.5, containing 0.1 m*M* EDTA. Disruption of the cells is achieved by sonication in an ice-cooled sonicator bath with 5 pulses of 15 sec and intervals of 30 sec between each pulse. If intact subcellular fractions are to be preserved, the homogenization buffer should be supplemented with 0.32 *M* sucrose and homogenization with a glass or Teflon homogenizer is preferred. Preparation of membrane-bound sialidases may require the addition of detergent if a lipid substrate is used. In most cases, 0.2% Triton X-100 (final concentration) is sufficient for enzyme solubilization from cells or subcellular fractions. Preparation of subcellular fractions follows standard procedures as described elsewhere. (References for the preparation of sialidases from rat erythrocytes or human fibroblasts and a brief description of their enzymatic properties can be found in Table I.[11–15]) The homogenized cells or subcellular fractions are incubated on ice for 15 min with an equal volume of 2-fold concentrated assay buffer without substrate prior to the assay reaction. Insoluble precipitates are then removed by centrifugation at 14,000*g* for 10 min at 4°. The supernatant is used for the initiation of the enzymatic reaction by the addition of a suitable substrate as described below.

Enzyme Preparation from Tissues

CYTOSOLIC SIALIDASES.[16–18] A frozen or freshly prepared tissue sample is homogenized in three volumes of 0.25 *M* sucrose containing 1 m*M* EDTA by the use of a glass–Teflon homogenizer. If the sample size is sufficiently large, a Waring blender running three cycles of 30 sec can be used. Crude

[11] J. Sagawa, T. Miyagi, and S. Tsuiki, *J. Biochem.* **107,** 452 (1990).
[12] M. Ziegler, V. Sury, and G. Bach, *Eur. J. Biochem.* **183,** 455 (1989).
[13] M. Lieser, E. Harms, H. Kern, G. Bach, and M. Cantz, *Biochem. J.* **260,** 69 (1989).
[14] V. Chigorno, G. Cardace, M. Pitto, S. Sonnino, R. Ghidoni, and G. Tettamanti, *Anal. Biochem.* **153,** 283 (1986).
[15] L. Caimi, A. Lombardo, A. Preti, U. Wiesmann, and G. Tettamanti, *Biochim. Biophys. Acta* **571,** 137 (1979).
[16] T. Miyagi and S. Tsuiki, *Eur. J. Biochem.* **141,** 75 (1984).
[17] T. Miyagi and S. Tsuiki, *J. Biol. Chem.* **260,** 6710 (1985).
[18] K. Dairaku, T. Miyagi, A. Wakui, and S. Tsuiki, *Biochem. Int.* **13,** 741 (1986).

cell debris is removed by centrifugation at 1000*g* for 10 min at 4°. Insoluble and membranous materials are removed from the remaining supernatant by centrifugation at 78,000*g* for 70 min at 4°. The resulting supernatant is used for determination of enzyme activity or further purification. In brief, the cytosolic sialidase from rat muscle or brain is first precipitated with ammonium sulfate at a concentration of 40–60% saturation. The ammonium sulfate precipitated fraction is redissolved, desalted, and then subjected to chromatography on a DEAE-cellulose column.[16–18] The cytosolic sialidase shows a broad specificity for gangliosides, synthetic substrates (such as 4-MU-SA), sialyllactose, and sialoglycoproteins. Among the various ganglioside substrates, the preferred substrates are GM3, GD1a, and GD1b (Table I).

LYSOSOMAL SIALIDASES.[12,14,19–25] The lysosomal enzyme is prepared by differential or density-gradient centrifugation starting with fresh tissues homogenized with three volumes of 0.32 *M* sucrose in 1–5 m*M* potassium phosphate buffer, pH 7.0, supplemented with 0.1 m*M* EDTA. Gentle homogenization with an ice-cooled glass or Teflon homogenizer should be performed to maintain the integrity of subcellular organelles. The homogenate is centrifuged at 1000*g* for 10 min at 4° and the pellet washed twice with four volumes of the homogenization buffer. The three supernatants are combined and centrifuged at 17,500*g* for 60 min at 4°. The enriched lysosomal fraction is then resuspended with the homogenization buffer (4 ml of final suspension/5 g of tissue) and aliquots of 4 ml/centrifugation tube layered onto 36 ml of 27% Percoll in the homogenization buffer. Centrifugation of the self-forming density gradient is carried out at 20,000*g* for 90 min in a fixed-angle rotor. A fraction of 6 ml containing a highly enriched lysosomal preparation is collected from the bottom of the tube and used for further enzyme purification or determination of enzyme activity. The soluble enzyme can be released from the lumen of lysosomes by hypotonic disruption of the membrane preparation. As with cytosolic sialidases, the lysosomal enzyme activity can be precipitated with ammonium sulfate. The membrane-bound sialidase is solubilized with a detergent,

[19] M. Hiraiwa, Y. Uda, S. Tsuji, T. Miyatake, B. M. Martin, M., Tayama, J. S. O'Brien, and Y. Kishimoto, *Biochem. Biophys. Res. Com.* **177,** 1211 (1991).

[20] M. Hiraiwa, Y. Uda, M. Nishizawa, and T. Miyatake, *J. Biochem.* **101,** 1273 (1987).

[21] M. Hiraiwa, M. Nishizawa, Y. Uda, T. Nakajima, and T. Miyatake, *J. Biochem.* **103,** 86 (1988).

[22] A. Fiorilli, B. Venerando, C. Siniscalo, E. Monti, R. Brescani, L. Caimi, A. Preti, and G. Tettamanti, *J. Neurochem.* **53,** 672 (1989).

[23] J. C. Michalski, A. P. Corfield, and R. Schauer, *Hoppe Seyler's Z. Physiol. Chem.* **363,** 1097 (1982).

[24] J. C. Michalski, A. P. Corfield, and R. Schauer, *Hoppe Seyler's Z. Physiol. Chem.* **367,** 715 (1986).

[25] T. Miyagi, J. Sagawa, K. Konno, S. Handa, and S. Tsuiki, *J. Biochem.* **107,** 787 (1990a).

e.g., 0.2% Triton X-100. The solubilized enzymes are distinguished from membrane-bound species by their high activity toward 4-MU-SA and inhibition by Triton X-100 (Table I). The intralysosomal localization of the lumenal enzymes, however, is very often insufficiently analyzed. The soluble sialidase from human placenta is associated in a complex with β-galactosidase, carboxypeptidase, and intralysosomal proteins. One species of lysosomal membrane-bound sialidase (membrane-bound sialidase II) appears to be identical to a membrane-bound enzyme found in synaptic vesicles. It is noteworthy that the membrane-bound enzyme from human liver lysosomes retains its activity after polyacrylamide gel electrophoresis (Table I).

PLASMA MEMBRANE-BOUND SIALIDASES.[8–15,24–35] These enzymes can be prepared from the particulate membrane fractions by differential centrifugation or from light membrane fractions enriched by density-gradient centrifugation from cultivated cells or tissue. Solubilization can be achieved by 0.5% (w/v) sodium deoxycholate supplemented with 0.1% (w/v) Triton X-100 or 0.5% CHAPS (3-[(3-cholamidopropyl)dimethylammonio]-1-propane-sulfonate) in 25 m*M* potassium phosphate buffer, pH 6.8, containing 1 m*M* EDTA. The plasma membrane-bound sialidase from brain tissues can also be obtained from a synaptosomal or a myelin membrane fraction (see below).

The isolation of synaptosomes from mouse brain can be achieved as follows. A mouse cerebrum is homogenized in four volumes of isolation buffer consisting of 0.32 *M* sucrose in 10 m*M* HEPES, pH 7.4, supplemented with 1 m*M* EDTA. After centrifugation at 1300*g* for 3 min, the resulting pellet is resuspended in the same buffer and centrifuged again. The two supernatants are combined and centrifuged at 16,700*g* for 10 min at 4°. The pellet is then homogenized in isolation buffer supplemented with 12% (w/w) Ficoll (final concentration) and transferred to an ultracentrifugation tube. The solution is carefully overlaid with 7.5% Ficoll in isolation buffer followed by a layer of buffer without Ficoll. The gradient is centrifuged at 99,000*g* for 60 min. Synaptosomes are recovered from the interface between

[26] T. Miyagi, J. Sagawa, K. Konno, and S. Tsuiki, *J. Biochem.* **107,** 794 (1990b).
[27] W. T. Norton and S. E. Poduslo, *J. Neurochem.* **21,** 749 (1973).
[28] J. E. Haley, F. G. Samuels, and R. W. Ledeen, *Cell. Mol. Neurobiol.* **1,** 175 (1981).
[29] H. C. Yohe, R. I. Jacobson, and R. K. Yu, *J. Neurosci. Res.* **9,** 401 (1983).
[30] M. Saito and R. K. Yu, *J. Neurochem.* **47,** 632 (1986).
[31] H. C. Yohe, M. Saito, R. W. Ledeen, T. Kunishita, J. R. Sclafani, and R. K. Yu, *J. Neurochem.* **46,** 623 (1986).
[32] M. Saito and R. K. Yu, *J. Neurochem.* **58,** 83 (1992).
[33] M. Saito and R. K. Yu, *J. Neurosci. Res.* **36,** 127 (1993).
[34] M. Saito, Y. Tanaka, C.-P. Tang, R. K. Yu, and S. Ando, *J. Neurosci. Res.* **40,** 401 (1995).
[35] T. Miyagi and S. Tsuiki, *FEBS Lett.* **206,** 223 (1986).

7.5 and 12% Ficoll, whereas myelin membranes are collected from the interface between 0 and 7.5% Ficoll. For further enrichment the synaptosomes are disrupted by homogenization in 1 m*M* sodium phosphate buffer, pH 7.5, supplemented with 0.1 m*M* EDTA. The homogenate is incubated on ice for 60 min and then centrifuged at 22,000*g* for 20 min at 4°. The pellet is resuspended in isolation buffer and layered onto a discontinuous gradient consisting of 0.6, 0.8, and 1.0 *M* sucrose in 10 m*M* HEPES buffer, pH 7.0. After centrifugation at 82,000*g* for 60 min at 4°, a synaptic plasma membrane fraction can be collected from the interphase between 0.8 and 1.0 *M* sucrose. Most plasma membrane-bound enzymes are specific for gangliosides and show only low activity toward sialoglycoproteins, sialyllactose, or 4-MU-SA. Synaptosomal sialidase II and myelin-associated sialidase (see next section) from rat brain, however, cleave sialyllactose and sialoglycoproteins to a considerable extent (Table I).

MYELIN-ASSOCIATED SIALIDASE.[27–34] Myelin is isolated from approximately 9–15 rat brains by centrifugation on three discontinuous sucrose gradients. The rat brains are homogenized in 20 volumes (w/v, 3–5 brains per 100 ml) of 0.32 *M* sucrose in distilled water by a glass or Teflon homogenizer running 5 strokes with a loose pestle and 5–7 strokes with a tight pestle. About 30–35 ml of the homogenate is layered onto 25 ml of 0.85 *M* sucrose and the gradient is centrifuged at 78,000*g* for 30 min. The myelin layer formed at the 0.32–0.85 *M* sucrose interface is collected with a Pasteur pipette and resuspended in about 60 ml of distilled water. The suspension is centifuged at 78,000*g* for 15 min at 4° and the pelleted myelin subjected to osmotic shock by dispersion in 60 ml of distilled water. After centrifugation at 12,000*g* for 10 min at 4°, the cloudy supernatant is discarded and the sedimented myelin subjected to a second osmotic shock and centrifuged as described. The resulting myelin pellets, equivalent to 3–5 rat brains, are combined and resuspended in 0.32 *M* sucrose and centrifuged on 0.85 *M* sucrose as described. The myelin fraction is again collected from the sucrose interface, diluted with distilled water and the dispersion centrifuged at 12,000*g* for 10 min at 4°. Further purification is achieved by "floating up" the myelin in a third sucrose gradient. This time, the pelleted myelin is resuspended in 0.85 *M* sucrose and overlaid with 0.32 *M* sucrose. After centrifugation at 78,000*g* for 15 min, the myelin is again collected from the sucrose interface, washed with distilled water, and lyophilized or further processed by delipidation. Freshly prepared or lyophilized myelin homogenized in 25 ml of distilled water is mixed with 250 ml of cold acetone at −20°. After stirring for 10 min, the mixture is centrifuged at 4300*g* for 10 min. The supernatant is discarded and the pellet extracted with 125 ml of cold ethanol as described for acetone extraction. The delipidated myelin is then homogenized with 25 ml of 1% Triton X-100 in 10 m*M* Tris-HCl buffer, pH 7.5, using a glass–Teflon homogenizer. The homogenate is centrifuged at 27,000g for 30 min at 4°. The supernatant is removed and the remaining pellet homogenized with 37.5 ml of 1% Triton X-100 in 1 *M* ammonium acetate, pH 7.5 (adjusted with NH_4OH). The solution is centrifuged at 27,000g for 30 min and the resulting supernatant dialyzed against 10 m*M* Tris-HCl buffer, pH 7.5, overnight in a cold room. The precipitate formed during dialysis is recovered by centrifugation and the resulting pellet resuspended in 25 ml of distilled water. The sialidase activity can be stabilized by dialysis against 10 m*M* Tris-HCl buffer. It should be mentioned that the myelin-associated sialidase shows a remarkable stability against heat inactivation as reported in the references given in Table I.

NUCLEAR SIALIDASE. A ganglioside sialidase has recently been found in nuclear membranes prepared from rat brain.[36] A rat brain tissue sample is homogenized in eight volumes of 1.3 *M* sucrose in 10 m*M* Tris-HCl buffer, pH 7.5, supplemented with 1 m*M* $MgCl_2$ and 1 m*M* phenylmethanesulfonyl fluoride (PMSF). Unbroken debris is removed by filtration through four layers of nylon stocking mesh and the filtrate centrifuged at 1000g for 20 min. The pellet is resuspended in 1.3 *M* sucrose-Tris buffer and mixed with 3.5 volumes of 2.4 *M* sucrose-Tris buffer. The mixture is then centrifuged at 100,000g for 60 min to precipitate the nuclei. The pellet is resuspended in 0.32 *M* sucrose-Tris buffer and overlaid on 2.2 *M* sucrose-Tris buffer. After centrifugation at 100,000g for 60 min, the nuclei are recovered in the pellet, with other subcellular fractions remaining at the interface between 0.32 and 2.2 *M* sucrose. The nuclei are incubated with 500 kU DNase I per brain for 1 hr at 4° in 0.32 *M* sucrose-Tris buffer supplemented with 1 m*M* PMSF. The incubation mixture is then overlaid on 2.2 *M* sucrose-Tris buffer and centrifuged at 100,000g for 60 min at 4°. Highly purified nuclear membranes are enriched at the 0.32–2.2 *M* sucrose interface. The nuclear enzyme cleaves GM3, GD1a, and GD1b, whereas sialyllactitol is only poorly accepted as a substrate.

Preparation of Substrates and Determination of Enzyme Activity

Nonradioactive Assays

DETERMINATION OF FREE SIALIC ACID WITH PERIODIC/THIOBARBITURIC ACID.[15,37,38] An enzyme preparation, containing 200–500 μg of protein in 100 μl of 100 m*M* sodium acetate buffer, pH 4.4–5.2 (or as indicated by the pH optimum of enzyme activity in Table I), optionally supplemented with 0.2% Triton X-100, is mixed with 100 μl of 4 m*M* sialyllactose, 0.1%

[36] M. Saito, C. L. Fronda, and R. K. Yu, *J. Neurochem.* **66,** 2205 (1995).
[37] L. Warren, *J. Biol. Chem.* **234,** 1971 (1959).
[38] G. Reutter and R. Schauer, *Methods Enzymol.* **230,** 168 (1994).

colominic acid, or 0.4 m*M* ganglioside in sodium acetate buffer. The reaction mixture is incubated for 2 hr at 37° in an incubation shaker. The enzyme reaction is terminated by quick freezing. For determination of the α2,8-sialidase activity with colominic acid, the assay sample is thawed and neutralized with 1 *N* NaOH. The mixture is then applied to a 4-mm id × 40-mm column of Bio-Rad (Hercules, CA) AG1-X8 (chloride from, 100–200 mesh) and the gel washed with 5 ml of distilled water. The bound sialic acid is eluted with 2.5 ml of 0.1 *M* NaCl and the eluate lyophilized. The residue is dissolved in 200 μl of distilled water and the free sialic acid determined by the periodate/thiobarbituric acid method. For the removal of interfering substances, the assay samples from the determination of sialidase activity toward sialyllactose, sialoglycoproteins, or gangliosides are mixed with 0.8 ml of 50 m*M* H_2SO_4. The mixture is applied to 500 μl of Dowex 2-X8 (chloride form) filled into a glass wool-plugged Pasteur pipette. The column is washed with 2.0 ml of distilled water and then eluted with 1.0 ml of 2 *M* sodium acetate/acetic acid buffer, pH 4.6. An aliquot of 200 μl of the eluate or the sample from the determination of enzyme activity toward colominic acid is mixed with 100 μl of 200 m*M* sodium (meta)periodate ($NaIO_4$) in 9 *M* H_2SO_4. The mixture is shaken and incubated for 20 min at room temperature. Then, 1.0 ml of 10% sodium arsenite ($NaAsO_2$) in 0.5 *M* Na_2SO_4, 0.1 *M* H_2SO_4 is added and the mixture vigorously vortexed until the brown color disappears. After addition of 3.0 ml of 0.6% thiobarbituric acid in 0.5 *M* Na_2SO_4, the solution is incubated for 15 min on a boiling water bath and then chilled on ice for 5 min. The red color developed during boiling may fade and the solution become cloudy. This change does not affect the final reading. The solution is extracted by thoroughly mixing it with 3.0 ml of cyclohexane followed by centrifugation at 500*g* for 5 min. The upper organic phase is removed for spectrophotometric reading at wavelengths of 549 and 532 nm. The molar extinction coefficient of the reaction product is 61,000. The amount of sialic acid is calculated according to

$$\mu\text{moles sialic acid} = 0.090\text{OD}_{549} - 0.033\text{OD}_{532}$$

FLUOROMETRIC ASSAY USING 4-METHYLUMBELLIFERYL-*N*-ACETYLNEURAMINIC ACID (4-MU-SA) AS A SUBSTRATE.[39–41] An enzyme preparation containing 50–150 μg of protein in 100 μl of 100 m*M* sodium acetate buffer or 4 m*M* phosphate–citrate buffer, pH 4.4–5.2 (see Table I for pH

[39] M. Potier, L. Mameli, M. Bélise, L. Dallaire, and S. B. Melançon, *Anal. Biochem.* **94,** 287 (1979).

[40] R. W. Myers, R. T. Lee, Y. C. Lee, G. H. Thomas, L. W. Reynolds, and Y. Uchida, *Anal. Biochem.* **101,** 166 (1980).

[41] W. Berg, G. Gutschker-Gdaniec, and R. Schauer, *Anal. Biochem.* **145,** 339 (1985).

optimum and optional detergent supplements), is supplemented with 200 μg of bovine serum albumin and then incubated with 0.2–1.0 m*M* 4-MU-Neu5Ac for 30 min to 1 hr at 37°. The reaction is terminated by the addition of 2.9 ml of 100 m*M* 2-amino-2-methyl-1-propanol-HCl buffer, pH 10.3, and the fluorescence immediately read with 1/5 to 1/10 of reaction mixture (excitation wavelength of 365 nm; emission wavelength of 440 nm). Standards are set up in the same buffer with 4-methylumbelliferone. The fluorogenic assay can also be used for visualization of sialidase in analytical polyacrylamide gels.

Assays Using Radiolabeled Substrates

PREPARATION OF TRITIUM-LABELED *N*-ACETYLNEURAMINE LACTITOL (NL[^{3}H]ol).[42] A solution of 231 μmol of NaB[^{3}H]$_4$ (7.29 mCi) in 2.0 ml of 0.01 *N* NaOH is added to 67 μmol of α2,3/2,6-*N*-acetylneuramin-lactose (NL) and incubated with occasional shaking in a stoppered test tube for 2.5 hr at room temperature. If radiolabeling of a specific sialyllactose isomer (α2,3 or 2,6) is desired, commercially available preparations of partially purified NL (e.g., from bovine colostrum grade I) must be chromatographically purified. After incubation, the mixture is supplemented with 20 mg of $NaBH_4$ in 0.5 ml of 0.01 *N* NaOH and the reaction allowed to proceed for another 4 hr. The reaction is stopped by decomposing the excess $NaBH_4$ by the dropwise addition of a total amount of 1.0 ml of acetone in the cold. The solution is then applied to a Bio-Rad AG50-X2 (H^+) column (9 × 70 mm) and the NL[^{3}H]ol preparation eluted with 60 ml of distilled water. The effluent is evaporated to dryness under reduced pressure at a temperature under 30°. Borate is completely removed by five cycles of repeated coevaporation with the addition of methanol. Further purification of [^{3}H]NL is achieved by removal of "acid stable" radioactive impurities by anion exchange column chromatography on Bio-Rad AG 1-X2 (formate form) or Bio-Rad AG 1-X8 (acetate form) in the cold. The column is washed with distilled water and then eluted with 10 m*M* pyridinium formate (AG 1-X2), or stepwise with 25, 50, and 100 m*M* pyridinium acetate, pH 5.0, in fractions of 20 ml. Fractions showing a high tritium count are pooled and lyophilized. The syrupy residue is dissolved in 1.0 ml of distilled water and passed through a Bio-Rad AG 50-X2 (H^+ form) as described. The effluents and washings are neutralized with 0.05 *N* NaOH and lyophilized. Bhavanandan *et al.*[42] reported a yield of 45.3 mg of the potassium salt of NL[^{3}H]ol with a specific radioactivity of 0.73 mCi/mmol. The purity of the product can be analyzed by high-performance thin-layer chromatography (HPTLC) on silica-coated glass plates using ethanol/*n*-butanol/pyridine/

[42] V. P. Bhavanandan, A. K. Veh, and R. Carubelli, *Anal. Biochem.* **69,** 385 (1975).

H_2O/HOAc (100:10:30:3, by volume) as the running solvent or by descending paper chromatography on 3 MM Whatman paper for 40 hr with ethyl acetate/pyridine/acetic acid/H_2O (5:5:1:3, by volume). Each NL[^{3}H]ol isomer should yield a single spot containing a total amount of more than 99% of the radioactivity.

ASSAY PROCEDURE.[30,36] An enzyme preparation containing 100–500 μg of protein in 100 μl of 100 m*M* sodium acetate buffer, pH 4.4–5.2 (see Table I for optimal pH and supplementation with detergents), is supplemented with 40–200 nmol of NL[^{3}H]ol (10^5 dpm) and incubated for 1 hr at 37°. The assay reaction is terminated by the addition of 900 μl of methanol. Uncleaved substrate is separated from tritium-labeled lactitol by application of the reaction mixture to a small column or glass wool-plugged Pasteur pipette filled with 500 μl of DEAE-Sephadex (acetate form) equilibrated with CH_3OH/H_2O (9:1, by volume). The reaction tube is washed twice with 250 μl of 90% methanol and the washings also applied to the column. The pass-through of 1.0 ml is combined with the effluent obtained with 2 times each of 0.5 ml of 90% methanol and collected into a scintillation vial for radioactivity counting. Alternatively, the enzyme reaction can be stopped by quick chilling on ice followed by application to a glass wool-plugged Pasteur pipette filled with 250 μl of Dowex AG 1-X2 (formate form, 200 mesh). The column is washed 3 times with 200 μl each of distilled water and the effluents collected for liquid scintillation counting.

RADIOLABELING AND ACTIVITY ASSAYS USING GANGLIOSIDES AS SUBSTRATES. *Tritium-labeling at C-3 position of sphingosine residue.*[43] A solution of 20 mg of gangliosides (e.g., GM3 or GD1a) in 10 ml of $CHCl_3$/CH_3OH (2:1, by volume) is mixed with 10 ml of 6% Triton X-100 prepared in the same solvent. The solvent is evaporated to dryness at 37° under vacuum. The residue is carefully dissolved in 10 ml of 3.6% 2,3-dichloro-5,6-dicyanobenzoquinone (DQQ) in toluene previously dehydrated with metallic sodium. The solution is continuously stirred at 37° for 40 hr in a screw-capped tube. The mixture is evaporated to dryness at 37° and the dark brown residue suspended in 10 ml of acetone and briefly sonicated in an ultrasonic bath. The gangliosides are insoluble and can be sedimented by centrifugation at 1000*g* for 10 min. The supernatant containing Triton X-100 and DQQ is discarded. The extraction procedure is repeated until a clear precipitate is obtained. The oxidized gangliosides are dissolved in $CHCl_3/CH_3OH/H_2O$ (60:35:7, by volume) and further purified by chromatography on a 700- × 40-mm column of silica gel, previously equilibrated with solvent. The elution pattern is monitored by HPTLC on silica gel-coated glass plates developed in the solvent system $CHCl_3/CH_3OH$/0.2% aqueous $CaCl_2$ (60:35:8, by volume). Gangliosides are visualized by expo-

[43] L. Svennerholm, *Biochim. Biophys. Acta* **24,** 604 (1957).

sure to iodine vapors or by staining with resorcinol–HCl reagent.[43] The fractions containing oxidized gangliosides are evaporated to dryness and the residue dissolved in 6 ml of *n*-propanol/H_2O (8:2, by volume). The reaction product (about 12 mg of oxidized gangliosides) is stable at 0°–4° for at least 6 months. An aliquot containing 1 mg of oxidized ganglioside is adjusted to a volume of 3.5 ml with 80% propanol and the mixture transferred to a vial containing 0.5 mg of solid NaB^3H_4. The solution is then transferred to a screw-capped tube and the walls of the vial washed twice with 0.3 ml each of 80% *n*-propanol. The combined solution and washings are stirred continuously at room temperature for 20 min. Complete reduction is achieved by the addition of 2 mg of unlabeled $NaBH_4$. The mixture is stirred for another 40 min at room temperature, and then 0.1 ml of 0.1 *N* acetic acid is added (the final pH is about 6). The solution is evaporated to dryness and the residue dissolved in 5 ml of water. After extensive dialysis against distilled water overnight the labeled ganglioside is further purified by chromatography on a silica gel column as described. The elution pattern is again monitored by HPTLC and liquid scintillation counting. The radioactive ganglioside-containing fractions are combined and evaporated to dryness. The residue is dissolved in 80% *n*-propanol, and should remain stable at 0°–4° for at least 6 months. Ghidoni *et al.*[44] reported a specific radioactivity of 1.25–1.28 Ci/mmol for tritium-labeled GD1a.

Tritium-labeling of double bond in sphingosine moiety.[31,45] Labeling of gangliosides (e.g., GD1a or GM3) can be performed in different solvent systems. An amount of 5–10 μmol of GD1a, which is more water soluble than GM3, in 0.8 ml of distilled water is prepared in a Teflon-lined screw cap and mixed with 0.1 ml 1 *M* NaOH and 50–100 μmol (2.7–5.4 mg) of KB^3H_4 (170 mCi to 13.3 Ci/mmol). The tube is flushed with a gentle stream of nitrogen and then 0.1 ml of 25 mg/ml $PdCl_2$ (14 μmol) is carefully layered over the ganglioside solution. The tube is immediately capped and the reaction mixture vigorously stirred for 3 hr at room temperature. Methanol (2–3 ml) is added and palladium precipitates removed by centrifugation at 500*g* for 10 min. The supernatant is adjusted to a pH of 5 by the addition of dilute acetic acid. In case of further precipitation of palladium, centrifugation is repeated and the supernatant evaporated to dryness under vacuum. Labile tritium is removed by repeated evaporation after dissolution of the residue in distilled water. Sodium ions are removed by chromatography of the aqueous solution of labeled gangliosides on 2 ml of Dowex 50-X2 (H^+ form) filled into a glass wool-plugged Pasteur pipette. The effluent and

[44] R. Ghidoni, S. Sonnino, M. Masserini, P. Orlando, and G. Tettamanti, *J. Lipid Res.* **22,** 1286 (1981).

[45] G. Schwarzmann, *Biochim. Biophys. Acta* **529,** 106 (1978).

washings are evaporated to dryness under vacuum, with toluene. Boric acid is removed either by dialysis or by repeated coevaporation from a solution of the labeled ganglioside in methanol. A more elegant method is to add mannitol to the ganglioside solution and the resulting mannitoborate complex and salts removed by reversed-phase silica gel column chromatography designed for organic solvents (e.g., SepPak, Waters Associates, Milford, MA).[46] This latter method eliminates the extensive dialysis or repeated methanol evaporation to remove borate, thus reducing time and the volume of radioactive waste.

For labeling of GM3, which is somewhat less soluble in water, the use of an organic solvent is advisable. The sample, 100 mg, is dissolved at 50° in 3 ml of tetrahydrofuran in a Teflon-lined screw-capped vial. After cooling down to room temperature, 6.75 mg of KB^3H_4 and 22.3 mg of $PdCl_2$ (125 μmol each) are added and the tube flushed with a gentle stream of nitrogen. Hydrogenation is initiated by the addition of 0.25 ml of 1 *M* NaOH and continued for several hours at room temperature. At the end of the reaction, $CHCl_3/CH_3OH$ (2:1, by volume) and a few drops of dilute acetic acid are added until the palladium precipitates. The palladium is filtered off and the labeled GM3 is further purified as described above for GD1a. Schwarzmann[45] reported a typical specific radioactivity of 74 Ci/mmol [^{3}H]GD1a and 385 Ci/mmol [^{3}H]GM3 with an overall yield of about 95%.

Assay procedures.[8,13,14,22,29,31,34] An enzyme preparation containing 10–200 μg of protein in 100 μl of sodium acetate buffer, pH 4.4–5.2 (see Table I for optimal pH), supplemented with 5–50 nmol of [^{3}H]GM3 (10^5 dpm) is incubated for 1hr at 37°. Triton X-100 or Triton CF-54 can be added to the incubation mixture at a final concentration of 0.1% (see Table I). The enzyme reaction is terminated by the addition of 2 ml of $CHCl_3/CH_3OH$ (2:1, v/v) and 0.4 ml of H_2O, and then subjected to Fölch's partition. Tritium-labeled lactosylceramide is recovered in the lower phase, while salts and large parts of the radioactive substrate are partitioned to the upper phase. Any unreacted substrate is separated from tritium-labeled lactosylceramide by application of the reaction mixture to a DEAE-Sephadex column (1.0 ml of the acetate from in a glass wool-plugged Pasteur pipette) equilibrated with the same solvent. The pass-through, two washings of the reaction tubes with 0.5 ml of solvent, and a further effluent obtained with 2.0 ml of solvent are collected into a scintillation vial and evaporated under a gentle stream of nitrogen. The residue is dissolved with 200 μl of methanol with brief sonication. After addition of 5 ml of Aquasol, the radioactivity is determined by liquid scintillation counting. If [^{3}H]GD1a is used as a substrate, the product has to be analyzed by HPTLC followed by autoradiography.

[46] H. C. Yohe, *J. Lipid Res.* **35,** 2100 (1994).

Radiolabeling of the sialic acid residue by enzymatic transfer.[35,47] Brains from 10 rats (12 days of age) are homogenized with 9 volumes of 0.25 *M* sucrose containing 0.1% mercaptoethanol with a Teflon-glass homogenizer. The homogenate is then centrifuged at 1000*g* for 10 min at 4°. The supernatant is removed and further centrifuged at 20,000*g* for 30 min at 4° and the resulting pellet resuspended in 9 volumes of 50 m*M* sodium cacodylate buffer, pH 7.0, containing 0.1% mercaptoethanol. The suspension is centrifuged at 40,000*g* for 30 min and the supernatant again centrifuged as described. The final pellet is resuspended in 4 volumes of 0.25 *M* sucrose with 0.1% mercaptoethanol, and the resulting suspension used for the transferase reaction.

For the sialytransferase reaction, an aliquot of 1.0 ml of the enzyme suspension containing approximately 31 mg of protein is adjusted to a final volume of 2.5 ml of a reaction mixture containing 100 m*M* sodium cacodylate buffer, pH 6.0, 1 m*M* lactosylceramide, 0.1 m*M* CMP-*N*-[^{14}C] acetylneuraminic acid (50 μCi), 0.4% Triton CF-54, 0.2% Tween 80. The reaction is incubated for 24 hr at 37° and then terminated by the addition of 50 ml of $CHCl_3/CH_3OH$ (2:1, by volume). The mixture is filtrated and evaporated to dryness under a gentle stream of nitrogen. The residue is dissolved in 2.5 ml of $CHCl_3/CH_3OH/H_2O$ (60:30:4.5, by volume) and applied to a Sephadex G-25 superfine column (10 $\times$ 200 mm) equilibrated with the same solvent. The column is eluted with 30 ml of the same solvent and the effluent evaporated to dryness. The residue is taken up in $CHCl_3/CH_3OH$ (2:1, by volume) and streaked onto two silica gel-precoated glass plates for HPTLC. The plates are developed in $CHCl_3/CH_3OH$/0.2% aqueous $CaCl_2$ (55:45:10, by volume) and scanned with a radioscanner for detection of radioactivity. Sphingolipids are visualized by exposure to iodine vapor. The radioactive band corresponding to GM3 is scraped from the plate and the ganglioside eluted with $CHCl_3/CH_3OH/H_2O$ (10:10:1, by volume). The solvent is evaporated and the radioactive product rechromatographed on Sephadex G-25 as described before. A recovery of 9×10^8 cpm [^{14}C]GM3 with a specific radioactivity of 217 mCi/mmol has been reported by Tallmann *et al.*[47]

Tritium labeling of sialic acid residue by oxidation/reduction.[48] GD1a (50 mg) is dissolved in 50 ml of 100 m*M* sodium acetate buffer, pH 5.5, containing 0.15 *M* NaCl. While stirring on an ice–water bath, 5 ml of ice cold 10 m*M* sodium (meta)periodate ($NaIO_4$) is added to the solution. After incubation for 10 min, the oxidation reaction is stopped by the addition of

[47] J. F. Tallman, P. H. Fishman, and R. C. Henneberry, *Arch. Biochem. Biophys.* **182,** 556 (1977).

[48] R. W. Veh, A. P. Corfield, M. Sander, and R. Schauer, *Biochim. Biophys. Acta* **486,** 145 (1977).

2.0 ml of glycerol. The reaction mixture is then extensively dialyzed against 50 m*M* sodium phosphate buffer, pH 7.4, followed by dialysis against distilled water in the cold room. The dialyzate is lyophilized and the residue used for reduction with $NaB[^3H]_4$. Lyophilizate (25 mg) is dissolved in 25 ml of 50 m*M* sodium phosphate buffer, pH 7.4, and supplemented with 2.5 ml of 100 m*M* $NaB[^3H]_4$ (25 mCi) in 0.01 *N* NaOH. The reaction mixture is incubated for 30 min at room temperature followed by the addition of 2.5 ml of unlabeled $NaBH_4$ freshly prepared in 0.01 *N* NaOH. The reaction is allowed to proceed for another 30 min at room temperature. Then, 1.0 ml of 0.5 *M* sodium acetate buffer, pH 5.5, is added and the mixture incubated for 5 min. The reaction mixture is applied to a hydrophobic Sep-Pak C_{18} column (7 × 300 mm) and the gel washed with 50 ml of 0.15 *M* NaCl followed by 50 ml of 50% CH_3OH. The radiolabeled ganglioside is then eluted with 15 ml of CH_3OH followed by 25 ml of $CHCl_3/CH_3OH$ (1 : 1, by volume). The two eluates are combined and evaporated to dryness under a gentle stream of nitrogen. A typical yield of 24 mg $[^3H]$GD1a is found with a specific radioactivity of 2.7×10^7 dpm/μmol (17.1 mCi/mmol). On condition of mild periodate oxidation as described, the major compounds are gangliosides containing tritium-labeled sialic acid and its derivative C_8-sialic acid. Both labeled compounds are accepted as sialidase substrates.

Assay procedure.[12,26,35,38] An enzyme preparation containing 10–100 μg of protein in 100 μl of 100 m*M* sodium acetate buffer, pH 4.4–5.2 (see Table I for the pH optimum), supplemented with 0.1 mg of bovine serum albumin and 5–10 nmol of ^3H- or ^{14}C-labeled ganglioside (10^4 dpm) is incubated for 1 hr at 37°. Optionally, Triton X-100 or CF-54 can be added to the reaction mixture at a final concentration of 0.1% (see Table I). After the enzyme reaction is finished, the assay mixture is chilled on ice and mixed with 100 μl of 0.15 *M* NaCl in distilled water. The reaction mixture is applied to 0.5 ml of Sep-Pak C_{18} in a glass wool-plugged Pasteur pipette. Free $[^3H]$sialic acid is then eluted with three times 0.6 ml of 0.15 *M* NaCl and mixed with 15 ml of Aquasol for liquid scintillation counting.

Acknowledgments

Studies from the authors' laboratory were supported by a USPHS grant NS11853 to RKY.

Section III

Sphingolipid–Protein Interactions and Cellular Targets

[28] Effects of Sphingosine and Other Sphingolipids on Protein Kinase C

By E. R. SMITH, ALFRED H. MERRILL, JR., LINA M. OBEID, and YUSUF A. HANNUN

Introduction

The generation of lipid second messengers that target lipid-activated kinases, such as protein kinase C (PKC), is one of the prominent events in transmembrane cellular signal transduction. In the case of PKC, at least 11 distinct isozymes have been characterized and are classified based on their lipid-activation requirements.[1,2] The classical or conventional PKCs ($\alpha, \beta_{\rm I}, \beta_{\rm II}, \gamma$), the first group identified, require 1,2-*sn*-diacylglycerol (DAG) and calcium for activation. The new or novel PKCs [δ, ε, η(L), and θ] respond to DAG but are not activated by calcium. Neither DAG nor calcium activate, and phorbol esters cannot down-regulate, the atypical PKCs (aPKC), which consist of two known isoforms, ζ and $\lambda(\iota)$. PKD, also known as PKCμ, may represent another subfamily.[3] All isozymes require the negatively charged phosphatidylserine (PS) for activation. Early studies indicated that PKC enzymes changed subcellular location on activation, typically translocating to the plasma membrane from the soluble cell fraction. The resultant high-affinity membrane–kinase interaction effectively activated the enzyme only in the presence of specific second messengers (e.g., DAG, calcium).[4] It is now believed that in addition to lipid–kinase interactions, specific protein–kinase interactions may localize individual isoforms to discrete intracellular sites where phosphorylation of substrates occur.[5]

Protein kinase C isozymes collectively have been found to be involved in cell growth and proliferation, cell survival, protein synthesis and transcription, and multiple other physiologic processes.[2] Discrete patterns and cell specificity of PKC isoforms may determine the specific function in cellular actions(s).[6] However, given the scope of the enzyme family in cells,

[1] H. Hug and T. F. Sarre, *Biochem J.* **291,** 329 (1993).
[2] Y. Nishizuka, *FASEB J.* **9,** 484 (1995).
[3] M. M. Chou, W. Hou, J. Johnson, L. K. Graham, M. H. Lee, C.-S. Chen, A. C. Newton, B. S. Schaffhausen, and A. Toker, *Curr. Biol.* **8,** 1069 (1998).
[4] A. C. Newton, *Curr. Biol.* **6,** 806 (1996).
[5] D. Mochly-Rosen, *Science* **268,** 247 (1995).
[6] L. V. Dekker and P. J. Parker, *Trends Biochem. Sci.* **19,** 73 (1994).

METHODS IN ENZYMOLOGY, VOL. 312
0076-6879/00 $30.00

it is perhaps not surprising that the paths of PKC intersect that of another group of pleiotropic cell modulators, the sphingolipids. Historically, sphingosine (D-*erythro*-4-*trans*-sphinganine) was first identified as an inhibitor of protein kinase C in *in vitro* assays[7,8] and has been used extensively as an inhibitor in cell culture studies to evaluate PKC-dependent or -related activities.[9–11] Sphingosine inhibits PKC by acting as a competitive inhibitor of activation by DAG, phorbol esters,[7,8] and for some isozymes, calcium, and blocks activation by unsaturated fatty acids and other lipids.[12] The *N*-methyl derivative (*N,N*-dimethylsphingosine)[13] and lysosphingolipids[14] have also been shown to inhibit PKC similarly to sphingosine. At least three reports indicate that endogenous levels of sphingosine and structurally related long-chain bases (e.g., sphinganine) can interfere with PKC activation[15–17]; thus the total level of sphingoid bases has been proposed to establish a set point for enzyme activation.[14] Furthermore, although it was not originally believed to modulate PKC, ceramide (*N*-acylsphingosine) is known to participate in numerous growth suppressive actions, cell death, senescence, and differentiation,[18–20] and has been found to activate PKC-ζ,[21,22] and indirectly inactivate other PKC isoforms.[23]

[7] Y. A. Hannun, C. R. Loomis, A. H. Merrill, and R. M. Bell, *J. Biol. Chem.* **261,** 12604 (1986).

[8] A. H. Merrill, Jr., S. Nimkar, D. Menaldino, Y. A. Hannun, C. Loomis, R. M. Bell, S. R. Tyagi, J. D. Lambeth, V. L. Stevens, R. Hunter, and D. C. Liotta, *Biochemistry* **28,** 3138 (1989).

[9] W. A. Khan, S. W. Mascarella, A. H. Lewin, C. D. Wyrick, F. I. Carroll, and Y. A. Hannun, *Biochem. J.* **278,** 387 (1991).

[10] A. H. Merrill, Jr., D. C. Liotta, and R. E. Riley, *in* "Handbook of Lipid Research: Lipid Second Messengers" (R. M. Bell, J. H. Exton, and S. M. Prescott, eds.), Vol. 8, p. 205. Alan R. Liss, New York, 1997.

[11] A. H. Merrill, Jr. and V. L. Stevens, *Biochem. Biophys. Acta* **1010,** 131 (1989).

[12] S. El Touny, W. Khan, and Y. A. Hannun, *J. Biol. Chem.* **265,** 16437 (1991).

[13] Y. Igarashi, S. Hakomori, T. Toyokuni, B. Dean, S. Fujita, M. Sugimoto, T. Ogawa, K. El-Ghendy, and E. Racker, *Biochemistry* **28,** 6796 (1989).

[14] Y. Hannun and R. M. Bell, *Science* **243,** 500 (1989).

[15] E. R. Smith, P. L. Jones, J. M. Boss, and A. H. Merrill, Jr., *J. Biol. Chem.* **272,** 5640 (1997).

[16] C. Rodriguez-Lafrasse, R. Rousson, S. Valla, P. Antignac, P. Louisot, and M. T. Vanier, *Biochem. J.* **325,** 787 (1997).

[17] C. Huang, M. Dickman, G. Henderson, and C. Jones, *Cancer Res.* **55,** 1655 (1995).

[18] L. M. Obeid, C. M. Linardic, L. A. Karolak, and Y. A. Hannun, *Science* **259,** 1769 (1993).

[19] R. Kolesnick and Z. Fuks, *J. Exp. Med.* **181,** 1949 (1995).

[20] S. Jayadev, B. Liu, A. E. Bielawska, J. Y. Lee, F. Nazaire, M. Y. Pushkareva, L. M. Obeid, and Y. A. Hannun, *J. Biol. Chem.* **270,** 2047 (1995).

[21] Y. M. Wang, M. L. Seibenhener, M. L. Vandenplas, and M. W. Wooten, *J. Neurosci. Res.* **55,** 293 (1999).

[22] G. Müller, M. Ayoub, P. Storz, J. Rennecke, D. Fabbro, and K. Pfizenmaier, *EMBO J.* **14,** 1961 (1995).

[23] J. Y. Lee, Y. A. Hannun, and L. M. Obeid, *J. Biol. Chem.* **271,** 13169 (1996).

For sphingolipids to function as effectors of PKC, dose-dependent inhibition or activation by the sphingolipid(s) must occur over the same range and under the same conditions as the biologic response being monitored. Specific assays of PKC activation, localization, phosphorylation, and lipid–kinase binding are all useful parameters. Nevertheless, studies that use sphingosine or other sphingolipids as pharmacologic tools for evaluating the actions or for investigating their roles as endogenous regulators of PKC are subject to certain cautions in monitoring the efficacy and specificity, delivery and concentration, and even handling of experimental samples. This chapter discusses some of these factors in assessing sphingolipids, specifically sphingosine and ceramide, as modulators of PKC.

Effectiveness and Specificity of Interaction

Structure and Stereospecificity

The value of sphingosine and ceramide as potential effectors of PKC in cells is modulated by structural and stereo considerations. These features can serve both experimental and control purposes.

The structural and stereospecificity of inhibition of PKC by sphingosine have been examined using multiple sphingosine analogs and long-chain amino bases.[8] Although the exact mechanism by which it inhibits PKC remains unknown, sphingosine is believed to localize in regions of acidic lipids (e.g., PS) and block enzyme binding or activity.[7,8,24] Inhibition requires a hydrophobic acyl chain, with a C_{18} sphingoid base optimal, and a positively charged amine. All stereoisomers of sphingosine inhibit with equal potency, and sphinganine, which lacks the 4,5-*trans* double bond in the long-chain base, is equally effective. These findings suggest, but do not prove, that nonspecific interactions account for the inhibitory effect.[7] Short-chain sphingosine analogs (e.g., C_{11}), which retain the free amine, do not inhibit PKC *in vitro* and in cells can be used as controls.[8]

Studies examining the role of ceramide as a lipid second messenger have shown both stereospecificity and structural selectivity. The amide-linked fatty acyl chain (C_{14} optimal) and the D-*erythro* isomer are necessary for mediating effects on cell growth, differentiation, and apoptosis. Moreover, the dihydro species (which lack the 4,5-*trans* double bond) are unable to mimic the biologic effects of ceramide on cell growth.[25,26] To our knowl-

[24] M. D. Bazzi and G. L. Nelsestuen, *Biochem. Biophys. Res. Commun.* **146,** 203 (1987).
[25] A. Bielawska, C. M. Linardic, and Y. A. Hannun, *J. Biol. Chem.* **267,** 18493 (1992).
[26] A. Bielawska, H. M. Crane, D. Liotta, L. M. Obeid, and Y. A. Hannun, *J. Biol. Chem.* **268,** 26226 (1993).

edge, no examination of these structural features has been explored on the role of ceramide binding and activation of PKC-ζ. Short-chain (C_2, C_6), C_{16}, and natural ceramides are equally potent and able to regulate PKC-ζ bifunctionally in U937 and PC12 cells,[21,22] with 5–25 n*M* causing maximum stimulation, and higher (>60 n*M*) concentrations resulting in less or no stimulation in U937 cells.[22] Nevertheless, C_6-dihydroceramide was not effective in causing inactivation of PKC-α, presumably indirectly via a ceramide-mediated protein phosphatase, in Molt-4 cells.[23]

Effective Concentration

The potency of sphingolipids in regulating PKC is primarily determined by their surface concentration.[27] Like other lipid molecules, this depends on their ability to partition into lipid bilayers, as well as the ratio of the sphingolipids relative to the total mass of lipids, which is directly related to the number of cells. In the absence of sphingolipid-binding proteins, sphingosine and natural long-chain ceramides will predominantly partition into cell membranes (>99%). Therefore, the molar concentration of sphingolipids in the incubation or culture medium may be misleading in predicting the effectiveness of the lipids in regulating PKC, if cell number or cell density is not considered. For example, by decreasing the number of cells or by increasing the concentration of sphingosine, the effective concentration of sphingosine in cell membranes will increase and, therefore, the dose range that apparently inhibits PKC will decrease. A concentration range between 1 and 10 μM per 10^6 cells is a useful starting point; however, key to these experiments is evaluating PKC responses over a range of concentrations, and for comparison between experiments, establishing and maintaining similar cell culture conditions.

Another major factor in appraising the effectiveness is the local concentration of the sphingolipids relative to PKC at specific membrane sites. This is especially critical in assessing whether endogenous sphingolipids may regulate PKC. Localization and translocation of PKC have been determined by cell fractionation,[28] immunofluorescence,[29,30] and electron microscopy.[31] Prior to activation, individual PKC isozymes appear to be restricted

[27] Y. A. Hannun, A. H. Merrill, Jr., and R. M. Bell, *Methods Enzymol.* **201,** 316 (1991).
[28] A. S. Kraft and W. B. Anderson, *Nature* **301,** 621 (1983).
[29] N. Shoji, P. R. Girard, G. J. Mazzei, W. R. Vogler, J. F. Kuo, *Biochem. Biophys. Res. Commun.* **135,** 1144 (1986).
[30] P. Sanchez, G. De Carcer, I. V. Sandoval, J. Moscat, and M. Diaz-Meco, *Mol. Cell. Biol.* **18,** 3069 (1998).
[31] M. Hagiwara, M. Sumi, N. Usuda, T. Nagata, and H. Hidako, *Arch. Biochem. Biophys.* **280,** 201 (1990).

to specific subcellular areas, and on activation, various isozymes translocate to specific sites within the same cell, which may include the plasma membrane, cytoskeletal elements, nuclei, and lysosomal-targeted endosomes.[5,30] This pattern of isozyme composition and distribution is cell type specific and should be determined for each cell or tissue, most easily by Western analysis using any of the many commercially available PKC isoform-specific antisera.

Determination of the sites of sphingosine and ceramide generation or the localized concentration of these lipids is more problematic. Exogenously supplied sphingolipids presumably localize initially in the plasma membrane, where they can potentially alter membrane-bound PKC activity or binding, then are metabolized or sorted to other membranes. The contribution of each action and over what time period obviously influences the overall regulation, and this most likely differs among tissues and cultured cell lines. Moreover, the intracellular locations of endogenous sphingosine and ceramide are not known with certainty and determinations are complicated by the fact that levels of these lipids are generally low in cells.[10]

There are clues as to where production of sphingosine and ceramide might take place. Both neutral and acidic sphingomyelin and ceramide hydrolases are present in cells, such that ceramide and sphingosine, respectively, could be produced at the plasma membrane and lysosome (or other acidic compartments). Generation of ceramide by agonist-stimulated turnover of sphingomyelin is believed to occur in the plasma membrane,[32,33] where it could potentially interact with PKC itself or PKC-regulatory proteins that are within or in proximity with membranes.[33] It is important to realize, however, that the lipids could move between intracellular membranes and/or concentrate in a specific area that differs from the site of production. For example, exogenously supplied fluorescent ceramides are sorted to and accumulate in the Golgi apparatus in many cells examined.[34–36] Whether the endogenously produced or exogenously supplied lipid could affect atypical PKC isoforms, PKC-ζ and PKC-λ/ι, which have been found anchored in lysosomal-targeted endosomes in HeLa cells,[30] is unclear. Thus, it is difficult to relate the results with exogenously added compounds to the same effects by endogenous sphingolipids even when their intracellular concentrations are known.[10]

[32] C. M. Linardic and Y. A. Hannun, *J. Biol. Chem.* **269,** 23530 (1994).
[33] Y. A. Hannun and C. M. Linardic, *Biochim. Biophys. Acta* **1154,** 223 (1993).
[34] N. G. Lipsky and R. E. Pagano, *J. Cell. Biol.* **100,** 27 (1985).
[35] N. G. Lipsky and R. E. Pagano, *Science* **228,** 745 (1985).
[36] N. G. Lipsky and R. E. Pagano, *Proc. Natl. Acad. Sci. U.S.A.* **80,** 2608 (1983).

Other Targets

The biologic targets of sphingosine have been found to be proteins involved in various signal transduction pathways and lipid metabolism[10,37] (reviewed elsewhere in this volume). The effects on PKC must be distinguished from those on other potential targets. For most of the isoforms regulated by DAG and/or phorbol ester, this has traditionally required comparing the dose responses of sphingosine inhibition of cellular PKC with the dose responses for the biologic response that is being analyzed. Inhibition by sphingosine is necessary, yet insufficient, to demonstrate that the response is mediated by PKC. Additional measurements must be used to verify PKC involvement. These include using other PKC-specific inhibitors to mimic the regulation (e.g., the isoquinoline sulfonamide derivative H7), nullifying the effects by specific activators (e.g., DAG and/or phorbol esters), using inactive structural analogs (described above), and losing the response following down-regulation of PKC by long-term application of phorbol esters.[38]

Proposed targets for ceramide include a ceramide-activated protein phosphatase[39,40] and protein kinase.[41,42] As with sphingosine, PKC-specific effects must be distinguished from other targets. Because atypical PKCs are neither translocated to the membrane fraction nor down-regulated by acute or long-term exposure to phorbol esters, phorbol ester binding is not useful for determining alterations in these isozymes. For dissecting the potential role of ceramide in regulating these PKCs, particularly PKC-ζ, immunne complex kinase assays and the autophorylation state of the enzyme can be used.[20,21,43] Other activators of the enzyme, including phosphatidylinositol 3,4,5-trisphosphate[42] and arachidonic acid,[21] should mimic, if not duplicate, the effects of ceramide on PKC activation.

[37] T. Megidish, J. Copper, L. Zhang, H. Fu, and S. Hakomori, *J. Biol. Chem.* **273,** 21834 (1998).

[38] S. Solanki, T. J. Slaga, M. Callaham, and E. Huberman, *Proc. Natl. Acad. Sci. U.S.A.* **78,** 1722 (1981).

[39] R. T. Dobrowsky and Y. A. Hannun, *J. Biol. Chem.* **267,** 5048 (1992).

[40] R. A. Wolff, R. T. Dobrowsky, A. Bielawska, L. A. Obeid, and Y. A. Hannun, *J. Biol. Chem.* **269,** 19605 (1994).

[41] T. Goldkorn, K. A. Dressler, J. Muindi, N. S. Radin, J. Mendelsohn, D. Menaldino, D. Liotta, and R. N. Kolesnick, *J. Biol. Chem.* **266,** 16092 (1991).

[42] S. Mathias, K. A. Dressler, and R. N. Kolesnick, *Proc. Natl. Acad. Sci. U.S.A.* **88,** 10009 (1991).

[43] H. Nakanishi, K. A. Brewer, and J. H. Exton, *J. Biol. Chem.* **268,** 13 (1993).

Use of Sphingolipids in Cell Systems

Method of Delivery

The amphilic nature of sphingosine and other sphingolipids results in nonspecific disruption of membranes, but toxicity at lower concentrations is due to disruption of cell signaling. The toxicity is dependent on the vehicle used to deliver the sphingoid base. In general this is due to three factors: the stability of the sphingoid base monomer (and possibly aggregates) in the aqueous phase, the partition coefficient for the sphingoid base in the membranes versus aqueous phase (and the amount of membrane lipid to accommodate the sphingoid base), and the presence of proteins and other surfaces that bind the sphingoid base (fatty acid-free albumin is often used to deliver sphingoid bases since it helps "buffer" the cells from too rapid exposure to high concentrations of the sphingoid bases.) The use of sphingosine in dimethyl sulfoxide (DMSO) solution can result in significant cytotoxicity if these considerations are not taken into account.[44] Ethanol solutions can be used for short-term cellular studies with sphingosine such as in the examination of platelet responses and neutrophil activation that occur over a few minutes. For long-term cellular experiments, sphingosine should be delivered to the cells as a 1:1 complex with serum albumin. The concentrations of stock solutions should be verified since sphingoid bases are hydroscopic and estimates of amounts based on weighing can be off by as much as 2-fold.

Many of the studies examining the effects of ceramide have used cell-permeable ceramide analogs, because natural long-chain ceramides are insoluble in aqueous tissue culture media and therefore are difficult to deliver. Permeable ceramides generally contain a C_2, C_6, or C_8 acyl group conjugated to the amino nitrogen of the sphingosine backbone. These analogs are first reconstituted as an ethanolic stock solution and dissolved in the culture media to the appropriate concentration. Although C_2-ceramide is usually effective, the authors have noted that stock solutions sometimes decompose during storage (perhaps due to the presence of acid from its synthesis); hence, this can give misleading results.

For both sphingosine and ceramide, the presence of sphingolipid binding proteins in the cell culture medium will act as a buffer to delay partitioning into cell membranes. For example, in the absence of serum proteins, nearly complete uptake of sphingosine occurs within minutes; this changes to

[44] J. D. Lambeth, D. N. Burnham, and S. R. Tyagi, *J. Biol. Chem.* **263,** 3818 (1988).

several hours and incomplete uptake in the presence of serum proteins. Albumin can also bind ceramides[45] and modulate their delivery.

Metabolism

Sphingosine and ceramide are intermediates in biosynthetic and degradative pathways of sphingolipids. When added to cells, both can be metabolized to more complex sphingolipids or degraded, and the extent of metabolism differs among cell types. Sphingosine is readily taken up and metabolized to sphingosine phosphate in platelets[27] and macrophages,[46] whereas in HL-60 cells, sphingosine is mainly incorporated into *N*-acyl derivatives or complex sphingolipids,[47] with a half-life of several (~6) hours. Therefore, it may be necessary to add sphingosine repetitively to tissue culture cells to optimize inhibition of PKC, but the intervals range from minutes to greater than 12–24 hr. Short-chain ceramides appear to be metabolized much more slowly in Molt-4,[48] H29rev,[49] U937,[22] and Caco-2 cells[50] in culture, such that, depending on the period of examination, daily addition may not be as necessary.

Manipulation of Experimental Systems

In addition to the mechanistic procedures discussed above, a few other considerations should be borne in mind when analyzing sphingolipids as modulators of PKC. These factors can be loosely described as "handling" effects.

First, many cultured cells undergo large increases in free long-chain bases and derivatives when culture medium is replenished. We and others have shown in J774A.1 macrophages,[46,51] Swiss 3T3 cells,[52] NIH 3T3 fibroblasts, A431, and NG108-5 cells[53] that changing cells from conditioned culture medium to fresh medium induces a transient "burst" in sphinganine

[45] H. Schulze, C. Michel, and G. van Echten-Deckert, *Methods Enzymol.* **311,** 22 (2000).

[46] E. R. Smith and A. H. Merrill, Jr., *J. Biol. Chem.* **270,** 18749 (1995).

[47] A. H. Merrill, Jr., A. Serini, V. L. Stevens, Y. A. Hannun, R. M. Bell, and J. M. Kinkade, Jr., *J. Biol. Chem.* **261,** 12610 (1986).

[48] G. S. Dbabo, M. Y. Pushkareva, S. Jayadev, J. K. Schwarz, J. M. Horowitz, L. M. Obeid, and Y. A. Hannun, *Proc. Natl. Acad. Sci. U.S.A.* **92,** 1347 (1995).

[49] R. J. Veldman, K. Klappe, D. Hoekstra, and J. W. Kok, *Biochem. J.* **331,** 563 (1998).

[50] E. R. Smith and V. L. Stevens, unpublished results (1999).

[51] L. A. Warden, D. S. Menaldino, T. Wilson, D. C. Liotta, E. R. Smith, and A. H. Merrill, Jr., *J. Biol. Chem.* **274,** 33875 (1999).

[52] J. J. Schroeder, H. M. Crane, J. Xia, D. C. Liotta, and A. H. Merrill, Jr., *J. Biol. Chem.* **269,** 3475 (1994).

[53] Y. Lavie, J. K., Blustzahn, and M. Liscovitch, *Biochim. Biophys. Acta* **1220,** 323 (1994).

and sphingosine, the latter by removing inhibitors of sphingolipid turnover.[51] The levels of free long-chain bases that are achieved within 30–60 min of incubation of J774A.1 macrophages are comparable with the levels of exogenous long-chain bases that inhibit PKC in human neutrophils.[54] These levels of sphinganine and sphingosine appear to alter the distribution and activity of the major PKC isoform (PKC-δ) in the macrophage cells, and the simple routine manipulation dramatically reduces the absolute amount of other cPKCs in the cells.[15] Additionally, the change to new medium stimulates *de novo* ceramide synthesis and increases the relative proportion of long-chain base 1-phosphates, thereby generating other potential bioactive lipids.[46] These pertubations may be responsible for anomalous cellular behavior and kinase activity that have been noted after changing culture medium or during cell lysate preparations for analysis of PKC.[22] Thus, it is necessary to determine that perturbations in PKC and the response being analyzed are not overtly due to changes in endogenous sphingolipid metabolism.

Secondly, modulation of free sphingolipid levels has been observed in human neutrophils by a number of factors, including some that are known to activate PKC. Free sphingosine increases in isolated human neutrophils in a time- and temperature-dependent manner, and serum, plasma, and low-density lipoproteins all augment the increases in free sphingosine.[55] Thus it is conceivable that serum lipids as well as the speed and agility in handling cells can influence the analysis of PKC activation. Moreover, other agents can decrease cellular sphingosine levels, including the PKC activators PMA (12-*O*-tetradecanoylphorbol 13-acetate) and arachidonic acid. PMA apparently increases the conversion of sphingosine to ceramide. Thus, some treatments may alter the overall state of PKC activation by removing an endogenous inhibitor of some isoforms and raising the levels of endogenous activators of other isoforms. To evaluate whether these changes influence PKC in a given system, mass levels of the endogenous sphingolipids should be compared to the activation status of the PKC enzymes within the cell.

Solutions

Ethanolic Sphingosine Stock

A concentrated ethanolic stock of sphingosine is first made by dissolving powdered sphingosine in absolute or 95% ethanol to a final concentration

[54] E. Wilson, M. C. Olcott, R. M. Bell, A. H. Merrill, Jr., and J. D. Lambeth, *J. Biol. Chem.* **261,** 12616 (1986).

[55] E. Wilson, E. Wang, R. E. Mullins, D. J. Uhlinger, D. C. Liotta, D. J. Lambeth, and A. H. Merrill, Jr., *J. Biol. Chem.* **263,** 9304 (1988).

of 100 m*M*. Heating the mixture at 37° or in hot water expedites the solubilization. At this concentration, sphingosine remains stable for several months when stored at −20° and longer stability is obtained by storing the solution under nitrogen or argon gas. The ethanol solution can be used directly to deliver sphingosine to cells or to prepare the serum albumin complex (see below). A similar method can be used to prepare ceramide stock solutions.

Sphingosine: BSA Conjugate

A 1 m*M* solution of fatty acid free bovine serum albumin (BSA) is prepared in phosphate-buffered saline (PBS) or other balanced salt solution. To prepare a 1–2 m*M* solution of sphingosine in albumin at a 1 : 1 or 2 : 1 molar ratio, an appropriate amount of sphingosine in ethanol is injected into the albumin using a glass syringe. The solution will initially turn cloudy but sphingosine will redissolve by incubating the solution in a shaking water bath at 37° for 1 hr or until the solution becomes clear. Solution is also aided by sonication in a bath-type sonicator. Stock solutions of the sphingosine : albumin conjugate can be stored at −20° for up to several months.

Monitoring PKC and PKC-Mediated Responses to Sphingolipids

As stated previously, dose-dependent inhibition or activation of PKC by the sphingolipid(s) should occur over the same range and under the same conditions as the biologic response being monitored. Some assays for monitoring PKC and PKC-related responses are described below.

Phorbol Dibutyrate Binding to Intact Cells

Since sphingosine is a competitive inhibitor of DAG and phorbol esters in the regulatory domain of PKC (the C_1-domain), decreased phorbol dibutyrate binding can be used to monitor inhibition by sphingosine[7,47] (either added exogenously or by an increase in endogenous sphinoid base amounts[15,16]). The effects of sphingosine on phorbol 12,13-dibutyrate (PDBu) binding to intact cells are examined under conditions identical to those in which the other cellular responses are measured. Buffers, cell numbers, cell conditions, and concentration of [^{3}H]PDBu should be the same. [^{3}H]PDBu is usually added to the medium near the dissociation constant of ~20 n*M* (~5×10^5 dpm/incubation). After incubation for 30 min at 37°, the medium is aspirated, and each dish washed quickly with ice-cold Krebs–Ringer phosphate-buffered saline (0.5 m*M* $MgCl_2 \cdot 6H_2O$, 0.12 *M* NaCl, 0.7 m*M* Na_2HPO_4, 15 m*M* $NaHCO_3$, 11 m*M* glucose). Cells are treated with 2% Triton X-100 (0.75 ml per dish) for 30–60 min at 37°, and the solubilized mixture is transferred to scintillation vials, 4 ml of scintillation cocktail is added, and the radioactivity determined by liquid scintillation counting. Specific binding is determined by the difference between total binding and binding in the presence of excess (>1 μM) unlabeled PDBu.

Effects of Enzyme Activators

Competitive activators should nullify the effect of sphingosine on PKC. Phorbol esters (PMA, PDBu) can be used, as well as the cell-permeable analogs of DAG, such as *sn*-1,2-dioctanoylglycerol (di-C_8) and 1-oleoyl-2-acetylglycerol (OAG).[9]

Protein Kinase C Activity

Determination of subcellular localization and enzymatic activity permits an assessment of the level of the PKC activation or inhibition. Translocation of the enzyme from the cytosolic to particulate fraction has traditionally been used to estimate activation; however, some isoforms in specific cells are localized to the particulate fraction and are not translocated on activation. Even in these cases, distribution can be altered on sphingosine or sphingoid base treatment, such that an increase in the soluble fraction reflects enzyme inhibition. For example, in J774A.1 macrophages, PKC-δ is the major isoform and is localized in the particulate fraction in resting cells. Treatment with exogenous sphinganine or an increase in endogenous long-chain bases causes an increase in the amount of PKC-δ in the cytosol.[15] An example of how this is assayed follows.

After treatment of the cell with the sphingolipid of interest and incubation for the desired time, the cells are washed quickly with 3 ml of ice-cold PBS (0.12 *M* NaCl, 2.7 m*M* KCl, 1.2 m*M* K_2HPO_4, 1 *M* NaH_2PO_4), scraped into ice-cold buffer A [20 m*M* Tris, pH 7.5, 1 m*M* phenylmethylsulfonyl fluoride (PMSF), 2 m*M* EDTA, 0.5 m*M* EGTA, 5 m*M* dithiothreitol (DTT), 50 μg/ml leupeptin, 20 μg/ml aprotinin], and treated with 4 m*M* diisopropyl fluorophosphate for 20 min on ice. Subcellular fractions are obtained by ultracentrifugation at 100,000*g* for 1 hr at 4°.[56,57] The particulate (membrane) fraction is solubilized in buffer A containing 1% Nonidet P-40 (NP-40) by rotating for 1.5 hr at 4°. Ultracentrifugation is repeated, and the solubilized membrane fraction is recovered as the supernatant. Membrane and cytosolic fractions are supplemented with glycerol to a final concentration of 15% (w/v) to prevent loss of PKC.

[56] C. Borner, S. N. Guadagno, D. Fabbro, and I. B. Weinstein, *J. Biol. Chem.* **267,** 12892 (1992).
[57] D. J. Uhlinger and D. K. Perry, *Biochem. Biophys. Res. Commun.* **187,** 940 (1992).

Before enzymatic assay, PKC is partially purified over DEAE-Sephacel columns. Membrane and cytosolic fractions are passed over individual 0.5-ml DEAE-Sephacel columns equilibrated with buffer B (20 m*M* Tris, pH 7.5, 1 m*M* PMSF, 0.5 m*M* EDTA, 0.5 m*M* EGTA, 0.5 m*M* DTT, 50 μg/ml leupeptin, 20 μg/ml aprotinin). The columns are washed with approximately 5 column volumes of buffer B, and the bound enzyme is eluted with 1–1.5 ml of buffer B containing 0.25 *M* NaCl and immediately assayed for activity. PKC activity is determined by measuring the incorporation of ^{32}P into a substrate protein, for example, myelin basic protein or histone III-S protein 1.[15,56] Calcium-independent activity is measured in the presence of DAG, PS, and 5 m*M* EGTA, whereas calcium-dependent activity is measured in the presence of 0.4 m*M* $CaCl_2$, PS, DAG, but without EGTA. Values are normalized to the amount of protein per sample.

Immune Complex Kinase Assay

Another procedure for analyzing the effects of sphingolipids on PKC measures activity in immunoprecipitated enzyme samples.[22,58] This is especially advantageous when it is desired to examine effects of sphingolipids on one PKC isoform. Cells are treated with the sphingolipid of interest, harvested into conical tubes using ice-cold PBS, and centrifuged. Lysis buffer consisting of 137 m*M* NaCl, 20 m*M* Tris, pH 8.0, 1 m*M* $MgCl_2$, 0.1 m*M* $CaCl_2$, 1 m*M* Na_4OV_3, 10% glycerol, 0.1% Triton X-100, 1 m*M* PMSF, 10 μg/ml leupeptin, and 5 μg/ml aprotinin is added to the cell pellet. The samples are sonicated for 5 sec and rotated for 30 min at 4° to release the membrane-bound proteins. The protein concentration is determined and normalized per volume. A polyclonal primary PKC antibody is added at a ratio of 4 μg of antibody per mg of lysate. Samples are rotated for 3 hr at 4°, after which 20 μl of goat anti-rabbit IgG-agarose beads (equilibrated to 50% in wash buffer) are added in lysis buffer that does not contain glycerol or Triton X-100. The samples are rotated for an additional 1.5 hr at 4°, then spun 4 min to pellet the beads. The beads are washed 5 times in wash buffer and 2 times in activity assay buffer. After washing, activity assays are performed in 40 μl of assay buffer containing 5 μl myelin basic protein (0.2 μg/μl), and 5 μl [γ-^{32}P]ATP (5 μCi of [γ-^{32}P]ATP and 10 μ*M* ATP). The reactions are incubated for 10 min at 37° and stopped by adding sodium dodecyl sulfate(SDS) sample buffer. Samples are boiled for 2 min

[58] M. M. Chou, W. Hou, J. Johnson, L. K. Graham, M. H. Lee, C.-S. Chen, A. C. Newton, B. S. Schaffhausen, and A. Toker, *Curr. Biol.* **8,** 1069 (1998).

and resolved by SDS–PAGE. The amount of ^{32}P incorporated into the substrate myelin basic protein is quantified by autoradiography.

Protein Substrate Phosphorylation

A direct measure of sphingosine or other sphingolipid modulation of PKC activation is an alteration in the ability of PKC to phosphorylate protein substrates. Although for atypical PKCs, the targets are not characterized, specific PKC substrates (e.g., 47-kDa proteins) are found in some cells. For these assays, cells are first incubated in phosphate-free medium, then labeled with ortho[^{32}P]phosphate at 0.2 mCi/ml for 75 min at 37° (used for platelets or other freshly isolated cells) or 0.05 mCi/ml for 4 hr (many cell lines in culture).[7,9,22,59] The cells are then washed free of radiolabel and treated with sphingolipids or under conditions that alter endogenous sphingolipid content. The reactions are stopped by lysing the cells in SDS sample buffer, and the phosphorylated proteins are analyzed by SDS–PAGE.

[59] V. L. Stevens, S. Nimkar, W. C. L. Jamison, D. C. Liotta, and A. H. Merrill, Jr., *Biochim. Biophys. Acta* **1051,** 37 (1990).

[29] Kinetic Analysis of Sphingoid Base Inhibition of Yeast Phosphatidate Phosphatase

By WEN-I. WU and GEORGE M. CARMAN

Introduction

During the past decade sphingolipids have been intensively studied because of their roles in cell signaling. Metabolites of sphingolipids, such as sphingoid bases, are particularly interesting because of their involvement in modulating multiple signaling pathways involved in protein kinases and phospholipases, and the intracellular levels of calcium, inositol trisphosphate, and cAMP.[1–3] One way to dissect the interactions between sphingoid bases and these metabolic pathways is to examine the direct effects of sphingoid bases on enzymes involved in the pathways through systematic kinetic analyses *in vitro.*

Phosphatidate phosphatase (3-*sn*-phosphatidate phosphohydrolase, EC 3.1.3.4) catalyzes the dephosphorylation of phosphatidate yielding diacyl-

[1] S. Spiegel and A. H. Merrill, *FASEB J.* **10,** 1388 (1996).
[2] M. A. Beaven, *Curr. Biol.* **6,** 798 (1996).
[3] Y. Igarashi, *J. Biochem.* (*Tokyo*) **122,** 1080 (1997).

0076-6879/00 $30.00

glycerol and P_i.[4] The enzyme plays an important role in regulating lipid synthesis in *Saccharomyces cerevisiae*[5,6] and in higher eukaryotic organisms.[7–9] Phosphatidate phosphatase is also involved in cell signaling mechanisms as part of the phospholipase D-phosphatidate phosphatase pathway for generating the lipid signaling molecule diacylglycerol.[10] The sphingoid bases sphingosine, phytosphingosine, and sphinganine have been shown to inhibit Mg^{2+}-dependent phosphatidate phosphatase from *S. cerevisiae*.[11] These positively charged lipids also inhibit phosphatidate phosphatase from higher eukaryotic cells.[12,13] In this chapter we discuss methods to study the kinetics of inhibition of the Mg^{2+}-dependent phosphatidate phosphatase from *S. cerevisiae* by sphingoid bases.

Preparation of Enzymes

The availability of purified preparations of phosphatidate phosphatase permits the examination of the effects of sphingoid bases on phosphatidate phosphatase activity under well-defined conditions. Two membrane-associated (45- and 104-kDa) forms of the Mg^{2+}-dependent phosphatidate phosphatase have been purified and characterized from *S. cerevisiae*.[14,15] Methods for the purification of these enzymes have been described in this series.[16,17]

Preparation of Radioactive Phosphatidate

[^{32}P]Phosphatidate is synthesized from diacylglycerol and [γ-^{32}P]ATP by the reaction of *Escherichia coli* diacylglycerol kinase[18] as described by

[4] S. W. Smith, S. B. Weiss, and E. P. Kennedy, *J. Biol. Chem.* **228,** 915 (1957).

[5] G. M. Carman and S. A. Henry, *Annu. Rev. Biochem.* **58,** 635 (1989).

[6] G. M. Carman and G. M. Zeimetz, *J. Biol. Chem.* **271,** 13293 (1996).

[7] D. N. Brindley, *Prog. Lipid Res.* **23,** 115 (1984).

[8] E. P. Kennedy, *in* "Lipids and Membranes: Past, Present and Future (J. A. F. Op den Kamp, B. Roelofsen, and K. W. A. Wirtz, eds.), pp. 171–206. Elsevier Science Publishers B.V., Amsterdam, 1986.

[9] T. Munnik, R. F. Irvine, and A. Musgrave, *Biochim. Biophys. Acta* **1389,** 222 (1998).

[10] J. H. Exton, *J. Biol. Chem.* **265,** 1 (1990).

[11] W.-I. Wu, Y.-P. Lin, E. Wang, A. H. Merrill, Jr., and G. M. Carman, *J. Biol. Chem.* **268,** 13830 (1993).

[12] T. J. Mullmann, M. I. Siegel, R. W. Egan, and M. M. Billah, *J. Biol. Chem.* **266,** 2013 (1991).

[13] Y. Lavie, O. Piterman, and M. Liscovitch, *FEBS Lett.* **277,** 7 (1990).

[14] Y.-P. Lin and G. M. Carman, *J. Biol. Chem.* **264,** 8641 (1989).

[15] K. R. Morlock, J. J. McLaughlin, Y.-P. Lin, and G. M. Carman, *J. Biol. Chem.* **266,** 3586 (1991).

[16] G. M. Carman and Y.-P. Lin, *Methods Enzymol.* **197,** 548 (1991).

[17] G. M. Carman and J. J. Quinlan, *Methods Enzymol.* **209,** 219 (1992).

[18] J. P. Walsh and R. M. Bell, *J. Biol. Chem.* **261,** 6239 (1986).

Morlock *et al.*[19] The reaction mixture contains 50 mM imidazole hydrochloride buffer (pH 6.6), 50 mM octyl-β-D-glucopyranoside, 0.25 mM diacylglycerol, 1 mM cardiolipin, 50 μCi [γ-^{32}P]ATP, 5 mM ATP, 50 mM NaCl, 12.5 mM $MgCl_2$, 1 mM EGTA, 10 mM 2-mercaptoethanol, and 0.04 unit of *E. coli* diacylglycerol kinase in a final volume of 0.1 ml. After incubation at 30° for 40 min, lipids are extracted from the reaction mixture by the addition of 0.5 ml acidic methanol (0.1 N HCl in methanol), 1 ml chloroform, and 1.5 ml 1 M $MgCl_2$. The chloroform phase is washed once with acidic methanol and 1 M $MgCl_2$. The radioactive phosphatidate is then purified by thin-layer chromatography (TLC) using the solvent system chloroform/methanol/water (65 : 25 : 4, v/v). The radioactive phosphatidate is extracted from the silica gel with chloroform/methanol/0.1 N HCl (1 : 2 : 0.8, v/v). Methanol and water are added to form a two-phase system. The choloroform phase containing the radioactive phosphatidate is adjusted to pH 7 with 1 N NH_4OH in methanol and then dried *in vacuo.* A Triton X-100/phosphatidate-mixed micelle solution (10 and 1 mM, respectively) is then added to the dried radioactive phosphatidate to a final radioactive specific activity of 10,000–20,000 cpm/nmol. The shelf-life of the radioactive substrate is about 3 weeks at −20°.

Preparation of Sphingoid Bases

Sphingoid bases (e.g., sphingosine, phytosphingosine, and sphinganine) can be purchased commercially as a dry powder. The dry sphingoid bases should be stored at −20° in a desiccated environment. We routinely prepare sphingoid bases solutions in a mixture of chloroform/methanol (9 : 1, v/v) just prior to each experiment.

Rational for Use of Triton X-100/Lipid-Mixed Micelles

The effects of sphingoid bases on phosphatidate phosphatase activity are examined using Triton X-100/lipid-mixed micelles. The nonionic detergent Triton X-100 is required to elicit a maximum turnover for phosphatidate phosphatase activity *in vitro.*[14,20] The function of Triton X-100 in the assay system is to form a uniform mixed micelle with the substrate phosphatidate.[20] The Triton X-100 micelle serves as a catalytically inert matrix in which phosphatidate is dispersed, preventing a high local concentration of phosphatidate at the active site.[20] In addition, this micelle system permits the analysis of phosphatidate phosphatase activity in an environment that

[19] K. R. Morlock, Y.-P. Lin, and G. M. Carman, *J. Bacteriol.* **170,** 3561 (1988).
[20] Y.-P. Lin and G. M. Carman, *J. Biol. Chem.* **265,** 166 (1990).

mimics the physiologic surface of the membrane.[21] In Triton X-100/phospholipid-mixed micelles, phosphatidate phosphatase activity follows surface dilution kinetics[21] where activity is dependent on both the bulk and surface concentrations of phosphatidate.[20] In experiments to examine the effects of sphingoid bases on phosphatidate phosphatase activity, the enzyme should be measured under conditions where activity is only dependent on the surface concentration of phosphatidate (i.e., molar concentrations of phosphatidate ≥ 0.1 m*M*).[20] The concentrations of phosphatidate and sphingoid bases are expressed as surface concentrations (in mol%) as opposed to bulk concentrations since these lipids form uniform mixed micelles with Triton X-100.[20,22]

Preparation of Triton X-100/Lipid-Mixed Micelles

Lipids (phosphatidate and sphingoid bases) in chloroform/methanol are transferred to a test tube, and solvent is removed *in vacuo* for 40 min. Triton X-100/lipid-mixed micelles are prepared by adding various amounts of a 5% (w/v) solution of Triton X-100 to the dried lipids. After the addition of Triton X-100, the mixture is vortexed. The surface concentration of lipids in mixed micelles is varied by the addition of Triton X-100. The total lipid concentration in Triton X-100/lipid-mixed micelles should not exceed 18 mol % to ensure that the structure of the mixed micelles is similar to the structure of pure Triton X-100.[23,24] The uniformity of the Triton X-100/lipid-mixed micelles is determined by gel filtration chromatography.[20] The mole percent of a lipid in a mixed micelle is calculated using the formula

$$\text{Mol\%}_{\text{lipid}} = [\text{lipid (bulk)}]/([\text{lipid (bulk)}] + [\text{Triton X-100}]) \times 100$$

Phosphatidate Phosphatase Assay

Phosphatidate phosphatase activity is measured by following the release of water-soluble [^{32}P]P$_i$ from chloroform-soluble [^{32}P]phosphatidate (20,000 cpm/nmol) at 30°.[16] The standard reaction mixture contains 50 m*M* Tris–maleate buffer (pH 7.0), 10 m*M* 2-mercaptoethanol, 2 m*M* $MgCl_2$, 1 m*M* Triton X-100, 0.1 m*M* phosphatidate, and an appropriate amount of pure enzyme in a total volume of 0.1 ml. The amount of enzyme used and the time of the reaction should be controlled so that activity is linear with time

[21] G. M. Carman, R. A. Deems, and E. A. Dennis, *J. Biol. Chem.* **270,** 18711 (1995).
[22] Y. A. Hannun, C. R. Loomis, A. H. Merrill, Jr., and R. M. Bell, *J. Biol. Chem.* **261,** 12604 (1986).
[23] D. Lichtenberg, R. J. Robson, and E. A. Dennis, *Biochim. Biophys. Acta* **737,** 285 (1983).
[24] R. J. Robson and E. A. Dennis, *Acct. Chem. Res.* **16,** 251 (1983).

and protein. One unit of phosphatidate phosphatase activity is defined as the amount of enzyme that catalyzes the formation of 1 μmol of product/min. Specific activity is defined as units/milligram of protein.

Analysis of Kinetic Data

Kinetic data are analyzed according to the Michaelis–Menten and Hill equations using the EZ-FIT Enzyme Kinetic Model Fitting computer program or a comparable program.

Inhibition of Phosphatidate Phosphatase Activity by Sphingoid Bases and Structural Requirements for Inhibition

The effects of sphingoid bases on phosphatidate phosphatase activities are examined using a subsaturating concentration (3 mol%) of phosphatidate.[11] By using a phosphatidate concentration near its K_m value for the phosphatases, the inhibitory effects of the sphingoid bases on activity can be observed more readily.[11] It is important to carry out a dose–response curve for the inhibition of activity by sphingoid bases. This information can be used for the design of kinetic-experiments to determine the mechanism of inhibition. For example, the activities of the 45- and 104-kDa forms of Mg^{2+}-dependent phosphatidate phosphatase are inhibited by sphingosine, phytosphingosine, and sphinganine in a dose-dependent manner between the concentrations of 0 and 15 mol%.[11]

The structural requirements for sphingoid base inhibition can be determined by examining the effects of compounds similar in structure to sphingoid bases on enzyme activity.[25] For example, stearylamine (a long-chain primary amine) and psychosine (sphingosine with a galactose on its hydroxyl group) inhibit the activities of the 45- and 104-kDa forms of Mg^{2+}-dependent phosphatidate phosphatase.[11] However, octylamine, oleic acid, and stearyl alcohol do not inhibit the phosphatidate phosphatases.[11] By conducting these experiments, it has been determined that a free amino group and a long-chain hydrocarbon are required for inhibition. The primary hydroxyl group on the sphingoid base does not appear to be critical for inhibition since both stearylamine and psychosine inhibit the phosphatidate phosphatase enzymes.[11] Psychosine is a less potent inhibitor of the phosphatases when compared with stearylamine and sphingoid bases.[11] This is presumably due to the bulky galactosyl group on the molecule, which could sterically hinder the interaction of the enzyme with the amino group on the sphingoid base molecule.[25]

[25] A. H. Merrill, Jr., S. Nimkar, D. Menaldino, Y. A. Hannun, C. Loomis, R. M. Bell, S. R. Tyagi, J. D. Lambeth, V. L. Stevens, R. Hunter, and D. C. Liotta, *Biochemistry* **28,** 3138 (1989).

Kinetic Analysis of Sphingoid Base Inhibition

A detailed kinetic analysis is performed to examine the mechanism of sphingoid base inhibition on phosphatidate phosphatase activity. The 104-kDa Mg^{2+}-dependent phosphatidate phosphatase has been used as a representative of both forms of the enzyme since the inhibition of both phosphatases by sphingoid bases is similar.[11] The phosphatidate dependence of

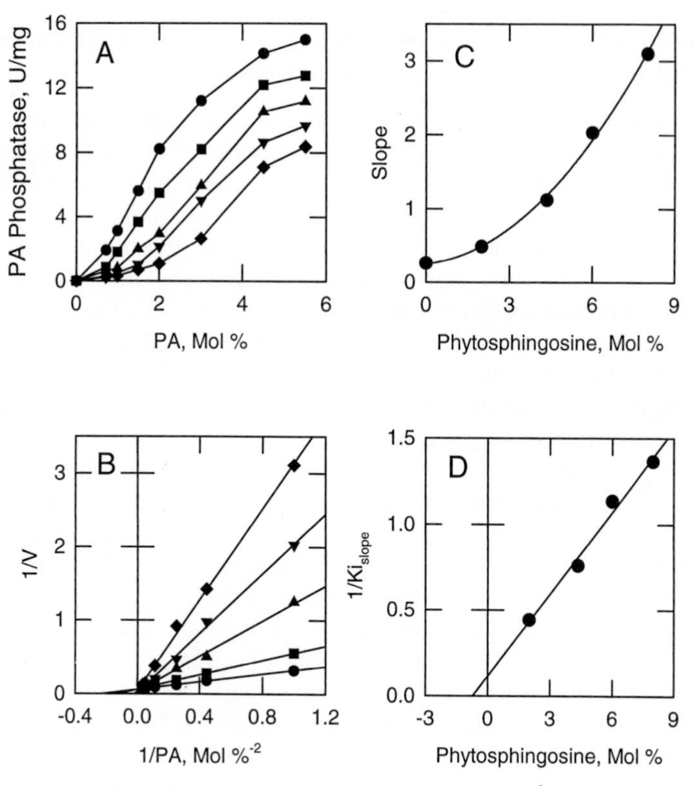

FIG. 1. Effect of phytosphingosine on the kinetics of 104-kDa Mg^{2+}-dependent phosphatidate phosphatase activity with respect to the surface concentration of phosphatidate. (A) 104-kDa phosphatidate phosphatase activity was measured as a function of the surface concentration (mol%) of phosphatidate (PA) (bulk concentration of 0.1 m*M*) at set surface concentrations (mol%) of phytosphingosine: 0 (●), 2 (■), 4.35 (▲), 6 (▼) 8 (◆). (B) Reciprocal plot of the data in part (A) where the phosphatidate concentration was raised to the Hill number of 2. (C) Replot of the slopes obtained from part (B) versus the phytosphingosine surface concentration. (D) Replot of the $1/K_i$ slope obtained from part (B) versus the phytosphingosine surface concentration. The lines drawn in (B) and (D) were a result of a least-squares analysis of the data. [Data from W.-I. Wu, Y.-P. Lin, E. Wang, A. H. Merrill, Jr., and G. M. Carman, *J. Biol. Chem.* **268,** 13830 (1993).]

TABLE I
KINETIC CONSTANTS FOR SPHINGOID BASES AND THEIR CELLULAR CONCENTRATIONS[a]

Sphingoid base	aK_i value (mol%)	Cellular concentration (mol%)
Phytosphingosine	0.4	0.16
Sphinganine	0.2	0.53
Sphingosine	1.5	ND[b]

[a] Data from W.-I. Wu, Y.-P. Lin, E. Wang, A. H. Merrill, Jr., and G. M. Carman, *J. Biol. Chem.* **268,** 13830 (1993).
[b] ND, not detected.

activity is examined in the absence and presence of several fixed concentrations of a sphingoid base. The concentrations of sphingoid base to be used in kinetic experiments is determined from dose–response curves of inhibition as discussed above. As an example, Fig. 1 shows the kinetics of inhibition by phytosphingosine, the yeast counterpart of sphingosine.[26] In the absence of phytosphingosine, the dependence of phosphatidate phosphatase activity on phosphatidate is cooperative with a Hill number of 1.92.[11] This indicates that the phosphatase has two phosphatidate binding sites. Phosphatidate phosphatase activity is inhibited by phytosphingosine in a dose-dependent manner at each phosphatidate concentration (Fig. 1A). Furthermore, the enzyme shows sigmoidal kinetics toward phosphatidate at each phytosphingosine concentration. The Hill number at each phytosphingosine concentration does not vary significantly from the value of 2. To examine the mechanism of inhibition, the data are transformed to a double reciprocal plot where the phosphatidate concentration is raised to the Hill number of 2.[27] Figure 1B shows that phytosphingosine does not affect the apparent V_{max} value for the enzyme but does cause an increase in the apparent K_m for phosphatidate. These results are consistent with phytosphingosine being a competitive inhibitor of phosphatidate phosphatase.[27] A replot of the slopes from the double reciprocal plot shown in Fig. 1B versus the phytosphingosine concentration results in a parabolic curve (Fig. 1C). This is indicative of parabolic competitive inhibition where two molecules of sphingoid base contribute to the exclusion of phosphatidate from the enzyme.[27]

[26] R. L. Lester and R. C. Dickson, *Adv. Lipid Res.* **26,** 253 (1993).
[27] I. H. Segel, "Enzyme Kinetics. Behavior and Analysis of Rapid Equilibrium and Steady-State Enzyme Systems." John Wiley and Sons, New York, 1975.

The K_i value for phytosphingosine cannot be determined from Fig. 1C because the curve is not linear. Furthermore, calculations of K_i values at each phytosphingosine concentration determined from the slopes of Fig. 1B are not true K_i values.[27] Because phytosphingosine is a parabolic competitive inhibitor, the K_i values calculated from Fig. 1B are actually K_i slope values.[27] K_i slope is not the K_i for the dissociation of an enzyme–inhibitor complex. Instead the K_i slope is a more complex function of K_i, which varies with the inhibitor concentration.[27] The K_i slope for each phytosphingosine concentration is calculated from the data in Fig. 1B using Eq. (1)[27]:

$$\text{Slope}[\text{I}] = \{1 + ([\text{I}]/K_i \text{ slope})\}\, K_m/V_{max} \tag{1}$$

where

$$K_i \text{ slope} = K_i^2/([\text{I}] + 2\,aK_i) \tag{2}$$

Equation (2) takes into account two inhibitor sites and the term a in the aK_i constant is an interaction factor for the inhibitor.[27] Equation (2) predicts that a plot of $1/K_i$ slope versus the inhibitor concentration should result in a straight line where the intercept of the inhibitor axis is equal to $-2aK_i$.[27] When $1/K_i$ slope is plotted versus the phytosphingosine concentration, a straight line is obtained (Fig. 1D). This analysis has also been conducted for sphingosine and sphinganine.[11] The aK_i values for phytosphingosine, sphingosine, and sphinganine have been determined from data such as that shown in Fig. 1D and are summarized in Table I.

Determination of the aK_i values for sphingoid bases is useful for comparison of the relative potency of sphingoid base inhibitors. Based on aK_i values, sphinganine is the most potent inhibitor of the phosphatase (Table I). Moreover, these inhibitor constants are useful for comparison with the cellular concentrations of sphingoid bases. The aK_i values for phytosphingosine and sphinganine are in the range of their cellular concentrations (Table I). Thus, phosphatidate phosphatase activity could be sensitive to changes in the phytosphingosine and/or sphinganine concentrations in the cell.

Acknowledgments

This work was supported in part by United States Public Health Service, National Institutes of Health grant GM-28140, New Jersey State funds, and the Charles and Johanna Busch Memorial Fund.

[30] Assays of Sphingosine-Dependent Kinase for 14-3-3 Protein

By TAMAR MEGIDISH, AKIKAZU HAMAGUCHI, KAZUHISA IWABUCHI, and SEN-ITIROH HAKOMORI

Introduction

A class of protein kinases termed *sphingosine-dependent kinases* (SDKs) is present in cytosolic and membrane fractions of various types of cells. Activities of SDKs are detectable only in the presence of sphingosine (Sph) or *N,N*-dimethyl-Sph (DMS), but not in the presence of ceramide, sphingosine 1-phosphate, diacylglycerol, or 14 other sphingolipids and phospholipids so far tested.[1] The SDK-activating effect of Sph and DMS is not due to their status as cationic amphiphiles, because other cationic amphiphiles such as stearylamine and hexadecyltrimethylammonium bromide did not activate SDKs.

One SDK, termed SDK1, specifically phosphorylates adapter/chaperone 14-3-3 protein, which is associated with a large variety of transducer molecules (some are listed in Table I) and modulates signal transduction. SDK1 is activated strongly by D-*erythro*-Sph and D-*erythro*-DMS, to a much lesser extent by L-*threo*-Sph and L-*threo*-DMS, and is unaffected by D-*erythro-N,N,N*-trimethyl-Sph.

Analysis of the phosphorylation site of 14-3-3 by SDK1 has been identified as Ser (S) next to Trp (W) in the common sequence as below, in three isoforms of 14-3-3[2]:

β (S60) YKNVVGARRS$\underline{\text{S}}$WRVISSIEQ
η (S59) YKNVVGARRS$\underline{\text{S}}$WRVISSIEQ
ζ (S58) YKNVVGARRS$\underline{\text{S}}$WRVVSSIEQ

Because this sequence is located at the longest helix (helix 3), which faces helix 1 of the counterpart 14-3-3 to form dimer, phosphate at S60 (for β isoform), S59 (η), or S58 (ζ) may interact with Arg-18 of helix 1. This Ser phosphorylation stabilizes the dimer, although dimer formation can occur without it.[2] Thus, SDK1-dependent phosphorylation may enhance interac-

[1] T. Megidish, T. White, K. Takio, K. Titani, Y. Igarashi, and S. Hakomori, *Biochem. Biophys. Res. Commun.* **216,** 739 (1995).
[2] T. Megidish, J. Cooper, L. Zheng, H. Fu and S. Hakomori, *J. Biol. Chem.* **278,** 21834 (1998).

0076-6879/00 $30.00

TABLE I
ASSOCIATION OF 14-3-3 WITH SIGNAL TRANSDUCERS, AND ITS MODULATORY EFFECT

Functional association with the key signaling molecules
PKC[6,7]; Raf[8–12]; cell cycle control[13] and cdc25 phosphatase[14]; PI3 kinase[15]; Bcr and Bcr-Abl[16]; KSR[17]; IGF-1 and insulin receptor substrate 1[18]
Significance of phosphorylation in 14-3-3
Raf association[19,20]; PKC inhibition[6]
Preferential association with phosphorylated targets
BAD[21]; tryptophan hydroxylase[22]; Cb1[23]; Raf-1 and Bcr[24]

tion of 14-3-3 with Raf, and perhaps with many other signal transducers as listed in Table I.

This chapter describes determination of SDK1 activity in mouse fibroblast BALB/c 3T3 A31 cells. We found that in order to observe good SDK1 activity, cells should be treated with 12-*O*-tetradecanoylphorbol 13-acetate (TPA) in serum-free conditions, followed by extraction and separation of SDK1 fraction by Q-Sepharose column chromatography. SDK1 activity is always found in cytosolic but not in membrane-bound fraction.

Cell Culture and Stimulation with TPA

Confluent fibroblasts treated with TPA under serum-free conditions are used for purification of SDK1 from cytosolic extracts, since intracellular Sph level is increased whereas PKC activity is reduced under these conditions.[3,4] Mouse BALB/c 3T3 A31 cells (American Type Culture Collection, Rockville, MD) are cultured in Dulbecco's modified Eagle's medium (DMEM) supplemented with 10% fetal bovine serum (FBS), in a humidified atmosphere containing 10% (v/v) CO_2 and 90% (v/v) air, at 37°. Number of passages is limited to five. Cells (5×10^5) are plated on 150-mm dishes, cultured for 6–8 days until confluence, washed 3× with serum-free DMEM, incubated in the same medium for 3 hr and then in DMEM containing 200 n*M* TPA in 0.01% dimethyl sulfoxide (DMSO) for 3–4 hr. Next, cells are washed with ice-cold phosphate-buffered saline (PBS) without Ca^{2+} and Mg^{2+}, harvested, and kept frozen at −80° until utilized for enzyme purification.

Separation of SDK1-Containing Fraction by Chromatography on Q-Sepharose

All procedures are carried out at 0–4°, and buffers are filtered through membrane and degassed under vacuum immediately prior to column chro-

[3] A. Rodriguez-Pena, and E. Rozengurt, *Biochem. Biophys. Res. Commun.* **120,** 1053 (1984).
[4] Y. Lavie, J. K. Blusztajn, and M. Liscovitch, *Biochim. Biophys. Acta* **1220,** 323 (1994).

matography. The following buffer systems are used. Buffer A: 20 m*M* Tris buffer (pH 8.5), 1 m*M* EDTA, 10 m*M* dithiothreitol (DTT), 10 m*M* NaF, and 0.1 mg/ml 4-(2-aminoethyl)-benzenesulfonyl fluoride (AEBSF). Buffer B: 20 m*M* Tris buffer (pH 8.5), 1 m*M* EDTA, 10 m*M* 2-mercaptoethanol. Homogenizing buffer: 20 m*M* Tris (pH 7.5), 0.5 m*M* EDTA, 0.5 m*M* EGTA, 12 m*M* NaF, 10 m*M* 2-mercaptoethanol, 10 μg/ml leupeptin, 10 μg/ml aprotinin, 10 μg/ml trypsin inhibitor, 10 μg/ml pepstatin, 0.5 mg/ml AEBSF, 5% glycerol.

Preparation of Cell Extract

Aliquots of packed frozen cells (2.5 ml) are thawed on ice and suspended in 45 ml homogenizing buffer. Homogenization is performed by 40 strokes in an ice-cooled, tight-fitting Dounce homogenizer. The homogenate is centrifuged at 500*g* for 10 min, and the supernatant centrifuged at 100,000*g* for 60 min. The pellet is saved as a source for other types of SDK (e.g., those that phosphorylate calreticulin or protein disulfide isomerase).[5] Since all SDK1 activity is present in the supernatant (cytosol), the cytosolic fraction is subjected to further separation of SDK1 fraction.

[5] T. Megidish, K. Takio, K. Titani, K. Iwabuchi, A. Hamaguchi, Y. Igarashi, and S. Hakomori, *Biochemistry,* **38,** 3369 (1999).
[6] A. Aitken, *Trends Biochem. Sci.* (*TIBS*) **20,** 95 (1995).
[7] N. Meller, Y.-C. Liu, T. L. Collins, N. Bonnefoy-Berard, G. Baier, N. Isakov, and A. Altman, *Mol. Cell. Biol.* **16,** 5782 (1996).
[8] E. Freed, M. Symons, S. G. MacDonald, F. McCormick, and R. Ruggieri, *Science* **265,** 1713 (1994).
[9] H. Fu, K. Xia, D. C. Pallas, C. Cui, K. Conroy, R. P. Narsimhan, H. Mamon, R. J. Collier, and T. M. Roberts, *Science* **266,** 126 (1994).
[10] B. Yamamori, S. Kuroda, K. Shimizu, K. Fukui, T. Ohtsuka, and Y. Takai, *J. Biol. Chem.* **270,** 11723 (1995).
[11] D. Morrison, *Science* **266,** 56 (1994).
[12] P. D. Burbelo and A. Hall, *Curr. Biol.* **5,** 95 (1995).
[13] J. C. Ford, F. Al-Khodairy, E. Fotou, K. S. Sheldrick, D. J. Griffiths, and A. M. Carr, *Science* **265,** 533 (1994).
[14] D. S. Conklin, K. Galaktionov, and D. Beach, *Proc. Natl. Acad. Sci. U.S.A.* **92,** 7892 (1995).
[15] N. Bonnefoy-Berard, Y.-C. Liu, M. von Willebrand, A. Sung, C. Elly, T. Mustelin, H. Yoshida, K. Ishizaka, and A. Altman, *Proc. Natl. Acad. Sci. U.S.A.* **92,** 10142 (1995).
[16] G. W. Reuther, H. Fu, L. D. Cripe, R. J. Collier, and A. M. Pendergast, *Science* **266,** 129 (1994).
[17] H. Xing, K. Kornfeld, and A. J. Muslin, *Curr. Biol.* **7,** 294 (1997).
[18] A. Craparo, R. Freund, and T. A. Gustafson, *J. Biol. Chem.* **272,** 11663 (1997).
[19] C. Rommel, G. Radziwill, J. Lovric, J. Noeldeke, T. Heinicke, D. Jones, A. Aitken, and K. Moelling, *Oncogene* **12,** 609 (1996).
[20] T. Dubois, C. Rommel, S. Howell, U. Steinhussen, Y. Soneji, N. Morrice, K. Moelling, and A. Aitken, *J. Biol. Chem.* **272,** 28882 (1997).
[21] J. Zha, H. Harada, E. Yang, J. Jockel, and S. J. Korsmeyer, *Cell* **87,** 619 (1996).

A

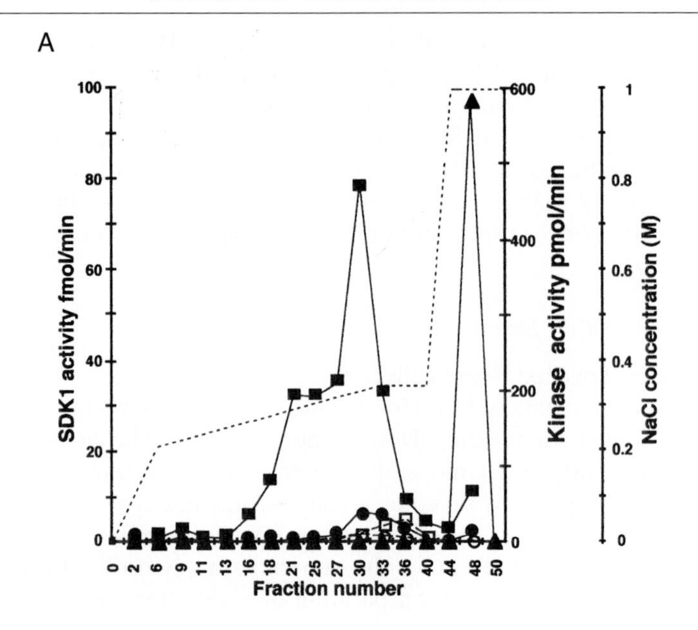

B

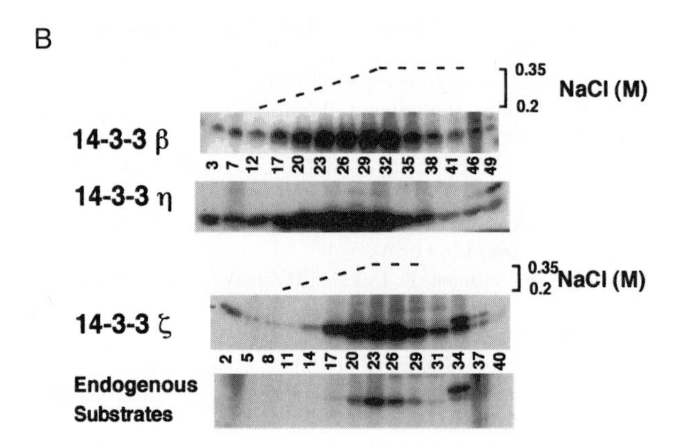

FIG. 1. Sequential purification and chromatographic separation of SDK1 from PKC and CKII. (A) Separation pattern of kinase activities for SDK1, PKC, and CKII in BALB/c 3T3 A31 cells through Q-Sepharose ion-exchange chromatography. Cytosolic fraction was loaded on Q-Sepharose column (1 × 10.5 cm), washed with buffer A, and eluted at a flow rate of 0.5 ml/min with 0–1 *M* NaCl with a combination of linear and stepwise gradients. Aliquots (5 μl) of fractions were analyzed for SDK1, PKC, and CKII activities. Left ordinate, SDK1 activity (fmol/min). Near right ordinate, PKC and CKII activities (pmol/min). Dotted line, change in NaCl concentration (*M*; far right ordinate) during gradient elution. ■, SDK1 activity tested with 120 n*M* (0.1 μg/30 μl) 14-3-3 β as substrate and 100 μ*M* DMS in 0.5% octyl-β-glucoside as lipid activator. □, SDK1 activity toward 28-kDa protein identified as endogenous 14-3-3, tested without addition of 14-3-3 substrate. ▲, CKII activity. ●, PKC activity tested in the

Q-Sepharose (QS) Column Chromatography

Cytosolic fraction (45 ml) is mixed with 2.5 ml 10% Triton X-100, agitated at 4° for 1 hr, immediately loaded on a Q-Sepharose column (1 × 10.5 cm) preequilibrated with buffer A, and washed with 4 volumes of buffer A. To separate SDK1 activity from other kinases, proteins are then eluted by a combination of stepwise changes and linear gradient changes of NaCl concentration (0–1.0 *M*), as indicated by dotted line in Fig. 1A. The linear gradients are 0–0.2 *M* NaCl for 15 min, 0.2–0.35 *M* for 85 min, constant at 0.35 *M* for 33 min, 0.35–1 *M* for 24 min, and finally constant at 1 *M* for 18 min. Elution is performed at a flow rate of 0.5 ml/min, and 1.5-ml fractions are collected and subjected to standard SDK1 assay. Fractions containing SDK1 activity are pooled and dialyzed 4–6 hr against 4 l buffer B.

Standard SDK1 Activity Assay

Sph and DMS both activate SDK1; however, DMS is used for detection of SDK1 activity since it cannot be metabolized to either ceramide or sphingosine 1-phosphate. SDK1 activity during purification is estimated by standard assay utilizing 0.1 μg 14-3-3 β and 50–100 μM DMS in 0.5% octyl-β-glucoside. SDK1 activity required Mg^{2+} with optimal concentration of 15 m*M* (Mn was ineffective), and the activity decreased as ionic strength increased. The total reaction volume of the standard assay system is 30 μl, i.e., 20 μl of substrate–kinase–sphingolipid mixture and 10 μl of ATP

[22] Y. Furukawa, N. Ikuta, S. Omata, T. Yamauchi, T. Isobe, and T. Ichimura, *Biochem. Biophys. Res. Commun.* **194,** 144 (1993).

[23] Y.-C. Liu, Y. Liu, C. Elly, H. Yoshida, S. Lipkowitz, and A. Altman, *J. Biol. Chem.* **272,** 9979 (1997).

[24] N. R., Michaud, J. R., Fabian, K. D., Mathes, and D. K. Morrison, *Mol. Cell. Biol.* **15,** 3390 (1995).

presence of MBP peptide substrate, PS/TPA, and Ca^{2+}. ○, PKC activity tested in the absence of PS/TPA and Ca^{2+}. Abscissa: selected chromatography fractions. (B) Elution profile of SDK1 activity from QS column, indicated by autoradiogram of SDS–PAGE gel. The cytosolic extract (step 1) was treated before fractionation with 0.5% octyl-β-glucoside instead of Triton X-100. The sample pH was increased to pH 8.5, and the sample was fractioned on a Q-Sepharose column equilibrated in buffer A containing 0.5% octyl-β-glucoside (pH 8.5). Phosphorylation by the indicated fractions was analyzed in the presence of 120 n*M* 14-3-3 β, η, and ζ, or without addition of exogenous 14-3-3 ("endogenous substrates"). Dotted lines shown above the gels indicate change in NaCl concentration (*M*) during elution.

solution. The mixture contains 5- or 10-μl kinase fraction eluted from column chromatography (containing 5.0–0.02 μg protein), 3 μl substrate solution containing 0.1 μg 14-3-3 in 50 m*M* Tris, 3 μl of 30 m*M* DTT, 6 or 1 μl of 50 m*M* Tris buffer, pH 7.5, and 3 μl of DMS (1 m*M* DMS in 5% octyl-β-glucoside) or vehicle (5% octyl-β-glucoside) alone. The reaction is initiated by addition of 10 μl of ATP solution consisting of 75 μM ATP, 45 m*M* magnesium acetate, and 2.5 μCi of [γ-^{32}P]ATP in 50 m*M* Tris-HCl, pH 7.5. The reaction mixtures are incubated at 30° for 15 min. The reaction is stopped by addition of 10 μl of 4× concentrated Laemmli's SDS–PAGE sample buffer and heating at 100° for 3 min. Phosphorylated proteins are resolved on 12% SDS–PAGE, and gels were stained with Coomassie Brilliant Blue R-250 (0.1%) in 50% methanol and 10% acetic acid, destained, dried, and subjected to autoradiography for 1–24 hr. Control assays are made using reaction without DMS, 14-3-3 substrate, or kinase. The 28-kDa band is excised and the amount of ^{32}P incorporated into the 14-3-3 (28-kDa) substrate determined by scintillation counting. SDK1 activity is calculated from the amount of ^{32}P incorporated into the substrate, taking into account the specific activity of the radioisotope and reaction time. Values obtained from gel without phosphorylated protein (blank) are subtracted from each determination.

The substrate specificity of SDK1 toward 14-3-3 isoforms, 14-3-3 mutants, and other SDK substrates (PDI, calreticulin) is tested in the presence of 120 n*M* protein.

Simplified SDK1 Activity Assay

SDK1 activity in cytosol fraction of cell homogenate or in fractions separated by Q-Sepharose column chromatography as above can be determined conveniently by the following simplified method. Add 10 μl of sample solution (either cytosol fraction or chromatography fraction) to a mixture of 3 μl substrate solution containing 0.1 μg 14-3-3, 3 μl of 30 m*M* DTT, 1 μl of 50 m*M* Tris buffer (pH 7.5), and 3 μl of DMS or Sph solution (1 mg/ml in 5% octylglucose) (total volume 10 μl). As a control, use 3 μl of 5% octylglucose without DMS or Sph.

To initiate the SDK1 reaction, the above mixture is added with 1 μl of 750 μM ATP, 1 μl of 450 m*M* $MgCl_2$, 7.75 μl of 50 m*M* Tris buffer (pH 7.5), and 0.25 μl of [γ-^{32}P]ATP (total volume 30 μl). Incubate at 37° for 15 min. To the reaction mixture, add 470 μl of radioimmunoprecipitation assay (RIPA) buffer (see composition below) and 4 μl of anti-14-3-3 antibody conjugated to agarose gel, and leave on rotary shaker at 4° for 3 hr or overnight. Centrifuge at 3000*g* for 5 min, discard supernatant, and wash anti-14-3-3 antibody-agarose gel 3× with RIPA buffer. Determine ^{32}P activity in gel with scintillation counter.

Anti-14-3-3 antibody-agarose gel is available from Santa Cruz Biotechnology (Santa Cruz, CA). Composition of RIPA buffer is 30 m*M* HEPES, pH 7.4, 150 m*M* NaCl, 1% Nonidet P-40, 0.5% sodium deoxycholate, 0.1% sodium dodecyl sulfate, 5 m*M* EDTA, 1 m*M* $NaVO_4$, 50 m*M* NaF, 1 m*M* PMSF, 10% pepstatin A, 10 μg/ml leupepsin, and 10 μg/ml aprotinin.

[31] Synthesis and Use of Caged Sphingolipids

By Roderick H. Scott, Jamie Pollock, Ahmet Ayar, Nicola M. Thatcher, and Uri Zehavi

Introduction

Sphingolipids (SLs) play important roles in intercellular and intracellular signaling and have been found to mobilize Ca^{2+} from intracellular stores in a variety of cell types including neonatal rat cultured sensory neurons; two examples from our work are presented in this article.[1,2] We have used the whole-cell recording variant of the patch-clamp technique to introduce SLs, including caged precursors and other compounds of interest, to the intracellular environment. We have then recorded Ca^{2+}-activated currents as a physiological index of a rise in intracellular Ca^{2+} following sphingolipid-evoked mobilization of Ca^{2+} from intracellular stores. According to Kaplan, "caged compounds" are photosensitive, normally inert molecules that fragment following irradiation with near-UV light yielding biologically active molecules.[3,4] Intracellular flash photolysis of caged photolabile compounds has proven useful in the study of a wide variety of intracellular signaling pathways. For example, cellular events activated by guanosine triphosphate (GTP) and GTP analogs (acting via G proteins), inositol 1,4,5-trisphosphate (IP_3), adenosine triphosphate (ATP), cGMP, cAMP, cyclic ADPribose, NO, H^+, and Ca^{2+} have been investigated using caged pre-

[1] A. Ayar, U. Zehavi, D. Trentham, and R. H. Scott, *J. Physiol.* **497,** 113P (1996).

[2] A. Ayar, N. M. Thatcher, U. Zehavi, D. R. Trentham, and R. H. Scott, *Acta Biochim. Polonica* **45,** 311 (1998).

[3] J. H. Kaplan, B. Forbush, and J. F. Hoffmann, *Biochemistry* **17,** 1929 (1978).

[4] J. E. T. Corrie and D. R. Trentham, *in* "Bioorganic Photochemistry, Vol. 2 Biochemical Applications of Photochemical Switches" (H. Morrison, ed.), p. 243. Wiley, New York, 1993.

METHODS IN ENZYMOLOGY, VOL. 312
0076-6879/00 $30.00

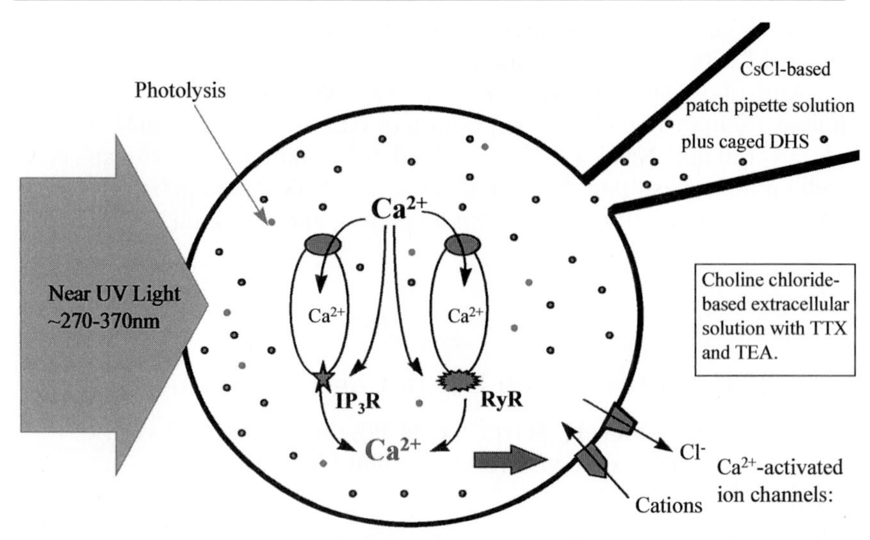

FIG. 1. Schematic diagram of experimental system, showing a cultured DRG neurone loaded, via the patch pipette solution, with caged DHS. Photolysis was achieved with intense near-UV light, and the photoreleased DHS mobilizes Ca^{2+} from intracellular stores, the rise in intracellular Ca^{2+} activates Ca^{2+}-activated currents, which can be recorded using the whole-cell variant of the patch-clamp method.

cursors or chelators.[5,6] Although not the only application of caged compounds, intracellular photorelease of molecules or ions of interest offers a number of benefits to researchers studying functional aspects of intracellular signaling. The properties of caged SLs that make them useful in the context of studying their potential roles as intracellular messengers include (1) rapid delivery of photoreleased sphingolipid close to possible sites of action; (2) protection of the species of sphingolipid from cellular metabolism prior to photolysis of the caged precursor; (3) potential for accurate kinetic studies because delays due to diffusion barriers can be minimized, and control and experimental data can be obtained from the same preparation; (4) intracellular concentration of photoreleased material can be estimated and varied by repeated photolysis or by varying the intensity and duration of irradiation; and (5) potential for application at the tissue, cellular, and subcellular levels. In this study we have combined intracellular flash photolysis of caged SLs with whole-cell patch-clamp electrophysiology (Fig. 1).

A few SLs, including glycosphingolipids (GSLs), were originally pro-

[5] A. M. Gurney *in* "Microelectrode Techniques. The Plymouth Workshop Handbook (D. Ogden, ed.), p. 389. The Company of Biologists Limited, Cambridge, 1994.

[6] G. Marriott, *Methods Enzymol.* **291,** 1 (1998).

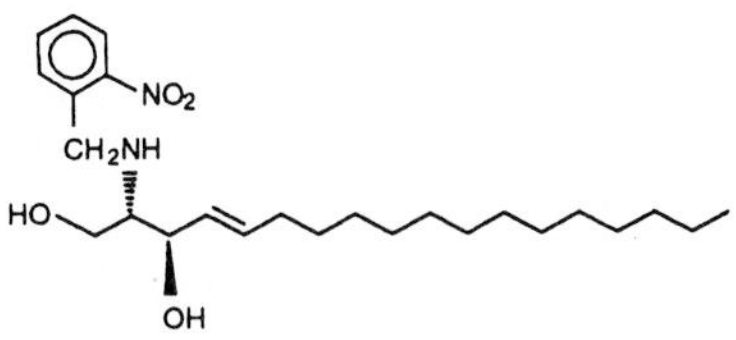

1

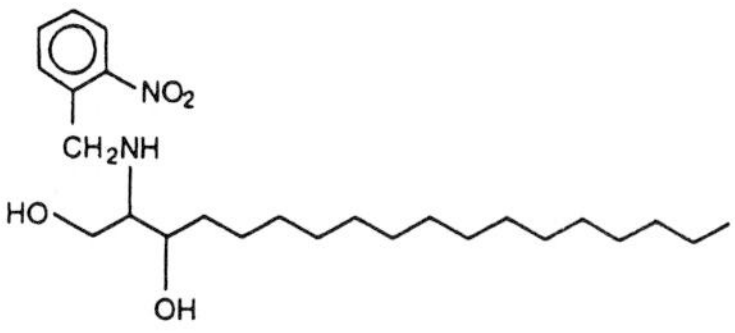

2

SCHEME 1. Structure of two 2-nitrobenzyl derivatives (compound **1**, caged Sph and compound **2**, caged DHS).

tected during synthesis as light-sensitive *N*-(4-carboxymethyl-2-nitrobenzyloxycarbonyl) derivatives and could have been considered as potentially caged SLs.[7,8] Converting the free amino function of *lyso*-SLs into *N*-(4-carboxymethyl-2-nitrobenzyloxycarbonyl) or *N*-(2-nitrobenzyl) derivatives afforded the first, intentionally synthesized, caged and potentially caged SLs where *N*-(4-carboxymethyl-2-nitrobenzyloxycarbonyl) derivatives could be readily subjected to saponification to the free carboxylic acids that could affect polarity and transport.[9,10] Of the potentially caged SLs prepared by us, two 2-nitrobenzyl derivatives (compound **1**, caged Sph, and compound **2**, caged DHS) are described here (Scheme 1). Additionally, sphingosine 1-phosphate can be caged as a derivatized 2-nitrobenzyl phosphate ester.[11]

Materials and Methods

General Synthesis of Caged Sphingolipids

Sphingosine sulfate ($Sph \cdot H_2SO_4$) can be purchased commercially or isolated, following hydrolysis from bovine brain[12–14]; DL-*erythro*-dihydrosphingosine (DHS) is purchased from Sigma (St. Louis, MO). Column chromatography is conducted on silica gel (Merck). Thin-layer chromatography (TLC) is performed on silica gel 60A KGF plates (Whatman, Clifton, NJ) or silica gel $60F_{254}$ sheets (Merck) and compounds are detected by viewing under UV light or by spraying with sulfuric acid and heating. Mass

[7] U. Zehavi, M. Herchman, S. Hakomori, and S. Köpper, *Glycoconj. J.* **7,** 219 (1990).
[8] U. Zehavi, M. Herchman, R. R. Schmidt, and T. Bär, *Glycoconj. J.* **7,** 229 (1990).
[9] U. Zehavi, *Chem. Phys. Lipids* **90,** 55 (1997).
[10] A. Tuchinsky and U. Zehavi, *Chem. Phys. Lipids* **92,** 91 (1998).
[11] L. Qiao, A. P. Kozikowski, A. Olivera, and S. Spiegel, *Bioorg, Medic. Chem. Lett.* **8,** 711 (1998).
[12] H. E. Carter, W. J. Haines, W. E. Ledyard, and W. P. Norris, *J. Biol. Chem.* **169,** 7 (1947).
[13] H. Thierfelder, *Hoppe-Seylers Z. Physiol. Chem.* **44,** 366 (1905).
[14] P. A. Levene and W. A. Jacob, *J. Biol. Chem.* **11,** 547 (1912).

spectra of the compounds are obtained using a VG platform at 3.5 kV with negative or positive ion electrospray. ^{1}H NMR spectra are recorded on a Bruker AM-400 or a Bruker AMX-400 (400 MHz) instrument in $CDCl_3$. UV absorbance is recorded on a Gilford 3AG linear transport instrument. The compounds synthesized are light-sensitive and must be kept in the dark.

2-N-(2-Nitrobenzylation) of Sphingosine Derivatives

Dry dimethylformamide (1 ml) 15 added to a mixture of 2-nitrobenzyl bromide (72 mg, 0.33 mmol) and a Sph derivative (0.33 mmol). The mixture is stirred well at 0° and 1,2,2,6,6-pentamethylpiperidine (200 μl in the case of Sph · H_2SO_4 and 66 μl in other cases) added. Stirring is continued for 1 hr at 0° and for 17 hr at room temperature. The solvent is evaporated *in vacuo* (30° bath) and the products applied to a silica gel column as described next.

Starting from Sph · H_2SO_4, the column (13 g silica gel, 1 cm in diameter) is washed first with petroleum ether/ethyl acetate, 2 : 1, v/v (60 ml) followed by petroleum ether/ethyl acetate, 1 : 1, v/v. Fractions 39–60 (5 ml/fraction) contained 2-*N*-(2-nitrobenzyl)sphingosine (compound **1**, pure by TLC, same solvent, 50 mg, 35% yield). MS: m/z = 435.3, calcd. 435.3, $[M + 1]^+$. ^{1}H NMR: δ 7.95 (d, 1H, $J_{3,4}$ 8.6 Hz, H-3 aromatic), 7.59–7.57 (m, 2H, aromatic), 7.46–7.41 (m, 1H, aromatic), 5.80–5.73 (m, 1H, olefinic), 5.46–5.40 (m, 1H, olefinic), 4.10 (app. d, 2H, J 1.5 Hz, benzylic), 3.70 (apparent doublet, 2H, J 6.7 Hz, H-1, H-1′), 2.63–2.66 (m, 1H), 2.02–2.07 (m, 1H), 1.37–1.30 (m, 2H, CH_2), 1.25 (m, 22H), 0.88 (t, 3H, J 7.0 Hz, CH_3-18).

Starting from DHS, the pure product, 2-*N*-(2-nitrobenzyl)-DL-*erythro*-dihydrosphingosine (compound **2**, 119 mg, 83% yield) elutes from the column and migrates on TLC like caged Sph (compound **1**). MS: m/z = 437.0, calculated. 437.3, $[M + 1]^+$. ^{1}H NMR: δ 7.94 (d, 1H, J 7.8 Hz, H-3 aromatic), 7.61–7.57 (m, 2H, aromatic), 7.46–7.41 (m, 1H, aromatic), 4.16 (d, 1H, J 13.8 Hz, benzylic), 4.06 (d, 1H, benzylic), 3.81–3.70 (m, 3H, H-1, H-1′, H-3), 2.63–2.60 (m, 1H, H-2), 1.49–1.45 (m, 2H, CH_2), 1.26 (m, 26H), 0.88 (t, 3H, J 6.8 Hz, CH_3-18).

Whole-Cell Recording

A use for the caged sphingolipids follows: Whole-cell patch-clamp[15] experiments with the compounds are conducted at room temperature on neonatal rat cultured dorsal root ganglion (DRG) neurons that have been in culture for at least 2 days. The recording solutions used in the voltage clamp experiments are designed to inhibit Na^+ and K^+ currents and isolate

[15] O. P. Hamill, A. Marty, E. Neher, B. Sakmann, and F. Sigworth, *Pflügers Arch.* **391,** 85 (1981).

voltage-activated Ca^{2+} currents and Ca^{2+}-activated nonselective cation and Cl^- currents. Low-resistance (3–10 MΩ) borosilicate glass patch pipettes are filled with CsCl-based patch pipette solution which contains (in m*M*) 140 CsCl, 0.1 $CaCl_2$, 1.1 or 20 EGTA, 2 $MgCl_2$, 2 ATP, 10 HEPES. Caged, *N*-(2-nitrobenzyl)dihydrosphingosine (100 μM), *N*-(2-nitrobenzyl)sphingosine (100 μM), or inert 2-(2-nitrobenzylamino)propanediol (100 μM) are included in the CsCl-based patch pipette solution with 0.3% (v/v) dimethylformamide (DMF) and 2 m*M* dithiothreitol (DTT). Some experiments are also carried out using patch pipette solution containing 10 μM ryanodine, 25 μM DHS with 0.4% DMF, dihydrosphingosine 1-phosphate (0.5–25 μM) with 0.4% DMF, or fumonisin B1 (500 n*M*) with 0.4% DMF. The extracellular bathing solution contains (in m*M*) 130 choline chloride, 2 $CaCl_2$, 3 KCl, 0.6 $MgCl_2$, 1 $NaHCO_3$, 10 HEPES, 5 glucose, 25 tetraethylammonium chloride, 0.0025 tetrodotoxin (Sigma, St. Louis, MO). Dantrolene (10 μM) is applied to the extracellular environment by low-pressure ejection from a blunt patch pipette.

Flash Photolysis

Intracellular photolysis of caged dihydrosphingosine, caged sphingosine, or 2-(2-nitrobenzylamino)propanediol is carried out after at least 5 min of equilibration in the whole-cell recording configuration. This period of equilibration allows for diffusion of the constituents of the patch pipette solution into the neuron. Photolysis is achieved using an XF-10 xenon flash lamp[16] (Hi-Tech Scientific, Salisbury, UK) with a UG11 bandpass filter. A 1-ms, 200-V flash of intense near-UV light has a power output of ~8 mJ mm^{-2} and gives about 5% photolysis of caged SLs in our system.[17] The mean quantum yield was 0.32 ($n = 6$) for the photolysis of 2-(2-nitrobenzylamino)propanediol, measured using HPLC reverse-phase chromatography and a mobile phase of 40% (v/v) methanol containing 100 m*M* potassium phosphate and 4 m*M* potassium acetate, pH 3.7.

Summary of Findings

Intracellular photorelease of 5–15 μM DHS evoked Ca^{2+}-activated inward currents from 88% ($n = 57$) of cultured DRG neurons studied. Photoreleased DHS gave rise to dose-dependent responses, which varied in amplitude and delay to activation following photolysis (Fig. 2A, B). No Ca^{2+}-activated currents were observed following intracellular flash pho-

[16] G. Rapp, *Methods Enzymol.* **291,** 202 (1998).
[17] K. P. M. Currie, J. F. Wootton, and R. H. Scott, *J. Physiol.* **482,** 291 (1995).

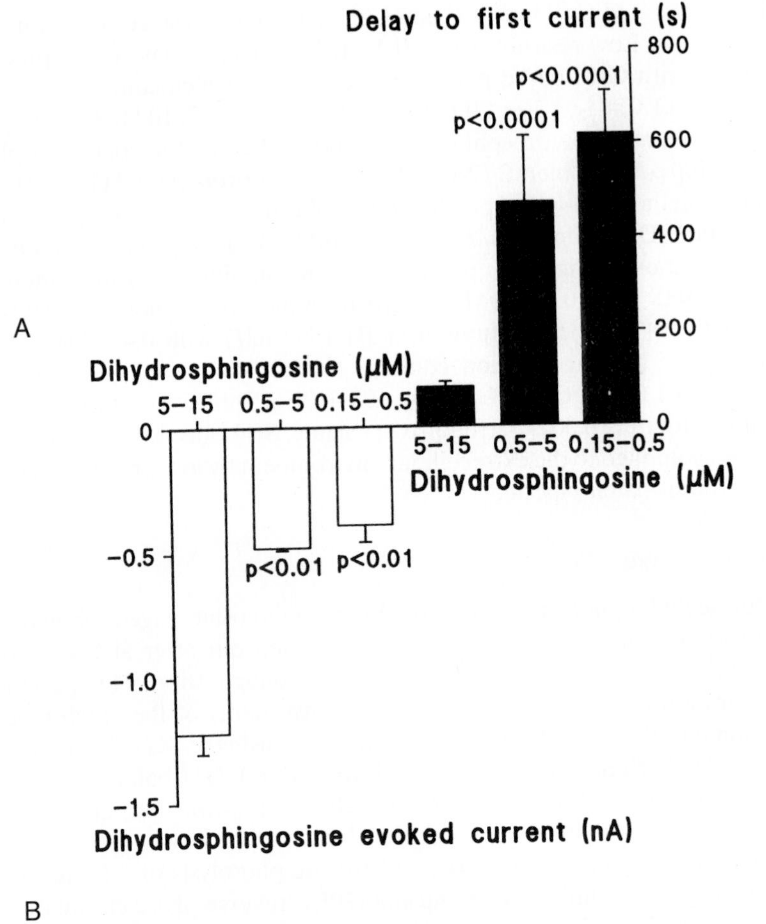
A
Delay to first current (s)
800
600
400
200
0
p<0.0001
p<0.0001
5–15 0.5–5 0.15–0.5
Dihydrosphingosine (μM)
Dihydrosphingosine (μM)
5–15 0.5–5 0.15–0.5
0
−0.5
−1.0
−1.5
p<0.01 p<0.01
Dihydrosphingosine evoked current (nA)

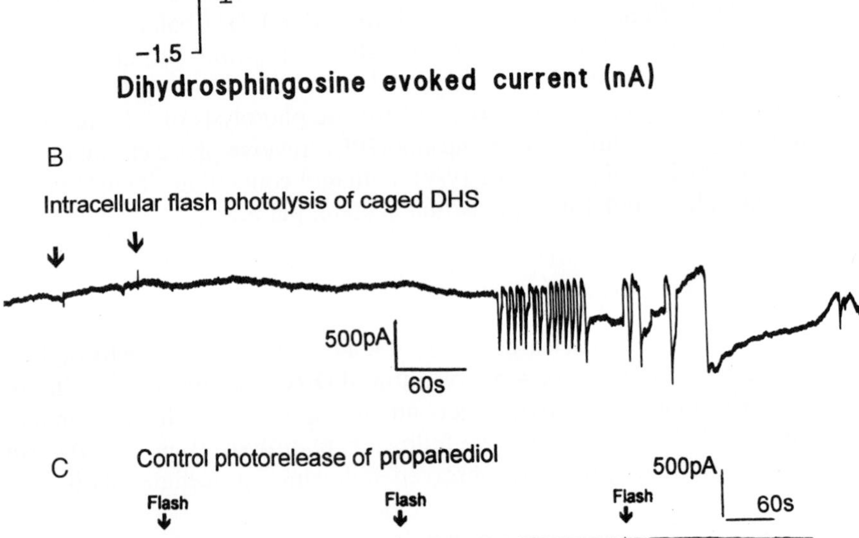
B
Intracellular flash photolysis of caged DHS
500pA
60s
C
Control photorelease of propanediol
Flash
Flash
Flash
500pA
60s

TABLE I
PROPERTIES OF Ca^{2+}-DEPENDENT CURRENTS EVOKED BY DIHYDROSPHINOGOSINE AND DIHYDROSPHINGOSINE 1-PHOSPHATE[a]

Compound	Concentration (μM)	Number of neurons	Percentage of neurons responding (%)	Peak amplitude of inward current (nA)	Delay to first response (sec)
Dihydrosphingosine	25	9	67	-1.15 ± 0.19	328 ± 114
Dihydrosphingosine 1-phosphate	25	9	56	-1.12 ± 0.34	169 ± 103
Dihydrosphingosine 1-phosphate	2.5	7	57	-0.84 ± 0.34	23 ± 23
Dihydrosphingosine 1-phosphate	0.5	10	30	-0.48 ± 0.17	75 ± 53

[a] Applied to intracellular environment via patch pipette solution.

tolysis of the inert compound, nitrobenzyl caged propanediol (Fig. 2C). This control indicated that the flashes of intense near-UV light and the by-product of photorelease, nitrosobenzaldehyde, did not evoke responses under our recording conditions. The inward currents were therefore activated as a result of a rise in intracellular sphingolipid concentration.

Experiments with uncaged SLs showed that DHS and dihydrosphingosine 1-phosphate also evoked inward Ca^{2+}-dependent currents in cultured DRG neurons (Table I). Comparing the delays to the first responses evoked by intracellular flash photolysis of 5–15 μM caged DHS with the delays produced by direct intracellular application of 25 μM DHS, it can be seen that the latter events were slower to develop by ~250 sec, an approximate 4-fold difference. This gives an approximation of the additional time due to diffusion barriers, before activation of DHS-evoked responses. The amplitudes of inward currents evoked by dihydrosphingosine 1-phosphate were dose dependent. However, as previously observed with a variety of Ca^{2+} mobilizing compounds including caffeine, cyclic ADPribose, the cytosolic sperm factor oscillogen and DHS the nature of Ca^{2+}-activated inward currents evoked by intracellular application of dihydrosphingosine

FIG. 2. (A) The relationship between the estimated concentration of photoreleased DHS and the amplitude of inward Ca^{2+}-activated currents, and the delay between intracellular photolysis of caged DHS and the development of the first response. (B) Inward current record obtained from a DRG neuron held at −90 mV showing responses to photoreleased DHS (5–15 μM). (C) Control record showing no responses to repeated intracellular flash photolysis of caged propanediol. Arrows show the point at which flash photolysis occurred.

1-phosphate was very variable. Dihydrosphingosine 1-phosphate responses varied from single transient currents to inward current oscillations (Fig. 3). Additionally, an anomaly was observed in the delay times, in that shorter delays were observed with concentrations of 2.5 and 0.5 μM compared with 25 μM dihydrosphingosine 1-phosphate. This finding may be indicative of inhibitory effects of higher concentrations of dihydrosphingosine 1-phosphate on metabolic processes or Ca^{2+} release mechanisms, which result in the slowed development of responses.

Several pieces of evidence suggest that intracellular application of DHS evoked mobilization of Ca^{2+} from intracellular stores. First, increasing the concentration of the Ca^{2+} chelator EGTA in the patch pipette solution

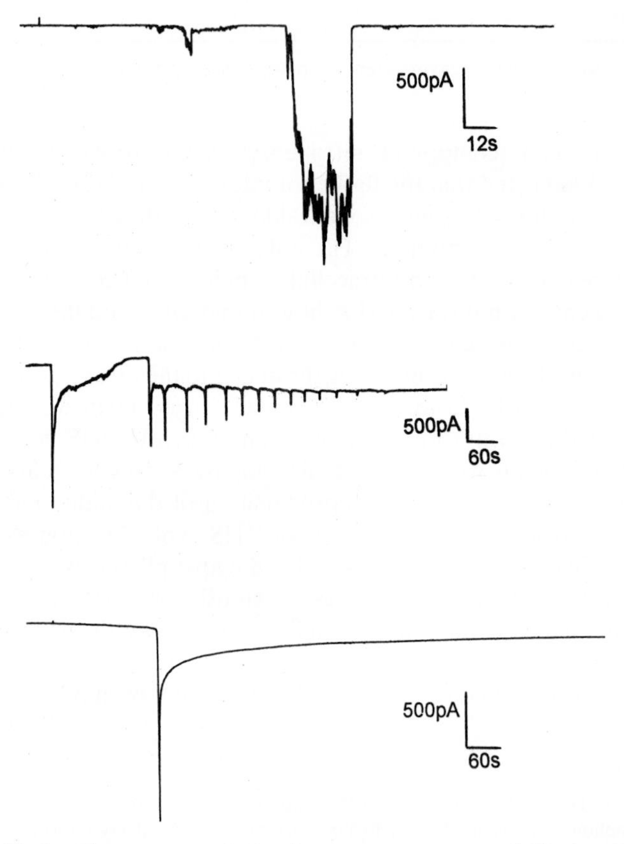

FIG. 3. Whole-cell current records showing responses to uncaged dihydrosphingosine 1-phosphate (25 μM) in three different cultured DRG neurons. The start of each record is the point at which the whole-cell recording configuration was achieved, and all neurons were voltage clamped at -90 mV.

from 1.1 to 20 mM attenuated DHS-activated responses. Second, Ca^{2+}-activated currents evoked by DHS still persisted when Ca^{2+} was removed from the extracellular environment. Third, the responses to intracellular photoreleased DHS (5–15 μM) were abolished by intracellular ryanodine (10 μM) and extracellular dantrolene (10 μM), which inhibit ryanodine receptor Ca^{2+} release channels on endoplasmic reticulum (ER) membranes.[1,2] Although a number of studies have linked sphingolipid-induced Ca^{2+} mobilization to IP_3-sensitive stores, in cultured DRG neurons no conclusive role for IP_3 was obtained. Intracellular heparin (100 μg/ml) failed to prevent DHS responses, but did significantly reduce the amplitude and increase the delay to activation of DHS-evoked currents ($n = 6$). This result was in part consistent with the finding that Ca^{2+} release within permeabilized DDT_1MF-2 cells evoked by sphingoid bases was not sensitive to heparin.[18]

Sphingolipids may act as intracellular messengers but may also diffuse out of the cell and bind to receptors on the cell membrane to activate different intracellular signaling cascades. To date, at least six distinct putative sphingolipid receptor subtypes have been identified. Several different classification systems, based on sphingolipid specificities and effector systems, have been used to describe these receptors (I, II, IIIa and IIIb, and EDG-1, EDG-2/*N*zg-1, EDG-3, EDG-4, H218/AGR16, and EDG-6).[19,20] The sphingolipid receptors are metabotropic receptors, which are linked, in many cases but not exclusively via pertussis toxin-sensitive G proteins, to effector signaling pathways. With this in mind we photoreleased DHS in cultured DRG neurons that had been pretreated with pertussis toxin (0.5 μg/ml) for 3 hr. Pertussis toxin pretreatment had no significant effect on the amplitude of inward currents evoked by DHS but significantly delayed their development (Fig. 4A). This could mean that a significant element of the DHS-evoked response involved activation of sphingolipid receptors on the cell surface and thus synergistic cross-talk between signaling pathways.

A further key issue for this type of functional study of SLs is whether DHS directly or indirectly evoked responses. Several indirect mechanisms are possible and these include metabolism to another sphingolipid and activation of distinct intracellular signaling pathways to generate a rise in intracellular Ca^{2+}. Intracellular photorelease of 5–15 μM Sph evoked similar responses to those currents activated by photorelease of DHS (Fig.

[18] T. K. Ghosh, J. Bian, and D. L. Gill, *J. Biol. Chem.* **269,** 22628 (1994).

[19] D. Meyer zu Heringdorf, C. J. van Koppen, and K. H. Jakobs, *FEBS Lett.* **410,** 34 (1997).

[20] S. Pyne, S. Rakhit, A.-M. Conway, A. McKie, P. Darroch, and N. Pyne, *Biochem. Soc. Trans.* **27,** 12 (1999).

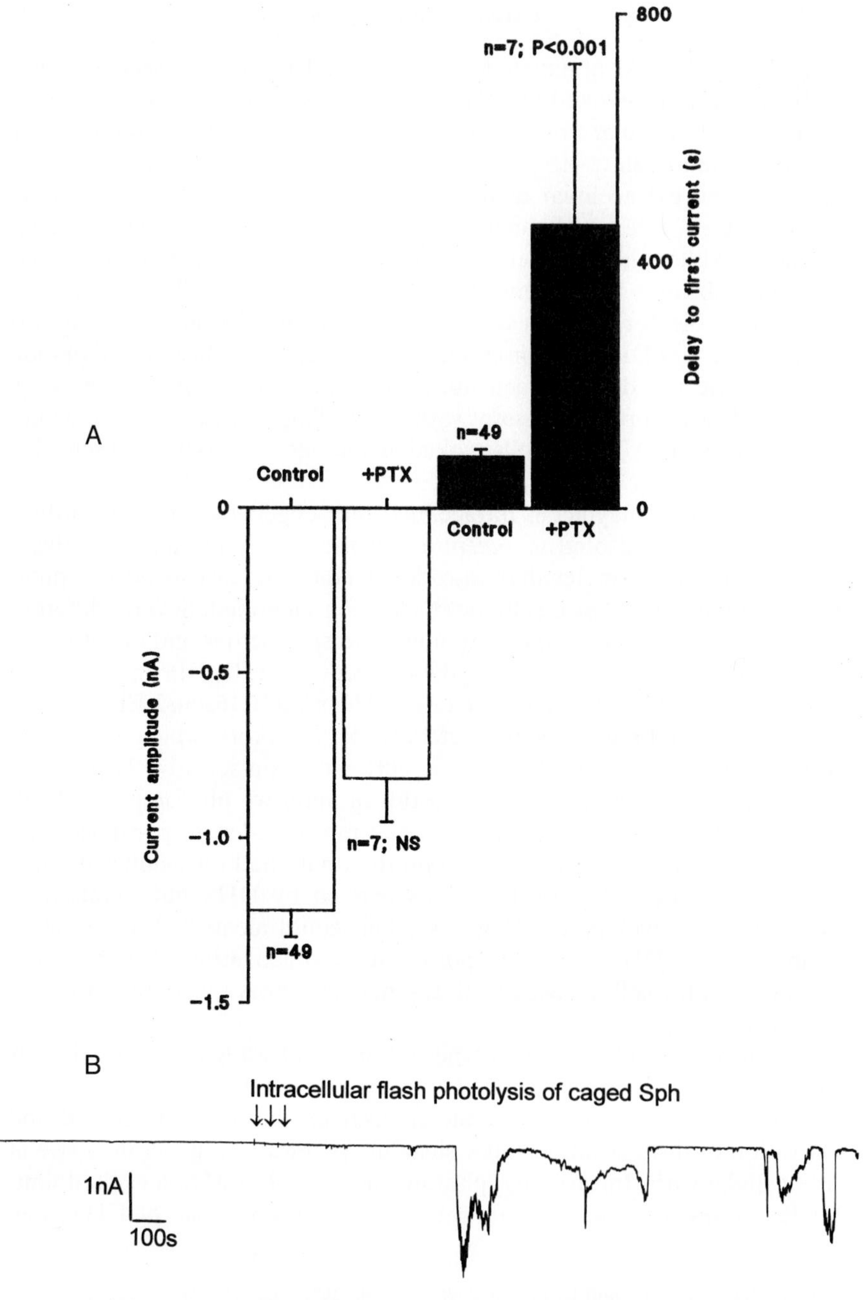

FIG. 4. (A) The influence of pertussis toxin (0.5 μg/ml) pretreatment for 3 hr on responses evoked by intracellular photolysis of caged DHS. (B) Inward current record obtained from a DRG neuron held at −90 mV showing responses to photoreleased Sph (15 μM).

4B). However, the delay to activation of the inward currents was more pronounced following intracellular photolysis of caged Sph compared with caged DHS. Intracellular photolysis of caged Sph evoked responses in 57% (n = 21) of cultured DRG neurons studied. The mean peak current amplitude and mean delay to activation of the Sph evoked events were -1.05 ± 0.21 nA and 276 ± 81 sec (n = 12), respectively. From these experiments it was far from clear which sphingolipid might be acting as an intracellular messenger to release Ca^{2+} from intracellular stores. Two approaches were taken to try to address the apparent lack of selectivity of the Ca^{2+} mobilization mechanism for a particular sphingolipid.

First, we explored the possibility that dihydrosphingosine may be converted to ceramide and then into Sph and possibly sphingosine 1-phosphate. This required investigation in cultured DRG neurons even though it may seem an unlikely prospect because DHS does not evoke Ca^{2+} responses in other preparations. Additionally, DHS has been used to competitively inhibit sphingosine kinase (K_i 5–18 μM) which generates sphingosine 1-phosphate. Many investigators suggest that sphingosine 1-phosphate is the primary candidate as a sphingolipid Ca^{2+}-mobilizing intracellular messenger.[21] There is evidence that fumonisin B1 isolated from fungal species including *Fusarium moniliforme* inhibits sphingolipid biosynthesis. This may occur as a result of close structural similarities between fumonisin B1 and SLs and thus inhibition of sphinganine *N*-acetyltransferase.[22,23] Intracellular fumonisin B1 (500 nM and 25 μM) was used in an attempt to modulate photoreleased DHS responses, and so implicate sphingolipid metabolism in Ca^{2+} mobilization from intracellular stores. However, in our studies both concentrations of intracellular fumonisin B1 alone evoked transient inward currents in about 50% of neurons (n = 19). These events were not significantly different from the inward currents activated by intracellular photorelease of DHS. Although the DHS-evoked currents appeared to persist with intracellular fumonisin B1, it was not possible to identify which compound was responsible for the activation of these responses (Fig. 5). Experiments were also carried out on neurons pretreated with extracellular fumonisin B1 (500 nM) for 24 hr; intracellular fumonisin B1 also evoked inward currents in these cells and responses also appeared to be activated by DHS. It was, however, noted that some neurons that appeared not to respond to fumonisin B1 did subsequently respond to intracellular photolysis of caged DHS. Nevertheless, the pharmacological effect of fumonisin B1 made a clear interpretation of the variable data very

[21] O. H. Choi, J.-H. Kim, and J.-P. Kinet, *Nature* **380,** 634 (1996).

[22] A. H. Merrill, Jr., D. C. Liotta, and R. Riley, *Trends Cell Biol.* **6,** 218 (1996).

[23] J. M. Yeung, H.-Y. Wang, and D. B. Prelusky, *Toxicol. Appl. Pharmacol.* **141,** 178 (1996).

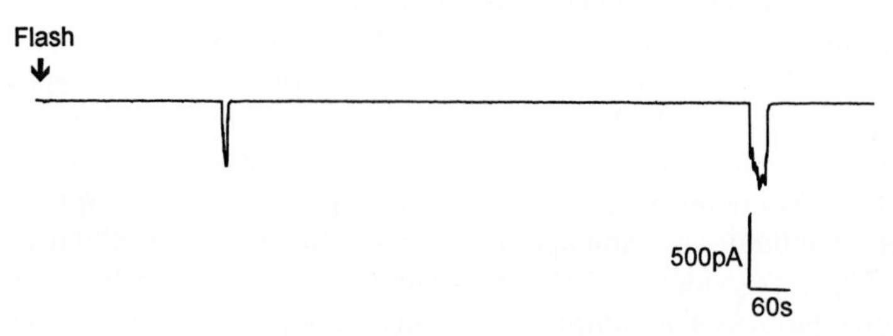

FIG. 5. Actions of intracellular fumonisin B1 (500 n*M*) on DRG neurons. (A) Intracellular fumonisin B1 alone evoked Ca^{2+}-activated currents. (B) Intracellular fumonisin B1 appeared not to prevent responses to photoreleased DHS. Both neurons were held at −90 mV.

difficult because we could not differentiate between DHS-evoked responses and the mimicking effects of the structurally closely related fungal toxin.

The second approach was to determine whether activation of the sphingolipid signaling pathway with a nonsphingolipid could generate release of Ca^{2+} from intracellular stores and activate inward currents. For these experiments we include 0.4 units/ml sphingomyelinase from *Bacillus cereus* (Sigma) in our patch pipette solution and applied it to the intracellular environment. Intracellular sphingomyelinase evoked Ca^{2+}-activated inward currents in 44% (n = 9) of cultured DRG neurons studied. The mean amplitude and delay time to the development of the first current were -1.16 ± 0.23 nA and 169 ± 60 sec (n = 4), respectively. It was also apparent that the responses to sphingomyelinase were influenced by entry of Ca^{2+} through voltage-activated channels on the cell membrane. If 20 high-voltage-activated Ca^{2+} currents were evoked by 100-ms step depolarizations, applied every 3 sec, from a holding potential of −90 to 0 mV, sphingomyelinase appeared to produce larger responses (Fig. 6). This may have related to the filling of intracellular Ca^{2+} stores, but may

A

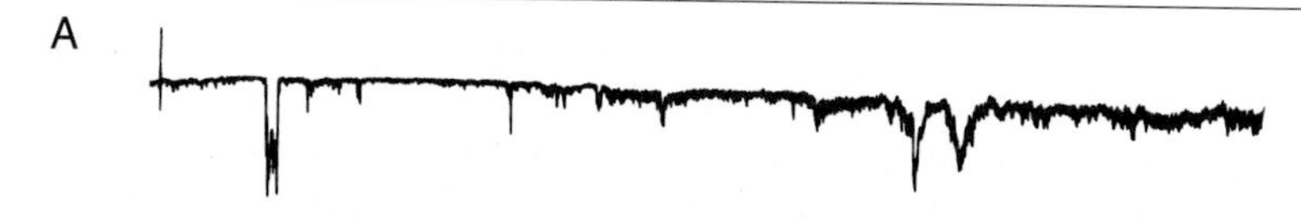

B

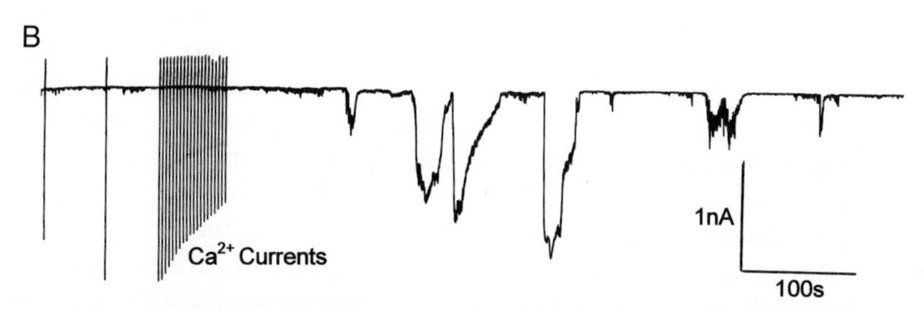

FIG. 6. Intracellular sphingomyelinase applied via the patch pipette solution evoked Ca^{2+}-dependent inward currents. (A) Current record of inward current activity recorded in the presence of 0.4 units/ml sphingomyelinase. (B) Current record showing enhancement of sphingomyelinase-evoked inward current activity by 20 high-voltage-activated inward currents. Both neurons were held at −90 mV.

also have been dependent on converging influences of Ca^{2+} entry and intracellular SLs on the set point for Ca^{2+}-induced Ca^{2+} release.

Conclusions and Future Goals

DHS, dihydrosphingosine 1-phosphate, and Sph applied to the intracellular environment evoked inward Ca^{2+}-activated currents in cultured DRG neurons. Intracellular sphingomyelinase mimicked these responses, which suggests that sphingolipid biosynthesis and metabolism may be of physiological importance rather than just a pharmacological curiosity in sensory neurons. One might envisage a metabotropic receptor mechanism coupled to sphingolipid signaling, similar to that reported in HEK-293 cells stably expressing human m2 and m3 muscarinic acetylcholine receptors.[24] Such a system might be important in the growth and development of neurons particularly because there is evidence that ceramide plays distinct roles during hippocampal neuron development.[25] The sphingolipid-evoked re-

[24] D. Meyer zu Heringdorf, H. Lass, R. Alemany, K. T. Laser, E. Neumann, C. Zhang, M. Schmidt, U. Rauen, K. H. Jakobs, and C. J. van Koppen, *EMBO J.* **17,** 2830 (1998).
[25] A. Schwartz and A. H. Futerman, *J. Neurosci.* **17,** 2929 (1997).

sponses appear to involve release of Ca^{2+} from ryanodine and dantrolene-sensitive intracellular stores. However, the involvement of putative sphingolipid receptors on the cell surface and activation of distinct signaling pathways in Ca^{2+} mobilization initiated by SLs has not been excluded. At this stage intracellular flash photolysis of different caged SLs and the measurement of the delays to the development of the responses have not yet made it possible to determine which SLs interact with the intracellular Ca^{2+} stores. The complexity of the responses and the apparent lack of discrimination between different SLs has, to date, prevented identification of the signaling molecule. Future experiments with caged sphingosine 1-phosphate may help shed light on this question. However, the complex nature of these Ca^{2+} release events may require additional approaches, particularly if another signaling pathway influenced by SLs is primarily responsible for the increase in intracellular Ca^{2+}. Fumonisin B1 could be a useful tool but in DRG neurons it failed to prevent responses to photoreleased DHS and actually activated responses alone. A further complexity may be bell-shaped dose–response relationships for sphingolipid-mediated events, with low concentrations of SLs evoking responses, which are attenuated by higher concentrations of SLs.

The present study illustrates how difficult it may be to identify the most appropriate concentrations of SLs to use, and the appropriate time course over which measurements need to be taken in single-cell functional studies. In conclusion, intracellular photolysis of caged SLs offers a useful approach to studying a signaling mechanism that mobilizes Ca^{2+} from ryanodine-sensitive intracellular stores. However, the present data have raised some critical questions regarding the relevance of SLs to the cellular physiology of sensory neurons and signaling events within their cell somas. Specifically, do any of the SLs directly interact with intracellular Ca^{2+} release mechanisms and are there roles for cross-talk between SLs and other intracellular signaling pathways that result in Ca^{2+} mobilization from intracellular stores?

Acknowledgments

The authors thank the MRC for support and Drs. David Trentham and Graeme Nixon for helpful discussion. Dr. Ahmet Ayar was supported by a scholarship from Firat University. Scheme 1 is reprinted from *Chemistry and Physics of Lipids,* **90,** Uri Zehvi, Synthesis of potentially caged sphingolipids, possible precursors of cellular modulators and second messengers, 55–61 (1997), with permission from Elsevier Science.

[32] Binding of Sphingosine 1-Phosphate to Cell Surface Receptors

By JAMES R. VAN BROCKLYN and SARAH SPIEGEL

Introduction

Sphingosine 1-phosphate (SPP) is a sphingolipid metabolite that has been shown to affect diverse biologic processes including cell proliferation, survival, motility, and differentiation.[1,2] SPP is capable of acting intracellularly as a second messenger in mitogenic signaling by platelet-derived growth factor[3] and in calcium release in response to ligation of the antibody receptors FcεR1[4] or FcγR1.[5] Furthermore, microinjection of SPP is mitogenic for Swiss 3T3 fibroblasts.[6] However, several other effects of SPP are mediated through cell surface receptors including platelet activation,[7] inhibition of melanoma cell motility,[8] activation of G_i protein-gated inward rectifying K^+ channels,[9] and Rho-dependent neurite retraction and cell rounding.[10,11] We have identified the G-protein-coupled receptor Edg-1 as a high-affinity receptor for SPP and have shown that signaling through Edg-1 led to morphogenetic differentiation of HEK293 cells.[12] In addition, two related G-protein-coupled receptors, Edg-3 and H218/Edg-5/AGR16,

[1] J. R. Van Brocklyn, O. Cuvillier, A. Olivera, and S. Spiegel, *J. Liposome Res.* **8,** 135 (1998).

[2] S. Spiegel, O. Cuvillier, L. C. Edsall, T. Kohama, R. Menzeleev, Z. Olah, A. Olivera, G. Pirianov, D. M. Thomas, Z. Tu, J. R. Van Brocklyn, and F. Wang, *Ann. N.Y. Acad. Sci.* **845,** 11 (1998).

[3] A. Olivera and S. Spiegel, *Nature* **365,** 557 (1993).

[4] O. Choi, H., J.-H. Kim, and J.-P. Kinet, *Nature* **380,** 634 (1996).

[5] A. Melendez, R. A. Floto, D. J. Gillooly, M. M. Harnett, and J. M. Allen, *J. Biol. Chem.* **273,** 9393 (1998).

[6] J. R. Van Brocklyn, M. J. Lee, R. Menzeleev, A. Olivera, L. Edsall, O. Cuvillier, D. M. Thomas, P. J. P. Coopman, S. Thangada, T. Hla, and S. Spiegel, *J. Cell Biol.* **142,** 229 (1998).

[7] Y. Yatomi, S. Yamamura, F. Ruan, and Y. Igarashi, *J. Biol. Chem.* **272,** 5291 (1997).

[8] S. Yamamura, Y. Yatomi, F. Ruan, E. A. Sweeney, S. Hakomori, and Y. Igarashi, *Biochemistry* **36,** 10751 (1997).

[9] C. J. van Koppen, D. Meyer zu Heringdorf, K. T. Laser, C. Zhang, K. H. Jakobs, M. Bünnemann, and L. Pott, *J. Biol. Chem.* **271,** 2082 (1996).

[10] F. R. Postma, K. Jalink, T. Hengeveld, and W. H. Moolenaar, *EMBO J.* **15,** 2388 (1996).

[11] K. Sato, H. Tomura, Y. Igarashi, M. Ui, and F. Okajima, *Biochem. Biophys. Res. Commun.* **240,** 329 (1997).

[12] M.-J. Lee, J. R. Van Brocklyn, S. Thangada, C. H. Liu, A. R. Hand, R. Menzeleev, S. Spiegel, and T. Hla, *Science* **279,** 1552 (1998).

0076-6879/00 $30.00

also bind SPP with nanomolar affinity.[13] H218, and to a lesser extent Edg-3, mediate cell rounding in response to SPP.[13]

Edg-1 was cloned as an immediate early gene induced during phorbol ester-stimulated differentiation of HUVEC cells, suggesting that it may play a role in differentiation of endothelial cells during angiogenesis.[14] In addition, Edg-1 is expressed in a variety of tissues with high levels of expression in the brain, heart, and spleen.[15] Expression is particularly high in the Purkinje cell layer of the cerebellum.[15] In the embryonic mouse, high levels of Edg-1 expression also exist in skeletal structures undergoing ossification.[15] Edg-3 is also expressed in a wide variety of tissues; however, high levels of expression were found in heart, placenta, liver, kidney, and pancreas with low levels in the brain.[16] H218 is expressed in the cardiovascular system[17] and in the brain during embryogenesis, where its expression is temporally regulated such that high levels of expression are found in neuronal cell bodies during early stages of differentiation and in axons during their outgrowth.[18] This led to the suggestion that H218 plays an important role in neuronal development and may steer axons by regulating their growth and inhibiting their extension.[18] Thus, SPP receptors exist in a wide variety of tissues and cells suggesting that many diverse biologic events may be regulated by SPP through Edg receptors. Because our data show that Edg-1, H218, and Edg-3 bind SPP with a high degree of specificity, we propose that these receptors should be renamed sphingosine 1-phosphate receptors SPPR1, SPPR2, and SPPR3, respectively.

Due to the dual actions of SPP, intracellularly and at the cell surface through SPPRs, it is important to characterize specific cell surface binding to cells in which SPP signaling is to be studied. However, accurate binding assays using lipophilic ligands are often problematic due to high nonspecific binding producing an unacceptably low signal-to-noise (*S/N*) ratio. This chapter details the assay used to characterize the binding of SPP to various SPPRs, which we recently developed to overcome these problems.

Preparation of [^{32}P]SPP

To reduce background due to nonspecific binding of SPP, it is important to use low concentrations of labeled SPP. Therefore, high specific activity

[13] J. R. Van Brocklyn, Z. Tu, L. Edsall, R. R. Schmidt, and S. Spiegel, *J. Biol. Chem.* **274,** 4626 (1999).

[14] T. Hla and T. Maciag, *J. Biol. Chem.* **265,** 9308 (1990).

[15] C. H. Liu and T. Hla, *Genomics* **43,** 15 (1997).

[16] F. Yamaguchi, M. Tokuda, O. Hatase, and S. Brenner, *Biochem. Biophys. Res. Commun.* **227,** 608 (1996).

[17] H. Okazaki, N. Ishizaka, T. Sakurai, K. Kurokawa, K. Goto, M. Kumada, and Y. Takuwa, *Biochem. Biophys. Res. Commun.* **190,** 1104 (1993).

[18] A. J. MacLennan, L. Marks, A. A. Gaskin, and N. Lee, *Neuroscience* **79,** 217 (1997).

is desirable. SPP can be labeled to high specific activity by taking advantage of the enzymatic phosphorylation of sphingosine catalyzed by sphingosine kinase. We have used sphingosine kinase from two sources, partially purified rat kidney enzyme[19] and recombinant sphingosine kinase expressed in HEK293 cells.[20] Recombinant kinase gives a greater yield. HEK293 cells transiently transfected with sphingosine kinase SPHK1a[20] are grown for 2 days. Cytosolic extracts are then prepared by freeze–thawing the cells as described[21] and protein concentration is determined. The phosphorylation reaction is performed using 1 ml sphingosine kinase buffer [0.1 *M* Tris HCl pH 7.4, 1 m*M* 2-mercaptoethanol, 1 m*M* EDTA, 10 m*M* $MgCl_2$, 1 m*M* $NaVO_3$, 15 m*M* NaF, 10 μg/ml leupeptin and aprotinin, 1 m*M* phenylmethylsulfonyl fluoride (PMSF), 0.5 m*M* 4-deoxypyridoxine, 40 m*M* β-glycerophosphate, and 20% (v/v) glycerol] to which is added 50 μM sphingosine, 100 μg protein from recombinant kinase extracts, and 300 μCi [γ-^{32}P]ATP (3000 Ci/mmol). The reaction mix is incubated for 1 hr at 37°. Further details regarding phosphorylation of sphingosine by sphingosine kinase and analysis of the product have been previously described.[21,21a] The reaction is stopped by addition of 20 μl 1*N* HCl, and [^{32}P]SPP is then extracted into the organic phase by addition of 1.6 ml $CHCl_3$/methanol/concentrated HCl (100/200/1, v/v), 1 ml 2 *M* KCl, and 1 ml $CHCl_3$, vortexing, and brief centrifugation. [^{32}P]SPP is then isolated by basic extraction into the aqueous phase after adding 2 ml methanol, 1 ml $CHCl_3$, 2 ml 2 *M* KCl, and 100 μl NH_4OH. The aqueous phase is transferred to a new tube and 3 ml $CHCl_3$ and 200 μl concentrated HCl are added to back-extract SPP into the organic phase.

Binding Assay

Tissue culture grade fetal bovine serum commonly contains high nanomolar levels of SPP.[22] Therefore, cells for binding assays should be washed thoroughly with phosphate-buffered saline (PBS) to remove serum. The cells are then harvested by scraping, pelleted, and resuspended in binding buffer (20 m*M* Tris-HCl, pH 7.4, 100 m*M* NaCl, 2 m*M* deoxypyridoxine, 15 m*M* NaF, 0.2 m*M* phenylmethylsulfonyl flouride (PMSF), and 1 μg/ml aprotinin and leupeptin) containing 4 mg/ml fatty acid-free bovine serum albumin (BSA).

[19] A. Olivera, T. Kohama, Z. Tu, S. Milstien, and S. Spiegel, *J. Biol. Chem.* **273,** 12576 (1998).

[20] T. Kohama, A. Olivera, L. Edsall, M. M. Nagiec, R. Dickson, and S. Spiegel, *J. Biol. Chem.* **273,** 23722 (1998).

[21] A. Olivera and S. Spiegel, *in* "Methods in Molecular Biology (Bird, I. M., ed.), Vol. 105, p. 233. Humana Press Inc., Totawa, New Jersey, 1998.

[22] L. C. Edsall and S. Spiegel, *Anal. Biochem* **272,** 80 (1999).

[^{32}P]SPP is dried under nitrogen in a glass tube and redissolved in binding buffer plus 4 mg/ml fatty acid-free BSA at 10 times the final desired concentration by vortexing and sonication. The inclusion of BSA in the buffer is necessary to solubilize SPP. Without BSA, most of the SPP will adhere to the glass tube and not be dissolved.

The K_D values for binding of SPP to SPPR1 (Edg-1), SPPR3 (Edg-3), and SPPR2 (H218) are 8, 23, and 27 n*M*, respectively, and binding typically saturates between 100 and 300 n*M*. Thus, final concentrations of [^{32}P]SPP should range from approximately 0.5 to 500 n*M*. For initial assessment of whether cells possess specific binding sites for SPP, a low concentration (approximately 1 n*M*) is recommended.

Cells (10^5 per sample) are then placed in microcentrifuge tubes on ice and the volume is brought to 180 μl by addition of binding buffer plus 4 mg/ml fatty acid-free BSA with or without 10 μM unlabeled SPP. [^{32}P]SPP (20 μl) is added and tubes are vortexed. Binding is allowed to continue for 30 min on ice and binding reaches equilibrium.[12] Cells are then pelleted by centrifugation in a microcentrifuge for 1 min at 3000*g* at 4° and washed twice with binding buffer plus 0.4 mg/ml fatty acid-free BSA. Binding is quantitated by liquid scintillation counting of an aliquot of resuspended cells. Specific binding is defined as binding in the absence of unlabeled SPP minus binding in the presence of excess unlabeled SPP.

Specificity of SPP binding can be easily assessed using a variety of structurally related lipids as unlabeled competitors. It is recommended that this be done using 1 n*M* [^{32}P]SPP and a concentration range of 100 n*M* to 1 μM competitor. Competitors can be used at concentrations higher than 20 μM; however, these concentrations are not recommended because they often lead to inconsistent, and in some cases, anomalous results in that they can actually cause increased binding of [^{32}P]SPP. Although the reason for this is uncertain, it is possible that at high concentrations micellar effects may cause increased nonspecific binding. A typical result for binding of [^{32}P]SPP to HEK293 cells overexpressing various SPPRs is shown in Fig. 1. Binding is clearly competed by unlabeled SPP or dihydro-SPP, which lacks the 4-*trans* double bond, but is unaffected by several related lipids.

Factors Affecting Binding of SPP to SPPR Expressing Cells

A common problem associated with binding experiments using lipid ligands is high nonspecific binding due to absorption of the hydrophobic ligand. The procedure outlined above is designed to minimize this effect. This section discusses several factors that can affect nonspecific binding.

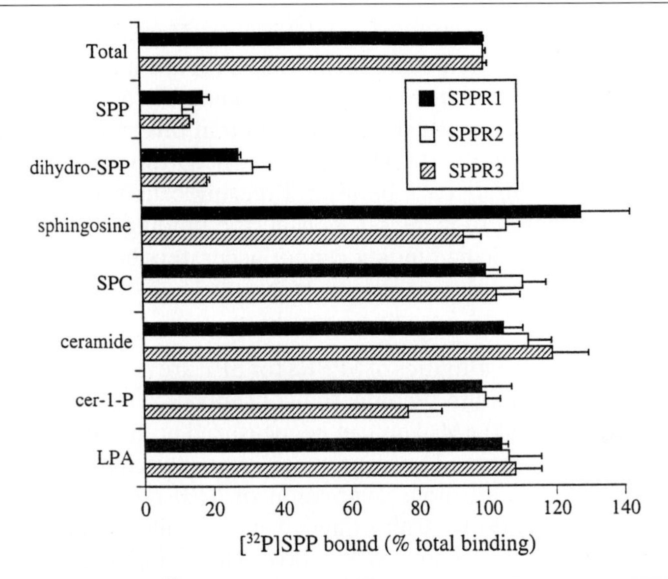

FIG. 1. Competition of [^{32}P]SPP binding to SPPR-expressing HEK293 cells. HEK293 cells stably expressing SPPR1 (Edg-1) or HEK 293 cells transiently transfected with plasmids encoding SPPR2 (H218) or SPPR3 (Edg-3) were incubated with 1 n*M* [^{32}P]SPP in the absence (total) or in the presence of 1 μM of the indicated lipids, and the amount of bound [^{32}P]SPP was determined as described. dihydroSPP, Sphinganine 1-phosphate; SPC, sphingosylphosphorylcholine; ceramide, C_2-ceramide; cer-1-P, C_8-ceramide 1-phosphate; LPA, lysophosphatidic acid. Results are means ± SD of triplicate determinations.

Temperature

It is vital that SPP binding experiments be performed on ice. Receptor binding measurements are often performed at 4° to prevent receptor internalization following ligand binding. Although this also applies to binding of SPP to SPPRs receptors, the most important reason to keep samples on ice is to reduce nonspecific absorption of SPP to cell membranes. When cells are incubated with exogenously added SPP at 37°, large amounts of SPP are taken up. In addition, when binding experiments are performed at room temperature, no specific binding can be detected, although a large amount of nonspecific binding is observed.

Cells vs. Membranes

Ligand binding experiments are often performed using membrane preparations from cells expressing the receptor of interest. Although this procedure can be used for SPP binding, it will result in greatly increased nonspecific binding. Using HEK293 cells stably expressing SPPR1, nonspecific binding is approximately 10% of total binding. However, when similar experiments are performed using membrane preparations from the same cells, nonspecific binding increases to approximately 50% total binding. Although specific binding can still be measured in this case, the results are less consistent, particularly at higher ligand concentrations. It is therefore recommended that whole cells be used. Freezing cells for use in binding experiments also leads to increased nonspecific binding, in some cases to such an extent that specific binding cannot accurately be measured.

Amount of [^{32}P]SPP and Receptor Expression

As stated above, the affinities of SPPRs for SPP are in the low nanomolar range. To obtain maximum binding, the concentrations of SPP used must therefore reach 100–500 n*M*. This can be measured accurately for cells that overexpress SPP receptors. However, when characterizing binding to cells with low levels of endogenous expression of SPP receptors, the higher concentrations of [^{32}P]SPP result in higher nonspecific binding in relation to specific binding (i.e., a low *S*/*N* ratio). Therefore, it is desirable to quantitate binding affinity using cells transfected with the receptor and expressing high receptor levels. Nevertheless, specific binding can readily be measured in cells expressing low levels of endogenous SPP receptors by using low nanomolar levels of [^{32}P]SPP.

Conclusions

SPP mediates diverse biologic effects by acting as both a second messenger and as an extracellular mediator through SPPRs. Therefore, to understand the molecular mechanisms that mediate the pleitropic effects of SPP, it is necessary to determine, for each cell type used, the possible contribution of SPP receptors. The assay detailed above is a relatively simple method to assess whether specific binding sites for SPP exist on the cell surface and to characterize the affinity and specficity of binding to such receptors.

Acknowledgments

Supported by research grants RO1 CA61774 and R01CA61774 from the National Institutes of Health and BE-275 from the American Cancer Society. Dr. James R. Van Brocklyn was supported by Postdoctoral Fellowship GM19209. We thank Dr. S. Milstien and the members of the Spiegel laboratory for helpful suggestions and discussions.

[33] Use of Short-Chain Ceramides

By CHIARA LUBERTO and YUSUF A. HANNUN

Introduction

Studies during the 1990s have pointed toward important roles for ceramide as a bioactive lipid mediator and have resulted in the hypothesis that ceramide is a key regulator of the eukaryotic stress response.[1,2] At the heart of this hypothesis are the fully documented changes in ceramide metabolism in response to various inducers of stress, which include cytokines (such as tumor necrosis factor α, TNFα), chemotherapeutic agents, heat, and radiation. These agents activate one or more pathways of ceramide formation, including various sphingomyelinases and the *de novo* pathway, that lead to substantial accumulation in endogenous levels of ceramide. In turn, ceramide has been shown to regulate many cellular functions including cell differentiation, cell senescence, and apoptosis through the regulation of several key components of the pathways such as caspases, the retinoblastoma gene product, bcl-2, bad, phospholipase D, and protein kinase C (PKC).[3]

One important tool in probing the cellular function of ceramide has come from the use of short-chain ceramides. Although these ceramides have provided a wealth of information, especially on candidate cellular activities of ceramide, their use is not straightforward and should be evaluated with careful consideration of their physical and biochemical properties and functions. This chapter presents a discussion on the reasons to use these analogs (the why), what situations to use them in (the when), and the best approaches to use them (the how).

Why Use Short-Chain Ceramides?

One of the most challenging aspects of ceramide-directed research has been the attempt to define and recapitulate cellular functions of ceramide. In theory, three general approaches could be employed in the investigation of cellular functions of any candidate second messenger and/or bioactive molecule: (1) deliver the molecule to cells; (2) use pharmacologic regulators of enzymes of metabolism of that molecule; and (3) use molecular biology

[1] Y. A. Hannun, *Science* **274,** 1855 (1996).
[2] S. Mathias, L. A. Pena, and R. Kolesnick, *Biochem. J.* **335,** 465 (1998).
[3] Y. A. Hannun and C. Luberto, *Trends Cell Biol.,* in press (2000).

0076-6879/00 $30.00

tools (such as expression of antisense and overexpression of sense constructs of enzymes that regulate the formation or metabolism of that molecule). In the earlier studies on ceramide, options (2) and (3) were not available due to the lack of the necessary molecular, biochemical, and pharmacologic foundations for the study of ceramide metabolism. Therefore, the initial studies aimed at exploiting the exogenous application of this lipid. However, the very high hydrophobicity of natural ceramides (Fig. 1) results in very poor solubility, thus preventing their simple delivery to cells. One of the major approaches to overcome this problem has been the development of ceramide analogs that are shortened either in the sphingoid backbone or in the fatty acid moiety (i.e., C_2- and C_6-ceramide, carrying an acetyl and hexanoyl group at the 2-amino position, respectively). These short-chain ceramides are quite soluble. In addition, they manifest a number of favor-

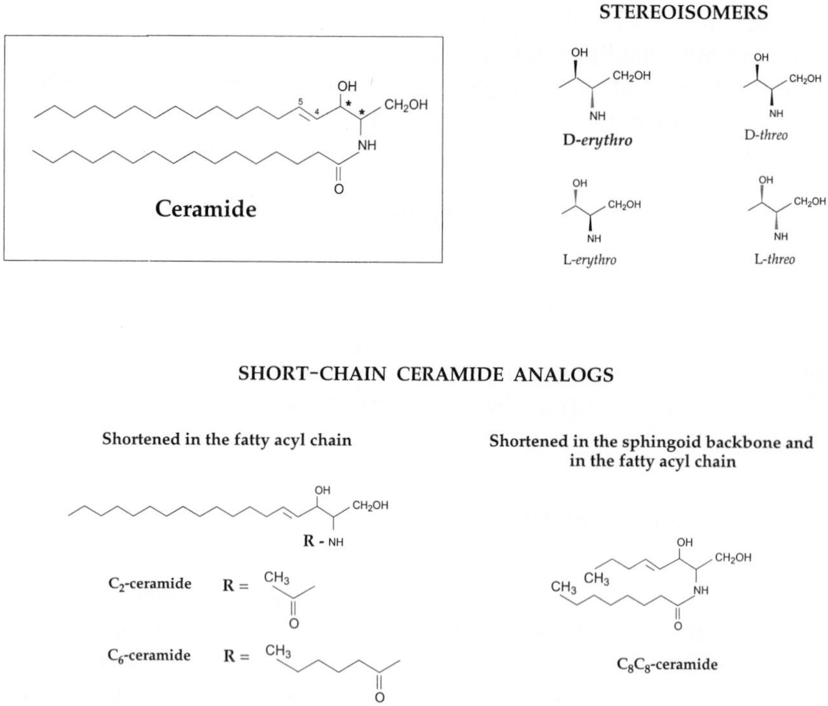

STEREOISOMERS

OH CH₂OH NH **D-*erythro*** OH CH₂OH NH D-*threo*

OH CH₂OH NH L-*erythro* OH CH₂OH NH L-*threo*

OH CH₂OH NH O 5 4 **Ceramide**

SHORT-CHAIN CERAMIDE ANALOGS

Shortened in the fatty acyl chain

OH CH₂OH R - NH

C_2-ceramide R = CH_3 O

C_6-ceramide R = CH_3 O

Shortened in the sphingoid backbone and in the fatty acyl chain

OH CH₂OH CH_3 CH_3 NH O

C_8C_8-ceramide

FIG. 1. Chemical structures of ceramide stereoisomers and short-chain analogs. The chemical structure of ceramide is characterized by the presence of a sphingoid backbone and a fatty acid chain. Asterisks indicate the chiral carbon atoms (C-2 and C-3) that allow the existence of four stereoisomers. The D-*erythro* form (indicated in bold) is the naturally occurring one.

able features: (1) they retain the basic stereochemistry and chemical features of ceramide with the sole exception of having shorter alkyl and/or acyl hydrocarbon chains (Fig. 1); (2) they are recognized by enzymes of ceramide metabolism[4]; and (3) they allow the development of structural analogs that permit a careful evaluation of structure function, a necessary goal in the understanding of the functions of ceramide and its metabolites (as well as that of any bioactive molecule) (Fig. 1). For these reasons, these compounds have become very useful in cellular studies.

When Is the Use of Short-Chain Ceramides Informative?

The use of short-chain ceramides continues to present an important and valuable source of information at different levels. First, these analogs provided the first direct evidence of the biologic functions that are affected/regulated by ceramide. Functions such as differentiation, growth arrest, and apoptosis are induced or regulated as a consequence of treatments with C_2-, C_6-, or C_8-ceramides in a variety of cell types.[3,4]

Second, the use of short-chain analogs allows investigators to proceed from these general cellular effects to more direct and indirect downstream molecular targets. Examples include ceramide-activated protein phosphatases (CAPPs)[5,6] and ceramide-activated protein kinase (CAPK/KSR)[7] as direct *in vitro* targets of ceramide, and caspases,[8] protein kinase C,[9] the retinoblastoma gene product (Rb),[6] bad,[10] bcl-2[11] and cyclin-dependent kinases as indirect *in vivo* targets.

Third, the use of radioactive and fluorescent ceramides, labeled in the fatty acid moiety or in the sphingoid backbone, makes it possible to follow the metabolic fate of this lipid.[12] This has been extremely informative over the years in delineating the complex network of sphingolipid metabolism. The fact that ceramide is intracellularly metabolized raises the question of

[4] R. Ghidoni, G. Sala, and A. Giuliani, *Biochim. Biophys. Acta* **1439,** 17 (1999).

[5] R. T. Dobrowsky, C. Kamibayashi, M. C. Mumby, and Y. A. Hannun, *J. Biol. Chem.* **268,** 15523 (1993).

[6] K. Kishikawa, C. Chalfant, D. Perry, A. Bielawska, and Y. A. Hannun, *J. Biol. Chem.* **274,** 21335 (1999).

[7] Y. H. Zhang, B. Yao, S. Delikat, S. Bayoumy, X. H. Lin, S. Basu, M. McGinley, P. Y. Chan-Hui, H. Lichenstein, and R. Kolesnick, *Cell* **89,** 63 (1997).

[8] G. S. Dbaibo, D. K. Perry, C. J. Gamard, R. Platt, G. G. Poirier, L. M. Obied, and Y. A. Hannun, *J. Exp. Med.* **185,** 481 (1997).

[9] J. Y. Lee, Y. A. Hannun, and L. M. Obeid, *J. Biol. Chem.* **271,** 13169 (1996).

[10] S. Basu, S. Bayoumy, Y. Zhang, J. Lozano, and R. Kolesnick, *J. Biol. Chem.* **273,** 30419 (1998).

[11] P. P. Ruvolo, D. Xingming, I. Takahiko, K. C. Boyd, and W. S. May, *J. Biol. Chem.* **274,** 20296 (1999).

[12] A. G. Rosenwald and R. E. Pagano, *Adv. Lipid Res.* **26,** 101 (1993).

whether effects observed after a treatment with short-chain analogs are due to the ceramide itself or to a metabolite. Given the fact that many sphingolipids exert biologic functions (e.g., sphingosine, sphingosine 1-phosphate, cerebrosides, gangliosides, and ceramide 1-phosphate) and that these lipids are metabolically connected, awareness of the extent of ceramide metabolism in the experimental model in use becomes a necessity.[3]

Finally, the observation that short-chain ceramide analogs are metabolized *in vivo* into sphingomyelin, sphingosine, and cerebrosides indicates that these compounds are recognized by the respective enzymes and therefore represent suitable substrates for *in vitro* and *in vivo* enzymatic assays and for the study of cellular metabolism of sphingolipids.

How Should Short-Chain Ceramides Be Used?

Whereas the benefits from the use of ceramide analogs are obvious, at the same time their use requires awareness of possible shortcomings that may lead to artifactual results or inappropriate conclusions. Following are some general considerations in optimizing the use of these analogs.

Quality/Identity of Short-Chain Ceramides

When handling ceramides, as is true for any compound, it is good practice to check their quality and identity. Lipids in particular are susceptible to degradation/oxidation, especially when in solution. Also the vehicle itself may be a source of impurities, which might accelerate this process. Therefore, it is advisable to check the purity of the compound, for instance by running thin-layer chromatography (TLC) in one or two different solvent systems. In this way the presence of unexpected and undesired by-products is avoided. In the same way, the identity of the compound needs to be checked. The most likely source of bias is represented by contamination with other stereoisomers. Separation of *erythro* vs. *threo* can be achieved by TLC analysis in diethyl ether : methanol (9 : 1, v/v for C_2-ceramide and 31 : 1, v/v for C_6-ceramide),[13] while separation of *D* vs. *L* enantiomers can be resolved by circular dicroism (CD) spectroscopy, chiral high-performance liquid chromatography (HPLC)columns, or chemical derivatization.

Delivery

Ceramide analogs can be dissolved in ethanol or dimethyl sulfoxide (DMSO) up to a concentration of 30 m*M* and higher (the diastereoisomers present slightly different solubility) and in this form can be easily delivered

[13] N. D. Ridgway, *Biochim. Biophys. Acta* **1256,** 39 (1994).

to cells. An alternative method of delivering short-chain ceramides to cells, which is less acute and perhaps more controlled, is in the form of complex with fatty acid-free bovine serum albumin (BSA).

Preparation of Lipid/BSA Complexes

In most cases complexes of BSA and sphingolipids contain equimolar amounts of the two components, even though other ratios may be used. For the preparation of 0.5 ml of approximately 1 m*M* complex, 500 nmol of the lipid suspended in 25 μl of ethanol is incubated with 475 μl of a 1 m*M* fatty acid-free BSA solution prepared in a suitable buffer (i.e., 50 m*M* Tris, pH 7.4) for 20–30 min at 30°. At this time, the solution should appear clear. If not, 2 cycles of 30 sec of sonication may be performed. If the presence of 1% (v/v) ethanol in the complex interferes with the future employment of the complexes, dialysis at 4° for 6–18 hr may be performed against the same buffer used during the preparation.

Uptake and Relevant Concentrations

The actual cellular concentrations following treatment of cells with a lipid applied extracellularly are determined by a number of factors such as solubility, uptake, and cell density. For example, because of the ability of BSA to complex lipids, the concentration of serum in the medium highly affects the cellular uptake of a given hydrophobic or amphiphilic molecule. Thus treatment with 10 μM D-*erythro*-C_2-ceramide in the presence of 10% serum is not equivalent to a treatment of 10 μM D-*erythro*-C_2-ceramide in the presence of 2% serum or in the absence of serum. In the latter case more "uncomplexed" ceramide is free to diffuse through the membranes, and initial uptake will be significantly higher. Moreover, treatment with 10 μM D-*erythro*-C_2-ceramide at a certain cell density is not equivalent to the same treatment when applied to twice that density of cells. This arises because the relevant concentration of a hydrophobic lipid of interest is the membrane-partitioned amount of that lipid, which is the amount of lipid of interest present in the membranes divided by the total amount of membrane lipids. Doubling the number of cells is equal to doubling the amount of membrane lipids, which leads to half of the relevant concentration (if uptake remains the same).

Measuring Uptake

Uptake is represented by the amount of molecule that associates with the cells. In the case of a treatment with short-chain ceramides (C_2-, C_6-, C_8-), during the first few minutes of incubation (up to 20 min depending on cell density, serum, analog concentration), the cells receive the molecule without significantly metabolizing it. During this time frame, variation of the molecule's content in the medium accounts with good approximation for the actual uptake. Once treated, collect the cells by centrifugation (600*g* at 4° for 5 min) and save the medium. Wash the cell pellet with an appropriate volume of phosphate-buffered saline (PBS) to remove the analog that is only loosely associated with the outside of the plasma membrane. After a second centrifugation, pool the medium and the PBS and determine the radioactivity or fluorescence (depending on the labeling method) present in the mix. Additionally, uptake can be measured by following the presence of the compound in the cells. In this case, after the wash, resuspend the cell pellet with fresh PBS (briefly sonicate if necessary) and measure the associated radioactivity or fluorescence. During a more extended incubation time, the compound reaches higher cellular concentrations and becomes subject to significant metabolism. Most of the metabolites of the short-chain ceramides are sufficiently soluble that they may be lost in the medium. The greater the metabolism, the higher the levels of metabolites that will be present in the medium. In this case a measure of the radioactivity associated with the medium would account not only for the analog still present but also for the metabolites. Therefore, TLC analysis of the medium becomes necessary in order to accurately measure the level of the analog still present. A significant part of the medium has to be subjected to lipid extraction; for example, if 10 ml of medium is used during the incubation, at least 3 ml should be used for analysis. The medium can be extracted according to the protocol of Bligh and Dyer (methanol : chloroform : water, 2 : 2 : 1.8).[14] In this case ceramide and most of the metabolites will partition in the organic phase (lower phase), whereas the more complex glycosphingolipids will partition in the upper phase. TLC analysis can be performed as indicated in the metabolism section (below).

Potency

A major consideration in the use of short-chain ceramides is the determination of an "acceptable" range of concentrations that will maximize the physiologic relevance of the observed effects and minimize any nonspecific effects. A reasonable reference would be using these compounds at a concentration within one or two orders of magnitude higher than the intracellular ceramide levels reached during agonist stimulation. This rule of thumb takes into account two major factors that tend to attenuate the physiologic effectiveness of short-chain ceramides: (1) short-chain ceramides are less

[14] E. G. Bligh and W. J. Dyer, *Can. J. Biochem. Physiol.* **37,** 911 (1959).

potent in activating CAPP *in vitro* than long-chain ceramides[5,15]; therefore, cellular effects mediated by CAPP will require higher concentrations of short-chain ceramides; and (2) it is hypothesized that natural ceramides are produced in specific compartments and therefore tend to achieve higher local concentrations, whereas the short-chain ceramides would partition into most membranes (this probably applies more to C_2- than to C_6-ceramide, which may favor the Golgi apparatus).[12,16]

Based on these considerations, we suggest that in practice a treatment of 2 million cells in 10 ml with 100 nmol of C_6-ceramide (10 μM), when a 50% uptake is observed, achieves concentrations of ceramide within this range. On the other hand, effects observed using concentrations equal to or higher than 40 μM need to be diligently documented with extensive specificity studies, and concentrations as high as 100 μM may result in a detergent-like effect that will disrupt the membranes. Moreover, given the many results observed with short-chain ceramides, it is clear that many effects are observed in the 1–10 μM range whereas other effects, and under similar culture conditions, require 40–100 μM. It seems to us that if the lower concentrations are bioactive, then the higher concentrations raise concerns as to whether they reflect effects of endogenous ceramides. The higher concentrations may indicate the need for further metabolism, cross-reaction with another metabolite (e.g., if ceramide is 10 times less active than sphingosine or ceramide phosphate on a given target or response, then the higher concentrations may be needed for ceramide to mimic the effects of sphingosine, ceramide phosphate, or other metabolites or analogs).

Metabolism

As already discussed, studies with short-chain ceramides must be accompanied by metabolic studies in order to determine whether any by-product has been formed under the working experimental conditions, and eventually it becomes important to discriminate between the direct effects of ceramide and any biologic effects exerted by subsequent metabolites.

Ceramide metabolism can be studied using different approaches. First, the use of labeled ceramides (radioactive/fluorescent) provides a direct way of tracing the formation of any metabolite carrying the label. A second approach would be the inhibition of specific metabolic pathways in order

[15] C. Chalfant, K. Kishikawa, M. C. Mumby, C. Kamibayashi, A. Bielawska, and Y. A. Hannun, *J. Biol. Chem.* **274,** 20313 (1999).

[16] R. E. van Meer and I. L. van Genderen, *Subcell. Biochem.* **23,** 1 (1994).

to avoid accumulation of one or more metabolites (Table I). By now, a few inhibitors are available including (1) D-*threo*-1-phenyl-2-decanoylamino-3-morpholino-1-propanol (D-t-PDMP) and 1-phenyl-2-palmitoylamino-3-pyrrolidino-1-propanol (P4) (5–20 μM),[17,18] which block the cerebroside/glycolipid pathway through inhibition of glucosylceramide synthase; (2) fumonisin B1 (50–100 μM), an inhibitor of ceramide synthase[19]; (3) D-MAPP (D-*erythro*-*N*-myristoyl-2-amino-1-phenylpropanol) (10–20 μM), an inhibitor of alkaline ceramidase,[20] and NOE (*N*-oleoylethanolamine) (50 μM), an inhibitor of the acid ceramidase[21,22]; and (4) the tricyclodecan-9-yl xanthogenate, D609 (10–50 μM), an inhibitor of sphingomyelin synthase[23] (but also widely used as a PC-PLC[24] and PLD inhibitor[25] without direct documentation). Specific inhibition of metabolic routes is potentially also achieved by antisense techniques. Unfortunately the only cloned enzymes involved in the metabolic networks of ceramide are glucosylceramide synthase and acid sphingomyelinase; thus our tools in this area are still very limited.

Lipid Extraction

When harvesting cells after incubation with short-chain analogs, it is important to consider that each cellular wash may enhance the loss of compounds and/or metabolites. Therefore, washing buffers or media need to be combined with the medium from the incubation. In particular, adherent cells can be harvested by scraping them directly in methanol instead of isotonic buffer. This will minimize the loss. Cells can be extracted according to the protocol of Bligh and Dyer[14] as indicated under the uptake section. An alternative method is represented by the one-step total lipid extraction (methanol : chloroform : water, 1 : 1 : 0.1) in which the lipids can

[17] A. Abe, N. S. Radin, J. A. Shayman, L. L. Wotring, R. E. Zipkin, R. Sivakumar, J. M. Ruggieri, K. G. Carson, and B. Ganem, *J. Lipid Res.* **36,** 611 (1995).

[18] L. Lee, A. Abe, and J. A. Shayman, *J. Biol. Chem.* **274,** 14662 (1999).

[19] A. H. Merrill, Jr., E.-M. Schmelz, D. L. Dillehay, S. Spiegel, J. A. Shayman, J. J. Schroeder, R. T. Riley, K. A. Voss, and E. Wang, *Toxicol. Appl. Pharmacol.* **142,** 208 (1997).

[20] A. Bielawska, M. S. Greenberg, D. K. Perry, S. Jayadev, J. A. Shayman, C. McKay, and Y. A. Hannun, *J. Biol. Chem.* **271,** 12646 (1996).

[21] D. A. Wiesner, J. P. Kilkus, A. R. Gottschalk, J. Quintáns, and G. Dawson, *J. Biol. Chem.* **272,** 9868 (1997).

[22] K. Pahan, F. G. Sheikh, M. Khan, A. M. S. Namboodiri, and I. Singh, *J. Biol. Chem.* **273,** 2591 (1998).

[23] C. Luberto and Y. A. Hannun, *J. Biol. Chem.* **273,** 14550 (1998).

[24] S. Schutze, K. Potthof, T. Machleidt, D. Berkovic, K. Wiegmann, and M. Kroenke, *Cell* **71,** 765 (1992).

[25] M. C. M. van Dijk, F. J. G. Muriana, J. de Widt, H. Hilkmann, and V. Blitterswijk, *J. Biol. Chem.* **272,** 11011 (1997).

TABLE I

MOST COMMONLY USED INHIBITORS FOR SOME OF THE ENZYMES REGULATING SPHINGOLIPID METABOLISM[a]

Enzymes	Reaction	Inhibitors	Refs.
SPT	serine + palmitoyl-CoA → 3-ketosphinganine	ISP-1 (0.01–2 μM)	b,c
		L-CS (50–100 μM)	d
Cer Synthase	dihydrosphingosine + fatty acyl-CoA → dihydroceramide sphingosine + fatty acyl-CoA → ceramide	FB1 (30–100 μM)	e
Alkaline CDase	ceramide → sphingosine + fatty acid	D-e-MAPP (5–20 μM)	f,g
Acid CDase	ceramide → sphingosine + fatty acid	NOE (50 μM)	h,i
Sph kinase	sphingosine + ATP → sphingosine-1P	N,N-DMS (1–10 μM)	j,k
GlcCer Synthase	ceramide + UDP-glucose → glucosylceramide	D-t-PDMP (5–20 μM)	l,m
		D-t-P4 (1–5 μM)	
LacCer Synthase	glucosylceramide + UDP-galactose → lactosylceramide	D-t-PDMP (5–20 μM)	l

[a] The concentrations indicated refer mainly to mammalian cellular studies. Cdase, ceramidase; Cer, ceramide; D-MAPP, D-*erythro-N*-myristoyl-2-amino-1-phenylpropanol; D-t-P4, 1-phenyl-2-hexadecanoylamino-3-pyrrolidino-1-propanol; D-t-P-DMP, 1-phenyl-2-decanoylamino-3-morpholino-1-propanol; FB1, fumonisin B1; GlcCer, glucosylceramide; ISP-1 also called myriocin; LacCer, lactosylceramide; L-CS, L-cycloserin; *N,N*-DMS, *N,N*-dimethylsphingosine; NOE, *N*-oleoylethanolamine; Sph, sphingosine; SPT, serine palmitoyltransferase.

[b] Y. Miyake, Y. Kozutsumi, S. Nakamura, T. Fujita, and T. Kawasaki, *Biochem. Biophys. Res. Comm.* **211,** 396 (1995).

[c] S. Furuya, J. Mitoma, A. Makino, and Y. Hirabayashi, *J. Neurochem.* **71,** 366 (1998).

[d] K. S. Sundaram and M. J. Lev, *Nurochem.* **42,** 577 (1984).

[e] A. H. Merrill, Jr., E.-M. Schmelz, D. L. Dillehay, S. Spiegel, J. A. Shayman, J. J. Schroeder, R. T. Riley, K. A. Voss, and E. Wang, *Toxicol. Appl. Pharmacol.* **142,** 208 (1997).

[f] A. Bielawska, M. S. Greenberg, D. Perry, S. Jayadev, J. A. Shayman, C. McKay, and Y. A. Hannun, *J. Biol. Chem.* **271,** 12646 (1996).

[g] Y. Y. Liu, T. Y. Han, A. E. Giuliano, S. Ichikawa, Y. Hirabayashi, and M. Cabot, *Exp. Cell. Res.* **252,** 464 (1999).

[h] D. A. Wiesner, J. P. Killkus, A. R. Gottschalk, J. Quintáns, and G. Dawson, *J. Biol. Chem.* **272,** 9868 (1997).

[i] K. Pahan, F. G. Sheikh, M. Khan, A. M. S. Namboodiri, and I. Singh, *J. Biol. Chem.* **273,** 2591 (1998).

[j] N. Auge, M. Nikolova-Karakashian, S. Carpentier, S. Parthasarathy, A. Negre-Salvayre, R. Salvayre, A. H. Merrill, and T. Levade, *J. Biol. Chem.* **274,** 21533 (1999).

[k] L. Edsall, J. R. Van Brocklyn, O. Cuvillier, B. Kleuser, and S. Spiegel, *Biochemistry* **37,** 12892 (1998).

[l] A. Abe, N. S. Radin, J. A. Shayman, L. L. Wotring, R. E. Zipkin, R. Sivakumar, J. M. Ruggieri, K. G. Carson, and B. Ganem, *J. Lipid Res.* **36,** 611 (1995).

[m] L. Lee, A. Abe, and J. A. Shayman, *J. Biol. Chem.* **274,** 14662 (1999).

be extracted in a monophase. In this case, after overnight incubation in the extracting mixture at 4°, the suspension is centrifuged at 2000g at 4° for 5 min and the supernatant is dried down. The lipid extract is then resuspended in a chloroform:methanol (2:1) mixture and subjected to thin-layer chromatography for analysis.

TLC Analysis

An appropriate TLC solvent system for analysis of metabolism of the short-chain ceramide analogs is the mixture chloroform:methanol:15 mM $CaCl_2$ (60:35:8).[26] In this solvent system ceramides, cerebrosides, and sphingomyelins are well separated. Some R_f values 0.72 for C_2-ceramide, 0.8 for C_6-ceramide, 0.55 for C_2-glucosylceramide, 0.6 for C_6-glucosylceramide, 0.25 for C_2-sphingomyelin, and 0.3 for C_6-sphingomyelin.

Specificity

Because the short-chain ceramides (with the exception of C_2-ceramide)[27,28] are not natural compounds, it is important to determine if the observed effects after cellular treatments are specific for ceramide and if they are biologically relevant. At least three tests can be used to confirm specificity of action. (1) Comparison of ceramide-induced effects vs. dihydroceramide. Dihydroceramide is the natural precursor for ceramide during *de novo* synthesis and differs from ceramide only in the absence of the *trans* double bond in the 4,5 position (Fig. 1). (2) Comparison of D-*erythro*-ceramide-induced effects vs. unnatural stereoisomers. The chemical structure of ceramide is characterized by the presence of two chiral centers, which determine the existence of four conformational stereoisomers: D-*erythro,* D-*threo,* L-*erythro,* and L-*threo* (Fig. 1). Because ceramide is naturally present only in the D-*erythro* conformation, it is particularly useful to compare the activity of the natural stereoisomer versus the others. (3) Comparison of short-chain ceramide analogs with short-chain diacylglycerol (DAG) analogs. Ceramide and DAG are structurally rather similar and demonstrate similar biophysical properties (such as formation of lamellar phases). If ceramide analogs act through nonspecific physicochemical mechanisms, one would expect a similar effect when using DAG analogs; however, sice DAGs activate PKC, whereas ceramides activate CAPPs, their more meaningful cellular activities tend to be opposite to each other.

The comparison of DAG and ceramide is also conceptually revealing.

[26] A. H. Futerman, B. Stieger, A. L. Hubbard, and R. E. Pagano, *J. Biol. Chem.* **265,** 8650 (1990).
[27] T. C. Lee, M. C. Ou, K. Shinozaki, B. Malone, and F. Snyder, *J. Biol. Chem.* **271,** 209 (1996).
[28] A. Abe, J. A. Shayman, and N. S. Radin, *J. Biol. Chem.* **271,** 14383 (1996).

(1) Both DAG and ceramide are hydrophobic neutral lipids that provide the key structural bases for the glycerolipids and sphingolipids, respectively. (2) Ceramide and DAG levels are comparable (ranging from a ratio of ceramide/DAG of 0.1–0.3 to 3–10 in different cells and in response to different stimulations). (3) Both molecules are recognized by protein targets stereospecifically: protein kinase C recognizes *sn*-1,2-DAG[29] and CAPP recognizes D-*erythro*-ceramides.[5,6] (4) A solid body of evidence supports the bioactive functions of DAG through PKC and a developing set of observations is beginning to support a role for CAPP in mediating ceramide effects (see below). (5) Exogenous short-chain DAGs are active in stimulating PKC in cells at concentrations usually in the 1–30 μM range.[30] These are very similar to the concentrations required for bioactivity of short-chain ceramides. This suggest to us that these are "legitimate" and reasonable concentrations of short-chain ceramides, expecially since the endogenous levels of ceramide and DAG are very similar.

Selectivity of Target and Response

In vitro, ceramides show selectivity of target action (e.g., in our hands they activate PP1 and PP2A phosphatases but not other phosphatases or kinases). In cells, ceramides treatments induce many of the biologic responses that can be observed with those agents that induce formation of endogenous ceramide (e.g., TNF, chemotherapeutic agents, heat stress, radiation).[3] Importantly, short-chain ceramides show selectivity of targets and responses at the level of the cell. For example, in TNF models, ceramide induces activation of caspases but not NF-κB,[31] a key transcription factor that may mediate an "antiapoptotic" response. Ceramides will cause inactivation of PKCα[9] but do not inactivate other kinases, such as the MAP kinase. Indeed, sphingosine appears to be a more consistent activator of the MAP kinase.[32]

Mechanism of Action

A final and critical consideration in the use of short-chain ceramides is the determination of their mechanism of action. Current studies show at least 2 or 3 direct targets for ceramide action including CAPPs,[5,6] CAPK,[7]

[29] M. J. O. Wakelam, *Biochim. Biophys. Acta* **1436,** 117 (1998).

[30] M. J. Jones and A. W. Murray, *J. Biol. Chem.* **270,** 5007 (1995).

[31] C. J. Gamard, G. S. Dbaibo, B. Liu, L. M. Obeid, and Y. A. Hannun, *J. Biol. Chem.* **272,** 16474 (1997).

[32] S. Spiegel, O. Cuvillier, L. C. Edsall, T. Kohama, R. Menzeleev, Z. Olah, A. Oliviera, G. Pirianov, D. M. Thomas, Z. Tu, J. R. Van Brocklyn, and F. Wang, *Ann. N.Y. Acad. Sci.* **845,** 11 (1998).

and PKC ζ.[33] If the shorter chain ceramides act specifically through one of these targets, then this resolves any issue concerning nonspecific action. For the two identified CAPPs (protein phosphatase 1 and protein phosphatase 2A) there are a number of potent and specific pharmacologic as well as protein inhibitors that allow a clear determination if the actions of ceramides are through these targets. Unfortunately, there are no known inhibitors yet for either CAPK or PKC ζ. Indeed, a growing body of evidence relates effects of ceramides to activation of CAPPs. These include c-*myc* down-regulation,[34] dephosphorylation of c-jun,[35] dephosphorylation of PKCα,[9] dephosphorylation of bcl-2,[11] and inactivation of PKC/Akt.[36]

Examples: Use of Short-Chain Ceramides in Cell Biology

There are several reported uses of ceramide that satisfy many, if not all, of the above considerations. One example of the efficacy of cellular treatments with short-chain ceramide analogs is represented by the effect that C_6-ceramide exerts on cyclin-dependent kinases (CDKs) in WI38 human lung fibroblasts. The CDKs regulate the progression of the eukaryotic cell cycle, and one of their established substrates is the retinoblastoma (Rb) gene protein. In WI38 cells, ceramide induces Rb dephosphorylation and thus cell cycle arrest. In line with these observations, C_6-ceramide exerts an inhibitory effect on CDK2 activity starting at concentrations as low as 3 μM and reaching its maximum at 10 μM (*relevant concentrations*). This inhibitory effect is limited to CDK2 and in a much lesser extent to CDC2, but not to CDK4 (*target selectivity*). Moreover, these effects are not observed when dihydroceramide or DAG is employed (*specificity of lipid action*). Finally, these effects of ceramide are inhibited by calyculin A, implicating a serine/threonine phosphatase in the mechanism of action (*mechanistic considerations*).

Supporting and Corroborating Lines of Investigation

Careful adherence to basic principles in using any pharmacologic agents (including recombinant material) is essential for appropriate interpretation of results. This is particularly compounded in the case of bioactive lipids,

[33] G. Muller, M. Ayoub, P. Storz, J. Rennecke, D. Fabbro, and K. Pfizenmaier, *EMBO J.* **14,** 1961 (1995).

[34] R. A. Wolff, R. T. Dobrowsky, A. Bielawska, L. M. Obeid, and Y. A. Hannun, *J. Biol. Chem.* **269,** 19605 (1994).

[35] J. G. Reyes, I. G. Robayna, P. S. Delgado, I. H. González, J. Q. Aguiar, F. E. Rosas, L. F. Fanjul, and C. M. R. De Galarreta, *J. Biol. Chem.* **271,** 21375 (1996).

[36] H. Zhou, S. A. Summers, M. J. Birnbaum, and R. N. Pittman, *J. Biol. Chem.* **273,** 16568 (1998).

which tend to "live" in membranes, and therefore raise concerns about either structural effects or nonspecific perturbations. In the case of ceramides, we suggest that close attention to the above criteria will minimize artifacts and will result in more reproducible results that allow progress in the study of ceramide-mediated biology. The study of ceramide metabolism and pharmacology has begun to generate additional tools to evaluate physiologic functions of this molecule. These are discussed next.

Use of Long-Chain Ceramides in Cell Experiments

The finding that permeability/solubility restrictions of the long-chain ceramides can be overcome by delivering the molecules as a 2% (v/v) dodecane solution in ethanol has resulted in a new breakthrough.[37] The application of the natural ceramide (C_{18}) at low nanomolar concentrations resulted in clear induction of apoptosis in U937 promonocytic cells. This observation confirms the results already acquired through the use of short-chain analogs, reducing the active concentrations to a range comparable to cellular ceramide levels.

Use of Inhibitors of Enzymes of Ceramide Metabolism

As discussed above, both D-MAPP and P4 (or PDMP) inhibit enzymes of ceramide clearance and therefore lead to accumulation of ceramide (Table I). In both cases, this is accompained by similar responses to those seen with the short-chain analogs (such as growth arrest, apoptosis, differentiation), demonstrating that the short-chain ceramides mimic the action of endogenous ceramide.[1]

Use of Recombinant Enzymes of Ceramide and Sphingolipid Metabolism

Finally, the induction of increased natural ceramide levels in a physiologic site through the overexpression of ceramide-producing enzymes (i.e., bacterial sphingomyelinase) recapitulates many, if not all, of the responses observed during treatment with short-chain ceramides (caspases activation, Rb dephosphorylation, growth arrest, DNA fragmentation).[38] Reciprocally, overexpression of ceramide clearing enzymes (most notably glucosylceramide synthase) attenuates the levels of ceramide in response to several cytotoxic agents[39] (and also TNF[40]). This results in amelioration and protec-

[37] L. Ji, G. Zhang, S. Uematsu, Y. Akahori, and Y. Hirabayashi, *FEBS Lett.* **358,** 211 (1995).
[38] P. Zhang, B. Liu, G. M. Jenkins, Y. A. Hannun, and L. M. Obeid, *J. Biol. Chem.* **272,** 9609 (1997).
[39] Y.-Y. Liu, T.-Y. Han, A. E. Giuliano, and M. C. Cabot, *J. Biol. Chem.* **274,** 1140 (1999).
[40] Y. Y. Liu, T. Y. Han, A. E. Giuliano, S. Ichikawa, Y. Hirabayashi, and M. C. Cabot, *Exp. Cell. Res.* **252,** 464 (1999).

tion from the cytotoxic effects of these molecules, strongly supporting a role for endogenous ceramide in these responses.

Conclusions

Taken together, these observations provide evidence of the important informative power of short-chain ceramide analogs. More specific inhibitors of the enzymes involved in the sphingolipid metabolism are on the way, as well as the enzymes themselves; this will be the future way to go.

Acknowledgments

This work was supported in part by NIH grant GM 43285. We thank David K. Perry for critical reading of this manuscript.

[34] Analysis of Ceramide-Activated Protein Phosphatases

By Charles E. Chalfant, Katsuya Kishikawa, Alicja Bielawska, and Yusuf A. Hannun

Introduction

The sphingomyelin cycle has emerged as a major player in the signaling pathways of apoptosis, cell differentiation, and cell cycle arrest. Ceramide is a product of the sphingomyelin cycle, and several lines of evidence point to an important role for ceramide in stress responses and growth suppression. First, the inducers of the sphingomyelin cycle relate to the common theme of growth suppression, differentiation, stress responses, inducers of cell damage, and apoptosis. These include tumor necrosis factor α (TNFα) γ-interferon, 1,α25-dihydroxyvitamin D_3, interleukin 1 (IL-1), ultraviolet light, heat, chemotherapeutic agents, FAS antigen, and nerve growth factor.[1–9] Second, the addition of exogenous ceramide or the en-

[1] T. Okazaki, R. M. Bell, and Y. A. Hannun, *J. Biol. Chem.* **264,** 19076 (1989).

[2] M.-Y. Kim, C. Linardic, L. Obeid, and Y. Hannun, *J. Biol. Chem.* **266,** 484 (1991).

[3] L. R. Ballou, C. P. Chao, M. A. Holness, S. C. Barker, and R. Raghow, *J. Biol. Chem.* **267,** 20044 (1992).

[4] M. Liscovitch, *Trends Biochem. Sci.* **17,** 393 (1992).

[5] Y. A. Hannun, *J. Biol. Chem.* **269,** 3125 (1994).

[6] J. Quintans, J. Kilkus, C. L. McShan, A. R. Gottshalk, and G. Dawson, *Biochem. Biophys. Res. Commun.* **202,** 710 (1994).

 0076-6879/00 $30.00

hancement of cellular levels of ceramide induces cell differentiation, cell cycle arrest, apoptosis, or cell senescence in various cell types.[4,10,11] Third, mechanistic insight relates the action of ceramide to key regulators of growth such as the retinoblastoma gene product (Rb), caspases, Bcl-2, and p53.[12–18] Fourth, studies in yeast have demonstrated an essential role for sphingolipids in many stress responses where ceramide may function in the adaptation to heat.[19] Finally, studies with knockout mice lacking acid sphingomyelinase or with fumonisin B1, an inhibitor of ceramide synthesis, have disclosed necessary roles for ceramide in growth regulation.[20,21]

These emerging roles of ceramide have necessitated a mechanistic understanding of ceramide action. The search for potential direct targets for ceramide action has led to the identification of two candidate ceramide-regulated enzymes, one of which is a ceramide-activated protein kinase (CAPK), and the other is a ceramide-activated protein phosphatase (CAPP).[22,23] Originally, CAPP was identified as a member of the 2A class of serine/threonine phosphatases (PP2A).[22–24] This was based on studies

---

[7] M. G. Cifone, R. De Maria, P. Roncaioli, M. R. Rippo, M. Azuma, L. L. Lanier, A. Santioni, and R. Testi, *J. Exp. Med.* **180,** 1547 (1994).

[8] J. C. Strum, G. W. Small, S. B. Pauig, and L. W. Daniel, *J. Biol. Chem.* **269,** 15493 (1994).

[9] A. Haimovitz-Friedman, C.-C. Kan, and D. Ehleiter, *J. Exp. Med.* **180,** 525 (1994).

[10] T. Okazaki, A. Bielawska, R. M. Bell, and Y. A. Hannun, *J. Biol. Chem.* **265,** 15823 (1990).

[11] L. M. Obeid, C. M. Linardic, L. A. Karolak, and Y. A. Hannun, *Science* **259,** 1769 (1993).

[12] C. J. Gamard, G. S. Dbaibo, B. Liu, L. M. Obeid, and Y. A. Hannun, *J. Biol. Chem.* **272,** 16474 (1997).

[13] G. S. Dbaibo, M. Y. Pushkareva, S. Jayadev, J. K. Schwartz, J. M. Horowitz, L. M. Obeid, and Y. A. Hannun, *Proc. Natl. Acad. Sci. U.S.A.* **92,** 1347 (1995).

[14] K. Cain, S. H. Inayat-Hussein, C. Couet, and G. M. Cohen, *Biochem. J.* **314,** 27 (1996).

[15] G. S. Dbaibo, M. Y. Pushkareva, R. A. Rachid, N. Alter, M. J. Smyth, L. M. Obeid, and Y. A. Hannun, *J. Clin. Invest.* **102,** 329 (1998).

[16] M. J. Smyth, D. K. Perry, J. Zhang, G. G. Poirier, Y. A. Hannun, and L. M. Obeid, *Biochem. J.* **316,** 25 (1996).

[17] N. Mizushima, R. Koike, H. Kohsaka, *et al., FEBS Lett.* **395,** 267 (1996).

[18] J. Zhang, N. Alter, J. C. Reed, C. Borner, L. M. Obeid, and Y. A. Hannun, *Proc. Natl. Acad. Sci. U.S.A.* **93,** 5325 (1996).

[19] G. M. Jenkins, A. Richards, T. Wahl, C. Mao, L. M. Obeid, and Y. A. Hannun, *J. Biol. Chem.* **272,** 32566 (1997).

[20] P. Santana, L. A. Pena, A. Haimovitz-Friedman, *et al., Cell* **86,** 189 (1996).

[21] A. H. Merrill, Jr., E-M. Schmelz, D. L. Dillehay, S. Spiegel, J. A. Shayman, J. J. Schroeder, R. T. Riley, K. A. Voss, and E. Wang, *Toxicol. Appl. Pharmacol.* **142,** 208 (1997).

[22] R. T. Dobrowsky, C. Kamibayashi, M. C. Mumby, and Y. A. Hannun, *J. Biol. Chem.* **268,** 15523 (1993).

[23] R. A. Wolff, R. T. Dobrowsky, A. Bielawska, L. M. Obeid, and Y. A. Hannun, *J. Biol. Chem.* **269,** 19605 (1994).

[24] K. Kishikawa, J. Y. Lee, C. E. Chalfant, A. Bielawska, S. H. Galadari, L. M. Obeid, and Y. A. Hannun, submitted for publication (1998).

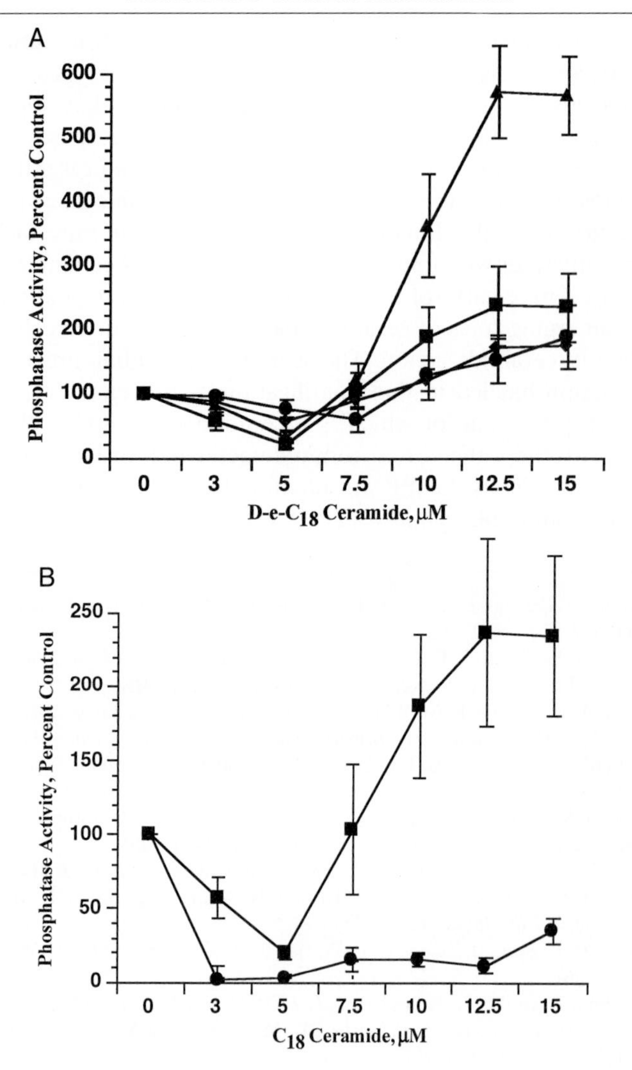

FIG. 1. (A) Activation of PP1αc, -γc, PP2Ac, and PP2A (AB′C) by long-chain ceramides *in vitro.* D-*erythro*-C_{18}-ceramide was solubilized at 37° in a 2% dodecane/98% ethanol solution and immediately added to the assay reactions at concentrations of 0, 3, 5, 7.5, 10, 12.5, and 15 μM (final dodecane concentration 0.02%, final ethanol concentration 0.98%). Enzyme preparations were then added and reactions were initiated by the addition of [^{32}P]MBP. Reactions were incubated at 30° for 20 min, terminated, and quantitated as described in this chapter. Results are expressed as percent of initial activity in the absence of D-*erythro*-C_{18}-ceramide. Data are mean ± S.E. of at least triplicate experiments reproduced on at least three separate occasions. The different phosphatases are designated: (▲) PP2A, (■) PP1αc, (◆) PP2Ac, and (●) PP1γc. (B) Effects of ceramide stereoisomers on protein phosphatases. Activation of PP1αc by long-chain ceramides is stereospecific. C_{18}-ceramide stereoisomers

using purified and reconstituted PP2A.[22,23] However, recently, we have also demonstrated that protein phosphatase-1 (PP1) is a target for ceramide and functions as a novel CAPP.[24] The specificity for CAPP activation *in vitro* closely resembled the specificity for various cellular activities of ceramide such as apoptosis.[22,23] Possible downstream targets for CAPP include c-Myc, c-Jun, and Rb, with the latter of these two serving as possible direct substrates.[3,5,24–28] These results indicate an important role for phosphatases in imparting apoptosis and growth inhibition. In this chapter, we describe the *in vitro* action of long-chain ceramides on these two CAPP phosphatases.

Assay Methods

Principle and Definition of Unit

For the following assays, myelin basic protein (MBP) is used as a substrate for PP1 and PP2A. For the purposes of defining phosphatase activity units, PP-1α and -γ are assayed by measuring the dephosphorylation of PNPP (although a variety of other substrates can be used). One unit of activity is defined as the amount of enzyme that will hydrolyze 1.0 nmol of PNPP per minute at 30°, pH 7.0. PP2A trimer and catalytic subunit are assayed by measuring the dephosphorylation of ^{32}P-labeled phosphorylase (although a variety of other substrates can be used). One unit of activity is defined as the amount of enzyme that will hydrolyze 1.0 nmol of phosphorylase per minute at 30°C, pH 7.0. In the studies described here, we use ^{32}P-labeled myelin basic protein (MBP) as a substrate for PP1 and PP2A.

[25] J. G. Reyes, I. G. Robayna, P. S. Delgado, I. H. Gonzalez, J. Q. Aguiar, F. E. Rosas, L. F. Fanjul, and C. M. R. Galarreta, *J. Biol. Chem.* **271,** 21375 (1996).
[26] J. Lozano, E. Berra, M. M. Municio, *et al., J. Biol. Chem.* **269,** 19200 (1994).
[27] M. E. Venable, G. C. Blobe, and L. M. Obeid, *J. Biol. Chem.* **269,** 26040 (1994).
[28] A. Gomez-Munoz, A. Martin, L. O'Brien, and D. N. Brindley, *J. Biol. Chem.* **269,** 8937 (1994).

---

were solubilized at 37° in a 2% dodecane/98% ethanol solution and immediately added to the assay reactions at concentrations of 0, 3, 5, 7.5, 10, 12.5, and 15 μM (final dodecane concentration 0.02%, final ethanol concentration 0.98%). Enzyme preparations were rapidly added and reactions were initiated by the addition of [^{32}P]MBP. Reactions were incubated at 30° for 20 min, terminated, and quantitated as described in the Materials and Methods section. Results are expressed as percent of initial activity in the absence of C_{18}-ceramide. Data are mean ± S.E. of at least triplicate experiments reproduced on at least three separate occasions. Stereoisomers are designated: (■) D-*erythro*-C_{18}-ceramide and (●) L-*erythro*-C_{18}-ceramide.

Reagents

MBP purified from bovine brain, adenosine triphosphate (ATP), protein kinase A (PKA) purified from bovine heart, and 2-mercaptoethanol are obtained from Sigma (St. Louis, MO). MBP is solubilized on ice with buffer A to a concentration of 25 mg/ml. The MBP solution is aliquoted at 40 μl per 1.5-ml microcentrifuge tube (1 mg MBP/reaction tube) and stored at $-20°$ until use. ATP is solubilized on ice with buffer A to a concentration of 40 m*M*. The 40 m*M*. ATP solution is aliquoted and stored at $-20°$. PKA is stored at $-20°$ receipt from the company and solubilized to a final concentration of 0.05 mg/ml using 6 mg/ml dithiothreitol (DTT) for 10 min just prior to use. 2-Mercaptoethanol is stored at room temperature after receipt from the company and diluted in water to a final concentration of 200 n*M* just prior to use. DTT is obtained from Bachem and stored at $-20°$ after dilution in water to 0.5 *M*. Dodecane is obtained from Aldrich Chemical Company (Milwaukee, WI). [γ-^{32}P]ATP (3000 Ci/mmol) is obtained from NEN Life Science. C_{18}-ceramides are synthesized as described.[29]

PP2A trimer (AB′C) and catalytic subunit (purified from bovine heart) were obtained from Drs. Craig Kamibayashi and Marc Mumby (Department of Pharmacology, University of Texas Health Science Center, Dallas, TX). Recombinant human PP1γc and rabbit PP1αc purified from *Escherichia coli* is obtained from Calbiochem-Novabiochem Corporation (La Jolla, CA).

Buffers

Buffer A: 50 m*M* Tris-HCl, pH 7.4
Buffer B: 50 m*M* Tris-HCl, pH 7.4, 150 m*M* KCl, and 0.1 m*M* dithiothreitol (DTT)

Procedures

Substrate Labeling

Myelin Basic Protein. Reactions are carried out in 1.5-ml screw-cap polypropylene microcentrifuge tubes. The vial already containing 40 μl MBP (25 mg/ml) is thawed on ice. Added in order to the reaction vial are:

10 μl 500 m*M* Tris-HCl, pH 7.4
50 μl 90 m*M* $MgCl_2$−1 m*M* ATP solution
5 μl 500 m*M* DTT

[29] A. Bielawska, C. M. Linardic, and Y. A. Hannun, *J. Biol. Chem.* **267,** 18493 (1992).

25 μl 200 mM 2-Mercaptoethanol
2 μl [γ-^{32}P]ATP
100 μl PKA
H_2O is added to a 500-μl reaction volume (~268 μl)

Final reaction concentrations of all components are 50 mM Tris-HCl, pH 7.4, 9 mM $MgCl_2$, 0.1 mM ATP, 5 mM DTT, 10 mM 2-mercaptoethanol, 2 μCi [γ-^{32}P]ATP, and 125 U of protein kinase A. After the components are mixed, the reaction is incubated at 37° for 2 hr. Ice-cold 100% trichloroacetic acid (TCA) is then added (0.17 ml), mixed, and incubated on ice for 30 min. The reaction is then centrifuged for 10 min at 13,000g in a Sorvall MC12 microcentrifuge, and the supernatant is then removed. The pellet is washed twice with −20° acetone (1.0 ml), dried, and reconstituted in buffer A.

HISTONE. 2 mg of histone is labeled as described for MBP.

Solubilization of Long-Chain Ceramides

Long-chain ceramides are solubilized in 100% ethanol at a concentration of 10 mM by incubation at 37° for 1–3 hr and frequent vortexing. Long-chain ceramides are then diluted to a 5 mM ceramide/2% dodecane/98% ethonol stock solution using a 4% dodecane/96% ethanol. This stock solution is then repeatedly chilled on ice and solubilized at 37° 5–7 times. The solution is then is stored at −20° until needed. To make dilutions or use the stock solution directly in the assays, the solution is incubated at 37° for at least 10 min with frequent vortexing immediately prior to use. Any dilutions made from the stock solution must also be incubated at 37° for 10 min with frequent vortexing immediately prior to use.

Phosphatase Assays

PROTEIN PHOSPHATASE-1 αc. Reactions are carried out in 13- × 100-mm borosilicate glass tubes. After ceramides are solubilized, they are added to tubes containing 50 mM Tris-HCl, pH 7.4, with the ethanol concentration not exceeding 1%. Stock enzyme is diluted to 1 U/ml in buffer A and 10 mU is added to each tube. Components are mixed well and preincubated for 10 min at 30°. The reaction is initiated with 0.005 ml of ^{32}P-labeled myelin basic protein (MBP) (1 mg/ml) in buffer A. After 20 min at 30°, the assay is terminated by the addition of 1 mM KH_2PO_4 in 1 N H_2SO_4 (0.1 ml). The released phosphate is then complexed with molybdate by the addition of 2% ammonium molybdate (0.3 ml), and the reaction is allowed to stand 10 min at ambient temperature. The phosphate : molybdate complex is then extracted from nonhydrolyzed [^{32}P]MBP by the addition of isobutanol : toluene (1 : 1) (1.0 ml) and by vortexing each reaction 10 sec. The reactions are then centrifuged at 1000*g* for 10 min and an aliquot of the upper organic phase is removed (0.5 ml), mixed with scintillant, and counted. A control incubation is performed in which the phosphatase is replaced by buffer A. Under these conditions, D-*erythro*-C_{18}-ceramide activates PP1αc (237% of control, EC_{50} 8.4 μM) (Fig. 1A).

PROTEIN PHOSPHATASE-1γc. PP1γc is assayed as described for PP1αc. The enzyme is diluted to 1 U/ml in buffer A and added at 10 mU per reaction. Under these conditions, D-*erythro*-C_{18}-ceramide activates PP1γc (181% of control, EC_{50} 11.25 μM) (Fig. 1A).

PROTEIN PHOSPHATASE 2A. PP2A (AB′C) is assayed as described for PP1α_c. The trimer is diluted to 1 U/ml in buffer A and 10 mU are added to each reaction. Under these conditions, D-*erythro*-C_{18}-ceramide activates PP2A (582% of control, EC_{50} 9.15 μM) (Fig. 1A).

PROTEIN PHOSPHATASE 2Ac. PP2Ac is assayed as described for PP-1αc. The enzyme is diluted to 1 U/ml and 10 mU are added to each reaction.

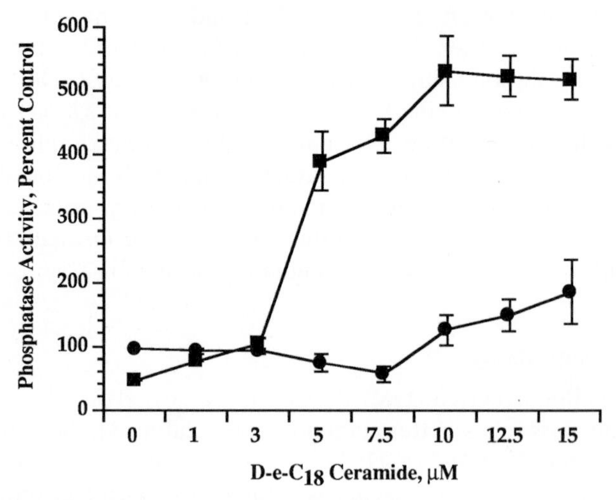

FIG. 2. Effect of physiological salt on ceramide-activated protein phosphatase. Dose response of PP1αc under physiologic salt conditions. Ceramide effects on PP1αc were assayed in (■) physiologic salt buffer (50 mM Tris-HCl, pH 7.4, 150 mM KCl, and 0.1 mM DTT) or (●) without the addition of salt (50 mM Tris-HCl, pH 7.4, and 0.1 mM DTT). Solubilized D-*erythro*-C_{18}-ceramide was added to the assay reactions at concentrations of 0, 5, 10, and 15 μM (final dodecane concentration 0.02%, final ethanol concentration 0.98%). PP1αc reactions were initiated by the addition of [^{32}P]MBP. Reactions were incubated at 30° for 20 min, terminated, and quantitated as described in this chapter. Results are expressed as percent of initial activity in the absence of D-*erythro*-C_{18}-ceramide and cations. Data are mean ± S.E. of at least triplicate experiments reproduced on at least three separate occasions.

Under these conditions, D-*erythro*-C_{18}-ceramide activates PP2Ac (172% of control, EC_{50} 11 μM) (Fig. 1A).

STEREOSPECIFICITY. Using the above conditions, activation of these phosphatases is stereospecific. This is demonstrated by the inability of L-*erythro*-C_{18}-ceramide, the mirror image of D-*erythro*-C_{18}-ceramide, to activate any of the phosphatases examined (Fig. 1B).

Effects of Salt, Reducing Agents, and Cations on Ceramide Responsiveness

ADDITION OF Mn^{2+} TO PHOSPHATASE ASSAY. Prebinding the enzymes with 1 mM $MnCl_2$ increases both basal activity and ceramide responsiveness, but the fold change in ceramide responsiveness is not affected (data not shown). Other cations such as Zn^{2+}, Fe^{2+}, and Fe^{3+} are inhibitory to the phosphatases and ceramide responsiveness.

Reducing agents such as glutathione and dithiothreitol are inhibitory to the phosphatases at levels as low as 1 mM. On the other hand, the addition of a very small amount of DTT (0.1 mM) actually enhances ceramide responsiveness without affecting basal activity. These reducing agents affect only the basal activity and do not affect the total ceramide activation. For example, in the presence of 10 mM glutathione, PP1αc is activated by D-*erythro*-C_{18}-ceramide to 423 fmoles P_i released/minute compared to acti-

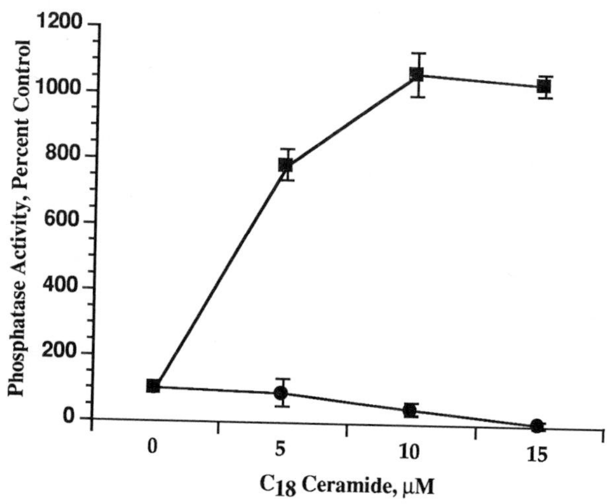

FIG. 3. Activation of PP1αc in the presence of physiologic salt is stereospecific. PP1αc was assayed as described in this chapter. Results are expressed as percent of initial activity in the absence of C_{18} ceramide and cations. Data are mean ± S.E. of at least triplicate experiments reproduced on at least three separate occasions. Data are presented as % control with (■) designating D-*erythro*-C_{18}-ceramide and (●) designating L-*erythro*-C_{18}-ceramide.

vation in the absence of glutathione of 441 fmoles P_i released/minute. Although the total activation is not significantly changed, the decrease in basal activity increases the fold response from 237% of control to 3115% of control.

Addition of physiologic salt such as 150 m*M* KCl significantly enhances ceramide activation of the phosphatases. D-*erythro*-C_{18}-ceramide can now activate the phosphatases at high concentrations without being solubilized with dodecane. Therefore the salt concentration may have an effect on ceramide solubility. D-*erythro*-C_{18}-ceramide solubilized in dodecane now will activate PP1αc at a lower dose (1–3 μM) and to a greater extent (~525% of control in the absence of salt) (Fig. 2). Stereospecificity is still retained and remarkable under these conditions as L-*erythro*-C_{18}-ceramide does not activate PP1αc (Fig. 3). PP1γc, PP2A (AB′C), and PP2Ac are also more responsive to D-*erythro*-C_{18}-ceramide in the presence of salt, demonstrating lower EC_{50} values and higher fold stimulations. Stereospecificity is also retained for these three phosphatases.

Optimal Conditions for Assaying Protein Phosphatases for Long-Chain Ceramide Responsiveness

The optimal conditions for assaying protein phosphatases for stereospecific ceramide responsiveness are 50 m*M* Tris-HCl, pH 7.4, 150 m*M* KCl, and 0.1 *M* DTT (buffer B). The enzyme should be prebound with 1 m*M* $MnCl_2$ for 20 min prior to addition to the reaction tube (Fig. 3).

Acknowledgments

This work was supported in part by the National Institutes of Health grant GM43825 and by the Department of Defense grant AIBS 512. We thank Dr. David Perry for careful review of the manuscript.

[35] Use of Affinity Chromatography and TID-Ceramide Photoaffinity Labeling for Detection of Ceramide-Binding Proteins

By STEFAN SCHÜTZE, MARC WICKEL, MICHAEL HEINRICH, SUPANDI WINOTO-MORBACH, THOMAS WEBER, JOSEF BRUNNER, and MARTIN KRÖNKE

Introduction

The neutral lipid ceramide has been recognized as a common intracellular second messenger for various exogenous stimuli including cytokines like tumor necrosis factor (TNF), interferon-γ, interleukin-1 (IL-1), nerve growth factor (NGF), or CD28 and CD95 (Fas/APO-1) receptor triggering, ionizing radiation, glucocorticoids, anticancer drugs, and serum deprivation.[1–3] The biologic responses to ceramide range from induction of proliferation and differentiation to cell-cycle arrest. The most prominent role of ceramide is the induction of apoptosis in various cell types. Ceramide is produced either by *degnovo* synthesis involving ceramide synthetase located in the Golgi, by hydrolysis of sphingomyelin engagin sphingomyelinases (SMase), or by deglycosylation of glucosylceramide. Two distinct SMases, a plasma membrane-bound N-SMase and an endolysosomal A-SMase are activated in response to ligation of the p55 TNF receptor (TR55), interleukin-1 receptor 1 (IL-1 R1) and the Fas receptor. Each type of SMase generates the second messenger ceramide, however with different kinetics and, most importantly, at different intracellular locations.

The identification of direct targets of ceramide is crucial to an understanding of the role of ceramide in cell signaling. Several enzymes have been described recently that can be activated by ceramide, including a ceramide-activated protein kinase (CAPK)[4] that has been suggested to be related to the kinase suppressor of Ras (KSR).[5] Further proteins that have been described in the literature as being ceramide responsive include

[1] B. Liu, L. M. Obeid, and Y. A. Hannun, *Semi. Cell Develop. Biol.* **8,** 311 (1997).

[2] S. Spiegel and A. H. Merrill, Jr., *FASEB J.* **10,** 1388 (1996).

[3] D. K. Perry and Y. A. Hannun, *Biochim. Biophys. Acta* **1436,** 233 (1998).

[4] S. Mathias, K. A. Dressler, and R. Kolesnick, *Proc. Natl. Acad. Sci. U.S.A.* **88,** 10009 (1991).

[5] Y. Zhang, B. Yao, S. Delikat, S. Bayoumy, X.-H. Lin, S. Basu, M. McGinley, P.-Y. Chan-Hui, H. Lichenstein, and R. Kolesnick, *Cell* **98,** 63 (1997).

0076-6879/00 $30.00

protein phosphatase 2A (CAPP),[6] protein kinase C ξ,[7] α, and δ,[8] protein kinase Raf-1,[9,10] the stress-activated/c-*jun* N-terminal protein kinase (JNK),[11–13] and a CPP32 (caspase 3) like apoptotic protease.[14–19] However, many of the reported ceramide responses are apparently indirect, probably involving unknown intermediates. Evidence for a direct physical interaction of ceramide has been difficult to obtain. Most importantly, the specificity of ceramide binding and activation of the target proteins should be demonstrated with competitive ceramide isomers and unrelated lipids.

As a tool for the identification of ceramide binding proteins, we describe here the application of (1) affinity chromatography based on coupling sphingosine to Sepharose-beads and (2) photoaffinity labeling using a radioiodinated analog of ceramide, 3-trifluoromethyl-3-(*m*-[^{125}I]iodophenyl) diazirine ceramide ([^{125}I]TID-ceramide).

Ceramide-Affinity Chromatography

In order to screen for intracellular ceramide-binding proteins, we generated a ceramide affinity matrix by coupling D-*erythro*-sphingosine via an aminohexanoic acid spacer to activated CH Sepharose 4B, resulting in an immobilized short acyl chain (C_6-) D-*erythro*-ceramide. The alkyl chain remains exposed to the fluid phase to interact with a hydrophobic cavity

[6] R. T. Dobrowski, C. Kamibayashi, M. C. Mumby, and Y. Hannun, *J. Biol. Chem.* **268,** 15523 (1993).

[7] G. Müller, M. Ayoub, P. Storz, J. Rennecke, D. Fabbro, and K. Pfizenmaier, *EMBO J.* **14,** 1961 (1995).

[8] A. Huwiler, D. Fabbro, and J. Pfeilschifter, *Biochemistry* **37,** 14556 (1999).

[9] A. Huwiler, J. Brunner, R. Hummel, M. Vervoodeldonk, S. Stabel, H. van den Bosch, and J. Pfeilschifter, *Proc. Natl. Acad. Sci. U.S.A.* **93,** 6959 (1996).

[10] G. Müller, P. Storz, S. Bourteele, H. Döppler, and K. Pfizenmaier, *EMBO J.* **17,** 732 (1998).

[11] J. K. Westwick, A. E. Bielawska, G. Dbaibo, Y. A. Hannun, and D. A. Brenner, *J. Biol. Chem.* **270,** 22689 (1995).

[12] C. Huang, W. Ma, M. Ding, G. T. Bowden, and Z. Dong, *J. Biol. Chem.* **272,** 27753 (1997).

[13] S. Schütze, T. Machleidt, D. Adam, R. Schwandner, K. Wiegmann, M.-L. Kruse, M. Heinrich, M. Wickel, and M. Krönke, *J. Biol. Chem.* **274,** 10203 (1999).

[14] M. J. Smyth, D. K. Perry, J. Zhang, G. G. Poirier, Y. Hannun, L. M. Obeid, *Biochem. J.* **316,** 25 (1996).

[15] L. Monney, R. Olivier, I. Otter, B. Jansen, G. G. Poirier, and C. Borner, *Eur. J. Biochem.* **251,** 295 (1998).

[16] J. Zhang, N. Alter, J. C. Reed, C. Borner, L. M. Obeid, and Y. Hannun, *Proc. Natl. Acad. Sci. U.S.A.* **93,** 5325 (1996).

[17] S. Yoshimura, Y. Banno, S. Nakashima, K. Takenaka, H. Sakai, Y. Nishimura, N. Sakai, S. Shimizu, Y. Eguchi, Y. Tsujimoto, and Y. Nozawa, *J. Biol. Chem.* **273,** 6921 (1998).

[18] N. Mizushima, R. Koike, H. Kohsaka, Y. Kushi, S. Handa, H. Yagita, and N. Miyasaka, *FEBS Lett.* **395,** 267 (1996).

[19] R. Anjum, A. M. Ali, Z. Begum, J. Vanaja, and A. Khar, *FEBS Lett.* **439,** 81 (1998).

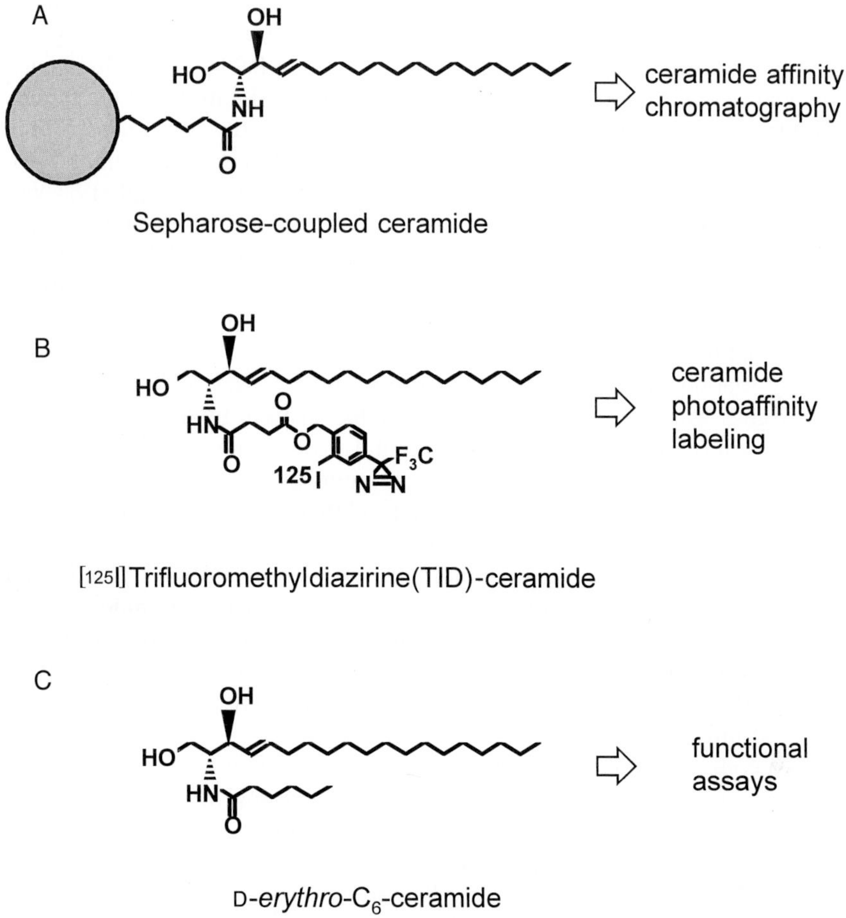

A

Sepharose-coupled ceramide → ceramide affinity chromatography

B

[125I]Trifluoromethyldiazirine(TID)-ceramide → ceramide photoaffinity labeling

C

D-*erythro*-C_6-ceramide → functional assays

FIG. 1. Identification of ceramide binding proteins by ceramide affinity chromatography, [^{125}I]TID-ceramide photoaffinity labeling, and synthetic C_6-ceramide. The structures and uses of the Sepharose-coupled D-*erytho*-sphingosine (A), the photoreactive ceramide analog [^{125}I]TID-ceramide (B), and D-*erythro*-C_6-ceramide (C) are shown.

of a signaling protein (Fig. 1A). The alkyl chain in ceramide appears to be critically involved in binding to proteins.[20,21]

Coupling of amino group-containing ligands to CH Sepharose 4B is normally carried out in aqueous buffer (see instructions by Pharmacia).

[20] M. Krönke, *Cytokine Growth Factor Rev.* **8,** 103 (1997).

[21] M. Heinrich, M. Wickel, W. Schneider-Brachert, C. Sandberg, J. Gahr, R. Schwandner, T. Weber, J. Brunner, M. Krönke, and S. Schütze, *EMBO J.* **18,** 5252 (1999).

Because sphingosine is barely soluble in water, we established nonaqueous coupling conditions. To do this, a related radiolabeled compound, *N*-([^{14}C]-6-aminohexanoyl)-D-*erythro*-sphingosine (4.0 mg; 0.288 × 10^6 dpm), was incubated with ca. 600 μl (corresponding to approximately 200 mg freeze-dried material) of activated CH Sepharose 4B in anhydrous tetrahydrofuran (THF) in the presence of 10 μl of triethylamine. After incubation at 4° for 3 hr on a wheel, 50 μl of ethanolamine was added and incubation was continued for 2 hr. The gel was then transferred on a filter and washed with THF (2 × 2 ml), methanol (1 × 2 ml), and 1 m*M* HCl (1 × 2 ml). Scintillation counting revealed that under the above coupling conditions at least 86% of the amino group-containing ligand became covalently bound to the Sepharose matrix.

By loading the ceramide-Sepharose affinity matrix with lysates from U937 cells followed by elution of the bound protein by an excess of ceramide, approximately 10% of the original protein applied to the matrix was recovered. Among them, several proteins are selectively enriched in the affinity eluates as compared to the total proteins prior to affinity chromatography.[22]

The specificity of ceramide binding is revealed by elution of the bound protein by D-*erythro*-ceramide versus D-*erythro*-dihydroceramide and the application of various unrelated lipids like diacylglycerol or free fatty acid. As an additional control for ceramide specificity, the dihydro form of D-*erythro*-dihydrosphingosine (sphinganine) is coupled to Sepharose and used as an affinity matrix.

To confirm the ceramide-binding characteristics of a ceramide-affinity purified protein, additional experiments are necessary. First, the proteins that were eluted from the D-*erythro*-ceramide affinity matrix are reprobed for their capacity to bind [^{125}I]TID ceramide (Fig. 1B; see section below on [^{125}I]TID-cermide photoaffinity labeling).

In addition, the physiologic significance of ceramide binding on the functional properties of the identified protein has to be tested, employing D-*erythro*-C_6-ceramide (Fig. 1C) and D-*erythro*-C_6-dihydroceramide as well as other unrelated lipids as controls.

Ceramide-binding proteins eluted from the affinity column can then be identified either by immunoblot screening using selected antibodies against presumed candidate targets, or by microsequencing of the affinity-purified proteins and data bank searches for homologs. Following this approach, we recently identified the aspartatic protease cathepsin D as a novel ceramide target.[21]

[22] M. Wickel, M. Heinrich, J. Brunner, T. Weber, M. Krönke, and S. Schütze, *Biochem. Soc. Trans.* **27,** 1 (1999).

Preparation of Affinity Matrix

Three grams of activated CH-Sepharose 4B (Amersham Pharmacia Biotech, SE-75184) is resuspended in 1 m*M* HCl for 2 hr at room temperature to obtain approximately 10 ml of swollen gel. Subsequently, activated CH-Sepharose 4B is washed (5 min at 500*g*) in increasing concentrations of THF: 30% THF, 70% THF, and 100% THF to provide anhydrous coupling conditions. Then 50 mg D-*erythro*-sphingosine (or D-*erythro*-dihydrosphingosine for negative control) is added in 6 ml anhydrous THF and coupled to the activated CH-Sepharose 4B following addition of 100 μl *N*-ethylmorpholine. The reaction between the *N*-hydroxysuccinimide ester in the activated CH-Sepharose 4B and the amine in sphingosine result in the formation of an amide bond and release of *N*-hydroxysuccinimide. After overnight incubation at 4° on a wheel, the affinity matrix is washed 3 times in 100% THF at 4°. Residual active sites are then blocked by ethanolamine (2 hr at room temperature). The affinity matrix is finally washed 3 times in 100% THF and 3 times in H_2O and stored in the presence of 0.1% sodium azide.

The affinity matrix is then transferred to an appropriate column, depending on the amount of protein to be analyzed. As an example, for 10 ml of packed gel, we used an XK 16/20 column (Amersham Pharmacia Biotech), 16 mm in diameter and 200 mm long. Prior to use, the affinity column is washed with 50 ml 1M potassium thiocyanate (KSCN), 1% Triton X-100 (v/v), followed with 100 ml of equilibration buffer (40 m*M* HEPES, pH 7.4, 150 m*M* KCl, 0.075% Triton X-100).

Preparation of Cell Extracts

2×10^7 cells are homogenized in 1 ml buffer H [150 m*M* KCl, 5 m*M* NaF, 1 m*M* phenylmethylsulfonyl fluoride (PMSF), 20 μM pepstatin, 20 μM leupeptin, and 20 μM antipain in 40 m*M* HEPES, pH 7.4] by passing the cells through a 28-gauge needle followed by 3× sonication for 10 sec at 4°. For analyzing total cellular proteins, Triton X-100 is added (final concentration 0.075%, v/v) and lysates are cleared by centrifugation (5 min, 1000*g*). The supernatant containing cytosolic as well as Triton X-100 soluble membrane proteins can then be applied to the column. If separation of cytosolic and membrane proteins is desired prior to chromatography, the cell homogenates are not supplemented by Triton X-100 but membranes are first separated from the lysates by centrifugation for 1 hr at 100,000*g*. The supernatant containing cytosolic proteins is saved and the pellet resuspended in buffer H and then adjusted to 0.075% (v/v) Triton X-100. Membrane proteins are solubilized by stirring for 30 min at 4°. After centrifugation for 1 hr at 100,000*g*, membrane protein is collected from the supernatants. Cytosolic and membrane proteins are then applied separately to the ceramide affinity columns.

Loading and Washing of Affinity Column

The volume in which the sample is applied is not critical, provided that the total capacity of the matrix is not exceeded. Best results were obtained at a protein-to-gel ratio of 0.6 mg protein/ml affinity matrix. The protein in homogenization buffer H (5 m*M* NaF, 1 m*M* PMSF, 20 μM pepstatin, 20 μM leupeptin, and 20 μM antipain in 40 m*M* HEPES, pH 7.4) supplemented with 0.075% Triton X-100, is loaded in an upward direction at a flow rate of 0.1 ml/min by cycling overnight at 4°.

After sample application, the column is washed with 7–10 volumes of the column with buffer H containing 0.075% Triton X-100 at 4° to remove unbound and nonspecifically adsorbed material. The choice of flow rate depends on the stability of protein absorption on the affinity matrix. Generally, the flow rate should be slow (0.5 ml/min). Efficiency of the washing is monitored by determining the protein content within the fractions (in the presence of Triton X-100, the BCA assay is used).

Elution of Affinity Matrix-Bound Protein

The principle of desorption from the affinity matrix is to change the binding equilibrium for the absorbed material from the stationary to the mobile phase. Because the binding between ceramide and protein is dominated by hydrophobic interactions, Triton X-100 as detergent is used. Specificity of the elution of ceramide-binding protein is obtained by competitive binding with an excess of soluble lipids contained within the elution buffer at room temperature. Best results were obtained by using an elution buffer consisting of 40 m*M* HEPES, pH 7.4, 0.075% Triton X-100, supplemented with 100 μM D-*erythro*-ceramide at a flow rate of 1.0 ml/min. To test for specificity of ceramide elution, various related or nonrelated lipids like D-*erythro*-dihydroceramide, D-*erythro*-sphingosine, D-*erythro*-dihydrosphingosine, free fatty acids, galactocerebrosides, sulfatides, or diacylglycerol are used instead of D-*erythro*-ceramide. The samples are concentrated by centrifugation using Centrex UF 10K MWCO filters.

The lipids used for competition are disolved in ethanol/tolinol (v/v). After evaporation of the solvents, the lipids were sonificated in 40 m*M* HEPES, pH 7.4, and 0.075% Triton X-100.

Regeneration and Equilibration

The affinity medium can be regenerated for reuse by washing the gel with 5 column volumes of high concentration detergent (1M KSCN, 1%

Triton X-100) followed by equilibration in 40 m*M* HEPES, pH 7.4, and 0.075% Triton X-100.

Analysis of Efficiency of Ceramide-Affinity Chromatography

The efficiency of the chromatography is monitored by comparing the protein pattern of (1) the original lysates, (2) the proteins that did not associate to the affinity matrix (flow-through), (3) the fraction of proteins washed off the matrix with excess of binding buffer, (4) the proteins that elute from the affinity column after addition of excess of competitor lipids, and (5) the protein that remains bound to the matrix after specific elution. Equal amounts of protein are loaded on SDS–polyacrylamide gels and separated proteins are analyzed by silver staining or by immunoblotting using specific antibodies against presumed target proteins.

[^{125}I]TID-Ceramide Photoaffinity Labeling

Photocross-linking is a powerful method for analyzing (transient) interactions between biologic molecules. A small photolabile group within one of the interacting molecules is converted by UV light activation into a highly reactive species, which then forms stable chemical connections between the interacting partners. A photoactivatable group that combines favorable properties is the 3-trifluoromethyl-3-phenyldiazirine (TPD) group. The TPD group is stable under a wide range of chemical and physical conditions including heat (80°), strong bases and acids, oxidizing conditions, and mild reducing agents. It can handeled without special precautions other than avoiding exposure to sunlight or UV. Photolysis occurs at long-wave UV (365 nm), i.e., under mild conditions that do not damage biologic structures. Photolysis of the TPD group results in formation of highly reactive carbene capable of reacting with the full range of functional groups occurring in biomolecules, including paraffinic CH bonds. The first TPD-based reagent that found widespread application was 3-trifluoromethyl-3-(*m*-[^{125}I]iodophenyl)diazirine ([^{125}I]TID).[23] Based on its highly apolar character, [^{125}I]TID was used to label membrane-integrated proteins. Now a variety of lipids containing the TID functionality can be obtained. The tin-containing precursor of these reagents can be converted by a simple, final step to a radioiodinated molecule with very high specific radioactivity (>2000 Ci/mmol).[24] The long-chain lipid [^{125}I]TID-PC/16 turned out to be an ideal reagent to investigate vectorial systems involving more than one membrane,

[23] J. Brunner, *Annu. Rev. Biochem.* **62,** 483 (1993).
[24] T. Weber and J. Brunner, *J. Am Chem. Soc.* **117,** 3084 (1995).

since it does not undergo rapid exchange between membranes.[25] Recently, analogs of lipid second messengers have been used as photoaffinity probes to identify proteins involved in signaling processes. For example, using a ^{32}P-labeled diazirine lysophosphatidic acid ([^{32}P]TID-LPA), the lysophosphatidic acid (LPA) receptor was identified.[26]

We employed [^{125}I]TID-ceramide to investigate the specificity of ceramide binding to the endolysosomal aspartic protease cathepsin D, one of the ceramide-target proteins that eluted from ceramide affinity chromatography.[21] [^{125}I]TID-ceramide was successfully used before to identify the protein kinase raf-1 and PKC-α and PKC-δ as ceramide targets.[8,9]

Radiolabeling of Tin Precursor of TID Ceramide

N-[3-[[[2-(Tributylstannyl)-4-[3-(trifluoromethyl)-3*H*-diazirin-3-yl] benzyl]oxy]carbonyl]propanoyl]-D-*erythro*-sphingosine (tin-ceramide) representing the tin precursor of the iodinated TID-ceramide, is synthesized and purified by HPLC as described.[24] Radiolabeling is performed by incubating 200 nmol tin-ceramide in 20 μl acetic acid and 5 mCi of Na^{125}I. The reaction is initiated by addition of 5 μl of peracetic acid (32% in acetic acid). After 2 min the reaction is quenced by the addition of 0.1 ml of 10% $NaHSO_3$ followed by neutralization with 70 μl of 5 *M* NaOH. Radiolabeled TID-ceramide is extracted with chloroform/methanol (2:1, total volume 300 μl). The organic phase is concentrated by a stream of N_2 and the residue subjected to TLC using as solvent dichloromethane/methanol (95:5). [^{125}I]TID-ceramide was identified by autoradiography, and the main radioactive band, comigrating with unlabeled TID-ceramide as standard, extracted with 0.2 ml of THF. After evaporation of the solvent, the residue was dissolved in toluene/ethanol (1:1) and stored at −20°.

Photoaffinity Labeling

For labeling of total cellular proteins, 10^5 cells in 100 μl serum-free medium are incubated with 50 n*M* [^{125}I]TID-ceramide [200 μCi/ml, dissolved in dimethyl sulfoxide (DMSO)] for 1 hr at 37° to allow for uptake of the radiolabeled lipid. Shorter incubation times (5 min) have also been reported.[8,9] Cells are then either kept nonphotolyzed as control or are subjected to UV photolysis for 2 min, 100 W, at 365 nm using a UVP Model B100A Black Ray Lamp (Herolab, Wiesloch, Germany). Cells are then

[25] T. Weber, G. Paesold, C. Galli, R. Mischler, G. Semenza, and J. Brunner, *Biol. Chem.* **269,** 18353 (1994).

[26] R. van der Bend, J. Brunner, K. Jalink, E. J. van Corven, H. W. Moolenaar, and W. J. van Blitterswijk, *EMBO J.* **11,** 2495 (1992).

washed and resuspended in solubilization buffer containing 40 m*M* HEPES, pH 7.4, 150 m*M* KCl, 5 m*M* NaF, 1 m*M* PMSF; 20 μM pepstatin, 20 μM leupeptin, and 20 μM antipain, 0.075% (v/v) Triton X-100 and homogenized by passing the cells through a 28-gauge needle. Lysates are cleared by centrifugation (5 min, 1000*g*) and the radiolabeled proteins analyzed after separation on SDS–PAGE by autoradiography using a phosphoimager. If a known protein is to be analyzed, the radiolabeled lysates can be used for immunoprecipitation employing specific antibodies against the target protein as descibed.[8,9,21] Buffer conditions for immunoprecipitation have to be adapted to the specific requirements of the respective antibodies. In some cases, [^{125}I]TID-ceramide cross-linked proteins lost their properties to become immunoprecipitated, probably due to changes in the antigen structure caused by the ceramide cross-linking.

In case of a poor uptake of [^{125}I]TID-ceramide in whole cells, cellular proteins can alternatively be photocross-linked using cell homogenates as starting material. The specificity of ceramide binding to target proteins is assayed by competition analysis employing [^{125}]TID-ceramide and increasing amounts of unlabeled D-*erythro*-ceramide or various related or nonrelated lipids like D-*erythro*-dihydroceramide, D-*erythro*-sphingosine, D-*erythro*-dihydrosphingosine, free fatty acids, galactocerebrosides, sulfatides, or diacylglycerol.

For competition analysis using a purified target protein, 1 μCi [^{125}I]TID-ceramide (0.5 pmol) is diluted in 5 μl DMSO. The competitor lipids (0.5–5000 pmol) are dissolved with the DMSO solution containing the [^{125}I]TID-ceramide. Then 10 μl solubilization buffer (40 m*M* HEPES, pH 7.4, 150 m*M* KCl, 0.075% Triton X-100) is added to each sample and sonication performed for 30 min at 4° in an ultrasonication water bath. The target protein, 2 μg for each competition sample, is suspended in 10 μl solubilization buffer and added to the lipid micelles. The protein/lipid micelles are incubated for 30 min on a shaker at room temperature. Photocross-linking is performed by UV-irradiation as described above and radiolabeled protein analyzed by autoradiography after separation by SDS–PAGE. Importantly, since the TID group itself may influence the binding to proteins based on its own hydrophobicity, a competition assay using nonradioactive TID is mandatory to ensure ceramide specificity of the [^{125}I]TID-ceramide-binding proteins.

Employing [^{125}I]TID-ceramide as a cross-linker, it should also be possible to map the ceramide-binding sites within a given ceramide target: photocross-linked native protein can be cleaved by cyanogenbromide or by tryptic digestion, resulting in the generation of several distinct peptides, including species that contain the radiolabeled ceramide. These peptides then can be further sequenced. Recombinant proteins carrying point mutations within the binding motive can be employed to further evaluate the structural requirements within the target protein for ceramide binding.

Acknowledgments

We thank K. Bernado for helpful discussions. This study was supported by a grant from the Deutsche Forschungsgemeinschaft (SFB 415) and by the Swiss National Science Foundation.

[36] Lectin-Mediated Cell Adhesion to Immobilized Glycosphingolipids

By BRIAN E. COLLINS, LYNDA J.-S. YANG, and RONALD L. SCHNAAR

Introduction

Carbohydrate-binding proteins (lectins) on one cell surface bind specifically to complementary target oligosaccharides on an apposing cell to initiate cell–cell recognition and adhesion.[1–3] The target glycoconjugates may be glycolipids, glycoproteins, or both, depending on the cell–cell interaction under investigation. The ability to purify or synthesize glycosphingolipids with well-defined oligosaccharide structures makes them excellent tools for investigating lectin recognition.

In nature, lectin–carbohydrate interactions are typically multivalent: an array of lectins on one cell binds to a complementary array of oligosaccharides on an apposing cell.[4,5] Assays that rely on monovalent lectin–ligand interactions may not be sufficiently sensitive to detect such interactions, which may have low "single-site" affinity but high multivalent affinity. Under physiologic conditions, multivalent interactions may also enhance recognition specificity, since modest differences in "single-site" affinity are magnified by multivalency to give high aggregate specificity. Accordingly, a high-throughput assay incorporating multivalent recognition and control of detachment force (adhesion strength) to measure adhesion of lectin-transfected cells to glycosphingolipids on an apposing surface was devel-

[1] N. Sharon and H. Lis, *Science* **246,** 227 (1989).
[2] L. A. Lasky, *Annu. Rev. Biochem.* **64,** 113 (1995).
[3] S. Kelm, R. Schauer, and P. R. Crocker, *Glycoconj. J.* **13,** 913 (1996).
[4] R. T. Lee and Y. C. Lee, *in* "Neoglycoconjugates: Preparation and Applications" (Y. C. Lee and R. T. Lee, eds.), p. 23. Academic Press, San Diego, 1994.
[5] W. I. Weis and K. Drickamer, *Annu. Rev. Biochem.* **65,** 441 (1996).

0076-6879/00 $30.00

oped to simulate *in vivo* multivalent recognition by cell surface lectins.[6] The assay consists of four steps: (1) preparation of lectin-transfected cells, (2) preparation of glycosphingolipid-coated microwells, (3) adhesion of lectin-transfected cells to glycosphingolipids, and (4) quantification of cell adhesion.

Lectin-Transfected Cells

Of the many cell systems that have been used for transient or stable expression of foreign genes, two of the more popular were tested and found to have sufficient cell surface lectin expression and low background binding necessary to study glycosphingolipid-directed cell adhesion: transiently transfected COS-1 cells, and stably transfected Chinese hamster ovary (CHO) cells.[7,8] Although both systems are useful for the study of glycosphingolipid-directed adhesion, transient transfection has proven most versatile. It is readily applied to most any vertebrate lectin cloned into an appropriate expression vector, results in high cell surface lectin expression, and transfected cells are ready for use within days without the need for further selection or cloning.[9,10]

COS-1 cells, simian virus 40 (SV40)-transformed African green monkey fibroblasts, are available from American Type Culture Collection (CRL-1650, Rockville, MD). High-level expression is obtained using lectins cloned into various plasmids carrying the SV40 origin of replication, such as the pcDNA1 family of vectors available from Invitrogen (Carlsbad, CA). The following transfection protocol[11] was optimized both for high transfection efficiency and high cell viability.

Reagents

Chloroquine (Sigma Chem. Co., St. Louis, MO), 100 m*M* in water, stable for 6 months at −20°

DEAE-dextran (Pharmacia, Piscataway, NJ), 4 mg/ml in phosphate-buffered saline (PBS), stable for 1 month at 4°

[6] L. J. S. Yang, C. B. Zeller, and R. L. Schnaar, *Anal. Biochem.* **236,** 161 (1996).

[7] L. J. S. Yang, C. B. Zeller, N. L. Shaper, M. Kiso, A. Hasegawa, R. E. Shapiro, and R. L. Schnaar, *Proc. Natl. Acad. Sci. U.S.A.* **93,** 814 (1996).

[8] B. E. Collins, L. J. S. Yang, G. Mukhopadhyay, M. T. Filbin, M. Kiso, A. Hasegawa, and R. L. Schnaar, *J. Biol. Chem.* **272,** 1248 (1997).

[9] M. Tiemeyer, B. K. Brandley, M. Ishihara, S. J. Swiedler, J. Greene, G. W. Hoyle, and R. L. Hill, *J. Biol. Chem.* **267,** 12252 (1992).

[10] B. E. Collins, M. Kiso, A. Hasegawa, M. B. Tropak, J. C. Roder, P. R. Crocker, and R. L. Schnaar, *J. Biol. Chem.* **272,** 16889 (1997).

[11] B. Seed and A. Aruffo, *Proc. Natl. Acad. Sci. U.S.A.* **84,** 3365 (1987).

Expression plasmid, 30 μg/ml DNA in sterile PBS
DMSO (Sigma), sterile solution of 10% (v/v) dimethyl sulfoxide in PBS, stable for 1 month at 4°

Procedure

COS-1 cells are maintained in Dulbecco's modified Eagle's medium (DMEM) supplemented with 10% fetal bovine serum (FBS). Twelve to 24 hr prior to transfection, cells (which had been passaged not less than 48 hr previously) are replated in 10-cm tissue culture dishes at 10^6 cells/dish, taking care to distribute them evenly, since cell clumping decreases transfection efficiency. The subsequent transfection procedure is performed rapidly to avoid undue changes in the pH of the medium. The medium is removed from the plates and replaced with 4 ml of transfection medium (DMEM supplemented with 2.5% FBS, 0.1 μ*M* chloroquine, 40 μg/ml DEAE-dextran). Expression plasmid in PBS (100 μl) is added dropwise in a spiral pattern, with the pipette tip touching the surface of the medium to avoid fluid shear. The plate is then gently agitated, without swirling, to ensure even plasmid distribution, and returned to the incubator for 3–3.5 hr (see *Notes*).

The transfection is stopped by removing the transfection medium and adding 5 ml of 10% DMSO (Sigma, tissue culture grade) in PBS. After 3 min, the DMSO solution is replaced with DMEM supplemented with 10% FBS and 20 μg/ml gentamicin, and the plate returned to the incubator for 48–72 hr to allow lectin expression.

Notes. Transfection is monitored by observing cell morphology starting 2 hr 45 min after addition of the plasmid, and every 10 min thereafter. As the transfection proceeds, cells become rounded with small clear intracellular vesicles. When a large majority of the cells have such vesicles (typically 3–3.5 hr), the transfection is stopped (see above). The transfection has gone too far if a significant percentage (>10%) of the cells is no longer attached to the plate.

Although lectin expression increases with incubation posttransfection, cell viability declines as plasmid replication and expression continues. Different plasmids vary in replication and expression rates, resulting in different optimal posttransfection times, which should be determined empirically (typically 48–72 hr).

Glycosphingolipid-Coated Microwells

To simulate a membrane monolayer, glycosphingolipids are stably coadsorbed with phosphatidylcholine and cholesterol onto the bottom of

96-well polystyrene multiwell plates. This is accomplished by adding the lipids in ethanol–water (1:1, v/v) and allowing a portion of the ethanol to evaporate. This results in the lipids adsorbing onto the plastic surface, where they are stable in aqueous solutions.[12]

Materials and Reagents

Polystyrene 96-well flat-bottom plates, Corning Costar Serocluster, (Corning, NY)
Phosphatidylcholine (100 μM, egg lecithin, Matreya) and cholesterol (400 μM, Matreya) in 100% ethanol
DMEM/HEPES, DMEM (Life Tech) supplemented with 10 mM HEPES and 4.6 mM sodium bicarbonate, pH 7.4

Procedure

Multiwell plates are prewashed with butanol, rinsed with ethanol, then air-dried prior to use to remove surface agents used in manufacture. Glycosphingolipids are evaporated from storage solvents and dissolved at the desired concentrations (up to 12 μM) in 100% ethanol containing 1 μM phosphatidylcholine and 4 μM cholesterol (a 100-fold dilution from stock). An equal volume of water is added, and 50 μl is placed into each well, which is then incubated uncovered for 90 min to allow partial evaporation and efficient lipid adsorption. Plates are subsequently washed three times with water to remove unadsorbed lipids and water is left in the wells prior to preblocking (up to 3 hr). Preblocking to reduce nonspecific cell adhesion is accomplished by adding 100 μl/well of DMEM/HEPES containing 1–25 mg/ml bovine serum albumin (BSA) and incubating at 37° for 10 min (see *Notes*).

Notes. Glycosphingolipid adsorption is quantified by recovering adsorbed lipids from the wells with butanol and analyzing by thin-layer chromatography, resulting in ≈40% adsorption efficiency for gangliosides, and higher for neutral glycosphingolipids.

Conditions for preblocking are optimized for each plasmid and cell type used. Additionally, significant variations in blocking with BSA type, manufacturer, and batch have been observed. It is advisable to test binding using various BSA batches and concentrations in the range of 1–25 mg/ml. Blocking should be performed at 37° (no blocking was observed at 4°).

Compared with PBS, the use of DMEM/HEPES throughout the assay results in increased cell viability.

[12] C. C. Blackburn, P. Swank-Hill, and R. L. Schnaar, *J. Biol. Chem.* **261,** 2873 (1986).

Adhesion of Lectin-Transfected Cells to Glycosphingolipids

Lectin-transfected cells are recovered from culture plates and incubated on glycosphingolipid-adsorbed wells to allow adhesion to occur. Carbohydrate-mediated cell adhesion, while highly specific, may be relatively sensitive to fluid sheer. Accordingly, nonadherent cells are removed using mild and controlled centrifugal detachment with the aid of a custom-designed centrifugation chamber.[6] Finally, adherent cells are recovered for quantification (see next section).

Procedure

Lectin-transfected cells are recovered from culture dishes by replacing their medium with ice cold hypertonic PBS/EDTA (0.31 *M* NaCl, 16 m*M* Na_2HPO_4, 54 m*M* KCl, 29 m*M* KH_2PO_4, 1 m*M* EDTA, pH 7.4) and incubating for 10–15 min at ambient temperature. The cells are readily removed by gentle trituration, immediately diluted 1:1 with DMEM/HEPES, collected by centrifugation (200*g*, 3 min, 4°), and resuspended in PBS supplemented with 2 mg/ml BSA.

Sialic acid residues on the COS cells may attenuate adhesion via steric inhibition or by specific binding to an expressed lectin.[8,13,14] To eliminate these possibilities and enhance glycosphingolipid-directed cell adhesion, cells are pretreated with *Vibrio cholerae* neuraminidase. After collection, 10^7 cells in 1 ml of PBS supplemented with 2 mg/ml BSA are placed in a 1.5-ml microcentrifuge tube, and 10 mU of *V. cholerae* neuraminidase (Roche Molecular Biochemicals, Indianapolis, IN) are added. After 1.5 hr at 37° on an end-over-end rotator, the cells are washed three times with PBS supplemented with 2 mg/ml BSA by microcentrifugation (5,000*g*, 15 sec, ambient temperature), then are resuspended at 2.5×10^5 cells/ml in DMEM/HEPES containing the same optimal BSA concentration used for plate blocking (see above). Viability through the neuraminidase pretreatment is >90% and, if stored on ice in the above medium, the cells are stable for up to 8 hr.

Cell suspension (200 μl, 50,000 cells) is added to each glycosphingolipid-adsorbed well (which still contains 100 μl of the blocking solution, see above). The plate is placed on a level surface at 4° for 10 min to allow cells to settle, and then moved to a 37° incubator for 45 min.

To avoid fluid sheer, the plate is carefully immersed upright in a vat of PBS (at ambient temperature), inverted (while immersed), and placed (inverted) in an immersed custom-designed Plexiglas box, which is sealed

[13] S. Braesch-Andersen and I. Stamenkovic, *J. Biol. Chem.* **269,** 11783 (1994).
[14] S. D. Freeman, S. Kelm, E. K. Barber, and P. R. Crocker, *Blood* **85,** 2005 (1995).

with a gasket to exclude air (Fig. 1). The inverted plate in its fluid-filled chamber is placed in a centrifuge carrier and centrifuged at 100*g* for 10 min at 4° to remove nonadherent cells, and the chamber is returned to the vat of PBS. While immersed, the plate is gently removed from the chamber but kept immersed (in the inverted orientation) for 5 min to allow any floating cells to settle away. The plate is then righted (while immersed), removed from the vat, and excess surface PBS removed by aspiration. At this point, all wells are full of PBS ($\approx$ 320 μl/well) and only adherent cells remain on the well bottoms. The cells are stable for 2 hr at 4° in this buffer.

Notes. Although Dulbecco's PBS supplemented with 1 m*M* EDTA is sufficient to remove transfected cells from their culture dishes, the use of hypertonic PBS/EDTA results in much higher cell viability and greatly improves cell adhesion.

Cell number in the adhesion assay is adjusted to minimize cell–cell interactions but maintain a high signal for quantification. For COS cells, 3.5–5 $\times$ 10^4 cells/well is optimal, whereas for smaller CHO cells, 7.5–10 $\times$ 10^4 cells/well is preferable.

The centrifugal detachment force, which can be varied from 1 to 250*g*, is optimized for each plasmid and cell system to give the maximum adhesion/background ratio. For a 1*g* detachment force, the plate can simply be immersed and maintained in an inverted orientation in a vat of PBS. Although this simple procedure does not require special equipment, background adhesion using 1*g* detachment is often too high to allow its routine use.

Outer perimeter rows and columns on the multiwell plate are more susceptible to inadvertent shear forces during the procedure, and should only be used after the investigator is experienced with the assay.

Quantification of Cell Adhesion

Cells that remain adherent after detachment (see above) are quantified by lysing them and measuring the activity of a released housekeeping enzyme, lactate dehydrogenase (LDH). This assay is proportional to cell number over a wide range, and is readily performed in a 96-well format using a kinetic plate reader. In this way, a single investigator can collect >500 cell adhesion data points in a single day.

Reagents

Triton X-100, 10% (v/v) in PBS

Assay buffer, pyruvate (4.7 m*M*, 0.5 mg/ml) and NADH (0.7 m*M*, 0.5 mg/ml) in PBS, prepared fresh daily

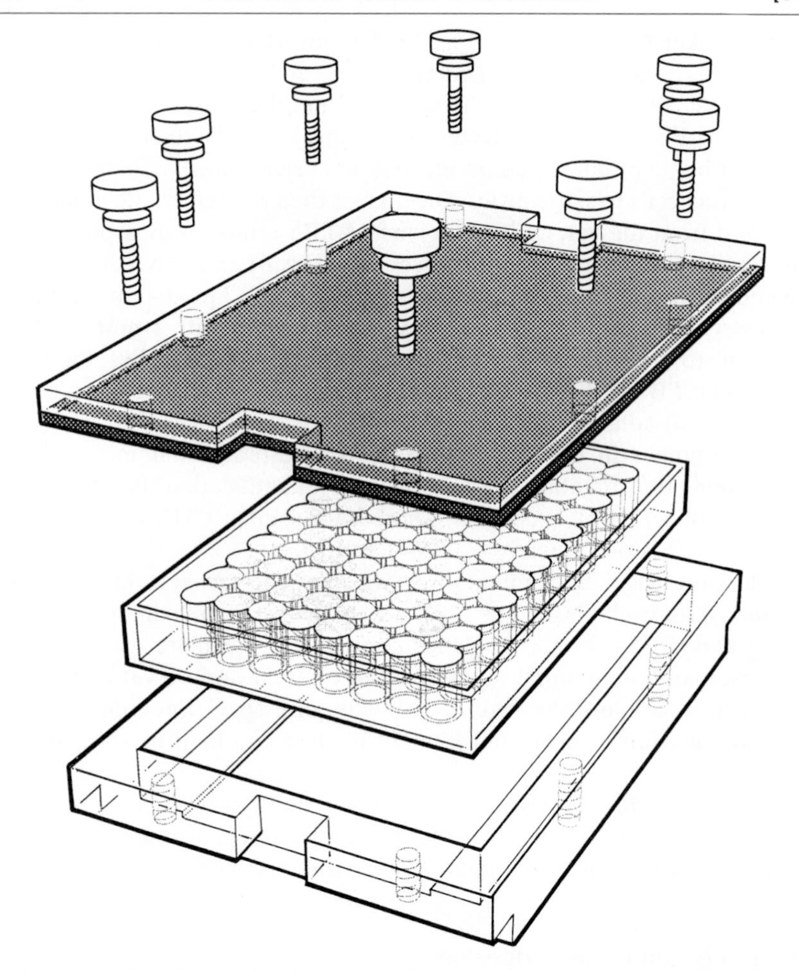

FIG. 1. Plexiglas adhesion chamber. The adhesion chamber was designed to allow detachment of nonadherent cells from the bottom of a standard 96-well plate using controlled centrifugal detachment forces. The chamber was designed specifically to fit in a Dupont–Sorvall microplate carrier (with thin metal plate spacers removed) for use with an H-1000B rotor in a RT6000 tabletop centrifuge. Adaptation to other carriers may require modification. The internal chamber is 129 mm (length) × 88 mm (width) × 15 mm (depth) with 8 mm (width) × 1 mm (depth) spacers affixed to the bottom along each long edge. Overall outside dimensions are 155 mm (length) × 106 mm (width) × 19 mm (depth) with end gaps cut to accommodate the plate carrier. The 5-mm-thick Plexiglas top has an attached foam rubber gasket and holes that line up with tapped holes in the chamber to allow it to be affixed using knurled-head polypropylene screws.

Procedure

After completion of the adhesion assay, 10 μl of 10% Triton X-100 in PBS is added to each well, the cells allowed to lyse for 10 min, and then triturated thoroughly with a multichannel micropipettor (LDH is stable in this solution for 1 hr). A portion of each cell lysate (80 μl) is transferred to a fresh multiwell plate and 120 μl of assay buffer added. The decrease in absorbance at 340 nm as a function of time is measured simultaneously in each well using a multiwell kinetic plate reader. For comparison, a known number of cells (typically 50,000) is added to an empty well after completion of the adhesion procedure, lysed, and an aliquot analyzed as above. This allows calculation of the adherent cell number from the kinetic data.

General Considerations

To ensure that differences in adhesion to different glycosphingolipids are not caused by variations in adsorption efficiency, stability, or metabolism

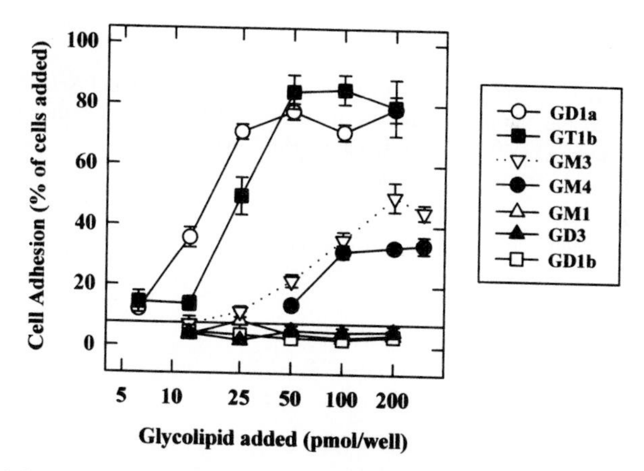

FIG. 2. MAG-mediated cell adhesion to gangliosides: structural specificity. COS cells were transiently transfected with a plasmid encoding full-length MAG, a member of the "siglec" family of vertebrate lectins.[3,16] After 48 hr of culture to allow MAG expression, cells were collected, pretreated with neuraminidase, and placed in wells previously adsorbed with an artificial membrane monolayer containing phosphatidylcholine, cholesterol, and the indicated gangliosides. After incubation (37°, 45 min) to allow adhesion to proceed, nonadherent cells were removed by centrifugation and adherent cells were quantified enzymatically. Adhesion is expressed relative to the total number of cells added to each well, and represents the mean ± S.E. (bars) of 3–103 replicate determinations. Background adhesion (7.5 ± 0.6%), represented by a horizontal line, was determined on wells adsorbed with phosphatidylcholine and cholesterol without gangliosides. Ganglioside nomenclature is that of Svennerholm.[17] (Reproduced with permission from Collins *et al.*[10])

of the lipids by cells, reextraction and quantification of the immobilized glycolipid after a control incubation are performed. Glycolipids to be tested are adsorbed, wells washed and preblocked, and 200 μl of medium containing 50,000 cells/well of nontransfected (nonadherent) COS cells are added. The wells are incubated and subjected to centrifugal detachment as for an adhesion experiment, except after removal of nonadherent cells the wells are washed briefly with water and the adsorbed glycolipids recovered with butanol. The recovered glycolipids are then qualitatively and quantitatively analyzed by TLC.[15]

Microscopic examination of wells prior to lysis is used to confirm that kinetic LDH quantification correlates with the appearance of bound cells. Cell viability during the assay is measured by microscopic examination and by measuring LDH activity released into the medium prior to immersion and detachment.

Differential adhesion of COS-1 expressing myelin-associated glycoprotein (MAG) to various gangliosides is shown in Fig. 2, as an example of application of the above assay.[10]

Acknowledgments

This work was supported by research grants from the National Science Foundation (IBN-9631745) and the National Institutes of Health (NS37096), and National Institutes of Health Training grants GM07626 (B.E.C.) and GM07309 (L.J.-S.Y.).

[15] R. L. Schnaar and L. K. Needham, *Methods Enzymol.* **230,** 371 (1994).

[16] P. R. Crocker, E. A. Clark, M. Filbin, S. Gordon, Y. Jones, J. H. Kehrl, S. Kelm, N. Le Douarin, L. Powell, J. Roder, R. L. Schnaar, D. C. Sgroi, K. Stamenkovic, R. Schauer, M. Schachner, T. K. van den Berg, P. A. van der Merwe, S. M. Watt, and A. Varki, *Glycobiology* **8,** v (1998).

[17] L. Svennerholm, *Prog. Brain Res.* **101,** xi (1994).

[37] Analysis of Glycolipid-Dependent Cell Adhesion Based on Carbohydrate–Carbohydrate Interaction

By KAZUKO HANDA, NAOYA KOJIMA, and SEN-ITIROH HAKOMORI

I. Introduction

Cell adhesion based on carbohydrate-carbohydrate interaction has been found typically in the following systems*: (1) homotypic adhesion of mouse embroyonal carcinoma F9 cells, mediated by Le^x-to-Le^x (Le^x = Galβ4[Fucα3]GlcNAcβ3Galβ4Glcβ1Cer interaction in the presence of bivalent cation[1,2]; (2) homotypic adhesion of rat basophilic leukemia (RBL) cells, mediated exclusively by Le^x-to-Le^x glycosphingolipid (GSL) interaction[3]; (3) heterotypic adhesion of mouse melanoma B16 cells to mouse lymphoma L5178 cells[4] or to endothelial cells,[5,6] mediated by interaction between GM3 (NeuAcα3Galβ4Glcβ1Cer) and Gg_3 (GalNAcβ4Galβ4Glc-β1Cer), or GM3 and LacCer (Galβ4Glcβ1Cer), respectively; and (4) homotypic adhesion of human embryonal carcinoma 2102 cells, mediated by heterotypic GSL interaction between Gb_4 (globoside, GalNAcβ3Galα4Gal-β4Glcβ1Cer) and GalGb_4, or Gb_4 and nLc_4. This interaction occurs even in the absence of divalent cations.[7]

Evidence that these cell adhesion systems are based on carbohydrate–carbohydrate interaction is as follows: (1) Adhesion occurs between cells expressing a defined GSL or carbohydrate epitope, but not cells lacking the epitope. (2) Exogenous addition of GSL enhances or induces GSL-dependent cell adhesion. (3) GSL liposomes or oligosaccharides of GSLs inhibit cell adhesion mediated by those GSLs, i.e., adhesion of cell a to

* Glycosphingolipids are abbreviated according to the recommendations of the IUPAC-IUB Commission on Biochemical Nomenclature (*Lipids* **12,** 455–463, 1977); however, the suffix -OseCer is omitted. Gangliosides are abbreviated according to Svennerholm (*J. Lipid Res.* **5,** 145–155, 1964) and IUPAC (1977).

[1] I. Eggens, B. A. Fenderson, T. Toyokuni, B. Dean, M. R. Stroud, and S. Hakomori, *J. Biol. Chem.* **264,** 9476 (1989).

[2] N. Kojima, B. A. Fenderson, M. R. Stroud, R. I. Goldberg, R. Habermann, T. Toyokuni, and S. Hakomori, *Glycoconj. J.* **11,** 238 (1994).

[3] M. Boubelík, D. Floryk, J. Bohata, L. Dráberová, J. Macák, F. Smíd, and P. Dráber, *Glycobiology* **8,** 139 (1998).

[4] N. Kojima, and S. Hakomori, *J. Biol. Chem.* **264,** 20159 (1989).

[5] N. Kojima and S. Hakomori, *J. Biol. Chem.* **266,** 17552 (1991).

[6] N. Kojima, M. Shiota, Y. Sadahira, K. Handa, and S. Hakomori, *J. Biol. Chem.* **267,** 17264 (1992).

[7] S. Yu, D. A. Withers, and S. Hakomori, *J. Biol. Chem.* **273,** 2517 (1998).

0076-6879/00 $30.00

cell b mediated by GSL a and b, respectively, can be inhibited by liposome of GSL a, liposome of GSL b, oligosaccharide a′ derived from GSL a, or oligosaccharide b′ derived from GSL b. (4) Interaction of GSLs involved in cell adhesion can be observed directly by various methods, e.g., (a) binding of GSL liposomes to GSL coated on plastic (polystyrene) surface[1,5,8]; (b) specific interaction of radiolabeled multivalent oligosaccharide by affinity chromatography with GSL affixed to a C_{18} column[2,5]; (c) molecular force measurements between two solid phases coated with GSLs[9]; and, (d) measurement of the physical state of GSL monolayers at the water surface during interaction with oligosaccharide–polystyrene conjugates.[10]

In this chapter we present a few examples of analysis of GSL-dependent cell adhesion, summarizing evidence that such adhesion is based on carbohydrate–carbohydrate interaction. A similar series of experiments could be conducted to analyze other carbohydrates thought to be involved in carbohydrate–carbohydrate interactions.

II. GM3-Dependent Adhesion of Mouse B16 Melanoma Cells to Mouse or Human Endothelial Cells

Mouse B16 melanoma cells, expressing a high level of GM3, display clear adhesion to nonactivated human or mouse endothelial cells (ECs). This adhesion is based on interaction of GM3 (expressed on B16 cells) with LacCer, Gg_3, or Gb_4 expressed on ECs. The adhesion system is independent of (vascular cell adhesion molecule) VCAM-integrin-based or selectin-based adhesion, but is implicated as intiating the metastatic process of B16 melanoma cells.

In the following sections, we describe (1) GM3-dependent adhesion of B16 cells to plates coated with LacCer, Gg_3, or other GSLs; (2) adhesion of GM3 liposomes to plates coated with LacCer, Gg_3, or other GSLs; and (3) adhesion of B16 cells to nonactivated human or mouse ECs, and its inhibition by LacCer, Gg_3, GM3 liposomes, or oligosaccharides of these GSLs.

GM3-Dependent Adhesion of B16 Melanoma Cells to Plates Coated with LacCer, Gg₃, or Other GSLs, as Compared to Fibronectin-Coated Plates

Preparation of Solid-Phase GSL Coated on Polystyrene Plastic Surface. A 50% aqueous ethonal solution of the GSL (50 $\mu g/\mu l$) is appropriately diluted and placed in each well of a 96-well flat-bottom culture plate

[8] K. M. Koshy and J. M. Boggs, *J. Biol. Chem.* **271,** 3496 (1996).

[9] Z. W. Yu, T. L. Calvert, and D. Leckband, *Biochemistry* **37,** 1540 (1998).

[10] K. Matsuma, H. Kitakouji, A. Tsuchida, N. Sawada, H. Isida, M. Kiso, and K. Kobayashi, *Chem. Lett.,* **1998,** 1293 (1998).

(Microtest III, Falcon, Lincoln Park, NJ), and dried at 37°. The quantity of GSL is indicated on the abscissa of Figs. 1D and 1E. GSL-coated solid phase is then incubated with 1% bovine serum albumin (BSA) for 1 hr at room temperature to block nonspecific binding, and washed twice with TBS (10 mM Tris-HCl, ph 8.0 150 mM NaCl) or PBS (phosphate-buffered saline, 8.1 m*M* Na_2HPO_4, 1.5 m*M* KH_2PO_4, 2.7 m*M* KCl, 137 m*M* NaCl, pH 7.4) before use. As a control, fibronectin (FN) is coated by placing FN solution in PBS at various concentrations as indicated on the abscissa of Fig. 1B and incubating at room temperature overnight.

Preparation of Culture of B16 Melanoma Variants. B16 melanoma variants with different degrees of metastatic potential, in the order BL6 $\cong$ F10 $>$ F1 $\gg$ WA4, show GM3 expression in the same relative order (Fig. 1A). Each of these variants is cultured in 10 ml of RPMI (Roswell Park Memorial Institute Media) 1640 containing 10% fetal calf serum (FCS) in 25-cm^2 Corning flasks and labeled with [^{3}H]thymidine (2 mCi/ml) for 16 hr. For measurement of cell adhesion, labeled cells are detached by treatment in 0.02% EDTA (ethylenediaminetetraacetic acid) at 37° for 5 min. Detached cells are collected and washed 2× with serum-free medium.

B16 Cell Adhesion to GSL- or FN-Coated Plates. Cells are suspended in serum-free RPMI medium at a density of 2×10^5 cells/ml, and 100-μl aliquots of suspension are placed in each well of 96-well plates coated with GSL or with FN as above. Plates are added with cells, centrifuged at 100*g* for 1 min, and incubated at 37° for appropriate times (30 min for the experiment shown in Figs. 1B, 1D, and 1E; 10–60 min for Fig. 1C). Cells are washed by gentle aspiration using a very thin capillary tip (Eppendorf, 200 Ultra Micro Tips, distributed through Brinkman Instruments Inc., Westbury, NY). PBS (100-μl aliquots) is added to each well and gently aspirated as above. This process is repeated once more. Remaining adherent cells are collected by cell harvester and radiolabel quantified using a scintillation counter.

Adhesion of GM3 Liposomes to Plates Coated with LacCer, Gg_3, or Other GSLs

Preparation of Single-Compartment GSL Liposomes with [^{14}C]Cholesterol. Cholesterol, synthetic dimyrisoyl-L-α-phosphatidylcholine, and purified GSLs are mixed in a weight ratio of 25 : 25 : 12.5 μg in chloroform–methanol (2 : 1). This solution is added with [^{14}C]cholesterol (55,000 cpm/μg) and evaporated to dryness under nitrogen. The original specific activity of [26-^{14}C]cholesterol (Dupont-New England Nuclear, Boston, MA) added is 53.0 mCi/mmol. The dried residue of the lipid mixture containing [^{14}C]cholesterol is dissolved in 100 μl of diethyl ether and injected by Hamilton syringe into 1.9 ml of prewarmed (60°–65°) PBS with or without

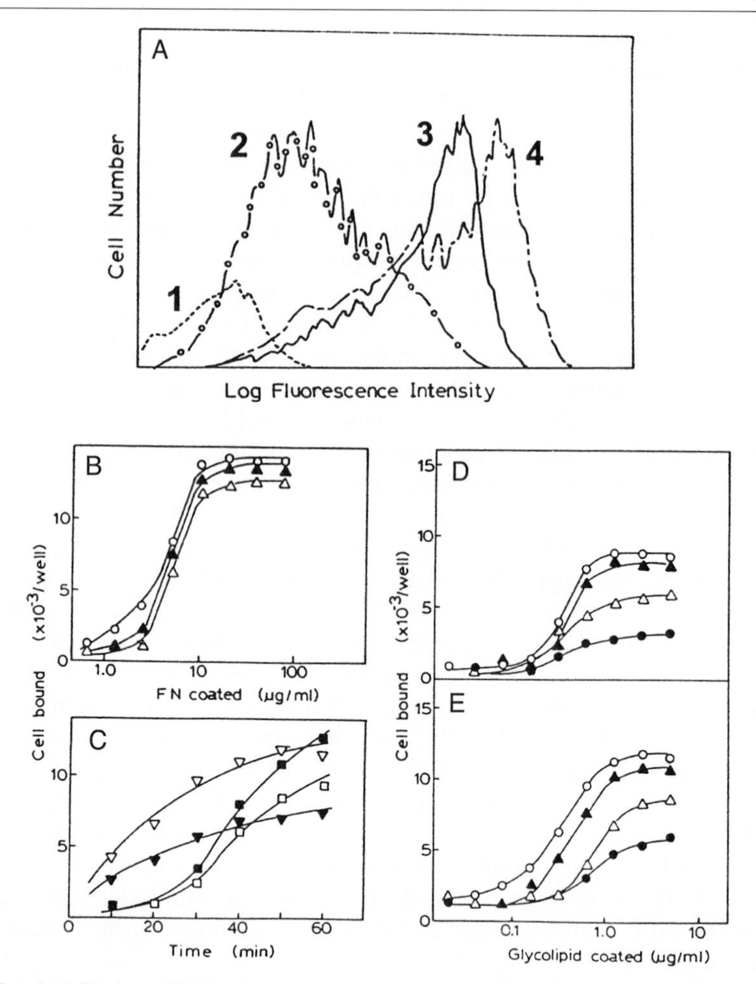

FIG. 1. Adhesion of B16 variants to plates coated with fibronectin (FN), LacCer, or Gg_3. (A) Degree of GM3 expression in B16 melanoma variants with different degrees of metastatic potential, analyzed by flow cytometry with anti-GM3 monoclonal antibody (MAb) DH2. Line 1, control. Line 2, WA4. Line 3, F1. Line 4, F10 (BL6 shows essentially the same pattern as F10). (B, D, and E) The 96-well flat-bottom assay plates were coated with various quantities (see abscissa) of FN (B), LacCer (D), or Gg_3 (E). [^{3}H]Thymidine-labeled B16 variant cells were placed (2×10^4 cells/well) on plates, centrifuged (100*g*, 1 min), and incubated in serum-free RPMI 1640 medium at 37° for 30 min. Ordinate, number of adherent cells as estimated from radioactivity count. ○, BL6; ▲, F10; △, F1; ●, WA4. (C) Time course of BL6 adhesion determined for wells coated with 1 μg/well of Gg_3 (▽) or LacCer (▼), or 5 μg/ml of FN (□) or LN (■). Note that adhesion via LacCer or Gg_3 occurs much more rapidly than that via FN or LN. Cell adhesion experiments were repeated 3–5 times. Points shown represent mean values.

divalent cation (see below). The PBS solution is strongly agitated with a vortex mixer during lipid/ether injection. Ether in the liposome suspension is removed by rotary evaporation for 15 min at ~30°. In the majority of experiments, PBS containing 0.9 m*M* $CaCl_2$, 0.5 m*M* $MgSO_4$, and 0.1 m*M* $MnCl_2$ is used; in a few experiments, $MnCl_2$ or all three of these divalent cations are omitted from the PBS.

Alternatively, a mixture of lipids in ether solution as described above is rotary-evaporated in a small flask, and the lipid film inside the flask sonicated with 2 ml of PBS containing divalent cation as above in a Branson model 5200 sonic bath for 2 hr. Binding of liposomes prepared by these two different methods gives essentially the same results.

In almost all experiments, freshly prepared liposomes are used. Alternatively, liposomes are refrigerated for a few days to a week at 7°. No difference in experimental results is observed between freshly prepared vs. 1-week-old liposome preparations. Immediately before use, liposome suspensions are vigorously agitated using a vortex mixer.

Binding of Liposomes of GSL-Coated Solid Phase. To each well of 96-well culture plate coated with GSL, 100 μl (25,000 cpm) of liposome containing GSL and [^{14}C]cholesterol is added and incubated at room temperature on a rotary shaker for 16 hr in PBS containing 1 m*M* $CaCl_2$ and 0.5 m*M* $MgCl_2$. Each well is slowly and carefully washed with the same PBS under slow suction using a very thin capillary tip as above; remaining liposome on solid phase is extracted twice with 2-propanol–hexane–water (55:25:20), and radioactivity counted. No radioactivity is observed in the wells after extraction. Results are shown in Fig. 2.

Adhesion of B16 Melanoma Cells to Nonactivated Human or Mouse Endothelial Cells, and Its Inhibition by LacCer, Gg_3 or GM3 Liposomes

Preparation of EC Monolayer. Flat-bottom plates with 96 or 48 wells (purchased respectively from Falcon, Lincoln Park, NJ, and Costar, Cambridge, MA) were pretreated with 0.5% gelatin at 37° for 1 hr. To each well a suspension of human umbilical vein ECs (Cell Systems, Kirkland, WA) or mouse splenic EC line SPE-1[11] (8×10^4 cells/ml) is placed and cultured until confluency. The culture medium used for human umbilical vein ECs was RPMI with 10% FCS, with addition of the special growth factor provided by Cell Systems, Kirkland, WA. SPE-1 cells are grown in RPMI supplemented with 10% FCS.

Adhesion of B16 Cells to EC Monolayer. Medium in each confluent EC culture is removed by aspiration, ECs were washed with PBS, and

[11] Y. Sadahira, T. Yoshino, and N. Kojima, *In Vitro Cell. Dev. Biol.* **30A,** 648 (1994).

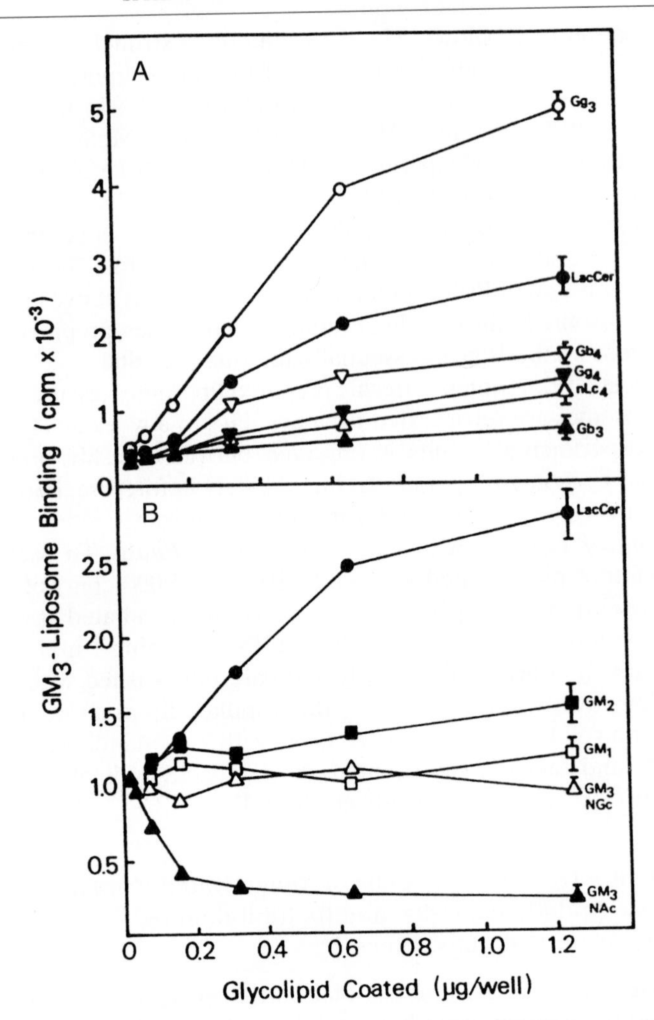

FIG. 2. Interaction of GM3-liposome with solid phase coated with various GSLs. GM3-liposome labeled with [^{14}C]cholesterol was prepared and incubated in plastic wells coated with various GSLs as described in the text. (A) Binding of GM3-liposome on solid phase coated with Gg_3 (○), LacCer (●), Gb_4 (▽), Gg_4 (▼), nLc_4 (paragloboside, Galβ4GlcNAcβ3Galβ4Glcβ1Cer) (△), and Gb_3 (▲). (B) Binding on solid phase coated with LacCer (●), GM2 (■), GM1 (□), NGcGM3 (△), and NAcGM3 (▲).

[^{3}H]thymidine-labeled B16 cells in PBS, prepared as described in the preceding section, are added to EC culture (without activation). The number of cells added per each well for 96-well plates is ~1×10^5; the number of cells added per each well for 48-well plates is ~3×10^5. Plates are incubated at 37° for 30 min. After incubation, cells are washed 2× with PBS, and adherent cells are detached by trypsinization, collected, and counted by scintillation counter.

Adhesion of B16 cells to nonactivated ECs as above is compared with adhesion to activated ECs. Activation of human umbilical vein ECs is achieved by treatment with 5 units/ml of human recombinant interleukin-1β (IL-1β) (Boehringer Mannheim Biochemicals, Indianapolis, IN; 1000 U/ml) for at least 4 hr at 37°. Activation of mouse SPE-1 cells is induced by 10 ng/ml recombinant tumor necrosis factor α (TNF-α).[11] Following activation of ECs, B16 cells are added and adhesion processed as above. Results are shown in Fig. 3.

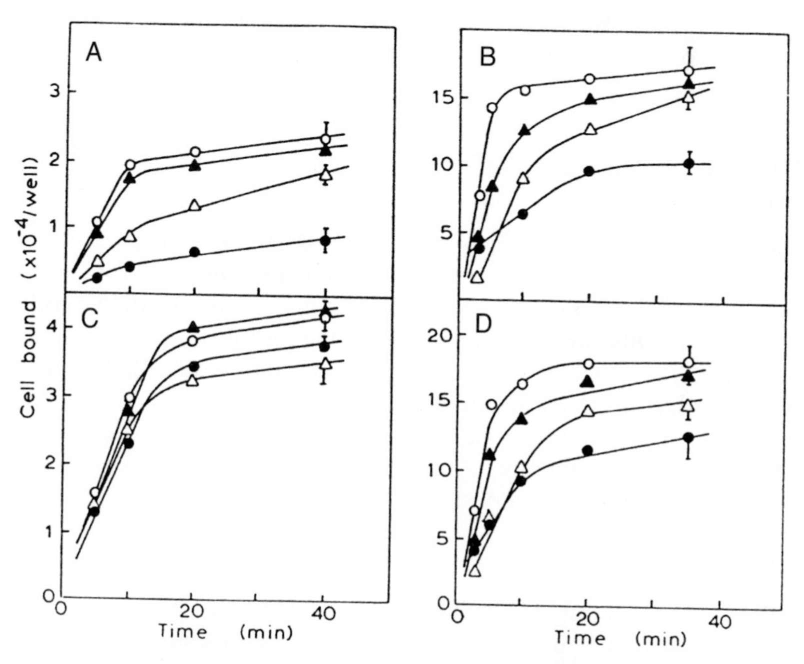

FIG. 3. Adhesion of B16 variants to nonstimulated and stimulated ECs. (A) Adhesion to nonstimulated human umbilical vein ECs. (B) Adhesion to nonstimulated mouse SPE-1 cells. (C) Adhesion to IL-1-stimulated human ECs. (D) Adhesion to TNF-α stimulated mouse SPE-1 cells. In all cases, ECs were cultured on 48-well plates precoated with gelatin. [^{3}H]Thymidine-labeled B16 variant cells were placed (5×10^4 cells/well) on plates, and incubated in serum-free RPMI 1640 medium at 37° for various times as shown on the abscissa. Ordinate, number of adherent cells as estimated from radioactivity count. ○, BL6; ▲, F10; △, F1; ●, WA4 cells. Note that the order of degree of adhesion to nonactivated human and mouse ECs is BL6 > F10 > F1 ≫ WA4, which is the same as the order of GM3 expression (Fig. 1A). Adhesion was enhanced on activation of human ECs by IL-1. However, activation of mouse ECs by TNF-α had essentially no effect on adhesion.

Inhibition of Cell Adhesion by Oligosaccharides and GSL Liposomes. BL6 cells are harvested with 0.02% EDTA, washed with PBS, and suspended in PBS at a concentration of 1×10^6 cells/ml. Methyl-β-D-lactoside and lactose are dissolved in PBS at a concentration of 200 mM. Liposomes (1 ml) are made from 500 nmol cholesterol, 500 nmol dipalmitoyl-PC, and 200 nmol GSL in PBS as described previously.[4] In this case, GSL concentration was 200 μM. Then 100 μl of 200 mM oligosaccharide or liposomes containing 200 μM of GSL is diluted twofold with 100 μl PBS. Next, 100 μl of BL6 cell suspension (2×10^5 cells) is added to 100 μl of oligosaccharide solution or liposome suspension, and incubated for 30 min at 37°. After incubation, mixtures of cells with oligosaccharide or liposome are placed on plates coated with LacCer (1 μg/well), human ECs, or mouse ECs, as described above. After incubation at 37° for 30 min, wells are washed and adherent cells collected and counted as described above. Results are shown in Fig. 4.

III. Gb_4-Dependent Adhesion of Human Embryonal Carcinoma 2102 Cells

Homotypic adhesion (autoaggregation) of 2102 cells is regarded as a model of the compaction process of human or primate embryonic cells at preimplantation "morula" stage. This adhesion process is mediated cooperatively by two adhesion systems: one involving carbohydrate–carbohydrate interaction, the other based on E-cadherin ("uvomorulin"). The former system is based on interaction of the major GSL, globoside (Gb_4), expressed on one cell with $GalGb_4$ or nLc_4 expressed on the counterpart cell surface; Ref. 7 and references therein). An example of an experiment demonstrating Gb_4-dependent adhesion is described below.

Cell Adhesion Process

2102 cells are cultured in Dulbecco's modified Eagle's medium (DMEM) containing 10% FCS, detached from confluent culture in 0.5 mM EDTA, collected, and suspended in 1% BSA in DMEM at an approximate density of 1×10^6 cells/ml. Aliquots (100 μl; 10^5 cells) of cell suspension are placed in each well of 96-well plates coated with various GSLs (Gb_4, Gg_3 GM1, GM2, Le^a, Le^b) at various quantities (0.1–1.5 nmol per well), and incubated for 30–120 min at 37° in a CO_2 incubator.

Washing Process

Each plate is filled with PBS containing Ca^{2+} and Mg^{2+}, placed in a large container filled with PBS, gently turned upside down, and allowed to

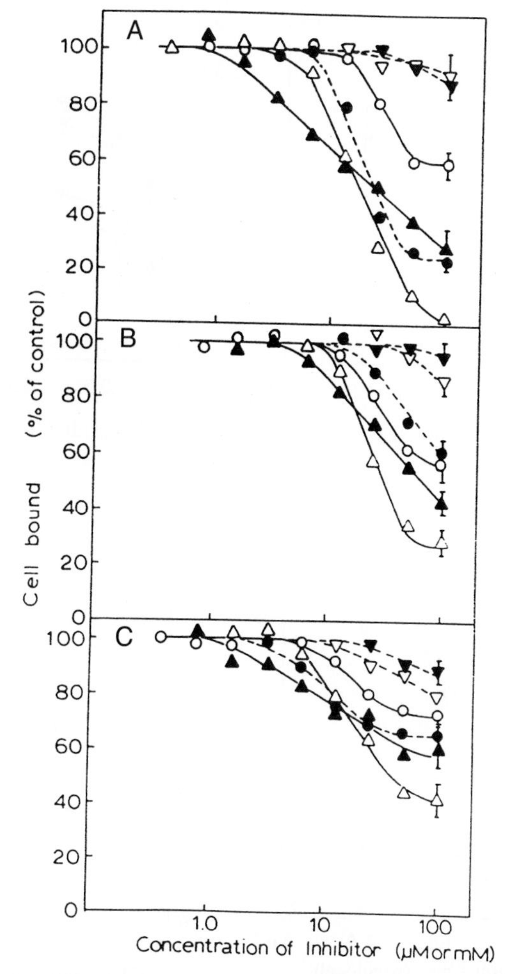

FIG. 4. Effect of oligosaccharides and GSLs on adhesion of B16/BL6 cells to LacCer-coated plates or to ECs. Labeled BL6 cells (5×10^5/ml) were incubated 30 min with various concentrations of lactose (▽), methyl-β-lactoside (●), methyl-β-N-acetyllactosaminide (▼), LacCer (△), Gg_3 (▲), or GM3 (○) as indicated on the abscissa. GSLs were added as liposomes, which were made from 500 nmol of cholesterol, 500 nmol of PC, and 200 nmol of GSL in 1 ml PBS. After incubation, 100 μl of each cell suspension (5×10^4 cells/well) was added to each well of 96-well flat-bottom plates coated with 1 μg/well of LacCer, or with confluent cultured ECs. After incubation (30 min for plates coated with LacCer or human ECs; 10 min for plates coated with mouse ECs), wells were washed and remaining adherent cells counted. (A) LacCer-coated plates. (B) Human ECs. (C) Mouse splenic EC line SPE-1. Concentrations on abscissa are expressed as μM for GSLs (solid lines) and as mM for oligosaccharides (dashed lines).

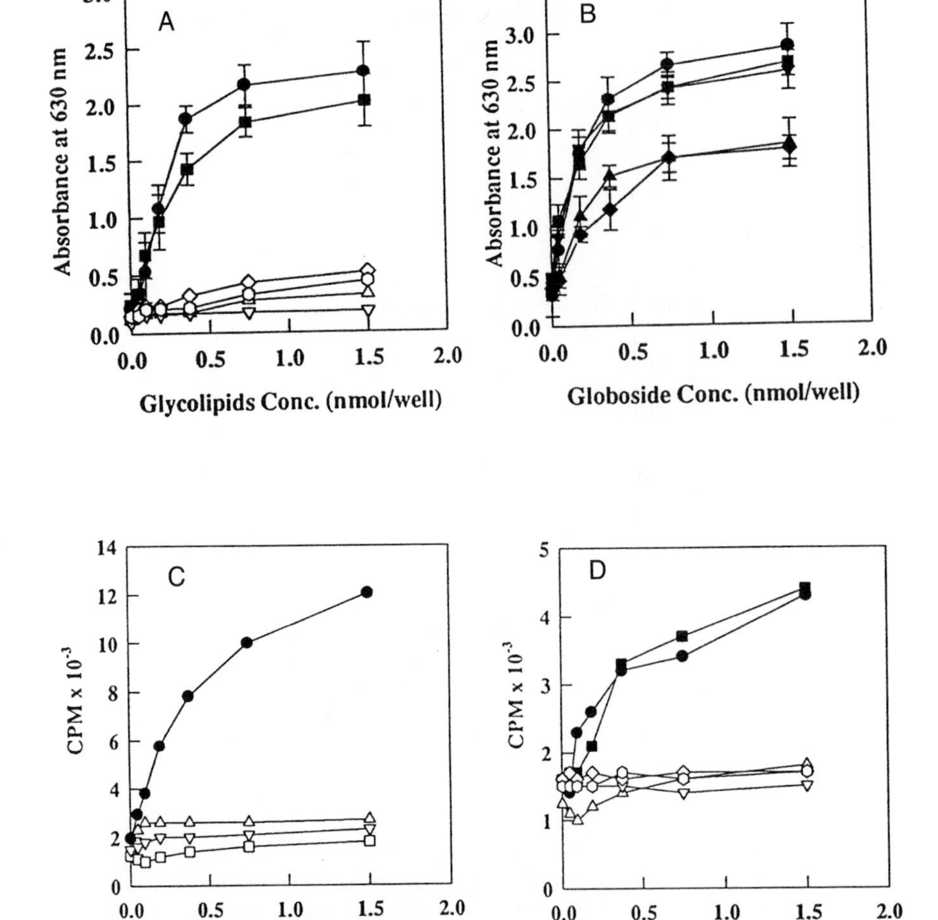

FIG. 5. Binding of 2102 cells or Gb_4 liposomes to plates coated with various GSLs, and the effect of anti-GSL antibodies on Gb_4-dependent 2102 cell adhesion. (A) 2102 cell binding to GSL-coated plates. 2102 cells (1×10^5/well) were incubated with Gg_3 (●), Gb_4 (■), GM1 (△), GM3 (▽), Le^a (○), and nLc_3 (◇) coated wells (0–1.5 nmol GSL/well). Adherent cells were stained with 0.1% Toluidine Blue O and measured at 630 nm. (B) Effect of anti-GSL antibodies on Gb_4-dependent adhesion. 2102 cells (1×10^5) were incubated in control medium or in medium containing MAbs 2D4 (anti-Gg_3) (5 μg/ml), 9G7 (anti-Gb_4) (20 μg/ml), MC631 (anti-SSEA-3) (20 μg/mL), or 1B2 (anti-nLc_4) (2 μg/ml) for 1 hr at 37°. Cells were washed and plated in wells coated with various quantities of Gb_4 (abscissa). ●, Control (no treatment of cells). ▼, 2D4. ■, 9G7. ◆, MC631. ▲, 1B2. Adherent cells were stained and measured as in part (A). Each point represents the mean of triplicate experiments. (C) Binding of Gb_4-liposomes to plates coated with nLc_4 (●), sialosyl-nLc_4 (▽), Lc_3 (□), and GM1 (△).

float with gentle shaking for 15–20 min. Nonadherent cells sedimented out of the plate. The plate is then turned right side up, removed from the container, and PBS in each well is aspirated gently using a very thin capillary tip.

Measuring of Adherent Cells

Adherent cells remaining after the above procedure are fixed with 150 μl of 3.7% paraformaldehyde for 30 min at room temperature, and stained with 0.1% Toluidene Blue O for 30 min. Cells remaining after washing with PBS are lysed in 10% acetic acid, and absorbance is measured at 630 nm.

Human embryonal carcinoma 2102 cells show clear adhesion to plates coated with Gb_4 or Gg_3, but not various other GSLs (Fig. 5A). The adhesion to Gb_4-coated plate is inhibited by preincubation of cells with anti-GalGb_4 MAb MC631 and anti-nLc_4 MAb 1B2, but not by preincubation with anti-Gg_3 MAb $2D_4$ or anti-Gb_4 MAb 9G7 (Fig. 5B). When Gb_4-coated plate is preincubated with anti-Gb_4 9G7, 2102 cell adhesion is inhibited in a dose-dependent manner (data not shown).

These results suggest that 2102 cell adhesion is mediated by interaction of Gb_4 (expressed on one cell) with nLc_4, GalGb_4, or Le^x (expressed on counterpart cell), i.e., Gb_4 is capable of binding to nLc_4 GalGb_4, or Le^x, but not to Gb_4. This is confirmed by binding of Gb_4 liposomes to plates coated with nLc_4 (Fig. 5C), GalGb_4, or Le^x (Fig. 5D), but not Gb_4 or various other GSLs tested. Each of these GSL epitopes is expressed in 2102 cells, at concentrations that vary during differentiation.[12] Although Gb_4 interacts strongly with Gg_3, Gg_3 is not expressed in 2102 cells[12] and therefore does not represent a physiologic epitope.

Comments

1. Methodological problems associated with study of cell adhesion or liposome binding to GSL-coated plates are (a) conditions for binding and

[12] C. Chen, B. A. Fenderson, P. W. Andrews, and S. Hakomori, *Biochemistry* **28,** 2229 (1989).

(D) Binding of Gb_4-liposomes to plates coated with GalGb_4 (●), Le^x (■), GM3 (○), Gb_4 (△), and Gb_3 (▽). Each point represents the mean of triplicate experiments. Standard variation is ~1% of value. Similar results were obtained in three separate experiments; one is shown here. Gb_4-liposomes labeled with [^{3}H]cholesterol and 96-well plates coated with various GSLs were prepared as described in Section II. Wells were coated with GSLs in various amounts (0–1.5 nmol/well). The amount of radioactivity in Gb_4-liposome bound to GSL-coated plate was measured by scintillation counter.

(b) conditions for washing (detach nonbinding cells or nonbinding liposomes). In the studies described in Section II, cell binding was performed by weak centrifugation (100*g*, 1 min) followed by incubation at 37°. Washing was performed by gentle aspiration using a special thin capillary tip. Adhesion varied depending on force of aspiration. More consistent results were obtained using the washing procedure described in Section III, in which washing/detachment of cells was performed by placing the plate upside down in a PBS-filled container and waiting for 20 min, i.e., detachment by a force of 1*g*. This procedure was applied successfully for study of selectin-dependent adhesion.[13]

2. Many types of GSL-dependent adhesion based on carbohydrate–carbohydrate interaction require a divalent cation, as described in Section II, i.e., GM3-dependent adhesion to LacCer or Gg_3 (melanoma cell adhesion to ECs). However, some types of GSL-dependent adhesion such as Gb_4-dependent adhesion (described in Section III) do not require divalent cations.

3. Liposome binding to GSL-coated polystyrene plate is enhanced greatly when the plate is coated with GSL–phosphatidylcholine–cholesterol mixture. However, the background value also becomes very high and obscures experimental variation, so such methods are not advisable.

4. The process of binding of cells to GSL-coated plate is much more rapid than integrin-dependent adhesion to fibronectin- or laminin-coated plate, as shown typically in Fig. 1C. This point is particularly important because cell adhesion based on carbohydrate–carbohydrate interaction may be the initial event leading to stronger adhesion based on protein adhesion receptors such as integrins.

Acknowledgment

This study was supported by National Cancer Institute Outstanding Investigator grant CA42505 (to S.H.). We thank Dr. Stephen Anderson for scientific editing and preparation of the manuscript.

[13] K. Handa, M. R. Stroud, and S. Hakomori, *Biochemistry* **36,** 12412 (1997).

[38] Analysis of Interactions between Glycosphingolipids and Microbial Toxins

By CLIFFORD A. LINGWOOD, BETH BOYD, and ANITA NUTIKKA

Introduction

Although several bacterial toxins have been reported to bind glycolipids on host cells, only two toxins exclusively utilize glycosphingolipids as their means of cell targeting and entry. These are cholera toxin (CT) (and its *Escherichia coli* homolog heat-labile toxin) and verotoxin (VT) from *E. coli* (and its *Shigella* homolog Shiga toxin, Stx), which bind, respectively, to the ganglioside GM1[1,2] and the neutral glycolipid Gb_3-globotriaosylceramide.[3] The recognition of glycolipids and their subsequent use in intracellular toxin trafficking provide some unusual features not seen for protein (or glycoprotein) receptors.

The distinct components of cellular metabolism are maintained topologically isolated by a complex series of intracellular membranes comprising the subcellular organelles. These membranes themselves have disparate features that allow distinct function. Temporary association of cytosolic components with membranes can regulate function and result in geographic separation.[4] Although transmembrane proteins may take advantage of this intracellular membrane network for sorting,[5] the distal components of proteins, which may serve as extracellular cell surface receptors, are unlikely to be responsive to compositional changes in the bilayer makeup. Lipids in general, and glycolipids specifically, however, being structural components of the membrane bilayer, are strongly influenced by the characteristic features of the membranes that contain them. This may provide an additional advantage to bacterial toxins which utilize cell surface glycolipids as their host cell receptors[6] in that such ligands may be able to sample, not only receptor distribution but also the receptor's bilayer microenvironment. This adds another dimension to receptor recognition.

[1] J. Moss, P. H. Fishman, V. C. Maganiello, M. Vaughan, and R. O. Brady, *Proc. Natl. Acad. Sci. U.S.A.* **73,** 1034 (1976).
[2] J. Holmgren, *Prog Brain Res.* **101,** 163 (1994).
[3] C. A. Lingwood, *Adv. Lipid Res.* **25,** 189 (1993).
[4] P. Wedegaertner, P. Wilson, and H. Bourne, *J. Biol. Chem.* **270,** 503 (1995).
[5] P. A. Gleeson, R. D. Teasdale, and J. Burke, *Glycoconj. J.* **11,** 381 (1994).
[6] K. Karlsson, *Chem. Phys. Lipids* **42,** 153 (1986).

METHODS IN ENZYMOLOGY, VOL. 312

0076-6879/00 $30.00

Assay of Glycolipid Binding

TLC Overlay

Glycolipid ligand binding is conveniently assayed by thin-layer chromatography (TLC) overlay procedures in which a cell lipid extract is separated on TLC by partition chromatography and then selective binding of the ligand to one or several of the separated species is directly performed on the plate. The procedure was first developed for cholera toxin[7] and involved the use of polyisobutylmethacrylate, supposedly to reorient the carbohydrate from interacting with the silica gel by which it was separated, to be available to the aqueous phase for ligand binding. Subsequent studies have shown that with appropriate selection of TLC material, the inclusion of this plasticizer is unnecessary.[8] Indeed it may cause binding artifacts.[9] The TLC overlay procedure has been widely used for the study of anti-carbohydrate antibody binding specificity, eukaryotic cell adhesion,[10] lectins[11] and carbohydrate binding proteins,[12] antibodies,[13] and intact pathogenic bacteria[14] and their isolated adhesins.[15] Glycolipid and glycoprotein (as neoglycolipid[16]) binding specificity has been analyzed by this methodology.

The advantages of this procedure are many, but include its rapidity and lack of ambiguity. Its resolving power is as good as the TLC solvent system used to separate the glycolipid mixtures. Combinations of this procedure have been developed such that the structure of unknown species to which a given ligand binds can be identified by mass spectroscopy straight from the TLC plate.[17] The drawback of the procedure is that it does not necessarily reflect the physiologic presentation of the glycolipid in cells. In this regard, the lipid moiety of the glycolipid can play a pivotal role in determining binding and this role can alter according to the manner in which the glycolipid is presented, i.e., on a TLC plate, in artificial membranes, or

[7] J. L. Magnani, D. F. Smith, and V. Ginsburg, *Anal. Biochem.* **109,** 399 (1980).
[8] C. A. Lingwood, *Can. J. Biochem. Cell Biol.* **63,** 1077 (1985).
[9] S. C. K. Yui and C. A. Lingwood, *Anal. Biochem.* **202,** 188 (1992).
[10] P. Swank-Hill, L. Needham, and R. Schnaar, *Anal. Biochem.* **163,** 27 (1987).
[11] R. W. Loveless, U. Holmskov, and T. Feizi, *Immunology* **85,** 651 (1995).
[12] D. D. Roberts, C. N. Rao, L. A. Liotta, H. R. Gralnick, and V. Ginsburg, *J. Biol. Chem.* **261,** 15 (1986).
[13] E. Nudelman, R. Kannagi, S. Hakomori, M. Parsons, M. Lipinski, J. Wiels, M. Fellows, and T. Tursz, *Science* **220,** 509 (1983).
[14] K.-A. Karlsson and N. Stromberg, *Methods Enzymol.* **138,** 220 (1987).
[15] C. A. Lingwood, G. Wasfy, H. Han, and M. Huesca, *Infect. Immun.* **61,** 2474 (1993).
[16] T. Feizi, M. S. Stoll, C.-T. Yuen, W. Chai, and A. M. Lawson, *Methods Enzymol.* **230,** 484 (1994).
[17] T. Taki, D. Ishikawa, S. Handa, and T. Kasama, *Anal. Biochem.* **225,** 24 (1995).

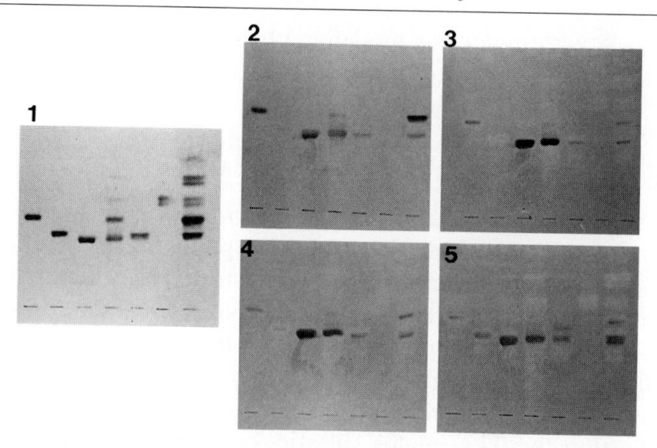

FIG. 1. Glycolipid binding specificity of verotoxins. (1) Orcinol chemical detection of glycolipids, (2) VT1 binding, (3) VT2 binding, (4) GT3 (a VT1-like mutant of VT2e [Tyrrell, 1992 #88) binding, (5) VT2e binding. Lane 1, Gb_3; lane 2, Gb_4; lane 3, amino Gb_4; lane 4, Gb_3 + amino Gb_4; lane 5, "hydroxylated" Gb_4; lane 6, lactosyl ceramide; lane 7, human renal neutral GSL. Note that some kidney extracts (as in lane 7) contain a Gb_4 subspecies that is recognized by VT1. This species was purified by extensive silica gel chromatography and partially characterized by 2D nuclear magnetic resonance (not shown) as being hydroxylated in the sphingosine. This purified Gb_4 subspecies was run in lane 5.

within a cell plasma membrane.[18,19] The degree to which the lipid moiety influences binding is likely dependent on the proximity of the sugar and carbohydrate. Thus for ligands that bind to distal sugars of long-chain glycolipids, this effect is likely to be minimal, whereas those that bind short-chain sugar-containing glycolipids, the effect can be dramatic.[19]

Experimental

SPECIFICITY. Verotoxins in fact comprise a family of toxins VT1, VT2, VT2c, and VT2e.[3] The glycolipid binding specificity varies in that VT2e, the pig edema disease toxin, binds Gb_4 in addition to Gb_3.[20] Although Gb_4 is not bound, removal of the *N*-acetyl group from the terminal sugar of Gb_4 generates amino-Gb_4, which is recognized well by all verotoxins.[21] This binding specificity (Fig. 1) determined by TLC overlay was found to

[18] N. Strömberg, P.-G. Nyholm, I. Pascher, and S. Normark, *Proc Natl. Acad. Sci. U.S.A.* **88,** 9340 (1991).

[19] B. Boyd, Z. Zhiuyan, G. Magnusson, and C. A. Lingwood, *Eur. J. Biochem.* **223,** 873 (1994).

[20] S. DeGrandis, H. Law, J. Brunton, C. Gyles, and C. A. Lingwood, *J. Biol. Chem.* **264,** 12520 (1989).

[21] P. Nyholm, G. Magnusson, Z. Zheng, R. Norel, B.-B. B. and L. C. A., *Chem. Biol.* **3,** 263 (1996).

correlate well with molecular modeling of the binding[21] using the coordinates of the crystal structure of the VT1 B subunit pentamer.

Procedure

1. Lay down lipid tracks (typically 0.5–5 μg of pure lipids; 20–50 μg of cell extract per lane) on the TLC plate. Since the migration of GSLs in whole-cell extracts may be distorted by other lipids present, this can be avoided by saponification of the extract and separation of the lower phase or the glycolipid fraction can be separated by silica gel chromatography.[22] Run two identical plates: one for VT binding, one for orcinol chemical detection.
2. Develop the TLC plates (plastic- or aluminum-backed SILG) in solvent [because some batches of plastic SILG plates blister in chloroform : methanol : water; 65 : 25 : 4 (v/v/v), use hexane : 2-propanol : water, 10 : 10 : 1].
3. Allow the TLC plate to dry completely before prewetting with water by letting the water run up the TLC plate as a solvent, which minimizes air being trapped.
4. After the TLC plate has been wetted, block with 1% gelatin from 45 min to overnight at 37°.
5. Rinse off traces of gelatin with 50 m*M* TBS (50 m*M* Tris buffer, pH 7.4, with 0.9% NaCl).
6. Incubate the TLC plate with purified VT1 diluted to a concentration of 1 μg per 15 ml in TBS (or about 50–60 ng/ml) for 45 min minimum at room temperature with gentle shaking. Rinse with TBS, 3× by hand vigorously for about 5 sec each and again (twice) on a shaking table for 3–5 min each.
7. Incubate with monoclonal anti-VT1 (1–2 μg/ml) diluted in TBS for 45 min minimum at room temperature with shaking. Rinse as before.
8. Incubate with goat anti-mouse IgG horseradish peroxidase conjugate (diluted 1 : 2000 in TBS) for 60 min at room temperature with shaking. Rinse as before.
9. Develop using a 3 mg/ml solution of 4-chloro-1-naphthol in methanol freshly mixed with 5 volumes of TBS and 1 : 1000 dilution of 30% H_2O_2. Develop until sufficient color has appeared (usually 1 to 10 min).
10. Stop with extensive washing under tapwater.

[22] B. Boyd and C. A. Lingwood, *Nephron.* **51,** 207 (1989).

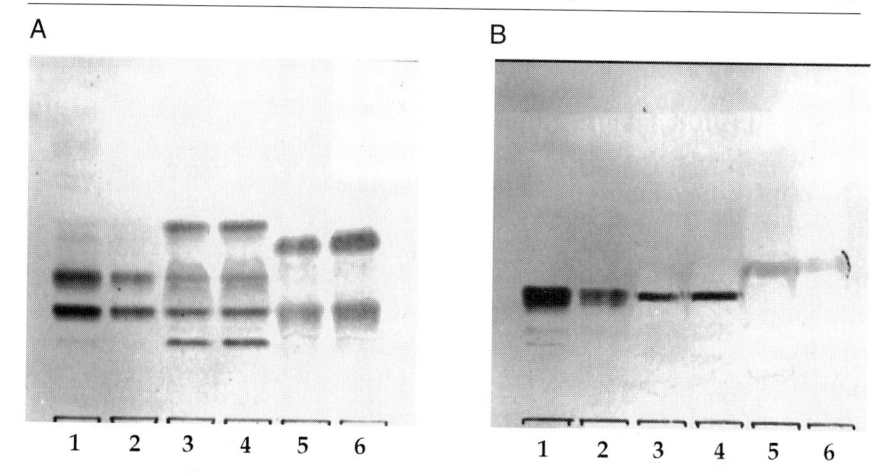

FIG. 2. Verotoxin binds Gb_3 in human platelets. (A) Orcinol spray, (B) VT1 binding. Lanes 1, 2, human renal Gb_3 and Gb_4 standard (1 mg/ml: 10 and 5 μl); lanes 3, 4, platelet neutral glycolipid fraction (5 and 10 μl); lanes 5, 6, platelet phospholipid fraction (5 and 10 μl). (In the phospholipid fraction—lanes 5, 6—some spillover of Gb_3 is seen whose migration is altered by the phospholipid content.)

Gb_3 IN PLATELETS. Verotoxin is strongly implicated in the etiology of hemolytic uremic syndrome (HUS).[23] One of the cardinal symptoms of HUS is thrombocytopenia.[24] We have therefore used TLC overlay to determine whether platelets contain the verotoxin receptor Gb_3. Platelets were extracted with organic solvent and the glycolipid and phospholipid fractions isolated by silica gel chromatography as previously.[22] Two aliquots were tested for VT1 binding by TLC overlay as described above (Fig. 2).

The finding that human platelets contain Gb_3 opens the possibility that VT may play a more direct role in the loss of platelets than previously suspected.

Microtiter Plate Glycolipid Binding Assays

Glycolipids can be immobilized on the plastic surface of microtiter plates in the presence of cholesterol and lecithin to form an artificial bilayer structure. This format of presentation more closely approaches that of a natural membrane. The effect of the chain length and unsaturation of the lipid moiety of the glycolipid on binding is more marked under such conditions than by determination of binding by TLC overlay.[25] Using radio-

[23] B. S. Kaplan, T. G. Cleary, and T. G. Obrig, *Pediatr. Nephrol.* **4,** 276 (1990).
[24] M. A. Karmali, *Clin. Microbiol. Rev.* **2,** 15 (1989).
[25] A. Kiarash, B. Boyd, and C. A. Lingwood, *J. Biol. Chem.* **269,** 11138 (1994).

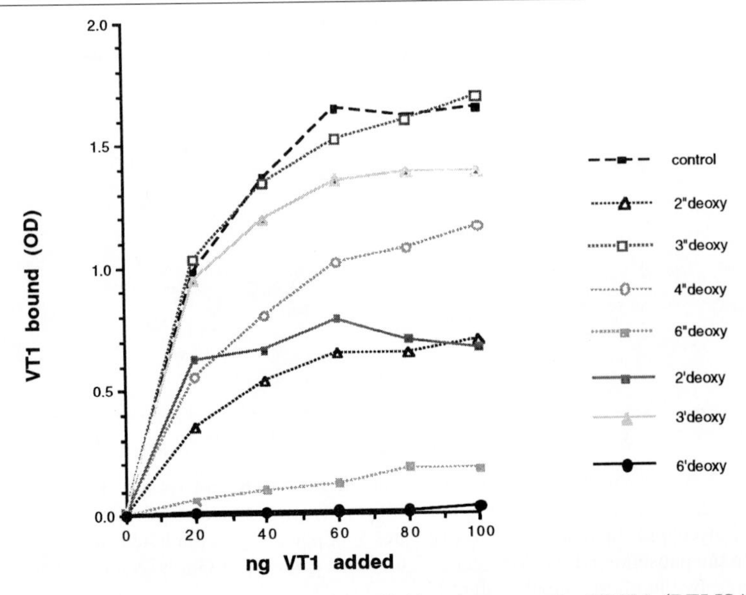

FIG. 3. VT1 binding deoxyl galabiosylglycolipids using a receptor ELISA (RELISA).

labeled ligand, this assay can be quantitative and Scatchard analysis performed to measure the kinetic parameters of binding. However, while the dissociation constant and number of binding sites can be calculated, determination of the stoichiometry of binding cannot because the relative availability of the Gb_3 on the surface bound (as opposed, for example, to that sequestered on the other surface of the bilayer) cannot be determined. The acyl chain length of the phosphatidylcholine can also effect glycolipid binding, indicating the influence of the membrane environment on glycolipid headgroup presentation for binding.[26] For the most part, short-chain hydrocarbon phospholipids promote headgroup recognition, whereas longer chain species decrease glycolipid headgroup availability for binding.

The confounding effect of such phospholipid modulation of receptor binding, though likely of relevance to the physiologic recognition of glycolipids in natural membranes, can be avoided by performance of binding to pure glycolipid immobilized on the microtiter plate. Analysis of verotoxin 1 binding to deoxy-Gb_3 analogs (in which each hydroxyl of the terminal galabiose was removed in turn) by such a receptor ELISA (Fig. 3, RELISA),

[26] S. Arab and C. A. Lingwood, *Glycoconj. J.* **13,** 159 (1996).

compares very well with a similar study using the [^{125}I]VT microtiter binding assay in the presence of cholesterol and lecithin.[21]

Procedure

This technique is very useful for indirectly quantitating binding of proteins to purified glycolipids. Due to the microtiter plate format, the binding of a series of glycolipid and/or protein concentrations can be simultaneously assayed and compared. Once conditions are standardized, the assay is highly reproducible. The major drawback of the technique is the variation observed between ELISA plate brands (see below). Glycolipid and GSL-binding protein concentrations must be individually determined for each brand of plate; some plates cause nonphysiologic GSL binding. For this reason, the RELISA should be used with caution to identify GSL binding specificities; the TLC overlay is the most robust for this.

PREPARATION OF PLATES. Lipids are coated in microtiter plate wells by evaporation from solution in ethanol. Glycolipid aliquots are measured from stock solutions in dichloromethane/methanol into screw-capped glass tubes, and the solvent removed under a stream of nitrogen. Ethanol is added to the lipid residue, and a uniform glycolipid solution is prepared by immersing the tube for 30 sec in a bath sonicator followed by 1 min of vortexing. Initially, glycolipid concentrations in the range of 0–2000 ng/well are screened for receptor activity. The volume of ethanolic glycolipid added per well is usually 50 or 100 μl. The glycolipid solution is pipetted into the microtiter plate wells and the ethanol is allowed to evaporate at room temperature, usually overnight. Once dry, the plates are stored in a desiccator at 4° and are stable in this form for several weeks. The remainder of the assay is similar to typical ELISAs with the exception that use of Tween 20, a mild detergent often used to reduce nonspecific interactions, is avoided since it may solubilize the immobilized glycolipids.

1. Blocking. The wells are blocked by addition of 200 μl/well of 0.2% (w/v) BSA in 10 m*M* sodium phosphate, 150 m*M* NaCl, pH 7.4 (BSA–PBS). The plate is covered with Parafilm in this and all subsequent incubations, which are carried out at room temperature by either static or with gentle agitation. Blocking is carried out for 1–2 hr, then the wells are washed twice with 150 μl/well of 0.2% BSA–PBS.
2. Binding. Dilutions of verotoxin are prepared in BSA–PBS to give a range of 0–1000 ng/well. The solution is dispensed into the wells (usually the same volume used for glycolipid coating is used here) and incubated for 1 hr.

3. Primary antibody. The wells are emptied and washed three times for 1 min each with 200 μl of 0.2% BSA–PBS. After the last wash, a 1 μg/ml solution of monoclonal anti-VT1 PH1[27] in 0.2% BSA–PBS is added to the wells and incubated for 1 hr.
4. Secondary antibody. Washing is performed as mentioned above and a 1:2000 dilution of goat anti-mouse IgG-horseradish peroxidase conjugate in 0.2% BSA–PBS is added to the wells and incubated for 1 hr. During the last 10 min the substrate solution is prepared; 2,2′-azinobis(3-ethylbenzthiazoline-6-sulfonic acid) diammonium salt (ABTS) is dissolved in 0.08 *M* citric acid, 0.1 *M* disodium phosphate, pH 7.4, at a concentration of 0.5 mg/ml. Then 3 μl of 30% H_2O_2 is added per 10 ml of solution. The solution is protected from strong light until use.
5. Color development. The wells are emptied and washed three times with 200 μl of 0.2% BSA–PBS for 3 min each. Finally, the wells are washed once with PBS, then 100 μl/well of ABTS solution is added. The plate is tapped gently then placed in the dark. After sufficient color has developed (usually 20–40 min), the absorbance of each well at 410 nm is determined using an ELISA plate reader.

This RELISA is quick, convenient, and allows quantitative comparison. However, because the bound ligand is detected immunologically, the assay does not permit kinetic analysis of the binding data. The assay does allow comparison of the binding to structural homologs of the receptor and as such is useful for verification of modeling predictions[21] and for screening for inhibitors of binding.

Caution, however, must be exercised in the choice of microtiter plate used to immobilize the glycolipid. Comparison of the VT1 binding specificity to Gb_3 and Gb_4 for typical commercially available plates is shown in Fig. 4. It can be seen that for only some plates is the distinction between Gb_3 and Gb_4 binding adequately maintained. This suggests that in associating with different types of plastics, the carbohydrate conformation of glycolipids can be altered, in this case to allow verotoxin binding to the penultimate galabiose moiety of Gb_4, normally not available for binding.

Retrograde Intracellular Glycolipid Traffic

The interaction of both CT and VT with their glycolipid receptors provides useful tools in cell biology. Indeed, susceptibility to modulation by CT treatment is considered diagnostic for a $G_s\alpha$-dependent pathway.[28]

[27] J. Boulanger, M. Petric, C. A. Lingwood, H. Law, M. Roscoe, and M. Karmali, *J. Clin. Micro.* **28,** 2830 (1990).

[28] J. Exton, *Ann. Rev. Pharmacol. Toxicol.* **36** (1996).

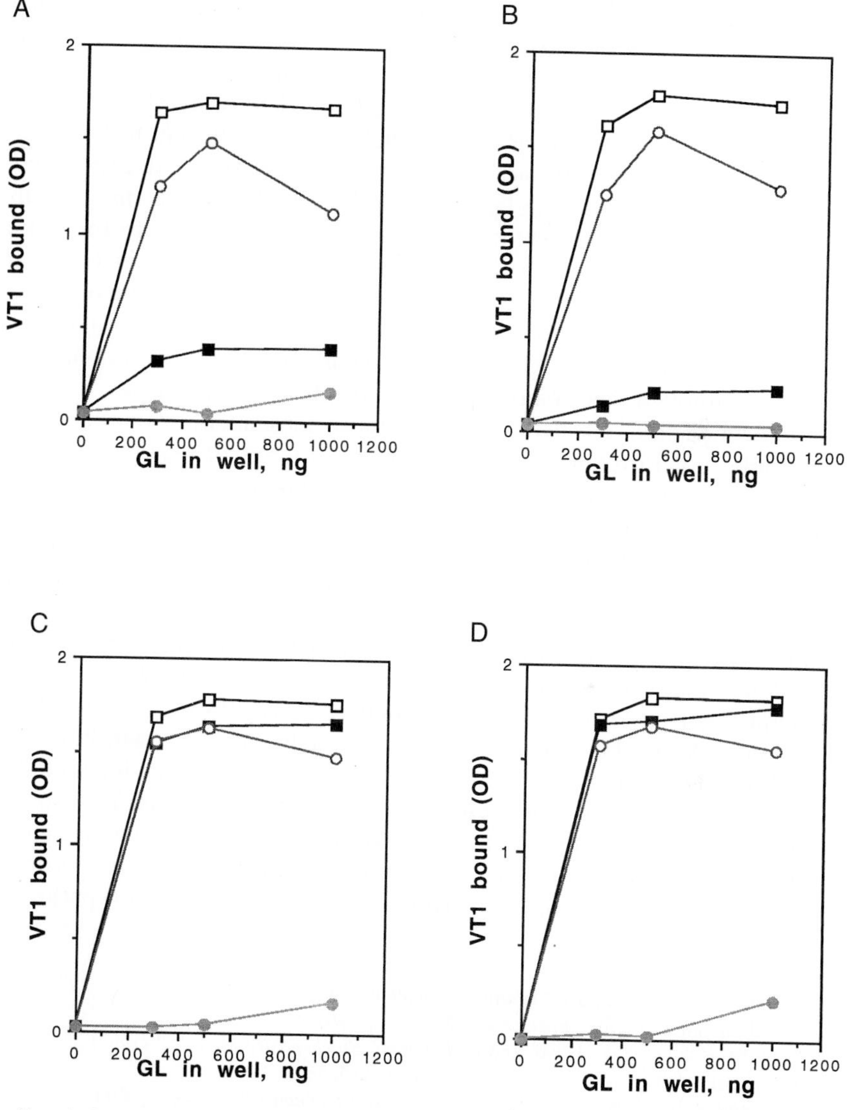

FIG. 4. Comparison of different commercial microtiter plates for the VT1 RELISA. (A) Corning, (B) Costar, (C) Nunc, (D) Immulon 96-well plates. Open squares, Gb_3; closed squares, Gb_4; open circles, globotriaosyl sulfonolipid [Boyd, 1994 #1446]; closed circles, 6″ deoxy globotriaosyl sulfonolipid [Nyholm, 1996 #1735]. Only Corning and Costar plates distinguish between Gb_3 and Gb_4 (similar results were found for Evergreen plates—not shown). The plates that show spurious high Gb_4 binding are "treated" by the manufacturer to increase adsorbance.

Cholera toxin and verotoxin are unique among bacterial toxins in that they recognize a single glycolipid species that mediates their eukaryotic cell targeting. The CT receptor GM1 is a five-sugar, sialic acid-containing glycolipid, and binding of cholera toxin is dependent on the terminal 2-3-linked sialic acid to GalNacβ1-4Gal. Cholera toxin will bind to the lipid-free GM1 ganglioside oligosaccharide,[29,30] whereas verotoxin does not bind, or binds very poorly, to the lipid-free oligosaccharide of the three-sugar-containing Gb_3.[19,31] This is likely due, as mentioned above, to the more distal nature of the binding site relative to the lipid moiety in GM1, as opposed to Gb_3 recognition by verotoxin. However, although the oligosaccharide alone is bound by CT, a lipid moiety is necessary for receptor function in cells. Thus neoglycoproteins decorated with the GM1 oligosaccharide, although they bind cholera toxin effectively, will not undergo receptor-mediated internalization and induce subsequent cytopathology.[32]

While cholera toxin is less discriminant, in terms of the nature of the lipid moiety, than is verotoxin, differential cytotoxicity is found for GM1 oligosaccharide linked to different classes of lipid species.[33] However, the precise intracellular routing of such neoglycolipids receptors and its relationship to cytotoxicity has not been investigated. Initial studies on the intracellular routing of cholera toxin to mediate cytopathology via activation of adenylate cyclase on the cytosolic leaflet of the plasma membrane concluded that cholera toxin was like other bacterial toxins in that the A subunit was translocated from the lumen of lysosomes to the cytosol for G protein ADP-ribosylation. More recent studies have shown, however, that both cholera toxin[34,35] and verotoxin[36,37] share a distinct intracellular trafficking pathway in that these toxins are transported from endosomes to the Golgi via a retrograde transport pathway.[38] Thus these toxins enter cells via the biosynthetic secretory pathway rather than the endosome/lysosomal degradatory route. Such retrograde glycolipid transport pathways

[29] P. Fishman, J. Moss, and J. Osborne, *Biochemistry* **17,** 711 (1978).

[30] A. Schön and E. Freire, *Biochemistry* **28,** 5019 (1989).

[31] P. M. St. Hilaire, M. K. Boyd, and E. J. Toone, *Biochemistry* **33,** 14452 (1994).

[32] T. Pacuszka and P. H. Fishman, *J. Biol. Chem.* **265,** 7673 (1990).

[33] T. Pacuszka, R. M. Bradley, and P. H. Fishman, *Biochemistry* **30,** 2563 (1991).

[34] M. P. Nambier, T. Oda, C. Chen, Y. Kuwazuru, and H. C. Wu, *J. Cell Physiol.* **154,** 222 (1993).

[35] A. Sofer and A. Futerman, *J. Biol. Chem.* **270,** 12117 (1995).

[36] K. Sandvig, O. Garred, K. Prydz, J. Kozlov, S. Hansen, and B. van Deurs, *Nature* **358,** 510 (1992).

[37] A. A. Khine and C. A. Lingwood, *J. Cell Physiol.* **161,** 319 (1994).

[38] K. Sandvig and B. van Deurs, *Physiol. Rev.* **76,** 949 (1996).

may underlie a role for these toxin receptor glycolipids in signal transduction.[39,40]

Use of Toxin Conjugates to Monitor Glycolipid Retrograde Transport within Cells

To monitor toxin glycolipid interactions after internalization into cells and intracellular trafficking, it is necessary to have a method for the visualization of the toxin-bound glycolipid. Such a method is also useful in determining the presence of the receptor glycolipid in frozen tissue sections. Cholera toxin (CT) has been used as a method for localizing GM1 ganglioside[41] and peroxidase-labeled Shiga toxin has been used to localize Gb_3[42] at the electron microscope (EM) level. We have been unable to label verotoxin with peroxidase without loss of receptor binding and cytotoxicity. We have however, been successful in fluorescently labeling the B subunit or holotoxin with fluorescein isothiocyanate (FITC) without effect on cytotoxic specific activity.[37] Similar fluorescent labeling of the CT B subunit is also effective.[43] The fluorescent toxin can be monitored, by either regular or confocal fluorescence microscopy, during intracellular trafficking once the toxin is internalizd into susceptible cells in culture. Such studies have shown that for cells sensitive to verotoxin, the toxin is internalized by a pH-independent[43] retrograde transport pathway to the Golgi. In highly VT-sensitive cells, retrograde trafficking of the VT/Gb_3 complex to the endoplasmic reticulum (ER) and nuclear envelope is seen.[44] Unlike CT, VT does not contain an ER retention sequence. The ER/nuclear targeting of VT is correlated with the presence of shorter chain fatty acid-containing Gb_3 isoforms, which are deficient in cells that target the toxin to the Golgi. Thus, the intracellular traffic of VT is likely a consequence of isoform restricted glycolipid transport.

A comparison of the FITC-VT B intracellular labeling of SF539 astrocy-

[39] A. A. Wolf, M. G. Jobling. S. Wimer-Mackin, M. Ferguson-Maltzman, J. L. Madara, R. K. Holmes, and W. I. Lencer, *J. Cell Biol.* **141,** 917 (1998).

[40] A. A. Khine, M. Firtel, and C. A. Lingwood, *J. Cell Physiol.* **176,** 281 (1998).

[41] H. A. Hansson, J. Holmgren, and L. Svennerholm, *Proc. Natl. Acad. Sci. U.S.A.* **74,** 3782 (1977).

[42] K. Sandvig, K. Prydz, M. Ryd, and B. van Deurs, *J. Cell Biol.* **113,** 553 (1991).

[43] F. Schapiro, C. A. Lingwood, W. Furuya, and S. Grinstein, *Am. J. Physiol.* **274,** 319 (1998).

[44] S. Arab and C. Lingwood, *J. Cell Physiol.* **177,** 646 (1998).

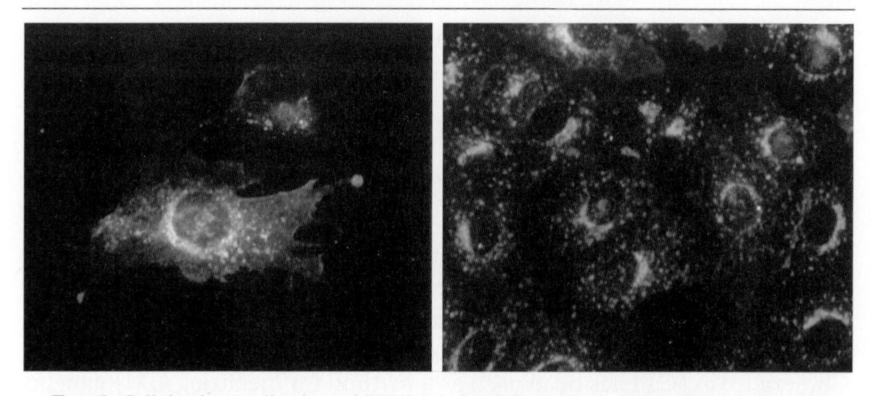

FIG. 5. Cellular internalization of FITC-labeled VT B. Once internalized by receptor-mediated endocytosis, VT B accumulates around the nucleus (ER and nuclear membrane) in SF539 astrocytoma cells (left-hand side) and in a juxtanuclear (Golgi) area in Vero cells (right-hand side).

toma cells, in which the ER/nuclear envelope is targeted, and Vero cells, in which the juxtanuclear Golgi is targeted, is shown in Fig. 5.

This retrograde Gb_3 transport pathway has been used to advantage to measure the acidic pH of the Golgi (since FITC fluorescence is pH sensitive) in living cells using FITC-VT1 B.[45] Similar measurement of the neutral pH of the ER was achieved[46] using a B subunit in which a C-terminal ER retention sequence had been engineered.[47]

Procedure

LABELING OF VT

1. Dissolve 1 mg VT (or VT B subunit)[48] in 0.1 ml of 0.1 *M* sodium bicarbonate (pH 9.0).
2. Dissolve the FITC in dimethyl sulfoxide (DMSO) at 10 mg/ml immediately before starting the reaction. Briefly sonicate.
3. While vortexing the VT solution, slowly add 5–10 μl FITC.
4. Incubate the reaction for 1 hr at room temperature with continuous stirring.

[45] J. H. Kim, C. A. Lingwood, D. B. Williams, W. Furuya, M. F. Manolson, and S. Grinstein, *J. Cell Biol.* **134,** 1387 (1996).

[46] J. H. Kim, L. Johannes, B. Goud, C. Autony, C. A. Lingwood, R. Daneman, and S. Grinstein, *Proc. Natl. Acad. Sci. U.S.A.* **95,** 2997 (1998).

[47] L. Johannes, D. Tenza, C. Anthoy, and B. Goud, *J. Biol. Chem.* **272,** 19554 (1997).

[48] M. Petric, M. A. Karmali, S. Richardson, and R. Cheung, *FEMS Microbiol. Lett.* **41,** 63 (1987).

5. Stop the reaction by adding 0.1 ml of freshly prepared 1.5 *M* hydroxylamine (pH 8.5) to a final concentration of 0.15 *M*.
6. Incubate for an additional 1 hr at room temperature.
7. Separate the conjugate from unreacted FITC and hydroxylamine on a gel filtration column (10 × 300 mm) equilibrated with PBS. The excluded fraction corresponding to the first fluorescent band to elute will be the VT conjugate.
8. Store at −70° or 4° in 0.01% sodium azide.

INTRACELLULAR STAINING WITH FITC-VT B

1. Cells are grown overnight on glass coverslips and incubated for 1 hr at 37° in the presence of 5 μg/ml of FITC-VT1 B.
2. Following extensive washing with 50 m*M* PBS, cells are mounted with DABCO and examined under the fluorescence microscope.

This Gb_3 trafficking pathway from the cell surface to the nucleus is also used by CD19 following ligation at the cell surface.[40] CD19 is a B-cell surface marker with an N-terminal extracellular domain with amino acid sequence similarity to VT B.[49]

Use of Toxin Conjugates to Monitor Tissue Distribution of Toxin Receptor Glycolipid

FITC labeled toxin was initially used to localize Gb_3 in the frozen sections of human kidney and showed that the receptor was present primarily in the glomerulus of pediatric renal tissue but absent from such structures in adults.[50] Thus receptor distribution during renal ontogeny reflects the high incidence of hemolytic uremic syndrome in the pediatric as opposed to adult population.[51] It can be seen in Fig. 6 that the toxin binds specifically to the membranes of cells within the glomerulus of a frozen renal section from a pediatric (1 yr, black male) minimum lesion nephrotic syndrome biopsy.

Procedure

VT STAINING OF FROZEN TISSUE SECTIONS

1. Let frozen sections (5 μm) air dry for 30–45 min.
2. Block with 1–2% normal goat serum for 30 min at room temperature. Rinse section with PBS.

[49] M. D. Maloney and C. A. Lingwood, *J. Exp. Med.* **180,** 191 (1994).
[50] C. A. Lingwood, *Nephron.* **66,** 21 (1994).
[51] P. Rowe, E. Orrbine, G. Wells, and P. McLaine, *J. Pediat.* **119,** 218 (1991).

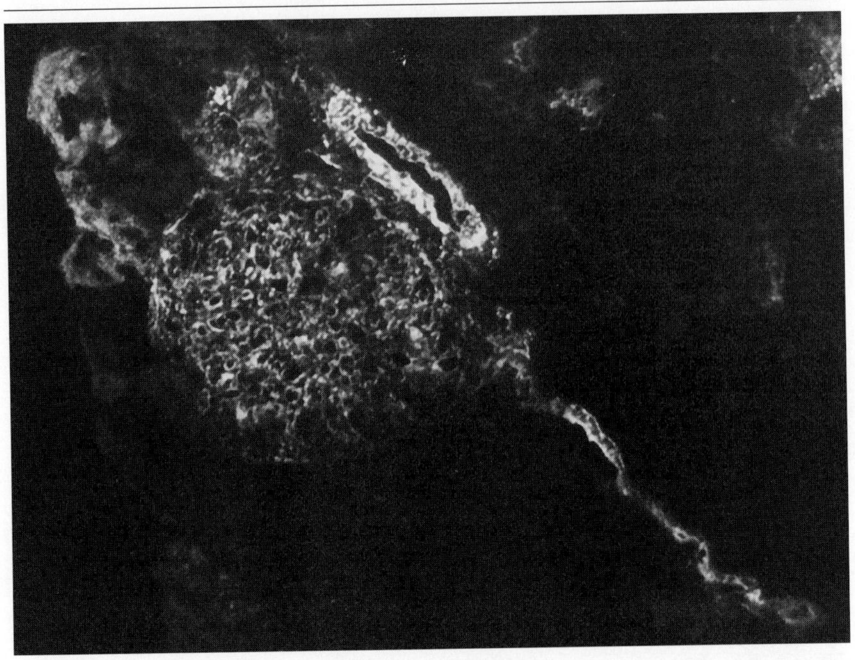

FIG. 6. FITC B labeling of the pediatric renal glomerulus. Tubular structures within the glomerulus are stained. A blood vessel adjacent to and a blood vessel leading into (or out of) the glomerulus are also stained.

3. Incubate section with 1 μg/ml of FITC-VT (holotoxin or B subunit) diluted into 1% with normal goat serum in PBS for 1 hr (in a humid chamber) protected from light. Rinse slides several times with PBS.
4. Fix sections with 2% formaldehyde in PBS for 20 min at room temperature. Rinse with PBS prior to mounting.
5. Section can be mounted using commercially available mounting media, e.g., DABCO.[37]

Subsequent studies have used FITC-B staining of frozen sections to confirm the up-regulation of Gb_3 in ovarian carcinoma.[52]

Toxin/glycolipid receptor interaction, particularly VT/Gb_3, provides a novel model of protein carbohydrate interaction, whereby the lipid plays an important role in sugar recognition. The interaction is a probe for the study of intracellular retrograde transport, glycolipid signal transduction,

[52] S. Arab, E. Russel, W. Chapman, B. Rosen, and C. Lingwood, *Oncol. Res.* **9,** 553 (1997).

and tissue receptor distribution, in relation to clinical pathology and may be increased in specific human neoplasias.

Acknowledgments

Studies in my laboratory have been supported by MRC grant MT13073 and NIH grant R01 DK52098.

[39] Oxidation of Aglycone of Glycosphingolipids: Serine and Ceramide Acid Precursors for Soluble Glycoconjugates

By MURUGESAPILLAI MYLVAGANAM and CLIFFORD A. LINGWOOD

Introduction

Synthetic glycoconjugates have been used as model compounds to study glycolipid–protein-mediated cellular events. Prevalent use of synthetic ligands is in part attributed to their solubility in water, whereas natural glycolipids tend to aggregate and thereby cause considerable difficulties in studying their interactions with proteins. Many soluble glycoconjugates have been contrived and tested.[1,2] Synthesis of glycoconjugates from hexose monomers requires intricate synthetic manipulations such as glycal assembly[3] and polymerization of glycosyl monomers.[4,5] For glycobiologists, efficient glycoconjugate protocols, particularly those that would enable the transformation of microgram quantities of natural glycolipids into neoglycoconjugates, would be very useful. Our strategy was to synthesize an "oligosaccharide" precursor from natural glycosphingolipids (GSLs) by trimming the hydrocarbon chains, i.e., the acyl chain and the sphingosine chain, by base hydrolysis and oxidative cleavage of the double bond, respectively. For the hydrolysis of acyl chains of GSLs, fine-tuned protocols based on alcoholic alkaline hydroxides are available.[6] However, there are only a few

[1] G.-J. Boons, "Carbohydrate Chemistry." Blackie Academic & Professional, 1998.
[2] E. E. Simanek, G. J. McGarvey, J. A. Jablonowski, and C.-H. Wong, *Chem. Rev.* **98,** 833 (1998).
[3] P. H. Seeberger, M. T. Bilodeau, and S. J. Danishefsky, *Aldrichimica Acta* **30,** 75 (1997).
[4] R. Roy, *Trends Glycosci. Glycobiotechnol.* **8,** 79 (1996).
[5] A. Furstner, *Top. Organometal. Chem.* **1,** 199 (1998).
[6] T. Taketomi, A. Harqa, K. Uemura, H. Kurahashi, and E. Sugiyama, *Biochem. Biophys. Res. Commun.* **224,** 462 (1996).

0076-6879/00 $30.00

reports on the oxidative cleavage of the sphingosine chain. The method described here enables efficient oxidative cleavage of the sphingosine double bond giving products containing a three-carbon-based "serine acid" or a four-carbon-based "ceramide acid" attached to the reducing end of the glycone (Scheme 1).[7,8]

Oxidation of Sphingosine Double Bond of Glycolipids

The sphingosine double bond of GSLs can be oxidatively cleaved to give aldehydes (glycosyl aldehydes) or carboxylic acids (glycosyl acids, Scheme 2). Affinity matrices having covalently immobilized GSL ligands can be synthesized by coupling glycosyl acids to an amino matrix. Such affinity columns are used in the purification of antibodies,[9] glycosylhydrolases, and transferases.[10] Also, the glycosyl acid or the aldehyde can be coupled to proteins or other aglycones to yield immunogenic neoglycoproteins[11] or high-affinity glycoconjugates.[7,8]

Treating peracetylated GSL precursors with OsO_4–$NaIO_4$[12] or natural GSLs with O_3 followed by nonoxidative degradation[11] of the ozonide gives the glycosyl aldehydes. These aldehydes are then treated with base to free the oligosaccharide,[12] or reductively coupled to proteins such as keyhole limpet hemocyanin or bovine serum albumin (BSA).[11] Reacting peracetylated GSLs with $KMnO_4$ crown ether complex[13] or with $KMnO_4$–acetone[14] or with O_3 followed by oxidative degradation of the ozonide[9] gives the glycosyl acid. Unlike the aldehydes, glycosyl acids are stable under basic conditions and can be coupled to various aglycones by exploiting methodologies developed for the synthesis of peptides. The method described in this review is based on the $KMnO_4$–$NaIO_4$[7,15] or $KMnO_4$–$NaIO_4$–K_2CO_3[8] catalytic systems, and whereas the former conditions give "glycosylceramide acids," the latter give "glycosylserine acids" (Scheme 3). Oxidation of deacyl GSLs gives such acids free of hydrocarbon chains and are referred to as "serine" and "ceramide" oligosaccharides.

[7] M. Mylvaganam and C. A. Lingwood, *J. Biol. Chem.*, **274,** 20725 (1999).
[8] M. Mylvaganam, L.-J. Meng, and C. A. Lingwood, *Biochemistry*, **38,** 10885 (1999).
[9] R. Laine, G. Yogeeswaren, and S. Hakomori, *J. Biol. Chem.* **14,** 4460 (1974).
[10] N. Taniguchi, K. Yanagisawa, A. Makita, and M. Naiki, *J. Biol. Chem.* **260,** 4908 (1985).
[11] F. Helling, A. Shang, M. Calves, S. Zhang, S. Ren, R. K. Yu, H. F. Oettgen, and P. O. Livingston, *Cancer Res.* **54,** 197 (1994).
[12] S.-I. Hakomori and B. Siddiqui, *Methods Enzymol.* **28,** 156 (1972).
[13] W. W. J. Young, R. A. Laine, and S. Hakomori, *J. Lipid Res.* **20,** 275 (1979).
[14] D. L. MacDonald, L. Pat, and S. Hakomori, *J. Lipid Res.* **21,** 642 (1980).
[15] R. U. Lemieux and E. von Rudloff, *Can. J. Chem.* **33,** 1701 (1955).

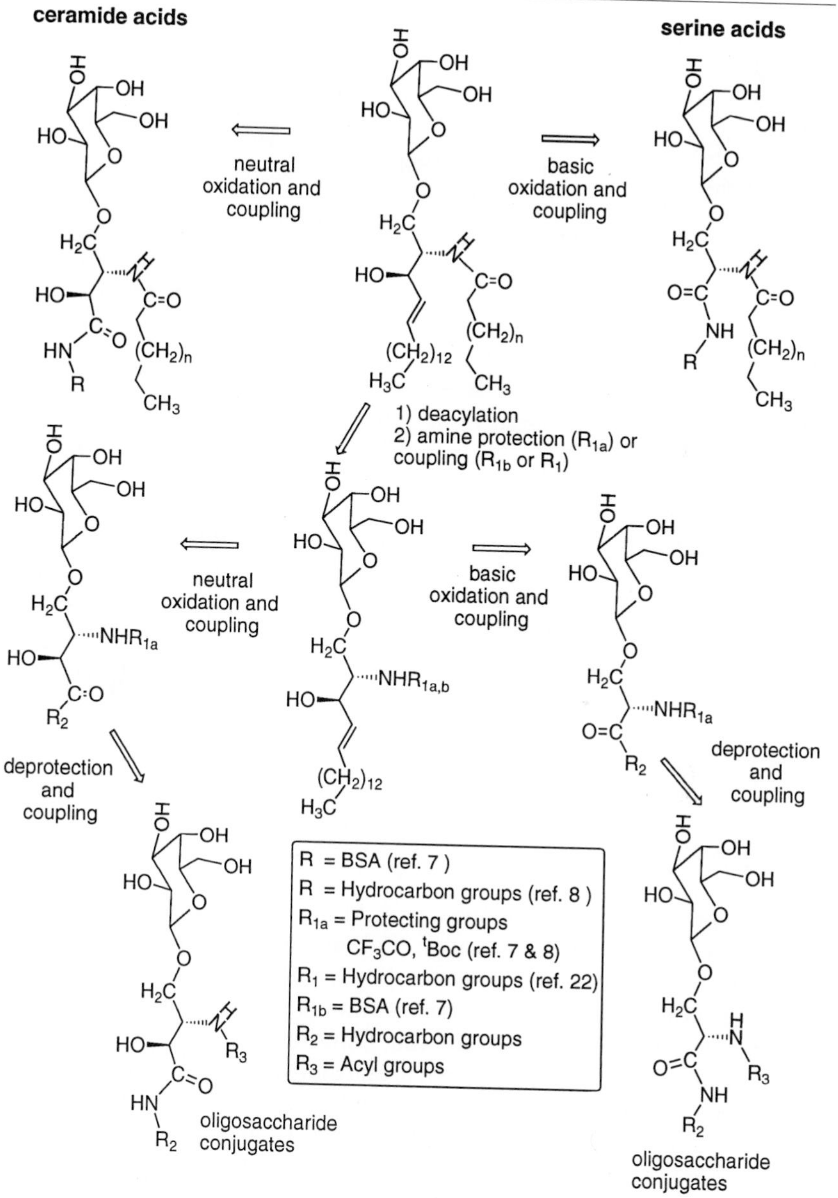

ceramide acids

serine acids

neutral oxidation and coupling

basic oxidation and coupling

1) deacylation
2) amine protection (R_{1a}) or coupling (R_{1b} or R_1)

neutral oxidation and coupling

basic oxidation and coupling

deprotection and coupling

deprotection and coupling

R = BSA (ref. 7)
R = Hydrocarbon groups (ref. 8)
R_{1a} = Protecting groups
CF_3CO, ^{t}Boc (ref. 7 & 8)
R_1 = Hydrocarbon groups (ref. 22)
R_{1b} = BSA (ref. 7)
R_2 = Hydrocarbon groups
R_3 = Acyl groups

oligosaccharide conjugates

oligosaccharide conjugates

SCHEME 1.

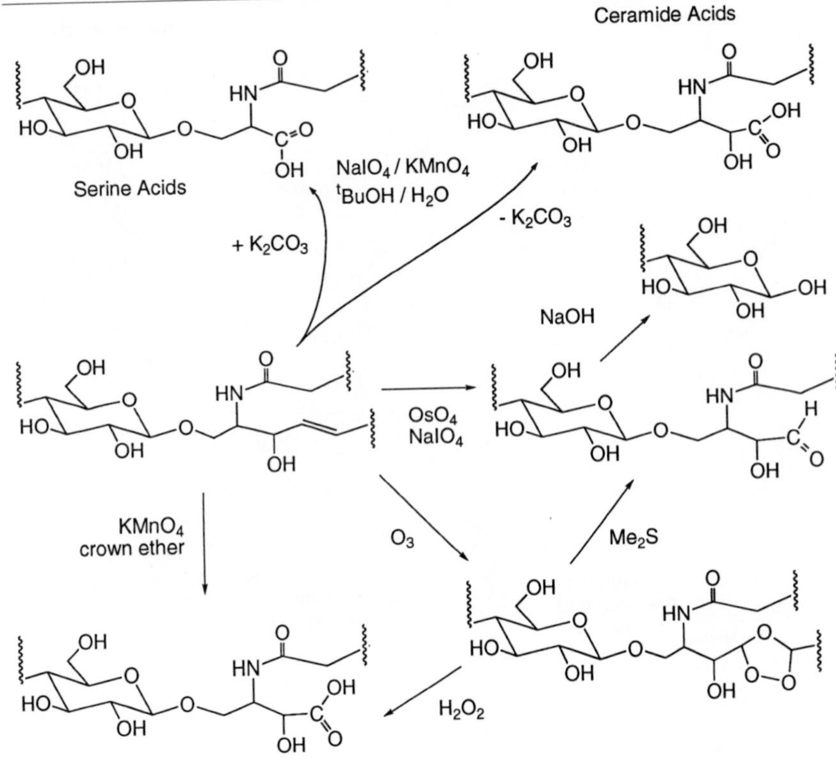

SCHEME 2.

Materials

Solvents. Dichloromethane (DCM), *tert*-butanol, 2-propanol, 1,2-dichloroethane (DCE), pyridine (Py), diethyl ether, methanol, chloroform ($CHCl_3$), and acetonitrile can be purchased from either Caledon (Georgetown, Ontario) or Aldrich (Milwaukee, WI) and ethanol from Commercial Alcohols Inc. (Brampton, Ontario).

Reagents. 0.5 *N* H_2SO_4 solution, acetic anhydride, 2-aminoadamantane ($AdaNH_2$), and 1-(trifluoroacetyl)imidazole (imidazole-TFa) are from Aldrich; ANALAR $KMnO_4$, ANALAR $NaHSO_3$, and 30% H_2O_2 are from BDH; *N-tert*-butoxycarbonyloxyimino-2-phenylacetonitrile (Boc-NO), triethylamine, 3-hydroxy-1,2,3-benzotriazin-4(3*H*)-one (HODBT), 1-ethyl-3-(3-dimethylaminopropyl)carbodiimide (EDAC), and 2,2′-azino-bis(3-ethylbenzthiazoline-6-sulfonic acid) (ABTS) are from Sigma (St. Louis, MO).

Chromatographic Materials. Silica gel 60 (40–63 μm or 230–400 mesh) and aluminum-backed nanosilica plates (alugram NanoSIL GI UV_{254},

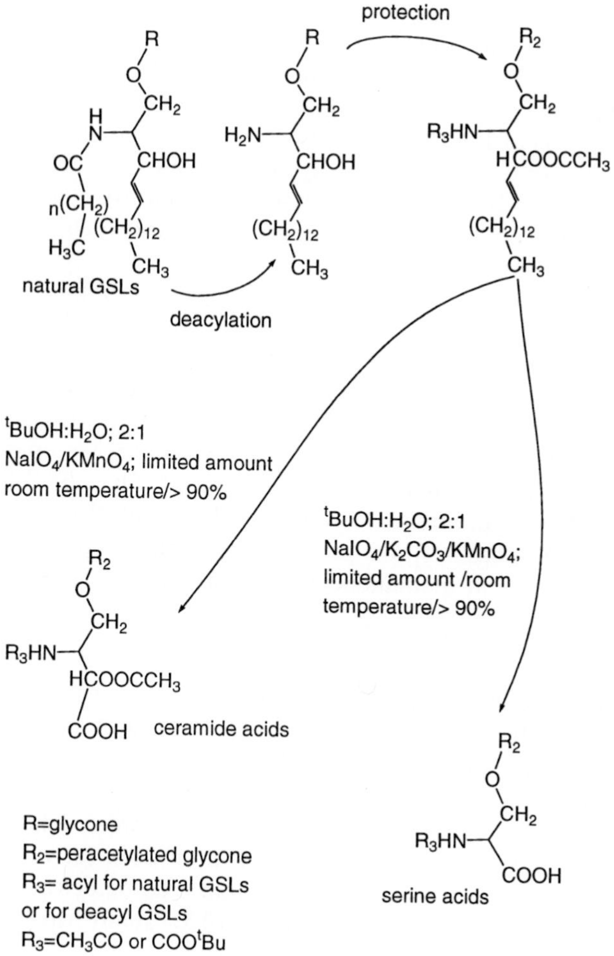

protection

natural GSLs

deacylation

$^tBuOH:H_2O$; 2:1
$NaIO_4/KMnO_4$; limited amount
room temperature/> 90%

$^tBuOH:H_2O$; 2:1
$NaIO_4/K_2CO_3/KMnO_4$;
limited amount /room
temperature/> 90%

ceramide acids

serine acids

R=glycone
R_2=peracetylated glycone
R_3= acyl for natural GSLs
or for deacyl GSLs
$R_3=CH_3CO$ or COO^tBu

SCHEME 3.

Macherey & Nagel) are supplied by Caledon. Hydroxyapatite is available from Bio-Rad (Hercules, CA) and molecular sieves (4 Å) are from Fisher.

Proteins and Antibodies. Goat anti-human IgG horseradish peroxidase conjugate (GAH), goat anti-rabbit IgG horseradish peroxidase conjugate (GAR), and goat anti-mouse IgG horseradish peroxidase conjugate (GAM) are from Bio-Rad. Bovine serum albumin (BSA) (99%, essentially fatty acid-free) is from Sigma. Recombinant gp120 (rgp120, 1 mg/ml in TBS, strain-LAV, baculovirus expression system) can be purchased from Protein

Sciences Corporation (Meriden, CT) and rabbit anti-gp120 polyclonal antibody [IgG, 1 mg/ml in phosphate-buffered saline (PBS)] is from Immuno Diagnostics, Inc. (Bedford, MA). Human sera from HIV patients containing anti-gp120 antibodies is obtained clinically (in our case, as a gift from Dr. S. Read, Division of Infectious Disease at the Hospital for Sick Children). Untreated, 96-well, flat-bottom, polystyrene plates are from Evergreen Scientific (supplied by Dia Med, Mississauga, Ontario). Verotoxin (VT)[16] is purified. Centricon-30 centrifugal concentrators are purchased from Amicon (Danvers, MA).

Glycolipids. Lactosylceramide (LC), globotriaosylceramide (Gb_3C), and globotetraosylceramide (Gb_4C) are purified from human kidney[17]; Forssmann antigen (Gb_5C) is purified from sheep blood[18]; and monosialylganglioside (GM1C) is purified from bovine brain.[19] Galactosylceramide (GalC), glucosylceramide (GlcC), and sulfogalactosylceramide (SGC) are purchased from Sigma. Gangliotetraosylceramide (Gg_4C) is prepared by acid hydrolysis of GM1C with 1 *M* acetic acid at 80° for 1 hr.[20] Deacylated derivatives [deacylglobotriaosylceramide (Gb_3S) and deacylgalactosylceramide (GalS, phychosine)] are prepared by saponification of Gb_3C and GalC at 102° with 1 *M* methanolic NaOH for 3 hr.[18]

Methods

TLC Analysis. Plate specifications: height, 85 mm; lane with, 8 mm; lane placement, 10 mm from bottom; solvent height, 5 mm from bottom; terminate running, 5 mm from top. Usually 1–3 μg of glycolipid species are analyzed.

After being developed with solvent, the plates are dried, sprayed with orcinol (0.5% in 3 *M* H_2SO_4), and heated in an oven (at 120–130°) until the plates turn pale yellow-orange (3–5 min). To detect amines (deacyl GSLs) plates are sprayed with ninhydrin (0.5% ninhydrin in 1:4 acetic acid:ethanol) and heated at 120° (3–5 min).

SDS–PAGE Analysis of Glycoproteins. During gel electrophoresis, neoglycoproteins tend to run as diffused bands when analyzed using 10% acrylamide (30% acrylamide containing 28:2; acrylamide:*N,N′*-methylenebisacrylamine) under standard conditions.[21] Samples are prepared under

[16] M. Petric, M. A. Karmali, S. Richardson, and R. Cheung, *FEMS Microbiol. Lett.* **41,** 63 (1987).
[17] B. Boyd and C. A. Lingwood, *Nephron.* **51,** 207 (1989).
[18] S. Head, K. Ramotar, and C. A. Lingwood, *Infect. Immun.* **58,** 1532 (1990).
[19] T. Yamakawa, R. Irie, and M. Iwanaga, *J. Biochem.* **48,** 490 (1960).
[20] C. A. Lingwood and A. Nutikka, *J. Cell Physiol.* **146,** 258 (1991).
[21] U. K. Laemmeli, *Nature* **227,** 680 (1970).

reducing conditions and the electrophoresis is run at 25 mA per gel, in a Bio-Rad Mini-PROTEAN II cell apparatus. Gels are run for an additional 15 min after the exit of the running front. Transfer to nitrocellulose is conducted by usual methods[22] using 100 mV for a period of 90 min.

Synthesis of Peracetylated Derivatives

Obtaining a clean peracetylated GSL precursor is crucial to the accomplishment of an efficient oxidation. It was observed that pure pyridine gave peracetyated glycolipids with fewer impurities. Ideally, freshly distilled pyridine predried over calcium hydride gives the best results; however, a new batch of commercially available pyridine stored over 4-Å activated molecular sieves for 12 hr in a nitrogen atmosphere gives few impurities during peracetylation.

To a dried sample of GSL or dGSL, a mixture of 1:1 acetic anhydride and pyridine (to a final concentration of 1 mg/ml) is added and stirred at 37° for 4 hr. In the case of dGSLs, treatment with acetic anhydride will result in the acetylation of the amine function along with peracetylation of hydroxyl groups. If a different N-substitution is desired, it should be incorporated prior to peracetylation. After the addition of acetic anhydride/pyridine mixture, displace the air inside the reaction container by purging with nitrogen before closing the container to reduce the amount of impurities formed.

Purification of Peracetylated Glycolipids. Dried material from above (3 mg) is dissolved in DCE (1 ml) and loaded on to a silica column (0.5 × 5 cm of silica gel in DCE) and eluted with DCM : methanol (25 ml : *Y*) where *Y* is varied from 100 μl in increments of 100 μl (for each solvent composition six 4-ml fractions are collected). The mobility of peracetylated derivatives during column chromatography varies significantly, with small changes in the activity of silica gel. Concomitant changes to the polarity of the eluent may be necessary.

The purity of the product (important for the oxidation to proceed smoothly) is checked by TLC (DCE : 2-propanol; 80 : 15) where initially the plates are developed in an I_2 chamber and the position of the band is marked with a pencil. The adsorbed I_2 is removed by heating the plate in an oven (120°) and then the glyco moiety is visualized with orcinol. All the peracetyaled GSLs and dGSLs give a single orcinol positive band that should coincide with the pencil mark.

[22] H. Towbin, T. Staehelin, and J. Gordon, *Proc. Nat. Acad. Sci. U.S.A.* **76,** 4350 (1979).

Synthesis of N-Protected dGSLs

Two types of amine protecting groups, *N*-trifluoroacetyl (*N*-TFa) and *N-tert*-butoxycarbonyl (*N-tert*-Boc) groups are used. After oxidation, both of these N-protecting groups can be removed at an appropriate stage of the synthesis.

N-Trifluoroacetyl dGSLs

As an example, synthesis of *N*-TFa derivative of deacylgalactosylceramide (GalS) is described. Dry GalS (3 mg, 6 μmol) is dissolved in chloroform and added to a solution of imidazole-TFa (approximately 1 *M* in DCM) in three portions (5 μl each, total of 15 μmol) at 30-min intervals, and stirred for an additional hour at room temperature. The reaction is monitored by TLC using $CHCl_3$: methanol : aqueous 0.88% KCl (70 : 30 : 2) as the solvent system. The nearest band to the dGSL corresponds to the *N*-trifluoroacetyl derivative. Once all the dGSL is consumed, ammoniacal ethanol (200 μL of 2 *M* NH_3 in ethanol) is added and the mixture dried under a stream of N_2.

Additional products are occasionally formed from acylation of some of the hydroxyl groups. If the TLC of the crude product shows additional bands, the dried material is treated with triethylamine : methanol : H_2O (2 : 6 : 10) (1 mg of total glycolipd/ml of amine solution) and monitored by TLC using $CHCl_3$: methanol : aqueous 0.88% KCl (70 : 30 : 2) every 2 hr. Once all the bands collapse to a single band, the reaction mixture is dried.

Purification. The dried product from above is redissolved in $CHCl_3$, loaded onto a silica column (0.5 × 6 cm of silica gel in $CHCl_3$) and eluted with $CHCl_3$: MeOH (98 : 2) (batch elution, 15 ml) and then with $CHCl_3$: methanol : H_2O (80 : 20 : 2) (ten 3-ml fractions). Solvent systems for mono- and disaccharide-containing GSLs are $CHCl_3$: methanol : aqueous 0.88% KCl (60 : 30 : 2) and for three or more sugar-containing glycolipids, $CHCl_3$: methanol : aqueous 0.88% KCl (60 : 35 : 8). The estimated yield was >90%.

N-tert-Butoxycarbonyl dGSLs

To a solution of dGSLs (about 1 mg, 1–2 μmol) in methanol (1 ml) is added solid Boc-NO (1 mg, 4 μmol) and stirred at 60° for 2 hr. The reaction mixture is dried under a stream of N_2, and the residue dissolved in $CHCl_3$ (1 ml) and loaded onto a silica column (0.5 × 5 cm of silica gel in $CHCl_3$). Excess reagents are eluted with $CHCl_3$: methanol (98 : 2) (batch elution, 15 ml) and the product is eluted as follows: for *tert*-Boc derivative of GalS, $CHCl_3$: methanol : H_2O (90 : 15 : 1), and for the *tert*-Boc derivative of deacylglobotriaosylceramide (Gb_3S), $CHCl_3$: methanol : H_2O (80 : 20 : 2).

TLC analysis of GalS-*tert*-Boc derivative was in $CHCl_3$: methanol : aqueous 0.88% KCl (80 : 20 : 2) and for Gb_3S-*tert*-Boc in $CHCl_3$: methanol : aqueous 0.88% KCl (65 : 25 : 4). The estimated yield is >90%.

Oxidation Reactions

All hydroxyl and amine groups have to be protected prior to oxidation. The oxidized product, i.e., the peracetylated glycosyl acid, is isolated from the inorganic oxidants by extractions with ethyl ether. The product, being peracetylated, is expected to partition into the ether phase. However, it was observed that some dGSLs (for example, *N*-acetyl derivatives as opposed to N-TFa derivatives) and charged GSLs mostly remain in the aqueous phase and a modified workup procedure must be employed.

Reagents. A 2:1 mixture of *tert*-butanol:H_2O and solutions of $NaIO_4$ (0.4 *M*), K_2CO_3 (0.25 *M*), and $KMnO_4$ (0.05 *M*). Quenching solution: A 5:1 mixture of 0.25 *M* $NaHSO_3$ solution and 0.5 N H_2SO_4.

Oxidation. Peracetylated glycolipid, $GSL(OAc)_n$ (0.5 mg) or peracetylated deacylglycolipid, $dGSL(OAc)_n$ (0.3 mg) is dried in a Kimax (15- × 100-mm) culture tube. Depending on the size (molecular weight) of the glycolipid, the amounts correspond to 1–0.3 μmol.

During the synthesis of ceramide acids, the dried precursor is dissolved in aqueous *tert*-butanol (500 μl) and solutions of $NaIO_4$ (30 μl, 10 μmol) and $KMnO_4$ (15 μl, 0.75 μmol) are added in the given sequence. A clear purple mixture is obtained.

During the synthesis of serine acids, the dried precursor is dissolved in aqueous *tert*-butanol (500 μl) and solutions of $NaIO_4$ (30 μl, 10 μmol), K_2CO_3 (40 μl, 10 μmol), and $KMnO_4$ (15 μl, 0.75 μmol) are added in the given sequence. After the addition of K_2CO_3, a cloudy solution is obtained, which turns purple on the addition of $KMnO_4$. The progress of the reaction is monitored by analyzing aliquots of the reaction mixture (10 μl) by TLC: solvent systems to analyze the oxidation of $GSL(OAc)_n$ and $dGSL(OAc)_n$ are $CHCl_3$: methanol : 0.88% aqueous KCl, 90 : 15 : 1 and 80 : 20 : 2, respectively. Oxidation of $GSL(OAc)_n$ gives multiple bands corresponding to glycosyl acids containing different fatty acids and dicarboxylic acids derived from the oxidation of glycolipids containing unsaturated fatty acids.

When pure precursors are used, catalytic regeneration of $KMnO_4$ proceeds smoothly until the reaction is terminated. Otherwise, the purple-colored reaction mixture diminishes in intensity with concomitant formation of MnO_2. In such cases, additional aliquots (5 μl) of $KMnO_4$ solution are added.

Workup. The reaction is quenched by the addition of 1 ml of water followed by 2 ml of quenching solution and the resulting colorless solution is extracted with diethyl ether (3 times, 5 ml each). Occasionally (due to insufficient quenching) the ether extract turns yellow, and in such cases, the combined extracts should be washed with 1 ml of quenching solution. The combined ether extract is washed (twice, 1 ml each) with water and dried under a stream of N_2 at 25° (higher temperatures tend to degrade some glycosyl acids). Residual water in the product is removed by adding absolute ethanol (1–2 ml) and drying under a stream of N_2.

Modified Workup. To isolate glycosyl acids derived from charged glycolipids (for example, SGC) and some deacyl glycolipids, a different workup procedure is employed. Although the alcoholic functions of these acids are also acetylated, the product partitions poorly into the diethyl ether phase. In such cases, the reaction is quenched with an excess of solid $NaHSO_3$ (50 mg). The colorless (occasionally pale yellow) suspension is dried on a rotary evaporator and extracted (three times, 5–7 ml) with $CHCl_3$: methanol : H_2O (80 : 20 : 2) and the salts are removed by passing the combined extract through a silica column (0.5 cm × 4 cm of silica gel in $CHCl_3$: methanol : H_2O; 80:20:2) and then dried.

This dried material (crude product) is free of inorganic reagents but contains impurities such as unoxidized dihydrosphingosine containing glycolipids (approximately 5%) and the carboxylic acid [$CH_3(CH_2)_{12}COOH$] from the cleavage of sphingosine chain. Further purification of this crude product depends on the type of glycoconjugate that will be synthesized from them.

Deprotecting Hydroxyl Groups. Peracetylated glycosyl serine and ceramide acids are deprotected by treating the dried crude product (0.5 mg) with triethylamine solution (1 ml of triethylamine : methanol : H_2O; 2 : 6 : 10) at 37° for 3 hr. The reaction can be monitored by TLC using the following solvent systems: for deprotected glycosylceramide acids containing mono and disaccharides, use $CHCl_3$: methanol : aqueous 0.88% KCl (65 : 25 : 4); for three or more sugar-containing acids, use $CHCl_3$: methanol : aqueous 0.88% KCl (60 : 40 : 9); and for glycosyl acids from deacyl GSLs, use $CHCl_3$: methanol : acetone : acetic acid : H_2O (5 : 5 : 5 : 2 : 2). Dried deprotected products are stored at −20°.

Deprotecting N-TFa Groups. At an appropriate stage of the synthesis, the trifluoroacetyl group is removed by treating with triethylamine solution (1 : 1; w/v of N-TFa precursor to triethylamine : methanol : H_2O; 2 : 6 : 10) at 37°. The reaction is monitored by TLC every 6 hr using $CHCl_3$: methanol : acetone : acetic acid : H_2O (5 : 5 : 5 : 2 : 2) as solvent system. The products are visualized on the plates using ninhydrin. Store dried product at −20°.

Deprotecting N-tert-Boc Groups: At an appropriate stage of the synthesis, the *N-tert*-Boc group is removed by treating the dried sample with pure formic acid (98%). Formic acid is added to give a final concentration of 2 mg of glycolipid precursor per ml of formic acid and the reaction is monitored by TLC every 4 hr using $CHCl_3$: methanol : acetone : acetic acid : H_2O (5 : 5 : 5 : 2 : 2) as the solvent system. On completion, 5 portions of methanol are added, the solvent evaporated, and the procedure repeated again to remove residual formic acid. The dried product is stored at −20°.

Synthesis of Glycoconjugates

Form the glycosyl acids, two classes of glycoconjugates have been synthesized. Coupling to a multivalent core such as BSA would give a highly soluble BSA(glycosyl)$_n$-type conjugate. Mass spectroscopic analysis of BSA conjugates derived from galactosylceramide acids showed that the neoglycoprotein could be formulated as BSA(glycosyl)$_9$.[7] However, coupling glycosyl ceramide acids derived from globotriaosylceramide to a core with a single amine function, such as an adamantane amine, gave a novel class of neohydrocarbon glycoconjugate that showed highly enhanced binding affinity for verotoxin. We presume these model compounds provide the rationale to design soluble monomeric glycolipid analogs.[23]

Neoglycoproteins

Essentially fatty acid-free BSA (Sigma) is further purified on a hydroxyapatite column[24] to remove the glycosylated globulins. We have observed that during conjugation, monomers and dimers of the BSA-conjugates are formed, and globulin impurities tend to interfere with the latter species.

To maximize the number of glycosyl units coupled to BSA, reactions are carried out with approximately 1 : 30 mole ratio (1 : 1 by weight) of BSA to glycosyl acid. The peracetylated glycosyl acids isolated by ether workup need no additional purification because the principle impurities (unoxidized glycolipids) are unlikely to interfere in coupling, and will be removed in the final stages of neoglycoprotein synthesis.

Prior to coupling, peracetylated glycosyl acids (isolated by ether extraction) are deprotected with triethylamine solution, as described under the oxidation protocol, and dried. Most of the hydrolyzed acetate is removed by dissolving the dried residue in acidic aqueous ethanol solution and removing the solvent with a rotary evaporator at 25°. Approximately two equivalents (i.e., with respect to acetate present, for example, 1 μmol of peraetylated galactosylceramide acid will give 6 μmol of acetate on deprotection) of acidic aqueous ethanol solution (10 m*M* HCl in 9 : 1, etha-

[23] M. Mylvaganam and C. Lingwood, *Biochem. Biophys. Res. Commun.* **257,** 391 (1999).
[24] L. H. Stanker, M. Vanderlaan, and H. Juarez-Salinas, *J. Immunol. Methods* **76,** 157 (1985).

nol : H_2O) is added. The dried sample is dissolved in methanol, transferred into a Kimax (15- × 100-mm) culture tube and dried.

Coupling. To the dried deprotected glycosyl acid (1 mg), a solution of BSA (1 ml of 1 mg/ml solution in 40 m*M* phosphate buffer, pH 7.4) and NHS (approximately 2 mg, 20 nmol) are added and the pH of the resulting mixture adjusted (with 0.1 *M* NaOH) between 7 and 8. EDAC is added to this mixture in three portions (total of 6 mg, 30 nmols) at 2-hr intervals and stirred at room temperature for 12 hours. Occasionally, the mixture is vortexed vigorously.

Purification. On standing overnight, the reaction mixture appears turbid. Phosphate buffer (3 ml of 40 m*M* phosphate buffer pH 7.4) is added, vortexed thoroughly, and the mixture centrifuged at 10,000 rpm for 5 min. The supernatant is transferred to a Centricon-30 that has been treated with 1% BSA (with spinning) and then with water. We have observed that some conjugates adsorb onto the Centricon membrane. Low molecular weight impurities, such as unoxidized dihydrosphingosine containing glycolipids, reagents etc., are removed by washing (3 cycles). During each cycle, the reaction mixture is concentrated to 0.5 ml and diluted with 2 ml of 40 m*M* phosphate buffer. The conjugates (0.5 mg/ml) are stored at −20°.

Neohydrocarbon Conjugates

The glycosyl acids isolated by ether workup need to be purified prior to further manipulations. After the synthesis of neohydrocarbon conjugates, both the product and impurities (i.e., the dihydrosphingosine containing glycolipids) show similar migration properties on silica gel. However, prior to coupling, the impurities, being nonpolar, have significantly higher mobility than the glycosyl acids and are therefore easily separated.

Oxidized material (0.5 mg) obtained from ether workup is dissolved in DCE (0.5 ml) and loaded on to a silica column (0.5 × 3 cm of silica gel in DCE) and eluted with DCM : methanol (25 ml : *Y*) where *Y* is varied in increments of 100 μl (for each solvent composition six 4-ml fractions are collected). This step-elution chromatography will remove the unoxidized dihydrosphingosine-containing species. The peracetylated glycosylserine or the ceramide acids are then eluted with methanol (3–5 ml) and dried under a stream of N_2.

Coupling. Dried peracetylated glycosyl acid is dissolved in basic acetonitrile (in 5 : 1, CH_3CN : triethylamine; 1 mg/ml). To this solution, $AdaNH_2$ (1 : 1 by weight), HODBT (1 : 1 by weight), and solid EDAC (1 mg) are added in the given order. Progress of the reaction is monitored by TLC ($CHCl_3$: methanol : aqueous 0.88% KCl; 90 : 15 : 1); the coupled products migrate closer to the solvent front. Once all the glycosyl acid is consumed, the reaction mixture is dried under a stream of N_2, dissolved in DCE, and purified on a mini silica column (for a 0.5-mg scale preparation, 0.5 × 2 cm of silica gel column), as described above (using DCM : methanol; 25 ml : *Y* solvent system).

Peracetylated neohydrocarbon conjugates are then deprotected with the triethylamine-methanol-H_2O system as described in the oxidation procedure. Most of the hydrolyzed acetate is removed by dissolving the dried residue in acidic aqueous ethanol solution (as described under neoglycoprotein synthesis) and dried on a rotary evaporator. Dried product is stored at −20°.

Ligand Binding

Depending on the physical properties of the glycoconjugates, they can be employed in different types of assays where qualitative and/or quantitative binding parameters are evaluated. For example soluble BSA-(glycosyl)$_n$ conjugates can be used to inhibit protein binding to glycolipid immobilized on a ELISA plate, incorporated into liposomes, or expressed on a cell membrane. Such data can provide information related to glycolipid organization in carbohydrate–protein interaction.

Neohydrocarbon Glycoconjugate Binding Assay. Methanolic solutions of natural glycolipids and neohydrocarbon glycoconjugates having *X* ng/50 μl (where *X* is 5, 10, 15, 20, 25, 30 and 35) are prepared. Add each test solution (50 μl/well in triplicate) to a 96-well plate and allow the solvent to evaporate overnight. Wells along the edges of the plate are excluded.

The wells are blocked with 0.2% BSA in PBS (BSA–PBS) for 1 hr at room temperature. The wells are washed once with BSA–PBS and 50 μl of VT_1 (verotoxin, 20 ng)[16] in BSA–PBS is added to each well. The plate is incubated at room temperature for 2 hr, washed three times with BSA–PBS (200 μl each) and incubated with 50 μl of PH1 (monoclonal anti VT_1 antibody, 50 ng per well)[25] in BSA–PBS for 1 hr. Plates are washed three times with BSA–PBS (200 μl each), and incubated with 50 μl of GAM peroxidase conjugate (1:2000 dilution in BSA–PBS) for 45 min. Throughout the assay, the edge wells are filled with BSA–PBS.

The wells are washed three times with BSA–PBS (200 μl each) and once with PBS (200 μl) and developed (100 μl of developer) for 10–20 min. The developer is composed of 5 mg of ABTS in 10 ml of citrate buffer (80 m*M* citric acid, 100 m*M* NaH_2PO_4, pH 4) containing 3 μl of H_2O_2.

[25] J. Boulanger, M. Petric, C. A. Lingwood, H. Law, M. Roscoe, and M. Karmali, *J. Clin. Micro.* **28,** 2830 (1990).

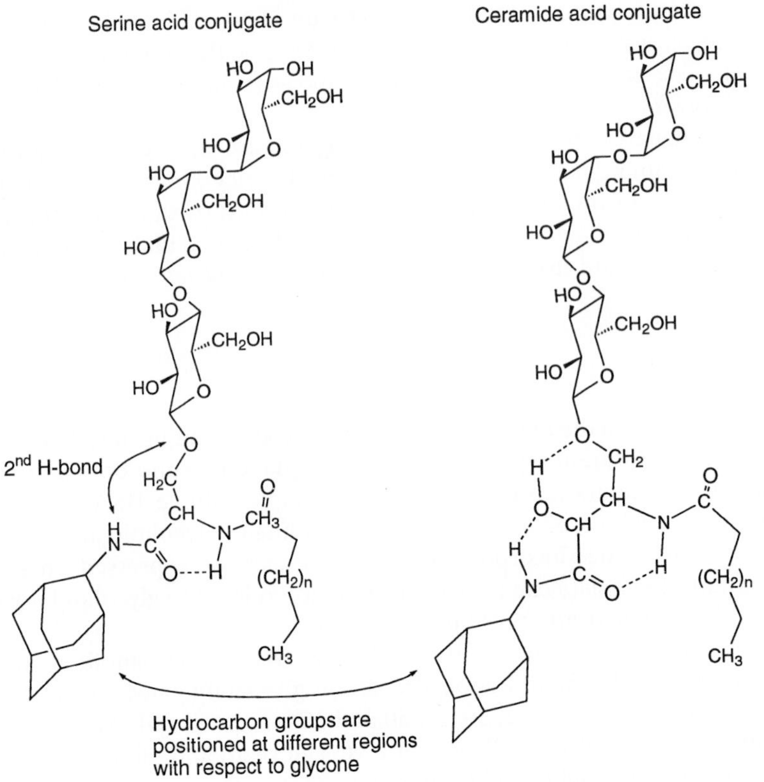

Serine acid conjugate

Ceramide acid conjugate

2nd H-bond

Hydrocarbon groups are positioned at different regions with respect to glycone

SCHEME 4.

Neohydrocarbon conjugates could also be assayed by using a TLC overlay method described by Lingwood *et al.*[26]

Neoglycoprotein Binding Assay

Neoglycoproteins derived from GalC, GlcC, LC, SGC, Gb_3C, and Gb_4C are adsorbed on to nitrocellulose membranes either directly or transferred after SDS–PAGE and tested for binding to HIV coat protein gp120 or to verotoxin. Direct adsorption is facilitated by placing the membrane on a filter paper placed inside a Büchner funnel, where the samples are applied under vacuum.

The membranes are blocked with 2.5% milk powder in TBS_{100} (10 m*M* TBS, 100 m*M* NaCl) for 1 hr at room temperature, rinsed 3 times with

[26] C. A. Lingwood, *Methods Enzymol.*

TBS_{100} and incubated with gp120 (1 μg/ml) in TBS_{50} (10 m*M* TBS, 50 m*M* NaCl) for 3 hr at room temperature. The blots are washed as above with TBS_{100} and incubated with human HIV serum (1 : 50) in 2.5% milk powder in TBS_{100} or rabbit anti-rgp120 (1 : 500) in TBS_{100} for 3 hr. After rinsing as above, blots are incubated with the secondary antibody (1 : 1000) in TBS_{100} for 45 min. Finally, the blots are rinsed as above and the binding visualized by treating with the developer (3 mg/ml solution of 4-chloro-1-naphthol in methanol mixed with 5 volumes of TBS_{100} and 1 : 1000 dilution of 30% H_2O_2.

For verotoxin binding, after adsorbing the neoglycoproteins and blocking as described above, the remainder of the procedure is identical to the TLC overlay assay as described by Lingwood *et al.*[26]

Summary

A new oxidation protocol for the cleavage of sphingosine double bonds is described. The procedure is applicable to both natural and deacyl glycolipids and can be applied to microgram quantities of precursors. Under neutral conditions, glycosyl ceramide acids are obtained and under basic conditions glycosyl serine acids are obtained. The glycosyl ceramide acid-based glycoconjugates—BSA–neoglycoprotein and adamantyl-neohydrocarbon—demonstrate the importance that an aglycone can play in carbohydrate–protein interaction. Studies with HIV coat protein gp120 and BSA–neoglycoprotein conjugates derived from galactosylceramide (GalC) showed that binding affinities of the conjugates depend on the manner in which the glycosyl unit is coupled to the protein. Deacyl-GalC conjugates, in which the glycosyl unit is coupled via the amine of the sphingosine, showed significantly lower affinity as compared to glycosylceramide acid conjugates. In the case of Gb_3–VT_1 binding, it was found that ceramide acid conjugates bound to VT_1 better than the serine acid conjugates. These studies show that the aglycone organization, particularly the region adjacent to the carbohydrate region (or in a membrane environment, the aglycone–glycone interface) modulate carbohydrate presentation. It is possible that in each of the conjugates described above, the interface region could have different hydrogen-bonding networks (see Scheme 4.) This, in turn, could influence the solvation and/or conformation of this region and thereby influence ligand binding.

Acknowledgment

This work was supported by Canadian MRC grant MT13073 and NIH grant R01 DK52098.

[40] Separation of Glycosphingolipid-Enriched Microdomains from Caveolar Membrane Characterized by Presence of Caveolin

By KAZUHISA IWABUCHI, KAZUKO HANDA, and SEN-ITIROH HAKOMORI

Introduction

The majority of glycosphingolipids (GSLs) and sphingomyelin are found, together with cholesterol and the phosphatidylglycosylinositol anchor, as components of a low-density, buoyant membrane fraction after cells are properly homogenized and subjected to sucrose or Ficoll density gradient centrifugation. Conveniently, the low-density membrane fraction can be separated after Dounce homogenization of cells in buffer containing 1% Triton X-100. This fraction (termed *detergent-insoluble membrane,* DIM) is also characterized by the presence of caveolin, a major scaffold membrane protein claimed to be specific for caveolae, which are invaginated structures of plasma membrane known for over 40 years and recently claimed to play an important role in endocytosis and signal transduction. In the current literature, GSLs and sphingomyelin, as well as certain signal transducer molecules such as Src family kinases, Rho, and Ras, are often considered to be common components of caveolae (for review, see Ref. 1).

We have found that the GSL-enriched fraction can be separated from the caveolin-containing fraction in DIM of mouse melanoma B16 cells by an immunoseparation technique using antibodies to GM3 (NeuAcα3Galβ4Glcβ1Cer) or to caveolin. Surprisingly, major signal transducer molecules (c-Src, Rho A, FAK) found originally in DIM were associated with GM3 and separated from the caveolin-containing fraction, which is characterized by the predominant presence of cholesterol, and presence of Ras, low level of sphingomyelin, but absence of GM3. The procedure for separation of GSL-enriched fraction from caveolin-containing fraction in B16 cells is described below for our analysis of B16 cells. The same methods can presumably be followed with other cell lines.

Preparation of Low-Density Detergent-Insoluble Membrane Fraction

Mouse melanoma B16 cells, grown in Dulbecco's modified Eagle's medium (DMEM) supplemented with 10% fetal calf serum (FCS), are har-

[1] R. G. Anderson, *Annual Rev. Biochem.* **67,** 199 (1998).

0076-6879/00 $30.00

vested in 0.02% ethylenediaminetetraacetic acid–phosphate-buffered saline (8.1 m*M* Na_2HPO_4, 1.5 m*M* KH_2PO_4, 2.7 m*M* KCl, 137 m*M* NaCl, pH 7.4) (EDTA–PBS), lysed, homogenized, and subjected to sucrose density gradient centrifugation to separate the low-density light-scattering membrane fraction by a modification of the method described previously.[2] Briefly, 1–5 $\times$ 10^7 cells are suspended in 1 ml of ice-cooled lysis buffer containing 1% Triton X-100, 10 m*M* Tris-HCl (pH 7.5), 150 m*M* NaCl, 5 m*M* EDTA, 75 units/ml aprotinin, and 1 m*M* phenylmethylsulfonyl fluoride (PMSF), and followed to stand for 20 min. When phosphyorylation assay is included, 1 m*M* Na_3VO_4 is added to inhibit phosphatase. The cell suspension is homogenized in a tight fitting Dounce homogenizer with 10 strokes in ice bath, and the lysate centrifuged for 5 min at 1300*g* to remove nuclei and large cellular debris. The supernatant fraction (postnuclear fraction) is subjected to sucrose density gradient centrifugation, i.e., the fraction is mixed with an equal volume of 85% sucrose (w/v) containing 10 m*M* Tris buffer (pH 7.5), 150 m*M* NaCl, 5 m*M* EDTA, with or without 1 m*M* Na_3VO_4 (Tris–NaCl–EDTA buffer).

The resulting diluent (2 ml) is placed at the bottom of a centrifuge tube, overlaid with 5.5 ml of 35% sucrose (w/v), and further overlaid with 5% sucrose in Tris-NaCl-EDTA buffer. Samples are centrifuged for 18 hr at 39,000 rpm (200,000*g* on average) at 4° in a SW41 swinging rotor (Beckman). A white light-scattering band under light illumination (presumably due to the Tyndall phenomenon of membranous fractions), located at the interface between 5 and 35% sucrose (fraction 5), is separated from other fractions, and termed DIM fraction. The entire procedure is performed at 0–4°. Protein content of each fraction is determined using a MicroBCA kit (Pierce Chemical Co., Rockford, IL).

Separation of DIM into GM3/Src-Containing and Caveolin-Containing Subfractions by Antibodies

A 1-ml aliquot of DIM fraction containing 25–30 μg protein is diluted 10$\times$ in IP buffer [50 m*M* Tris-HCl (pH 7.4), 150 m*M* NaCl, 2 m*M* NaF, 1 m*M* EDTA, 1 m*M* EGTA, 1 m*M* Na_3VO_4, 1 m*M* PMSF, 75 U/ml aprotinin, 1% Triton X-100], precleared by incubating with 50 μl protein G-Sepharose beads (Pharmacia, Piscataway, NJ), and placed in a rotary mixer at 4° for 2 hr. After centrifugation at 270*g* for 2 min, the supernatant (10 ml) is collected and added with purified anti-GM3 MAb DH2[3] at a final concentration of 1 μg/ml. In addition, 1-ml aliquots of DIM fraction are similarly

[2] W. Rodgers and J. K. Rose, *J. Cell Biol.* **135,** 1515 (1996).
[3] T. Dohi, G. Nores, and S. Hakomori, *Cancer Res.* **48,** 5680 (1988).

treated and added with anti-caveolin rabbit IgG (Transduction Laboratories, Lexington, KY), or control normal rabbit IgG or control mouse IgG (both from Santa Cruz Biotechnology, Santa Cruz, CA), at a final concentration of 1 μg/ml. Each mixture is placed overnight in a rotary mixer at 4°, then 50 μl of protein G-Sepharose beads are added, and placed again in a rotary mixer for 2 hr. The beads are washed 3× with IP buffer by centrifugation at 270*g* for 2 min.

The immunocomplexes adhering to the beads are treated differently depending on subsequent analysis. (1) For Western blot analysis, the washed beads are suspended with 100 μl sample buffer with 2-mercaptoethanol, heated to 95° for 3 min, centrifuged at 1000*g* for 2 min, and the supernatant subjected to SDS–PAGE. (2) For dot-blot analysis, bound materials on beads are eluted by 1 ml of 0.2 *M* glycine-hydrochloride (pH 2.3) containing 1% Triton X-100 with sonication in ice-cold water for 15 sec. After centrifugation at 270*g* for 2 min, the resultant supernatants are subjected to the assay. (3) For second immunoprecipitation analysis, immunoprecipitates on beads are eluted by mixing with 0.2 *M* glycine-hydrochloride (pH 2.3) containing 1% Triton X-100 as above and centrifugation for 270*g* for 2 min. The supernatants are collected and immediately neutralized with 1 *M* Tris-HCl (pH 10) to pH 7.4, then diluted 10× with IP buffer, and incubated with 1 μg/ml anti-caveolin IgG, anti-GM3 MAb DH2, or normal rabbit IgG in a rotary mixer at 4° overnight. The mixtures are added with 50 μl of protein G-Sepharose beads, and placed again in a rotary mixer for 2 hr. Beads are washed 3× with IP buffer, by centrifugation at 270*g* for 2 min, then suspended with 100 μl sample buffer with 2-mercaptoethanol, heated to 95° for 3 min, and centrifuged at 1000*g* for 2 min, and the supernatant subjected to SDS–PAGE.

Western blot analysis is performed by modification[4] of the method originally described.[5] Briefly, proteins separated on 5–15% linear gradient SDS–PAGE under reducing conditions are electroblotted onto polyvinylidene difluoride (PVDF) membranes. The membranes are incubated with anti-c-Src rabbit or goat IgG (Santa Cruz), anti-Rho A mouse MAb (Santa Cruz), anti-Ras rat MAb (Santa Cruz), or anti-FAK rabbit IgG (Santa Cruz) for 2 hr at room temperature, then incubated with HRP-labeled proper secondary antibody. The reacted proteins are detected by chemiluminescence method with Super-Signal-CL-HRP (Pierce).

Dot immunoblot analysis is performed using anti-GM3 MAb or anti-

[4] K. Iwabuchi, S. Yamamura, A. Prinetti, K. Handa, and S. Hakomori, *J. Biol. Chem.* **273,** 9130 (1998).

[5] H. Towbin, T. Staehelin, and J. Gordon, *Proc. Natl. Acad. Sci. U.S.A.* **76,** 4350 (1979).

caveolin antibody, by a modification of the method described previously.[6] A PVDF membrane (Immobilon-P, Millipore Corp., Bedford, MA) is soaked in fixation buffer (62.5 m*M* Tris-HCl, pH 8.9, containing 5% methanol) for at least 2 hr before assembly in Bio-Dot apparatus (Bio-Rad Laboratory, Cambridge, MA). The collected desorbed fraction is loaded onto PVDF membrane by aspiration. Next, the blotted membranes are washed in fixation buffer and dried. DIM diluted to various concentrations is also blotted and used as standards. After brief soaking in methanol, the membranes are blocked in 5% skim milk in TBS-T [10 m*M* Tris-HCl, pH 8.0, 150 m*M* NaCl containing 0.05% Tween 20 at room temperature for 3 hr, and then incubated with 1 μg/ml anti-GM3 MAb DH2, or rabbit anti-caveolin IgG in TBS-T containing 1% normal goat serum at room temperature for 2 hr. After extensive washing with TBS-T, the membranes are incubated with HRP-conjugated goat anti-mouse IgG (Southern Biotechnology Associates, Birmingham, AL) or HRP-conjugated goat anti-rabbit IgG (Santa Cruz) in TBS-T at room temperature for 1 hr. The membrane is washed 5 times in TBS-T, and developed using chemiluminescence method with Super-Signal-CL-HRP (Pierce).

The dot immunoblot analysis of DIM fraction separated by antibodies to c-Src, caveolin, hepatocyte growth factor/scatter factor receptor, and GM3 is shown in Fig. 1. Western blot analyses show that c-Src in DIM is associated with GM3 but not with caveolin (Fig. 2).

Lipid Composition of DIM Subfractions of Mouse Melanoma B16 Cells: GSL Signaling Domain and Caveolin-Containing Fraction

The membrane subfractions immunoseparated with antibodies are mixed with 1 ml methanol, sonicated for 15 min, added to 2 ml chloroform, and further sonicated for 5 min. To extract lipids from liquid samples, 10 volumes of methanol is added, and the mixture sonicated for 15 min in a water bath sonicator, then added to 20 volumes of chloroform, sonicated for 5 min, and centrifuged at 1500*g* at 4° for 15 min. The supernatants are collected, dried under N_2 stream, the residue is dissolved in methanol–water, 3:7, v/v). The solutions are applied to Bond Elut packed columns (1 ml, C_{18}, Analytichem International, Harbor, CA), washed and preequilibrated with methanol–water, 3:7. The columns are washed with 10 column volumes of distilled water, and bound materials are eluted with 2 ml chloroform–methanol, 2:1. The eluates are dried under N_2, and dissolved in ~20 μl chloroform–methanol, 2:1.

Three-dimensional TLC is carried out according to the method de-

[6] H. Towbin and J. Gordon, *J. Immunol. Methods* **72,** 313 (1984).

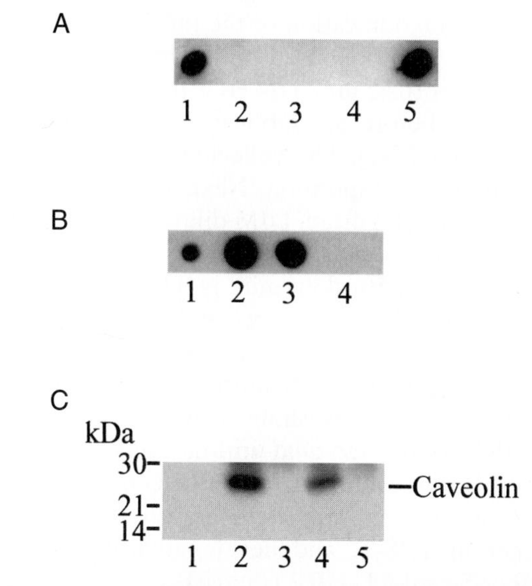

FIG. 1. Separation of DIM into two subfractions. (A) DIM was immunoprecipitated with anti-c-Src rabbit IgG (dot 1), anti-caveolin rabbit IgG (dot 2), anti-c-MET rabbit IgG (dot 3), rabbit normal IgG (dot 4), or DH2 (dot 5). GM3 content of each dot was determined by immunoblotting with MAb DH2. (B) DIM was incubated on anti-caveolin rabbit IgG- or DH2-coated dishes. The bound and unbound fractions were collected and the amount of caveolin in each fraction was determined using anti-caveolin rabbit IgG. Dot 1, unbound to anti-caveolin; dot 2, bound to anti-caveolin; dot 3, unbound to anti-GM3 DH2; dot 4, bound to anti-GM3 DH2. (C) DIM was immunoprecipitated with DH2 (lane 1) or anti-caveolin rabbit IgG (lane 2). The immunoprecipitate with anti-caveolin rabbit IgG was eluted and reimmunoprecipitated with DH2 (lane 3) or anti-caveolin rabbit IgG (lane 4). Lane 5, immunoprecipitate with normal rabbit IgG. Caveolin content was determined by Western blotting with anti-caveolin rabbit IgG. Locations of molecular markers are indicated at left.

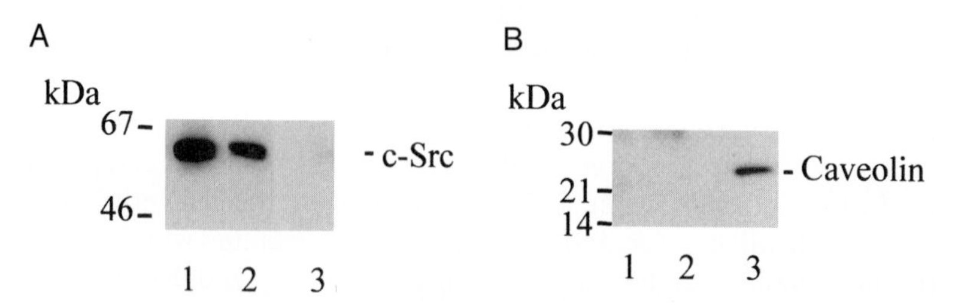

FIG. 2. c-Src is associated with GM3 but not with caveolin. DIM was immunoprecipitated with anti-c-Src rabbit IgG (lane 1), anti-GM3 MAb DH2 (lane 2), or anti-caveolin rabbit IgG (lane 3). Each immunoprecipitate was subjected to SDS–PAGE followed by Western blotting with anti-c-Src goat IgG (A) or anti-caveolin rabbit IgG (B).

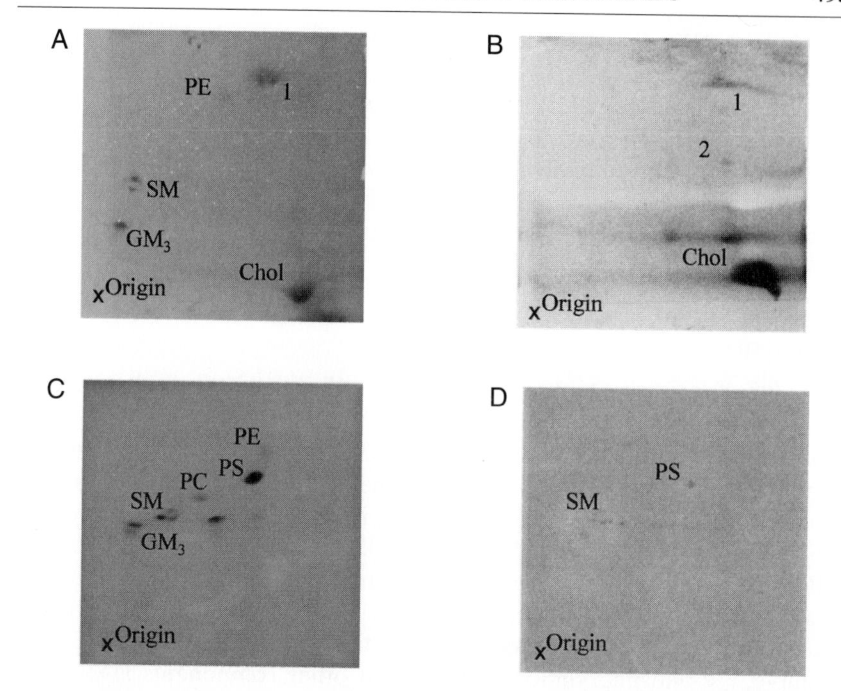

FIG. 3. Three-dimensional TLC pattern of lipid components. (A and C) Subfraction of DIM immunoprecipitated with anti-GM3 DH2. (B and D) Subfraction of DIM immunoprecipitated with anti-caveolin rabbit IgG. TLC plates in parts (A) and (B) were sprayed with primulin. In parts (C) and (D), DIM was prepared from cells metabolically labeled with $[^3H]$Ser, and TLC plates were subjected to autoradiography. Standard lipids were used to assign each spot. Spot 1 in (A) and (B) correspond to GlcCer, and spot 2 in (B), are not identified. PE, Phosphatidylethanolamine; PS, phosphatidylserine; PC, phosphatidylcholine; PI, phosphatidylinositol; Chol, cholesterol; SM, sphingomyelin.

scribed previously.[7] In brief, samples are spotted at the lower left-hand corner of a 10-cm × 10-cm high-performance thin-layer chromatography (HPTLC) plate (Merck, Rahway, NJ). The first chromatographic run is performed with chloroform/methanol/formic acid/water 65:25:8.9:1.1 (v/v/v/v). The second run is performed with chloroform/methanol/4.4 *N* ammonia, 50:40:10 (v/v/v) at rotation 90° from the first direction. The third run is performed with diethyl ether in a direction opposite to that of the second run. The plate is sprayed with 0.03% primuline (Aldrich Chemical Co., Milwaukee, WI) in acetone/water, 80:20 (v/v), and photographed under UV light. For autoradiography, TLC plates are exposed to Kodak

[7] K. Yokoyama, H. Nojiri, M. Suzuki, M. Setaka, A. Suzuki, and S. Nojima, *FEBS Lett.* **368,** 477 (1995).

(Rochester, NY) BioMax MS film at −80° with Kodak TranScreen-LE intensifying screen.

Lipid composition of DIM subfraction adsorbed on anti-GM3 antibody-Sepharose beads is shown in Fig. 3A, and composition of caveolar membrane adsorbed on anti-caveolin antibody is shown in Fig. 3B. [^{3}H]Ser metabolic labeling pattern of lipids (this labels sphingosine and phosphatidylserine) of the same subfractions as in panels A and B are shown in Figs. 3C and 3D, respectively.

Comments

Methods are described for preparation of low-density membrane fraction (herein termed *DIM*) using medium containing 1% Triton X-100, and separation of subfractions representing GM3-enriched microdomain and caveolae. The former subfraction contained GM3, sphingomyelin, c-Src, Rho A, and FAK. The latter subfraction, in contrast, contained caveolin, a high quantity of cholesterol, and essentially all Ras present in the original DIM fraction. The subfraction representing GM3-enriched microdomain was characterized by (1) maintenance of GM3-dependent adhesion and (2) susceptibility to being activated for signal transduction through GM3. For example, ^{32}P-phosphorylation of c-Src and other components is strongly enhanced on GM3-dependent adhesion or addition of anti-GM3 antibody.[8]

Acknowledgment

This study was supported by National Cancer Institute Outstanding Investigator Grant CA42505 (to S.H.). We thank Dr. Stephen Anderson for scientific editing and preparation of the manuscript.

[8] K. Iwabuchi, K. Handa, and S. Hakomori, *J. Biol. Chem.* **273,** 33766 (1998).

[41] Reconstitution of Sphingolipid–Cholesterol Plasma Membrane Microdomains for Studies of Virus–Glycolipid Interactions

By DJILAI HAMMACHE, GÉRARD PIÉRONI, MARC MARESCA, SERGE IVALDI, NOUARA YAHI, and JACQUES FANTINI

Introduction

Glycosphingolipids can serve as attachment sites for a number of pathogens including bacteria, parasites, or viruses.[1] In addition, these lipids may play an active role in viral fusion, e.g., by increasing the lateral membrane tension and/or the curvature of the membrane.[2] In this respect, galactosylceramide (GalCer) has been shown to promote membrane fusion of Semliki Forest virus,[3] while neutral glycolipids terminating in galactose were found to mediate myxovirus-induced membrane fusion.[4] The fusion of type 1 human immunodeficiency virus (HIV-1) with the plasma membrane of $CD4^+$ cells requires specific glycosphingolipids such as the monosialoganglioside GM3 (NeuAcα2-3Galβ1-4Glc-Cer) or the globotriaosylceramide Gb_3 (Galα1-4Galβ1-4Glc-Cer).[5,6] Moreover, GalCer is an alternative receptor allowing HIV-1 binding to neural and intestinal $CD4^-$ cells.[7,8]

In the outer leaflet of the plasma membrane, glycosphingolipids organize into moving platforms, or rafts, onto which specific proteins attach within the bilayer.[9] This lateral organization probably results from preferential packing of sphingolipids and cholesterol, resulting in the formation of membrane microdomains. To analyze glycosphingolipid–HIV-1 interactions, a reconstituted monolayer of purified glycosphingolipid is prepared at the air–water interface as model for a glycosphingolipid membrane microdo-

[1] K.-A. Karlsson, *Annu. Rev. Biochem.* **58,** 309 (1989).

[2] J. P. Monck and J. M. Fernandez, *Curr. Opin. Cell Biol.* **8,** 524 (1996).

[3] J. L. Nieva, R. Bron, J. Corver, and J. Wilschut, *EMBO J.* **13,** 2797 (1994).

[4] R. T. Huang, *Lipids* **18,** 489 (1983).

[5] D. Hammache, N. Yahi, G. Piéroni, F. Ariasi, C. Tamalet, and J. Fantini, *Biochem. Biophys. Res. Commun.* **246,** 117 (1998).

[6] A. Puri, P. Hug, K. Jernigan, J. Barchi, H.-Y. Kim, J. Hamilton, J. Wiels, G. J. Murray, R. O. Brady, and R. Blumenthal. *Proc. Natl. Acad. Sci. U.S.A.* **95,** 14435 (1998).

[7] J. M. Harouse, S. Bhat, S. L. Spitalnik, M. Laughlin, K. Stefano, D. H. Silberberg, and F. Gonzalez-Scarano, *Science* **253,** 320 (1991).

[8] N. Yahi, S. Baghdiguian, H. Moreau, and J. Fantini, *J. Virol.* **66,** 4848 (1992).

[9] K. Simons and E. Ikonen, *Nature* **387,** 569 (1997).

0076-6879/00 $30.00

main.[10–12] This technology is based on the physical properties of lipids. When spread on an air–water interface, the lipids automatically form a monomolecular film and thus imitate one leaflet of a membrane. Comparative physicochemical studies of monolayers and bilayers (i.e., liposomes) demonstrated the accuracy of lipid monolayers as a model for biologic membranes.[13] The surface tension of the monolayer is a function of lateral packing density of the lipid film. Thus, the insertion of a protein within the lipid film can be measured as an increase of surface pressure.[14] However, the surface pressure is not changed when the protein binds to the monolayer without inducing a lateral packing of lipid molecules.

To assess that the glycosphingolipid organize into a monomolecular film, it is necessary to measure the variations of surface pressure vs. apparent molecular area (i.e., isotherms). In these experiments, the lipid films are compressed at a constant rate (e.g., 10 mm/min) and the surface pressure is measured with a fully automated computerized Langmuir film balance (KSV, Oriel, Courtaboeuf, France). Symmetrical compression with two barriers is achieved with a stepper motor. All isotherms are run at least twice in the direction of increasing pressure. The high compressibility of the glycosphingolipids at all film pressures and the absence of discontinuities in their isotherms show that they exist in the liquid expanded state up to film collapse. In our experiments, glycosphingolipid monolayers are prepared at an initial pressure of 10 m*N*/m corresponding to a compressible film.[15]

Surface Pressure Measurements

The penetration of different molecules (drugs, peptides, proteins) into lipid monolayers can be easily determined with a specifically designed Langmuir film balance (e.g., the Microtrough S system from Kibron Inc., Finland). Like the KSV system, this apparatus is also convenient for the determination of isotherms. A computerized homemade system can also be set up with a precision balance, a sensor, and a computer (Fig. 1). This apparatus is ideal for routine studies of protein–lipid interactions. The balance (HM-202, 42 g/0.01 mg, AND, Oxford, UK), equipped with a standard built-in underhook, is driven by the Collect software (Labtronics Inc., Guelph, Ontario, Canada). The trough compartment is a Teflon tank

[10] T. Sato, T. Serizawa, and Y. Okahata, *Biochim. Biophys. Acta* **1285,** 14 (1996).

[11] W. Curatolo, *Biochim. Biophys. Acta* **906,** 111 (1987).

[12] D. Hammache, G. Piéroni, N. Yahi, O. Delézay, N. Koch, H. Lafont, C. Tamalet, and Jacques Fantihi, *J. Biol. Chem.* **273,** 7967 (1998).

[13] R. C. MacDonald and S. A. Simon, *Proc. Natl. Acad. Sci. U.S.A.* **84,** 4089 (1987).

[14] B. Maggio, *Prog. Biophys. Mol. Biol.* **62,** 55 (1994).

[15] B. Maggio, F. A. Cumar, and R. Caputto, *Biochem. J.* **171,** 559 (1978).

FIG. 1. Experimental device for measuring the surface tension at the air–water interface. The Langmuir film balance consists of a computerized balance equipped with a standard built-in underhook. The trough compartment is a Teflon tank placed on a magnetic stirrer (2) inside a homemade Plexiglas chamber. The sensor is a platinum plate (1). The balance (3) is connected to a computer (4) and the data analyzed with the Collect software.

placed on a magnetic stirrer inside a homemade Plexiglas chamber. The sensor is a platinum plate (length 1.962 cm). A bunsen burner is necessary for cleaning the sensor by flaming.

For the 1.962-cm plate, an apparent mass decrease of 4 mg corresponds to a surface pressure increase of 1 m*N*/m. After being dissolved in a mixture of hexane : chloroform : ethanol, 11 : 5 : 4 (v : v : v),[16] lipids are spread inside a Teflon tank. In our experiments, the subphases are 10 ml of pure water obtained by filtration through a Milli-Q water purification system (Millipore, Saint-Quentin, France). When the barrier of the film balance is run across a subphase of pure water, no change in surface tension is observed. As a further precaution against the introduction of surface active contaminant with the pure water, the surface layer of water is removed with a Pasteur pipette connected to a water-driven aspirator after the barrier had been run across the subphase. This compression cleaning cycle is repeated twice before films of lipids were spread. Each run is performed with a fresh film and subphase.

[16] D. S. Johnston and D. Chapman, *Biochim. Biophys. Acta* **937,** 10 (1988).

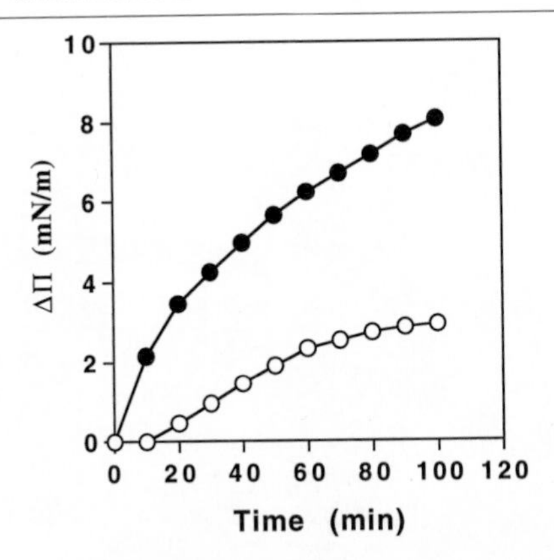

FIG. 2. Measurement of the kinetics of interaction between HIV-1 gp120 and GalCer monolayers. Monolayers of GalCer–HFA (●) or GalCer–NFA (○) are prepared at an initial surface pressure of 10 m*N*/m and this figure shows the variations induced by the addition of HIV-1 gp120 (1.5 n*M*) underneath the monolayer.

Kinetics Studies

To illustrate the use of this system to measure glycolipid interactions, the interaction of HIV-1 surface envelope glycoprotein gp120 with monolayers of GalCer was followed by measurement of the surface pressure increase as a function of time in a constant area setup. In these experiments, the viral glycoprotein was incubated with a monolayer of either GalCer–HFA (i.e., with a α-hydroxyl group in the fatty acid of the ceramide, HFA) or GalCer–NFA (i.e., without the α-hydroxyl group, NFA). The recombinant preparation of HIV-1 gp120 (IIIB isolate) was free of contaminant proteins as demonstrated by SDS–PAGE analysis.[12] The kinetics of the interaction demonstrate the preference of gp120 for GalCer–HFA vs. GalCer–NFA (Fig. 2).

To investigate further the lipid specificity of the penetration process, experiments were performed with GalCer–HFA, GalCer-NFA, GlcCer, ceramide–HFA and ceramide–NFA (Fig. 2). In these studies, the increase in surface pressure ($\Delta\Pi$) caused by penetration of the monolayer by the added component was measured as a function of initial surface pressure of the monolayer. The underlying idea is that lipid monolayers show decreased compressibility with increasing surface pressure, so that $\Delta\Pi$ is expected to decrease as the initial surface pressure of the monolayer increases.[17] Under

[17] J. D. Lear and M. Rafalski, in "Viral Fusion Mechanisms" (J. Bentz, ed.), p. 55. CRC Press, Boca Raton, Florida, 1993.

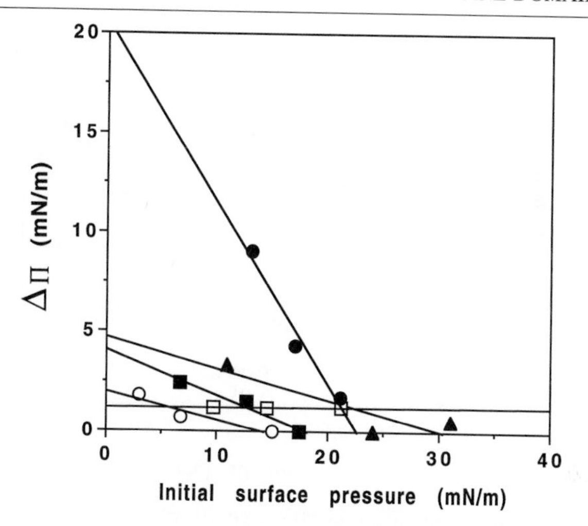

FIG. 3. Example of the maximal surface pressure increase (ΔΠ) reached after injection of HIV-1 gp120 (1.5 n*M*) under different glycolipid films at various initial surface pressures. GalCer–HFA (●), GalCer–NFA (○), Cer–HFA (■) Cer–NFA (□), GlcCer (▲).

these conditions, the data in Fig. 3 show that gp120 interacts specifically with monolayers of GalCer–HFA but not of GalCer–NFA. The specificity of the recombinant glycoprotein for GalCer–HFA was further demonstrated by the lack of penetration of monolayers of GluCer or ceramide.

Reconstitution of Cholesterol-GalCer Microdomains

Monolayers consisting of various amounts of GalCer–HFA in cholesterol can be prepared as models for cholesterol–GalCer plasma membrane microdomains. As shown in Fig. 4, a nonlinear relationship between GalCer concentration and gp120 binding was observed. Moreover, a threshold level of 40% GalCer–HFA in the monolayer was necessary to detect the interaction. These data are in agreement with those of Long *et al.*[18] who studied the binding of HIV-1 gp120 to GalCer-containing liposomes. These results show that the mere presence of GalCer in a model membrane (monolayer or bilayer) is not sufficient for GalCer binding. Rather, a critical concentration of GalCer molecules must be attained for efficient gp120 binding. This may explain why $CD4^-$ cells expressing low levels of GalCer on their cell surface are not infectable by HIV-1.[19]

[18] D. Long, J. F. Berson, D. G. Cook, and R. W. Doms, *J. Virol.* **68,** 5890 (1994).

[19] J. Fantini, D. G. Cook, N. Nathanson, S. L. Spitalnik, and F. Gonzalez-Scarano, *Proc. Natl. Acad. Sci. U.S.A.* **90,** 2700 (1993).

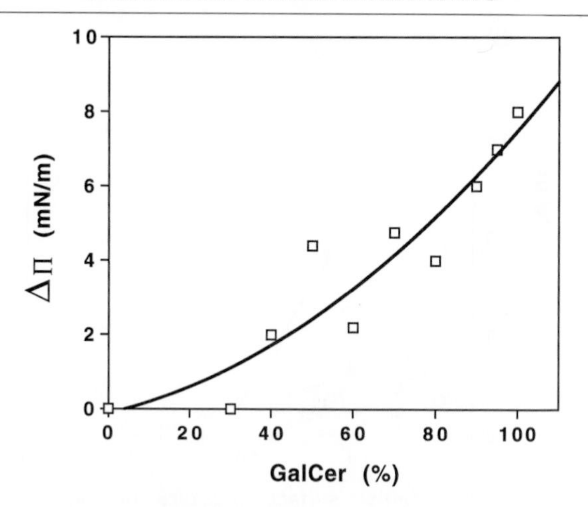

FIG. 4. Measurement of interaction of HIV-1 gp120 with reconstituted microdomains of GalCer and cholesterol. Cholesterol/GalCer–HFA monolayers with the indicated percentage of GalCer–HFA are prepared at various initial surface pressure (Πi) values (range 9–12 mN/m). The surface pressure increase (ΔΠ) induced by the addition of HIV-1 gp120 (2.5 n*M*) is measured and the value obtained for an initial surface pressure of 10 m*N*/m extrapolated from ΔΠ/Πi plots (see text).

Mapping of GalCer-Binding Site on gp120

To determine the domain of gp120 involved in GalCer recognition, the viral glycoprotein is preincubated with antibodies against distinct epitopes of the protein. The anti-gp120 antibodies are purified and stored at −80° in phosphate-buffered saline (PBS) : glycerol (1 : 1) without bovine serum albumin (BSA). The absence of contaminant proteins in the antibodies preparation is very important since serum albumin, for instance, may interact with virtually any lipid monolayer. The immune complexes are then added underneath a GalCer–HFA monolayer. The data summarized in Table I show that both the C2 and V3 domains of gp120 are important for GalCer binding, in agreement with previous studies.[20,21]

Interaction of CD4 with Glycosphingolipid Microdomains

In the plasma membrane of human T lymphocytes, CD4 and GM3 are colocalized in the same detergent-insoluble microdomain,[22] suggesting that

[20] S. Bhat, R. V. Mettus, E. P. Reddy, K. E. Ugen, V. Srikathan, W. V. Williams, and D. B. Weiner, *AIDS Res. Hum. Retroviruses* **9,** 175 (1993).

[21] D. G. Cook, J. Fantini, S. L. Spitalnik, and F. Gonzalez-Scarano, *Virology* **201,** 206 (1994).

[22] M. Sorice, I. Parolini, T. Sansolini, T. Garofalo, V. Dolo, M. Sargiacomo, T. Tai, C. Peschle, M. R. Torrisi, and A. Pavan, *J. Lipid. Res.* **38,** 969 (1997).

TABLE I
EFFECT OF ANTI-gp120 ANTIBODIES ON INTERACTION BETWEEN HIV-1 gp120 AND GalCer–HFA[a]

Antibody	ΔΠ (m*N*/m)	Inhibition (%)
None	4.975	0
Anti-V3 (F5)	2.825	43
Anti-V3 (110H)	1.325	73
Anti-C2 (110E)	1.575	68
Anti-Cter (D7326)	4.875	2

[a] HIV-1 gp120 (1.5 n*M*) was preincubated for 30 min at 37° with the indicated antibody (5 μg/ml). The complex was then added underneath a monolayer of GalCer–HFA at an initial surface pressure of 10–12 m*N*/m. The variations in surface pressure were measured after 40 min. The F5, 110H, and 110E antibodies were kindly provided by F. Traincard and J. C. Mazié (Institut Pasteur, Paris). The D7326 antibody was purchased from Aalto Bioreagents (Dublin, Ireland).

CD4 may bind to the glycosphingolipid. This type of interaction (i.e., the insertion of recombinant soluble CD4 in lipid monolayers) can also be followed by measurement of the surface pressure increase in a constant area setup. As shown in Fig. 5, the addition of CD4 under a GM3 monolayer results in a maximal pressure increase of 11.8 m*N*/m. To investigate the specificity of the penetration process, the increase in surface pressure (ΔΠ) caused by the addition of CD4 under the GM3 monolayer is measured at various initial surface pressures. As shown in the inset, the compressibility of the monolayer gradually decreases as the initial pressure of the monolayer increases. The influence of the initial surface pressure on the compressibility of the monolayer demonstrates the high specificity of the interaction, as previously established for several other lipids and ligands.[14]

The soluble form of CD4 used in these illustrations is a 4-domain CD4 lacking the transmembrane domain of the protein (kindly provided by the Medical Research Council, UK). Therefore, to rule out the possibility that the surface pressure increase might be due to an impurity in the recombinant preparation, CD4 can be depleted with anti-CD4 antibodies (MT-151, Boehringer-Mannheim, Mannheim, Germany) and protein A-Sepharose (Sigma). Briefly, 100 μl of protein-A Sepharose is incubated with 20 μl of the MT-151 monoclonal antibody (MAb) (0.2 mg/ml) for 1 hr at room temperature with gentle stirring. The protein A-Sepharose antibody complexes are rinsed 5 times with 500 μl of PBS (centrifugation in an Eppendorf

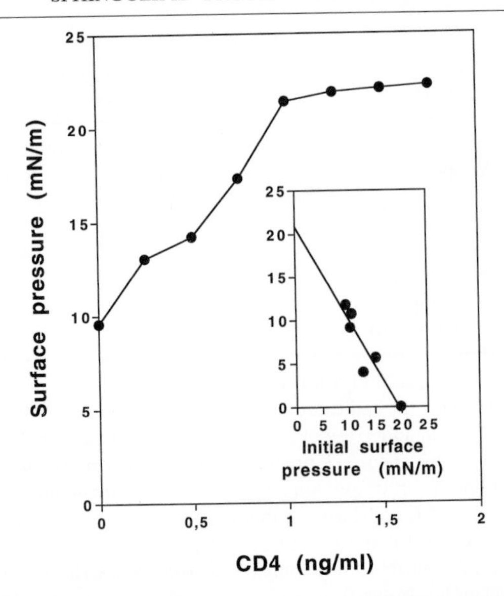

FIG. 5. Variations in surface pressure of a monolayer of human GM3 after injection of soluble recombinant CD4. *Inset:* Maximal surface pressure increase reached after injection of soluble recombinant CD4 (1 ng/ml) under a monomolecular film of human GM3 at various initial surface pressures.

tube, 7000 rpm, 30 sec). The pellet is resuspended with 8 μl of recombinant soluble CD4 (0.5 mg/ml) and 32 μl of PBS. After 1 hr at room temperature, the tubes are centrifugated and the CD4-depleted supernatant used immediately. This supernatant does not react with the GM3 monolayer.

For other controls, one establishes that the activity of CD4 is not altered by immunoprecipitation with control anti-HLA antibodies; also, CD4 can be preincubated with 1.85 mg/ml of 3′-sialyllactose [the oligosaccharide corresponding to the sugar moiety of GM3 (Oxford GlycoSystems, Abington, UK)], which prevents the increase in the surface pressure of the GM3 monolayer. Taken together, these analyses can suggest that a protein (CD4) is binding to a glycosphingolipid (GM3) through a specific interaction with its sugar moiety.

Reconstitution of HIV-1 Fusion Complex in Glycosphingolipid Microdomain

We have demonstrated that the binding of HIV-1 gp120 with CD4 complexed with GM3 demasks the V3 domain of gp120, allowing secondary interactions between V3 amino acids and the GM3 patch.[5] The sequential interaction of CD4 and gp120 with the glycosphingolipid microdomain induces a biphasic increase of the surface pressure. The first response corresponds to the insertion of CD4 in the patch, and the second one to the CD4-induced penetration of gp120, through its V3 domain, in the glycosphingolipid monolayer. In a typical experiment (Fig. 6), recombinant soluble CD4 (0.5 ng/ml) is added first under a monolayer of glycosphingolipid (Gb_3 in this case). After reaching a plateau value, gp120 is added at a concentration of 1.85 n*M* and the variations of surface pressure measured from this second input.

Taken together, these data confirm that reconstituted glycosphingolipid microdomains are recognized by gp120.[12] CD4 interacts with either GM3 or Gb_3, and induces glycosphingolipid-dependent interactions with gp120 from selected HIV-1 isolates. These secondary interactions are specifically abrogated with 3′-sialyllactose and monoclonal antibodies against the V3 loop, suggesting that charged V3 amino acid residues interact with the oligosaccharide moiety of the glycosphingolipid (Ref. 5 and data not shown).

Glycosphingolipid Microdomains as Preferential Sites of Formation of HIV-1 Fusion Complex

The entry of HIV-1 into $CD4^+$ cells requires the sequential interaction of the viral surface envelope glycoprotein gp120 with the CD4 receptor

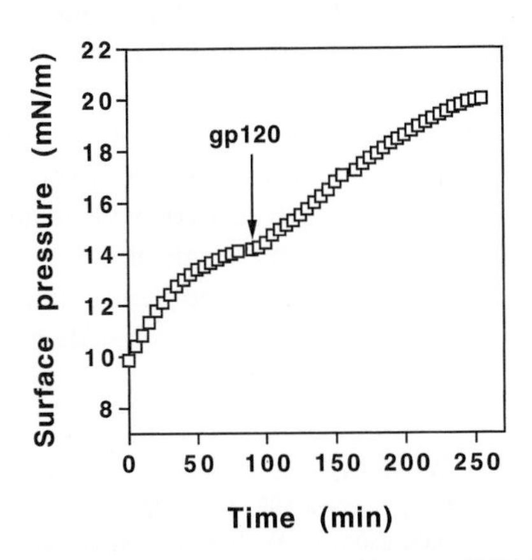

FIG. 6. Measurement of CD4-induced interactions between HIV-1 gp120 with a reconstituted microdomain of Gb_3. At time 0, recombinant CD4 (0.5 ng/ml) is added under a monolayer of Gb_3 prepared at an initial surface pressure of 10 m*N*/m. The first increase in surface pressure is due to the interaction of CD4 with the monolayer of Gb_3. After reaching a plateau value, HIV-1 gp120 is added at a concentration of 1.85 n*M*. Secondary gp120–Gb3 interactions are evidenced by a second phase of surface pressure increase.

and a coreceptor (or a fusion cofactor) on the cell surface.[23] The coreceptors identified so far for human (HIV) and simian (SIV) immunodeficiency viruses include chemokine receptors (mainly CXCR4, CCR5, CCR3, CCR2b) and a series of orphan receptors including virus-encoded receptors all belonging to the family of seven-transmembrane domains receptors.[24]

Following a primary interaction with CD4, a conformational change in gp120 renders cryptic regions of the viral glycoprotein (including the V3 domain[25] available for secondary interactions with either CXCR4 or CCR5.[24] Because seven-transmembrane domains receptors are almost flush with the cell membrane, binding of gp120 to the coreceptor is necessary to move the viral spike close to the target membrane.[23] Finally, gp120–coreceptor interactions trigger additional conformational changes in the HIV-1 envelope glycoprotein trimer that lead to exposure of the fusion peptide at the N terminus of the transmembrane glycoprotein gp41.[23] Because coreceptors are important determinants of virus tropism, HIV-1 isolates are functionally classified with respect to their ability to use a given coreceptor[26]: for instance, viruses using CXCR4 but not CCR5 are referred to as X4, whereas isolates using CCR5 but not CXCR4 are called R5. Dual-tropic viruses able to use either CXCR4 or CCR5 are referred to as R5X4.

The presence of CD4 in glycosphingolipid microdomains suggests a specific role for these lipids in the HIV-1 fusion process. Recent data show that in the presence of CD4, gp120 from HIV-1(IIIB) (a T-cell line adapted X4 isolate) interacts preferentially with Gb_3, whereas gp120 from HIV-1 (89.6) (a dual-tropic, primary R5X4 isolate) interacts exclusively with GM3.[27] Therefore, HIV-1 isolates might select glycosphingolipids of their choice in addition to the chemokine receptors to promote fusion. GM3 and Gb_3 are highly expressed in human macrophages but are also present in $CD4^+$ lymphocytes and T-cell lines.[28] These glycosphingolipids share a common structural feature, i.e., a free hydroxyl group in position 4 of a terminal galactose residue. Based on the inhibitory effect of 3′-sialyllactose on GM3–CD4 interaction, one can hypothesize that CD4 binds to this structural determinant borne by the oligosaccharide part of both GM3 and

[23] J. Binley and J. P. Moore, *Nature* **387,** 346 (1997).

[24] P. R. Clapham and R. A. Weiss, *Nature* **388,** 230 (1997).

[25] P. D. Kwong, R. Wyatt, J. Robinson, R. W. Sweet, J. Sodroski, and W. A. Hendrickson, *Nature* **393,** 648 (1998).

[26] E. A. Berger, R. W. Doms, E. M. Fenyo, B. T. M. Korber, D. R. Littman, J. P. Moore, Q. J. Sattentau, H. Schuitemaker, and J. Sodroski, *Nature* **391,** 240 (1998).

[27] D. Hammache, N. Yahi, M. Maresca, G. Pièroni, and J. Fantini, *J. Virol.* **73,** 5244 (1999).

[28] J. Fantini, C. Tamalet, D. Hammache, C. Tourrès, N. Duclos, and N. Yahi. *J. AIDS Hum. Retrovirol.* **19,** 221 (1998).

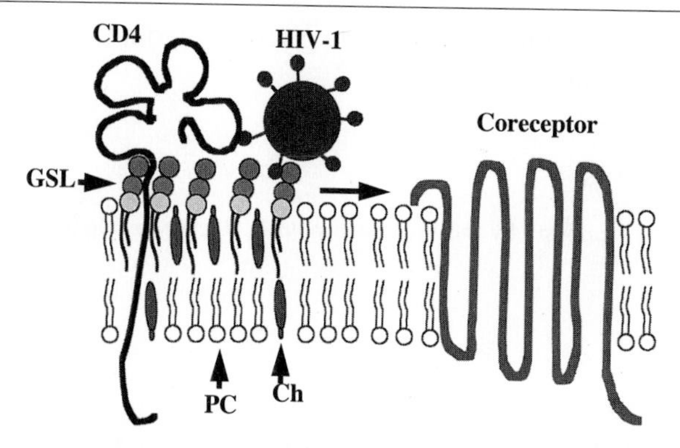

FIG. 7. Plasma membrane glycosphingolipid microdomains as preferential sites of formation of the HIV-1 fusion complex. In the plasma membrane of $CD4^+$ cells, CD4 is present in glycosphingolipid-enriched microdomains but is not associated with HIV-1 coreceptors. Once bound to CD4, the viral particle is conveyed to an appropriate coreceptor by the glycosphingolipid raft, which moves freely in the external leaflet of the plasma membrane. Ch, cholesterol; GSL, glycosphingolipid; PC, phosphatidylcholine.

and Gb_3. In the model shown in Fig. 7, the glycosphingolipid interacts with domain 4 of CD4, which is plausible given its proximity to the plasma membrane.

Glycosphingolipids recognized by both CD4 and gp120 may induce the formation of a trimolecular complex CD4–glycosphingolipid–gp 120.[5] The role of the glycosphingolipid in this multimolecular organization could be to facilitate the migration of the CD4–gp 120 complex to an appropriate coreceptor (e.g., CCR5 or CXCR4), since CD4 and these coreceptors are not physically associated in absence of HIV-1.[29] By moving freely in the external leaflet of the plasma membrane, the glycosphingolipid patch may behave as a raft dragging the CD4 receptor and taking aboard the viral particle (Fig. 7). The binding of the virion to the raft is stabilized by secondary interactions between the polar heads of glycosphingolipid molecules and the V3 loop of gp120.[5] The raft may then float on the cell surface until finding an adequate coreceptor that can displace the glycosphingolipid–V3 loop interactions to its own profit, resulting in the initiation of the fusion process. In the absence of any available coreceptor, the glycosphingolipid may eventually allow the conformational change of gp41, as

[29] C. L. Lapham, J. Ouyang, B. Chandrasekhar, N. Y. Nguyen, D. S. Dimitrov, and H. Golding, *Science* **274,** 602 (1996).

may be the case for human erythrocyte glycosphingolipids transferred into murine cells expressing human CD4.[6]

Acknowledgments

This work was supported by the Fondation pour la Recherche Médicale (SIDACTION grant).

[42] Analysis of Ceramides Present in Glycosylphosphatidylinositol Anchored Proteins of *Saccharomyces cerevisiae*

By ISABELLE GUILLAS, MARTINE PFEFFERLI, and ANDREAS CONZELMANN

Principle

Glycosylphosphatidylinositol (GPI) anchored proteins are found on the cell surfaces of virtually all eukaryotic organisms.[1] The genome of *Saccharomyces cerevisiae* contains about 70 open reading frames encoding proteins that would be predicted to be anchored by a GPI anchor.[2,3] The presence of a GPI anchor has been confirmed in several yeast proteins, and the detailed structures of the GPI anchor of a pool of GPI proteins as well as of individual proteins such as Gas1p or cAMP binding protein have been reported.[4,5] The carbohydrate core structure linking the C-terminal end of yeast GPI proteins to a lipid moiety is the same as in all other eukaryotic organisms, namely, protein-CO-NH$(CH_2)_2$-PO_4-6Manα1-2Manα1-6Manα1-4GlcNα1-6*myo*-inositol-PO_4-lipid (Fig. 1). The lipid moiety consists of either a ceramide or a mono- or diacylglycerol.

Ceramides are found on the majority of yeast anchors; they mainly consist of C18:0- or C20:0-phytosphingosine (PHS) and a C26:0 fatty acid. In contrast, the lipid moiety of Gas1p, an abundant GPI protein, is sensitive to PI-PLC and to mild alkaline hydrolysis, and contains C26:0 fatty acids.[5] Thus, yeast GPI proteins are made with two different kinds

[1] M. J. McConville and M. A. Ferguson, *Biochem. J.* **294,** 305 (1993).
[2] K. Hamada, S. Fukuchi, M. Arisawa, M. Baba, and K. Kitada, *Mol. Gen. Genet.* **258,** 53 (1998).
[3] L. H. Caro, H. Tettelin, J. H. Vossen, A. F. Ram, H. van den Ende, and F. M. Klis, *Yeast* **13,** 1477 (1997).
[4] G. Muller, K. Schubert, F. Fiedler, and W. Bandlow, *J. Biol. Chem.* **267,** 25337 (1992).
[5] C. Fankhauser, S. W. Homans, J. E. Thomas Oates, M. J. McConville, C. Desponds, A. Conzelmann, and M. A. Ferguson, *J. Biol. Chem.* **268,** 26365 (1993).

0076-6879/00 $30.00

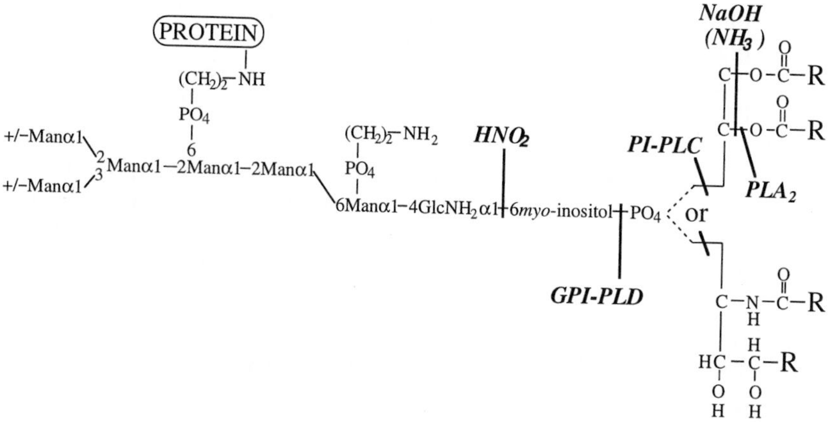

FIG. 1. Structural variants of the yeast GPI anchor. Relevant cleavage procedures are indicated. Anchors can contain an optional fifth mannose residue linked either to α1,2 or α1,3. This residue is added to GPI proteins in the Golgi [C. Fankhauser, S. W. Homans, J. E. Thomas Oates, M. J. McConville, C. Desponds, A. Conzelmann, and M. A. Ferguson, *J. Biol. Chem.* **268,** 26365 (1993); G. Sipos, A. Puoti, and A. Conzelmann, *J. Biol. Chem.* **270,** 19709 (1995)]. They also can bear a second ethanolaminephosphate on the α1,4-linked mannose (unpublished observation). GPI-PLD, GPI-specific phospholipase D; PI-PLC, phosphatidylinositol-specific phospholipase C.

of lipids, both of which contain very long fatty acids. In both types of lipids, the C26 : 0 fatty acid may be hydroxylated on C_2. The complete GPI lipids that are transferred onto newly synthesized proteins shortly after their translocation into the endoplasmic reticulum (ER), however, contain only diacylglycerol as a lipid moiety.[6,7] This suggests that lipids of GPI anchors are exchanged by a process called *anchor remodeling*. Data so far indicate that three types of enzymes are intimately involved in the elaboration of the mature lipid moiety of GPI proteins: (1) enzymes exchanging diacylglycerol for ceramide, (2) enzymes exchanging C_{16} or C_{18} for C26 : 0 fatty acids, and (3) enzymes responsible for the presence of monohydroxylated C_{26} fatty acids, the latter being localized in the Golgi apparatus.[7,8]

The functional significance of lipid remodeling is not clear, in part because there are no mutants that are deficient specifically in these processes and the enzymes. Because the majority of GPI anchored proteins of *S. cerevisiae* eventually lose part of the anchor, including the lipid moiety,

[6] A. Conzelmann, A. Puoti, R. L. Lester, and C. Desponds, *EMBO J.* **11,** 457 (1992).
[7] G. Sipos, F. Reggiori, C. Vionnet, and A. Conzelmann, *EMBO J.* **16,** 3494 (1997).
[8] F. Reggiori, E. Canivenc-Gansel, and A. Conzelmann, *EMBO J.* **16,** 3506 (1997).

and become covalently attached to the β1,6-glucans of the cell wall,[9–11] it may be that the lipid remodeling of GPI proteins contributes to the correct transport and sorting of GPI proteins rather than altering their function per se. The study of remodeling events requires methods that can analyze the lipid moieties of individual proteins at different stages of maturation. The methods described here rely on procedures that allow for rapid, specific, and efficient metabolic radiolabeling of the lipid moiety of GPI proteins.

The analytical methods that provide structural information on the labeled anchor lipids have been developed in other labs. The procedure described here is useful for biosynthetic studies using secretory mutants or for kinetic studies as recently described,[6–8,12] but results obtained with these methods will eventually have to be corroborated by mass spectrometry.

Overall Approach

A typical experiment will follow the various steps denoted in Fig. 2. Metabolic labeling with *myo*-[2-^{3}H]inositol ([^{3}H]Ins) specifically labels GPI proteins but not other proteins.[6] Metabolic labeling with [4,5-^{3}H]dihydrosphingosine ([^{3}H]DHS) is also specific for GPI proteins and labels the ceramide-containing GPI proteins but not the ones that contain a diacylglycerol, although part of the [^{3}H]DHS is degraded to [^{3}H]palmitaldehyde and is subsequently used for the biosynthesis of phosphatidic acid and phospholipids.[13,14] Metabolic labeling with [^{3}H]palmitate or [^{3}H]myristate leads to extensive and almost specific labeling of GPI proteins, but curiously, after [^{3}H]palmitate labeling, the protein-associated label has been found almost exclusively in the C_{18}-phytosphingosine component of GPI anchors containing ceramide.[6] Note that the labeled anchor lipids isolated after labeling cells with [^{3}H]DHS or ^{3}H-labeled fatty acids may not be representative of the lipids that are normally added to GPI proteins. Thus, after labeling with [^{3}H]DHS, it has been noted that about 15% of anchor ceramides contained [^{3}H]DHS; the others contained [^{3}H]phytosphingosine. When analyzed by mass spectrometry, the proportion of DHS-containing

[9] C.-F. Lu, R. C. Montijn, J. L. Brown, F. Klis, J. Kurjan, H. Bussey, and P. N. Lipke, *J. Cell Biol.* **128,** 333 (1995).

[10] H. de Nobel and P. N. Lipke, *Trends Cell Biol.* **4,** 42 (1994).

[11] R. Kollar, B. B. Reinhold, E. Petrakova, H. J. Yeh, G. Ashwell, J. Drgonova, J. C. Kapteyn, F. M. Klis, and E. Cabib, *J. Biol. Chem.* **272,** 17762 (1997).

[12] F. Reggiori and A. Conzelmann, *J. Biol. Chem.* **273,** 30550 (1998).

[13] C. Mao, M. Wadleigh, G. M. Jenkins, Y. A. Hannun, and L. M. Obeid, *J. Biol. Chem.* **272,** 28690 (1997).

[14] J. D. Saba, F. Nara, A. Bielawska, S. Garrett, and Y. A. Hannun, *J. Biol. Chem.* **272,** 26087 (1997).

Isolation of lipid moieties of radiolabeled GPI proteins

Yeast cells, metabolically labeled with tritiated inositol or dihydrosphingosine

↓

Mechanical disruption of cells in chloroform–methanol using glass beads

↓

Extensive delipidation of proteins and cell walls with organic solvents → Labeled lipids (→ Desalting of anchor lipids by butanol extraction)

↓

Solubilization of proteins by boiling in SDS

↓ (two branches)

Dilution into Triton X-100, affinity chromatography of proteins on concanavalin A-Sepharose or immunoprecipitation | SDS–PAGE, fluorography, isolation of bands of interest

↓

Elution of bound material in form of amino acids and peptides using a protease

↓

Purification of GPI anchor peptides on octyl-Sepharose

↓

Liberation of lipid moieties of anchors using nitrous acid deamination, PI-PLC, GPI-PLD, strong acid hydrolysis, etc.

↓

Desalting of anchor lipids by butanol extraction

↓

Thin-layer chromatography

↓

Detection and quantitation of different anchor lipids by two-dimensional radioscanning → Recovery of lipids and further fragmentation by partial deacylation with methanolic NH_3, PLA_2, strong acid hydrolysis etc. (→ Desalting of anchor lipids by butanol extraction); → Fluorography

FIG. 2. Outline of a typical experiment investigating the lipids of GPI anchors. The recovery of total anchor lipids typically amounts to 0.2–0.5% of the [^{3}H]Ins initially added to cells.

anchor ceramides amounted to only 5%.[5,8] No bias is expected if anchors are labeled with [^{3}H]Ins.

Extensive extraction with organic solvents will not achieve a complete delipidation of proteins; further delipidation steps (e.g., affinity chromatography on concanavalin A-Sepharose) must also be used. All of the [^{3}H]Ins-labeled proteins of *S. cerevisiae* are retained on concanavalin A-Sepharose, which indicates that all of the major GPI proteins are glycoproteins. Alternatively, proteins may be further delipidated by immunoprecipitation or by gel electrophoresis. In all cases, GPI peptides can conveniently be recovered using protease digestion as long as the protease is not contaminated with enzymes that would hydrolyze the anchor components under study. To analyze anchor lipids, we use pronase, because it does not contain phospholipase activity, which would remove the labeled inositol from the lipid. The subsequent chromatography on octyl-Sepharose removes amino acids, hydrophilic glycopeptides, and detergent.

Experimental Procedures

Radiolabeling of Cells with Myo-[2-^{3}H]Inositol or [4,5-^{3}H]Dihydrosphingosine and Preparation of Delipidated Proteins

[^{3}H]Ins and [^{3}H]DHS are available at specific activities of 15–20 Ci/mmol and 30–60 Ci/mmol, respectively, commercially (e.g., Anawa Trading SA, Wangen, Switzerland). Catalytic reduction of C_{18}-sphingosine with tritium gas can also be performed (in our case, by NEN-Du Pont, Les Ulis, France) to yield C_{18}-[4, 5-^{3}H]DHS that is purified by preparative thin-layer chromatography (TLC) as recently described.[15] Cells are grown in synthetic minimal medium containing 2% glucose,[16] but inositol is omitted from the vitamin supplement, and the medium is supplemented with 1% casein hydrolyzate (casamino acids, GibcoBRL, Paisley, Scotland). Inositol must be omitted to induce the inositol transporter Itr1p.[17–19] Casamino acids contain small amounts of inositol that do not interfere with the induction of the transporter, but better incorporation of [^{3}H]Ins is achieved if cells are labeled in glucose containing minimal medium supplemented with a mixture of pure amino acids (i.e., SC medium).[16]

Labeling is performed by resuspending exponentially growing cells in fresh medium at about $1–2 \times 10^8$ cells/ml and adding 40–150 μCi of [^{3}H]Ins or 100 μCi of [^{3}H]DHS/ml of medium from concentrated stocks kept in water or methanol, respectively. With both radiotracers, the bulk of radioactivity is incorporated into cellular lipids within 10–30 min and fresh medium can be added if labeling is to be continued. Note that under the indicated conditions, the incorporation of these radiotracers into GPI

[15] F. Reggiori, E. Canivenc-Gensel, and A. Conzelmann, *Methods Mol. Biol.* **116,** 91 (1998).
[16] F. Sherman, *Methods Enzymol.* **194,** 3 (1991).
[17] K. Lai and P. McGraw, *J. Biol. Chem.* **269,** 2245 (1994).
[18] J. Nikawa, K. Hosaka, and S. Yamashita, *Mol. Microbiol.* **10,** 955 (1993).
[19] K. Lai, C. P. Bolognese, S. Swift, and P. McGraw, *J. Biol. Chem.* **270,** 2525 (1995).

proteins continues even while the labeled lipids are being chased by lipids made by cells from unlabeled, endogenously synthesized Ins and DHS. To stop the incorporation of radiolabel into proteins, protein synthesis inhibitors such as cycloheximide must be used. Note, however, that [^{3}H]DHS continues to be incorporated into mature GPI proteins for as long as 60 min after protein synthesis has been arrested, apparently because the ceramides of mature proteins continue to be exchanged.[8]

In contrast, [^{3}H]Ins incorporation into proteins is rapidly arrested by cycloheximide.[20] Typically, we label 0.25–1 ml of cell suspension for 120 min in a shaking water bath. At the end, cells can be washed in water but we prefer to process them immediately after centrifugation in a screw-topped Eppendorf tube. The cell pellet is resuspended in 500 μl of chloroform–methanol (1:1, v/v) by sonication and vortexing, and 200 μl of acid washed glass beads is added. Cells are broken by repeated vortexing (5 times for 1 min each). After sedimentation of glass beads the supernatant is transferred to a fresh tube, the glass beads are rinsed twice with 400 μl of chloroform–methanol–water (10:5:3, v/v), and the supernatants are pooled with the first supernatant. Precipitated proteins and cell wall fragments are then sedimented at 15,000*g* for 5 min at 4°. The supernatant contains lipids that can serve as a reference for the cell's free (i.e., not protein bound) lipids and is stored at −20°.

The protein pellet is further delipidated by resuspension in 0.5 ml chloroform–methanol–water (10:10:3), sonicated in a bath-type sonicator, centrifuged, and supernatant removed. This is done four times. Proteins are then resuspended in 0.5 ml ethanol–water–diethyl ether–pyridine–ammonia (15:15:5:1:0.018 v/v) and incubated at 37° for 15 min[21] and the solvent is removed in a Speed-Vac evaporator. This treatment may solubilize some labeled lipids since it seems to render the subsequent delipidation steps more efficient. The protein pellet is then delipidated once more with chloroform–methanol–water (10:10:3). The dried pellet is boiled for 5 min in solubilization buffer (80 m*M* Tris-HCl, pH 6.8, 2% sodium dodecyl sulfate (SDS), 5% 2-mercaptoethanol). When starting with 1–2 × 10^8 cells, about 200 μl of this buffer must be used.

Affinity Chromatography of Proteins on Concanavalin A-Sepharose and Generation of Anchor Peptides

All of the following procedures will be described for a sample starting with 1–2 × 10^8 cells. To remove insoluble material, add 1 ml of concanavalin

[20] A. Conzelmann, C. Fankhauser, and C. Desponds, *EMBO J.* **9,** 653 (1990).
[21] B. A. Hanson and R. L. Lester, *J. Lipid. Res.* **21,** 309 (1980).

A (ConA)-Sepharose buffer (50 mM Tris-HCl, 1% Triton X-100, 150 mM NaCl, 1 mM $CaCl_2$, 1 mM $MgCl_2$, pH 7.4), mix, and centrifuge at 15,000g for 5 min. Transfer the supernatant to a 15-ml plastic tube. To avoid denaturation of ConA by SDS, a 20-fold excess of Triton X-100 should be added. Thus, 8 ml of ConA-Sepharose buffer is added to neutralize the SDS of 200 μl of solubilization buffer. ConA-Sepharose (Pharmacia, Uppsala, Sweden) is sedimented two times in ConA-Sepharose buffer by centrifugation at 200g for 5 min to remove free ConA before use. The washed ConA-Sepharose is added to the sample (100 μl of packed beads for material from 1–2 $\times$ 10^8 cells) and the sample is incubated at room temperature for 2 hr on a rotating wheel.

The Sepharose is then sedimented and washed by repeated resuspension in large volumes of ConA-Sepharose buffer. Beads are left standing a few minutes to allow for lipid exchange, and sedimentation is terminated by centrifugation for 3 min at 12g. Washing the Sepharose in this manner should continue until the radioactivity in the supernatant becomes constant (at approximately 300–500 cpm/ml). Sepharose may be washed once more with ConA-Sepharose buffer containing Triton X-100 at 0.02% to reduce the amount of detergent carried over into subsequent steps. At this stage proteins may be eluted intact by incubation with 1 M α-methylmannoside or by boiling in SDS sample buffer.[22]

Several of the procedures for liberating the GPI anchor lipid moieties can be done on the intact proteins.[8] However, results seem to be more quantitative when GPI anchor peptides are prepared using protease first. For this, ConA-Sepharose is resuspended in pronase buffer (100 mM Tris-HCl, 1 mM $CaCl_2$, 0.02% Triton X-100, 10 mM NaN_3, 20 μg/ml gentamycin, pH 8.0; 0.5 ml/0.1 ml of Sepharose) and treated for 16 hr with 0.5 mg of pronase at 37°. Pronase should be preincubated for 30 min to digest contaminating hydrolases. Samples are boiled for 5 min, centrifuged, and the supernatant is transferred to a fresh tube. Concanavalin A-Sepharose beads are rinsed 2 times with 0.4 ml of 0.1 M ammonium acetate, 5% 1-propanol and these washes are added to the pronase eluate.

Purification of GPI Anchor Peptides on Octyl-Sepharose

This method is commonly used to purify GPI anchor peptides and has been described elsewhere in this series.[23] Our procedure is similar, however, we use 1 ml of octyl-Sepharose per column, adjust the pH of the sample first to 4.5 by adding 50-μl aliquots of 10% acetic acid, reapply the sample twice to ensure complete adsorption, and elute columns either with a gradi-

[22] U. K. Laemmli, *Nature* **227,** 680 (1970).
[23] P. Schneider and M. A. J. Ferguson, *Methods Enzymol.* **250,** 614 (1995).

ent of propanol or with 5% 1-propanol, then 3 ml of 25% 1-propanol, then 4 ml of 50% 1-propanol. When using a gradient, we generally obtain three peaks of radioactivity: the first, small peak of uncharacterized material contains the bulk of detergents (SDS and Triton X-100); the two other peaks eluting between 30 and 40% propanol are not well separated and correspond to GPI peptides, which contain the same lipid moieties and the same number of mannose residues on their GPI anchor but which may differ by the number of residual amino acids. When using stepwise elution with 25 and 50% 1-propanol, the anchor peptides are found in the first 1.5 ml of the 50% 1-propanol eluate.

Liberation of Lipid Moieties

In all procedures it is important to perform an incubation control that documents that the lipids derive from the cleavage of GPI anchor peptides and are not simply due to incomplete delipidation of the anchor peptide preparation under study (e.g., see Fig. 3, lane 3). Nitrous acid (HNO_2) deamination liberates phosphatidylinositol (PI) or inositolphosphorylceramide (IPC) and is the method of choice for [^{3}H]Ins-labeled samples (Fig. 1). Nitrous acid treatment is carried out exactly as previously de-

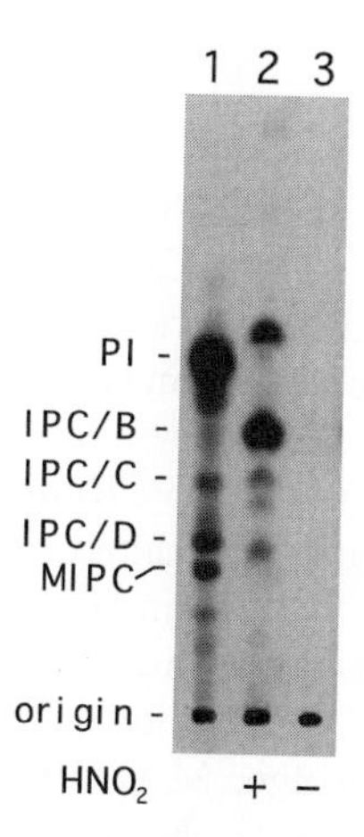

FIG. 3. [^{3}H]Ins-labeled GPI anchor lipids of W303 wild-type cells. 10^8 cells were labeled for 2 hr with 40 μCi [^{3}H]Ins. Lipids were extracted and desalted by butanol–water partitioning. Proteins were completely delipidated by affinity chromatography over ConA-Sepharose and their lipid moieties were liberated using HNO_2. Free lipids (lane 1) and liberated anchor lipids (lane 2) were separated by TLC using solvent (55:45:5). One-half of the material was control incubated without HNO_2 (lane 3). The fluorogram was exposed during 4 weeks. IPC/B, IPC/C, and IPC/D denote subclasses of IPC. MIPC is mannosylated IPC.

scribed.[24] If cells have been labeled with [^{3}H]DHS, enzymes such as GPI-PLD and PI-PLC will also liberate radiolabeled lipid moieties. For GPI-PLD treatment, the anchor peptides are resuspended in 100 μl of GPI-PLD buffer (50 m*M* Tris-HCl, 10 m*M* NaCl, 2.6 m*M* $CaCl_2$, 20% 1-propanol, pH 4.5), split into two equal aliquots, and incubated with or without 0.5 U of GPI-PLD[25] at 37° overnight. PI-PLC treatment is carried out in the same way but PI-PLC buffer (20 m*M* Tris-HCl, 0.2 m*M* EDTA, 20% 1-propanol) and 0.05 U of PI-PLC (ICN, Costa Mesa, CA) are used. For treatment with PLA_2 from bee venom (Sigma, St. Louis, MO), anchor peptides or anchor lipids are resuspended in 30 μl PLA_2 buffer (25 m*M* Tris-HCl, pH 7.2, 2 m*M* $CaCl_2$, 0.1% sodium deoxycholate) and incubated with or without 2 U of PLA_2 for 5 hr at 37°.

After HNO_2 or enzyme treatments, the liberated lipids are desalted by partitioning between butanol and water. For this, 400 μl of water saturated 1-butanol is added at the end of the incubation. After vortexing and centrifugation at 15,000*g* for 1 min, the upper, butanol phase is removed. After a second, identical butanol extraction, the pooled butanol phases are back extracted 2 or 3 times with 500 μl of water.

To free the long-chain bases one can use strong acid hydrolysis. For this, peptides are resuspended in 90% methanol, 1*M* HCl and are incubated at 80° for 16 hr. They are then dried in the Speed-Vac and residual HCl is flash evaporated twice with 200 μl dry methanol.

Thin-Layer Chromatography

Dried lipids are taken up in chloroform-methanol (1 : 1) and spotted on 0.2-mm-thick silica gel plates (20 × 20 cm, Merck, Darmstadt, Germany, silica gel 60). Ceramides, diacylglycerol, and long-chain bases are well separated using the solvent chloroform–methanol–2*M* ammonium hydroxide (40 : 10 : 1); PIs and IPCs are well separated by using the solvent chloroform–methanol–0.25% KCl (55 : 45 : 5). Radioactive bands can be localized and quantitated by one- and two-dimensional radioscanning (LB2842, Berthold AG, Regensdorf, Switzerland). At this stage, interesting bands can be scraped and eluted from the silica for further enzymatic and chemical treatments as depicted in Fig. 2. Alternatively, plates are sprayed with EN^{3}HANCE (NEN, Cambridge, MA) and exposed to film (X-OMAT; Eastman Kodak Co., Rochester, NY) at −80°. A typical example of the latter is shown in Fig. 3. Note that the main IPC of GPI anchors is IPC/B, a species that represents only a minor fraction of the cell's free IPC (Fig. 3, lane 2). Note also that delipidation of anchor peptides ap-

[24] M. L. Güther, W. J. Masterson, and M. A. Ferguson, *J. Biol. Chem.* **269,** 18694 (1994).
[25] M. A. Davitz, J. Hom, and S. Schenkman, *J. Biol. Chem.* **264,** 13760 (1989).

pears to have been complete since no lipids are visible if HNO_2 is omitted (Fig. 3, lane 3).

Acknowledgments

I.G. is supported by grant 3100-049664.96/1 from the Swiss National Foundation for Scientific Research. M.P. works for the EUROFAN2 project and is supported by OFES contract 97.0537-2.

[43] Preparation of Functionalized Lipid Tubules for Electron Crystallography of Macromolecules

By ELIZABETH M. WILSON-KUBALEK

Introduction

Sample preparation still remains the critical step for obtaining high-resolution structural information about biological macromolecules. The methodology for preparing two-dimensional (2D) crystals is maturing rapidly both for membrane proteins and for soluble proteins that have been ordered on lipid layers (reviewed in Refs. 1 and 2). The latter method, introduced by Uzgiris and Kornberg,[3] is particularly versatile due to the possibility of modifying the lipid substrate to facilitate protein binding. Binding and crystallization of the protein at the lipid surface can be achieved by the use of specific ligand-derivatized headgroups[3] or by nonspecific, electrostatic interactions with charged lipids.[4,5] The method has been further generalized through the introduction of Ni^{2+}-chelating moieties to the headgroups of lipids, expanding the application of the lipid layer crystallization method to a large variety of histidine-tagged proteins.[6–9] Streptavidin crys-

[1] B. K. Jap, M. Zulauf, T. Scheybani, A. Hefti, W. Baumeister, U. Aebi, and A. Engel, *Ultramicroscopy* **46,** 45 (1992).
[2] W. Chiu, A. J. Avila-Sakar, and M. F. Schmid, *Adv. Biophys.* **34,** 161 (1997).
[3] E. E. Uzgiris and R. D. Kornberg, *Nature* **301,** 125 (1983).
[4] S. A. Darst, H. O., Ribi, D. W. Pierce, and R. D. Kornberg, *J. Mol. Biol.* **203,** 269 (1998).
[5] S. A. Darst, E. W. Kubalek, A. M. Edwards, and R. D. Kornberg, *J. Mol. Biol.* **221,** 347 (1991).
[6] E. W. Kubalek, S. F. J. Le Grice, and P. O. Brown, *J. Struct. Biol.* **113,** 117 (1994).
[7] E. Barklis, J. McDermott, S. Wilkens, E. Schabtach, M. F. Schmid, S. Fuller, S. Karanjia, Z. Love, R. Jones, Y. Rui, X. Zhao, and D. Thompson, *EMBO J.* **16,** 1199 (1997).
[8] C. Venien-Bryan, F. Balavoine, B. Toussaint, C. Mioskowski, E. A. Hewat, B. Helme, and P. M. Vignais, *J. Mol. Biol.* **274,** 687 (1997).
[9] N. Bischler, F. Balavoine, P. Milkereit, H. Tschochner, C. Mioskowski, and P. Schultz, *Biophys. J.* **74,** 1522 (1998).

0076-6879/00 $30.00

tals grown on biotinylated lipid layers by this method diffract to <3-Å resolution.[10,11] However, obtaining a high-resolution structure requires a tilt series reconstruction and the collection of electron diffraction data. This is a time-consuming task, and the limits on tilting in the microscope cause the resolution in the final three-dimensional (3D) map perpendicular to the plane of the crystal to be significantly poorer than the resolution in the plane of the crystal. 3D maps from helical specimens do not suffer from these limitations; tilting is not necessary because all required molecular views are present in a single image, and helical analysis and averaging allow calculation of 3D maps with isotropic resolution and high signal-to-noise ratios.[12,13]

The method described here has combined the principles of the lipid layer crystallization methodology with the propensity of certain lipids to form nanoscale tubules[14] to develop a general approach for helical crystallization of macromolecules.[15] Hydrated mixtures of the tubule forming glycolipid galactosylceramide (GalCer) and derivatized lipids or charged lipids form unilamellar nanotubules[15] (Fig. 1A). By doping the GalCer with the nickel lipid, 1,2-dioleoyl-*sn*-glycero-3-[(*N*-(5-amino-1-carboxypentyl)iminodiacetic acid)succinyl] (nickel salt) (DOGS-NTA-Ni) helical arrays of two histidine-tagged Fabs were formed.[15] Similarly, doping with a biotinylated lipid allowed cyrstallization of streptavidin.[15] In two cases, the proteins g-actin and RNA polymerase, which have the affinity for positively charged surfaces, formed helical arrays on 1,2-dioleoyl-*sn*-glycero-3-ethylphosphocholine (DO-ethyl-PC)-doped tubules.[15] A third protein, annexin V, which has the opposite affinity, formed a helical array on 1,2-dioleoyl-*sn*-glycero-3-[phospho-L-serine] (DOPS)-doped tubules.[15] Electron micrographs (EM) of the arrays were suitable for helical image analysis and 3D reconstruction.[15] The generality of this method may allow a wide variety of proteins to be crystallized on lipid nanotubes under physiological conditions.

Lipids

The glycolipid galactosylceramide (GalCer) can be purchased from Sigma Chemical Co. (St. Louis, MO). The nickel lipid DOGS-NTA-

[10] E. W. Kubalek, R. D. Kornberg, and S. A. Darst, *Ultramicroscopy* **35,** 295 (1991).
[11] A. J. Avila-Sakar and W. Chiu, *Biophys. J.* **70,** 57 (1996).
[12] D. G. Morgan, and D. DeRosier, *Ultramicroscopy* **46,** 263 (1992).
[13] M. Whittaker, B. O. Carragher, and R. A. Milligan, *Ultramicroscopy* **58,** 245 (1995).
[14] V. S. Kulkarni, W. H. Anderson, and R. E. Brown, *Biophys, J.* **69,** 1976 (1995).
[15] E. W. Kubalek, R. E. Brown, H. Celia, and R. A. Milligan, *Proc. Natl. Acad. Sci. U.S.A.* **95,** 8040 (1998).

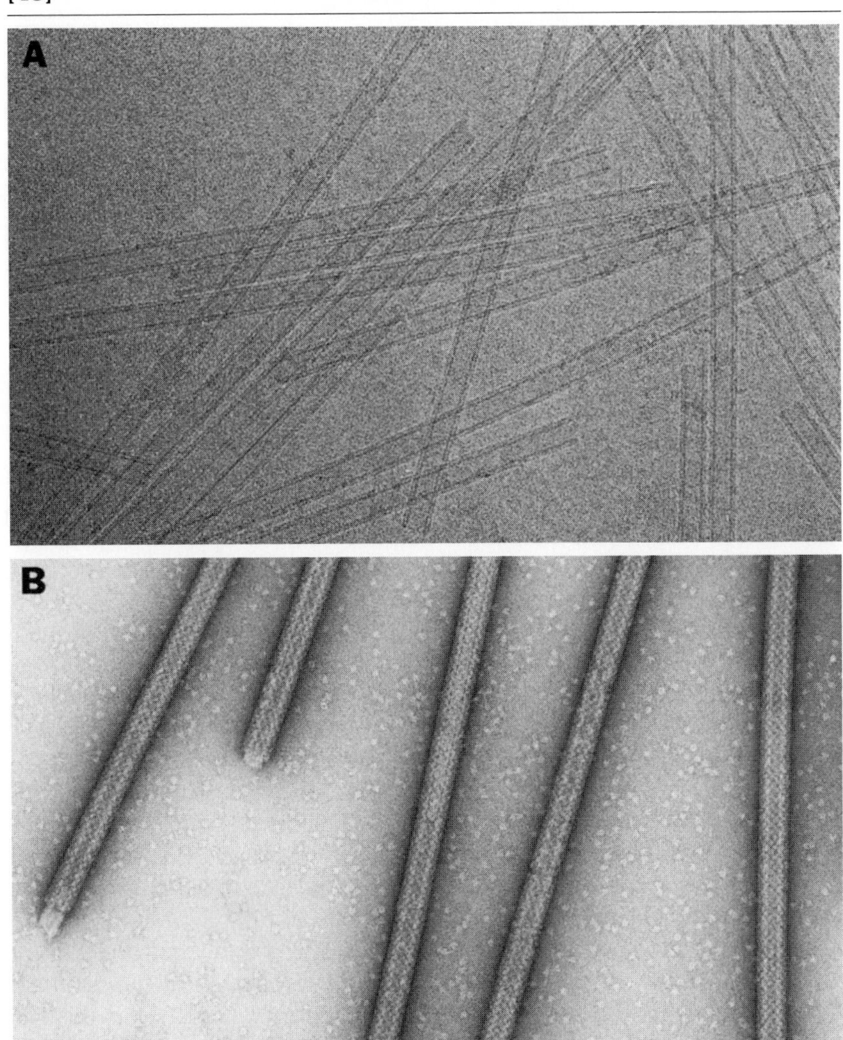

FIG. 1. (A) Unilamellar lipid tubules doped with nickel lipid and preserved in vitreous ice over a holey carbon film; 50% GalCer, 50% DOGS-NTA-Ni (w/w). Magnification: ×144,000. (B) Helical arrays of RNA polymerase on lipid tubules doped with DO-ethyl-PC (positive surface charge); 90% GalCer, 10% DO-ethyl-PC. A low background of unbound single protein molecules of RNA polymerase. Magnification: ×87,500.

Ni, 1,2-dioleoyl-*sn*-glycero-3-[(N(5-amino-1-carboxypentyl) iminodiacetic acid) succinyl] (nickel salt) that binds histidine-tagged proteins can be purchased from Avanti Polar lipids Inc. (Alabaster, AL). Charged lipids that have been used include DO-ethyl-PC, 1,2-dioleoyl-*sn*-glycero-3-ethylphosphocholine; DOPS, 1,2-dioleoyl-*sn*-glycero-3-[phospho-L-serine]; DOTAP, 1,2-dioleoyl-3-trimethylammonium-propane; and can also be purchased from Avanti. Biotinylated lipid, *N*-[(6-(biotinoyl)amino)hexanoyl] dipalmitoyl-L-α-phosphatidylethanolamine, can be obtained from Molecular Probes (Eugene, OR).

Procedure

Tubule Preparation

Resuspend GalCer with pure grade chloroform/methanol (50:50) and store in 100-μl aliquots at 10 mg/ml in a $-20°$ freezer. Use glass syringes with Teflon plugs when handling organic solvents or lipids in organic solvents. Derivatized lipids and charged lipids are usually stored in chloroform. Use clean glass micropipettes or syringes to mix lipids in the organic solvents. When lipid tubules are required, aliquots (5–50 μl) of GalCer in chloroform/methanol (50:50) at 1–10 mg/ml can be mixed with derivatized lipids, such as nickel lipid (DOGS-NTA-Ni), or charged lipids, such as DO-ethyl-PC, at 1–10 mg/ml, at ratios varying from 10 to 80% (w/w) and 10–50% (w/w), respectively. The lipids in the organic solutions should be mixed in small glass vials that can hold 1–2 ml and have Teflon screw caps. The lipid mixes in the organic solvents are then dried down under a gentle stream of either pure argon or nitrogen gas. A clear or an opaque residue forms on the bottom of the glass vial. Resuspend dried lipid mix in a buffer of choice to a final concentration of 0.5–2 mg/ml total lipid.

The lipid mix now in a buffer is sonicated in a water bath (ultrasonic cleaner, e.g., Branson 2510, Danbury, CT) for 1–6 min at room temperature. The sonicated mixture is usually slightly cloudy; if a visible pellet still remains on the glass further sonication may be necessary. The total yield of tubules can be variable, depending on the lipids used as dopants. In practical terms, when 5 μl of a tubule preparation is pipetted onto a 400-mesh EM grid and subsequently stained, hundreds of tubules should be visible on each grid square. Tubule length can vary considerably; however, tubule diameter should be more consistent, ~27 nm in most cases (Fig. 1A).

Buffers

It is best to prepare the tubules at concentrations between 0.5 and 2.0 mg/ml in a simple buffer such as 10 m*M* TrisCl, pH 7.0, or 10 m*M* HEPES, pH 7.0. It is possible to dilute the tubules into an appropriate multicomponent buffer for crystallization trials. Incorporating the nickel lipid, DOGS-NTA-Ni into the tubes is possible in a variety of buffers from pH 5.5 to 8.0. Buffer concentration in the range from 10 to 100 m*M* does not affect tube formation. Addition of 100 m*M* NaCl helps prevent tubules from aggregating. Sonication in buffers containing divalent cations such as Mg^{2+} and Ca^{2+} at 5 m*M* perturbs the formation of the tubules. However sonication in 1 *M* urea does not adversely affect tubule formation.

Crystallization Trials

Protein concentrations in the range from 50 to 500 μg/ml have been used successfully. In cases where the protein concentration is low, the use of small volumes of concentrated tubules is recommended to minimize dilution effects. In the upper ranges of protein concentration, successful preparations have been made by the mixing of equal volumes of protein sample and tubule solutions. All manipulations can be carried out in small volumes, typically 10–30 μl in microfuge tubes. Further, samples may be prepared under a wide range of temperatures. Helical arrays have been successfully prepared at 4° and 37°. Since the precise conditions for the preparation of satisfactory samples is largely dependent on the proteins and lipids involved, the search for these conditions is empirical. RNA polymerase, for example, formed helical arrays on tubules that were doped with positively charged DO-ethyl-PC, but did not form ordered arrays with tubules doped with the positively charged DOTAP.

It is important to frequently monitor the progress of one's attempts by negative staining. To stain these samples, first apply 5 μl of protein/tubule mixture to carbon-coated glow-discharged treated EM grids. In cases where the ionic strength or added salts of the buffer are high, a single 5-μl droplet of low ionic strength wash is recommended. Staining is achieved with 1–2% uranyl acetate.

When extensive ordering of the protein molecules is observed on many tubules, by negative staining methods, as in Fig. 1B, one can further study the specimen in vitreous ice over a holey carbon film. This technique helps to preserve the helical integrity of the specimen.

Acknowledgments

I am grateful to Seth Darst (Rockefeller) for providing the *Escherichia coli* RNA polymerase. The original work described here was supported in part by grants from the National Institutes of Health (AR39155, AR44278, and GM52468 to Ronald A. Milligan).

Section IV

Sphingolipid Transport and Trafficking

[44] Applications of BODIPY–Sphingolipid Analogs to Study Lipid Traffic and Metabolism in Cells

By RICHARD E. PAGANO, RIKIO WATANABE, CHRISTINE WHEATLEY, and MICHEL DOMINGUEZ

Introduction

Sphingolipids labeled with the fluorophore, boron dipyrromethene difluoride (BODIPY, Molecular Probes, Eugene, OR), represent a novel class of fluorescent lipid analogs for studying lipid metabolism and traffic in animal cells. The fluorescence emission of these analogs shifts from green to red wavelengths with increasing concentrations in membranes,[1] providing a means to observe and quantify the distribution of a particular analog (and its metabolites) within living cells by fluorescence microscopy. These analogs have been used to study ceramide accumulation at the Golgi complex and lipid transport along the secretory and endocytic pathways, and to screen for mutants in the secretory pathway by fluorescence activated cell sorting (FACS).[1–4] In this chapter we highlight two new uses of BODIPY–sphingolipid (SL) analogs to study lipid transport and metabolism in cells. In the first section, techniques for quantifying the concentration of BODIPY–SL analogs in various intracellular compartments of the living cell, based on the concentration-dependent spectral properties of these probes, are presented. In the second section, a novel screening method is described that employs BODIPY–ceramide to identify mutants in lipid metabolism.

Measurements of Concentration-Dependent Spectral Shift of BODIPY–Lipids in Living Cells

General Overview

The generalized structure of a BODIPY–SL is shown in Fig. 1A. A number of BODIPY–SL analogs are now available from Molecular Probes. These include analogs of ceramide (Cer), sphingomyelin (SM), glucosylceramide (GlcCer), lactosylceramide (LacCer), and GM1 ganglioside (GM1)

[1] R. E. Pagano, O. C. Martin, H. C. Kang, and R. P. Haugland, *J. Cell Biol.* **113,** 1267 (1991).
[2] O. C. Martin and R. E. Pagano, *J. Cell Biol.* **125,** 769 (1994).
[3] C.-S. Chen, O. C. Martin, and R. E. Pagano, *Biophys. J.* **72,** 37 (1997).
[4] N. T. Ktistakis, C.-Y. Kao, R.-H. Wang, and M. G. Roth, *Mol. Biol. Cell* **6,** 135 (1995).

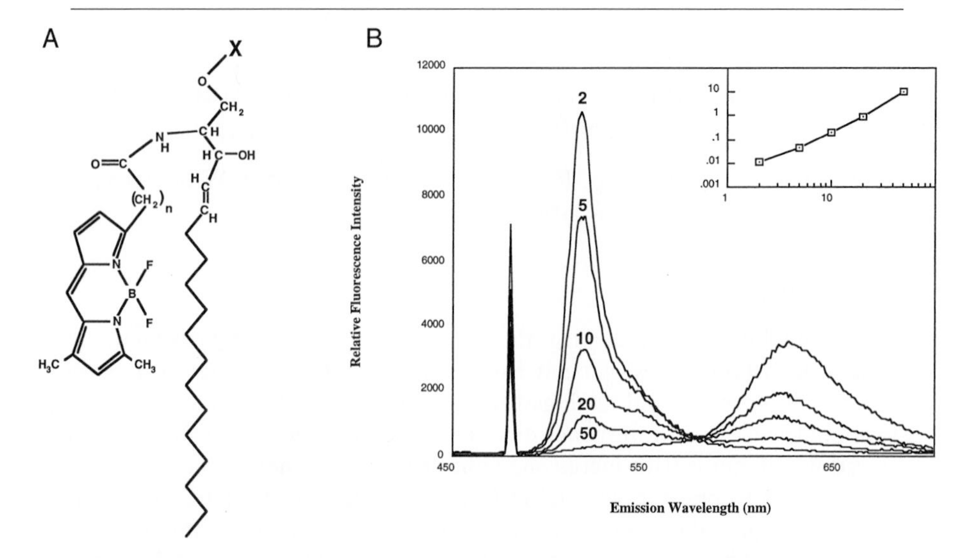

FIG. 1. Structure of BODIPY–SLs and their spectral properties in lipid vesicles. (A) General structure of a BODIPY–SL showing the sphingosine backbone, BODIPY-fatty acid, and polar headgroup, "X." For C_5-DMB-SL analogs, $n = 4$. (B) Fluorescence emission spectra of lipid vesicles formed from phosphatidycholine and 2, 5, 10, 20, or 50 mol % BODIPY-Cer (λ_{ex} = 480 nm). Note the decrease in green fluorescence at 515 nm and the increase in red fluorescence at 620 nm with increasing concentration of the BODIPY–lipid. *Inset:* Ratio of relative fluorescence intensities at 620 and 515 nm vs. mol % fluorescent Cer. Virtually identical spectra are obtained using other BODIPY-labeled sphingolipids (data not shown). (Adapted from Pagano *et al.*[1] and reproduced by permission from The Rockefeller University Press.)

containing a BODIPY–fatty acid. Although some BODIPY–SL analogs are available with different chain-length fatty acids or with various substitutions on the BODIPY moiety to modify its spectral properties, we find the five-carbon fatty acid, 5-(5,7-dimethyl-BODIPY)-1-pentanoic acid (C_5-DMB-), to be ideally suited for most applications. In addition to the commercially available C_5-DMB-SL analogs noted above, synthesis of others is readily achieved by *N*-acylation of the corresponding "lyso-SL" using the *N*-hydroxysuccinimide ester of C_5-DMB-fatty acid (Molecular Probes) as described.[5] Using this procedure, we have synthesized C_5-DMB-galactosylceramide (GalCer), -LacCer[2], and -sulfatide, as well as several stereoisomers of C_5-DMB-Cer and -SM (unpublished). The BODIPY–SL analogs are readily purified by preparative thin-layer chromatography (TLC) using chloroform/methanol/15 m*M* $CaCl_2$ (60 : 35 : 8, v/v/v) as the developing solvent. Concentrated stock solutions of the BODIPY–SL are prepared

[5] G. Schwarzmann and K. Sandhoff, *Methods Enzymol.* **138,** 319 (1987).

in chloroform–methanol (19:1, v/v) and are stored at −20°. Lipids are periodically checked for purity and repurified when necessary.

The rates of spontaneous transfer for some of the C_5-DMB-SLs in artificial lipid vesicles (liposomes) have been measured.[6] Although these rates are significantly slower than those seen for the corresponding analogs labeled with NBD (7-nitrobenz-2-oxa-1,3-diazol-4-yl)hexanoic acid,[7,8] they are sufficient to promote integration of the SL analogs into cell membranes during low-temperature incubations (see below). Furthermore, the BODIPY fluorophore is a particularly attractive alternative to NBD because it has a higher fluorescence yield and is more photostable than NBD.[9] In addition, the BODIPY fluorophore is less polar than NBD and therefore is probably better anchored in the membrane bilayer than NBD, which "loops back" to the membrane/water interface.[10,11]

Figure 1B shows the fluorescence emission spectra of a series of lipid vesicles comprised of phosphatidylcholine and containing various concentrations of C_5-DMB-Cer. At low mole fractions, a single fluorescence emission peak at ~515 nm is observed. However, with increasing mole fractions the fluorescence intensity at this wavelength decreases while a second peak of fluorescence appears at ~620 nm. The ratio of fluorescence intensities at 620 and 515 nm increases approximately linearly from 2 to 50 mol % C_5-DMB-Cer (Fig. 1, inset). Virtually identical results have been obtained using various molecular species of phosphatidylcholine (or crude lipid extracts) and using vesicles containing other C_5-DMB-SL analogs in place of C_5-DMB-Cer, demonstrating that the shift in fluorescence emission is not limited to the Cer analog. Additional control experiments have demonstrated that the spectral properties of C_5-DMB-lipids are not sensitive to pH, membrane potential, or membrane curvature.[3,12,13]

The concentration-dependent spectral properties are also readily observed in cell membranes. For example, when cultured fibroblasts are incubated with C_5-DMB-SM at 4°, washed, and observed under the fluorescence microscope (λ_{ex} = 450–490 nm; $\lambda_{em} \geq$ 520 nm), prominent labeling of the plasma membrane is observed. However, the *color* of the fluorescence emission varies from green to red/orange, as the concentration of C_5-DMB-

[6] J. Bai and R. E. Pagano, *Biochemistry* **36,** 8840 (1997).
[7] R. E. Pagano and R. G. Sleight, *Science* **229,** 1051 (1985).
[8] R. E. Pagano and O. C. Martin, *Biochemistry* **27,** 4439 (1988).
[9] I. D. Johnson, H. C. Kang, and R. P. Haugland, *Anal. Biochem.* **198,** 228 (1991).
[10] A. Chattopadhyay and E. London, *Biochemistry* **26,** 39 (1987).
[11] D. E. Wolf, A. P. Winiski, A. E. Ting, K. M. Bocian, and R. E. Pagano, *Biochemistry* **31,** 2865 (1992).
[12] J. Karolin, L. B.-A. Johansson, L. Strandberg, and T. Ny, *J. Am. Chem. Soc.* **116,** 7801 (1994).
[13] H. C. Kang and R. P. Haugland, *Proc. SPIE* **1063,** 68 (1989).

SM used during the initial incubation is increased.[3] To date, the concentration-dependent spectral properties of C_5-DMB-lipids have been used (1) to study accumulation of C_5-DMB-Cer at the Golgi apparatus, (2) to study subpopulations of endosomes formed during the initial stages of endocytosis, (3) to study lipid transport and accumulation in the lysosomes of mucolipidosis type IV cells, and (4) to screen for lipid accumulation in the lysosomes of certain lipid storage diseases.[1,3,14,15] These phenomena can be studied by quantitative fluorescence microscopy as outlined below.

Basic Methods

Preparation of BSA Complexes. For most studies, it is very convenient to use a bovine serum albumin (BSA) complex[16] of the BODIPY–SL as a vehicle for introduction of the fluorescent lipids into cells. These complexes are prepared as follows. A 500-nmol aliquot of BODIPY–SL is transferred to a glass tube and dried under nitrogen and then *in vacuo* for several hours. The fluorescent lipid is then dissolved in 400 μl of ethanol and, while vortex mixing, injected into a small volume of buffer (typically 0.5–1 ml) containing an equimolar amount of defatted BSA (DF-BSA). The resulting complex is dialyzed overnight against several changes of buffer. [For some applications we centrifuge the BODIPY–SL/DF-BSA complex in a tabletop ultracentrifuge (30 min at 100,000*g*) to remove any aggregates that may be present.] The final concentration of the BODIPY–SL/DF-BSA is determined by measuring the relative fluorescence intensity (λ_{ex} = 505 nm; λ_{em} = 515 nm) of an aliquot against a known standard and the complex is then stored at 4° or −20° under nitrogen.

Incubation of BODIPY–SL with Cells. In selecting a particular BODIPY–SL for an experiment, it is important to recognize that various C_5-DMB-SL analogs are transported differently within cells and thus different analogs are used for different applications: (1) C_5-DMB-Cer is useful for labeling the Golgi apparatus and for studying SL transport along the secretory pathway; (2) C_5-DMB-SM, -LacCer, -GalCer, -GM1, or -sulfatide label the plasma membrane of most cell types at low temperature, and on warming cells, serve as markers for endocytosis; and (3) C_5-DMB-GlcCer labels the plasma membrane of cells at low temperature, but is internalized by a combination of endocytic and nonendocytic mechanisms.[1–3,15]

In a typical experiment, we incubate cells with 1–2 μM BODIPY–SL/DF-BSA for 30 min at 4°. Incubations are generally performed in a serum-

[14] C. Chen, G. Bach, and R. E. Pagano, *Proc. Natl. Acad. Sci. U.S.A.* **95,** 6373 (1998).

[15] R. E. Pagano and C. Chen, *Ann. N.Y. Acad. Sci.* **845,** 152 (1998).

[16] R. E. Pagano and O. C. Martin, *in* "Cell Biology: A Laboratory Handbook" (J. E. Celis, ed.), Vol. 2, 2nd ed., p. 507. Academic Press, New York, 1998.

free medium. (In the presence of serum, the BODIPY–SL partitions into the large excess of serum lipoproteins, thereby reducing the effective concentration of BODIPY–SL available for transfer into cell membranes.) Following the low-temperature incubation, the cells are washed and warmed to 37° for various times, prior to observation under the fluorescence microscope or to lipid extraction and analysis.

Transport to Endosomes and Lysosomes. Immediately following incubation of various cell types with C_5-DMB-SM, -LacCer, -GalCer, -GM1, or -sulfatide at low temperature, prominent labeling of the plasma membrane is observed. To observe subsequent transport of these analogs into "very early" endosomes, cells are briefly (7 sec to 2 min) warmed to 37° to allow internalization to occur. Endocytosis is stopped by chilling the cells and adding a "cocktail" of inhibitors.[3] The specimens are then "back-exchanged" to remove any fluorescent lipid present at the plasma membrane so that the internalized lipid can be readily observed. For back-exchange of BODIPY–SLs, cells are typically incubated with 5% DF-BSA at 11° (6 changes; 10 min each) which results in removal of ≥95% of the BODIPY–SL analogs from the surface of most cell types. Transport to lysosomes is studied in a similar manner except for the following: (1) Cells are preincubated overnight with 1–2 mg/ml Cascade blue dextran (10 kDa, lysine fixable; Molecular Probes) in culture medium for ≥24 hr at 37°, washed, and further incubated ("chased") in complete medium for 2 hr at 37° to label the lysosomes.[17] (2) The cells are incubated with BODIPY–SL as described above and "chased" for relatively long periods of time (0.5–2 hr) at 37° to allow transport of the lipid probe to the lysosomes.

Quantitative Studies

For quantitative studies we use an Olympus IX70 fluorescence microscope equipped with filter packs that allow the specimens to be excited at 450–490 nm and viewed at "green" (520–560 nm), "green + red" ($\lambda_{em} \geq$ 520 nm), or "red" ($\lambda_{em} \geq$ 590 nm) wavelengths. For specimens that are also labeled with Cascade blue fluorescent dextran (see below), samples are excited at 365 nm and fluorescence is observed at ≥420 nm. A 12-bit, cooled CCD camera (ORCA100; Hamamatsu Photonics Systems, Bridgewater, NJ) is also attached to the microscope for image acquisition. In a typical experiment, several low-light-level images of a given cell are acquired at different emission wavelengths, along with corresponding background images of blank areas of the culture dish at these wavelengths. All images are corrected for background prior to further analysis. Quantitative

[17] M. Koval and R. E. Pagano, *J. Cell Biol.* **111,** 429 (1990).

image analysis is performed using the Metamorph image processing program (Universal Imaging, Media, PA).

Early Endosomes. When cells labeled with C_5-DMB-SM, -LacCer, -GalCer, -GM1, or -sulfatide are briefly chased at 37° as described above and observed under the fluorescence microscope in the "green + red" channel, a large number of variously colored endosomes are seen within single cells. These range in color from red/orange (often at the periphery of the cell, to green (more widely scattered through the cytoplasm). The ratio of red-to-green fluorescence (R/G) of individual endosomes is quantified as follows. Samples are excited with blue light and low-light-level images of a given cell are acquired at "green" and "red" wavelengths. (With some microscopes it is necessary to first align the green and red images; however, we can detect no movement of the images using our Olympus IX70 microscope when changing from one filter set to another.) A region of a single cell containing at least 100 endosomes is selected, and a threshold level is then determined such that all the endosomes are included. A binary mask, defined by that threshold value, is then used to isolate the endosomes within that region so that the intensities of the individual endosomes in the corresponding red and green images of the cell can be measured.

This procedure is repeated on images collected from five or six different cells in the same (or identically treated) culture dishes. The resulting data are then used to compute either an average R/G value or a histogram of R/G values. A typical result for 30-sec endosomes in human skin fibroblasts labeled with C_5-DMB-SM is shown in Fig. 2. The R/G values can also be

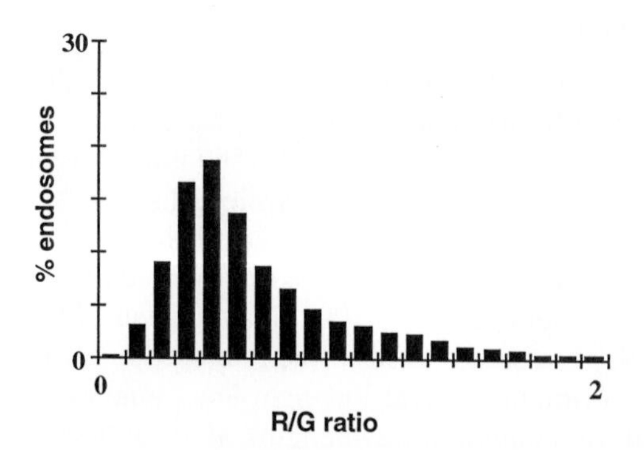

FIG. 2. Measurement of BODIPY–lipid transport into early endosomes. Human skin fibroblasts were incubated with C_5-DMB-SM at 4°, washed, and warmed to 37° for 30 sec. The cells were then back-exchanged and the R/G ratio was determined for ≥500 endosomes in five or six different cells by image analysis (see text). The R/G values were then plotted as a function of the percent of the total number of endosomes.

used to estimate the approximate molar density of the BODIPY–SL analog in the endosomes by comparing these values to an appropriate standard curve. This curve is generated by quantifying fluorescence images of a series of giant vesicles formed from cellular or synthetic lipids and various known amounts of a BODIPY–SL.[1,3]

Lysosomes. The procedure for quantifying the R/G ratio in lysosomes is somewhat more complicated than for endosomes because after long periods of chase the BODIPY–SL is present in both endosomes and lysosomes. In these studies, cells are preincubated overnight with blue dextran to prelabel the lysosome compartment.[17] Following pulse labeling with the fluorescent lipid, three low-light-level images of a given cell corresponding to blue dextran and the "green" and "red" components of BODIPY fluorescence, are then acquired. The punctate fluorescent structures in a blue dextran image (several hundred for a single human skin fibroblast) are used to create a binary mask corresponding to the lysosomes as described above (Fig. 3A). A second binary mask corresponding to the lipid labeled endosomes and lysosomes is created using the "green" image of the same cell (Fig. 3B). The overlapping region of these two masks (Fig. 3C) are then used to identify regions of the green and red images corresponding to the lysosomes, and the intensities (and areas) of these structures are then quantified. Intensity values are also corrected for autofluorescence of the lysosomes at green and red wavelengths measured in cells treated with blue dextran alone. The relative amount of BODIPY–lipid in the lysosomes of a given cell is then calculated from the ratio of red to green fluorescence in the lipid-labeled lysosomes times the area of those structures. We have

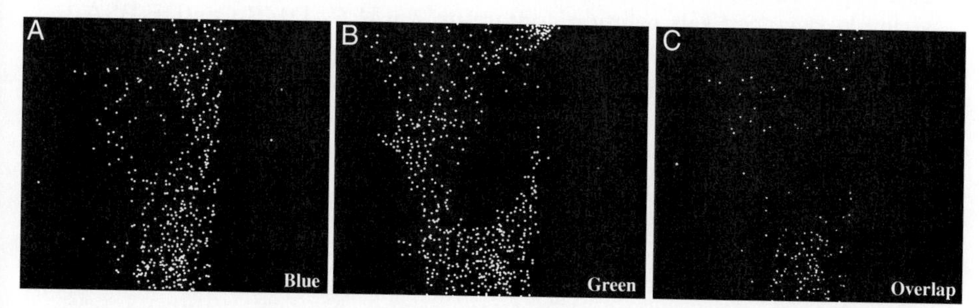

Fig. 3. Binary masks used to quantify BODIPY–lipid transport into lysosomes. Human skin fibroblasts were incubated overnight with Cascade blue fluorescent dextran to label the lysosomes, and subsequently pulse labeled with C_5-DMB-LacCer. Fluorescence images of the blue dextran and of the BODIPY–LacCer (at green wavelengths) were then used to construct binary images corresponding to (A) the lysosomes and (B) endosomes + lysosomes. The overlapping region of these two binary masks (C) was then used for quantifying lipid accumulation in the lysosomes. (See text for details.)

previously used this procedure to compare the rates of accumulation of C_5-DMB-LacCer in the lysosomes of normal human skin fibroblasts vs. mucolipidosis type IV cells and showed that there was a defect in the mutant cell type that affects membrane sorting and/or late steps of endocytosis.[14] This methodology is also being used in our laboratory to quantify the accumulation of BODIPY–SL analogs in various lipid storage diseases.[18]

Assay of Sphingomyelin Synthase Activity on Replica Colonies

In this section we present a novel method developed for screening for sphingomyelin synthase (SMS) activity using C_5-DMB-Cer and replica cell colonies. The method was originally developed to screen mutagenized Chinese hamster ovary (CHO) cells grown on polyester (PE) disks for clones in which SMS activity was greatly reduced or absent compared to wild type. (The reader is referred to previous publications for details on the replica plating of animal cells.[19,20]) However, this method can be easily adapted to replica colonies of bacteria or yeast on filter paper.

Step I. Incubation of Replica Colonies on PE Disks or Filter Paper with C_5-DMB-Cer

In a typical experiment, a PE disk (~7.8 cm in diameter) containing several hundred colonies of cells, ranging in size from 1 to 2 mm in diameter, is used. Pending identification of clone(s) of interest on the PE disk, the corresponding master plate is held at 33° in a CO_2 incubator while the PE disk is placed in a 100-mm-diameter tissue culture dish and covered with a simple balanced salt solution containing 2 μM C_5-DMB-Cer/DF-BSA. (We typically use a HEPES-buffered minimal essential media for this purpose.) During this incubation period, a portion of the C_5-DMB-Cer will be metabolized to C_5-DMB-SM, however little or no metabolism to C_5-DMB-GlcCer will take place because C_5-DMB-Cer is a poor substrate for glucosylceramide synthase.[1] Thus, following incubation of the PE disks with C_5-DMB-Cer, each of the wild-type cell colonies should contain both C_5-DMB-Cer and -SM, while the mutant clone of interest should only contain C_5-DMB-Cer. In addition, there should be a "high background" of C_5-

[18] V. Puri, R. Watanabe, M. Dominguez, X. Sun, C. L. Wheatley, D. L. Marks, and R. E. Pagano, *Nature Cell Biol.* **1,** 386 (1999).

[19] C. R. H. Raetz, M. M. Wermuth, T. M. McIntyre, J. D. Esko, and D. C. Wing, *Proc. Natl. Acad. Sci. U.S.A.* **79,** 3223 (1982).

[20] J. D. Esko, *Methods Cell Biol.* **32,** 387 (1989).

DMB-Cer on the remaining areas of the disks, which contain no cell colonies (Fig. 4A).

Step II. Transfer of C_5-DMB-Lipids from PE Disk to TLC Plate

In this step, the fluorescent lipids are extracted from the PE disk and transferred to a TLC plate in a single step. This procedure is carried out by placing the PE disk on a TLC plate, overlaying it with three pieces of Whatman (Clifton, NJ) filter paper saturated with chloroform–methanol (2/1, v/v), and applying strong pressure for several minutes by means of a glass plate placed on top of the filter papers (Fig. 4B). The solvent composition for this transfer was chosen by trial and error to maximize extraction of BODIPY fluorescence from the colonies without causing migration of the fluorescent SM analog once it is transferred from the colony on the PE disk to the corresponding position on the TLC plate. Observation of the TLC plate under UV light immediately after this step revealed a bright uniform disk of fluorescence corresponding in size and shape to the original PE disk (not shown). Presumably this fluorescence was due to the large excess of C_5-DMB-Cer that was present in all areas of the PE disk.

Step III. Separation of C_5-DMB-Cer and -SM by TLC

Following transfer of the fluorescent lipids, the TLC plate was developed in chloroform–methanol (85 : 15, v/v). [This developing solvent was chosen by observing the migration of TLC standards of C_5-DMB-SM and -Cer in various solvent systems. The indicated solvent mixture results in migration of all the fluorescent ceramide toward the solvent front, while no movement of C_5-DMB-SM is detected (Fig. 4C).] A typical result is shown in Fig. 5. Comparison of the array of fluorescent "dots" with a Coomassie Blue stained PE disk showed perfect coincidence of the blue colonies on the PE disk with the fluorescent spots on the TLC plate (not shown). To verify that the fluorescence present in the "dot-like" pattern was due to C_5-DMB-SM, the silica gel corresponding to the "dot-like" region was scraped from the TLC plate and the fluorescent lipids were extracted with chloroform–methanol. TLC analysis of this extract confirmed that only C_5-DMB-SM was present.

We attempted to use this procedure to screen mutagenized CHO-K1 cells for temperature-sensitive mutants in SM synthase activity but no SMS-deficient clones were detected. This method is currently being used to screen for SM synthase using bacterial and yeast colonies transformed with mammalian cDNA libraries. For these cell types, replicas are made on Whatman paper 50 and the replica filters are incubated with 2–4 μM C_5-DMB-Cer. Transfer to TLC plates and separation of C_5-DMB-lipids by

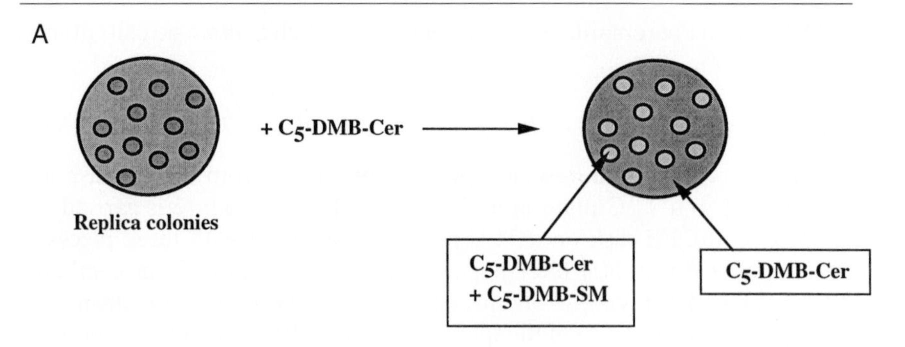
A
+ C_5-DMB-Cer
Replica colonies
C_5-DMB-Cer
+ C_5-DMB-SM
C_5-DMB-Cer

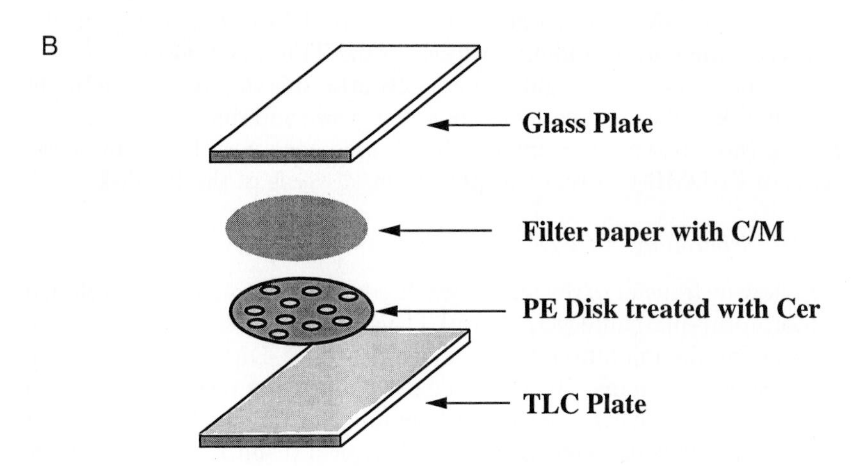
B
Glass Plate
Filter paper with C/M
PE Disk treated with Cer
TLC Plate

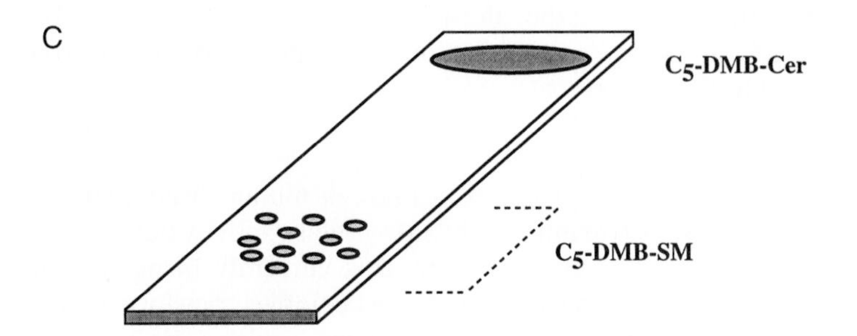
C
C_5-DMB-Cer
C_5-DMB-SM

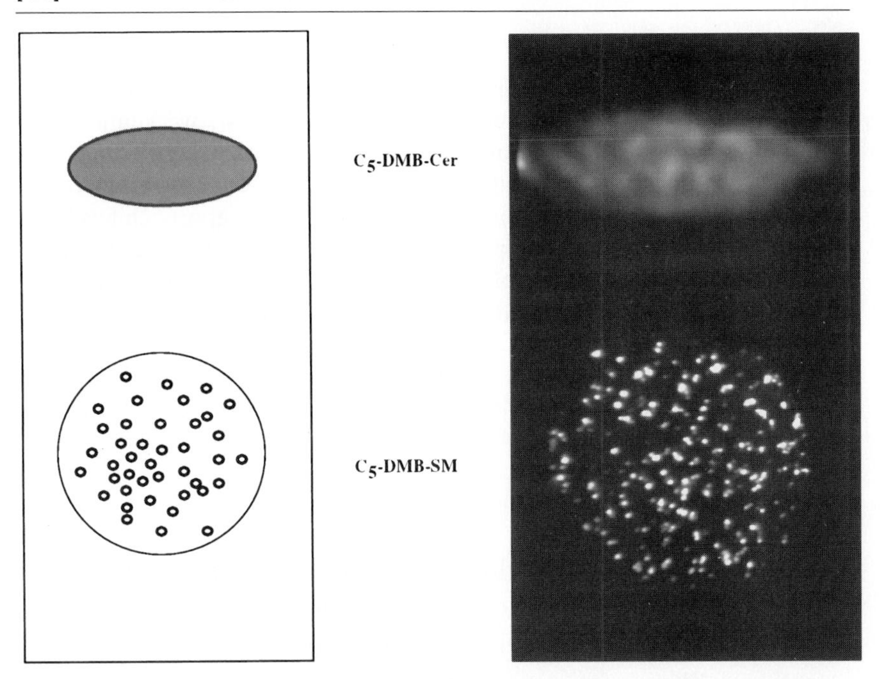

FIG. 5. Photograph of a TLC plate showing distribution of C_5-DMB-Cer and -SM from an experiment using CHO-K1 fibroblasts grown on PE disks. Accompanying diagram shows a schematic representation of the TLC plate after development.

TLC is performed as described above for PE disks. In the case of yeast, we found that C_5-DMB-Cer is metabolized to fluorescent inositol-P-ceramide (IPC) (unpublished observation), raising the possibility that this method may also be useful in finding mutants in the yeast *Saccharomyces cerevisiae* sphingolipid biosynthetic pathway.

FIG. 4. Use of C_5-DMB-Cer to screen for mutants in lipid metabolism. (A) Replica cell colonies grown on polyester (PE) disks are incubated with C_5-DMB-Cer. (B) Following this incubation, the fluorescent lipids present on the PE disks are transferred to a 10 × 20-cm TLC plate by overlaying the PE disk with filter paper saturated with chloroform–methanol (2:1, v/v). This results in efficient transfer of the C_5-DMB-lipids to the TLC plate. (C) The TLC plate is then chromatographed in chloroform–methanol (85:15, v/v). In this solvent system, the excess C_5-DMB-Cer migrates toward the solvent front, while the C_5-DMB-SM associated with the individual cell colonies remains in a position corresponding to the cell colony from which it was transfered. (See text for details.)

Conclusions

In this chapter we have highlighted two uses of BODIPY–SL analogs to study lipid transport and metabolism in cells. First, we summarized methodology, which takes advantage of the concentration-dependent spectral shift of BODIPY–SL analogs to estimate the concentration of an analog (and its metabolites) in living cells. To date, this approach has been particularly useful in studying lipid transport along the endosome/lysosome pathway. However, future applications may allow investigators to study "lipid sorting" of SL analogs along the secretory pathway following their *de novo* synthesis from C_5-DMB-Cer. Second, we presented a novel method exploiting some of the properties of C_5-DMB-Cer to screen for mutants in SL metabolism. This methodology may be useful in both eukaryotic and prokaryotic systems.

[45] Using Biotinylated Gangliosides to Study Their Distribution and Traffic in Cells by Immunoelectron Microscopy

By GÜNTER SCHWARZMANN, ALEXANDER VON COBURG, and WIEBKE MÖBIUS

Introduction

Gangliosides and other sphingolipids are predominantly located in the outer leaflet of the plasma membrane of vertebrate cells, where they are assumed to serve regulatory functions in cell growth and differentiation, and to have a role in cell adhesion and recognition as well as signal transduction.[1–3] In addition, available evidence suggests that gangliosides are components of lipid microdomains at the cell surface that hosts many glycosylphosphatidylinositol (GPI)-anchored proteins and may function as a platform for signal transduction.[4,5] Furthermore, these microdomains or lipid rafts may be important for sorting of membrane components in endocytosis. Transport of gangliosides, and sphingolipids in general, between different

[1] S. Hakomori, *J. Biol. Chem.* **265,** 18713 (1990).

[2] K. Iwabuchi, S. Yamamura, A. Prinetti, K. Handa, and S. Hakomori, *J. Biol. Chem.* **273,** 9130 (1998).

[3] I. Flores, A. C. Martinez, Y. A. Hannun, and I. Merida, *J. Immunol.* **160,** 3528 (1998).

[4] N. M. Hooper, *Curr. Biol.* **8,** R114 (1998).

[5] T. Harder and K. Simons, *Eur. J. Immunol.* **29,** 556 (1999).

0076-6879/00 $30.00

organelles in the biosynthetic as well as endocytic pathway is believed to be mediated primarily by a vesicular mechanism.[6]

Morphologic studies on the intracellular distribution and trafficking of membrane lipids have made use of fluorescent lipid analogs[7–10] and were confined to the limited resolution of light microscopy. For higher resolution electron microscope (EM) studies utilizing cholera toxin as a probe for the localization of ganglioside GM1[11,12] or antibodies against Forssman glycolipid[13] have been performed. These studies offer important insights into the distribution of these endogenous glycosphingolipids. For transport studies, however, labeled ganglioside derivatives are needed that can be discriminated from the endogenous lipids. Because it is possible to incorporate exogenous gangliosides into the plasma membrane of cultured cells,[14–18] suitably labeled ganglioside analogs seem to be advantageous for the study of endocytic membrane flow.

The first part of this article describes the synthesis of biotin-tagged GM1 (biotin-GM1) derivatives[19] with different acyl chain length that prove effective in the study of their distribution in the plasma membrane and intracellular transport along the endocytic pathway by electron microscopy using gold-conjugated anti-biotin antibodies.[20]

The following section outlines the procedures involved in the incorporation of biotin-GM1 in cultured tissue and its immunoelectron microscopic

[6] G. Schwarzmann and K. Sandhoff, *Biochemistry* **29,** 10865 (1990).

[7] G. van Meer, *Annu. Rev. Cell Biol.* **5,** 247 (1989).

[8] R. E. Pagano, *Curr. Opin. Cell Biol.* **2,** 652 (1990).

[9] U. Pütz and G. Schwarzmann, *Eur. J. Cell Biol.* **68,** 113 (1995).

[10] A. Sofer, G. Schwarzmann, and A. H. Futerman, *J. Cell Sci.* **109,** 2111 (1996).

[11] R. G. Parton, *J. Histochem. Cytochem.* **42,** 155 (1994).

[12] H. A. Hansson, J. Holmgren, and L. Svennerholm *Proc. Natl. Acad. Sci. U.S.A.* **74,** 3782 (1977).

[13] I. L. van Genderen, G. van Meer, J. W. Slot, H. J. Geuze, and W. F. Voorhout, *J. Cell Biol.* **115,** 1009 (1991).

[14] R. Callies, G. Schwarzmann, K. Radsak, R. Siegert, and H. Wiegandt, *Eur. J. Biochem.* **80,** 425 (1977).

[15] K. Radsak, G. Schwarzmann, and H. Wiegandt, *Hoppe-Seyler's Z. Physiol. Chem.* **363,** 263 (1982).

[16] G. Schwarzmann, P. Hoffmann-Bleihauer, J. Schubert, K. Sandhoff, and D. Marsh, *Biochemistry* **22,** 5041 (1983).

[17] G. Schwarzmann, D. Marsh, V. Herzog, and K. Sandhoff, *in* "Gangliosides and Modulation of Neuronal Function" (H. Rahmann, ed.), Vol. H7, p. 217. NATO ASI Series, Berlin, Heidelberg, Springer, 1987.

[18] H. E. Saqr, D. K. Pearl, and A. J. Yates, *J. Neurochem.* **61,** 395 (1991).

[19] B. Albrecht, G. Pohlentz, K. Sandhoff, and G. Schwarzmann, *Chem. Phys. Lipids* **86,** 37 (1997).

[20] W. Möbius, V. Herzog, K. Sandhoff, and G. Schwarzmann, *J. Histochem. Cytochem.* **47,** 1005 (1999).

localization. Special emphasis is given to the description of embedding techniques and the preparation of cryosections.

Materials and General Methods

Materials

N,N'-Dicyclohexylcarbodiimide and *N*-hydroxysuccinimide are obtained from Fluka (Buchs, Switzerland). LiChroprep RP-18, silica gel Si 60 (15–40 μm), prepacked silica gel chromatography columns Lobar, thin-layer chromatography (TLC) plates (silica gel Si 60) and *p*-dimethylamino-cinnamaldehyde, hydroquinone, osmium tetroxide, proteinase K (EC 3.4.21.14), silver lactate, tungstophosphoric acid, and uranyl acetate are from E. Merck (Darmstadt, Germany). Glutaraldehyde (EM grade, 25%), methylcellulose, polyvinylpyrrolidone (PVP 10,000), and defatted bovine serum albumin (BSA) are purchased from Sigma-Aldrich Chemie (Steinheim, Germany). [1-^{14}C]Octadecanoic acid (2146 MBq/mmol) is bought from Amersham Life Science (Braunschweig, Germany). Sodium [1-^{14}C]octanoate (2109 MBq/mmol), neutral, acidic, and basic alumina W 200 are from ICN Biomedicals (Eschwege, Germany). D-Biotinoyl-ε-amino-caproic acid *N*-succinimidyl ester, GM1-β-galactosidase (EC 3.2.1.23), and *Vibrio cholerae* sialidase (EC 3.2.1.18) are purchased from Boehringer Mannheim (Mannheim, Germany). *N*-2-Hydroxyethylpiperazine-*N'*-2-ethansulfonic acid (HEPES) is from Calbiochem (La Jolla, CA). Trypsin (EC 3.4.21.4) and Dulbecco's modified Eagle's medium (DMEM) is obtained from Gibco (Eggenstein, Germany). Fetal calf serum (FCS) is from Cytogen (Lohmar, Germany). VISKING dialysis tubing type 20/32 is obtained from Serva (Heidelberg, Germany). Anti-biotin antibody gold conjugates (10 nm and ultra-small gold particles) and gum arabic are obtained from Aurion (Wageningen, The Netherlands). Poly/Bed 812, nadic methyl anhydride, dodecenylsuccinic anhydride and 2,4,6-tri(dimethylamino-methyl)phenol are purchased from Polyscience (Warrington, PA). LR Gold and benzil are obtained from Agar Scientific (Essex, UK). BEEM capsules and EM grids are from W. Plannet GmbH (Wetzlar, Germany). β-Hexosaminidases A and B (EC 3.2.1.52) are prepared from human placenta.[21] All solvents are analytical grade and are obtained from either E. Merck (Darmstadt, Germany) or Riedel-de Haën (Seelze, Germany). All other chemicals are of the highest purity available.

[21] B. Liessem, G. J. Glombitza, F. Knoll, J. Lehmann, J. Kellermann, F. Lottspeich, and K. Sandhoff, *J. Biol. Chem.* **270,** 23693 (1995).

Methods

Solvents. Some solvents are prepared dry and/or argon-saturated prior to use. Anhydrous ethyl acetate is obtained by refining the analytical grade solvent over neutral alumina. *N,N*-Dimethylformamide (DMF) is purified and freed from basic decomposition products by passing, under argon, over acidic alumina.

Thin-Layer Chromatography. Column chromatographic elution profiles and the progress of reactions are routinely followed by TLC in tanks (Camag, Muttenz, Switzerland) with vapor saturation. R_f values are determined from 20-cm plates that are developed for 16–18 cm. Detection on TLC plates of compounds is accomplished by spraying the plates with a mixture of acetic acid/sulfuric acid/anisaldehyde (500:10:2, v/v/v),[22] followed by heating for 10 min at 120° or, alternatively, by dipping the plates into a solution of ceric ammonium nitrate $(NH_4)_2Ce(NO_2)_6$ in 20% sulfuric acid followed by heating with a heat gun. This procedure yields blue spots on a slightly yellowish background after cooling. Reaction intermediates with a free amino group are also detected by spraying with 1% ninhydrin in *n*-butanol prior to heating for 20 min at 120°. For the recognition of products containing a biotin residue, plates are sprayed with a 1:1 mixture of ethanolic sulfuric acid (1.5 ml sulfuric acid/100 ml ethanol) and ethanolic *p*-dimethylaminocinnamaldehyde (200 mg/100 ml ethanol). Biotin residues yield red colored spots at room temperature.[23] Radioactive products are localized and quantified by the use of a Fuji BAS 1000 Bio Imaging analyzer (Raytest, Pforzheim, Germany) or by radioscanning using an automatic TLC linear analyzer (Tracemaster 40, Berthold, Wildbad, Germany). Radioactive bands on TLC plates are visualized also by exposure to X-ray sensitive film (Kodak Biomax, Kodak, Rochester, NY).

Quantification of Radioactivity. Radioactivity is determined by a liquid scintillation counter Tri-Carb 1900 (Canberra Packard, Frankfurt, Germany), using Ultima gold (Canberra Packard) as scintillation cocktail.

Removal of Salts and Other Polar Compounds from Reaction Products. Sphingosine-containing reaction products are freed of salts and other polar and water-soluble materials of low molecular weight by short dialysis in the cold or by reversed-phase chromatography similar to the procedure described.[24] Briefly, Pasteur pipettes containing a small cotton plug are packed with a 1:1 slurry of LiChroprep RP-18 in methanol and washed successively with methanol, chloroform/methanol (1:1, v/v), methanol, and water prior to use. Lipophilic reaction products (less than 2 mg) are ad-

[22] E. Stahl and U. Kaltenbach, *J. Chromatogr.* **5,** 351 (1961).
[23] D. B. McCormick and J. A. Roth, *Anal. Biochem.* **34,** 226 (1970).
[24] M. Williams and R. McCluer, *J. Neurochem.* **35,** 266 (1980).

sorbed onto these minicolumns from a solution in methanol/water (1:3 to 2:3, v/v) and then eluted with methanol after the nonadsorbed salts and other polar materials are washed out. Sep-Pak cartridges (Waters Division of Millipore Corporation, Milford, MA) can be used instead of self-prepared minicolumns. If more material (up to 100 mg) has to be desalted, LiChroprep RP-18 can be filled in bigger columns; alternatively, Lobar Fertigsäulen LiChroprep RP-18 size A can be used.

Column Chromatography. Separation of reaction products and purification of the target compounds are achieved by medium-pressure column chromatography using heavy-walled glass columns (with adjustable column adapters) with up to 100-cm lengths and having various inner diameters (0.6–1.2 cm) (Latek GmbH, Eppelheim, Germany) and a high-pressure pump (Latek P 400) or any other suitable columns and pumps. The flow rate of solvents is usually equivalent to 2 bed volumes per hour. Fractions of 2–8 ml, depending on the column size, are collected. For chromatography silica gel Si 60 in glass columns or Lobar Fertigsäulen of size A and B with LiChroprep Si 60 are used. The latter are more expensive than self-packed columns but have a more reproducible performance.

High-Performance Liquid Chromatography (HPLC). For final purification biotin-GM1 derivatives are subjected to HPLC on columns (1×25 cm) containing reversed-phase material (ProSep C_{18}, with 5-μm mean particle diameter; Latek, Eppelheim, Germany).

Fast Atom Bombardment Mass Spectra (FABMS). FABMS spectra are recorded in the positive-ion mode on a ZAB HF instrument (VG Analytical, Manchester, UK).[25] Data shown are nominal masses.

Matrix-Assisted Laser Desorption Ionization Time-of-Flight Mass Spectra (MALDI-TOF-MS). MALDI-TOF-MS is performed in the positive-ion mode with a Micromass TofSpec E (Micromass UK Limited, Manchester, UK).

500-MHz ^1H Nuclear Magnetic Resonance (NMR) Spectroscopy. A Bruker AMX 500 NMR spectrometer (Bruker, Karlsruhe, Germany) is used for ^{1}H NMR measurements. Chemical shifts (δ) are indicated in ppm relative to internal tetramethylsilane.

Ganglioside GM1. Gangliosides are extracted from bovine brain with mixtures of chloroform/methanol/0.1 *M* potassium chloride (8:4:3, v/v/v) and partitioned as described.[26] Prior to purification of monosialoganglioside GM1, the isolated gangliosides are treated with sialidase as described[27] to convert oligosialogangliosides GD1a, GD1b, GT1b, and GQ1b into GM1.

[25] H. Egge and J. Peter-Katalinic, *Mass Spectrom. Rev.* **6,** 331 (1987).
[26] J. Folch, M. B. Lees, and G. H. Sloan Stanley, *J. Biol. Chem.* **226,** 497 (1957).
[27] R. Kuhn, H. Wiegandt, and H. Egge, *Angew. Chem.* **73,** 580 (1961).

During this hydrolysis, all sialyl residues are split but for the one that is linked to position 3 of the inner galactose moiety.[27] Final purification of GM1 is achieved by medium-pressure chromatography on columns (1.2 or 2.5 × 100 cm or Lobar Fertigsäulen size B) of LiChroprep Si 60 using mixtures of chloroform, methanol, and water of increasing polarity or linear gradients of 2-propanol/*n*-hexane/water (55 : 45 : 5 to 55 : 25 : 15, v/v/v). Alternatively, GM1 can be obtained from Sigma or Matreya or Calbiochem.

All preparations of GM1 from bovine brain contain both C_{18}- and C_{20}-sphingosine-containing GM1 species in almost similar proportions.

N-Succinimidyl Ester of Labeled Fatty Acids. For the preparation of radiocarbon labeled biotin-GM1 the specific radioactivity of the commercially available fatty acids is reduced to 185 MBq/mmol by diluting with the appropriate amount of unlabeled fatty acid. Fatty acid *N*-succinimidyl ester is then made by the method of Lapidot *et al.* with modifications as described.[28,29] Typically, the labeled fatty acids (53 μmol) are dissolved in dry ethyl acetate (2 ml) and a slight molar excess of solid powdered *N*-hydroxysuccinimide (6.9 mg, 60 μmol) is added. An equimolar amount of solid *N,N'*-dicyclohexylcarbodiimide (11 mg, 53 μmol) is then added and the mixture is stirred under argon for 2 days at 30°.

N,N'-Dicyclohexylurea formed during the reaction is removed by centrifugation for 5 min at 2000*g* at 20°, washed once with dry ethyl acetate, and the clear supernatants are combined and stored under argon at −20° until use. In case of sodium [1-^{14}C]octanoate, the free acid is generated by dissolving the salt (8 μmol, 1.33 mg, 17.2 MBq) in 10 m*M* hydrogen chloride (1 ml). The free [1-^{14}C]octanoic acid is then extracted twice with 2 ml *n*-hexane. The organic phase is passed through a small column containing powdered anhydrous sodium sulfate (300 mg) to remove traces of water. Hexane is then evaporated in a gentle stream of nitrogen with no heating under a well-vented hood. Note that under this condition less than 0.1% of radioactivity is lost, if at all.[30] The remaining oily residue is immediately dissolved in dry ethyl acetate containing the appropriate amount of unlabeled octanoic acid for the conversion of [1-^{14}C]octanoic acid into its *N*-succinimidoyl derivative (185 MBq/mmol). The yield of active *N*-succinimidoyl derivatives depends on traces of water in the reagents and may vary from 70 to 90%.

Alternatively, sodium [1-^{14}C]octanoate can be dried azeotropically from its commercially available ethanolic solution in a nitrogen jet following the addition of a small volume of benzene and mixed with the respective amount

[28] Y. Lapidot, S. Rappaport, and Y. Wolman, *J. Lipid Res.* **8,** 142 (1967).
[29] B. Albrecht, U. Pütz, and G. Schwarzmann, *Carbohydr. Res.* **276,** 289 (1995).
[30] G. Schwarzmann, *Methods Enzymol.* **311,** 601 (2000).

of unlabeled free octanoic acid prior to being taken up in ethyl acetate. In this case the amount of the acid, corresponding to the salt form, may not be converted into the *N*-succinimidoyl derivative. This more simple procedure is applicable when the specific radioactivity is reduced to 15% or less.

Cell Culture. Monolayer cultures of human skin fibroblasts obtained from biopsy of a male infant were cultured in DMEM containing 10% fetal calf serum (FCS). The cells were grown at 37° in a water-saturated atmosphere of 5% CO_2 in air. For morphologic studies, cultures were seeded with 7×10^5 cells in 25-cm^2 tissue culture flasks. The cultures were then grown for 1–2 days prior to use.

Biotinylated Ganglioside GM1

Principle

This section describes the synthesis of two biotin-tagged derivatives of GM1 as outlined in Fig. 1. The intermediate deacylated ganglioside GM1 (monodeacetyllyso-GM1) containing two amino groups is obtained by vigorous alkaline hydrolysis of ganglioside GM1 with potassium hydroxide in methanol at 100°. By this procedure the fatty acyl and the acetyl group of the sialic acid residue are completely removed with very little hydrolysis of the acetamido group of the *N*-acetylgalactosaminyl moiety of GM1.[31] The amino group of the sphingosine residue is more basic than those of the neuraminic acid moiety. Thus, selective acylation of the free amino group of the sphingosine moiety of deacylated GM1 can be achieved with the *N*-succinimidyl ester of either [1-^{14}C]octadecanoic acid or [1-^{14}C]octanoic acid. Thereafter, the free amino group of the neuraminyl group of the resulting monodeacetylated GM1 derivatives is acylated with an excess of D-biotinoyl-ε-aminocaproic acid *N*-succinimidyl ester yielding the desired biotinylated gangliosides.

Procedures

Deacylated GM1. In a heavy-walled Teflon-lined screw-capped flask (Pyrex) ganglioside GM1 (78 mg, 50 μmol) is suspended in freshly prepared 0.8 *M* potassium hydroxide in dry methanol (15 ml) that had been saturated with argon. After flushing with argon, the flask is tightly sealed and the suspension is stirred at 100°. After a few minutes, the ganglioside is completely dissolved and the solution turns yellow during the course of hydroly-

[31] G. Schwarzmann and K. Sandhoff, *Methods Enzymol.* **138,** 319 (1987).

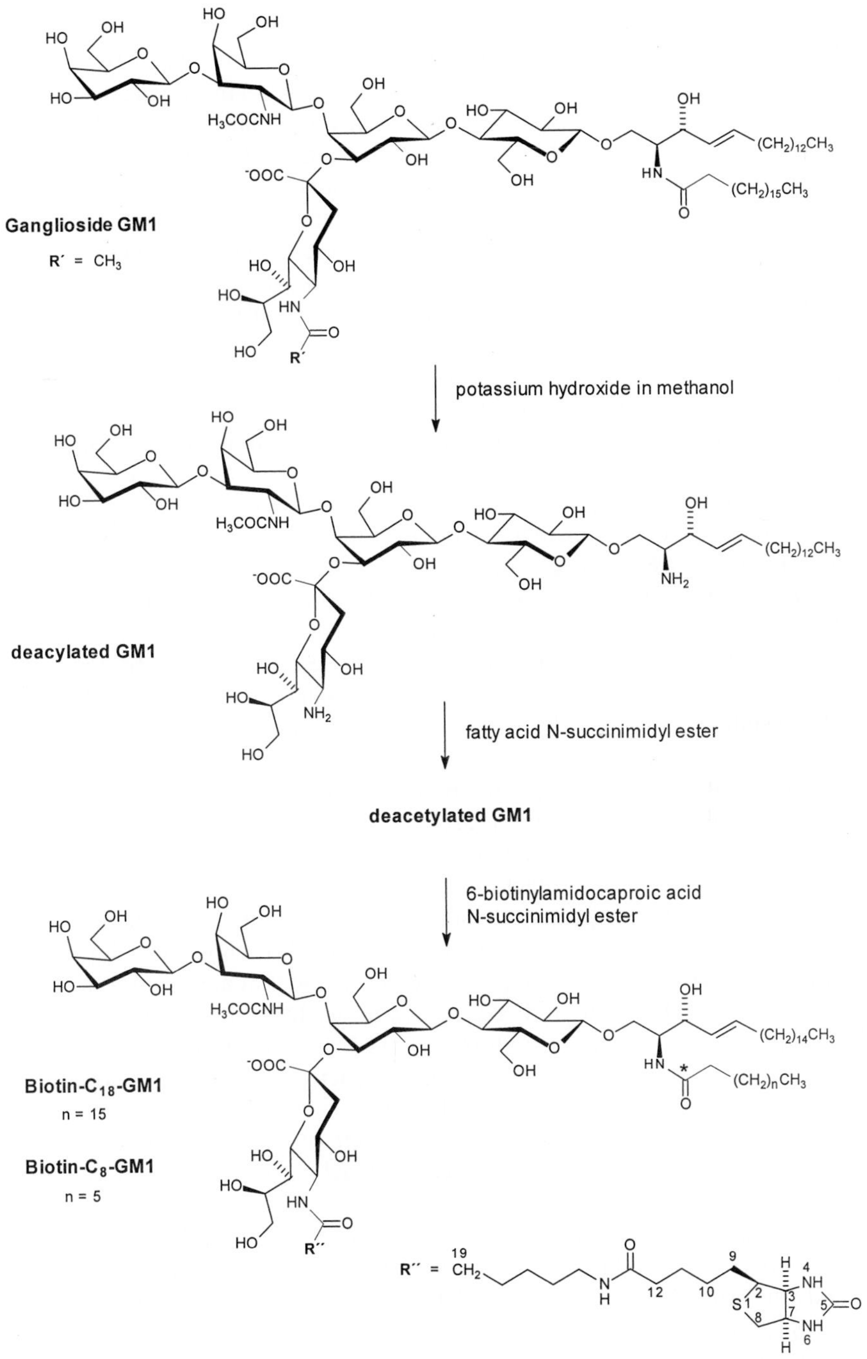

Ganglioside GM1

R′ = CH_3

↓ potassium hydroxide in methanol

deacylated GM1

↓ fatty acid N-succinimidyl ester

deacetylated GM1

↓ 6-biotinylamidocaproic acid N-succinimidyl ester

Biotin-C_{18}-GM1

n = 15

Biotin-C_8-GM1

n = 5

FIG. 1. Synthetic scheme for the preparation of biotin-GM1. Asterisk denotes position of radiocarbon.

sis. After 20 hr the solution is cooled to 20° and neutralized by careful addition of acetic acid. Following removal of methanol in a nitrogen stream, the wet residue is suspended in water (3 ml) and dialyzed at 10° for 6 hr against distilled water (10 liters), with two changes. During dialysis, some material sediments in the dialysis bag and is isolated by centrifugation. The pellet as well as the clear supernatant containing the deacylated GM1 and fatty acids in, however, different proportions are lyophilized separately. To facilitate dissolution of the raw products, the freeze-dried pellet is treated with methanol (0.8 ml) followed by chloroform (0.6 ml), whereas the freeze-dried supernatant is first treated with 2.5 *M* ammonia (0.18 ml) followed by chloroform (0.6 ml) under vigorous stirring. Both mixtures are then combined under vigorous stirring to yield an almost clear solution.

The solution is subjected to column chromatography on LiChroprep Si 60 (1.2 × 100 cm) using chloroform/methanol/2.5 *M* ammonia (60:40:9, v/v/v) as the mobile phase. The elution profile is monitored by TLC in the same solvent system. All fractions containing a ninhydrin- as well as anisaldehyde-positive single major spot of the desired product (R_f 0.07) are pooled and lyophilized from water following removal of the solvents. The yield of deacylated GM1 is in the range of 44–47 mg (35–37.5 μmol, 65–75%). Minor reaction products are totally deacylated GM1 (dideacetyl-lyso-GM1, roughly 5%, R_f 0.03) and GM1 lacking merely the acetyl group of the sialic acid residue (monodeacetyl-GM1, about 3%, R_f 0.16).

Deacetylated GM1. A small amount of deacylated GM1 (12.5 mg, 10 μmol) dissolved in dimethylformamide (DMF, 0.3 ml) is, after the addition of *N,N*-diisopropylethylamine (0.02 ml), dried in a stream of nitrogen to remove any volatile amino and imino compounds that may otherwise interfere with the *N*-acylation reaction. The residue is then redissolved in DMF (0.2 ml) and *N,N*-diisopropylethylamine (0.01 ml) and treated at 30° for 48 hr under argon with an equimolar amount of the *N*-succinimidoyl derivative of [1-^{14}C]octadecanoic acid or [1-^{14}C]octanoic acid that are dried from their solution in ethyl acetate and redissolved in dry DMF (0.15 ml).

The reaction is then monitored by TLC in chloroform/methanol/2.5 *N* NH3 (60:40:9, v/v/v). One major radioactive spot (R_f 0.16 and 0.13, respectively, depending on the acyl chain length) should be obtained. If TLC shows that more than a trace amount of deacylated GM1 is still present, a small amount of the *N*-succinimidoyl derivative of [1-^{14}C]octadecanoic acid or [1-^{14}C]octanoic acid, respectively, is added and the acylation reaction is continued. After almost all deacylated GM1 has been converted into labeled deacetylated GM1 the solvents are removed in a nitrogen jet and the crude product is purified by column chromatography on silica gel Si 60 (Lobar Fertigsäulen size A) using chloroform/methanol/2.5 *M* ammonia (60:40:9, v/v/v) as the mobile phase. The elution profile is monitored by TLC in the same solvent system. All fractions containing a ninhydrin-positive and radioactive single major spot of the desired product (R_f 0.16 or 0.13, respectively, depending on the acyl chain length) are pooled and lyophilized from water following removal of the solvents.

The yield of deacetylated GM1 derivatives is in the range of 9–10.5 mg and 8.2–9.5 mg, respectively, depending on the acyl chain length (6–7 μmol, 60–70%). FABMS: ($C_{71}H_{129}N_3O_{30}$ for the C_{18}-sphingosine analog of the octadecanoyl derivative, 1504.81 g/mol), $[M+Na^+]$ at m/z 1526; ($C_{73}H_{133}N_3O_{30}$ for the C_{20}-sphingosine analog of the octadecanoyl derivative, 1532.86 g/mol), $[M + Na^+]$ at m/z 1554; ($C_{61}H_{109}N_3O_{30}$ for the C_{18}-sphingosine analog of the octanoyl derivative, 1364.54 g/mol), $[M + Na^+]$ at m/z 1386; ($C_{63}H_{113}N_3O_{30}$ for the C_{20}-sphingosine analog of the octanoyl derivative, 1392.59 g/mol), $[M+Na^+]$ at m/z 1414.

Note that in this microscale preparation small amounts of amine-like substances may quench the yield by reacting with the *N*-succinimidoyl derivatives of fatty acids. This unwanted side reaction can be minimized by eliminating volatile amines and imines or ammonia in a nitrogen jet in the presence of a less volatile tertiary amine such as *N,N*-diisopropylethylamine.[29]

Biotinylated GM1 (Biotin-C_{18}-GM1 and Biotin-C_8-GM1). For the preparation of biotin-GM1 derivatives, the respective radioactive deacetylated GM1 derivatives (5 μmol) are dissolved in DMF (0.2 ml) and *N,N*-diisopropylethylamine (0.01 ml). The solvents are removed in a nitrogen jet to expel volatile and reactive amino and imino compounds and the residues are redissolved in DMF (0.2 ml) and *N,N*-diisopropylethylamine (0.01 ml) prior to the addition of solid D-biotinoyl-ε-aminocaproic acid *N*-succinimidyl ester (3.6 mg, 10 μmol). The reaction mixtures are stirred, under argon, for 24 hr at 40°. After drying in a stream of nitrogen, the residues are dissolved in methanol/water (85 : 15, v/v, 1.5 ml) and purified by reversed-phase HPLC in methanol/water (85 : 15, v/v). Fractions of 4 ml are collected and monitored for radioactivity. The pure products (about 3 μmol each, 60% with respect to deacetylated GM1 derivatives) should yield one single biotin-positive spot on TLC plates (R_f 0.21 and 0.15, chloroform/methanol/acetic acid/water, 60 : 35 : 6 : 2, v/v/v/v, for biotin-C_{18}-GM1 and biotin-C_8-GM1, respectively) that, in addition, is also detected by dipping the plate in ceric ammonium nitrate in 20% sulfuric acid followed by heating. The solvent mixture is reduced in a flash evaporator and the remaining material is freeze dried. The pure compounds (biotin-C_{18}-GM1 and biotin-C_8-GM1) are stored in methanol under argon at −20° until use.

Alternatively, the biotin-GM1 derivatives can be purified by preparative TLC using chloroform/methanol/2.5 *M* ammonia (60 : 40 : 9, v/v/v) as the mobile phase (R_f 0.31 for the octadecanoyl derivative of biotin-GM1).

Briefly, the dried reaction products are applied to TLC plates as a streak and the plates are developed in a tank with vapor saturation. The band containing the desired product is scraped after wetting the plate with a spray of methanol/water (1 : 1, v/v) and biotin-GM1 is extracted with chloroform/methanol (1 : 1, v/v). Final purification is achieved with LiChroprep RP-18. This procedure gives a yield of only about 40%. Also, it warrants mention that the sulfur atom in the biotin heterocycle is prone to easy oxidation by air, yielding mainly the respective sulfoxide; further oxidation gives rise to the corresponding sulfone.[19]

Characterization of Biotin-C_{18}-GM1 and Biotin-C_8-GM1. The biotinylated derivatives are best characterized by NMR spectroscopy and mass spectrometry. The following data are obtained: ^{1}H NMR (500 MHZ, $CDCl_3$/CD_3OD, 1 : 1, v/v): sphingosine unit: δ 5.71 (m, 1 H, H-5), δ 5.69 (m, 1 H, H-3), δ 5.44 (dd, 1 H, $J_{3,4}$ 7.5 Hz, H-4), δ 4.33 (m, 1 H, H-2), δ 2.93 (dd, 1 H, $J_{1a,1b}$ 7.8 Hz, $J_{1a,2}$ 5.1 Hz, H-1), δ 2.02 (m, 2 H, H-6), δ 1.32-1.23 (m, 26 H, 13 CH_2), δ 0.88 (t, 3 H, $J_{19,20}$ 7.1 Hz, CH_3); octanoic acid unit: δ 2.29 (m, 2 H, H-2), δ 1.32-1.23 (m, 10 H, 5 CH_2), δ 0.88 (t, 3 H, $J_{7,8}$ 7.1 Hz, CH_3); glucose (I) unit: δ 4.28 (d, 1 H, $J_{1,2}$ 7.8 Hz, H-1); galactose (II) unit: δ 4.39 (d, 1 H, $J_{1,2}$ 7.9 Hz, H-1); galactose (IV) unit: δ 4.35 (m, 1 H, H-1); galactose *N*-acetyl unit: δ 4.77 (d, 1 H, $J_{1,2}$ 8.3 Hz, H-1), δ 1,97 (s, 3 H, CH_3); biotin residue (for designation of atoms of this residue see Fig. 1): δ 4.54 (m, 2 H, H-8), δ 3.18 (m, 2 H, H-3/H-7), δ 2.21 (t, 2 H, $J_{11,12}$ 6.7 Hz, H-6), δ 2.17 (t, 2 H, $J_{19,20}$ 7.6 Hz, H-19). FABMS: ($C_{89}H_{158}N_6O_{33}S$ for the C_{20}-sphingosine homolog of the octadecanoyl derivative, 1872.31 g/mol), [M + Na^+] at *m/z* 1893; ($C_{79}H_{138}N_6O_{33}S$ for the C_{20}-sphingosine homolog of the octanoyl derivative, 1732.04 g/mol), [M + Na^+] at *m/z* 1753. MALDI-TOF-MS (data shown represent the monoisotopic mass of the most abundant molecular ion): ($C_{89}H_{158}N_6O_{33}S$ for the C_{20}-sphingosine homolog of the octadecanoyl derivative, 1871.06 Da), [M + Na^+] at *Da/e* 1894.3; ($C_{79}H_{138}N_6O_{33}S$ for the C_{20}-sphingosine homolog of the octanoyl derivative, 1730.90 Da), [M + Na^+] at *Da/e* 1753.9.

Enzymatic Degradation of Biotin-GM1 Derivatives with GM1-β-Galactosidase, β-Hexosaminidases and Sialidase. For further characterization of the target products and for the test of susceptibility to the action of lysosomal glycohydrolases the biotin-GM1 derivatives are sequentially subjected to the action of GM1-β-galactosidase, β-hexosaminidases, and sialidase. Briefly, 4 nmol each of the octadecanoyl and octanoyl derivative of biotin-GM1 are dissolved in a total volume of 0.2 ml of 200 m*M* sodium citrate buffer, pH 4.3, containing 2 m*M* sodium taurodeoxycholate and 50 μg bovine serum albumin (BSA). As a control one-quarter (50 μl) of the resulting mixture is diluted with water (50 μl) and incubated for 3 days at 37° (assay without enzymes). The remainder is treated with 17 pKatal

GM1-β-galactosidase overnight at 37°. One-third is then removed from this mixture (assay with GM1-β-galactosidase) and to the remaining mixture (100 μl) 17,000 pKatal β-hexosaminidases is added. After an incubation for 24 hr at 37° half of this assay mixture is removed (assay with GM1-β-galactosidase and β-hexosaminidases).

Following the addition of 1300 pKatal sialidase the remainder is incubated for 24 hr at 37° (assay with GM1-β-galactosidase, β-hexosaminidases and sialidase). All assays are stopped with methanol (50 μl) and the resulting mixtures are desalted using minicolumns filled with LiChroprep RP 18 (see above). The lipid extracts are then analyzed by TLC using chloroform/methanol/acetic acid/water (60:35:6:2, v/v/v/v) as the mobile phase, and the radioactive bands are visualized on an X-ray sensitive film. The results of such an experiment with the octanoyl derivative of biotin-GM1 are shown in Fig. 2. Note that sialidase does not split the modified sialyl residue. This is due to steric hindrance caused by the bulky and spacer linked biotin moiety. This is important in view of immunolabeling of biotin-gangliosides endocytosed by cells.

Incorporation of Biotinylated GM1 into Cultured Fibroblasts

Principle

In the case of an exogenous antigen such as biotinylated GM1, it is necessary to incorporate into cells sufficient amounts for detection on ultrathin sections with gold conjugated antibodies. It is known that gangliosides form micelles in aqueous media above their critical micellar concentration of 10^{-9} *M* or less and [32,33] when incubated with cells, ganglioside micelles adhere to the cell surface. Single ganglioside molecules dissociate from the adhering micelles and insert into the plasma membrane in a slow process. Therefore, for the incorporation of a sufficient amount of biotin-C_{18}-GM1 into cells, prolonged incubation with a micellar solution of biotin-C_{18}-GM1 at 37° is required.

Derivatives of glycosphingolipids with a short acyl chain moiety in place of the natural long-chain residue (such as biotin-C_8-GM1) form 1:1 complexes with defatted BSA that allow insertion of these lipids into plasma

[32] S. Formisano, M. L. Johnson, G. Lee, S. M. Aloj, and E. Edelhoch, *Biochemistry* **18,** 1119 (1979).

[33] W. Mraz, G. Schwarzmann, J. Sattler, T. Momoi, B. Seemann, and H. Wiegandt, *Hoppe Seyler's Z. Physiol. Chem.* **361,** 177 (1980).

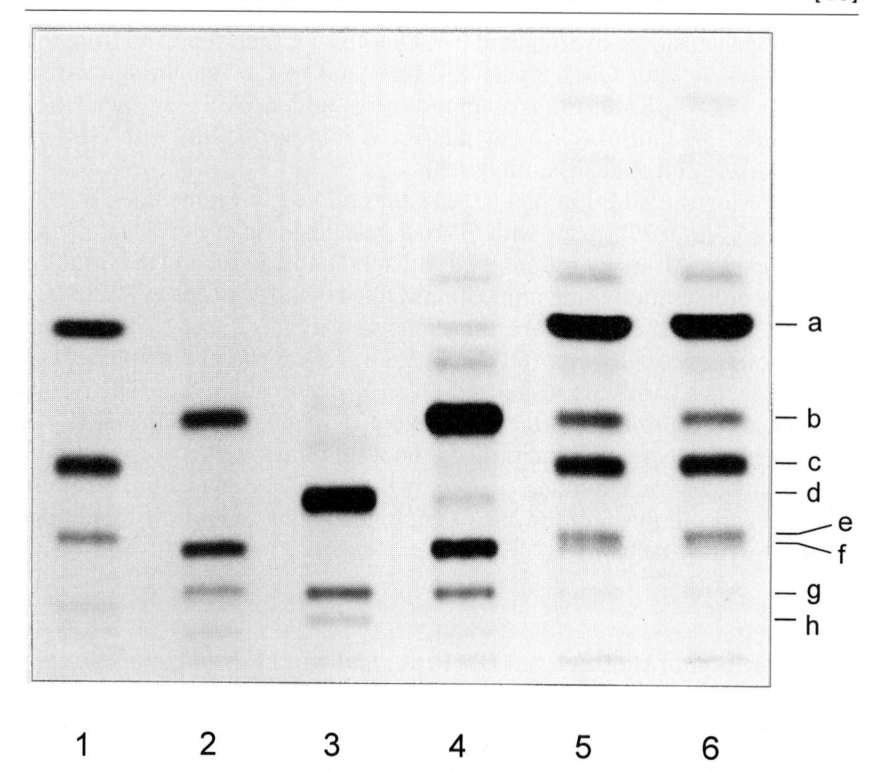

FIG. 2. Sequential hydrolysis of biotin-C_8-GM1 by GM1-β-galactosidase, β-hexosaminidase A and sialidase. Biotin-C_8-GM1 was treated sequentially with glycohydrolases as described in the text. Lane 1, biotin-C_8-GM3; lane 2, biotin-C_8-GM2; lane 3, biotin-C_8-GM1 after treatment without enzyme; lane 4, biotin-C_8-GM1 after treatment with GM1-β-galactosidase; lane 5, biotin-C_8-GM1 after treatment with GM1-β-galactosidase and subsequent treatment with β-hexosaminidase; lane 6, biotin-C_8-GM1 after subsequent treatment with GM1-β-galactosidase, β-hexosaminidase A, and sialidase. (a) biotin-C_8-GM3, (b) biotin-C_8-GM2, (c) biotin-C_8-GM3-sulfoxide, (d) biotin-C_8-GM1, (e) biotin-C_8-GM3-sulfone, (f) biotin-C_8-GM2-sulfoxide, (g) biotin-C_8-GM1-sulfoxide, biotin-C_8-GM2-sulfone, and (h) biotin-C_8-GM1-sulfone.

membranes even under temperature block of endocytosis.[34,35] Because of its short acyl chain, biotin-C_8-GM1 dissociates easily from the lipid–BSA complex and inserts rapidly into the plasma membrane at 4°. This is not possible for glycosphingolipids carrying a long acyl chain. However, if

[34] R. E. Pagano, *in* "Methods in Cell Biology: Fluorescence Microscopy of Living Cells in Culture" (Y.-L. Wang and D. L. Taylor, eds.), Vol. 29, p. 75. Academic Press, San Diego, 1989.

[35] G. Schwarzmann, P. Hofmann, U. Pütz, and B. Albrecht, *J. Biol. Chem.* **270,** 21271 (1995).

incubation is performed at 37°, both procedures of application of gangliosides are suitable. In this case, the amount of biotin-C_8-GM1 incorporated by confluent fibroblasts exceeds the amount of incorporated C_{18}-biotin-GM1 by a factor of 1.6, when both derivatives are applied at the same concentration of 10 μM for 48 hr at 37°.

Procedures

Incubation of Cultured Fibroblasts with Biotin-C_{18}-GM1. To prepare incubation media, an aliquot of the biotin-C_{18}-GM1 stock solution (100 nmol) in methanol is transferred into a test tube and dried under a stream of nitrogen. The remainder is dissolved in 20 μl ethanol. Then 10 ml of Dulbecco's modified Eagle's medium (DMEM) containing 0.3% FCS are added. The amount of FCS in the culture media is limited to 0.3% to minimize the inhibition of ganglioside uptake by serum proteins without impairing cell viability that otherwise may be reduced in serum-free media.[36] After stirring and sonification at 37° the final concentration of biotin-C_{18}-GM1 can be controlled by the amount of radioactivity in the media. It warrants mention that the amount of biotin-C_{18}-GM1 taken up by cells varies with the concentration of the derivative, the incubation time and the cell type. According to our experience, a concentration of 10 μM biotin-C_{18}-GM1 in the culture media is sufficient for metabolic as well as morphologic studies. When applied at this concentration to cultured confluent human fibroblasts in 25-cm^2 culture flasks for different time periods between 24 and 120 hr at 37°, maximal incorporation was reached after 72 hr yielding around 3 nmol biotin-C_{18}-GM1 per cell pellet following cell harvest by treatment with trypsin/EDTA (1%/0.002% in PBS).[15]

Incubation of Cultured Fibroblasts with Biotin-C_8-GM1. An aliquot of the stock solution of biotin-C_8-GM1 (100 nmol) in methanol is dried under a stream of nitrogen. The dried lipid is dissolved by adding first 20 μl of ethanol and then 1 ml of 10^{-4} M defatted BSA (7 mg/ml) in either HMEM if incubation with cells is performed at 4° or DMEM if cells are incubated with the lipid–BSA complex at 37°. After vigorous stirring, the resulting clear solution is diluted with 9 ml of HMEM or DMEM, respectively, to yield a 10 μM lipid–BSA complex.

Localization of Biotin-C_{18}-GM1 by Immunoelectron Microscopy

The localization of lipid antigens such as biotinylated gangliosides by immunoelectron microscopy is hampered by the fact that well-defined fixa-

[36] S. Sonderfeld, E. Conzelmann, G. Schwarzmann, J. Burg, U. Hinrichs, and K. Sandhoff, *Eur. J. Biochem.* **149,** 247 (1985).

tives for lipids do not exist. Therefore, the preservation of the membrane structure is of eminent importance and lipid extraction or redistribution during sample preparation and immunolabeling should be avoided. Different approaches are described to meet this requirement for the localization of biotin-C_{18}-GM1: (1) preembedding labeling of fixed cell monolayers by applying single-step labeling using antibodies coupled to ultra-small gold particles combined with silver enhancement according to Danscher,[37] (2) postembedding labeling after dehydration at low temperature and embedding in a hydrophilic resin, and (3) labeling of cryosections.

Localization of Biotin-C_{18}-GM1 by Preembedding Labeling

Principle

Because of the possibility that lipids may be lost on dehydration and embedding, preembedding labeling is a feasible procedure to circumvent a possible delocalization of biotin-C_{18}-GM1 during sample processing. In principle, preembedding labeling is a problematic procedure for high-resolution immunocytochemistry, especially for the detection of intracellular antigens. The most crucial step is the permeabilization of the cells to render the antigens accessible to antibodies. For the localization of membrane antigens such as biotin-C_{18}-GM1 it is evident that all membrane active substances such as detergents, mostly used for permeabilization, must be strictly avoided. Permeabilization is achieved by the fixation itself and the subsequent treatment of fixed monolayer cells with 1% sodium borohydride. Sodium borohydride is normally applied at a low concentration (0.05–0.1%) to reduce remaining aldehydes after fixation before immunolabeling, and to block autofluorescence for light microscopy studies.[38,39] According to our experience 1% sodium borohydride in PBS sufficiently permeabilizes cell membranes thus allowing antibody penetration for detection of intracellular biotin-C_{18}-GM1. The mechanism of permeabilization by sodium borohydride is not fully understood but it is possible that the cells are permeabilized physically by hydrogen bubbles that emerge inside the fixed cell from the reaction of sodium borohydride with protons.

For the localization of biotin-C_{18}-GM1, it is advantageous to use antibiotin antibody gold conjugates for single-step labeling. An efficient detection of intracellular antigens by preembedding labeling was facilitated by

[37] G. Danscher, *Histochemistry* **71,** 81 (1981).

[38] K. Weber, P. C. Rathke and M. Osborn, *Proc. Natl. Acad. Sci. U.S.A.* **75,** 1820 (1978).

[39] R. Bacallao and E. H. K. Stelzer, *in* "Methods in Cell Biology: Vesicular Transport" (A. M. Tartakoff, ed.), Vol. 31A, p. 437. Academic Press, San Diego, 1989.

the introduction of ultra-small gold conjugates in combination with silver enhancement.[37] Due to their reduced size antibody conjugates coupled to ultra small gold particles penetrate into cells like nonconjugated antibodies. For visualization the gold particles are enlarged by silver enhancement.

Figure 3a shows an example for the localization of biotin-C_{18}-GM1 by preembedding labeling followed by the silver enhancement procedure. This demonstrates that the localization of intracellular antigens is possible when applying this procedure. However, cells treated with the preembedding labeling protocol show a poor preservation of their ultrastructure. In this case, a poorly preserved ultrastructure must be accepted if antibody penetration has to be achieved. Another disadvantage of preembedding labeling is the low sensitivity and resolution of the label. In spite of this, the preembedding labeling study is important to serve as a reference for the localization of biotin-C_{18}-GM1 by alternative procedures that could cause a redistribution of biotin-C_{18}-GM1 during sample preparation and, therefore, could result in a distorted localization.

Procedures

Incubation of Cells. Confluent fibroblasts grown as monolayers in 8-cm^2 petri dishes are incubated with biotin-C_{18}-GM1 at a concentration of 10 μM in DMEM supplemented with 0.3% FCS at 37° for 72 hr. For controls, cells are incubated in the same medium lacking biotin-C_{18}-GM1.

Fixation of Cells. After removing the incubation medium, cells are washed three times with PBS and fixed with a mixture of 1% formaldehyde and 0.5% glutaraldehyde in 0.2 *M* HEPES, pH 7.2, for at least 1 hr. It is also possible to store the cells in this fixative at 4° until further processing.

Preparation of Fixative. Based on the description of Robertson *et al.*[40] an 8% formaldehyde stock solution in 0.2 *M* HEPES is prepared from depolymerized paraformaldehyde as follows: 8 g of paraformaldehyde is suspended in 50 ml water and, after addition of some droplets of 6 *M* sodium hydroxide, stirred in a closed bottle at 60° until the solution is clear. This formaldehyde solution is then diluted 1:1 with 0.4 *M* HEPES, pH 7.2, filtered through a 0.2-μm filter, and stored in aliquots at −20°. Before use the formaldehyde solution is warmed to 37° and agitated until the solution is clear. Glutaraldehyde is available as a 25% solution (EM grade) and should be stored at −20°. To obtain 20 ml of the final fixative, 2.5 ml of the 8% formaldehyde solution and 400 μl of the glutaraldehyde solution are diluted with 17.1 ml 0.2 *M* HEPES, pH 7.2.

[40] J. D. Robertson, T. S. Bodenheimer, and D. E. Stage, *J. Cell Biol.* **19,** 159 (1963).

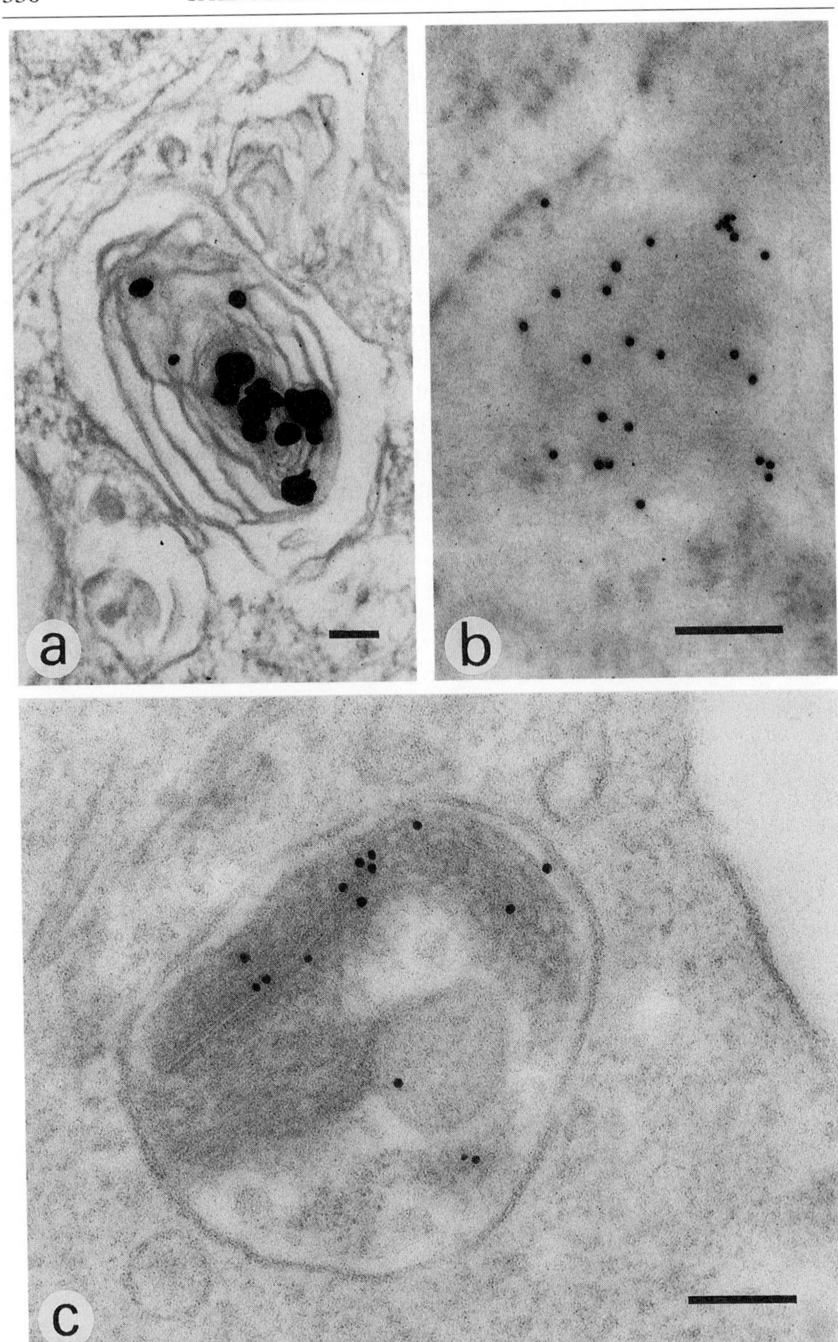
a
b
c

Treatment of Cells with Sodium Borohydride. Sodium borohydride solutions are not stable at pH 7. Therefore, cells are treated with freshly prepared sodium borohyride solutions two times for 10 min at room temperature. Then the cells are washed 3 times for 5 min with PBS.

Immunolabeling. All steps including blocking, immunolabeling, and subsequent washings for removal of unbound antibodies should be performed under constant shaking. Cells are blocked with 5% BSA and 0.2% cold water fish skin gelatin (CWFS-gelatin) in PBS for 1 hr at room temperature. Cells are then incubated overnight with a goat anti-biotin antibody conjugated to ultra-small gold particles diluted 1:100 with PBS containing 5% BSA at 4°. After immunolabeling, cells are washed three times for 10 min with PBS containing 5% BSA and six times for 5 min with PBS at room temperature. After washing the monolayer cells are fixed with 2% glutaraldehyde in 0.2 *M* HEPES, pH 7.2, for 10 min at room temperature.

Epon Embedding. To produce hard blocks for thin sectioning the Epon resin is prepared as follows: Poly/Bed 812 (60.16 g), nadic methyl anhydride (40.87 g) and dodecenylsuccinic anhydride (18.61 g) are stirred carefully for 10 min to avoid the production of air bubbles. Then 2,4,6-tri(dimethylaminomethyl)phenol (1.67 g) is added and the mixture is stirred for 20 min. The ready mixture is stored at −20° until use. It is advantageous to fill the resin into capped disposable 10-ml plastic syringes and to remove all air before capping the syringe. During the embedding procedure the resin can be easily dispensed directly from the syringe.

Monolayer cells are postfixed with 1% osmium tetroxide in 0.2 *M* HEPES, pH 7.2, for 20 min at room temperature in the dark. After osmium fixation cells are washed three times with water. For dehydration, the cells are washed two times for 10 min with 30, 50, and 75% ethanol at room temperature. At this stage cells are stained *en bloc* with 1.5% uranyl acetate

---

FIG. 3. Localization of biotin-C_{18}-GM1 by preembedding labeling and postembedding labeling of LR Gold sections. Confluent human fibroblasts were incubated for 72 hr with 10 μM biotin-C_{18}-GM1. Sections were examined and photographed with a Phillips CM 120 electron microscope (Phillips, Eindhoven, The Netherlands). (a) Fixed monolayer cells were treated with 1% sodium borohydride and immunolabeled with goat anti-biotin antibodies coupled to ultra-small gold particles. After embedding in Epon and sectioning, the ultra-small gold particles were enlarged on Epon sections by the silver enhancement procedure,[37] resulting in silver precipitates of irregular size and shape that indicate the localization of anti-biotin antibodies over the internal membranes of a multilamellar organelle. (b and c) Immunolabeling of LR Gold sections with anti-biotin antibodies coupled to 10-nm gold particles. In (b) postfixation with osmium tetroxide was omitted and, therefore, membranes lack any contrast. In comparison to this, excellent contrast is obtained by postfixation with osmium tetroxide (c). Bar: 100 nm.

and 1.5% tungstophosphoric acid in 75% ethanol for 1 hr at room temperature in the dark. This staining solution should be filtered through a 0.2-μm filter before use to remove undissolved salt. After several washes with 100% ethanol the monolayer cells are infiltrated with Epon/ethanol (1:1 and 3:1, v/v) for 30 min. Note that Epon mixes with ethanol under vigorous stirring. This makes it possible to leave out washing steps with propylene oxide and, therefore, keeps the petri dish intact and the monolayer cells attached to the substrate. For further infiltration the Epon/ethanol mixture is replaced by 100% Epon. After several changes most of the Epon should be removed from the dish, leaving a very thin layer of the resin. Then Epon-filed BEEM capsules are placed upside down on the monolayer. Polymerization of the resin is achieved by incubation for 24 hr at 40° followed by 24 hr at 65°. Epon-embedded monolayer cells can be removed very easily from the petri dish by taking off the Epon-filled BEEM capsule. The smooth block face contains the monolayer cells ready for sectioning parallel to the substrate.

Silver Enhancement. For silver enhancement on thin sections nickel grids are recommended. If copper grids are used, they should be covered with a supporting film to avoid a reaction of the developer with the grid. EM grids with Epon sections are placed on droplets of water on a piece of Parafilm and then transferred onto droplets of the developer solution. The developer according to Danscher[37] contains the following ingredients:

A: protecting colloid:	33% gum arabic in water (w/v)
B: citrate buffer:	2.55 g citric acid and 2.35 g sodium citrate (dihydrate) are dissolved in 10 ml water (pH 3.8)
C: reducing agent:	hydroquinone: 0.85 g in 15 ml water
D: silver ion supply:	silver lactate: 0.11 g in 15 ml water

Silver lactate is highly sensitive to light. Solutions should therefore be carefully protected from light. All solutions with the exception of solution B can be stored in aliquots at −20° until use. The developer is prepared by mixing 6 parts of A with 1 part of B and 1.5 parts of C. To the resulting mixture 1.5 parts of D are added immediately before use. The silver enhancement should be performed in the dark. Because the silver grains grow with time, the development should be stopped after different time points starting with 15 min. The reaction is stopped by washing with water. After silver enhancement Epon sections are counterstained with uranyl acetate.

Localization of Biotin-C_{18}-GM1 in LR Gold Embedded Samples

Principle

Hydrophilic acrylic resins such as LR Gold, LR White, or Lowicryl K4M are widely used for immunocytochemical applications. Because of their low viscosity (8 centipoise at 25°) the resins rapidly infiltrate into cells and tissues. Acrylic resins are more hydrophilic than epoxy resins and, therefore, tolerate residual water in the sample. We used LR Gold because it is a single component embedding media of low toxicity that polymerizes by illumination with UV light at low temperature (−20°) in the presence of a radical forming compound. Therefore, a progressive lowering of temperature (PLT) protocol for sample dehydration and embedding is applied. Both, the ability to polymerize at low temperature and the hydrophilicity of the resin are important to maintain maximal immunoreactivity and to minimize lipid extraction.

Note that we tested whether or not fixation with osmium tetroxide prior to embedding enhances the retention of biotin-C_{18}-GM1 in the sample. Figure 3b shows a labeled section of a nonosmicated sample, whereas Fig. 3c shows a section of an osmium tetroxide-postfixed sample after immunolabeling with anti-biotin gold conjugates (10 nm). Because the labeling is always found close to organelles it can be concluded that redistribution of biotin-C_{18}-GM1 is negligible in both cases. This observation is in agreement with the finding that a same amount of cell-associated biotin-C_{18}-GM1 (10–15%) is extracted during dehydration irrespective of osmium tetroxide postfixation. The osmium tetroxide-fixed samples, however, show better contrast than the nonosmicated samples. Alternatively, good contrast may also be obtained by postfixation of the tissue with uranyl acetate prior to embedding in LR Gold and by staining of the labeled thin sections with 2% osmium tetroxide followed by a treatment with lead citrate.[41]

Procedures

Incubation with Biotin-C_{18}-GM1, Fixation, and Postfixation. Human skin fibroblasts are incubated with biotin-C_{18}-GM1 at a concentration of 10 μM for 72 hr at 37° as described above. For a control, cells are incubated with media lacking biotin-C_{18}-GM1. After careful washes with PBS the cells are detached from the substratum with proteinase K (0.05 mg/ ml in PBS) for 3 min on ice.[42] Detached cells are pelleted in 0.5-ml Eppendorf caps (1

[41] M. A. Berryman and R. D. Rodewald, *J. Histochem. Cytochem.* **38,** 159 (1990).
[42] J. Green, G. Griffiths, D. Louvard, P. Quinn, and G. Warren, *J. Mol. Biol.* **152,** 663 (1981).

min at 1000*g*) and fixed with 4% formaldehyde in 0.2 *M* HEPES, pH 7.2. Samples were postfixed with 1% osmium tetroxide in water for 15 min at 4° in the dark. In control samples, the postfixation step is omitted. After postfixation the samples are washed with water.

Progressive Lowering of Temperature (PLT) Embedding. Fixed cells are embedded in LR Gold following a protocol suggested by Polyscience (Warrington, PA) with minor modifications:

2 × 20 min	50% ethanol, 0°
2 × 30 min	70% ethanol, −20°
2 × 30 min	90% ethanol, −20°
30 min	LR Gold/ethanol 1 : 1, −20°
60 min	LR Gold/ethanol 7 : 3, −20°
60 min	LR Gold pure, −20°
60 min	LR Gold + 0.5% benzil (UV initiator), −20°
Overnight	LR Gold + 0.5% benzil, −20°

The Eppendorf caps are completely filled with the LR Gold/benzil solution and capped tightly to exclude air, because oxygen could interfere with the polymerization reaction. Polymerization is achieved by illumination with UV light for 48 hr at −20° and another 48 hr at room temperature. For polymerization at low temperature, the UV lamp is placed at the bottom of a deep freezer at a distance of about 15 cm below the samples. This arrangement makes sure that the pellets are well illuminated and that polymerization starts there. The dark color of osmium tetroxide-fixed cell pellets could interfere with UV exposure and, therefore, prevent polymerization. This, however, is never a problem with small pellets. After the Eppendorf caps are removed, polymerized samples can be cured further by exposure to sunlight. Loss of smell is a good indication of complete polymerization.

Note that optimal infiltration is achieved when the pellets are carefully resuspended at each incubation step. After each step the cells are pelleted by short and fast centrifugation to minimize loss of material. The best way is to use a cooled Eppendorf centrifuge with a swing-out rotor for Eppendorf caps. All infiltration steps at −20° are best performed in a common deep freezer.

Sectioning and Immunolabeling. LR Gold blocks are easy to section by standard techniques. Because of the hydrophilic nature of LR Gold, it is helpful to lower the water level in the knife-boat to prevent water from leaping onto the block face. Because LR Gold sections are less stable under the electron beam than Epon sections on noncoated grids, the former should be collected on Formvar-coated grids. For immunolabeling a long sheet of Parafilm is attached to a glass plate by spreading some water between the film and the plate. Droplets of the various solutions used during immunolabeling are positioned on the Parafilm sheet. For blocking and washing steps, large droplets (50–100 μl) are useful. In the case of rare antibody solutions, the size of a droplet can be reduced to 10 μl. All steps are carried out at room temperature. All solutions are filtered through a 0.2-μm filter before use. Antibodies are diluted with filtered block solution. Immunolabeling is performed as outlined below:

1. The grids are floated with the sections facing down on droplets of blocking solution composed of 1% BSA and 1% CWFS-gelatin in 0.2 *M* HEPES, pH 7.2, and incubated for 15 min to block nonspecific binding sites.
2. Grids are transferred to droplets containing the goat anti-biotin gold conjugate diluted 1 : 100 with blocking solution and incubated for 1 hr.
3. After labeling the grids are washed 5 × 5 min in 0.2 *M* HEPES. Then the labeled sections are fixed with 2% glutaraldehyde in 0.2 *M* HEPES. After 5 × 5 min washing with water, sections are stained in 2% uranyl acetate in water for 10 min.
4. After 3 × 5 min washes with water the grids are air-dried.

Note that the specificity of the anti-biotin antibody can be tested easily in the control experiments by omitting the incubation with biotin-C_{18}-GM1.

Localization of Biotin-C_{18}-GM1 on Ultrathin Cryosections

Principle

The thawed frozen section technique, introduced by Tokuyasu,[43] using chemically fixed and cryoprotected material provides a major technical advance for immunocytochemistry. The advantages over other methods include a high sensitivity for immunolabeling because antigens are not masked by any resin and, therefore, are better accessible for antibodies. Furthermore, the processing of samples is fast and denaturation of antigens is minimized. However, cryosections are sensitive objects because of the lack of a supporting resin and, therefore, structural distortion and material extraction often occur, limiting the usefulness of this method. This is of particular importance in the case of cellular structures that are less amenable to chemical fixation such as lipid-rich organelles. The step of thawing and transfer of the sections to the grid is most critical for the structural preserva-

[43] K. T. Tokuyasu, *J. Cell Biol.* **57,** 551 (1973).

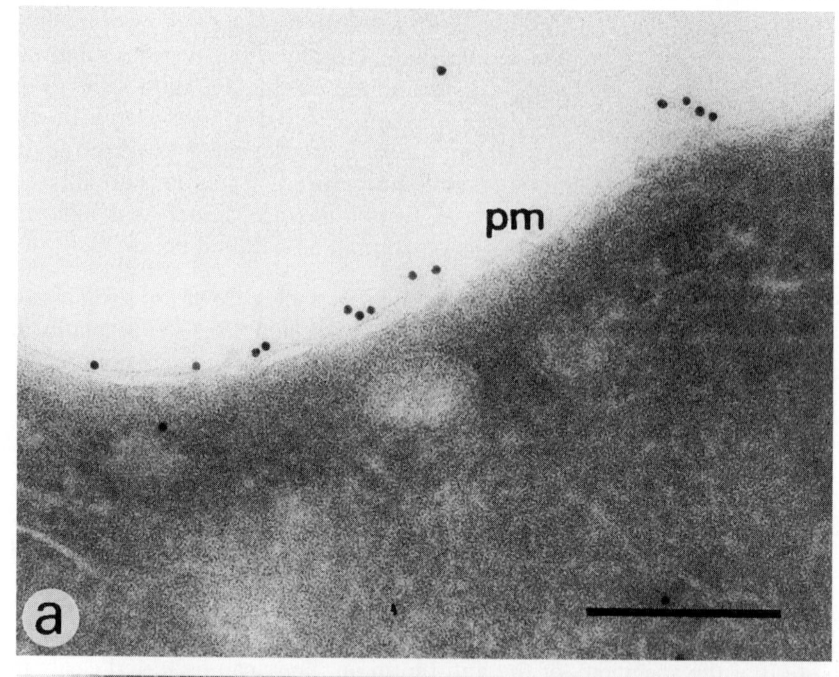
pm
a

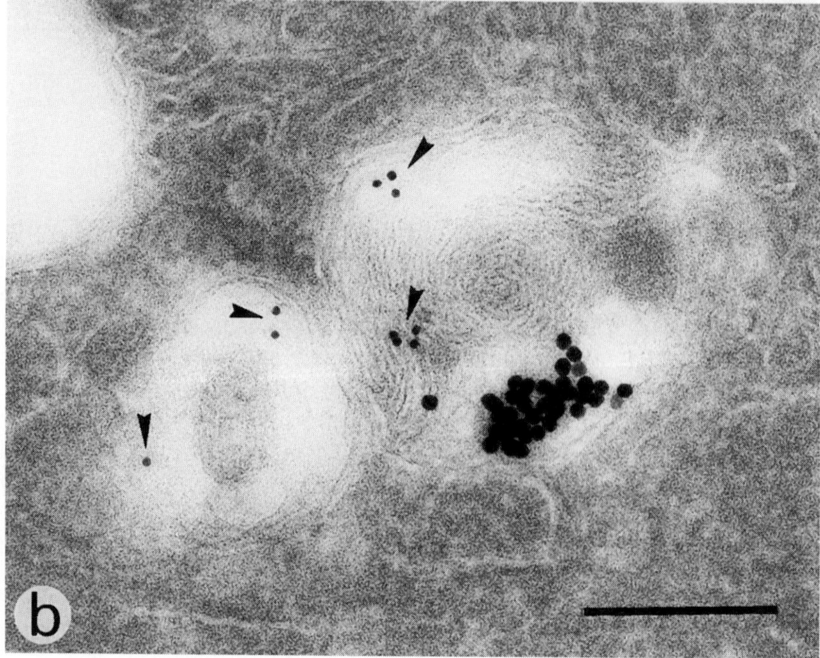
b

tion and damage occurs mainly at this stage and not during immunolabeling.[44] Substitution of the conventional sucrose transfer medium by either a mixture of methylcellulose and sucrose (MS) or methylcellulose containing uranyl acetate (MU) remarkedly improves the structural integrity of cryosections. This modification of the section pickup protocol is an important improvement for the localization of biotin-C_{18}-GM1 on cryosections. Figure 4 shows the localization of biotin-C_{18}-GM1 at the plasma membrane of human fibroblasts (Fig. 4a) and over the internal membranes of a multilamellar organelle that was preloaded with BSA-gold (large gold particles, prepared as described by Handley[45]) before incubation of the cells with biotin-C_{18}-GM1 (Fig. 4b). The sections were picked up with MU and labeled with anti-biotin gold (10 nm). As visible in Fig. 4b, the multilamellar lipid structures are well preserved. In contrast, Fig. 5 shows cryosections that were picked up with MS (see also next section). In this case internal membrane structures of lipid-rich organelles are still visible but less preserved, whereas the conventional sucrose pickup procedure resulted in a complete loss of lipid-rich structures, leaving empty vacuoles in the cells.

It warrants mention that in some samples precipitates over the cells occurred with the MU pickup that did not appear with MS pickup. As another disadvantage of the MU pickup the labeling density is often diminished. This limits the MU pickup procedure, but it is always a good idea to test both pickup solutions.

Procedures

Incubation of Cells with Biotin-C_{18}-GM1. Human skin fibroblasts are incubated with biotin-C_{18}-GM1 at a concentration of 10 μM for 48 hr at 37° and harvested with proteinase K as described above. For controls,

[44] W. Liou, H. J. Geuze, and J. W. Slot, *Histochem. Cell Biol.* **106,** 41 (1996).

[45] D. A. Handley, *in* "Colloidal Gold: Principles, Methods and Applications" (M. A. Hayat, ed.), Vol. 1, p. 13. Academic Press, San Diego, 1989.

FIG. 4. Localization of biotin-C_{18}-GM1 on cryosections. Confluent human fibroblasts were incubated for 48 hr with 10 μM biotin-C_{18}-GM1. Cryosections were picked up with MU and labeled with anti-biotin antibodies coupled to 10-nm gold particles. Biotin-C_{18}-GM1 is detectable at the plasma membrane (a) and over the internal membranes of multilamellar organelles (b) as indicated by arrowheads. These organelles were preloaded with BSA-gold (16 nm) by incubating the cells with BSA-gold overnight followed by a chase period of 4 hr at 37° in medium lacking BSA-gold. Therefore, these organelles are very likely late endosomes or lysosomes. The internal membrane structures are well preserved. pm, Plasma membrane. Bar: 200 nm.

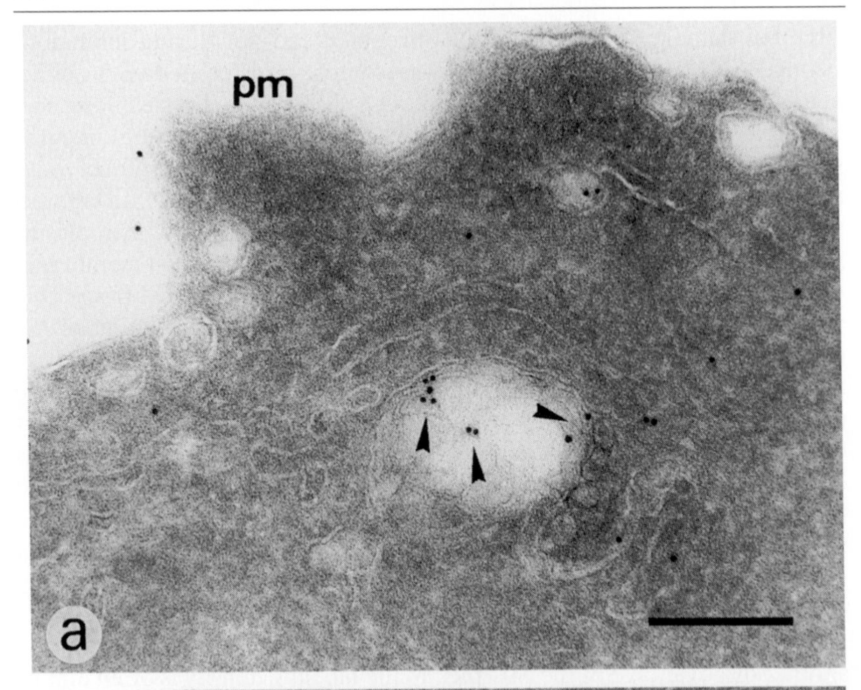
pm
a

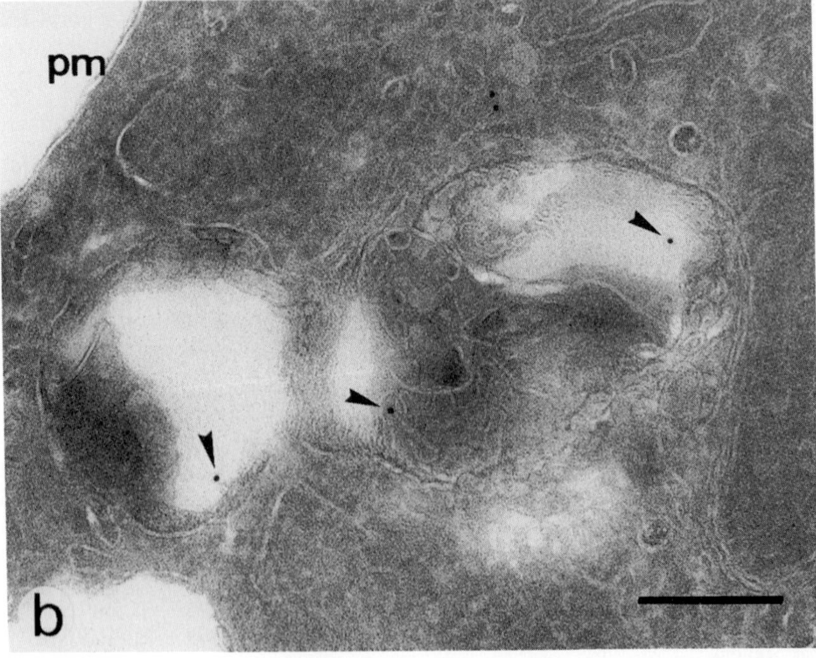
pm
b

cells are incubated with media lacking biotin-C_{18}-GM1. Detached cells are pelleted in 1.5-ml Eppendorf caps (1 min at 1000*g*).

Fixation of Cells and Subsequent Infiltration with Cryoprotectant. Cell pellets are fixed with 4% formaldehyde and 0.1% glutaraldehyde in 0.2 *M* HEPES, pH 7.2. Glutaraldehyde is included in the fixative to prevent the pellets from dispersion during infiltration with the cryoprotectant. Pellets are infiltrated with 50% polyvinylpyrrolidone/1.15 *M* sucrose overnight on a rotator at 4°. For preparation of the cryoprotectant the protocol by Tokuyasu[46] with modifications by Hans Stukenbrok is used:

1. Dissolve 0.58 g sodium hydrogencarbonate and 20 mg sodium azide in 20 ml water, then add 10 ml of 0.1 *M* phosphate buffer, pH 7.4.
2. To this solution add 100 g of polyvinylpyrrolidone (MW 10,000) and stir in a water bath at 60° until the solution becomes transparent.
3. Add 100 ml of 2.3 *M* sucrose in 0.1 *M* phosphate buffer to the solution of polyvinylpyrrolidone and continue stirring in a water bath until the components are mixed well resulting in a solution of high viscosity that can be stored at 4°. After infiltration the specimens are mounted to specimen stubs and frozen in liquid nitrogen.

Note that the replacement of 0.1 *M* phosphate buffer by 0.1 *M* HEPES, pH 7.2, avoids the formation of precipitates during MU pickup.

Preparation of 2% Solution of Methylcellulose (25 Centipoise, 25°). In an Erlenmeyer flask, 98 ml water is heated to 95° and the Erlenmeyer flask is removed from the hot plate before 2 g of methylcellulose is added.[47] After mixing, the dispersion is put on ice and stirred slowly overnight at 4°. Then the resulting dispersion is kept for 3 days at 4°. Before use the dispersion is centrifuged at 300,000*g* for 65 min at 4°. The supernatant is a clear solution of methylcellulose that is stored in the cold. The 1:1 mixture of 2% methylcellulose and 2.3 *M* sucrose in water (MS) should be prepared in the cold by gentle mixing. The MU pickup solution is made of 900 μl

[46] K. T. Tokuyasu, *Histochem. J.* **21,** 163 (1989).
[47] K. T. Tokuyasu, *J. Ultrastruct. Res.* **63,** 287 (1978).

FIG. 5. Localization of biotin-C_{18}-GM1 and biotin-C_8-GM1 on cryosections. Confluent human fibroblasts are incubated either with 10 μM biotin-C_{18}-GM1 (a) or 10 μM biotin-C_8-GM1 (b) for 48 hr at 37° and processed as above. Cryosections were picked up with MS and labeled with anti-biotin antibodies coupled to 10-nm gold. The internal membrane structures of the labeled organelles are still visible but less preserved than after MU pickup (for comparison, see Fig. 4). The labeling density in (a) is much higher than in (b; see arrow heads). pm, Plasma membrane. Bar: 200 nm.

of a 2% solution of methylcellulose and 100 μl of a filtered 4% solution of uranyl acetate.

Cryosectioning. Ultrathin cryosections are prepared using an ultracryomicrotome (Leica Ultracut UCT, Leica, Bensheim, Germany) equipped with an antistatic device (Diatome, Biel, Switzerland) and a diamond knife (Diatome, Biel, Switzerland or Drukker International, Cuijk, The Netherlands). Before sectioning it is necessary to trim a cutting surface area of roughly 0.3 × 0.5 mm. The block is oriented with the short side parallel to the knife edge and cutting is performed at −120° at a speed of 0.8–10 mm/s. The section thickness is set to 50 nm. Sections are picked up with either a 1:1 mixture of 2% methylcellulose and 2.3 *M* sucrose in water (MS) or with 1.8% methylcellulose containing 0.4% uranyl acetate (MU) and transferred to Formvar-coated grids.

Immunolabeling. Before immunolabeling the sections are washed with water. Immunolabeling is performed as described above. After the final contrasting step in 2% uranyl acetate the grids are washed 3 × 5 min with water, then transferred to droplets of MU on ice and incubated for 10 min. To dry the sections, tungsten loops of 3.5 mm diameter are used to loop out the grid. The grid is placed in the middle of the loop and excess fluid is removed by touching the surface of a filter paper. This procedure results in the so-called "positive–negative staining."[48]

Use of Biotin-C_8-GM1 for Studies of Ganglioside Transport in Cells: Applications and Limitations

Principle

Membrane lipids are removed from the plasma membrane by endocytosis and are delivered to lysosomes or are recycled back to the plasma membrane. To analyze these processes, time course studies of endocytosis of a defined lipid appear to be promising. For such studies it is necessary to insert a defined lipid such as biotinylated GM1 into the plasma membrane under temperature block of endocytosis. Incubation of cells with a micellar solution of biotin-C_{18}-GM1 at 4° leads to only a negligible amount of incorporated ganglioside molecules, insufficient for their detection by immunoelectron microscopy. For practical reasons, it is only possible to insert into the plasma membrane at 4° "semitruncated" biotinylated GM1 derivatives, e.g., biotin-C_8-GM1, bearing a short acyl chain instead of a long acyl

[48] G. Griffiths, R. Brands, B. Burke, D. Louvard, and G. Warren, *J. Cell Biol.* **95,** 781 (1982).

chain as found in natural gangliosides. Such "semitruncated" lipids easily form complexes with defatted BSA and readily dissociate from BSA, facilitating their insertion into the plasma membrane even at 4°. In contrast, long acyl chain lipids form tight complexes with defatted BSA and their dissociation rate is too small to serve as a vehicle for ganglioside incorporation into plasma membranes. However, "semitruncated" lipids have a high off-rate[49] and, in contrast to lipids with long acyl chains, can be back-exchanged easily from membranes using an excess of either defatted BSA or acceptor membranes such as liposomes. Therefore, particular care must be taken to prevent back-exchange of membrane-inserted biotin-C_8-GM1 molecules in the course of sample preparation and immunolabeling for immunoelectron microscopy.

Extraction of plasma membrane-inserted biotin-C_8-GM1 is not observed when PBS, 2% methylcellulose, or 1% CWFS-gelatine in 0.2 *M* HEPES is used as a medium. However, a significant back-exchange (more than 50%) of biotin-C_8-GM1 from intact plasma membranes is caused by anti-biotin antibodies coupled to 10-nm gold particles. This extraction of short acyl chain biotin-GM1 by antibody gold conjugates obviously takes place also during immunolabeling of cryosections as shown by their scarce labeling in Fig. 5b as compared to Fig. 5a, where long acyl chain biotin-GM1 was detected by the same anti-biotin antibody gold conjugate using the same labeling protocol. In this case incorporation was performed at 37° for 48 hr with 10 μM biotin-GM1 of either chain length for comparison. Even though about 50% more of the short-chain derivative than of the long-chain analog is incorporated, much less labeling in the case of C_8-biotin-GM1 is obtained. This cannot be understood if we assume that the back-exchange phenomenon for sectioned membranes (cryosections) is the same as for intact plasma membranes. It may be that the short acyl chain biotin-GM1 is more easily extracted from cryosections than from intact membranes. Whatever the reason may be it is clear that biotin-GM1 bearing a C_8 or a shorter acyl chain is not appropriate for time course studies by immunoelectron microscopy. The use of a biotin-GM1 with a decanoyl or dodecanoyl chain may help to overcome this problem. These chain lengths will still allow insertion into the plasma membrane at low temperature and will render biotin-GM1 less extractable.

[49] R. E. Brown, F. A. Stephenson, T. Markello, Y. Barenholz, and T. E. Thompson, *Chem. Phys. Lipids* **38,** 79 (1985).

Acknowledgments

Support for this work was provided by the Deutsche Forschungsgemeinschaft SFB 284. We thank Dr. B. Liessem for a sample of β-hexosaminidases. The mass spectrometric analyses were performed by Dr. G. Pohlentz (FABMS) and J. Hörnschemeyer (MALDI-TOF-MS), hereby gratefully acknowledged. The skillful technical assistance of P. Hofmann is gratefully acknowledged.

[46] Assays for Transmembrane Movement of Sphingolipids

By DAN J. SILLENCE, RENÉ J. RAGGERS, and GERRIT VAN MEER

Introduction

It is well established that the lipids in mammalian cells, like membrane proteins, are not evenly distributed over the different cellular membranes and that each organelle has a specific lipid composition. The plasma membrane and the organelles with which it is in direct contact via recycling membrane vesicles, like endosomes and trans-Golgi network, are enriched in glycosphingolipids, sphingomyelin (SM), and cholesterol when compared to the endoplasmic reticulum (ER), mitochondria, and peroxisomes. Glycosphingolipids can form a protective barrier between the cell and its environment. In addition, glycosphingolipids are thought to be important for cell–cell and cell–substratum interactions. Some act as receptors for bacteria and viruses. Both glycosphingolipids and SM are believed to be involved in cell signaling.[1–3] However, whereas so far glycosphingolipids fulfill all functions on the outside of the cell, in the exoplasmic leaflet of the plasma membrane bilayer, SM signaling reportedly involves hydrolysis by a sphingomyelinase on the cytosolic aspect of the plasma membrane.[4] In the Golgi, glycosphingolipids, SM, and cholesterol are thought to form domains. These domains may sort membrane proteins by lateral clustering in the lumenal leaflet[5] or by increasing the plasma membrane's bilayer thickness, in which case they may be present on either side of the membrane.[6] The identification

[1] C. M. Linardic and Y. A. Hannun, *J. Biol. Chem.* **269,** 23530 (1994).
[2] S.-i. Hakomori and Y. Igarashi, *J. Biochem.* **118,** 1091 (1995).
[3] N. Andrieu, R. Salvayre, and T. Levade, *Eur. J. Biochem.* **236,** 738 (1996).
[4] S. Tomiuk, K. Hofmann, M. Nix, M. Zumbansen, and W. Stoffel, *Proc. Natl. Acad. Sci. U.S.A.* **95,** 3638 (1998).
[5] K. Simons and E. Ikonen, *Nature* **387,** 569 (1997).
[6] S. Munro, *EMBO J.* **14,** 4695 (1995).

0076-6879/00 $30.00

in the cytosol of a transfer protein for glycosphingolipids,[7,8] galectins of unknown function,[9] and annexins with lectin properties[10] support a physiologic relevance for pools of sphingolipids on the cytosolic surface of cellular membranes. To understand these functions more fully, the topology of the lipid molecules, i.e., their distribution across the lipid bilayer, also termed *transmembrane* distribution or *sidedness,* must be appreciated.

Sphingomyelin and all glycosphingolipids containing more than one carbohydrate are synthesized at the lumenal surface of the Golgi, which is topologically equivalent to the outer leaflet of the plasma membrane. They can be transported in the lumenal leaflet of carrier vesicles and there is no need for them to move across membranes to reach the surface of the cell.[11–15] However, evidence has been provided for a SM pool on the cytosolic surface of the plasma membrane,[1,3] while the presence of cytosolic pools of higher glycosphingolipids has been claimed (see Chan and Liu[16]). The monohexosylsphingolipids, glucosylceramide (GlcCer) and galactosylceramide (GalCer), are special cases. GlcCer is the only glycosphingolipid synthesized on a cytosolic leaflet, in the early Golgi.[15,17–19] While GlcCer is required for protein sorting in the cytosolic leaflet (Sprong *et al.,* unpublished), it is also utilized for the synthesis of higher glycosphingolipids in the lumenal leaflet. GlcCer must therefore "translocate" across the lipid bilayer toward the exoplasmic leaflet of the Golgi membrane[14,15] or another membrane of the vacuolar system. GalCer is synthesized in the lumen of the ER[20] and seems to translocate readily to the cytosolic leaflet of the ER and back into the Golgi.[15] The transmembrane movement of sphingolipids may be subject to regulation.

In recent years, we have demonstrated that various short-chain analogs

[7] R. J. Metz and N. S. Radin, *J. Biol. Chem.* **255,** 4463 (1980).

[8] A. Abe, *J. Biol. Chem.* **265,** 9634 (1990).

[9] S. H. Barondes, D. N. W. Cooper, M. A. Gitt, and H. Leffler, *J. Biol. Chem.* **269,** 20807 (1994).

[10] K. Kojima, K. Yamamoto, T. Irimura, T. Osawa, H. Ogawa, and I. Matsumoto, *J. Biol. Chem.* **271,** 7679 (1996).

[11] A. H. Futerman, B. Stieger, A. L. Hubbard, and R. E. Pagano, *J. Biol. Chem.* **265,** 8650 (1990).

[12] D. Jeckel, A. Karrenbauer, R. Birk, R. R. Schmidt, and F. Wieland, *FEBS Lett.* **261,** 155 (1990).

[13] W. W. Young, Jr., M. S. Lutz, and W. A. Blackburn, *J. Biol. Chem.* **267,** 12011 (1992).

[14] H. Lannert, C. Bünning, D. Jeckel, and F. T. Wieland, *FEBS Lett.* **342,** 91 (1994).

[15] K. N. J. Burger, P. van der Bijl, and G. van Meer, *J. Cell Biol.* **133,** 15 (1996).

[16] K.-F. Chan and Y. Liu, *Glycobiology* **1,** 193 (1991).

[17] H. Coste, M. B. Martel, and R. Got, *Biochim. Biophys. Acta* **858,** 6 (1986).

[18] A. H. Futerman and R. E. Pagano, *Biochem. J.* **280,** 295 (1991).

[19] D. Jeckel, A. Karrenbauer, K. N. J. Burger, G. van Meer, and F. Wieland, *J. Cell Biol.* **117,** 259 (1992).

[20] H. Sprong, B. Kruithof, R. Leijendekker, J. W. Slot, G. van Meer, and P. van der Sluijs, *J. Biol. Chem.* **273,** 25880 (1998).

of GlcCer and SM can be translocated from the inner to the outer leaflet of the plasma membrane by the multidrug transporters MDR1 P-glycoprotein and MRP1.[21,22] Since the related MDR3 P-glycoprotein has been shown to be a translocator of the natural phospholipid phosphatidylcholine, MDR1, MRP1, and other ABC transporters are candidates for translocating natural GlcCer *in vivo*. Here, we compare available techniques to study the transmembrane distribution of sphingolipids, and describe the requirements that these assays should meet in order to be applicable to measurements of transbilayer mobility. Translocation activity of a particular protein can then be tested by transfection into a cellular assay system, or by reconstitution into model membranes.

General Aspects of Assays for Translocation of Sphingolipids across Membranes

To determine transmembrane movement of a sphingolipid, an assay first must be developed to measure its distribution across the bilayer at a specific moment in time. The following criteria can be applied to judge the usefulness of a particular approach: (1) The procedure must recognize the lipid of interest on a membrane surface. Since a certain lipid will only rarely be present exclusively on one side of the bilayer, the assay should preferably quantitate the amount of the lipid that is accessible, instead of merely showing that a lipid is or is not present on that surface. If possible, all molecules of the lipid present on that surface should be recognized: biologic membranes are not a homogenous dispersion of lipid molecules, and the reactivity of a particular lipid molecule may vary due to local phase separations and protein–lipid and lipid–lipid interactions. (2) The procedure must discriminate the lipids in the one leaflet from those in the opposite leaflet of the bilayer. (a) For this, the recognizing agent should not permeate the membrane. This necessitates that the recognition reaction be terminated before the lipid from the inaccessible leaflet becomes accessible to the reagent, for example, during lipid extraction and analysis. (b) Lipid should not move between the two bilayer leaflets during the assay. The assay must be significantly faster than the natural transbilayer movement, or it should be performed under conditions where this movement is inhibited, e.g., by reduced temperature. In addition, the procedure should not induce redistribution of lipids between the two bilayer leaflets, i.e., the assay should be noninvasive. (3) The amount of the lipid under study should not change

[21] A. van Helvoort, A. J. Smith, H. Sprong, I. Fritzsche, A. H. Schinkel, P. Borst, and G. van Meer, *Cell* **87,** 507 (1996).

[22] R. J. Raggers, A. van Helvoort, R. Evers, and G. van Meer, *J. Cell Sci.* **112,** 415 (1999).

during the assay due to processes independent of the assay, such as via uptake by exocytosis and endocytosis with hydrolysis and (re)synthesis. In cases where these strict criteria are not met but where accurate quantitation is possible, the lipid pool in the outer leaflet may be calculated from a kinetic analysis (see van Meer *et al.*[23,24]). It should be noted that the cell does not necessarily have to survive the assay as long as the plasma membrane remains intact.

To determine the rate of lipid translocation across a membrane ($t_{1/2}$ in biomembranes between seconds and hours), on-line measurements with a fast response would be ideal, but such methodology is not yet available for natural sphingolipids. Usually, the pool of a lipid in one leaflet is radiolabeled and its transbilayer distribution is determined at multiple time points. Rates of transbilayer movement can then be calculated from the values at the different time points. The transbilayer distribution of the lipid at a defined time point can be determined from its accessibility to chemical modification (which generally affects membrane proteins as well), enzymatic modification, noncovalent labeling by a binding protein, or exchangeability by a transfer protein. Lipid sidedness and translocation may also be studied by using analogs of the lipids of interest. Examples are fluorescent analogs, spin-labeled analogs, analogs with shortened acyl chains, or a combination of these. Their sideness may then be monitored in time by the techniques mentioned above or by noninvasive spectroscopic methods. Generally, results are more convincing when confirmed by independent methods.

Calculating Translocation of Sphingolipids across Membranes

Transbilayer Distribution

The simplest situation is encountered when the transbilayer distribution of a lipid can be studied in a system possessing a single membrane. Quantitation of the amount of the lipid that is accessible versus the amount that is shielded directly yields the percentage of the lipid that is present in the outer leaflet and the ratio of that lipid in the outer vs. the inner leaflet. This is the case for unilamellar liposomes, most membrane viruses, gram-positive bacteria, erythrocytes, and isolated cell organelles (except for mitochondria, nuclei, and chloroplasts). More complicated is the situation in eukaryotic cells where the unreacted lipid can be present not only in the

[23] G. van Meer, K. Simons, J. A. F. Op den Kamp, and L. L. M. van Deenen, *Biochemistry* **20,** 1974 (1981).

[24] G. van Meer and J. A. F. Op den Kamp, *J. Cell Biochem.* **19,** 193 (1982).

TABLE I
POTENTIAL METHODS FOR MEASURING TRANSBILAYER DISTRIBUTION[a]

Sphingolipid	Amino reagents	Chemical oxidation	Enzymatic hydrolysis/ modification	Antibodies toxins	Monomeric transfer
Lyso-sphingolipids[b]	30, 31				[c]
Sphingomyelin			44–47	88	73, 74
Glucosylceramide		3[a]	17		72
Galactosylcermide		43	43	89, 90	72
Lactosylceramide		(43)[d]	(43)[d]	91	72
Complex glycolipids		13, 42	56–66	60, 67	72

[a] Numbers refer to references.
[b] Methods applicable to *N*-acylated sphingolipids can in principle be applied to corresponding lyso forms.
[c] Spontaneous transfer.
[d] Terminal galactose of lactosylceramide should have properties similar to those of GalCer.

inner leaflet of the plasma membrane but also in intracellular membranes. The outside/inside ratio can only be calculated from the fraction of the lipid that is in the plasma membrane. These data are mostly unavailable or controversial.[25–27] Published calculations depend on the preparation of "pure" plasma membranes[28,28a] and their significance depends on how accurately contamination with other membranes was determined.

Transbilayer Mobility

Spontaneous transbilayer mobility is a chance process that can be described by a set of simple logarithmic formulas. A simple one-bilayer system can be considered a two-pool closed system, where measurement of one of the two pools over time allows a full description of the kinetics.[24] Admittedly, this approach neglects the potential presence of lateral domains, which may result in the presence of two lipid pools within one membrane leaflet with potentially different translocation properties. Of course, activities of translocators must be approached as enzyme activities. Characteriza-

[25] G. van Meer, *TIBS* **12,** 375 (1987).
[26] Y. Lange, M. H. Swaisgood, B. V. Ramos, and T. L. Steck, *J. Biol. Chem.* **264,** 3786 (1989).
[27] D. E. Warnock, C. Roberts, M. S. Lutz, W. A. Blackburn, W. W. Young, Jr., and J. U. Baenziger, *J. Biol. Chem.* **268,** 10145 (1993).
[28] M. Record, A. El Tamer, H. Chap, and L. Douste-Blazy, *Biochim. Biophys. Acta* **778,** 449 (1984).
[28a] J. A. Post, G. A. Langer, J. A. F. Op den Kamp, and A. J. Verkleij, *Biochim. Biophys. Acta* **943,** 256 (1988).

tion of such enzyme activities is hampered by the fact that kinetics applicable to soluble molecules cannot be applied directly to membrane-embedded substrates (see, e.g., Mosior and Newton[29]). The possibility that translocators are present in specialized lateral membrane domains further complicates kinetics characterization.

Chemical Modification (Table I)

Principle of Assay

A reagent is added to the outside of a closed structure where it reacts with a particular lipid in the outer leaflet but not with the same lipid in the inner leaflet. The properties of the lipid in the outer leaflet have now been changed, and the modified lipid is separated from the unmodified lipid and quantitated.

Amino Reagents

A typical example of this approach is the labeling of primary amino groups. After years of work on the organization of membrane proteins, labeling reagents such as formylmethionyl(sulfonyl) methylphosphate[30] and trinitrobenzenesulfonic acid (TNBS)[31] were used in the first studies on lipid asymmetry to label phosphatidylethanolamine. Only 20% of this lipid was found to reside in the outer leaflet of the erythrocyte membrane.[30,31] TNBS has been applied to nucleated cells,[32–35] in which cases less than 5% of the cellular phosphatidylethanolamine was accessible. Later studies made use of the fluorescent label fluorescamine.[36] The reactions can be stopped by lowering the pH and by adding an excess of free amino groups. Reacted phosphatidylethanolamine can be conveniently separated from phosphatidylethanolamine by thin-layer chromatography (TLC). The penetration of the label through the membrane can be reduced by low temperature, or the two pools can be discriminated kinetically.[23] Note that the physical

[29] M. Mosior and A. C. Newton, *Biochemistry* **37,** 17271 (1998).
[30] M. S. Bretscher, *J. Mol. Biol.* **71,** 523 (1972).
[31] S. E. Gordesky and G. V. Marinetti, *Biochem. Biophys. Res. Commun.* **50,** 1027 (1973).
[31a] D. J. Sillence, R. J. Raggers, D. C. Neville, D. J. Harvey, and G. van Meer, *J. Lipid Res.* **4,** 1252 (2000).
[32] R. G. Sleight and R. E. Pagano, *J. Biol. Chem.* **258,** 9050 (1983).
[33] T. Kobayashi and R. E. Pagano, *J. Biol. Chem.* **264,** 5966 (1989).
[34] J. E. Vance, E. J. Aasman, and R. Szarka, *J. Biol. Chem.* **266,** 8241 (1991).
[35] L. S. Kean, A. M. Grant, C. Angeletti, Y. Mahé, K. Kuchler, R. S. Fuller, and J. W. Nichols, *J. Cell Biol.* **138,** 255 (1997).
[36] R. D. Koynova and B. G. Tenchov, *Biochim. Biophys. Acta* **727,** 351 (1983).

properties of the reaction product are different from those of the original phosphatidylethanolamine and indeed the labeling procedure has been reported to disrupt the bilayer and lead to enhanced phospholipid transbilayer movement.[37]

Although only few sphingolipids contain a primary amino group, they may be interesting because of their special functions. This is especially true for lysosphingolipids, which lack the amide-linked fatty acid and thereby expose an amino group at the membrane–water interface. *In vitro,* a quantitative assay for sphingolipids has been based on the reactivity of this amino group with fluorescamine.[38,39] Besides lyso-SM (sphingosylphosphocholine) and lysoglycolipids, interesting molecules are the signaling molecules sphingosine 1-phosphate, *N,N*-dimethylsphingosine and free sphingosine/sphinganine.

Another sphingolipid with a free aminogroup is ceramide phosphoethanolamine, an SM analog of low abundance in mammalian cells.[40] Yet others are glycosylphosphatidylinositol (GPI) protein anchors in yeast, which obtain glucosamine by deacetylation of *N*-acetylglucosamine, and obtain a ceramide anchor by exchange of the original diacylglycerol.[41]

Chemical Oxidation

The gangliosides GM1, GM2, GM3, and GD1a have been efficiently detected by mild oxidation of their sialic acid with $NaIO_4$.[13,42] The periodate oxidation was carried out on the surface of intact cells, after which the gangliosides were isolated and reacted with dinitrophenylhydrazine. The DNP products of the oxidized gangliosides were then separated from the nonoxidized gangliosides by TLC. Because the oxidation is carried out at 0°, membrane transport does not affect the results. The assay has been used to study transport of radiolabeled newly synthesized gangliosides to the cell surface. Unfortunately, no control experiments have been carried out so far to test whether only sialic acids on the outside of the plasma membrane are oxidized. Because gangliosides are only minor components of cellular membranes, it is unlikely that their oxidation would affect the intactness of the plasma membrane.

Under different conditions (50 vs. 2 m*M* periodate and longer times)

[37] G. van Duijn, J. Luiken, A. J. Verkleij, and B. de Kruijff, *Biochim. Biophys. Acta* **863,** 193 (1986).

[38] M. Naoi, Y. C. Lee, and S. Roseman, *Anal. Biochem.* **58,** 571 (1974).

[39] T. J. Higgins, *J. Lipid Res.* **25,** 1007 (1984).

[40] M. Malgat, A. Maurice, and J. Baraud, *J. Lipid Res.* **27,** 251 (1986).

[41] A. Conzelmann, A. Puoti, R. L. Lester, and C. Desponds, *EMBO J.* **11,** 457 (1992).

[42] H. Miller-Podraza and P. H. Fishman, *Biochemistry* **21,** 3265 (1982).

galactosylceramide has also been oxidized. Essentially all galactosylceramide was oxidized in myelin, whereas only 50% was oxidizable in sonicated myelin lipids, suggesting that the majority of the galactosylceramide was situated in the external leaflet.[43] In contrast to the gangliosides, galactosylceramide is a major lipid in myelin and in apical membranes of some epithelial cells. Therefore, studies of transbilayer distribution using galactosylceramide oxidation require controls for the maintenance of the intactness of the membrane (see Galactose oxidase section below).

TEMPO Oxidation

Recently, we have developed a method that uses TEMPO nitroxides, which specifically oxidize primary hydroxyls and amines, such as the C-6 hydroxyl of the glucose moiety in GlcCer. This oxidation reaction relies on the generation of an active nitrosonium ion by *in situ* oxidation of the TEMPO free radical by OCl^-/Br^-. We noticed that the leakage of TEMPO through membranes can be essentially prevented by using the carboxy-TEMPO analog. Carboxy-TEMPO can be purchased from several commercial sources as a dry powder in the free acid form. Its solubility is greatly enhanced by conversion to the sodium salt. A typical experiment involves weighing out 10 mg of carboxy-TEMPO (15 m*M*) and sonicating in 830 μl of 80 m*M* $NaHCO_3$/60 m*M* NaOH, pH 13. After cooling on ice, the nitrosonium ion is generated by addition of NaBr (2 m*M*) followed by NaOCl (10 m*M*). This reaction mixture is then diluted to isosmolarity by the addition of 300 μl of water and 2 ml of 100 m*M* NaCl/60 m*M* $NaHCO_3$/2 m*M* KCl. 1 *M* NaOH is then added until the solution has a pH of 9.5. Butylated hydroxytoluene (100 μM) is then added to inhibit nonspecific lipid peroxidation. The oxidation product of GlcCer can be separated from native GlcCer by thin layer chromatography (TLC).[31a]

Enzymatic Modification

Principle of Assay

An enzyme is added to one side of the membrane where it modifies a substrate lipid. The enzyme may need a second substrate and/or cofactors. The enzyme modifies a particular lipid in the outer leaflet but not the inner leaflet, after which the enzymatic reaction is stopped, and the modified lipid product separated from the unmodified lipid and quantitated.

[43] C. Linington and M. G. Rumsby, *J. Neurochem.* **35,** 983 (1980).

Sphingomyelinase

Phospholipases are the prototypic enzymes for asymmetry studies.[44] In the case of the sphingolipids, sphingomyelinase (SMase from *Staphylococcus aureus* or *Bacillus cereus*) splits SM into the phosphocholine headgroup and ceramide backbone. The products can be easily separated from SM by lipid extraction and TLC. The assay can be performed at 37°[44] but also at 15°.[45,46] The latter has been convenient in studies on living cells, because it eliminates the effects of endocytosis and exocytosis during the assay. In most cases, the loss of SM is quantitated. It is taken to reflect the fraction of SM that was originally present on the side where SMase had been added. SMase has access to SM in all membranes studied so far, which is in contrast to some phospholipases that cannot hydrolyze substrate lipids in cellular plasma membranes. Of the erythrocyte SM, 73% was degraded by SMase alone. SMase has been used in combination with other phsopholipases.[44,47]

Sphingomyelinase added to the outside of a closed membrane will not cross the membrane and will only have access to the SM on the exposed surface. A second positive characteristic of the assay is that it can be conveniently stopped by chelating divalent cations. On the negative side, it should be realized that the assay is invasive because SMase changes the membrane under study. The membrane phospholipid SM is replaced by ceramide, a hydrophobic lipid without a polar headgroup that by itself does not form bilayers in aqueous environments. The potential problems for the asymmetry assay are illustrated by observations on the effects of generic phospholipases C on membranes. Phospholipases C cleave the headgroup from glycerophospholipids and thereby replace phospholipids by diacylglycerols. Extensive treatment of membranes resulted in so-called signet ring structures where the diacylglycerol accumulated in the hydrophobic membrane interior, and due to the removal of phospholipid from the surface, the membrane shrunk with a 50% reduction in surface area.[48] By definition this also implies a 50% reduction in the surface area of the inner bilayer leaflet. Half of the lipids from the inner leaflet must have moved across the membrane to the outer leaflet where they were then exposed to the phospholipase C. Accidentally, phospholipase C did not induce leaks in the membrane and hydrolysis stopped when 50% of the lipids had been

[44] A. J. Verkleij, R. F. A. Zwaal, B. Roelofsen, P. Comfurius, D. Kastelijn, and L. L. M. van Deenen, *Biochim. Biophys. Acta* **323,** 178 (1973).

[45] Y.-J. Shiao and J. E. Vance, *J. Biol. Chem.* **268,** 26085 (1993).

[46] A. van Helvoort, M. L. Giudici, M. Thielemans, and G. van Meer, *J. Cell Sci.* **110,** 75 (1997).

[47] D. Allan and C. M. Walklin, *Biochim. Biophys. Acta* **938,** 403 (1988).

[48] J. B. Finean and A. Martonosi, *Biochim. Bophys. Acta* **98,** 547 (1965).

degraded. These features have been taken as evidence that phospholipase C can be succesfully used in asymmetry studies, even in a study confirming the 50% reduction in surface area.[49] It is unclear why the intact phospholipids in the outer leaflet are no longer substrates for the phospholipase C, but this may be a matter of product inhibition. Still, clearly the assay does not meet the requirement that the agent (enzyme) should not have access to lipids in the other leaflet and should not induce redistribution of the lipids across the bilayer. Phospholipase C could be utilized in an assay for the distribution of lipids when the hydrolyzable lipid is only a minor fraction of the total membrane lipid.[50]

The dramatic effects observed for the generic phospholipase C apply to SMase. In erythrocytes, ceramide produced during the assay accumulated in the hydrophobic membrane interior where it gave rise to ceramide droplets.[44] The reduction in the amount of SM in the outer leaflet induced invagination of the erythrocyte membrane.[47] SMase induced ATP-independent endocytosis in living cells.[51] It is not clear whether the addition of SMase to cells has led to artifacts in previously published studies. Mammalian plasma membranes typically contain 15–25 mol % SM, and the higher the mole percentage of SM in a membrane, the higher the risk of membrane damage and artefactual redistribution.[52] In a membrane virus containing 25 mol % SM, a kinetic assay (at reduced temperature) yielded smaller numbers for the SM pool in the outer leaflet,[23] than an assay where the reaction was allowed to go to completion.[53]

A further problem with the use of SMase is that the ceramide produced can be utilized as a substrate by the enzyme SM synthase. SM synthase, which is present in Golgi membranes and also on the outside of plasma membranes,[46,54] transfers the phosphocholine headgroup from phosphatidylcholine to ceramide and thereby generates diacylglycerol. Whether or not this occurs can be assessed from the relative amounts of ceramide and phosphocholine produced. Finally, commercial SMase may be contaminated by phospholipase C.

[49] J. A. Higgins, *J. Cell Sci.* **53,** 211 (1982).

[50] R. Sundler, A. W. Alberts, and P. R. Vagelos, *J. Biol. Chem.* **253,** 5299 (1978).

[51] X. Zha, L. M. Pierini, P. L. Leopold, P. J. Skiba, I. Tabas, and F. R. Maxfield, *J. Cell Biol.* **140,** 39 (1998).

[52] G. Basáñez, M. B. Ruiz-Argüello, A. Alonso, F. M. Goñi, G. Karlsson, and K. Edwards, *Biophys. J.* **72,** 2630 (1997).

[53] D. Allan and P. Quinn, *Biochim. Biophys. Acta* **987,** 199 (1989).

[54] A. van Helvoort, W. van't Hof, T. Ritsema, A. Sandra, and G. van Meer, *J. Biol. Chem.* **269,** 1763 (1994).

Glucocerebrosidase and Endoglycoceramidase

Comparable to SMase, acid glucocerebrosidase (glucoceramidase) has been applied as a tool to study the topology of GlcCer, by hydrolysis to glucose and ceramide.[17] One difficulty is that the enzyme normally acts in the lumen of the lysosome, under conditions of low pH, and requires the presence of the activator protein saposin C.[55] The enzyme does not show any activity when added to the outside of cells (our unpublished observations). It may be that the addition of saposin C or of the glycolipid transfer protein (see below) suffices to activate the protein. Endoglycoceramidase or ceramide glycanase (*Rhodococcus*) cleaves the carbohydrate chain from ceramide, hydrolyzing most of the glycolipids on intact erythrocytes within 2 hr at 37° when used in the presence of an activator protein.[56] GlcCer and GalCer are not substrates.[57]

Neuraminidase

Neuraminidase (sialidase) removes terminal sialic acids from gangliosides. Using this technique it has been concluded that all GM3 was present on the membrane surface of the enveloped Sindbis virus.[58,59] In neuroblastoma cells, neuraminidase from *Vibrio cholerae* cleaved the sialic acid from most GM3, while it also removed the terminal sialic acid from GD1a to yield GM1a.[60] In contrast, in macrophages most GD1a but no GM3 was degraded by this enzyme.[61] Neuraminidase from *Clostridium perfringens* was used to study transport of newly synthesized GM3, GD3, and GT3 to the surface of retina cells.[62]

N-Deacylase

A *Pseudomonas N*-deacylase removes the acyl chain of a number of glycolipids and to a lesser degree SM.[63] The use of this enzyme may be preferable over using hydrolases that cleave off the headgroup, since this enzyme produces two membrane lipids, a lysosphingolipid and a free fatty

[55] W. Fürst and K. Sandhoff, *Biochim. Biophys. Acta* **1126,** 1 (1992).
[56] M. Ito, Y. Ikegami, and T. Yamagata, *Euro. J. Biochem.* **218,** 645 (1993).
[57] T. Yamagata and M. Ito, *in* "CRC Handbook of Endoglycosidases and Glycoamidases" (N. Takahashi and T. Muramatsu, eds.), p. 133. CRC Press, Boca Raton, Florida, 1992.
[58] W. Stoffel, R. Anderson, and J. Stahl, *Hoppe-Seylers Z. Physiol. Chem.* **356,** 1123 (1975).
[59] W. Stoffel and W. Sorgo, *Chem. Phys. Lipids* **17,** 324 (1976).
[60] H. Miller-Podraza, R. M. Bradley, and P. H. Fishman, *Biochemistry* **21,** 3260 (1982).
[61] L. J. Macala and H. C. Yohe, *Glycobiology* **5,** 67 (1995).
[62] V. M. Rosales Fritz and H. J. Maccioni, *J. Neurochem.* **65,** 1859 (1995).
[63] M. Ito, T. Kurita, and K. Kita, *J. Biol. Chem.* **270,** 24370 (1995).

acid, that in the absence of a scavenger like albumin do not affect the stability of the membrane.[44,50]

Galactose Oxidase

Galactose oxidase (*Dactylium dendroides*) shows specificity for galactose and *N*-acetylgalactosamine whose primary hydroxyls are oxidized to aldehydes. From studies where the oxidized lipids were subsequently reduced by NaB^3H_4, it was concluded that most GM3 and globoside were present in the outer leaflet of the erythrocyte membrane. Although lactosylceramide was a substrate in the presence of detergent, it was not oxidized in the intact membrane, probably because it does not protrude far enough from the membrane (or is not present on the cell surface).[64] In contrast to intact erythrocytes, inside-out vesicles showed very little labeling supporting the presence of GM3 and globoside in the outer leaflet of the erythrocyte membrane.[65] In a later study in which the fraction of globoside that was oxidized was determined by mass spectrometry, only up to two-thirds was oxidized, implying that its distribution across the erythrocyte membrane bilayer has not been unequivocally determined.[66] In intact macrophages GM1a was efficiently labeled by this procedure.[61] In myelin, approximately 50% of the galactosylceramide could be oxidized by galactose oxidase, whereas all was oxidized by periodate (see above), suggesting an accessibility problem in the outer leaflet.[43]

Noncovalent Protein Binding

Principle of Assay

A protein with a binding specificity for a particular glycolipid class is added to cells and its association is measured. Quantitation can be performed by using a labeled binding protein or a labeled secondary protein (like an antibody); alternatively, fluorescent labels can be studied by light microscopy whereas gold-labeled proteins can be visualized by electron microscopy (EM). So far, this approach has demonstrated the presence of essentially all glycolipids on the cell surface, the outer leaflet of the plasma membrane, but only rarely have binding proteins been applied to study the distribution across the bilayer.

[64] C. G. Gahmberg and S.-I. Hakomori, *J. Biol. Chem.* **248,** 4311 (1973).
[65] T. L. Steck and G. Dawson, *J. Biol. Chem.* **249,** 2135 (1974).
[66] A. Lampio, J. Finne, D. Homer, and C. G. Gahmberg, *Eur. J. Biochem.* **145,** 77 (1984).

Antibodies

It has become clear that many antibodies are directed against glycolipids. The prime example is the Forssman antigen,[67] a 5-sugar lipid of the globo series. Antibodies against dozens of glycolipids are now available, and many have been used to localize the antigenic substrate.

Toxins

As for the antibodies, it has become clear over the years that a number of toxins use glycolipids as receptors. The clearest example, cholera toxin, has been applied to study the organization of GM1.[60] From knowing the stoichiometry of binding, one toxin molecule to five GM1, careful quantitation assigned most GM1 in these cells to the cell surface. The toxin has since been used to show that GM1 on the surface is concentrated in caveolae.[68] A potential artifact in such studies is lateral redistribution of the glycolipid due to the oligomeric nature of the binding.

Lectins

By the method of fracture label, which permits cytochemical characterization of each of the two freeze-fracture faces (each bilayer leaflet) of membranes, it was demonstrated that the lectin concanavalin A labeled the glycosphingolipid lipophosphonoglycan selectively in the outer bilayer leaflet of the plasma membrane of *Acanthamoeba castellanii.*[69]

Lipid Exchange and Transfer Proteins

Principle of the Assay

Sphingolipids possessing only one fatty chain, like lyso-SM (sphingosylphosphocholine), lysoglycolipids, sphingosine 1-phosphate, *N,N*-dimethylsphingosine, and free sphingosine/sphinganine, can be depleted from an accessible surface by the addition of a lipid scavenger like serum albumin or liposomes. Because these lipids occur in only low mol %, it is unlikely that their depletion changes physical properties of the membrane.

For normal sphingolipids, monomeric diffusion between membranes is an exceedingly slow process with half-times on the order of days.[70] However, exchange can be accelerated by transfer proteins, soluble proteins that can

[67] J. Forssman, *Biochem. Zeitschr.* **37,** 78 (1911).
[68] R. G. Parton, *J. Histochem. Cytochem.* **42,** 155 (1994).
[69] B. Bloj and D. B. Zilversmit, *J. Biol. Chem.* **256,** 5988 (1981).
[70] M. C. Phillips, W. J. Johnson, and G. H. Rothblat, *Biochim. Biophys. Acta* **906,** 223 (1987).

bind and carry a single lipid molecule.[71] In the presence of liposomes that contain an excess of unlabeled lipid these proteins can mediate the complete replacement of radiolabeled lipid on the surface of a closed membrane (like an intact cell). Some transfer proteins bind tightly to the lipid, never leave the membrane surface in an empty state, and thus cannot cause net mass transfer of lipid. In such cases, the membrane structure is not perturbed and so only the lipid on the outer leaflet of the membrane is exchanged. The fraction of radiolabeled lipid that was lost reflects the fraction of that lipid in the external leaflet. Some transfer proteins can exist in an empty state in solution. They should be used with caution as they can generate net transfer of lipid mass between membranes, which changes their lipid composition and, potentially, organization. Note that for a transfer protein to generate net transfer there needs to be a difference in the energy state of the lipid between donor and acceptor membranes.

Glycolipid and SM Transfer Proteins

A transfer protein that can exchange glycolipids has been isolated and its amino acid sequence determined.[7,8] By a monomeric carrier mechanism it transfers many different types of glycosphingolipids (and to some extent SM) between membranes.[72] In a transfer assay it should be used with caution because it can exist in solution in an empty state. A transfer protein with a preference for SM has been reported,[73] while a phosphatidylinositol transfer protein isoform has also been found with a relatively high affinity for SM.[74]

Nonspecific Transfer Protein and Saposins

Both SM and glycosphingolipids are substrates for a nonspecific transfer protein.[69] *In vivo,* this protein, also called sterol carrier protein 2, is generated from a higher molecular weight precursor in the peroxisomes. Its function in lipid transfer is unclear.[75] Because the protein apparently does not carry monomeric lipids and because of its lack of specificity, it scrambles the lipid composition of accessible membrane surfaces and causes net trans-

[71] K. W. A. Wirtz, *Annu. Rev. Biochem.* **60,** 73 (1991).

[72] T. Sasaki and A. Abe, *Methods Enzymol.* **179,** 559 (1989).

[73] E. V. Dyatlovitskaya, N. G. Timofeeva, E. F. Yakimenko, L. I. Barsukov, G. I. Muzya, and L. D. Bergelson, *Eur. J. Biochem.* **123,** 311 (1982).

[74] K. J. de Vries, A. A. J. Heinrichs, E. Cunningham, F. Brunink, J. Westerman, P. J. Somerharju, S. Cockcroft, K. W. A. Wirtz, and G. T. Snoek, *Biochem. J.* **310,** 643 (1995).

[75] U. Seedorf, M. Raabe, P. Ellinghaus, F. Kannenberg, M. Fobker, T. Engel, S. Denis, F. Wouters, K. W. Wirtz, R. J. Wanders, N. Maeda, and G. Assmann, *Genes Develop.* **12,** 1189 (1998).

fer. It should be used with caution. The lysosomal saposins, which serve functions in presenting glycolipids to their hydrolytic enzymes for degradation, were originally characterized as lipid exchange proteins *in vitro*.[55]

Application of Analogs of Lipids of Interest

Analogs are different from their natural counterparts. Data on the behavior of analogs provide insights into the behavior of the natural lipids but cannot be extrapolated without independent confirmation. Clearly, the closer the resemblance in structure, the better the prediction. Besides the fact that lipid analogs will only be applied when convenient assays for the natural lipids are not available, an advantage is that the structure of the lipid analog is chemically defined. In addition, analogs sometimes allow approaches that are inherently impossible for the natural lipids (see below).

Short-Chain Lipids: Principle of Assay

Transfer rates of two-chain sphingolipids can be dramatically increased by replacing the fatty acyl moiety or the sphingoid base with shorter alkyl chains. Typically a C_4–C_6 chain is used that is labeled with ^3H or ^{14}C, with fluorescent moieties like NBD or BODIPY (Molecular Probes, Eugene, OR) or with a spin-label. For these lipid analogs, the off-rate from membranes is sufficiently high that they can be depleted from an accessible membrane surface by the proper scavenger, such as an excess of liposomes[76] or serum albumin.[77] Sometimes analogs have been used that carry two short chains, such as a truncated ceramide containing a C_8-sphingosine and a C_8 fatty acid.[19] Because these are highly water soluble, no scavenger is needed for depletion. These assays can be performed on ice, whereby one monitors the loss of the lipid after exogenous addition and an internalization incubation[78] or the appearance of a lipid after intracellular synthesis from a precursor, typically ceramide.[19,76,77] In addition, such lipids can be conveniently incorporated into membranes from BSA solutions or ethanolic solutions.

Short-Chain Sphingolipids Added to the Outside of Cells or Organelles

In many studies analogs of glycolipids, notably GlcCer, or SM have been added to cells that carry a short fluorescent fatty acid. Whereas origi-

[76] N. G. Lipsky and R. E. Pagano, *J. Cell Biol.* **100,** 27 (1985).

[77] G. van Meer, E. H. K. Stelzer, R. W. Wijnaendts-van-Resandt, and K. Simons, *J. Cell Biol.* **105,** 1623 (1987).

[78] O. C. Martin and R. E. Pagano, *J. Cell Biol.* **125,** 769 (1994).

most studies used C6-NBD, the new analogs containing a C5-BODIPY chain were later shown to have several advantages in terms of being more apolar and having superior fluorescent properties.[79] After insertion into the outer leaflet of the plasma membrane, evidence for translocation across the plasma membrane bilayer can obtained in either of two ways. First, after a time interval a fraction of the short-chain lipid can no longer be extracted ("back-exchanged") by BSA in the extracellular medium. Second, uptake can be assessed by fluorescence microscopy, whereby the fluorescence pattern can provide information on the mechanism of uptake. From such a study it has been concluded that C5-BODIPY-SM is taken up by endocytosis, whereas most C5-BODIPY-GlcCer enter the cell by translocation.[78] These two processes can be distinguished by performing the assay at reduced temperature or after energy depletion, which inhibits endocytosis.[78] Alternatively, translocation of the GlcCer analogs C6-NBD-GlcCer and C8C8-GlcCer across the membrane of isolated Golgi has been assessed by conversion to the analogous lactosylceramide, which, being in the Golgi lumen, was protected against BSA extraction.[14,15] Finally, the NBD moiety can be chemically quenched by dithionite,[80] which can be used to establish the transbilayer distribution.

Important data on sphingolipid translocation across the plasma membrane have been obtained in the opposite direction. For this, short-chain analogs of ceramide, C6-NBD-ceramide, or C_8/C_8-ceramide were added to cells. After uptake into the external leaflet of the plasma membrane, these lipids flipped across the plasma membrane, a process probably facilitated by the absence of a polar headgroup, as was evident from the fact that they were converted by enzymes in the cellular Golgi complex to GlcCer and SM. This uptake is independent of vesicular traffic as it continues below 15°. Because short-chain GlcCer is synthesized at the cytosolic surface of the Golgi and because it can freely exchange across aqueous phases, it has free access to the cytosolic surface of the plasma membrane. Translocation toward the outer leaflet can then be measured by accessibility to depletion by BSA in the extracellular medium. That this approach really works was demonstrated by our studies on cells transfected with various multidrug transporters, in which it was demonstrated that the MDR1 P-glycoprotein (but not MDR3) is capable of translocating various short-chain GlcCer analogs across the plasma membrane, whereas the multidrug resistance protein MRP1 could only translocate C6-NBD-GlcCer.[21,22] SM is synthesized in the Golgi lumen and does not have access to the cytosolic surface of the plasma membrane. Still, a slight modification of the assay made it

[79] R. E. Pagano, O. C. Martin, H. C. Kang, and R. P. Haugland, *J. Cell Biol.* **113,** 1267 (1991).
[80] J. McIntyre and R. G. Sleight, *Biochemistry* **30,** 11819 (1991).

possible to study translocation of short-chain SM as well. For this, cells were treated with brefeldin A, which results in mixing of the membranes of Golgi and ER and allows translocation of newly synthesized SM to their cytosolic surface.[21,22] Since natural GlcCer and SM do not freely diffuse through the cytosol at 15° this approach can only be applied to their analogs.

On-Line Measurements of Fluorescence

The monitoring of fluorescent lipids on line has great potential for studying kinetic properties of transbilayer movement. For example, as can be done with spin-labels, fluorescence can be quenched selectively on one side of the membrane. Quenching can be irreversible, such as the quenching of NBD fluorescence by dithionite,[80] which has been evaluated for its applicability in lipid asymmetry studies.[81,82] Quenching can be reversible, like quenching of NBD fluorescence by neighboring trinitrophenyl groups.[83] These can be introduced selectively into one leaflet of the bilayer by labeling with TNBS (see above). Reversible quenching can also be performed by the presence of a second fluorescent probe that accepts the fluorescence energy of the first probe and generates resonance energy transfer.[84] In addition, high concentrations of a fluorescent probe result in self-quenching or excimer formation[85] and dilution by transport can be measured. These techniques are potentially useful for the study of transbilayer movement of lipids. However, since these techniques have not been applied to sphingolipids, the issue is not discussed in detail here.

Short-Chain Spin-Labeled Sphingolipids

Sidedness and transbilayer redistribution of short-chain sphingolipids have also been approached by using spin labels, notably SM carrying a 4-doxyl pentanoyl chain (C5-doxyl).[86] After incorporation into the plasma membrane of erythrocytes from ethanolic solution, the amount of C5-doxyl-SM in the cytoplasmic leaflet can be determined at a specific time-point as the residual electron spin resonance (ESR) signal after reduction of the probe in the outer leaflet using sodium ascorbate. Reduction takes 5 min at 4°.[86] Alternatively, C5-doxyl-SM can be removed from the outer leaflet by back-exchange against BSA, after which the residual ESR signal is

[81] C. Balch, R. Morris, E. Brooks, and R. G. Sleight, *Chem. Phys. Lipids* **70,** 205 (1994).

[82] T. Pomorski, A. Herrmann, B. Zimmermann, A. Zachowski, and P. Muller, *Chem. Phys. Lipids* **77,** 139 (1995).

[83] A. J. Schroit and R. E. Pagano, *Cell* **23,** 105 (1981).

[84] J. W. Nichols and R. E. Pagano, *Biochemistry* **21,** 1720 (1982).

[85] R. C. Hresko, I. P. Sugar, Y. Barenholz, and T. E. Thompson, *Biophys. J.* **51,** 725 (1987).

[86] A. Zachowski, P. Fellman, and P. F. Devaux, *Biochim. Biophys. Acta* **815,** 510 (1985).

determined. This method has, for example, been used to study lipid redistribution from the cytosolic to the lumenal surface of microsomes.[87] The resolution of this technique depends on the time and temperature needed for the ESR measurement, which is typically minutes at room temperature. It should be established that the measured ESR signal is due to the original lipid and not a metabolic product.

Acknowledgments

The original work on which this paper was based has been supported by a Marie Curie Fellowship from the TMR program of the EC to DS, and by the Netherlands Foundation for Chemical Research (CW) with financial aid from the Netherlands Organization for Scientific Research (NWO) to R.R. and G.v.M.

[87] A. Herrmann, A. Zachowski, and P. F. Devaux, *Biochemistry* **29,** 2023 (1990).
[88] K. Hanada, T. Hara, M. Fukasawa, A. Yamaji, M. Umeda, and M. Nishijima, *J. Biol. Chem.* **273,** 33787 (1998).
[89] B. Ranscht, P. A. Clapshaw, J. Price, M. Noble, and W. Seifert, *Proc. Natl. Acad. Sci. U.S.A.* **79,** 2709 (1982).
[90] R. Bansal, A. E. Warrington, A. L. Gard, B. Ranscht, and S. E. Pfeiffer, *J. Neurosci. Res.* **24,** 548 (1989).
[91] F. W. Symington, W. A. Murray, S. I. Bearman, and S.-i. Hakomori, *J. Biol. Chem.* **262,** 11356 (1987).

Section V

Other Methods

[47] Compilation of Methods Published in Previous Volumes of *Methods in Enzymology*

By ALFRED H. MERRILL, JR.

Introduction

The *Methods in Enzymology* series has published many chapters on sphingolipids beyond the ones included in this volume. To facilitate location of these chapters, we list below the ones that we have been able to identify under the subheadings of the two volumes dedicated to sphingolipids (Vol. 311 and this volume).

Sphingolipid Metabolism

Enzymes of ceramide biosynthesis, Vol. 209, 427 (1992).[1]
Serine palmitoyltransferase, Vol. 311, 3 (2000).[2]
Assay of the *Saccharomyces cerevisiae* dihydrosphingosine C-4 hydroxylase, Vol. 311, 9 (2000).[3]
Ceramide synthease, Vol. 311, 15 (2000).[4]
Dihydroceramide desaturase, Vol. 311, 22 (2000).[5]
Assays for the biosynthesis of sphingomyelin (ceramide phosphocholine) and ceramide-phosphoethanolamine, Vol. 311, 31 (2000).[6]
Phospholipid exchange protein-dependent synthesis of sphingomyelin, Vol. 98, 596 (1983).[7]
Glucosylceramide synthase: assay and properties, Vol. 311, 42 (2000).[8]
Methods for studying glucosylceramide synthase, Vol. 311, 50 (2000).[9]
A novel assay method for the biosynthesis of galactosyl- and glucosylceramides, Vol. 72, 384 (1981).[10]

[1] A. H. Merrill, Jr., and E. Wang, *Methods Enzymol.* **209,** 427 (1992).
[2] R. C. Dickson, R. L. Lester and M. M. Nagiec, *Methods Enzymol.* **311,** 3 (2000).
[3] M. M. Grilley and J. Y. Takemoto, *Methods Enzymol.* **311,** 9 (2000).
[4] E. Wang and A. H. Merrill, Jr., *Methods Enzymol.* **311,** 15 (2000).
[5] H. Schulze, C. Michel, and G. van Echten-Deckert, *Methods Enzymol.* **311,** 22 (2000).
[6] M. Nikolova-Karakashian, *Methods Enzymol.* **311,** 31 (2000).
[7] D. R. Voelker and E. P. Kennedy, *Methods Enzymol.* **98,** 596 (1983).
[8] J. A. Shayman and A. Abe, *Methods Enzymol.* **311,** 42 (2000).
[9] P. Marks, P. Paul, Y. Kamisaka, and R. E. Pagano, *Methods Enzymol.* **311,** 50 (2000).
[10] E. Costantino-Ceccarini and A. Cestelli, *Methods Enzymol.* **72,** 384 (1981).

0076-6879/00 $30.00

Analysis of galactolipids and UDP-galactose: ceramide galactosyltransferase, Vol. 311, 59 (2000).[11]

UDP-galactose: ceramide galactosyltransferase from rat brain, Vol. 71, 521 (1981).[12]

Glycosyltransferase assay, Vol. 138, 567 (1987).[13]

Assay of lactosylceramide synthase and comments on its potential role in signal transduction, Vol. 311, 73 (2000).[14]

In vitro assays for enzymes of ganglioside synthesis, Vol. 311, 82 (2000).[15]

Analysis of sulfatide and enzymes of sulfatide metabolism, Vol. 311, 94 (2000).[16]

1-*O*-acylceramide synthase, Vol. 311, 105 (2000).[17]

N-*a*cetylation of sphingosine by platelet-activating factor: sphingosine transacetylase, Vol. 311, 117 (2000).[18]

Inositolphosphoryl ceramide synthase from yeast, Vol. 311, 123 (2000).[19]

Enzymes of sphingolipid metabolism in plants, Vol. 311, 130 (2000).[20]

Purification and characterization of recombinant human acid sphingomyelinase expressed in insect Sf21 cells, Vol. 311, 149 (2000).[21]

Human acid sphingomyelinase from human urine, Vol. 197, 536 (1991).[22]

Purification of neutral sphingomyelinase from human urine, Vol. 197, 540 (1991).[23]

Purification of rat brain membrane neutral sphingomyelinase, Vol. 311, 156 (2000).[24]

Sphingomyelinase assay using radiolabeled substrate, Vol. 311, 164 (2000).[25]

Robotic assay of sphingomyelinase activity for high-throughput screening, Vol. 311, 168 (2000).[26]

[11] H. Sprong, G. van Meer, and P. van der Sluijs, *Methods Enzymol.* **311,** 59 (2000).

[12] N. M. Neskovic, P. Mandel, and S. Gatt, *Methods Enzymol.* **71,** 521 (1981).

[13] B. E. Samuelsson, *Methods Enzymol.* **138,** 567 (1987).

[14] S. Chatterjee, *Methods Enzymol.* **311,** 73 (2000).

[15] G. Pohlentz, C. Kaes, and K. Sandhoff, *Methods Enzymol.* **311,** 82 (2000).

[16] F. B. Jungalwala, M. R. Natowicz, P. Chaturvedi, and D. S. Newburg, *Methods Enzymol.* **311,** 94 (2000).

[17] J. A. Shayman and A. Abe, *Methods Enzymol.* **311,** 105 (2000).

[18] T. C. Lee, *Methods Enzymol.* **311,** 117 (2000).

[19] A. S. Fischl, Y. Liu, A. Browdy, and A. E. Cremesti, *Methods Enzymol.* **311,** 123 (2000).

[20] D. V. Lynch, *Methods Enzymol.* **311,** 130 (2000).

[21] S. Lansmann, O. Bartelsen, and K. Sandhoff, *Methods Enzymol.* **311,** 149 (2000).

[22] L. E. Quintern and K. Sandhoff, *Methods Enzymol.* **197,** 536 (1991).

[23] S. Chatterjee and N. Ghosh, *Methods Enzymol.* **197,** 540 (1991).

[24] B. Liu and Y. A. Hannun, *Methods Enzymol.* **311,** 156 (2000).

[25] B. Liu and Y. A. Hannun, *Methods Enzymol.* **311,** 164 (2000).

[26] A. G. Barbone, A. C. Jackson, D. M. Ritchie, and D. C. Argentieri, *Methods Enzymol.* **311,** 168 (2000).

A high throughput sphingomyelinase assay, Vol. 311, 176 (2000).[27]

The action of pure phospholipases on native and ghost red cell membranes, Vol. 32, 131 (1974).[28]

Analysis of sphingomyelin hydrolysis in caveolar membranes, Vol. 311, 184 (2000).[29]

Protein activator (coglucosidase) for the hydrolysis of β-glucosides, Vol. 83, 596 (1982).[30]

Ceramidases, Vol. 311, 194 (2000).[31]

Purification of acid ceramidase from human placenta, Vol. 311, 201 (2000).[32]

Ceramide kinase, Vol. 311, 207 (2000).[33]

Assaying sphingosine kinase activity, Vol. 311, 215 (2000).[34]

Yeast sphingosine 1-phosphate phosphatases: assay, expression, deletion, purification, and cellular localization by GFP tagging, Vol. 311, 223 (2000).[35]

Analysis of ceramide 1-phosphate and sphingosine 1-phosphate phosphatase activities, Vol. 311, 233 (2000).[36]

Sphingosine 1-phosphate lyase, Vol. 311, 244 (2000).[37]

Activator proteins for lysosomal glycolipid hydrolysis, Vol. 138, 792 (1987).[38]

Sphingolipid hydrolases and activator proteins, Vol. 311, 255 (2000).[39]

Protein activators for the hydrolysis of GM1 and GM2 gangliosides, Vol. 83, 588 (1982).[40]

Ceramide trihexosidase from human placenta, Vol. 50, 533 (1978).[41]

Arylsulfatases A and B from human liver, Vol. 50, 537 (1978).[42]

Sphingolipid hydrolyzing enzymes in the gastrointestinal tract, Vol. 311, 276 (2000).[43]

[27] D. F. Hassler, R. M. Laethem, and G. K. Smith, *Methods Enzymol.* **311,** 176 (2000).

[28] B. Roelofsen, R. F. Zwaal, and C. B. Woodward, *Methods Enzymol.* **32,** 131 (1974).

[29] R. T. Dobrowsky and V. R. Gazula, *Methods Enzymol.* **311,** 184 (2000).

[30] N. S. Radin and S. L. Berent, *Methods Enzymol.* **83,** 596 (1982).

[31] M. Nikolova-Karakashian and A. H. Merrill, Jr., *Methods Enzymol.* **311,** 194 (2000).

[32] T. Linke, S. Lansmann, and K. Sandhoff, *Methods Enzymol.* **311,** 201 (2000).

[33] S. Bajjalieh and R. Batchelor, *Methods Enzymol.* **311,** 207 (2000).

[34] A. Olivera, K. D. Barlow, and S. Spiegel, *Methods Enzymol.* **311,** 215 (2000).

[35] C. Mao and L. Obeid, *Methods Enzymol.* **311,** 223 (2000).

[36] D. N. Brindley, J. Xu, R. Jasinska, and D. W. Waggoner, *Methods Enzymol.* **311,** 233 (2000).

[37] P. P. Van Veldhoven, *Methods Enzymol.* **311,** 244 (2000).

[38] E. Conzelmann and K. Sandhoff, *Methods Enzymol.* **138,** 792 (1987).

[39] U. Bierfreund, T. Kolter, and K. Sandhoff, *Methods Enzymol.* **311,** 255 (2000).

[40] S. C. Li and Y. T. Li, *Methods Enzymol.* **83,** 588 (1982).

[41] J. W. Kusiak, J. M. Quirk, and R. O. Brady, *Methods Enzymol.* **50,** 533 (1978).

[42] A. L. Fluharty and J. Edmond, *Methods Enzymol.* **50,** 537 (1978).

[43] R. D. Duan and A. Nilsson, *Methods Enzymol.* **311,** 276 (2000).

Properties of animal ceramide glycanases, Vol. 311, 287 (2000).[44]
Ceramide glycanase from leech, *Hirudo medicinalis,* and earthworm, *Lumbricus terrestris,* Vol. 179, 479 (1989).[45]
Ceramide glycanase from the leech *Macrobdella decora* and oligosaccharide-transferring activity, Vol. 242, 146 (1994).[46]
Endo-β-galactosidase from *Flavobacterium keratolyticus,* Vol. 83, 619 (1982).[47]
Endo-β-galactosidase from *Escherichia freundii,* Vol. 83, 610 (1982).[48]
Enzymatic *N*-deacylation of sphingolipids, Vol. 311, 297 (2000).[49]
Genetic approaches for studies of glycolipid synthetic enzymes, Vol. 311, 303 (2000).[50]
Use of yeast as a model system for studies of sphingolipid metabolism and signaling, Vol. 311, 319 (2000).[51]
Enzymic diagnosis of the genetic mucopolysaccharide storage disorders, Vol. 50, 439 (1978).[52]
Enzymic diagnosis of sphingolipidoses, Vol. 50, 456 (1978).[53]
Enzymatic diagnosis of sphingolipidoses, Vol. 138, 727 (1987).[54]
Assay of enzymes of lipid metabolism with colored and fluorescent derivatives of natural lipids, Vol. 72, 351 (1981).[55]

Inhibitors of Sphingolipid Metabolism

Isolation and characterization of novel inhibitors of sphingolipid synthesis: australifungin, viridiofungins, rustmicin, and khafrefungin, Vol. 311, 335 (2000).[56]
Fermentation, partial purification, and use of serine palmitoyltransferase inhibitors from *Isaria* (= *Cordyceps*) *sinclairii,* Vol. 311, 348 (2000).[57]
Isolation and characterization of fumonisins, Vol. 311, 361 (2000).[58]

[44] M. Basu, P. Kelly, M. Girzadas, Z. Li, and S. Basu, *Methods Enzymol.* **311,** 287 (2000).
[45] Y. T. Li and S. C. Li, *Methods Enzymol.* **179,** 479 (1989).
[46] Y. T. Li and S. C. Li, *Methods Enzymol.* **242,** 146 (1994).
[47] M. Kitamikado, M. Ito, and Y. T. Li, *Methods Enzymol.* **83,** 619 (1982).
[48] Y. T. Li, H. Nakagawa, M. Kitamikado, and S. C. Li, *Methods Enzymol.* **83,** 610 (1982).
[49] M. Ito, K. Kita, T. Kurita, N. Sueyoshi, and H. Izu, *Methods Enzymol.* **311,** 297 (2000).
[50] S. Ichikawa and Y. Hirabayashi, *Methods Enzymol.* **311,** 303 (2000).
[51] N. Chung and L. M. Obeid, *Methods Enzymol.* **311,** 319 (2000).
[52] C. W. Hall, I. Liebaers, P. Di Natale, and E. F. Neufeld, *Methods Enzymol.* **50,** 439 (1978).
[53] K. Suzuki, *Methods Enzymol.* **50,** 456 (1978).
[54] K. Suzuki, *Methods Enzymol.* **138,** 727 (1987).
[55] S. Gatt, *et al., Methods Enzymol.* **72,** 351 (1981).
[56] S. M. Mandala and G. H. Harris, *Methods Enzymol.* **311,** 335 (2000).
[57] R. T. Riley and R. D. Plattner, *Methods Enzymol.* **311,** 348 (2000).
[58] F. I. Meredith, *Methods Enzymol.* **311,** 361 (2000).

Inhibitors of glucosylceramide synthase, Vol. 311, 373 (2000).[59]
Inhibitors of cerebroside metabolism, Vol. 72, 673 (1981).[60]

Chemical and Enzymatic Syntheses

Glycosphingolipids: structure, biologic source, and properties, Vol. 179, 167 (1989).[61]
Synthesis of sphingosine and sphingoid bases, Vol. 311, 391 (2000).[62]
Synthesis of sphingosine, radiolabeled-sphingosine, 4-methyl-*cis*-sphingosine, and 1-amino derivatives of sphingosine via their azido derivatives, Vol. 311, 441 (2000).[63]
Total synthesis of sphingosine and its analogs, Vol. 311, 458 (2000).[64]
Lysogangliosides: synthesis and use in preparing labeled gangliosides, Vol. 138, 319 (1987).[65]
Radiolabeling of the sphingolipid backbone, Vol. 311, 480 (2000).[66]
Preparation of radiolabeled ceramides and phosphosphingolipids, Vol. 311, 499 (2000).[67]
Synthesis of key precursors of radiolabeled sphingolipids, Vol. 311, 518 (2000).[68]
Metabolic radiolabeling of glycoconjugates, Vol. 230, 16 (1994).[69]
The preparation of Tay-Sachs ganglioside specifically labeled in either the *N*-acetylneuraminosyl or *N*-acetylgalactosaminyl portion of the molecule, Vol. 35, 541 (1975).[70]
Practical synthesis of *N*-palmitoylsphingomyelin and *N*-palmitoyldihydrosphingomyelin, Vol. 311, 535 (2000).[71]
Chemical synthesis of choline-labeled lecithins and sphingomyelins, Vol. 35, 533 (1975).[72]
Synthesis and biologic activity of glycolipids, with a focus on gangliosides and sulfatide analogs, Vol. 311, 547 (2000).[73]

[59] J. A. Shayman, L. Lee, A. Abe, and L. Shu, *Methods Enzymol.* **311,** 373 (2000).
[60] N. S. Radin and R. R. Vunnam, *Methods Enzymol.* **72,** 673 (1981).
[61] C. L. Stults, C. C. Sweeley, and B. A. Macher, *Methods Enzymol.* **179,** 167 (1989).
[62] C. Curfman and D. Liotta, *Methods Enzymol.* **311,** 391 (2000).
[63] K. H. Jung and R. R. Schmidt, *Methods Enzymol.* **311,** 441 (2000).
[64] P. M. Koskinen and A. M. Koskinen, *Methods Enzymol.* **311,** 458 (2000).
[65] G. Schwarzmann and K. Sandhoff, *Methods Enzymol.* **138,** 319 (1987).
[66] A. Bielawska, Y. A. Hannum, and Z. Szulc, *Methods Enzymol.* **311,** 480 (2000).
[67] A. Bielawska and Y. A. Hannun, *Methods Enzymol.* **311,** 499 (2000).
[68] A. Bielawska, Z. Szulc, and Y. A. Hannun, *Methods Enzymol.* **311,** 518 (2000).
[69] A. Varki, *Methods Enzymol.* **230,** 16 (1994).
[70] J. F. Tallman, E. H. Kolodny, and R. O. Brady, *Methods Enzymol.* **35,** 541 (1975).
[71] A. S. Bushnev and D. C. Liotta, *Methods Enzymol.* **311,** 535 (2000).
[72] W. Stoffel, *Methods Enzymol.* **35,** 533 (1975).
[73] T. Ikami, H. Ishida, and M. Kiso, *Methods Enzymol.* **311,** 547 (2000).

Sphingolipid photoaffinity labels, Vol. 311, 568 (2000).[74]
Disaccharide units from complex carbohydrates of animals, Vol. 50, 272 (1978).[75]
Chemical synthesis of phospholipids and analogs of phospholipids containing carbon–phosphorus bonds, Vol. 35, 429 (1975).[76]
Chemical synthesis of glycosylamide and cerebroside analogs composed of carba sugars, Vol. 247, 136 (1994).[77]
Synthesis and characterization of metabolically stable sphingolipids, Vol. 311, 601 (2000).[78]
Synthesis of ganglioside analogs containing sulfur in place of oxygen at the linkage positions, Vol. 242, 183 (1994).[79]
Preparation of fluorescence-labeled neoglycolipids for ceramide glycanase assays, Vol. 278, 519 (1997).[80]
Ganglioside-based neoglycoproteins, Vol. 242, 17 (1994).[81]
Synthetic soluble analogs of glycolipids for studies of virus–glycolipid interactions, Vol. 311, 626 (2000).[82]
Synthesis of ganglioside GM3 and analogs containing modified sialic acids and ceramides, Vol. 242, 173 (1994).[83]
Synthesis and characterization of ganglioside GM3 derivatives: lyso-GM3, de-*N*-acetyl-GM3, and other compounds, Vol. 179, 242 (1989).[84]
Preparation of radioactive gangliosides, ^{3}H or ^{14}C isotopically labeled at oligosaccharide or ceramide moieties, Vol. 311, 639 (2000).[85]
Synthesis of sialyl Lewis X ganglioside and analogs, Vol. 242, 158 (1994).[86]
Fluorescent gangliosides, Vol. 138, 313 (1987).[87]
Sialic acid analogs and application for preparation of neoglycoconjugates, Vol. 247, 153 (1994).[88]

[74] F. Knoll, T. Kolter, and K. Sandhoff, *Methods Enzymol.* **311,** 568 (2000).
[75] G. Dawson, *Methods Enzymol.* **50,** 272 (1978).
[76] A. F. Rosenthal, *Methods Enzymol.* **35,** 429 (1975).
[77] S. Ogawa and H. Tsunoda, *Methods Enzymol.* **247,** 136 (1994).
[78] G. Schwarzmann, *Methods Enzymol.* **311,** 601 (2000).
[79] H. Ishida, M. Kiso, and A. Hasegawa, *Methods Enzymol.* **242,** 183 (1994).
[80] K. Matsuoka, S. I. Nishimura, and Y. C. Lee, *Methods Enzymol.* **278,** 519 (1997).
[81] J. A. Mahoney and R. L. Schnaar, *Methods Enzymol.* **242,** 17 (1994).
[82] J. Fantini, *Methods Enzymol.* **311,** 626 (2000).
[83] M. Kiso and A. Hasegawa, *Methods Enzymol.* **242,** 173 (1994).
[84] G. A. Nores, N. Hanai, S. B. Levery, H. L. Eaton, M. E. Salyan, and S. Hakomori, *Methods Enzymol.* **179,** 242 (1989).
[85] S. Sonnino, V. Chigorno, and G. Tettamanti, *Methods Enzymol.* **311,** 639 (2000).
[86] A. Hasegawa and M. Kiso, *Methods Enzymol.* **242,** 158 (1994).
[87] S. Spiegel, *Methods Enzymol.* **138,** 313 (1987).
[88] R. Brossmer and H. J. Gross, *Methods Enzymol.* **247,** 153 (1994).

Enzymatic synthesis of [^{14}C]ceramide, [^{14}C]glycosphingolipids, and ω-aminoceramide, Vol. 311, 682 (2000).[89]

Methods for Isolating and Analyzing Sphingolipids

Separation of sphingosine bases by chromatography on columns of silica gel, Vol. 35, 529 (1975).[90]
Inositol-containing sphingolids, Vol. 138, 186 (1987).[91]
Thin-layer chromatography of neutral glycosphingolipids, Vol. 35, 396 (1975).[92]
Thin-layer chromatography of neutral glycosphingolipids and gangliosides, Vol. 72, 185 (1981).[93]
Thin-layer chromatography of glycosphingolipids, Vol. 230, 371 (1994).[94]
Isolation and characterization of glycosphingolipid from animal cells and their membranes, Vol. 32, 345-67 (1974).[95]
DEAE-silica gel and DEAE-controlled porous glass as ion exchangers for the isolation of glycolipids, Vol. 72, 174 (1981).[96]
Isolation of glycosphingolipids, Vol. 230, 348 (1994).[97]
Use of antibodies and the primary enzyme immunoassay (PEIA) to study enzymes: the arylsulfatase A–anti-arylsulfatase A system, Vol. 73, 550 (1981).[98]
Isolation of blood group ABH-active glycolipids from human erythrocyte membranes, Vol. 50, 207 (1978).[99]
Isolation of poly(glycosyl)ceramides with A, B, H, and I blood-group activities, Vol. 50, 211 (1978).[100]
Glycosphingolipids: structure, biologic source, and properties, Vol. 50, 236 (1978).[101]
Gangliosides: structure, isolation, and analysis, Vol. 83, 139 (1982).[102]

[89] M. Ito, S. Mitsutake, M. Tani, and K. Kita, *Methods Enzymol.* **311,** 682 (2000).
[90] Y. Barenholz and S. Gatt, *Methods Enzymol.* **35,** 529 (1975).
[91] R. A. Laine and T. C. Hsieh, *Methods Enzymol.* **138,** 186 (1987).
[92] V. P. Skipski, *Methods Enzymol.* **35,** 396 (1975).
[93] S. K. Kundu, *Methods Enzymol.* **72,** 185 (1981).
[94] R. L. Schnaar and L. K. Needham, *Methods Enzymol.* **230,** 371 (1994).
[95] S. I. Hakomori and B. Siddiqui, *Methods Enzymol.* **32,** 345 (1974).
[96] S. K. Kundu, *Methods Enzymol.* **72,** 174 (1981).
[97] R. L. Schnaar, *Methods Enzymol.* **230,** 348 (1994).
[98] E. A. Neuwelt, *Methods Enzymol.* **73,** 550 (1981).
[99] S. I. Hakomori, *Methods Enzymol.* **50,** 207 (1978).
[100] J. Koscielak, H. Miller-Podraza, and E. Zdebska, *Methods Enzymol.* **50,** 211 (1978).
[101] B. A. Macher and C. C. Sweeley, *Methods Enzymol.* **50,** 236 (1978).
[102] R. W. Ledeen and R. K. Yu, *Methods Enzymol.* **83,** 139 (1982).

Isolation and purification of gangliosides from plasma, Vol. 138, 300 (1987).[103]

Short-bed, continuous development, thin-layer chromatography of glycosphingolipids, Vol. 138, 125 (1987).[104]

High-pressure liquid chromatography analysis of neutral glycosphingolipids: perbenzoylated mono-, di-, tri-, and tetraglycosylceramides, Vol. 138, 117 (1987).[105]

Isolation and purification of glycosphingolipids by high-performance liquid chromatography, Vol. 138, 3 (1987).[106]

High-performance liquid chromatography of membrane lipids: glycosphingolipids and phospholipids, Vol. 172, 538 (1989).[107]

Radioisotopic assay for galactocerebrosides and sulfatides, Vol. 83, 191 (1982).[108]

Isolation of specific sugar sequences using carbohydrate-binding proteins, Vol. 83, 241 (1982).[109]

Regulation of cholera toxin by temperature, pH, and osmolarity, Vol. 235, 517 (1994).[110]

Removal of water-soluble substances from ganglioside preparations, Vol. 35, 549 (1975).[111]

Methods for determination of gangliosides in retinas and rod outer segments, Vol. 81, 304 (1982).[112]

Analysis of glycosphingolipids by high-resolution proton nuclear magnetic resonance spectroscopy, Vol. 83, 69 (1982).[113]

Lectin affinity chromatography of glycolipids and glycolipid-derived oligosaccharides, Vol. 179, 30 (1989).[114]

Two-dimensional proton magnetic resonance spectroscopy, Vol. 179, 122 (1989).[115]

[103] S. Ladisch and B. Gillard, *Methods Enzymol.* **138,** 300 (1987).
[104] W. W. Young, Jr., and C. A. Borgman, *Methods Enzymol.* **138,** 125 (1987).
[105] M. D. Ullman and R. H. McCluer, *Methods Enzymol.* **138,** 117 (1987).
[106] R. Kannagi, K. Watanabe, and S. Hakomori, *Methods Enzymol.* **138,** 3 (1987).
[107] R. H. McCluer, M. D. Ullman, and F. B. Jungalwala, *Methods Enzymol.* **172,** 538 (1989).
[108] H. Singh and J. N. Kanfer, *Methods Enzymol.* **83,** 191 (1982).
[109] D. F. Smith, *Methods Enzymol.* **83,** 241 (1982).
[110] C. L. Gardel and J. J. Mekalanos, *Methods Enzymol.* **235,** 517 (1994).
[111] T. P. Carter and J. N. Kanfer, *Methods Enzymol.* **35,** 549 (1975).
[112] H. Dreyfuss, N. Virmaux-Colin, S. Harth, and P. Mandel, *Methods Enzymol.* **81,** 304 (1982).
[113] J. Dabrowski, P. Hanfland, and H. Egge, *Methods Enzymol.* **83,** 69 (1982).
[114] D. F. Smith and B. V. Torres, *Methods Enzymol.* **179,** 30 (1989).
[115] J. Dabrowski, *Methods Enzymol.* **179,** 122 (1989).

Computer simulations of nuclear Overhauser effect spectra of complex oligosaccharides, Vol. 240, 446 (1994).[116]

Direct chemical ionization mass spectrometry of carbohydrates, Vol. 138, 59 (1987).[117]

Mass spectrometry of mixtures of intact glycosphingolipids, Vol. 193, 623 (1990).[118]

Desorption mass spectrometry of glycosphingolipids, Vol. 193, 713 (1990).[119]

High-mass gas chromatography-mass spectrometry of permethylated oligosaccharides, Vol. 193, 733 (1990).[120]

Infrared and laser Raman spectroscopy, Vol. 32, 247 (1974).[121]

Glycolipids, Vol. 138, 575 (1987).[122]

Methods for Analyzing Aspects of Sphingolipid Metabolism in Intact Cells

Use of *N*-([1-^{14}C]hexanoyl)-D-*erythro*-sphingolipids to assay sphingolipid metabolism, Vol. 209, 437 (1992).[123]

Sphingolipid–Protein Interactions and Cellular Targets

Use of sphingosine as inhibitor of protein kinase C, Vol. 201, 316 (1991).[124]

Covalent attachment of glycolipids to solid supports and macromolecules, Vol. 50, 137 (1978).[125]

Glycolipid transfer protein from pig brain, Vol. 179, 559 (1989).[126]

Sulfatide-binding proteins, Vol. 138, 473 (1987).[127]

Cerebroside transfer protein, Vol. 98, 613 (1983).[128]

[116] C. A. Bush, *Methods Enzymol.* **240,** 446 (1994).

[117] V. N. Reinhold, *Methods Enzymol.* **138,** 59 (1987).

[118] B. E. Samuelsson, W. Pimlott, and K. A. Karlsson, *Methods Enzymol.* **193,** 623 (1990).

[119] J. Peter-Katalinic and H. Egge, *Methods Enzymol.* **193,** 713 (1990).

[120] G. C. Hansson and H. Karlsson, *Methods Enzymol.* **193,** 733 (1990).

[121] D. F. Hoelzl Wallach, and A. R. Oseroff, *Methods Enzymol.* **32,** 247 (1974).

[122] M. Basu, T. De, K. K. Das, J. W. Kyle, H. C. Chon, R. J. Schaeper, and S. Basu, *Methods Enzymol.* **138,** 575 (1987).

[123] A. H. Futerman and R. E. Pagano, *Methods Enzymol.* **209,** 437 (1992).

[124] Y. A. Hannun, A. H. Merrill, Jr., and R. M. Bell, *Methods Enzymol.* **201,** 316 (1991).

[125] W. W. Young, Jr., A. Laine, and S. I. Hakomori, *Methods Enzymol.* **50,** 137 (1978).

[126] T. Sasaki and A. Abe, *Methods Enzymol.* **179,** 559 (1989).

[127] D. D. Roberts, *Methods Enzymol.* **138,** 473 (1987).

[128] N. S. Radin and R. J. Metz, *Methods Enzymol.* **98,** 613 (1983).

Adhesion of eukaryotic cells to immobilized carbohydrates, Vol. 179, 542 (1989).[129]
Gangliosides that modulate membrane protein function, Vol. 179, 521 (1989).[130]
Ganglioside-agarose and cholera toxin, Vol. 34, 610 (1974).[131]
Replacement of glycosphingolipid ceramide residues by glycerolipid for microtiter plate assays, Vol. 242, 198 (1994).[132]

Sphingolipid Transport and Trafficking

Estimating sphingolipid metabolism and trafficking in cultured cells using radiolabeled compounds, Vol. 311, 656 (2000).[133]
Localization of ganglioside GM1 with biotinylated choleragen, Vol. 184, 405 (1990).[134]
Transfer of bulk markers from endoplasmic reticulum to plasma membrane, Vol. 219, 189 (1992).[135]

Other Methods

Purification of rat liver 5′-nucleotidase as a complex with sphingomyelin, Vol. 32, 368 (1974).[136]
Sphingolipid-dependent signaling in regulation of cytochrome P450 expression, Vol. 272, 381 (1996).[137]

The author apologizes for any other chapters that have been overlooked in preparing this list.

[129] R. L. Schnaar, B. K. Brandley, L. K. Needham, P. Swank-Hill, and C. C. Blackburn, *Methods Enzymol.* **179,** 542 (1989).
[130] Y. Igarashi, H. Nojiri, N. Hanai, and S. Hakomori, *Methods Enzymol.* **179,** 521 (1989).
[131] I. Parikh and P. Cuatrecasas, *Methods Enzymol.* **34,** 610 (1974).
[132] R. Roy, A. Romanowska, and F. O. Andersson, *Methods Enzymol.* **242,** 198 (1994).
[133] L. Riboni, P. Viani, and G. Tettamanti, *Methods Enzymol.* **311,** 656 (2000).
[134] H. Asou, *Methods Enzymol.* **184,** 405 (1990).
[135] F. Wieland, *Methods Enzymol.* **219,** 189 (1992).
[136] C. C. Widnell, *Methods Enzymol.* **32,** 368 (1974).
[137] E. T. Morgan, M. Nikolova-Karakashian, J. Q. Chen, and A. H. Merrill, Jr., *Methods Enzymol.* **272,** 381 (1996).

Author Index

Numbers in parentheses are footnote reference numbers and indicate that an author's work is referred to although the name is not cited in the text.

A

B

C

D

E

F

G

H

I

J

K

L

N

O

P

Q

R

S

T

U

V

W

X

Y

Z

Subject Index

A

B

C

H

M

N

O

P

R

S

ISBN 0-12-182213-3